AF743416

THE CENTRAL PARSECS OF THE GALAXY

COVER ILLUSTRATION: The cover illustration is a 12.5 micron image of the central parsec of the Galaxy obtained with the mid-infrared camera MIRLIN at Keck II in cooperation with JPL/Caltech, UCLA, and the Keck Observatory (*see Cotera et al., this volume, p. 240)*

A SERIES OF BOOKS ON RECENT DEVELOPMENTS IN ASTRONOMY AND ASTROPHYSICS

Managing Editor, D. Harold McNamara
Production Manager, Enid L. Livingston

A.S.P. CONFERENCE SERIES PUBLICATIONS COMMITTEE

Sallie Baliunas, Chair
Carol Ambruster
Catharine Garmany
Mark S. Giampapa
Kenneth Janes

© Copyright 1999 Astronomical Society of the Pacific
390 Ashton Avenue, San Francisco, California 94112

All rights reserved

Printed by Sheridan Books, Inc.

First published 1999

Library of Congress Catalog Card Number: 99-66053
ISBN 1-58381-013-7

Please contact proper address for information on:

PUBLISHING:
Managing Editor
PO Box 24463
211 KMB
Brigham Young University
Provo, UT 84602-4463 USA

Phone: 801-378-2298
Fax: 801-378-4049
E-mail: pasp@astro.byu.edu

ORDERING BOOKS:
Astronomical Society of the Pacific
CONFERENCE SERIES
390 Ashton Avenue
San Francisco, CA 94112 - 1722 USA

Phone: 415-337-2624
Fax: 415-337-5205
E-mail: catalog@aspsky.org
Web Site: http://www.aspsky.org

A SERIES OF BOOKS ON RECENT DEVELOPMENTS IN ASTRONOMY AND ASTROPHYSICS

Vol. 1-Progress and Opportunities in Southern Hemisphere Optical Astronomy: CTIO 25th Anniversary Symposium
ed. V. M. Blanco and M. M. Phillips ISBN 0-937707-18-X

Vol. 2-Proceedings of a Workshop on Optical Surveys for Quasars
ed. P. S. Osmer, A. C. Porter, R. F. Green, and C. B. Foltz ISBN 0-937707-19-8

Vol. 3-Fiber Optics in Astronomy
ed. S. C. Barden ISBN 0-937707-20-1

Vol. 4-The Extragalactic Distance Scale: Proceedings of the ASP 100th Anniversary Symposium
ed. S. van den Bergh and C. J. Pritchet ISBN 0-937707-21-X

Vol. 5-The Minnesota Lectures on Clusters of Galaxies and Large-Scale Structure
ed. J. M. Dickey ISBN 0-937707-22-8

Vol. 6-Synthesis Imaging in Radio Astronomy: A Collection of Lectures from the Third NRAO Synthesis Imaging Summer School
ed. R. A. Perley, F. R. Schwab, and A. H. Bridle ISBN 0-937707-23-6

Vol. 7-Properties of Hot Luminous Stars: Boulder-Munich Workshop
ed. C. D. Garmany ISBN 0-937707-24-4

Vol. 8-CCDs in Astronomy
ed. G. H. Jacoby ISBN 0-937707-25-2

Vol. 9-Cool Stars, Stellar Systems, and the Sun. Sixth Cambridge Workshop
ed. G. Wallerstein ISBN 0-937707-27-9

Vol. 10-Evolution of the Universe of Galaxies: Edwin Hubble Centennial Symposium
ed. R. G. Kron ISBN 0-937707-28-7

Vol. 11-Confrontation Between Stellar Pulsation and Evolution
ed. C. Cacciari and G. Clementini ISBN 0-937707-30-9

Vol. 12-The Evolution of the Interstellar Medium
ed. L. Blitz ISBN 0-937707-31-7

Vol. 13-The Formation and Evolution of Star Clusters
ed. K. Janes ISBN 0-937707-32-5

Vol. 14-Astrophysics with Infrared Arrays
ed. R. Elston ISBN 0-937707-33-3

Vol. 15-Large-Scale Structures and Peculiar Motions in the Universe
ed. D. W. Latham and L. A. N. da Costa ISBN 0-937707-34-1

Vol. 16-Proceedings of the 3rd Haystack Observatory Conference on Atoms, Ions and Molecules: New Results in Spectral Line Astrophysics
ed. A. D. Haschick and P. T. P. Ho ISBN 0-937707-35-X

Vol. 17-Light Pollution, Radio Interference, and Space Debris
ed. D. L. Crawford ISBN 0-937707-36-8

Vol. 18-The Interpretation of Modern Synthesis Observations of Spiral Galaxies
ed. N. Duric and P. C. Crane ISBN 0-937707-37-6

Vol. 19-Radio Interferometry: Theory, Techniques, and Applications, IAU Colloquium 131
ed. T. J. Cornwell and R. A. Perley ISBN 0-937707-38-4

Vol. 20-Frontiers of Stellar Evolution: 50th Anniversary McDonald Observatory (1939-1989)
ed. D. L. Lambert ISBN 0-937707-39-2

Vol. 21-The Space Distribution of Quasars
ed. D. Crampton ISBN 0-937707-40-6

Vol. 22-Nonisotropic and Variable Outflows from Stars
ed. L. Drissen, C. Leitherer, and A. Nota ISBN 0-937707-41-4

Vol. 23-Astronomical CCD Observing and Reduction Techniques
ed. S. B. Howell ISBN 0-937707-42-4

Vol. 24-Cosmology and Large-Scale Structure in the Universe
ed. R. R. de Carvalho ISBN 0-937707-43-0

Vol. 25-Astronomical Data Analysis Software and Systems I
ed. D. M. Worrall, C. Biemesderfer, and J. Barnes ISBN 0-937707-44-9

Vol. 26-Cool Stars, Stellar Systems, and the Sun, Seventh Cambridge Workshop
ed. M. S. Giampapa and J. A. Bookbinder ISBN 0-937707-45-7

Vol. 27-The Solar Cycle: Proceedings of the National Solar Observatory/Sacramento Peak 12th Summer Workshop
ed. K. L. Harvey ISBN 0-937707-46-5

Vol. 28-Automated Telescopes for Photometry and Imaging
ed. S. J. Adelman, R. J. Dukes, Jr., and C. J. Adelman ISBN 0-937707-47-3

Vol. 29-Viña Del Mar Workshop on Catacysmic Variable Stars
ed. N. Vogt ISBN 0-937707-48-1

Vol. 30-Variable Stars and Galaxies
ed. B. Warner ISBN 0-937707-49-X

Vol. 31-Relationships Between Active Galactic Nuclei and Starburst Galaxies
ed. A. V. Filippenko ISBN 0-937707-50-3

Vol. 32-Complementary Approaches to Double and Multiple Star Research, IAU Collouquium 135
ed. H. A. McAlister and W. I. Hartkopf ISBN 0-937707-51-1

Vol. 33-Research Amateur Astronomy
ed. S. J. Edberg ISBN 0-937707-52-X

Vol. 34-Robotic Telescopes in the 1990s
ed. A. V. Filippenko ISBN 0-937707-53-8

Vol. 35-Massive Stars: Their Lives in the Interstellar Medium
ed. J. P. Cassinelli and E. B. Churchwell ISBN 0-937707-54-6

Vol. 36-Planets Around Pulsars
ed. J. A. Phillips, S. E. Thorsett, and S. R. Kulkarni ISBN 0-937707-55-4

Vol. 37-Fiber Optics in Astronomy II
ed. P. M. Gray ISBN 0-937707-56-2

Vol. 38-New Frontiers in Binary Star Research: Pacific Rim Colloquium
ed. K. C. Leung and I.-S. Nha ISBN 0-937707-57-0

Vol. 39-The Minnesota Lectures on the Structure and Dynamics of the Milky Way
ed. Roberta M. Humphreys ISBN 0-937707-58-9

Vol. 40-Inside the Stars, IAU Colloquium 137
ed. Werner W. Weiss and Annie Baglin ISBN 0-937707-59-7

Vol. 41-Astronomical Infrared Spectroscopy: Future Observational Directions
ed. Sun Kwok ISBN 0-937707-60-0

Vol. 42-GONG 1992: Seismic Investigation of the Sun and Stars
ed. Timothy M. Brown ISBN 0-937707-61-9

Vol. 43-Sky Surveys: Protostars to Protogalaxies
ed. B. T. Soifer ISBN 0-937707-62-7

Vol. 44-Peculiar Versus Normal Phenomena in A-Type and Related Stars, IAU Colloquium 138
ed. M. M. Dworetsky, F. Castelli, and R. Faraggiana ISBN 0-937707-63-5

Vol. 45-Luminous High-Latitude Stars
ed. D. D. Sasselov ISBN 0-937707-64-3

Vol. 46-The Magnetic and Velocity Fields of Solar Active Regions, IAU Colloquium 141
ed. H. Zirin, G. Ai, and H. Wang ISBN 0-937707-65-1

Vol. 47-Third Decennial US-USSR Conference on SETI
ed. G. Seth Shostak ISBN 0-937707-66-X

Vol. 48-The Globular Cluster-Galaxy Connection
ed. Graeme H. Smith and Jean P. Brodie ISBN 0-937707-67-8

Vol. 49-Galaxy Evolution: The Milky Way Perspective
ed. Steven R. Majewski ISBN 0-937707-68-6

Vol. 50-Structure and Dynamics of Globular Clusters
ed. S. G. Djorgovski and G. Meylan ISBN 0-937707-69-4

Vol. 51-Observational Cosmology
ed. G. Chincarini, A. Iovino, T. Maccacaro, and D. Maccagni ISBN 0-937707-70-8

Vol. 52-Astronomical Data Analysis Software and Systems II
ed. R. J. Hanisch, R. J. V. Brissenden, and Jeannette Barnes ISBN 0-937707-71-6

Vol. 53-Blue Stragglers
ed. Rex A. Saffer ISBN 0-937707-72-4

Vol. 54-The First Stromlo Symposium: The Physics of Active Galaxies
ed. Geoffrey V. Bicknell, Michael A. Dopita, and Peter J. Quinn ISBN 0-937707-73-2

Vol. 55-Optical Astronomy from the Earth and Moon
ed. Diane M. Pyper and Ronald J. Angione ISBN 0-937707-74-0

Vol. 56-Interacting Binary Stars
ed. Allen W. Shafter ISBN 0-937707-75-9

Vol. 57-Stellar and Circumstellar Astrophysics
ed. George Wallerstein and Alberto Noriega-Crespo ISBN 0-937707-76-7

Vol. 58-The First Symposium on the Infrared Cirrus and Diffuse Interstellar Clouds
ed. Roc M. Cutri and William B. Latter ISBN 0-937707-77-5

Vol. 59-Astronomy with Millimeter and Submillimeter Wave Interferometry, IAU Colloquium 140
ed. M. Ishiguro and Wm. J. Welch ISBN 0-937707-78-3

Vol. 60-The MK Process at 50 Years: A Powerful Tool for Astrophysical Insight: A Workshop of the Vatican Observatory
ed. C. J. Corbally, R. O. Gray, and R. F. Garrison ISBN 0-937707-79-1

Vol. 61-Astronomical Data Analysis Software and Systems III
ed. Dennis R. Crabtree, R. J. Hanisch, and Jeannette Barnes ISBN 0-937707-80-5

Vol. 62-The Nature and Evolutionary Status of Herbig Ae / Be Stars
ed. P. S. Thé, M. R. Pérez, and E. P. J. van den Heuvel ISBN 0-937707-81-3

Vol. 63-Seventy-Five Years of Hirayama Asteroid Families: The Role of Collisions in the Solar System History
ed. Y. Kozai, R. P. Binzel, and T. Hirayama ISBN 0-937707-82-1

Vol. 64-Cool Stars, Stellar Systems, and the Sun, Eighth Cambridge Workshop
ed. Jean-Pierre Caillault ISBN 0-937707-83-X

Vol. 65-Clouds, Cores, and Low Mass Stars
ed. Dan P. Clemens and Richard Barvainis ISBN 0-937707-84-8

Vol. 66- Physics of the Gaseous and Stellar Disks of the Galaxy
ed. Ivan R. King ISBN 0-937707-85-6

Vol. 67-Unveiling Large-Scale Structures Behind the Milky Way
ed. C. Balkowski and R. C. Kraan-Korteweg ISBN 0-937707-86-4

Vol. 68-Solar Active Region Evolution: Comparing Models with Observations
ed. K. S. Balasubramaniam and George W. Simon ISBN 0-937707-87-2

Vol. 69-Reverberation Mapping of the Broad-Line Region in Active Galactic Nuclei
ed. P. M. Gondhalekar, K. Horne, and B. M. Peterson ISBN 0-937707-88-0

Vol. 70-Groups of Galaxies
ed. Otto G. Richter and Kirk Borne ISBN 0-937707-89-9

Vol. 71-Tridimensional Optical Spectroscopic Methods in Astrophysics, IAU Colloquium 149
ed. G. Comte and M. Marcelin ISBN 0-937707-90-2

Vol. 72-Millisecond Pulsars: A Decade of Surprise.
ed. A. A. Fruchter, M. Tavani, and D. C. Backer ISBN 0-937707-91-0

Vol. 73-Airborne Astronomy Symposium on the Galactic Ecosystem: From Gas to Stars to Dust
ed. M. R. Haas, J. A. Davidson, and E. F. Erickson ISBN 0-937707-92-9

Vol. 74-Progress in the Search for Extraterrestrial Life: 1993 Bioastronomy Symposium
ed. G. Seth Shostak ISBN 0-937707-93-7

Vol. 75-Multi-Feed Systems for Radio Telescopes
ed. D. T. Emerson and J. M. Payne ISBN 0-937707-94-5

Vol. 76-GONG '94: Helio- and Astero-Seismology from the Earth and Space
ed. Roger K. Ulrich, Edward J. Rhodes, Jr., and Werner Däppen ISBN 0-937707-95-3

Vol. 77-Astronomical Data Analysis Software and System IV
ed. R. A. Shaw, H. E. Payne, and J. J. E. Hayes ISBN 0-937707-96-1

Vol. 78-Astrophysical Applications of Powerful New Databases: Joint Discussion No. 16 of the 22nd General Assembly of the IAU
ed. S. J. Adelman and W. L. Wiese ISBN 0-937707-97-X

Vol. 79-Robotic Telescopes: Current Capabilities, Present Developments, and Future Prospects for Automated Astronomy
ed. Gregory W. Henry and Joel A. Eaton ISBN 0-937707-98-8

Vol. 80-The Physics of the Interstellar Medium and Intergalactic Medium
ed. A. Ferrara, C. F. McKee, C. Heiles, and P. R. Shapiro ISBN 0-937707-99-6

Vol. 81-Laboratory and Astronomical High Resolution Spectra
ed. A. J. Sauval, R. Blomme, and N. Grevesse ISBN 1-886733-01-5

Vol. 82-Very Long Baseline Interferometry and the VLBA
ed. J. A. Zensus, P. J. Diamond, and P. J. Napier ISBN 1-886733-02-3

Vol. 83-Astrophysical Applications of Stellar Pulsation. IAU Colloquium 155
ed. R. S. Stobie and P. A. Whitelock ISBN 1-886733-03-1

Vol. 84-The Future Utilisation of Schmidt Telescopes, IAU Colloquium 148
ed. Jessica Chapman, Russell Cannon, Sandra Harrison, and Bambang Hidayat ISBN 1-886733-05-8

Vol. 85-Cape Workshop on Magnetic Cataclysmic Variables
ed. D. A. H. Buckley and B. Warner ISBN 1-886733-06-6

Vol. 86-Fresh Views of Elliptical Galaxies
ed. Alberto Buzzoni, Alvio Renzini, and Alfonso Serrano ISBN 1-886733-07-4

Vol. 87-New Observing Modes for the Next Century
ed. Todd Boroson, John Davies, and Ian Robson ISBN 1-886733-08-2

Vol. 88- Clusters, Lensing, and the Future of the Universe
ed. Virginia Trimble and Andreas Reisenegger ISBN 1-886733-09-0

Vol. 89-Astronomy Education: Current Developments, Future Coordination
ed.John R. Percy ISBN 1-886733-10-4

Vol. 90-The Origins, Evolution, and Destinies of Binary Stars in Clusters
ed. E. F. Milone and J. C. Mermilliod ISBN 1-886733-11-2

Vol. 91-Barred Galaxies, IAU Colloquium 157
ed. R. Buta, D. A. Crocker, and B. G. Elmegreen ISBN 1-886733-12-0

Vol. 92-Formation of the Galactic Halo–Inside and Out
ed. H. L. Morrison and A. Sarajedini ISBN 1-886733-13-9

Vol. 93-Radio Emission from the Stars and the Sun
ed. A. R. Taylor and J. M. Paredes ISBN 1-886733-14-7

Vol. 94-Mapping, Measuring, and Modelling the Universe
ed. Peter Coles, Vincent Martinez, and Maria-Jesus Pons-Borderia ISBN 1-886733-15-5

Vol. 95-Solar Drivers of Interplanetary and Terrestrial Disturbances: Proceedings of 16th International Workshop, National Solar Observatory/Sacramento Peak
ed. K. S. Balasubramaniam, S. L. Keil, and R. N. Smartt ISBN 1-886733-16-3

Vol. 96- Hydrogen-Deficient Stars
ed. C. S. Jeffery and U. Heber ISBN 1-886733-17-1

Vol. 97-Polarimetry of the Interstellar Medium
ed.W. G. Roberge and D. C. B. Whittet ISBN 1-886733-18-X

Vol. 98-From Stars to Galaxies: The Impact of Stellar Physics on Galaxy Evolution
ed. Claus Leitherer, Uta Fritze-von Alvensleben, and John Huchra ISBN 1-886733-19-8

Vol. 99-Cosmic Abundances: Proceedings of the 6th Annual October Astrophysics Conference
ed. Stephen S. Holt and Geroge Sonneborn ISBN 1-886733-20-1

Vol. 100-Energy Transport in Radio Galaxies and Quasars
ed. P. E. Hardee, A. H. Bridle, and J. A. Zensus ISBN 1-886733-21-X

Vol. 101-Astronomical Data Analysis Software and Systems V
ed. George H. Jacoby and Jeannette Barnes ISSN 1080-7926

Vol. 102-The Galactic Center, 4th ESO/CTIO Workshop
ed. Roland Gredel ISBN 1-886733-22-8

Vol. 103-The Physics of Liners in View of Recent Observations
ed. M. Eracleous, A. Koratkar, C. Leitherer, and L. Ho ISBN 1-886733-23-6

Vol. 104-Physics, Chemistry, and Dynamics of Interplanetary Dust, IAU Colloquium 150
ed. Bo A. S. Gustafson and Martha S. Hanner ISBN 1-886733-24-4

Vol. 105-Pulsars: Problems and Progress, IAU Colloquium 160
ed. M. Bailes, S. Johnston, and M. A. Walker ISBN 1-886733-25-2

Vol. 106-Minnesota Lectures on Extragalactic Neutral Hydrogen
ed. Evan D. Skillman ISBN 1-886733-26-0

Vol. 107-Completing the Inventory of the Solar System: A Symposium held in conjuunction with the 106th Annual Meeting of the ASP
ed. Terrence W. Rettig and Joseph M. Hahn ISBN 1-886733-27-9

Vol. 108-M. A. S. S. Model Atmospheres and Spectrum Synthesis: 5th Vienna Workshop
ed. S. J. Adelman, F. Kupka, and W. W. Weiss ISBN 1-886733-28-7

Vol. 109-Cool Stars, Stellar Systems, and the Sun, Ninth Cambridge Workshop
ed. Roberto Pallavicini and Andrea K. Dupree ISBN 1-886733-29-5

Vol. 110-Blazar Continuum Variability
ed.H. R. Miller, J. R. Webb, and J. C. Noble ISBN 1-886733-30-9

Vol. 111-Magnetic Reconnection in the Solar Atmosphere: Proceedings of a Yohkoh Conference
ed. R. D. Bentley and J. T. Mariska ISBN 1-886733-31-7

Vol. 112-The History of the Milky Way and Its Satellite System
ed. A. Burkert, D. H. Hartmann, and S. R. Majewski ISBN 1-886733-32-5

Vol. 113-Emission Lines in Active Galaxies: New Methods and Techniques, IAU Colloquium 159
ed. B. M. Peterson, F. Z. Cheng, and A. S. Wilson ISBN 1-886733-33-3

Vol. 114-Young Galaxies and QSO Absorption-Line Systems
ed. Sueli M. Viegas, Ruth Gruenwald, and Reinaldo R. de Carvalho ISBN 1-886733-34-1

Vol. 115-Galactic and Cluster Cooling Flows
ed. Noam Soker ISBN 1-886733-35-X

Vol. 116-The Second Stromlo Symposium: The Nature of Elliptical Galaxies
ed. M. Arnaboldi, G. S. Da Costa, and P. Saha ISBN 1-886733-36-8

Vol. 117- Dark and Visible Matter in Galaxies
ed. Massimo Persic and Paolo Salucci ISBN 1-886733-37-6

Vol. 118-First Advances in Solar Physics Euroconference: Advances in the Physics of Sunspots
ed. B. Schmieder, J. C. del Toro Iniesta, and M. Vázquez ISBN 1-886733-38-4

Vol. 119-Planets Beyond the Solar System and the Next Generation of Space Missions
ed. David R. Soderblom ISBN 1-886733-39-2

Vol. 120-Luminous Blue Variables: Massive Stars in Transition
ed. Antonella Nota and Henny J. G. L. M. Lamers ISBN 1-886733-40-6

Vol. 121-Accretion Phenomena and Related Outflows, IAU Colloquium 163
ed. D. T. Wickramasinghe, G. V. Bicknell and L. Ferrario ISBN 1-886733-41-4

Vol. 122-From Stardust to Planetesimals: Symposium held as part of the 108th Annual Meeting of the ASP
ed. Yvonne J. Pendleton and A. G. G. M. Tielens ISBN 1-886733-42-2

Vol. 123-The 12th 'Kingston Meeting': Computational Astrophysics
ed. David A. Clarke and Michael J. West ISBN 1-886733-43-0

Vol. 124-Diffuse Infrared Radiation and the IRTS
ed. Haruyuki Okuda, Toshio Matsumoto, and Thomas L. Roellig ISBN 1-886733-44-9

Vol. 125- Astronomical Data Analysis Software and Systems VI
ed. Gareth Hunt and H. E. Payne ISBN 1-886733-45-7

Vol. 126-From Quantum Fluctuations to Cosmological Structures
ed. D. Vallis-Gabaud, M. A. Hendry, P. Molaro, and K. Chamcham ISBN 1-886733-46-5

Vol. 127-Proper Motions and Galactic Astronomy
ed. Roberta M. Humphreys ISBN 1-886733-47-3

Vol. 148-Origins
ed. Charles E. Woodward, J. Michael Shull, and Harley A. Thronson, Jr. ISBN 1-886733-68-6

Vol. 149-Solar System Formation and Evolution
ed. D. Lazzaro, R. Vieira Martins, S. Ferraz-Mello, J. Fernández, and C. Beaugé ISBN 1-886733-69-4

Vol. 150-New Perspectives on Solar Prominences, IAU Colloquium 167
ed. David Webb, David Rust, and Brigitte Schmieder ISBN 1-886733-70-8

Vol. 151-Cosmic Microwave Background and Large Scale Structure of the Universe
ed. Yong-Ik Byun and Kin-Wang Ng ISBN 1-886733-71-6

Vol. 152-Fiber Optics in Astronomy III
ed. S. Arribas, E. Mediavilla, and F. Watson ISBN 1-886733-72-4

Vol. 153-Library and Information Services in Astronomy III, (LISA III)
ed. Uta Grothkopf, Heinz Andernach, Sarah Stevens-Rayburn, and Monique Gomez ISBN 1-886733-73-2

Vol. 154-Cool Stars, Stellar Systems, and the Sun, Tenth Cambridge Workshop
ed. Robert A. Donahue and Jay A. Bookbinder ISBN 1-886733-74-0

Vol. 155-Second Advances in Solar Physics Euroconference: Three-Dimensional Structure of Solar Active Regions
ed. Costas E. Alissandrakis and Brigitte Schmieder ISBN 1-886733-75-9

Vol. 156-Highly Redshifted Radio Lines
ed. C. L. Carilli, S. J. E. Radford, K. M. Menten and G. I. Langston ISBN 1-886733-76-7

Vol. 157-Annapolis Workshop on Magnetic Cataclysmic Variables
ed. Coel Hellier and Koji Mukai ISBN 1-886733-77-5

Vol. 158-Solar and Stellar Activity: Similarities and Differences
ed. C. J. Butler and J. G. Doyle ISBN 1-886733-78-3

Vol. 159-BL Lac Phenomenon
ed. Leo O. Takalo and Aimo Sillanpää ISBN 1-886733-79-1

Vol. 160-Astrophysical Discs, An EC Summer School
ed. J. A. Sellwood and Jeremy Goodman ISBN 1-886733-80-5

Vol. 161-High Energy Processes in Accreting Black Holes
ed. Juri Poutanen and Roland Svensson ISBN 1-886733-81-3

Vol. 162-Quasars and Cosmology
ed. Gary Ferland and Jack Baldwin ISBN 1-886733-83-X

Vol. 163-Star Formation in Early Type Galaxies
ed. Jordi Cepa and Patricia Carral ISBN 1-886733-84-8

Vol. 164-Ultraviolet–Optical Space Astronomy Beyond HST
ed. Jon A. Morse, J. Michael Shull and Anne L. Kinney ISBN 1-886733-85-6

Vol. 165-The Third Stromlo Symposium: The Galactic Halo
ed. Brad K. Gibson, Tim S. Axelrod, and Mary E. Putman ISBN 1-886733-86-4

Vol. 166-Stromlo Workshop on High-Velocity Clouds
ed. Brad K. Gibson and Mary E. Putman ISBN 1-886733-87-2

Vol. 167-Harmonizing Cosmic Distance Scales in a Post-HIPPARCOS Era
ed. Daniel Egret and André Heck ISBN 1-886733-88-0

Vol. 168-New Perspectives on the Interstellar Medium
ed. A. R. Taylor, T. L. Landecker, and G. Joncas — ISBN 1-886733-89-9

Vol. 169-11th European Workshop on White Dwarfs
ed. J.-E. Solheim and E. G. Meistas — ISBN 1-886733-91-0

Vol. 170-The Low Surface Brightness Universe, IAU Colloquium 171
ed. J. I. Davies, C. Impey, and S. Phillipps — ISBN 1-886733-92-9

Vol. 171-LiBeB, Cosmic Rays, and Related X-and Gamma-Rays
ed. Reuven Ramaty, Elisabeth Vangioni-Flam,
Michel Cassé, and Keith Olive — ISBN 1-886733-93-7

Vol. 172-Astronomical Data Analysis Software and Systems VIII
ed. David M. Mehringer, Raymond L. Plante, and
Douglas A. Roberts — ISBN 1-886733-94-5

Vol. 173-Theory and Tests of Convection In Stellar Structure
ed. Álvaro Giménez, Edward F. Guinan, and Benjamín Montesinos — ISBN 1-886733-95-3

Vol. 174-Catching the Perfect Wave: Adaptive Optics and Interferometry for the 21st Century, A Symposium held as a part of the 110th Annual Meeting of the ASP
ed. Sergio R. Restaino, William Junor, and Nebojsa Duric — ISBN 1-886733-96-1

Vol. 175-Structure and Kinematics of Quasar Broad Line Regions
ed. C.M. Gaskell, W. N. Brandt, M. Dietrich, D. Dultzin-Hacyan,
and M. Eracleous — ISBN 1-886733-97-X

Vol. 176-Observational Cosmology: The Development of Galaxy Systems
ed. Giuliano Giuricin, Marino Mezzetti, and Paolo Salucci — ISBN 1-58381-000-5

Vol. 177-Astrophysics with Infrared Surveys: A Prelude to SIRTF
ed. Michael D. Bicay, Chas A. Beichman, Roc M. Cutri,
and Barry F. Madore — ISBN 1-58381-001-3

Vol. 178-Stellar Dynamos: Nonlinearity and Chaotic Flows
ed. Manuel Núñez and Antonio Ferriz-Mas — ISBN 1-58381-002-1

Vol. 179-Eta Carinae At The Millennium
ed. Jon A. Morse, Roberta M. Humphreys, and Augusto Damineli — ISBN 1-58381-003-X

Vol. 180-Synthesis Imaging in Radio Astronomy II
ed. G. B. Taylor, C. L. Carilli, and R. A. Perley — ISBN 1-58381-005-6

Vol. 181-Microwave Foregrounds
ed. Angelica de Oliveira-Costa and Max Tegmark — ISBN 1-58381-006-4

Vol. 182-Galaxy Dynamics, A Rutgers Symposium
ed. David Merritt, J. A. Sellwood, and Monica Valluri — ISBN 1-58381-007-2

Vol. 183-High Resolution Solar Physics: Theory, Observations and Techniques
ed. T. R. Rimmele, K. S. Balasubramaniam, and R. R. Radick — ISBN 1-58381-009-9

Vol. 184-Third Advances in Solar Physics Euroconference: Magnetic Fields and Oscillations
ed. B. Schmieder, A. Hofmann, and J. Staude — ISBN 1-58381-010-2

Vol. 185-Precise Stellar Radial Velocities, IAU Colloquium 170
ed. J. B. Hearnshaw and C. D. Scarfe ISBN 1-58381-011-0

Vol. 186-The Central Parsecs of the Galaxy
ed. Heino Falcke, Angela Cotera, Wolfgang J. Duschl, Fulvio Melia, and Marcia J. Rieke ISBN 1-58381-012-9

Book orders or inquiries concerning these volumes should be directed to the:

Astronomical Society of the Pacific Conference Series
390 Ashton Avenue
San Francisco, CA 94112-1722 USA

Phone: 415-337-2126
Fax: 415-337-5205
E-mail: catalog@aspsky.org
Web Site: http://www.aspsky.org

ASTRONOMICAL SOCIETY OF THE PACIFIC
CONFERENCE SERIES

Volume 186

THE CENTRAL PARSECS OF THE GALAXY
GALACTIC CENTER WORKSHOP

Proceedings of a meeting held at
Tucson, Arizona, USA
7-11 September, 1998

Edited by

Heino Falcke
Max-Planck-Institut für Radioastronomie, Bonn, Germany

Angela Cotera
Steward Observatory, The University of Arizona, Tucson, Arizona, USA
SETI Institute, Mountain View, California, USA

Wolfgang J. Duschl
Institut für Theoretische Astrophysik, Heidelberg, Germany
Sonderforschungsbereich 328, Evolution of Galaxies, Universität Heidelberg
Max-Planck-Institut für Radioastronomie, Bonn, Germany

Fulvio Melia
Physics Department and Steward Observatory, The University of Arizona
Tucson, Arizona, USA

and - 3 AUG 2004

Marcia J. Rieke
Steward Observatory, The University of Arizona, Tucson, Arizona, USA

~~524.64 GAL~~

523.1125
GAL

2247884

Table of Contents

Session III: How to Reconcile Models and Observations of Sgr A*?

Session IV: How are the Various Components of Sgr A West Related?

Session V: How, When, Where and Why are Stars Formed in the Galactic Center?

Session VI: What is the Metallicity as Determined from Old and New Stars?

Session VII: How Do the Various Components of the Inner 100 Parsecs Interact With Each Other?

Session VIII: What is the Strength and Influence of the Magnetic Field in the Galactic Center?

Session IX: What Would the Galactic Center Look Like from 1 Mpc?

Session X: What Can We Learn from High-Energy Emission in the Galactic Center?

Summary

The Central Parsecs – Preface

Galactic Center Workshop in Tucson, Sep. 7-11, 1999

Though largely shrouded by the intervening gas and dust, the Galactic Center is now actively being explored observationally at radio, sub-millimeter, infrared, X- and γ-ray wavelengths with unprecedented clarity and spectral resolution. This region of the Galaxy is attracting many astronomers and astrophysicists interested in learning about the physics of massive black hole accretion, magnetized gas dynamics and unusual stellar formation and evolution, among others. It is currently one of the most exciting fields to be entered by someone just starting out in research.

This ever growing interest in Galactic Center research has manifested itself in a series of conferences over the last few years, e.g., the 4th ESO/CTIO Workshop held in La Serena, Chile, in 1996, and the IAU Symposium 184 held in Kyoto, Japan, in 1997. These meetings have been extremly valuable in showing the wealth of observational data now becoming available. The Galactic Center Workshop 1998 "The Central Parsecs", whose proceedings are presented in this volume, was organized in response to the need to begin digesting this explosion of information.

We have organized this Galactic Center Workshop to provide a forum

- for *observers*, to rapidly present and disseminate the most recent observational results in this fast moving field;
- for *theorists*, to present and discuss the various alternative models and possibly to get closer to a consenus on at least some aspects of the theoretical understanding;
- for *both observers and theorists*, to organize collaborations and discuss common areas where they may find complementary strength in relevant observations and addressing the important questions.

The important role played by the direct scientific interaction among the participants at this Workshop is exemplified by the inclusion of transcriptions of the dedicated discussion sessions held at various points during this meeting. The discussion sessions were led by expert moderators. The session transcripts were edited only to the degree necessary to make them readable but still to keep their original liveliness. This approach of dedicated discussion sessions is somewhat unusual but was necessary for the reasons discussed above. The degree to which we have succeeded in achieving these goals is left to the discretion of the participants and the reader.

A conference of this kind would not have been possible without the enthusiastic participation of the attendees and the generous support from a number of sources. For financial support, we are grateful to Sonderforschungsbereich 328 at the University of Heidelberg, the Physics Department at The University of Arizona, the NICMOS collaboration, Steward Observatory, the Max-Planck-Institut für Radioastronomie, Bonn, and the Institut für Theoretische Astrophysik, Heidelberg. In addition, managing a conference of this scope requires the dedication and hard work of many people to prepare and run the meeting. For this we are deeply grateful to the Scientific Organizing Committee, the *Galactic Center Newsletter* and—in particular—the Local Organizing Committee.

Heidelberg, July 1999

Wolfgang J. Duschl and Fulvio Melia (for the Editors)

THE CENTRAL PARSECS
GC98
GALACTIC CENTER WORKSHOP
TUCSON, 7 - 11 SEPTEMBER 1998

Scientific Organizing Committee: Angela Cotera*, Wolfgang J. Duschl*, Heino Falcke*, Reinhard Genzel, W. Miller Goss, Fulvio Melia*, Peter G. Mezger, Mark Morris, Marcia Rieke, Chris Sellgren, Yoshiaki Sofue and Farhad Yusef-Zadeh.
(*co-chairs)

Local Organizing Committee: Angela Cotera (chair), Robert Coker, Sera Markoff, Fulvio Melia and Susan Stolovy.

Participant List

Anantharamaiah, Kuduvalli, National Radio Astronomy Observatory, P.O. Box O, Socorro, NM 87801, USA ⟨ anantha@aoc.nrao.edu ⟩

Antonucci, Robert, UCSB Physics Dept, Santa Barbara CA 93106, USA ⟨ antonucci@physics.ucsb.edu ⟩

Auer, Richard, Institut für Theoretische Astrophysik, Tiergartenstr. 15, 69121 Heidelberg, Germany ⟨ raa@platon.ita.uni-heidelberg.de ⟩

Backer, Donald, Astronomy Department, 601 Campbell Hall, University of California, Berkeley, CA 94720, USA ⟨ dbacker@astro.berkeley.edu ⟩

Beckert, Thomas, Inst. f. Theo. Astrophysik, Tiergartenstr. 15, D-69121 Heidelberg, Germany ⟨ tbec@ita.uni-heidelberg.de ⟩

Becklin, Eric, UCLA - Division of Astronomy, 405 Hilgard Ave, Los Angeles, CA 90095-1562, USA ⟨ becklin@astro.ucla.edu ⟩

Biermann, Peter L., Max-Planck-Institut für Radioastronomie, Auf dem Hügel 69, D-53121 Bonn, Germany ⟨ plbiermann@mpifr-bonn.mpg.de ⟩

Blum, Bob, c/o CTIO/AURA, casilla 603, La Serena, Chile ⟨ rblum@noao.edu ⟩

Bower, Geoffrey, Max Planck Institut für Radioastronomie, Auf dem Hügel 69, D 53121 Bonn, Germany ⟨ gbower@nrao.edu ⟩

Coil, Alison, 2619 B Etna St., Berkeley, CA 94704, USA ⟨ acoil@cfa.harvard.edu ⟩

Coker, Robert, Physics Dept, PAS #81, 1118 E. 4th St, University of Arizona, Tucson, AZ 85721, USA ⟨ rfc@physics.arizona.edu ⟩

Cotera, Angela, Steward Observatory, University of Arizona, Tucson, AZ 85721, USA ⟨ cotera@as.arizona.edu ⟩

Doeleman, Sheperd, Haystack Observatory, Off Route 40, Westford, Mass 0186, USA ⟨ dole@newton.haystack.edu ⟩

Donea, Alina-Catalina, Str. Cutitul de Argint 5, Bucharest, RO-75212, Romania ⟨ adonea@roastro.astro.ro ⟩

Duschl, Wolfgang J., Institut für Theoretische Astrophysik, Tiergartenstr. 15, D-69121 Heidelberg, Germany ⟨ wjd@ita.uni-heidelberg.de ⟩

Eckart, Andreas, MPE, Postfach 160, D 85740 Garching, Germany ⟨ eckart@mpe.mpg.de ⟩

Falcke, Heino, Max-Planck-Institut für Radioastronomie, Auf dem Hügel 69, D-53121 Bonn, Germany ⟨ hfalcke@mpifr-bonn.mpg.de ⟩

Fatuzzo, Marco, Department of Physics, 1118 4th Street, University of Arizona, Tucson, AZ 85721, USA ⟨ fatuzzo@newton.physics.arizona.edu ⟩

Figer, Donald, Division of Astronomy, UCLA, MSA9371, Los Angeles, CA 90095-1562, USA ⟨ figer@astro.ucla.edu ⟩

Genzel, Reinhard, Max-Planck-Institut für extraterrestrische Physik, Postfach 1603, 85740 Garching, Germany ⟨ genzel@mpe-garching.mpg.de ⟩

Gezari, Dan, Dan Gezari, NASA/Goddard, Code 685, Greenbelt, MD 20771, USA ⟨gezari@stars.gsfc.nasa.gov⟩

Ghez, Andrea, Dept. of Physics and Astronomy, Box 951562, UCLA, Los Angeles, CA 90095-1562, USA ⟨ghez@astro.ucla.edu⟩

Goss, Miller, National Radio Astronomy Observatory, P.O. Box O, Socorro, New Mexico 87801, USA ⟨mgoss@nrao.edu⟩

Jauch, Zeno, Institut für Theoretische Astrophysik, Tiergartenstr. 15, 69121 Heidelberg, Germany ⟨zeno@ita.uni-heidelberg.de⟩

Kassim, Namir, Code 7213, Naval Research Laboratory, Washington, DC 20375-5351, USA ⟨nkassim@shimmer.nrl.navy.mil⟩

Krichbaum, Thomas P., Max Planck Institut für Radioastronomie, Auf dem Hügel 69, D-53121 Bonn, Germany ⟨tkrichbaum@mpifr-bonn.mpg.de⟩

Lang, Cornelia, National Radio Astronomy Observatory, Box 0, Socorro, NM 87801, USA ⟨clang@astro.ucla.edu⟩

Lazio, Joseph, NRL, Code 7210, Washington, DC 20375-5351, USA ⟨lazio@rsd.nrl.navy.mil⟩

Maeda, Yoshitomo, Department of Astronomy and Astrophysics, 525 Davey Laboratory, Pennsylvania State University, University Park, PA 16802-6305, U.S.A. ⟨maeda@astro.psu.edu⟩

Mahadevan, Rohan, Institute of Astronomy,, University of Cambridge,, Madingley Road,, Cambridge, CB3 0HA, England ⟨rohan@ast.cam.ac.uk⟩

Markoff, Sera, Physics Department, PAS #81, University of Arizona, Tucson, AZ 85721, USA ⟨sera@physics.arizona.edu⟩

Melia, Fulvio, Steward Observatory, The University of Arizona, Tucson, AZ 85721, USA ⟨melia@physics.arizona.edu⟩

Mezger, Peter G., Max Planck Institut für Radioastronomie, Auf dem Hügel 69, 53121 Bonn, Germany ⟨pmezger@mpifr-bonn.mpg.de⟩

Najarro de la Parra, Francisco, Instituto de Estructura de la Materia, C/ Serrano 121, 28006 Madrid, Spain ⟨najarro@astro.iem.csic.es⟩

Narayan, Ramesh, Harvard-Smithsonian Center for Astrophysics, 60 Garden Street MS 51, Cambridge, MA 02138, USA ⟨rnarayan@cfa.harvard.edu⟩

Novak, Giles, Department of Physics and Astronomy, Northwestern University, 2145 Sheridan Rd., Evanston, IL 60208-3112, USA ⟨g-novak@nwu.edu⟩

Ott, Thomas, MPE, Postfach 1603, 85740 Garching, Germany ⟨ott@mpe.mpg.de⟩

Pankonin, Vernon, National Science Foundation, Suite 1045, 4201 Wilson Blvd, Arlington, VA 22230, USA ⟨vpankoni@nsf.gov⟩

Ramirez, Solange, Astronomy Department, 174 West 18th Avenue, Columbus, OH 43210, USA ⟨solange@astronomy.ohio-state.edu⟩

Reid, Mark, Harvard-Smithsonian CfA MS42, 60 Garden Street, Cambridge, MA 02138, USA ⟨reid@cfa.harvard.edu⟩

Rieke, George, Steward Observatory, University of Arizona, Tucson, AZ 85721, USA ⟨grieke@as.arizona.edu⟩

Rieke, Marcia, Steward Observatory, University of Arizona, Tucson, AZ 85721, USA ⟨mrieke@as.arizona.edu⟩

Roberts, Doug, Astronomy Department, 1002 West Green St., Urbana, IL 61801, USA ⟨dougr@ncsa.uiuc.edu⟩

Sakano, Masaaki, Department of Physics, Faculty of Science, Kyoto University, Sakyo, Kyoto 606-8502, Japan ⟨sakano@cr.scphys.kyoto-u.ac.jp⟩

Sanders, Robert H., Kapteyn Astronomical Institute, Postbus 800, 9700 AV Groningen, The Netherlands ⟨sanders@astro.rug.nl⟩

Sellgren, Kris, Astronomy Department, Ohio State University, 174 West 18th Av., Columbus, OH 43210, USA ⟨sellgren@astronomy.ohio-state.edu⟩

Serabyn, Gene, JPL 171-113, 4800 Oak Grove Dr., Pasadena, CA 91109, USA ⟨serabyn@tacos.caltech.edu⟩

Simpson, Janet, NASA/Ames Research Center, MS 245-6, Moffett Field, CA 94035-1000, USA ⟨simpson@cygnus.arc.nasa.gov⟩

Sjouwerman, Lorant, JIVE, Radiosterrenwacht Dwingeloo, Postbus 2, 7990 AA Dwingeloo, The Netherlands ⟨sjouwerm@jive.nfra.nl⟩

Sofue, Yoshiaki, Institute of Astronomy, University of Tokyo, Mitaka, Tokyo 181, Japan ⟨sofue@mtk.ioa.s.u-tokyo.ac.jp⟩

Stolovy, Susan, Steward Observatory, University of Arizona, Tucson, AZ 85721, USA ⟨sstolovy@as.arizona.edu⟩

Tanner, Angelle, 8731 Math Sciences, UCLA, Westwood, CA 90095, USA ⟨tanner@astro.ucla.edu⟩

Townes, Charles, University of California, Department of Physics, 366 Le Conte Hall #7300, Berkeley, CA 94720-7300, USA ⟨cht@ssl.berkeley.edu⟩

Tsuboi, Masato, Inst. of Astroph. and Planetary Science, Ibaraki University, 2-1 Bunkyo, Mito, 310-8512, Japan ⟨tsuboi@masato.sci.ibaraki.ac.jp⟩

Vollmer, Bernd, Institut für Theoretische Astrophysik, Tiergartenstr. 15, D-69121 Heidelberg, Germany ⟨vollmer@DAEC.obspm.fr⟩

Wardle, Mark, School of Physics, A28, University of Sydney, Sydney, NSW 2006, Australia ⟨wardle@physics.usyd.edu.au⟩

Winnberg, Anders, NRAO, P.O. Box 0, Socorro, NM 87801-0387, USA ⟨anders@oso.chalmers.se⟩

Yusef-Zadeh, Farhad, Northwestern University, Dept. Physics and Astronomy, 2145 Sheridan Road, Evanston, IL 60208, USA ⟨zadeh@nwu.edu⟩

Zensus, Anton, Max Planck Institut für Radioastronomie, Auf dem Hügel 69, 53121 Bonn, Germany ⟨azensus@mpifr-bonn.mpg.de⟩

Zhao, Jun-Hui, Harvard-Smithsonian Center for Astrophysics, 60 Garden St, MS 78, Cambridge, MA 02138, USA ⟨jzhao@cfa.harvard.edu⟩

Zylka, Robert, Max Planck Institut für Radioastronomie, Auf dem Hügel 69, 53121 Bonn, Germany ⟨rzylka@mpifr-bonn.mpg.de⟩

Photo: Alaina Levine, University of Arizona

Color Plates

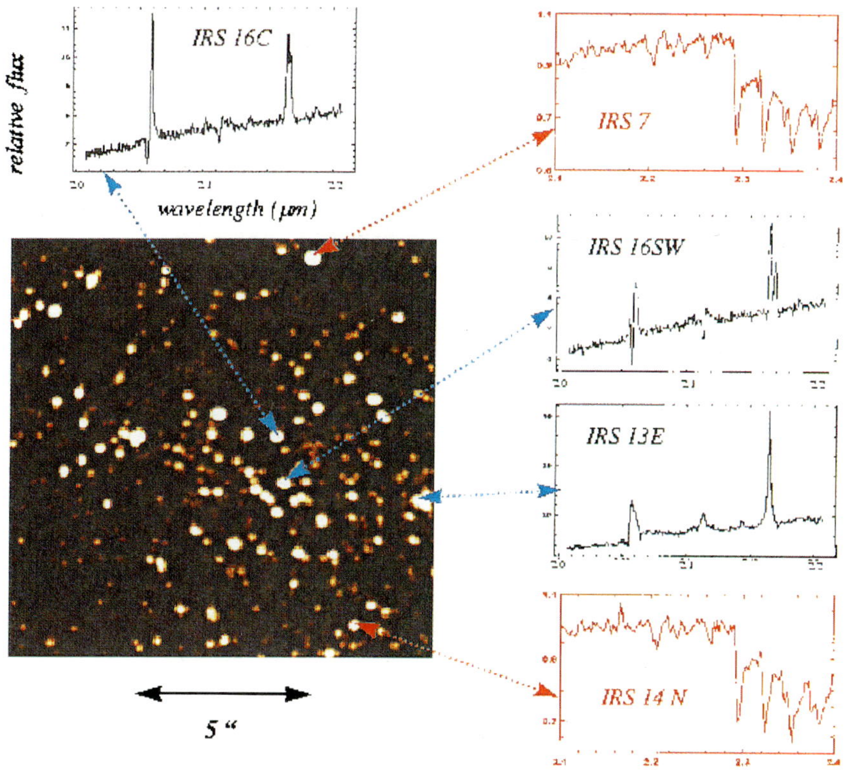

Plate I: False color near-infrared (2μm) image of the central 10" (0.39 pc) of the Galaxy, obtained with the MPE SHARP camera in 1994 on the ESO NTT (Eckart et al. 1995), along with a selection of stellar spectra obtained with the MPE 3D spectrometer (Genzel et al. 1996 and unpublished). The brightest stars are overexposed and are about 200 times brighter than the faintest stars. Stars with emission lines (from HeI and HI) are luminous and massive, blue supergiants. Stars with absorption bands (due to CO, NaI and CaI) are cool, red giants and supergiants. From these spectra radial velocities are obtained for individual stars. *(Figure 1 from Genzel & Eckart, this volume, p. 3)*

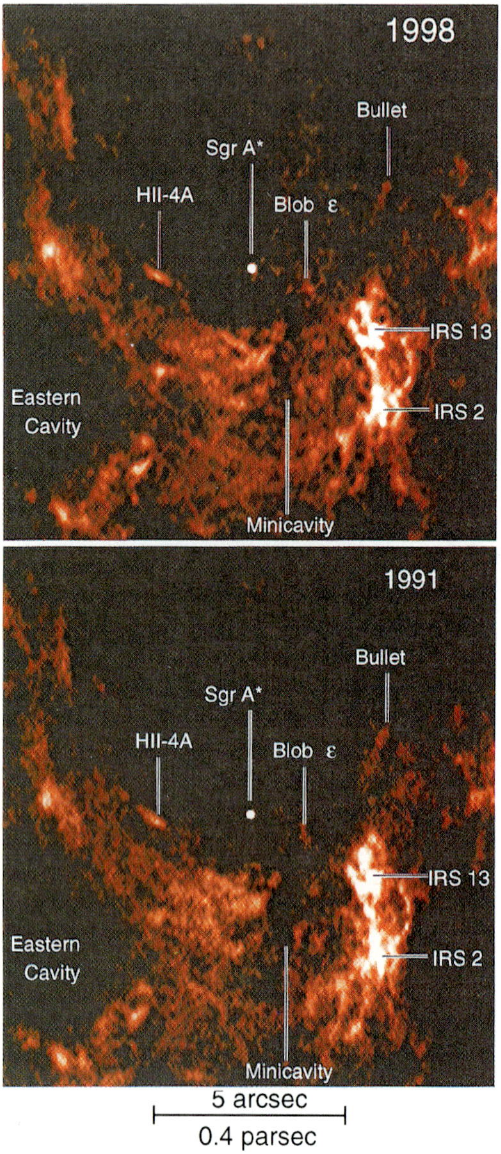

Plate II: Radio images of the ionized gas in the vicinity of Sgr A* observed with the VLA in its A and B configurations at 13 mm in 1998 (top) and 1991 (bottom). The FWHM beam is 0.10″. The images were made using maximum entropy method (MEM). The flux density of Sgr A* varies. The flux densities of Sgr A* on the observing dates of 1991JUN22, 1992JAN19, 98MAY08 and 1998SEP08 were 0.97±0.03, 1.08±0.03, 1.06±0.03, and 1.37±0.05 Jy, respectively. The peak intensity in both images was unified. *(Figure 1 from Zhao & Goss, this volume, p. 224)*

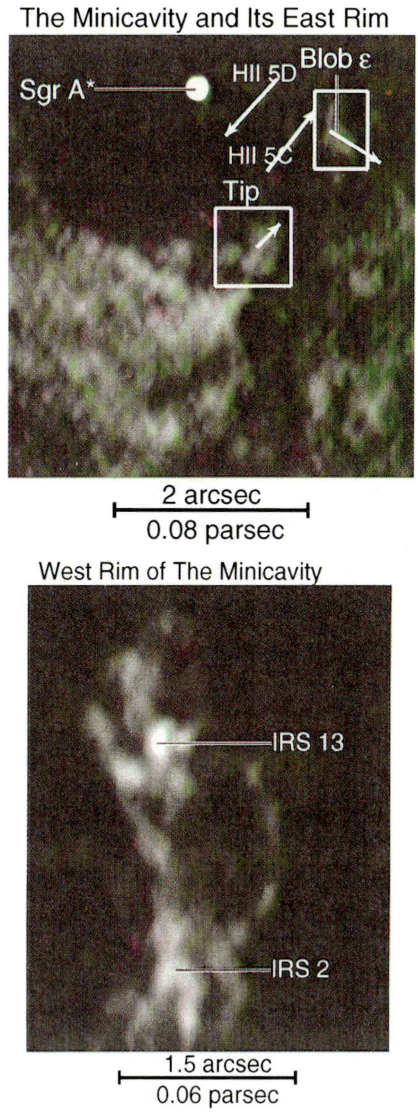

Plate III: Images (at 13 mm) at the two epochs are overlaid to illustrate proper motions of the ionized components. Purple is 1991 and green 1998. Top panel (Fig. 2a) is for the region east of the Minicavity including Blob ϵ and the Tip, and bottom panel (Fig. 2b) is for the region west of the Minicavity, IRS 13 and IRS 2. The FWHM beam is $0.10''$. The proper motion vectors in the top panel are on a relative scale with the value for HII 5c = 1160 km s^{-1}. *(Figure 2 from Zhao & Goss, this volume, p. 224)*

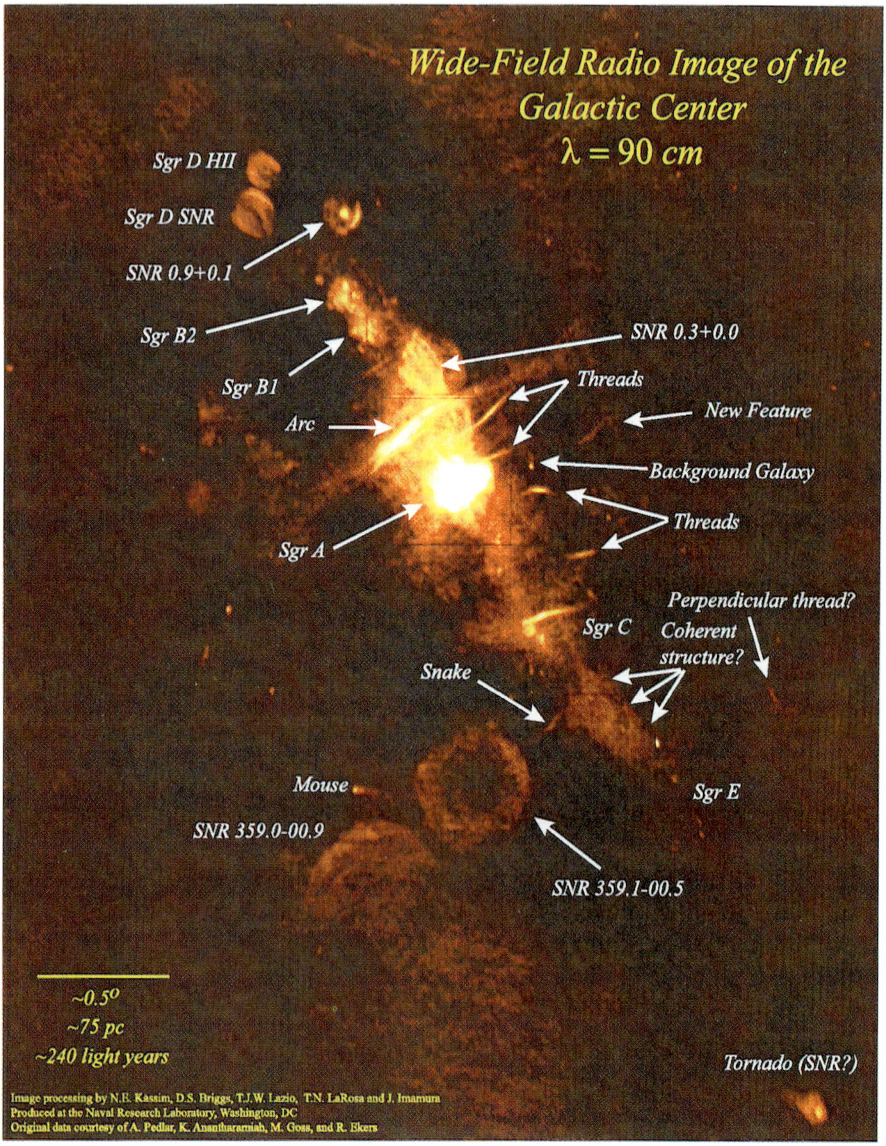

Plate IV: Full 330 MHz wide-field image of the Galactic Center at an angular resolution of approximately 45″. The original data were obtained by Pedlar et al. (1989) and Anantharamaiah et al. (1991) using the VLA in the B-, C-, and D configurations. The rms noise in the image is approximately 5 mJy beam^{-1}. The image has not been corrected for primary beam attenuation. *(Figure 3 from Kassim et al., this volume, p. 403)*

Session I

What is the Central Matter Distribution?

The Central Parsecs of the Galaxy
ASP Conference Series, Vol. 186, 1999
H. Falcke, A. Cotera, W.J. Duschl, F. Melia, M.J. Rieke, eds.

The Galactic Center Black Hole

Reinhard Genzel and Andreas Eckart

Max-Planck Institut für extraterrestrische Physik, Garching, FRG

Abstract. Over 6 years of high spatial resolution, near-infrared imaging and spectroscopy have made it possible to measure stellar velocities down to separations of less than five light days from the compact radio source Sgr A* at the dynamic center of the Milky Way. These confirmed measurements imply the presence of a central dark mass of 2.6×10^6 solar masses. Simple physical considerations show that this dark mass cannot consist of a stable cluster of stars, stellar remnants, substellar condensations or a degenerate gas of elementary particles. Momentum and energy balance arguments suggest that at least 10^3 to 10^5 solar masses must be associated with Sgr A* itself and likely is enclosed within less than 8 light minutes (equivalent to 15 Schwarzschild radii of a million solar mass black hole). Therefore the dark mass at the center of the Milky Way is currently the most compelling case for a massive black hole.

1. Introduction

Massive black holes are used to explain a large number of observational facts linked to 'active galactic nuclei'(AGNs, e.g. quasars) and their evolution in the early Universe. Amongst these are their high luminosity (up 10^{14} $L_\odot$) produced within their central light year, highly collimated relativistic jets, and rapidly varying X- and γ-ray emission. For a direct proof of the 'black hole'paradigm it is necessary, however, to determine the characteristic mass concentration and show the existence of an event horizon.

An unambiguous method for carrying out such a proof is the determination of the form of the gravitational potential from the velocity field of stars and gas orbiting the black hole candidate. Ground-based and Hubble Space Telescope observations of the Doppler shifts of spectral lines from gas and stars have indeed shown that many (and perhaps most) nearby galaxies have massive dark mass concentrations in their nuclei (e.g. Kormendy and Richstone 1995). With the exception of detailed radio Very-Long-Baseline-Interferometry (VLBI) observations of the galaxy NGC 4258 (Myoshi et al. 1995), none of these measurements have high enough resolution yet to prove that the central dark mass must be a black hole and could not be, for instance, a dense compact cluster of stellar remnants. In contrast the nucleus of the Milky Way at a distance of ~8 kpc is a thousand times closer than the nearest AGN and one hundred thousand times closer than the nearest quasar. It is thus a unique laboratory to test the black hole paradigm. The Galactic Center, however, is heavily extincted by dust (A_V~30) so that observations in the visible are not practical. With the advent of sensitive infrared detectors, high resolution images and imaging spectrometers it has recently become possible to study the stellar velocity field at unprecedented resolution and provide the best evidence yet for a massive black hole at the nucleus of a galaxy.

2. Early Evidence for a Compact Mass

The first indications for a central mass concentration in the Milky Way emerged in the late seventies from spectroscopic observations of a mid- infrared fine structure line of Ne^+ (Wollman et al. 1977, Lacy et al. 1979). These measurements showed unusually large Doppler shifts (±250 km/s) of ionized gas clouds in the central parsec, toward the maximum stellar density. As radio interferometric observations had discovered a compact, non-thermal radio source, Sgr A*, in the same region (Balick and Brown 1974, most recent measurements by Bower & Backer 1998 and Krichbaum et al. 1998), a plausible interpretation – in analogy to quasars – was that the large gas velocities indicate orbital motions in the vicinity of a million solar mass black hole, coincident with Sgr A* (Lynden-Bell and Rees 1971, Lacy, Hollenbach and Townes 1982). Further infrared (and radio) spectroscopic data taken by various groups in the eighties strengthened the gas dynamic evidence for this central mass concentration (e.g. Serabyn and Lacy 1985, Genzel and Townes 1987). These measurements were, however, not considered compelling by many researchers in the field. In addition to gravitational forces, gas may be affected by magnetic fields, radiation pressure, stellar winds and friction with other gas components. All these phenomena are known to be present in the Galactic Center thus making the interpretation uncertain. Beginning in the late eighties several groups began measuring radial velocities of late type, red giant and supergiants (e.g. Rieke and Rieke 1988, Sellgren et al. 1990, Krabbe et al. 1995, Haller et al. 1996). Some spectra are shown in Figure 1. These measurements confirm and strengthen the evidence for the presence of a 1 to 3×10^6 $M_\odot$ central mass that cannot be accounted for solely by the stellar cluster that is sampled by the near-infrared light.

3. Status in 1998: 6 Years of Proper Motion Observations

Two important developments have made it possible in the recent years to measure the stellar velocity field down to scales as small as 5 light days and thus, to be able to place decisive constraints on the nature of the central mass concentration. On the one hand this significant progress was possible due to the discovery (by Forrest et al. 1987, Allen et al. 1990 and Krabbe et al. 1991) of a compact cluster of hot, luminous emission line stars (the so called 'HeI stars'). that must have formed in the last few million years and now power the central parsec. They provide radial velocity measurements to a scale of $\sim 1''$ (0.04pc). Using the integral field imaging spectrometer, 3D, Genzel et al. (1996) have been able to determine radial velocities for about 220 of these stars (see also Krabbe et al. 1995, Haller et al. 1996).

On the other hand – as it turns out – the most important development has been the first measurement of stellar proper motions. Figure 2 shows a 0.15" resolution, 2μm image obtained with the MPE 'SHARP'camera on the 3.5m New

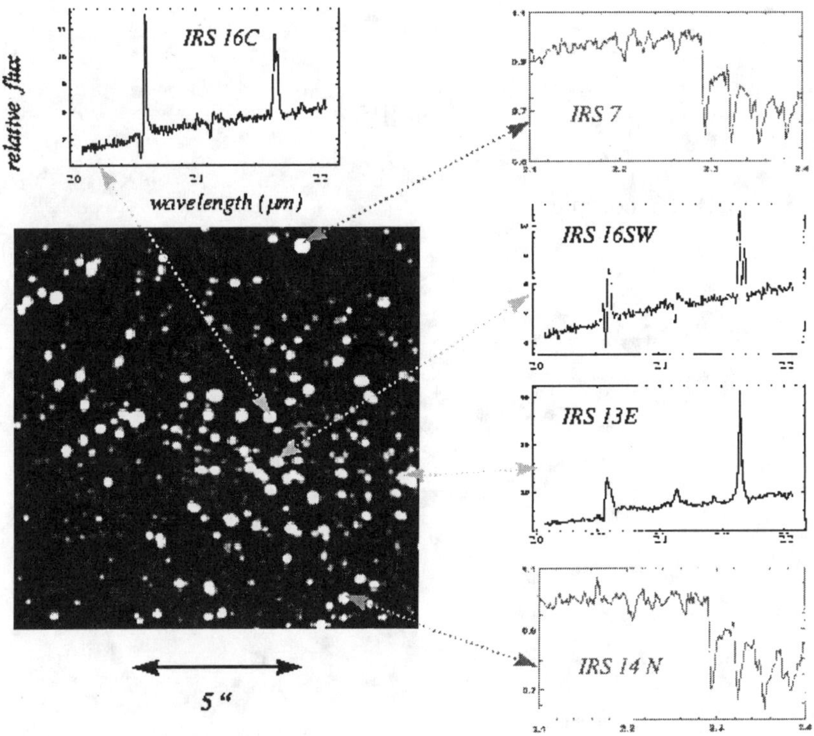

Figure 1. *(See Plate I for color version)* False color near-infrared (2μm) image of the central 10" (0.39 pc) of the Galaxy, obtained with the MPE SHARP camera in 1994 on the ESO NTT (Eckart et al. 1995), along with a selection of stellar spectra obtained with the MPE 3D spectrometer (Genzel et al. 1996 and unpublished). The brightest stars are overexposed and are about 200 times brighter than the faintest stars. Stars with emission lines (from HeI and HI) are luminous and massive, blue supergiants. Stars with absorption bands (due to CO, NaI and CaI) are cool, red giants and supergiants. From these spectra radial velocities are obtained for individual stars.

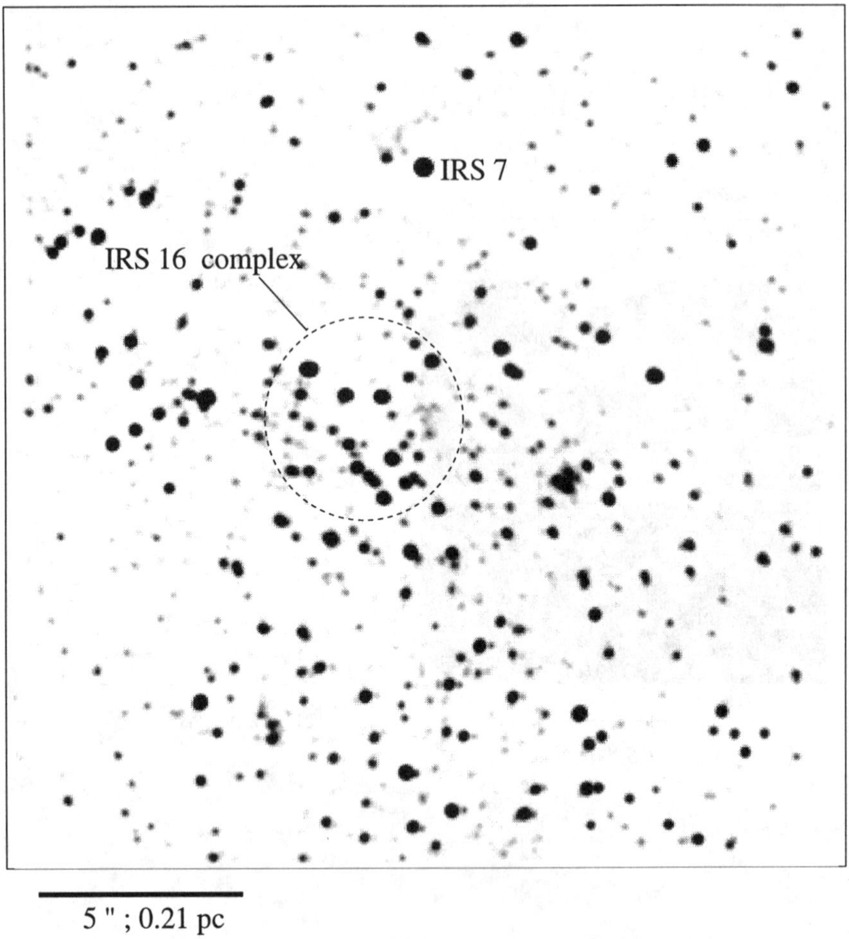

Figure 2. The central stellar cluster at a wavelength of 2.2 μm the near-infrared K-band. This is a mosaic of speckle image reconstructions from data taken with the SHARP speckle camera on the 3.5m New Technology Telescope (NTT) of the European Southern Observatory (ESO). The angular resolution is 0.15" FWHM. This image represents a dynamic range of almost a factor of 10^4: The strongest source (IRS 7) has a K-magnitude of 6.7 (1320 mJy) while the faintest sources visible in this image have K-magnitudes of about 16 (0.25 mJy). The location of Sgr A* close to the center of the image can be obtained via comparison to the next figure.

Technology Telescope (NTT) of the European Southern Observatory (ESO). The excellent resolution, image quality and high dynamic range of this images are the result of combining the shift-and-add speckle imaging technique with non-linear CLEANing techniques that allow to correct for the non-diffraction limited parts of the point spread function in these images. The resulting near-infrared maps reveal close to 10^3 stars in the central parsec, concentrated and centered on or very near the compact radio source Sgr A*. Combining about 70 independent high resolution images between 1992 and 1998, Eckart and Genzel (1996, 1997 and unpublished) and Genzel et al. (1997) have derived (relative) proper motions for about 70 stars. Figure 3 shows examples of the data obtained. The derived proper motion vectors (without error bars) are plotted for a number of stars on the high resolution image, assuming a Sun-Galactic Center distance of 8 kpc.

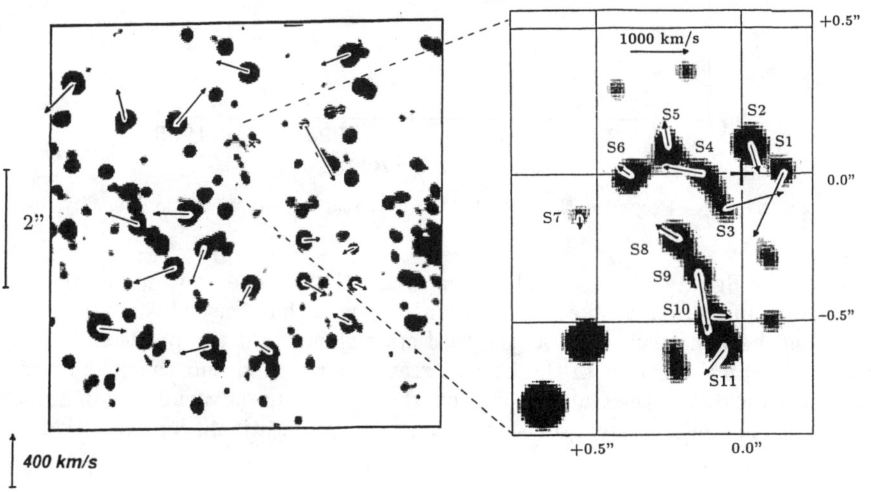

Figure 3. Proper motion vectors for in the central few arcseconds (left) including the vectors for the central sources close to Sgr A* (at the position of the cross) in April 1994 (right).

Of special interest is the immediate vicinity of Sgr A* (right side of Figure 3) where one finds a $\sim 1''$ diameter concentration of faint stars. Several of these stars in this so-called Sgr A* cluster show proper motions in excess of 1000 km/s, the fastest one (S1: v$\sim$1400 km/s; see Figure 4) is currently also one of the stars closest to Sgr A* ($\sim$0.13").

Stars further out have velocities of only a few hundred km/s. This finding is exactly what one would expect if Sgr A* were coincident with a large compact mass. Because of the obvious importance of these large motions and the substantial technical difficulties in deriving believable proper motions of faint stars in such a crowded environment, a confirmation of these results was critical. This has now happened. A group from the University of California, Los Angeles has used the 10m Keck telescope on Mauna Kea, Hawaii, and carried out yet higher resolution 2μm imagery of the central few arcseconds. Combining data from 4 epochs between 1995 and 1998 this group fully confirms the very high veloci-

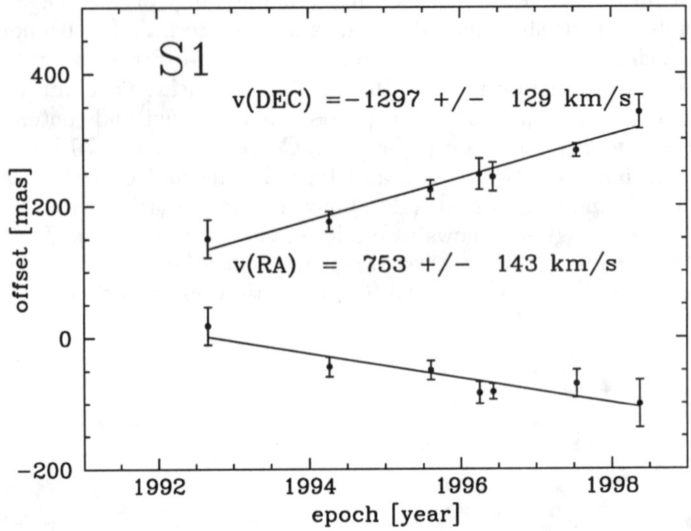

Figure 4. Proper motions of S1 showing the largest proper motion velocity.

ties of the Sgr A* cluster stars (Ghez et al. 1998). The continuous motion of the stars measured by both groups and the fact that they cluster around Sgr A*, and have similar colors as the He I stars shows that the positional changes present orbital motions in the central gravitational field and that the stars are actually located in the Galactic Center (see discussions on variability in a double or multiple stellar system, or due to a central gravitational lens in Eckart and Genzel 1996, 1997).

4. Derived Mass Distribution

A first analysis of the radial and proper motions shows that the projected velocity dispersions of a number of stars in a given annulus of projected radius p increase with $p^{-1/2}$ between p$\sim$1 pc and p$\sim$0.01 pc, as expected in the potential of a central point mass (a 'Kepler law'). The location of the largest stellar velocities (the dynamic center), the stellar density maximum, and the position of Sgr A* (now determined relative to the stars to 30 milli-arcseconds, Menten et al. 1997) all agree to within ±0.004pc (0.1", Ghez et al. 1998). Between $5'' \geq p \geq 1''$ – where both radial and proper motions, sometimes from the same stars, are available – the mean velocities in all three directions agree within the error bars. This means that anisotropy of the stellar orbits –caused, for instance by predominantly very elliptical orbits– does not play a significant role in the Galactic Center.

The final distribution of the enclosed mass as a function of true separation from Sgr A* is shown in Figure 5 and is the result of applying the 'Jeans'equation as well as projected mass estimators to all available stellar radial and proper motion data (Eckart and Genzel 1997, Genzel et al. 1997). Figure 5 shows

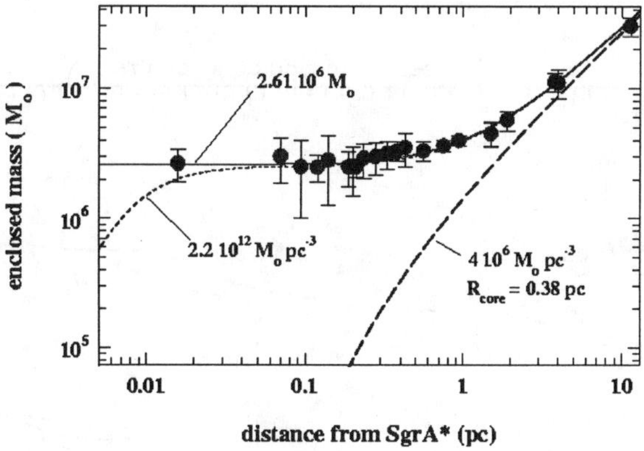

Figure 5. Mass modelling of the stellar proper and radial motions.

the modelling results with the addition of two points from gas kinematics at R=1.5 and 4 pc. Shown as filled circles with 1σ error bars are the various mass estimates listed in Table 2 of Genzel et al. (1997) and discussed in the text, assuming a Sun-Galactic Center distance of 8 kpc. The thick dashed curve represents the mass model for the (visible) nearly isothermal stellar cluster cluster (M/L(2μm)=2, R_{core}=0.38 pc, ρ(R=0)=4$\times 10^6$ $M_\odot pc^{-3}$, Genzel et al. 1996). The thin continuous curve is the sum of this stellar cluster, plus a point mass of 2.61 $[\pm 0.15_{stat}, \pm 0.35_{stat+sys}]\times 10^6$ $M_\odot$. The thin dotted curve is the sum of the visible stellar cluster, plus a α=5 Plummer model of a dark cluster of central density 2.2$\times 10^{12}$ $M_\odot pc^{-3}$ and R_o=0.0065 pc. It provides a χ^2 fit 1σ worse than the best fit with central density >7.5$\times 10^{13}$ $M_\odot pc^{-3}$. This density is about 500,000 times larger than that of the visible cluster. The central mass is 'dark', as it exhibits a mass to luminosity ratio of 100 or greater.

5. Spectra of the Stars Close to Sgr A*

Since the field around Sgr A* is very crowded most of the spectra are contaminated by other nearby stars. The central Sgr A* cluster, however, is located west of the IRS16 complex in a void of bright stars. The separation to any bright star is larger than 1″ and the contaminating effects of neighboring sources is comparatively small. About half of the Sgr A* cluster sources are distributed in north-south direction. This coincidence can be exploited to extract low resolution spectra of individual stars in this region.

In Figure 6 we show the spectra obtained for the sources S1, S2, S8 and S11. The resulting spectra are fairly flat and increase slowly in flux toward longer wavelength. CO band-head absorption is absent so S1, S2, S8 and S11 cannot be late type giant stars. While somewhat redder than the most prominent He I emission line stars in the IRS 16 complex their spectral characteristics are consistent with early type stars located in the central cluster (i.e. $A_K \sim 3$). There is also no indication for very strong line emission (or absorption) at the wavelength

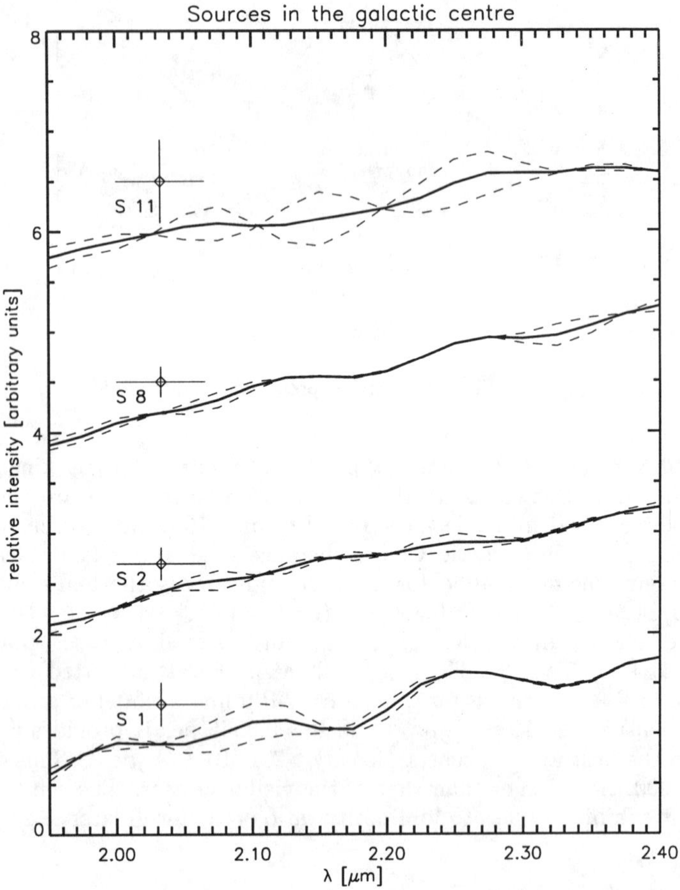

Figure 6. Extracted and calibrated speckle spectra of the sources S1, S2, S8 and S11 obtained from our speckle spectroscopy data as described in the text. The thick black lines represent the mean of two independent data sets show by the dashed lines. The spectral resolution of $\lambda/\Delta\lambda$=35 as well as the 1σ error are indicated by the cross.

of the He I and/or Brγ emission lines. This is not surprising, however, as the narrow He I/Brγ emission lines cannot be seen even in the classical He I emission line stars at the low spectral resolution. These results are fully consistent with the broad band JHK colors and narrow band filter data presented in Eckart et al. (1995). From this data one can conclude that the $m_K \sim 14.5$ sources in the central Sgr A* cluster are most likely moderately luminous (L$\sim$5,000 to 10,000 $L_\odot$) early type stars. If they are on the main sequence they would have to be O9 – B0.5 stars with masses of 15 to 20 $M_\odot$.

6. Possible Identification of Sgr A* in the Infrared

A comparison of all available data taken under the best seeing conditions allowed us to produce very deep K-band images of the central $3'' \times 3''$ of the Galactic Center at a very high angular resolution (Eckart et al. 1997). This is especially true for the April, June 1996 and in July 1997 observations that were carried out with a 50 mas and 25 mas pixel scale. The high signal to noise in the combined data sets resulted in reproducible images below the diffraction limit of the NTT at an angular resolution of 70 mas to 100 mas (FWHM). A comparison of the images reveals the existence of a $m_K \sim 15$ source (S12) between S1, S2 and S3 in the June 1996 and July 1997 epochs. This source was not detected in August 1992 and April 1994 nor in May 1998. For these epochs any source at that position would have been fainter than $m_K \sim 16.3$ (see Figure 7). Most importantly the location of S12 is coincident with the compact radio source Sgr A*, to within the ± 30 mas uncertainties of the radio-infrared reference frames (Menten et al. 1997). Despite this fact the proper motions of S12 are small when compared with the motions of the surrounding sources in the Sgr A* cluster (see Figure 8). In addition S12 appears to be a time variable source. The difference of 1.3 magnitudes corresponds to a minimum flux density variability of a factor of 3 to 4 (see Figure 9). The m_K=15.0 brightness of the newly found source translates into a dereddened flux density of 13 mJy (assuming $A_V = 30^m$ and $A_K = 0.11 \times A_V$). The upper limit to the flux density in April 1996 then corresponds to about 4 mJy. (Here it should be noted that S3 with an angular separation from Sgr A* of about $0.1''$ shows flux density variability as well. As indicated in the SHARP data the flux density of S3 decreased by almost one magnitude between 1994 and 1998.) There are at least two plausible possible interpretations of the data that could explain the nature of the newly found source S12:

- S12 is the infrared counterpart of the compact radio source SgrA*. Its variability could be explained by the theoretical models of black hole accretion (Melia 1992, Narayan, Yi and Mahadevan 1995, Narayan 1996, Coker and Melia, 1997).
- Alternatively the variability could be due to gravitational lensing. In this case the source would be a more distant central cluster member not intimately associated with Sgr A* itself. At a distance to the Galactic Center of 8 kpc and an enclosed mass of 2.6×10^6 $M_\odot$ the size of the Einstein ring within which significant flux density changes and apparent motions due to lensing occur is then of the order of only 10 mas. Depending on the

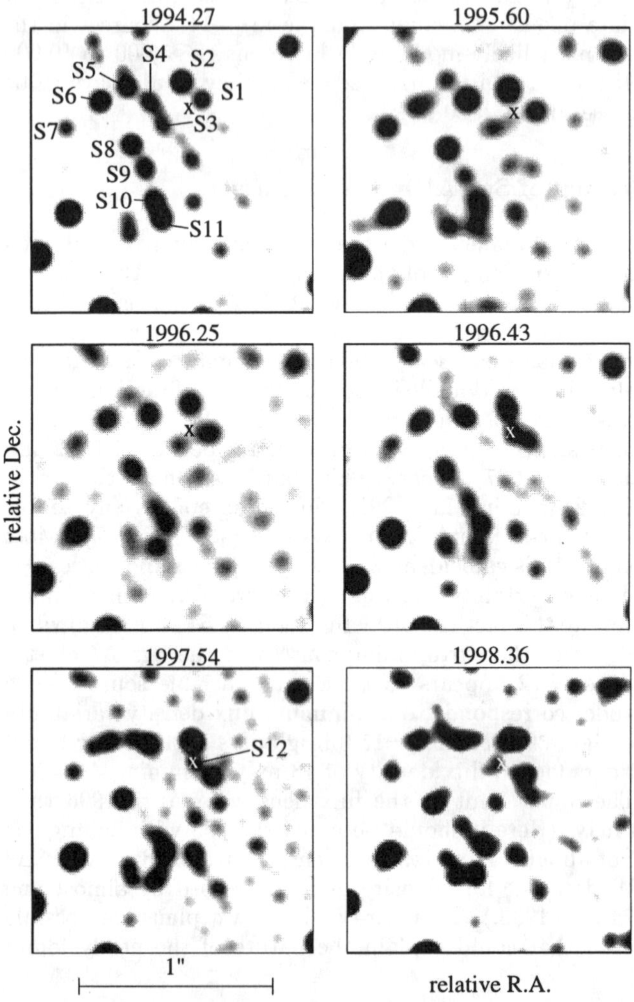

Figure 7. Grey scale representations of deep, super-resolution K-band maps (70 to 100 mas FWHM) of the central 3″. Each map contains up to about 5 hours of integration time with pixel scales of 25 and 50 milli-arcseconds. The 3σ noise floor is approximately at m_K 17. Sources S1 to S11 are indicated in the 1994.27 map. All maps except the 1994 one show evidence of a source (S12 as labeled in the 1997.54 map) right at the position of the compact radio source Sgr A* as marked by the "x".

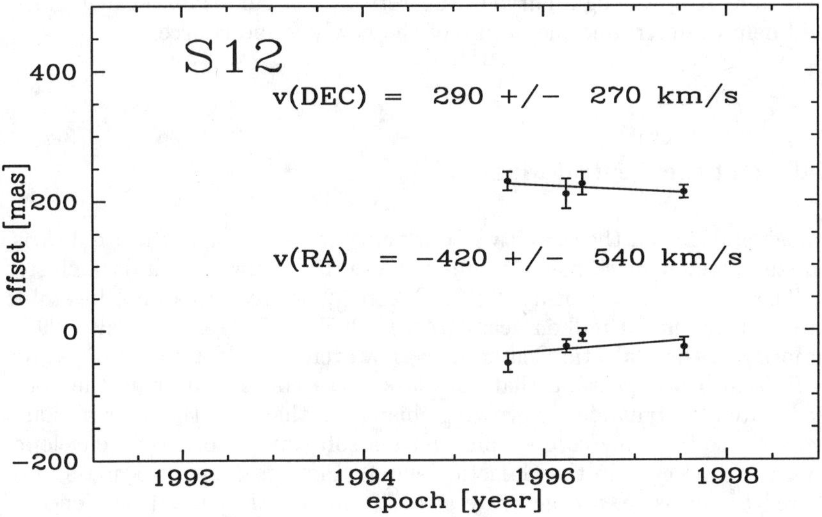

Figure 8. Proper motions of the possible Sgr A* counterpart S12.

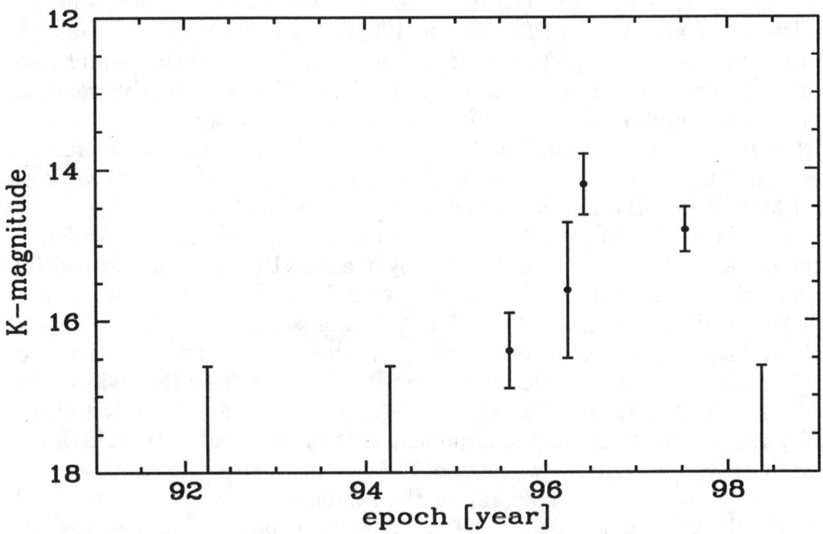

Figure 9. Variability of the K-band magnitude of the possible Sgr A* counter part S12.

transverse velocity of the source the duration of the corresponding lense effect will only be of the order of one year. Monitoring the flux density and comparing the light curve to theoretical light curves for lensing effects will help to determine the nature of the newly found source.

7. Nature of the Central Mass

Basic considerations on the stability of dark clusters composed of white dwarfs, neutron stars, stellar black holes or sub-stellar entities show that a dark cluster of mass 2.6×10^6 $M_\odot$ and density 2×10^{12} $M_\odot$ pc^{-3} or greater can not be stable for more than about 10 million years (Maoz 1995, 1997, Genzel et al. 1997). The majority of the Galactic Center stars, however, are older than 10^8 or 10^9 years. It is also not possible that the dark mass concentration is the core-collapsed state of a dynamically evolving cluster. In that case the distribution – while very dense in a tiny core – would have a soft, quasi-isothermal envelope, unlike what is observed in the Galactic Center. Finally, if the dark mass were conjectured to consist of a degenerate gas of fermions, the m^{-2} dependence of the Chandrasekhar mass on the mass m of the constituent particles requires that the mass of the fermions cannot be much larger than the electron mass. The only realistic configuration without net electric charge would then be a positron-electron plasma which would, however, rapidly decay through annihilation line radiation. Two further arguments strengthen the conclusion that the dark mass in the Galactic Center must, in fact, be a black hole. The first comes from the fact that Sgr A* itself is known from VLBI measurements to have a proper motion less than about 20 km/s (Backer 1996). In the very dense Galactic Center core the fast moving stars near Sgr A* and Sgr A* itself should have approximately the same kinetic energy. The large (factor 100) difference in observed motions and a balance of momentum and/or energy suggests that at least 10^3 to 10^5 solar masses must be associated with Sgr A* itself. If one further assumes that the mass of Sgr A* is at least as concentrated as its radio emission (radius 1.5×10^{13} cm, Backer 1996), the inferred density of Sgr A* then ranges between $10^{18.5}$ to $10^{20.5}$ $M_\odot$ pc^{-3}. These lower limits are only about 6 orders of magnitude smaller than the equivalent density of a 2.6×10^6 $M_\odot$ black hole within its Schwarzschild radius of $\sim10^{12}$ cm. The second argument is an inversion of the well known dilemma that if Sgr A* is a million solar mass black hole it is presently radiating at a rest mass energy to radiation conversion efficiency of 10^{-5} to 10^{-6}, considering the accretion of stellar wind gas from its environment (Melia 1992). The only possible way out – apart from very large amplitude variability in the accretion – is the argument that in pure radial (Bondi-Hoyle) or in low density, non-radial flows most of the rest mass energy of the accretion flow can be advected into the hole, rather than radiated away (Rees et al. 1982, Melia 1992, Narayan et al. 1995, 1997). This explanation then requires the existence of an event horizon and does not work with any configuration other than a black hole (Narayan et al. 1997). Taking all these arguments together it is hard to escape the conclusion that the core of the Milky Way in fact harbors a massive, but presently inactive central black hole.

8. Prospects

Imaging and spectroscopic studies of the Galactic Center will be continued. Diffraction limited spectroscopy using the upcoming large diameter telescopes such as the Keck telescope (10 m), the ESO Very Large Telescope (8.2 m), and the Large Binocular Telescope (LBT 2×8.4m) in connection with adaptive optics will allow measurement of stellar spectra of even fainter sources in the crowded central regions. These spectra will make it possible to study the distribution of the different stellar populations as a function of distance from the Galactic Center in more detail. It will help to determine the stellar evolution in such a dense stellar cluster harboring a massive black hole.

The telescopes mentioned above as well as interferometers like the planned ESO VLTI (Very Large Telescope Interferometer) will also be ideally suited to obtain more exact measurements of the relative stellar positions and the proper motions (Quirrenbach & Eckart 1995). These observations will allow derivation of proper motions for a larger number of stars and will therefore put the presented analysis on an even firmer statistical footing. In addition it will be possible to trace stars closer to the center approaching the relativistic regime. Also obviously a yet larger baseline in time is needed to obtain information on the exact orbits of the stars closest to SgrA*. The current data implies that the orbital time scales for the innermost stars range from several 10 to a few hundred years. In the next few years it should therefore be possible to measure the curvature of the stellar trajectories. A derivation of orbits will lead to a better understanding of the central cluster dynamics and will help to determine the amount of the enclosed dark mass and its compactness with a higher precession.

However, the stellar orbits in the central nuclear cluster will not be simple Keplerian ellipses. For the Galactic Center the potential is composed of that of an unresolved black hole mass and the potential of the extended stellar cluster. The latter contribution becomes increasingly dominant at separations beyond the stellar cluster core radius of about 0.3 to 0.4 pc. Therefore, relativistic periastron rotation and most of all the fact that the stellar cluster is resolved will give orbits in forms of a rosette. Depending on the eccentricity of the orbits the combination of both effects will result in periastron rotation in the range of $\sim 1''$ to several degrees for a single orbit, and will therefore be the dominant effect. For stars with small semi-major axes with orbits that stay within the central few arcseconds the observed periastron rotation will be a very precise measure of the compactness of the central mass. In addition to the compactness the observed orbits will result in an improved determination of the location of the dynamical center.

The required knowledge of the potential can in principal be obtained from the number density distribution of stars as a function of separation from SgrA*. However, with the current brightness limit at 2 μm of $m_K \sim 16 - 17$ for the groundbased experiments, the number density counts just start to probe the distribution of the major mass carrying population of stars. Here the deep integrations with the Hubble Space Telescope will help to determine the exact cluster morphology and potential.

References

Allen, D.A., Hyland, A.R. and Hillier, D.J. , MNRAS 244, 706 (1990)

Backer, D.C., in Unsolved Problems of the Milky Way, eds. L. Blitz and P. Teuben (Kluwer:Dordrecht), 193 (1996)

Balick, B. and Brown, R.L., Ap.J. 194, 265 (1974)

Bower, G.C., Backer, D.C., Ap.J. 496, L97 (1998)

Coker, R.F., Melia, F., Ap.J. 488, L149 (1997)

Eckart, A., Genzel, R., Hofmann, R., Sams, B. and Tacconi-Garman, L.E., Ap.J. 445, L26 (1995)

Eckart, A. and Genzel, R., NATURE 383, 415 (1996)

Eckart, A. and Genzel, R., MNRAS 284, 576 (1997)

Forrest, W.J., Shure, M.A., Pipher, J.L. and Woodward, C.A.in 'The Galactic Center', ed.D.Backer, AIP Conf. Proc 155, 153 (1987)

Genzel, R. and Townes, C.H. , Ann.Rev.Astr.Ap. 25, 377(1987)

Genzel, R., Thatte, N., Krabbe, A., Eckart, A., Kroker, H. and Tacconi-Garman, L.E. Ap.J.472, 153 (1996)

Genzel, R., Eckart, A., Ott, T. and Eisenhauer, F., MNRAS in press (1997)

Ghez, A., Klein, B., Morris, M. and Becklin, E., Ap.J. in press (1998)

Guesten, R., Genzel, R., Wright, M.C.H., Jaffe, D.T., Stutzki, J. and Harris, A.I., Ap.J. 472, 153 (1987)

Haller, J.W., Rieke, M.J., Rieke, G.H., Tamblyn, P., Close, L. and Melia, F., Ap.J. 456, 194 (1996)

Kormendy, J. and Richstone, D., Ann.Rev.Astr.Ap.33, 581 (1995)

Krabbe, A., Genzel, R., Drapatz, S. and Rotaciuc, V., Ap.J. 382, L19 (1991)

Krabbe, A. Genzel, R., Eckart, A., Najarro, F., Lutz, D. et al., Ap.J.Lett. 447, L95 (1995)

Krichbaum, T.P., et al. A&A 335 L106 (1998)

Lacy, J.H., Baas, F., Townes, C.H. and Geballe, T.R. Ap.J. 227, L17 (1979)

Lacy, J.H., Townes, C.H. and Hollenbach, D.J., Ap.J. 262, 120 (1982)

Lacy, J.H., Achtermann, J.M. and Serabyn, E., Ap.J. 380, L71 (1991)

Lynden-Bell, D. and Rees, M., MNRAS 152, 461 (1971)

Maoz, E. , Ap.J. 447, L91 (1995)

Maoz, E., 1997, IAU Symposium no. 184, The Central Regions of the Galaxy and Galaxies

Melia, F., Ap.J. 387, L25 (1992)

Menten, K.M., Reid, M., Eckart, A. and Genzel, R., Ap.J. 475, L111 (1997)

Myoshi,M, Moran, J.M., Hernstein, J., Greenhill, L., Nakai, N., Diamond, P. and Inoue, M., NATURE 373, 127 (1995)

Narayan, R., Yi, I., and Mahadevan, R., NATURE 374, 623 (1995)

Narayan, R., Mahadevan, R., Grindlay, J., Popham, R.G. and Gammie, C., Ap.J. 492, 554 (1997)

Rees, M., Phinney, E.S., Begelman, M.C. and Blandford, R.D., NATURE 295, 17 (1982)

Rieke, G.H. and Rieke, M.J., Ap.J. 330, L33 (1988)

Sellgren, K., McGinn, M.T., Becklin, E. and Hall, D.N.B., Ap.J. 359, 112 (1990)

Serabyn, E. and Lacy, J. , Ap.J. 293, 445 (1985)

Wollman, E.R., Geballe, T.R., Lacy, J.H., Townes, C.H. and Rank, D.M., Ap.J. 218, L103 (1977)

The Central Parsecs of the Galaxy
ASP Conference Series, Vol. 186, 1999
H. Falcke, A. Cotera, W.J. Duschl, F. Melia, M.J. Rieke, eds.

High Proper Motion Stars in the Vicinity of Sgr A*: Evidence for a Supermassive Black Hole at the Center of Our Galaxy

A. M. Ghez, M. Morris, E. E. Becklin

University of California, Dept. Physics & Astronomy, Los Angeles, CA 90095-1562

Abstract. Over a three year period (1995-8), we have conducted a diffraction-limited imaging study at 2.2 μm of the inner $6'' \times 6''$ of the Galaxy's central stellar cluster using the W. M. Keck 10-m telescope. Stars moving with velocities as large as 1,400 ± 100 km/sec have been detected. The peaks of both the stellar surface density and the velocity dispersion are consistent with the position of the unusual radio source and black hole candidate, Sgr A*, suggesting that Sgr A* is coincident (± $0\farcs1$) with the dynamical center of the Galaxy. As a function of distance from Sgr A*, the velocity dispersion displays a falloff well fit by Keplerian motion ($\sigma_v \sim r^{-0.5\pm0.1}$) about a central dark mass of $2.6(\pm0.2) \times 10^6 M_\odot$ confined to a volume of at most 10^{-6}pc^3, consistent with earlier results. Thus, independent of the presence of Sgr A*, the large inferred central density of at least 10^{12}M$_\odot$/pc^3, which exceeds the volume-averaged mass densities found at the center of any other galaxy, leads us to the conclusion that our Galaxy harbors a massive central black hole.

1. Introduction

More than a quarter century ago Lynden-Bell & Rees (1971), extrapolating from the idea that the highly energetic phenomena observed in very active galaxies are powered by massive central black holes, suggested that much less active galaxies may also harbor massive, though possibly dormant, central black holes. The proximity of the center of our own Milky Way, a mere 8 kpc (Reid 1993), makes it an ideal test case for exploring this hypothesis.

Early on, indirect support for a central black hole arose from the discovery of the unusual radio source Sgr A*; its non-thermal spectrum (e.g., Serabyn et al. 1997, Beckert et al. 1996), compact size (Rogers et al. 1994), and lack of detected motion (Backer 1994) led researchers to associate it with the putative black hole. Although many of Sgr A*'s characteristics are well matched by the expected environmental characteristics of an accreting black hole, direct dynamical evidence is necessary to conclusively demonstrate the presence of a central black hole.

2. Proper-Motion Study

In 1995 we initiated a diffraction-limited proper-motion study of the Galaxy's central stellar cluster using the W. M. Keck I 10-m telescope. Ghez et al. (1998) report the results from the 1995-1997 observations and here we review those results with an additional year's worth of data.

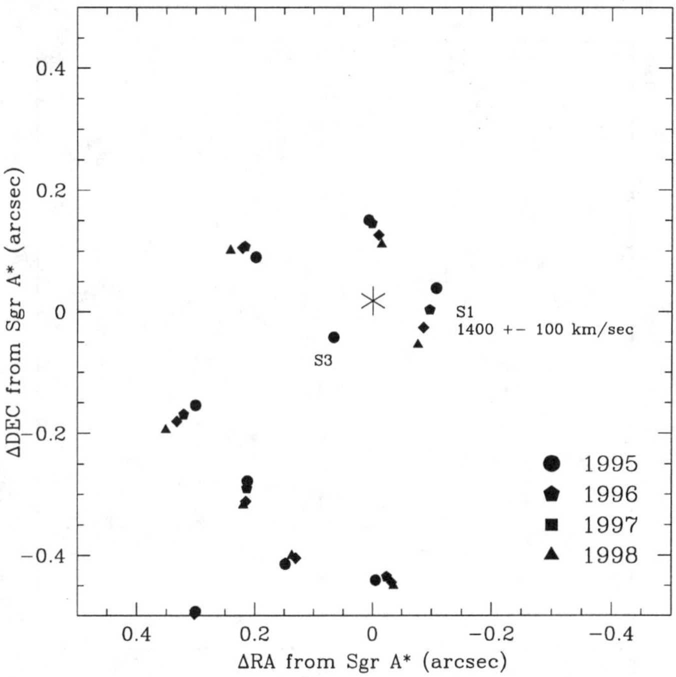

Figure 1. The Keck position measurements of stars within the one arcsec^2 region centered on the position of the radio source Sgr A* (starred point). Significant velocities, which reach 1400 ± 100 km/sec, are easily detected in this region. Each year's measurement is represented by a different symbol: 1995 by circles, 1996 by pentagons, 1997 by diamonds, and 1998 by triangles.

A diffraction-limited resolution of 0.″05 was obtained by summing a large number of very short exposure images aligned on the brightest speckle of IRS 16C. Thus, the resulting 6″×6″ K[2.2 μm]-band images have the highest angular resolution obtained at near-infrared wavelengths from ground or space and reveal a large population of faint stars. Using an unbiased approach for identifying stars, we selected a sample of ~90 stars with brightnesses ranging from K = 9 to 17 mag and two-dimensional velocities as large as 1,400 ± 100 km/sec to be included in the proper motion study. Figure 1 shows the positions for stars within a one arcsec^2 region centered on the nominal position of Sgr A*, where motion for several of the stars can be easily seen. Compared to earlier work (Eckart & Genzel 1997; Genzel et al. 1997), the source confusion is reduced by a factor of nine, the number of stars with proper motion measurement in the central 25 arcsec^2 of our Galaxy is doubled, and the accuracy of the velocity measurements in the central one arcsec^2 is improved by a factor of four.

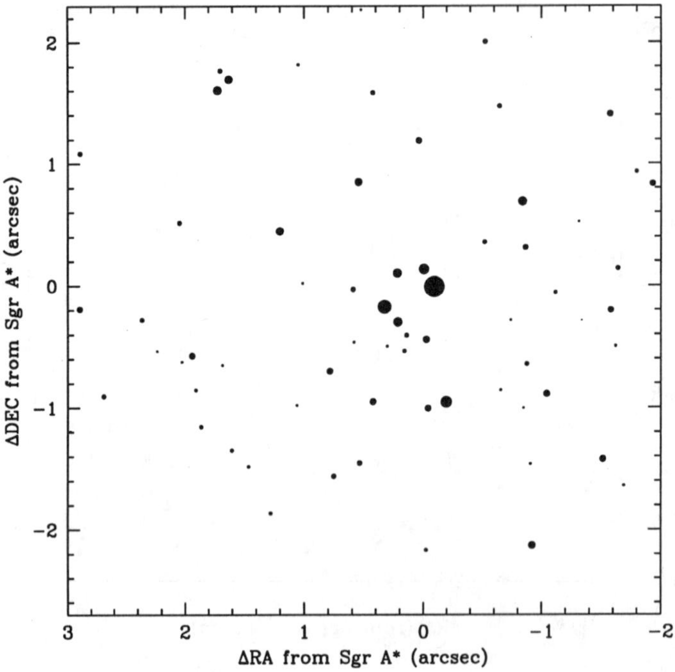

Figure 2. The positions of the stars that were unambiguously detected in all four years are displayed with point sizes scaled to the stars' velocities. A clear increase in the velocities is visible at the field center, where stars reach velocities of 1,400 km/sec.

In addition to our primary field, we also mapped a larger region with less sensitivity to faint sources to include two SiO maser sources – IRS 7 and IRS 10 (Menten et al. 1997) – thereby linking the infrared and radio reference frames and securing the position of Sgr A* to ± 30 mas. Although no 2.2 μm source is detected at the exact nominal Sgr A* location, it is coincident with the region containing the highest velocity stars (see Figure 2) as well as with the highest surface density of sources in our images. Based on the two-dimensional velocity dispersion, the dynamical center appears to be consistent with the position of

Sgr A* to within $\sim$ 100 mas. Analysis of the one-dimensional velocity dispersion clearly demonstrates the presence of $2.6(\pm0.2) \times 10^6 M_\odot$ interior to a radius of $\sim$ 0.015 pc, or densities in excess of $10^{12} M_\odot/pc^3$ (see Figure 3). This volume-averaged mass density exceeds that inferred so far for the center of any other galaxy. The high mass-to-light ratio and high density lead us to conclude that our Galaxy harbors a massive central black hole. Ghez et al. (1998) have argued that any plausible cluster of stars or stellar remnants having the mass and density required to account for these observations would be unstable on relatively short time scales, and is therefore not a viable alternative.

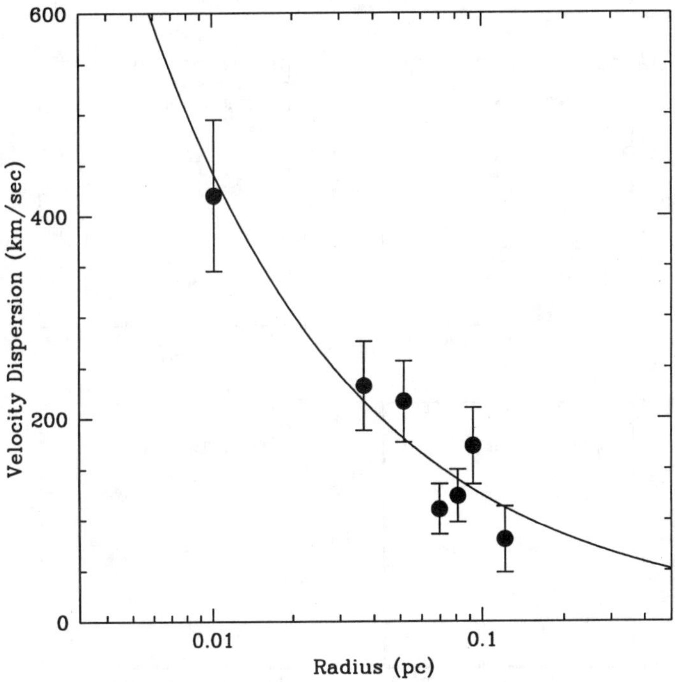

Figure 3. The projected stellar velocity dispersion as a function of projected distance from Sgr A* is consistent with Keplerian motion, which implies the gravitational field is dominated by mass within 0.01 pc.

3. The Mystery of S3

Of all the sources that pass our criteria for being tracked in the proper motion experiment, only one source, S3, is lost (see Figures 1&4). It appears only in our 1995 map as a K = 15 mag source and has no evident counterpart brighter than 16.5 in subsequent years. S3 also has the distinction of being the source closest to the nominal position of Sgr A*, separated by a mere 70 ± 30 mas. If one associates S3 with Sgr A* and Sgr A* with the black hole, this raises the possibility that S3 may represent a gravitational lensing event or an enhancement of the accretion rate onto the black hole. The former case has been explored by Wardle & Yusef-Zadeh (1992), and by Alexander & Sternberg (1999); the latter authors suggest that there is only a 1% probability of such a lensing event being detected. Other possibilities include the detection of a variable star or variable extinction. More intensive observing campaigns recently undertaken will address the presence and characteristics of variable sources within this region.

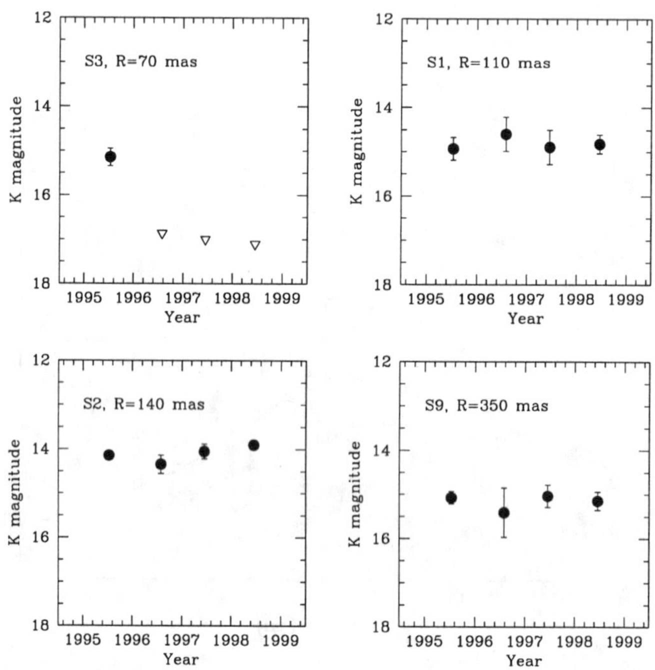

Figure 4. K-band light curves for several sources in the central one arcsec^2. The name of the star and its distance from the nominal position of Sgr A* is provided in the top left corner of each panel. S3 is only detected in 1995 (the triangles represent upper limits). For comparison, three other stars of comparable brightness and in similarly crowded regions are shown. This star is particularly intriguing as its position is closest to the nominal position of Sgr A*, the unusual radio source near the position of the black hole.

Acknowledgments. Support for this work was provided by the National Science Foundation and the Packard Foundation.

References

Alexander, T. and Sternberg, A. 1999, ApJ, in press

Backer, D.C. 1994, in IAU Symp. No. 169: Unsolved Problems in the Milky Way

Beckert, T., Duschl, W. J., Mezger, P. G., & Zylka, R. 1996, A&A, 307, 450

Eckart, A. and Genzel, R 1997, MNRAS, 284, 576

Genzel, R., Eckart, A., Ott, T. Eisenhauer, F. 1997, MNRAS, 291, 219

Ghez, A. M., Klein, B. L., Morris, M., and Becklin, E. E. 1998, ApJ, 509, 678

Lynden-Bell, D. & Rees, M. J. 1971, MNRAS, 152, 461

Menten, K. M., Reid, M. J., Eckart, A., Genzel, R. 1997, ApJ, 475 , L111

Reid, M. ARA&A, 31, 345

Rogers, A. E. E., Doeleman, S., Wright, M. C. H., Bower, G. C., Ba cker, D. C., Padin, S., Philips, J. A., Emerson, D. T., Greenhill, L., Moran, J. M., & Kellermann, K. I. 1994, ApJ, 434, L59

Serabyn, E, Carlstrom, J., Lay, O., Lis, D. C., Hunter, T. R., & Lacy, J. H. 1997, ApJ, 490, L77

Wardle, M., & Yusef-Zadeh, F. 1992, ApJ, 387, L65.

The Central Parsecs of the Galaxy
ASP Conference Series, Vol. 186, 1999
H. Falcke, A. Cotera, W.J. Duschl, F. Melia, M.J. Rieke, eds.

The Proper Motion and Mass of Sgr A*

M. J. Reid

Harvard–Smithsonian CfA

A. C. S. Readhead and R. C. Vermeulen

California Institute of Technology

R. N. Treuhaft

Jet Propulsion Laboratory

Abstract.

The *apparent* proper motion of Sgr A*, relative to extragalactic sources, is consistent with that expected from the Sun orbiting the center of the Galaxy. Using the VLBA we obtain a lower limit for the peculiar motion of Sgr A* of about 20 km s^{-1} , implying a lower limit for mass of Sgr A* of $\sim 10^3$ $M_\odot$. This rules out the possibility that Sgr A* is a stellar system, such as a contact binary, containing ~ 10 $M_\odot$ and radiating near its Eddington luminosity. A mass of more than $\sim 10^3$ $M_\odot$ and a luminosity $\lesssim 10^5$ $L_\odot$ indicates that Sgr A* is radiating at $\lesssim 0.1\%$ of its Eddington limit.

All observations are consistent with Sgr A* being a super-massive black hole. Since the lower limit for the mass of Sgr A* is only about 0.1% of the gravitational mass inferred from the stellar motions, one cannot claim from our observations alone that even a significant fraction of the dark mass must be in Sgr A*. However, alternative models of dark matter distributions are severely restricted by these observations.

1. Introduction

Recent measurements of proper motions of stars very close to Sgr A* strongly suggest that Sgr A* is a super-massive black hole (Eckart & Genzel 1997, Genzel et al. 1997, Ghez et al. 1999). Stellar speeds in excess of 1000 km s^{-1} have been seen for stars at a projected distance of 0.015 pc from the position of Sgr A* determined by Menten et al. (1997), implying a central mass of 2.6×10^6 $M_\odot$. While it is unlikely that this mass could take any stable form other than that of a super-massive black hole (Maoz 1998), a convincing case has yet to be made that the radio source Sgr A* is such a black hole for several reasons. Firstly, without an understanding of the nature of dark matter it is hard to make a case that an extreme density cannot exist. Secondly, the low apparent luminosity of Sgr A* ($\lesssim 10^5$ $L_\odot$) is difficult to explain. It could, for example, originate from a stellar system, such as a contact binary, radiating at its Eddington limit. Finally, even though lifetimes of dense stellar clusters are short, the possibility of a steady state resupplying of members has not been yet addressed.

Studies of the radio emission of Sgr A* itself can provide strong evidence, complementary to that of the stellar proper motions, for (or against) the supermassive black hole hypothesis for Sgr A*. VLBI studies at cm- and mm-wavelengths have revealed that the radio emission from Sgr A* comes from a region $\lesssim$ 5 AU. (eg, Rogers et al. 1994, Lo et al. 1998). However, without knowing the mass of Sgr A*, this is not necessarily a stringent limit. While this corresponds to $\lesssim$ 100 Schwarzschild radii for a 2.6×10^6 $M_\odot$ black hole, were Sgr A* to be a contact binary of $\sim$ 10 $M_\odot$, one might expect even smaller sizes.

Another approach to constraining the mass *directly associated* with Sgr A* is to measure its proper motion. Were Sgr A* to contain of order a solar mass, then even if placed precisely at the center of some dark matter distribution, random perturbations owing to close encounters with nearby stars would gradually move Sgr A* outward, and it would acquire a speed comparable to that of the nearby stars ($\gtrsim$ 1000 km s^{-1}). Thus, the detection of, or limits on, the proper motions Sgr A* are invaluable to gaining a better understanding of the nature of this object.

In 1991 we started a program with the Very Long Baseline Array (VLBA) of the National Radio Astronomy Observatory [1] (NRAO) to measure the apparent motion of Sgr A* . In principle, VLBA observations of Sgr A* , phase-referenced to extragalactic radio sources, can achieve an accuracy sufficient to detect secular motions of $\lesssim$ 1 km s^{-1} for a source at the distance of the Galactic Center. The detailed results of the first two years of observations are given by Reid et al. (1999) and only a brief summary will be presented here.

2. Observations

Our observations using the VLBA were conducted in the late-night, early-morning periods of 1995 March 4, 1996 March 20 and 31, and 1997 March 16 and 27. The observing frequency was 43.2 GHz and only the inner five VLBA stations were used, as baselines longer than about 1500 km heavily resolve the scatter broadened image of Sgr A* at 43 GHz (eg, Bower & Backer 1998). We observed Sgr A* and two extragalactic sources, J1745–283 and J1748-291, switching rapidly among them in order to form the phase-delay difference between an extragalactic source and Sgr A*.

Interferometer fringe amplitudes were calibrated in the standard manner for the VLBA. Phases were calibrated using the strong fringes obtained on Sgr A*. Images of the extragalactic sources, phase referenced to Sgr A*, were made and preliminary relative positions of the extragalactic sources were obtained from these images. More accurate relative positions were obtained by model fitting the relative phase-delay data, because the model fitting approach allowed us to solve for, and improve upon, the vertical atmospheric delay for each VLBA

[1]The National Radio Astronomy Observatory is operated by Associated Universities Inc., under cooperative agreement with the National Science Foundation.

antenna. We estimate that relative position errors are ≈ 0.0001 and ≈ 0.0004 arcsec in the East–West and North–South coordinates, respectively, for one day's observation.

The apparent motion of Sgr A* with respect to the radio source J1745–283 is shown in Fig. 1. This motion can be used to estimate the rotation rate of the Galaxy and any peculiar motion of the super-massive black hole candidate Sgr A*. The dashed line is the variance-weighted least-squares fit to the data, and the solid line denotes the orientation of the Galactic Plane. Clearly the apparent motion of Sgr A* is almost entirely in the Galactic Plane. Converting to galactic coordinates, yields an apparent motion of -0.0059 ± 0.0004 and $+0.0002 \pm 0.0003$ arcsec y^{-1} in galactic longitude and latitude, respectively. The apparent motion of Sgr A* in the plane of the Galaxy should be dominated by the effects of the orbit of the Sun around the Galactic Center, while the motion out of the plane should contain only small terms from the Z-component of the solar motion and the peculiar motion of Sgr A*.

3. Limits on the Peculiar Motion of Sgr A*

Assuming a distance of 8.0 ± 0.5 kpc (Reid 1993), the apparent angular motion of Sgr A* in the plane of the Galaxy translates to -225 ± 20 km s^{-1} . The uncertainty includes the effects of measurement errors and the uncertainty in R_0 . Assuming the peculiar motion of Sgr A*is small, and removing a 5.25 ± 0.62 km s^{-1} component of the solar motion in the direction of galactic rotation (Dehnen & Binney 1998) from the *reflex* of our apparent motion of Sgr A* , yields an estimate for Θ_0 of 219 ± 20 km s^{-1} . This value is consistent with most recent estimates of about 220 km s^{-1} (Kerr & Lynden-Bell 1986).

After removing the best current estimate of the *angular* motion of the Sun around the Galactic Center, determined from Hipparcos measurements (Feast & Whitelock 1997), our VLBA observations yield an estimate of the peculiar motion of Sgr A* of -1 ± 17 km s^{-1} towards positive galactic longitude. This estimate of the "in plane" motion of Sgr A* comes from differencing two *angular* motions, and, since this difference is small, the uncertainty in R_0 does not significantly affect this component of the peculiar motion of Sgr A*. Given the excellent agreement of our global and the Hipparcos local measures of the angular rotation rate of the Galaxy, and the lack of a detected peculiar motion for Sgr A*, it is likely that Sgr A* is at the dynamical center of the Galaxy.

Whereas the orbital motion of the Sun (around the Galactic Center) complicates estimates of the "in plane" component of the peculiar motion of Sgr A*, motions out of the plane are simpler to derive. One needs only to subtract the small Z-component of the solar motion from the observed motion of Sgr A* to estimate the out-of-plane component of the peculiar motion of Sgr A*. Removing 7.17 km s^{-1} z-component of the solar motion (Dehnen & Binney 1998) from our measured apparent motion of Sgr A* out of the plane of the Galaxy, we estimate the peculiar motion of Sgr A* is 16 ± 11 km s^{-1} toward the North Galactic Pole. The uncertainty is dominated by our proper motion measurements and can be greatly improved by future measurements.

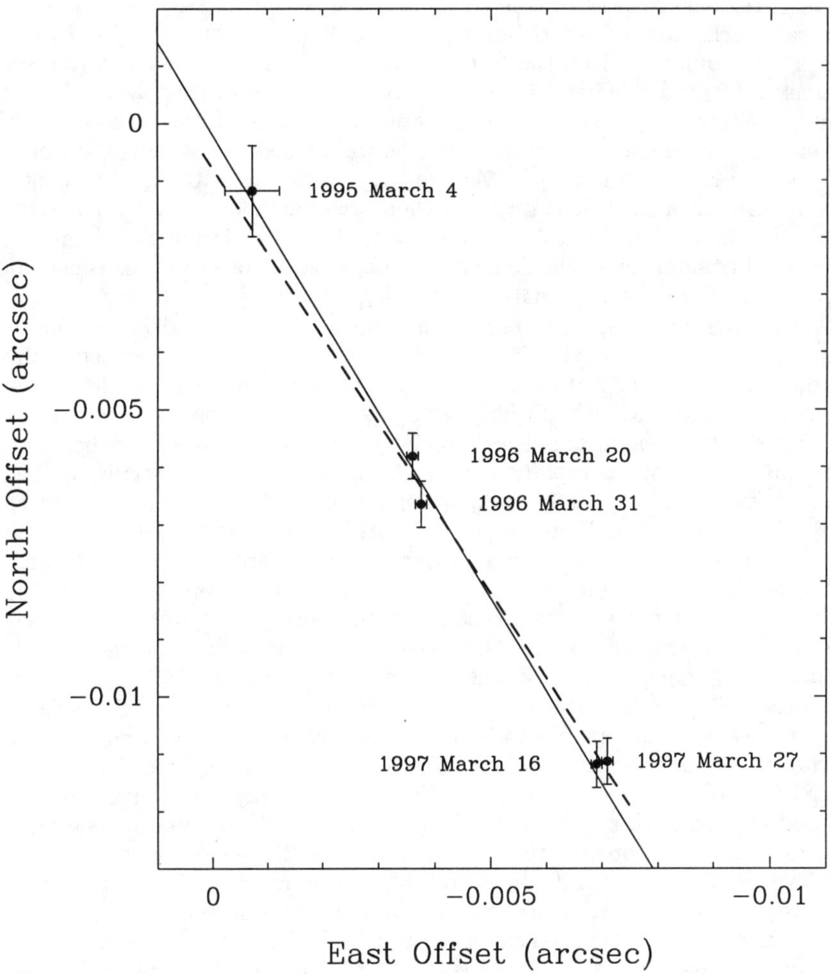

Figure 1. Position residuals of Sgr A* relative to J1745–283 on the plane of the sky. North is to the top and East to the left. Each measurement is indicated with the date of observation and $1-\sigma$ error bars. The dashed line is the variance-weighted best-fit proper motion, and the solid line gives the orientation of the Galactic plane.

4. Limits on the Mass of Sgr A*

Our estimate of a peculiar motion of Sgr A* provides an upper limit of about 20 km s^{-1} for motions both in and out of the galactic plane. Since stars in the inner-most regions of the central cluster move at speeds in excess of 1000 km s^{-1} , a central "dark mass" of approximately 2.6×10^6 $M_\odot$ contained within 0.015 pc of Sgr A* is required. Given the fact that independent measurements (Backer & Sramek 1999, and this paper) show that Sgr A* moves at least two orders of magnitude slower than its surrounding stars, Sgr A* must be much more massive than individual stars observed in the central cluster. Of course, the upper limits on the proper motion of Sgr A* are fully consistent with Sgr A* being a super-massive black hole. The question then arises as how to quantify a lower limit to the mass of Sgr A*, and assuming that Sgr A* is not the dominant gravitational component in the central 0.01 pc, how the observations constrain possible distributions of dark matter.

Neither the virial theorem, nor the assumption of equipartition of kinetic energy among Sgr A* and the dense central star cluster, leads to trustworthy limits on the mass of Sgr A*. Attempts to use the virial theorem to estimate the kinetic energy of a massive black hole requires differencing two large and uncertain quantities (the potential energy associated with the massive black hole and the kinetic energies of the rapidly moving stars) and will be essentially useless. Alternatively, the assumption of equipartition of kinetic energy is unlikely to be valid for the case of a small number of stars orbiting a massive black hole. The influence of a dominant central mass produces largely stable stellar orbits, much as the gravitational dominance of the Sun in our solar system leads to stable planetary orbits. In our solar system the kinetic energy of Jupiter exceeds that of the the Sun by a factor close to the inverse of the ratio of their masses. Hence, the Sun is not in equipartition of kinetic energy with the planets.

In order to evaluate better how an upper limit to the motion of Sgr A* can be used to provide a lower limit to its mass, we carried out N-body simulations of stars orbiting about a massive black hole. We used a simple, direct integration code (NBODY0) of Aarseth (1985), modified for our purposes. Initial simulations used 255 stars orbiting a 2.6×10^6 $M_\odot$ black hole. The stellar masses were chosen randomly to represent the upper end of a stellar mass function with a power law distribution from 20 down to 2 $M_\odot$. The number of stars and their masses used in this simulation are comparable to those observed within the central 0.5 arcsec or 4000 AU (Genzel et al. 1997). Stellar orbits were chosen by randomly assigning a distance from the black hole, uniformly distributed in the range 10 to 10,000 AU and starting with eccentricities of $\approx 20\%$.

The N-body simulations show quasi-random motions of the massive black hole. After relatively short periods of time ($\ll 10,000$ years) a "steady state" condition appeared to be reached. The motion of Sgr A* changed completely in all three coordinates on time scales $\ll 100$ years and was $\lesssim 0.1$ km s^{-1} in each coordinate. Thus the expected motion from a 2.6×10^6 $M_\odot$ black hole is well below our upper limits on Sgr A*'s peculiar motion.

While the simplest interpretation is that Sgr A* is a 2.6×10^6 $M_\odot$ black hole, our upper limit on any peculiar motion for Sgr A* currently is two to three orders of magnitude above its expected motion. Thus, it seems reasonable to investigate the possibility that the mass within a radius of 0.01 pc is not dominated by Sgr A*, but instead is in some form of "dark" matter. In this case, Sgr A* will react to the gravitational potential and orbit the center of mass of this dark matter.

Fig. 2 displays the enclosed mass versus radius relation for two mass models that are consistent with the stellar proper motions. These models yield flat enclosed-mass versus radius relations at radii of $\gtrsim 0.01$ pc where measurements exist. The most centrally condensed mass distribution is shown as the horizontal dash-dot line labeled "point mass". The least centrally condensed mass distribution plotted is for a Plummer density distribution, where density, ρ, is given by $\rho_0 \; (1 + (r/r_0)^2)^{-\alpha/2}$, for $\rho_0 = 6 \times 10^{11}$ $M_\odot$ pc^{-3}, $r_0 = 0.01$ pc, and $\alpha = 5$. This distribution is shown with a curved dash-dot line labeled "Plummer model". It is difficult to make a physically reasonable mass distribution that is significantly less centrally condensed than this and consistent with the stellar motion data. Thus, these two "extreme" model distributions approximately bound all allowed mass distributions.

Assuming that Sgr A* has a mass $\ll 10^6$ $M_\odot$, it will orbit about the center of mass of the system. Orbital speed for a body at a radius, r, from the center of mass is given by $V = \sqrt{GM_{encl}/r}$, where G is the gravitational constant and M_{encl} is the enclosed mass at that radius. Setting $V = 20$ km s^{-1} , our limit for the motion of Sgr A*, produces the sloping solid line in Fig. 2. Only enclosed masses below that line are permitted by our observations. For radii greater than about 3×10^{-5} pc (6 AU), this limit rules out a wide range of mass distributions, leaving only the *least* centrally condensed mass models as possibilities. For radii less than about 3×10^{-5} pc, the orbital motion of Sgr A* produces angular excursions less than 0.0008 arcsec. In this case, while the orbital speed might greatly exceed 20 km s^{-1} , we may have failed to detect these excursions owing to our poor temporal sampling and ≈ 0.0004 arcsec errors in the North–South direction. Thus, we do not extend the motion limit line below 3×10^{-5} pc, and at this radius we replace the motion limit with a vertical line in Fig. 2.

The proper motions of stars and the limits on Sgr A* combine to exclude almost all of "parameter space" for models of the density distribution of material in the inner 0.1 pc of the Galactic Center. The stellar motions tend to exclude "soft" gravitational potentials (i.e., the least centrally condensed mass distributions) and the motion limit for Sgr A* tends to exclude "hard" gravitational potentials. The point-like mass distribution in Fig. 2 essentially requires a super-massive black hole. Since we have assumed that this is **not** Sgr A*, we would be left with the question of why a low-mass Sgr A* radiates far more than a super-massive black hole in essentially the same environment. Other mass models with the extremely centrally condensed mass distributions require exceedingly high mass densities ($\gtrsim 10^{14}$ $M_\odot$ pc^{-3}) and theoretical arguments suggest that such models are unlikely to be stable for even 10^7 yr, regardless of the composition of the matter (Maoz 1998).

The hardest cases to exclude on either observational or theoretical grounds are given by the mass distributions that are the least centrally condensed, but still consistent with the stellar motion data. For this case, one can ask the question, at what mass will perturbations by stars in the central cluster lead to detectable motion of Sgr A*? In order to answer this question, we modified the N-body code described above to include a fixed gravitational potential ap-

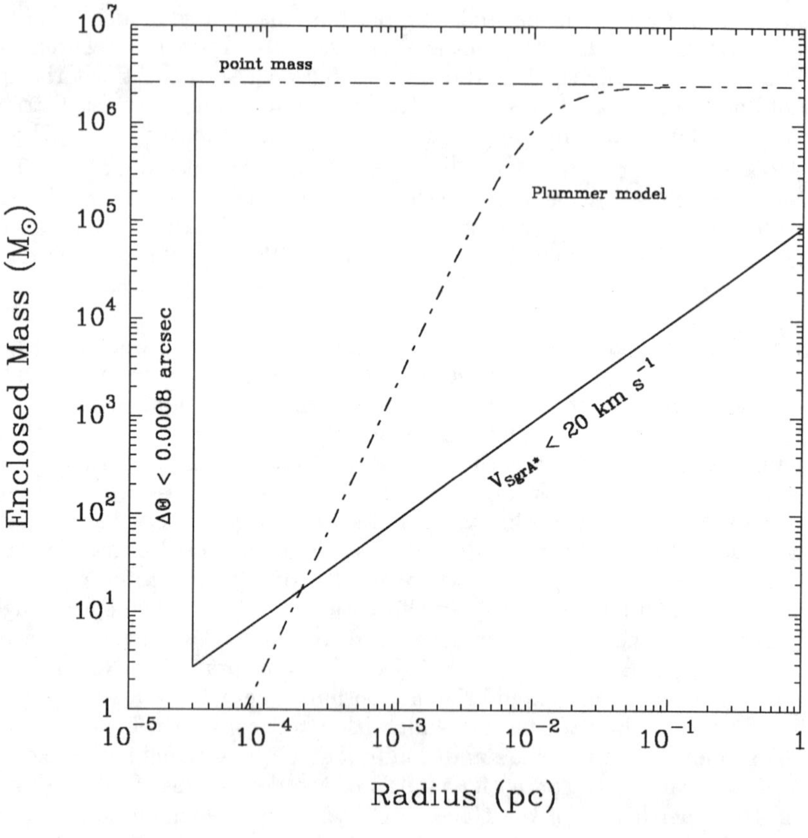

Figure 2. Enclosed mass versus radius for various model distributions of dark matter, assuming the mass of Sgr A* $\ll 10^6$ $M_\odot$. Models labeled "point mass" and "Plummer model" represent hard and "soft" gravitational potentials and approximately bound mass distributions that are consistent with stellar proper motion data. The Plummer model has a core density of 6.0×10^{11} $M_\odot$ pc^{-3} at a radius of 0.01 pc and $\alpha = 5$. The sloping solid line indicates the upper limit for enclosed mass based on the proper motion of Sgr A*. The vertical solid line at 3×10^{-5} pc indicates the upper limit in radius, where angular excursions of Sgr A* of < 0.0008 arcsec could be missed owing to insufficient astrometric accuracy.

propriate for a Plummer law mass distribution with $\alpha = 5$, $\rho_0 = 6 \times 10^{11}$ $M_\odot$ pc^{-3}, and $r_0 = 0.01$ pc ("Plummer model" in Fig. 2). Sgr A* was assigned a mass $\ll 10^6$ $M_\odot$ and the entire system, including 254 stars was allowed to evolve in time. For this gravitational potential there is little enclosed mass within $\sim 10^{-4}$ pc to bind Sgr A*. We found that when Sgr A*'s mass was less than $\approx 3,000$ $M_\odot$, Sgr A* gradually moved outward from the center of the gravitational potential and achieved orbital speeds in excess of 20 km s^{-1} . Thus, allowing for reasonable projections of the the velocity vector for Sgr A*, we conclude that the lack of detectable motion places a conservative lower limit of about 10^3 $M_\odot$ for the mass associated with Sgr A*.

References

Aarseth, S. 1985, in *Multiple Time Scales*, eds. J. U. Brackbill & B. I. Cohen, (Academic Press: New York), pp. 378–418

Backer, D. & Sramek, R.A. 1999, to be submitted to ApJ

Bower, G. C, & Backer, D. C. 1998, ApJ, 496, L97

Dehnen, W. & Binney, J. J. 1998, MNRAS, 298, 387

Eckart, A. & Genzel, R. 1997, MNRAS, 284, 576

Feast, M. & Whitelock, P. 1997, MNRAS, 291, 683

Genzel, R., Eckart, A. Ott, T. & Eisenhauer, F. 1997, MNRAS, 291, 219

Ghez, A. M., Klein, B. L., Morris, M. & Becklin, E. E. 1999, to appear in ApJ

Kerr, F. J. & Lynden-Bell, D. 1986, MNRAS, 221, 1023

Lo, K. Y., Shen, Z.-Q., Zhao, J.-H. & Ho, P. T. P. 1999, ApJ, 508, L61

Maoz, E. 1998, ApJ, 494, L181

Menten, K. M., Reid, M. J., Eckart, A. & Genzel, R. 1997, ApJ, 475, L111

Reid, M. J. 1993, ARA&A, 31, 345

Reid, M. J., Readhead, A. C. S., Vermeulen, R. C, & Treuhaft, R. N. 1999, to be submitted to ApJ

Rogers, A. E. E. et al. 1994, ApJ, 434, L59

The Central Parsecs of the Galaxy
ASP Conference Series, Vol. 186, 1999
H. Falcke, A. Cotera, W.J. Duschl, F. Melia, M.J. Rieke, eds.

NICMOS Search for the Tip of the Main Sequence

Marcia J. Rieke

Steward Observatory, University of Arizona, Tucson, AZ 85721

Abstract. Two of the strengths of the NICMOS (Near-Infrared Camera and Multi-Object Spectrometer) cameras on the Hubble Space Telescope are sensitivity and the stability of the point spread function. These characteristics permit a search for the late O to early B main sequence stars that should be present based on the existence of blue and red supergiants such as the IRS16 stars and IRS7. This paper presents progress reports on the search for these stars, and on the establishment of the NICMOS photometric system in support of such a search.

1. Introduction

Bright, post-main sequence stars were discovered some time ago at the Galactic Center: for example IRS7, identified as a red supergiant in 1982 (Lebofsky et al.) Good spectral types for the bluest stars have become available only recently (Tamblyn et al. 1996, Najarro et al. 1997, Hanson et al. 1998). The progenitor masses of these stars are difficult to assign with high accuracy but appear likely to lie in the range of 20 $M_{\odot}$ and up. This leads to an estimate of the tip of the main sequence lying at as bright a level as O6-O8, but given the uncertainties, it could lie at as faint a level as B2-B4. Modeling of the stellar population (e.g. Tamblyn and Rieke 1993, Genzel et al. 1996) supports an age of ~10 million years for the epoch of last star formation, but none of the models can reproduce the observed numbers of blue and red supergiants. This failure of the models may result from the known deficiencies in stellar evolution models where the post-main sequence phases are the most difficult to predict in detail. Langer and Maeder (1995) disscuss this issue in detail and provide examples of how the blue to red supergiant ratio from models does not match the data even in regions much easier to observe than the Galactic Center.

Measurement of the location of the tip of the main sequence in a color magnitude diagram, a classical technique for determining the age of a stellar population, would permit a greater understanding of the star formation process in the extreme environment present at the Galactic Center.

2. Observations

Four filters were used in observing the stars at the Galactic Center: F110M and F145M with Camera 1, and F160W and F222M with Camera 2. The effective wavelengths, bandwidths, zero-magnitude fluxes for the Johnson system and pixel sizes are listed in Table 1. All data were taken using non-destructive reads which permitted rejection of cosmic-rays in the time domain (MULTI-ACCUM readout mode). A total of three positions were observed: the center, a position 2′ away, and a position 2° away. While the F110M images were being acquired, Camera 2 was used to observe in F110W, F160W, and F222M thus providing another set of comparison fields. The F145M data came from a proper motion program and was taken at a different time.

Table 1. NICMOS Filter Properties

Filter	λ_{eff} (μm)	$\Delta\lambda$ (μm)	0-mag(Jy)	Pixel Size(")
F110M	1.100	0.20	1779	0.042
F145M	1.452	0.20	1369	0.042
F160W	1.593	0.41	1083	0.076
F222M	2.216	0.14	668	0.076

All images were dark subtracted using dark frames taken on-orbit. After dark subtraction, signals were extracted using a linear regression fit to the values for each pixel. All images were flat-fielded using on-orbit flat-fields generated using the internal NICMOS flat-field lamp. This lamp has been demonstrated to flatten images to level of less than 2% residual scatter by imaging a star in a 12×12 grid across the face of each camera and then measuring the flux as a function of position of the star. The data reduction procedure followed is essentially that of the "NICRED" package (McLeod 1997). The resulting image in the F145M filter is presented in Figure 1.

The Camera 1 images were produced using four field centers to cover the 19′×19′area covered by the Camera 2 images. Every position was observed using a four-position dither and final images were produced by shifting and adding the dithered images.

The resulting four images were then processed using DAOPHOT (Stetson 1987) as implemented in IRAF. Figure 2 shows a sample radial profile from the F145M point-spread function that was produced using DAOPHOT. The characteristic speckles in a NICMOS image with exposed stars were not mistakenly identified as separate objects as DAOPHOT was run in an iterative fashion with bright stars removed in the first pass and subsequent passes run using the output from the previous pass as input. The resulting star lists from each filter were merged by using the IRAF immatch package.

The fluxes from DAOPHOT were calibrated using observations of P330E which has been well-observed on the ground (Persson et al. 1998). Tranformation of the groundbased data for this star which is a solar analog to the NICMOS system was accomplished by normalizing a solar spectrum to the J, H, K fluxes for P330E and then integrating the resulting spectrum across the NICMOS filters. The magnitudes adopted for P330E are F110M=11.90, F145M=11.70, F160W=11.48, and F222M=11.44.

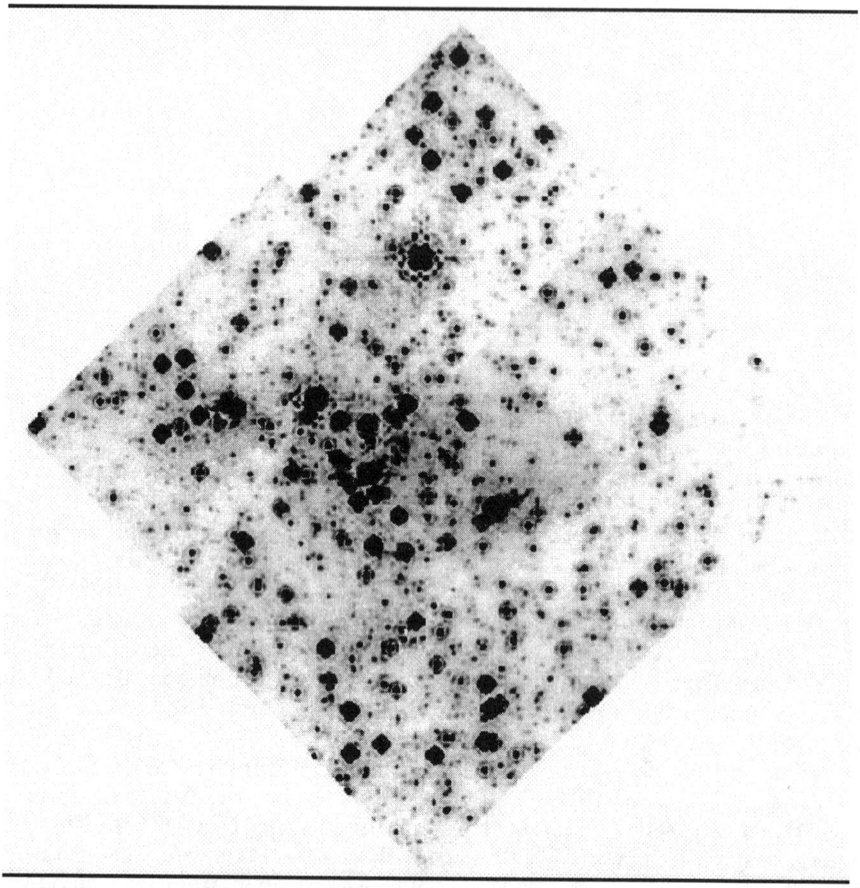

Figure 1. The central 20″ of the Galaxy at 1.45μm.

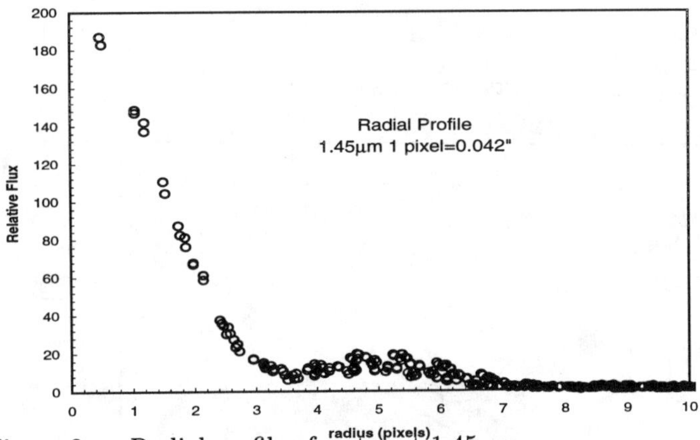

Figure 2. Radial profile of a star at $1.45\mu m$.

3. Extinction Law and Color Corrections

To achieve the goal of finding the tip of the main sequence, very careful dereddening is required. An error in dereddening of only 5% leads to an error in the J (1.25μm) values sufficiently large to wipe out any chance of distinguishing intrinsically blue stars from intrinsically red stars. In addition, any extinction law must be translated from the groundbased filter system to the NICMOS filter system. The extinction to the Galactic Center is high at about 30 magnitudes at V, equivalent to over 3 magnitudes at K. At J the extinction is nearly 8 magnitudes and it is even higher in the NICMOS F110M filter. The extremely red colors necessitate computing color corrections caused by the shift of flux towards the long-wavelength end of the filter as compared to the relatively blue-colored standard stars. Figures 3 and 4 illustrate corrections derived by reddening Kurucz stellar spectra obtained from the Space Telescope Science Institute Calibration Database with a law obtained by matching the NICMOS colors of the IRS16 sources with spectra in Tamblyn et al (1996). Synthetic photometry of the Kurucz models through groundbased JHK filters matches all but red giants and supergiants well, so the computed corrections are a useful starting point but need to be re-computed for late-type, post-main sequence stars. The extinction coefficients used are $A_{F110M} = 0.365$, $A_{F160W} = 0.176$, and $A_{F222M} = 0.105$. This is slightly different from interpolating the law presented in Rieke and Lebofsky (1985) which may result from the taking the color corrections into account and the uncertainties in the old law at I(0.9μm) which influence the interpolation to 1.10μm. The figures show the correction that must be added to compensate for the reddened colors.

4. Results

The stars detected at the Galactic Center at F110M, F160W, and F222M have been dereddened and color-corrected using the values presented in the previous section. For this initial attempt at dereddening, all objects with F110M-F160W

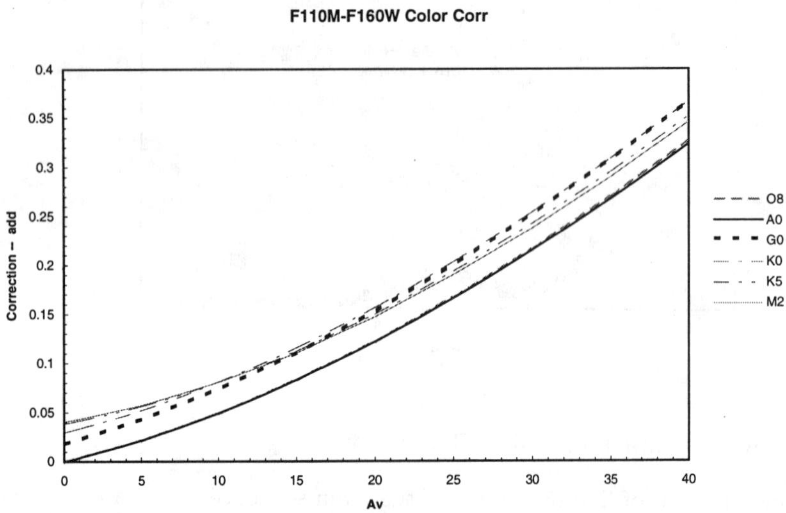

Figure 3. Correction to the observed F110M-F160W color as a function of A_V and spectral type.

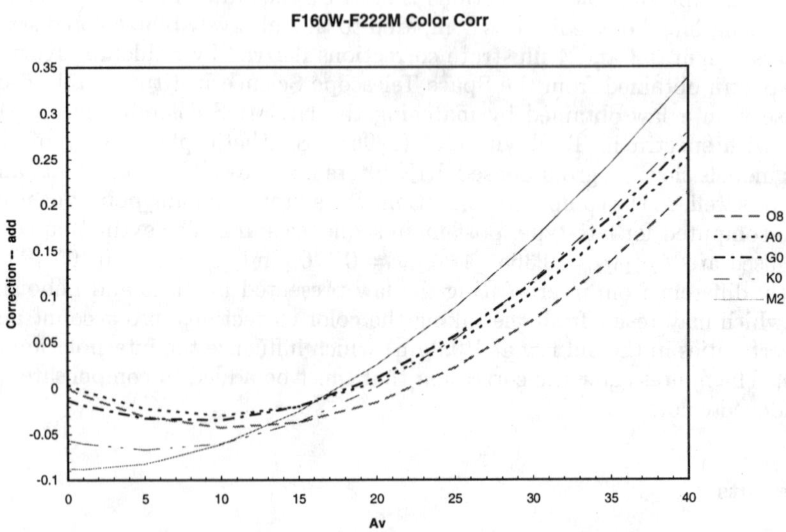

Figure 4. Correction to the observed F160W-F222M color

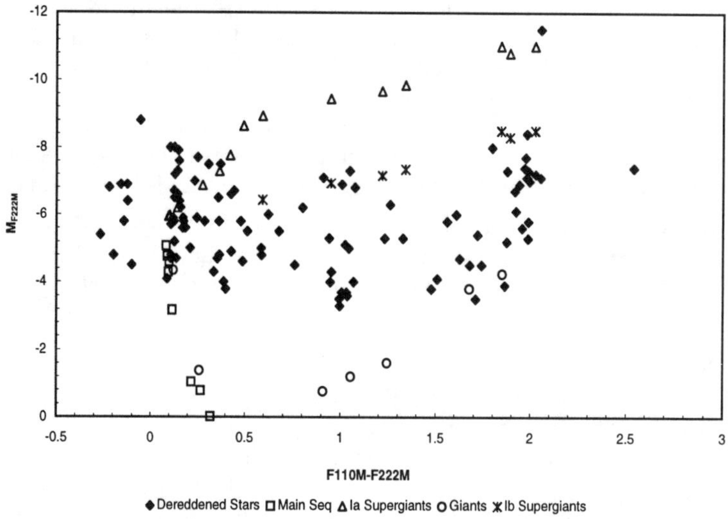

Figure 5. Preliminary color magnitude diagram for stars at the Galactic Center.

< 5.9 had color corrections for O8V stars applied while stars redder than this had corrections derived from M0III stars applied. This simplification may cause extra scatter in the colors. As a result of the relative exposure times and much higher extinction in F110M, the images from this filter set the limiting magnitude for the data set; nearly ten times as many stars were detected at F160W and F222M as are detected in the complete color set. Figure 5 shows the result of dereddening and color correcting the complete color set. To construct this color magnitude diagram, a distance of 8 Kpc has been assumed for the Galactic Center.

The average level of extinction for the stars in Figure 5 is $A_V = 30.5$; this level is biased by the detection limit at F110M which precludes objects, especially intrinsically red ones, from being detected at much higher extinction levels. The color-magnitude diagram still exhibits more scatter than is desirable and a more sophisticated routine for dereddening and color correcting that applies color corrections appropriate to the derived spectral type is needed. Some of the scatter may also reflect nature of the stars themselves. Stars with hot dust surrounding them, such as IRS1, IRS9, and IRS13, have been excluded from the diagram.

Also shown in Figure 5 are the locations of the Kurucz model stars. From this plot it appears likely that stars at a tip of the main sequence near O8V have been detected, but better dereddening is needed to make this result unequivocal. A quick comparison with other fields at $2'$ and $2°$ away from the Galactic Center confirms the existence of a population of bright, intrinsically blue stars at the Center.

The colors of the sources closest to the position of Sgr A* are of special interest as groundbased data is not yet able to measure these sources accurately at J and H. Most of these stars have colors similar to the IRS16 sources. For example, the observed colors of S1 in the nomenclature of Eckart & Genzel (1997) are $F110M - F160W = 5.68 \pm .12$, $F160W - F222M = 1.90 \pm .09$, and $F222M = 13.69 \pm .05$ as compared to $F110M - F160W = 5.58 \pm .02$, $F160W - F222M = 2.08 \pm .02$, and $F222M = 9.00 \pm .02$ for IRS16NE. Whether these stars are members of the population at the tip of the main seqeunce or have otherwise been modified by their environment awaits further study.

Acknowledgments. The National Aeronautics and Space Administration has funded this work as well as the development of the NICMOS instrument. George Rieke and Almudena Alonso-Herrero have given much help to the author.

References

Eckart, A., & Genzel, R. 1997, MNRAS, 284, 567

Genzel, R., Thatte, N., Krabbe, A., Kroker, H., & Tacconi-Garman, 1996, ApJ, 472, 153

Hanson, M. M., Rieke, G. H., & Tamblyn, P. 1998, American Astronomical Society, 193, 104.02

Langer, N. & Maeder, A. 1995, A&A, 295, 685

Lebofsky, M. J., Rieke, G. H., & Tokunaga, A. 1982, ApJ, 263, 736

McLeod, B., 1997, in 1997 HST Calibration Workshop, S. Casertano, Robert Jedrzejewski, Charles D. Keyes, & Mark Stevens, Baltimore: Space Telescope Science Institute, 281

Najarro, P., Krabbe, A., Genzel, R., Lutz, D., Kudritzki, R., & Hillier, D. 1997, A&A, 325, 700

Persson, E., Murphy, D. C., Krzeminski, W., Roth, M. & Rieke, M. J., 1998, AJ, 116, 2475

Rieke, G. H., & Lebofsky, M. J. 1985, ApJ, 288, 618

Stetson, P. B. 1987, PASP, 99, 191

Tamblyn, P. & Rieke, G. H. 1993, ApJ, 414, 573

Tamblyn, P., Rieke, G. H., Hanson, M. M., Close, L. M., McCarthy, D. W., & Rieke, M. J. 1996, ApJ, 456, 206

The Central Parsecs of the Galaxy
ASP Conference Series, Vol. 186, 1999
H. Falcke, A. Cotera, W.J. Duschl, F. Melia, M.J. Rieke, eds.

NICMOS Observations of the Central Parsec: Deep Imaging at 1.6 microns, Short-Term Variability, and Ionized Emission

S. R. Stolovy, D. W. McCarthy, F. Melia, G. Rieke, M. J. Rieke

Steward Observatory, University of Arizona, Tucson, AZ 85721

F. Yusef-Zadeh

Dearborn Observatory, Northwestern University

Abstract. A suite of diffraction-limited observations of the central parsec taken with Camera 1 (11″ FOV) of NICMOS is presented. The search for Sgr A* at 1.6 microns is discussed, as well as an exploration of possible short-term variability that could be expected from an inhomogeneity in the accretion flow. High-resolution NICMOS Pa alpha and [FeII] images illustrate the complex morphology of the ionized gas in the region.

1. Introduction

The proximity of our own Galactic Center (GC) at only $\sim$ 8 kpc provides a unique opportunity to test the theory of galactocentric black holes, as well as to study in detail the stellar and gaseous environment in the immediate vicinity of the proposed black hole. This article presents preliminary results of three independent NICMOS projects to study the central parsec of the Galaxy: 1) a search for a near-infrared counterpart to Sgr A* at 1.6 μm and its possible variability and a search for [SiVI] (PI: D. McCarthy); 2) mapping the distribution and proper motions of the ionized gas as imaged in Pa α in the central parsec (PI: S. Stolovy); and 3) mapping the distribution of [FeII] in the central parsec as a tracer of shocks or low-ionization regions (PI: F. Melia).

2. Sgr A* at 1.6 μm

The notion that the GC harbors a supermassive black hole was proposed as early as 1971 (Lynden-Bell and Rees). Sgr A*, a strong source of synchrotron radio emission, is located at the dynamical center of the Galaxy. Its lack of proper motion, combined with radial-velocity measurements of the surrounding gas and stars as well as near-infrared stellar proper motion measurements have led many to conclude that Sgr A* is associated with a dark object of mass $\sim 2.6 \times 10^6 M_\odot$ confined to a region smaller than 0.015 pc (e.g. Genzel et al. 1997, Ghez et al. 1998, Eckart and Genzel 1997). A supermassive black hole provides the most stable configuration for such a mass concentration. The evidence is convincing, but the true test of the existence of a black hole should lie in observational evidence for general relativistic effects, namely the existence of an event horizon. General relativistic calculations predict a modulation of flux from any inhomogeneity in accreting material according to a characteristic timescale. This timescale corresponds to the period of the innermost stable orbit, which is on the order of minutes, as discussed further below. An observed light curve, coupled with the near-IR flux estimates, will test general relativistic predictions and will further constrain accretion models. The near-IR fractional flux change of a gravitationally lensed enhancement in the accretion flow can be as high as $\sim$ 40%, depending on the model and geometry (Hollywood et al. 1995). The NICMOS images were taken not only to establish the flux level at the position of Sgr A* at 1.6 μm, but to analyze the images for short-term variability. The [SiVI] observations were planned in order to search for the high-excitation conditions of an AGN-like nucleus.

2.1. Observations and Data Reduction

Deep images of the central parsec at 1.6 μm were taken with camera 1 with the F160W filter ($\Delta\lambda \sim 0.4$ μm) on August 31, 1997 and August 19, 1998. Six 416-second exposures were taken during one orbit at each epoch, dithering each exposure by 5.5 pixels. The camera 1 plate scale is approximately 0.0431″. One limitation of HST is that continuous observations for several hours are not possible, as roughly half of each 90-minute orbit is unavailable for observations due to the interruption of occultation by the Earth.

All data presented in this article were reduced with the "calnica" program using improved flats and darks and an improved linearity correction algorithm as compared to what was provided by the pipeline from the Space Telescope Institute. In order to increase the time resolution, the non-destructive "multi-accum" readout sequence (totaling 416 seconds) was broken up and re-reduced to produce three, 128-second exposures. In this regime, non-linear behavior of pixels entering saturation cannot be properly corrected. Although this affected the bright IRS16 sources, the sources in the inner arcsec (the focus of this paper) remained linear in response and therefore the photometry is reliable. Cosmic ray hits have been corrected for or flagged from the statistics. In order to make mosaics or to compare photometry, all images were aligned with a cross-correlation technique ("xregister" in iraf) and then refined to an accuracy of 0.02 pixels (0.9 mas) by subtracting each image from a reference image and minimizing the residuals. Shifts and rotations were done with a cubic spline interpolation. On examination of the "jitter" files, the typical RMS deviation from perfect pointing was on the order of 3-4 mas, which dominates the alignment accuracy.

2.2. Analysis/Discussion

Where is Sgr A*? Since Sgr A* has no clear near-infrared counterpart, we must derive its position by aligning with other objects in the GC that are detected both in the radio and infrared. It is a lucky coincidence that there are several stars in the GC that have detected maser emission (Menten et al. 1997), thus providing a way to align the radio and infrared reference frames. The only stellar maser source that appears in the NICMOS field is IRS 7. We adopt the of fset of the SiO peak of IRS7 from Sgr A* to derive the IR position of Sgr A* (Sgr A* is 0.040″W and 5.583″ S of IRS 7; Menten et al. 1997). We assume that

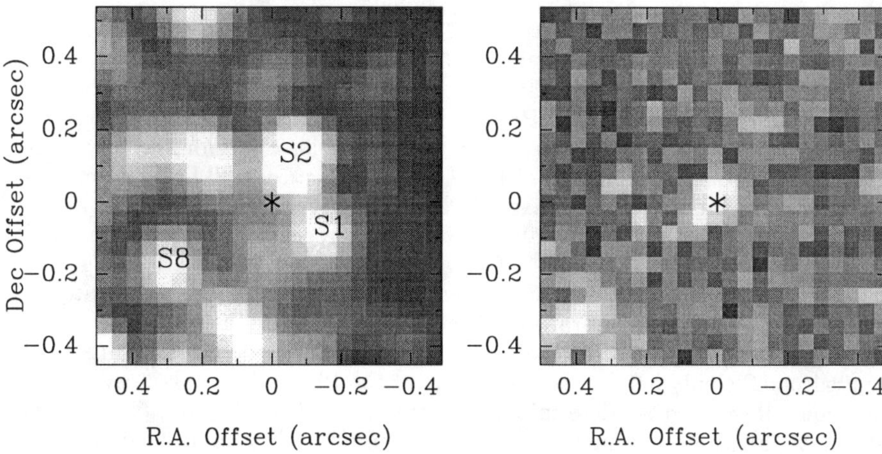

Figure 1. a). (left) The inner 1″ of the Galaxy at 1.6 μm, from the Aug. 1998 observations. The position of Sgr A* is marked with an asterisk. This image was constructed from a median of 15 aligned 128-sec exposures and is displayed in a linear grayscale. Several of the inner cluster "S" sources (nomenclature of Genzel et al. 1997) are indicated. b). (right) The ratio of the standard deviation per pixel for the 15 exposures divided by the square root of the image (a measure of the photon noise) in Figure 1a. The signal/noise in this ratio image at Sgr A* is $\sim$ 20. The photometric properties of Sgr A* as well as other objects located diagonally from lower left to upper right are contaminated somewhat by scattered light.

IRS 7 has no proper motion relative to Sgr A*, but this has not been clearly established.

Determining the Flux of Sgr A* at 1.6 μm Determining the near-IR flux at Sgr A* is complicated by severe crowding in the central 1″. There is also a diffuse, unresolved background contribution from faint bulge stars. In addition, overlapping diffraction patterns from the bright IRS16 stars contribute to a non-uniform background at the position of Sgr A*. The NICMOS observations will nevertheless provide the best estimate to date of the flux from Sgr A* at 1.6 μm in this very crowded region. NICMOS is also uniquely suited for a variability study due to the stability of the Point Spread Function (PSF). Although no definitive detection of a near-IR counterpart to Sgr A* has been made thus far, there is some evidence for a transient K$\sim$ 15 source (Genzel et al. 1997; Ghez et al. 1998). Accretion models have had to invoke advection dominated flows (ADAF) whereby the plasma disappears into the event horizon before it has a chance to radiate (e.g. Narayan et al. 1998) in order to explain the low near-infrared luminosity. The only previous observational campaign to look for such a variation is described as a 2σ signature in Close et al. (1995). However, those observations suffered from much poorer spatial resolution, effectively combining the flux of a dozen K $\sim$ 14-16 stars now resolved by NICMOS and Keck within the inner 0.5″ of Sgr A* (Ghez et al. 1998).

Various flux estimation techniques will be carefully explored elsewhere, including subtraction of nearby PSF's, deconvolution and modelling the background. For this paper, we present only a *maximum* value for the flux at Sgr A* from the combined median of the Aug. 1998 data set. This is determined by measuring the flux at the pixel closest to Sgr A*, assuming all of this flux is due to a point source, and scaling up by the percentage of flux in the central pixel of the PSF. Clearly this is an overestimate, due to the overlapping PSF's from nearby sources and the diffuse background component. A composite PSF was constructed (with the iraf task "psf") from three bright, isolated, and unsaturated stars in the 11″ field of view. The empirically determined flux in the central pixel is only 5.9% of the total flux of the PSF. Given these caveats, the maximum flux of Sgr A* determined with this method is 1.9×10^{-4} Jy. Note that although this is a maximum value, it is not an upper limit, as the signal/noise is $\sim$ 230 at the position of Sgr A*. The noise level is 2.6×10^{-7} Jy/pixel.

Another estimate for the flux was obtained by subtracting the composite PSF from the position at Sgr A* and determining the flux level where a "hole" appears as compared to the background level. The backgrounds level just outside the inner cluster (0.5″ from Sgr A*), excluding any clear point sources is about 4×10^{-6} Jy/pixel (2 mJy/arcsec^2). A change in flux is clearly evident with sub traction (or addition) of a source as faint as 5×10^{-6} Jy. This method further constrains the 1.6 μm flux at Sgr A* to less than 1.0×10^{-4} Jy. Adopting an extinction law of $A_{F160W}/A_v = 0.179$ and A_v=31.5 toward the Galactic Center (M. Rieke, private communication), the 1.6 μm correction is 5.64 magnitudes. The extinction-corrected flux at Sgr A* from this method is 18 mJy, with a 1.0 mJy uncertainty due to the dominant source of error of estimating the background level.

Variability: Theoretical Background and Preliminary Results Any density enhancement in an accretion flow will be magnified in a characteristic way as it passes behind the black hole from our vantage point (e.g. Hollywood et al. 1995). Quasi-periodic fluctuations will be seen on a similar timescale, even if the infall is not confined to a stable, flattened disk, as long as the accretion isn't purely radial. If the black hole is rotating, the relative orientation of the spin angular momentum vector of the black hole and of the accreting material will dictate the location of the innermost stable orbit (including the effects of frame dragging). For a maximally rotating black hole, this shortest distance would be 1 R_g, corresponding to a shortest timescale of 1.3 minutes; the longest timescale would correspond to 12 R_g, or 57 minutes (the period is given by $P = 2\pi(R_g/c)(R/R_g)^{1.5}$). Note that the Schwarzchild radius for the GC black hole is 7.7×10^{11} cm (twice the gravitational radius R_g of GM/c^2). Another possible variability mechanism could relate to a relativistic jet, which could vary on similar or longer timescales (H. Falcke, private communication; Falcke 1998).

Besides the proposed accretion processes, other sources of variability must be considered: can we distinguish between them? The high stellar density in the GC region is a blessing in disguise; although the severe crowding near the very center is problematic, the field provides a large statistical sample of stars (over 100 within 5″ of Sgr A*) that can be used as variability calibra-

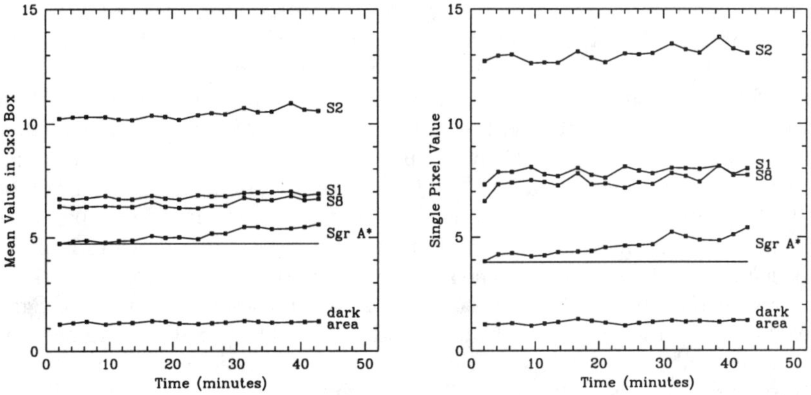

Figure 2. Preliminary light curves of several objects within 1″ of Sgr A* over one orbit from Aug. 1998. A straight line is plotted for reference at the initial flux value for Sgr A*. The dark area is located about 1″ NW of Sgr A*. The last 3-6 exposures are contaminated by additional scattered light in some objects. a). (left) Fluxes averaged over a 3x3 pixel box (0.13″). b). (right) Lightcurves for the same objects for the single pixel closest to the centroid of each source. The background noise per pixel per 128-sec exposure is about 0.1 y-axis units (1 unit corresponds to about 2.5 μJy/pixel).

tors, What kind of intrinsic variability might be expected in objects in the field of view? Fortunately, all of these expected variability sources have longer timescales than that of the accretion processes. Stellar opacity changes typically occur over months/years (e.g. long-period variables). Wind source fluctuation timescales would be on order 1 day; e.g., for the Wolf-Rayet star closest to Sgr A*, IRS16NW, using the stellar radius and wind velocity parameters from Najarro et al. (1997), the timescale would be of order $R_{star}/V_{wind\infty} \sim$ 18 hours. Other effects would be stellar rotations/sunspots/binaries (>days), and gravitational lensing of stars passing behind the black hole (weeks to years; see Alexander and Sternberg 1998). Lensing events may explain the disappearance of K-band sources possibly coincident with Sgr A* between observing runs separated by months. These sources reportedly reached K$\sim$15, but then faded 1.5-2 magnitudes to the detection limit (Ghez et al. 1998, Genzel et al. 1997). Lensing and accretion phenomena should be achromatic in the near-IR and will occur at the position of Sgr A* at our resolution (corresponding to $\sim$ 1000 AU).

The inner arcsec from Aug. 1998 is shown in Figure 1a, made from a median of the first 5 pointings (fifteen 128-second exposures). Figure 1b shows the standard deviation per pixel normalized by the square root of the median value of that pixel. A peak above this "noise" ratio does appear at the position of Sgr A* relative to neigboring regions, however the significance of this peak is uncertain until the effects of scattered light are removed. Scattered light contaminates the photometry of some sources in the last two of six pointings, and especially in the last pointing. The clumpy distribution of scattered light is seen streaking from bright sources when they illuminate the edge of the detector. An attempt is underway to remove the effects of scattered light by subtracting a composite light curve made from many pixels affected by scattered light. The Aug. 1997 data is not affected by scattered light at the position of Sgr A*, and shows no detectable variability; the source of variability may be intermittent.

Figure 2a shows Aug. 1998 light curves for various objects in the inner 1″ constructed from a mean in a 3x3 box (roughly corresponding to the FWHM of the PSF). Figure 2b shows the same thing but only for the single pixel closest to the centroid of the objects. A preliminary analysis of a composite lightcurve of pixels affected by scattered light indicates a sudden increase of about 10% in the countrate at the the 5th dither position, whereas the lightcurve of Sgr A* increases more gradually with time. Small (even 0.01 pixel) alignment errors can also contribute to flux variations. Further statistical tests, photometry, scattered-light models, a time-series analysis, and a discussion of the implications for accretion models will be treated elsewhere.

3. [SiVI] observations

[SiVI], among other coronal lines, has been detected in the Seyfert 2 galaxies NGC1068 and Circinus; a hard radiation field from the AGN, not from stars or shocks, is proposed as the excitation mechanism (Marconi et al. 1996). Such a "coronal" line req uires very high electron densities (n $\sim$ 10^8 cm^{-3}) and excitation (167 eV), which would not be expected from even the hottest stars in the GC, which have $T_{eff} \sim 30,000$ K. One orbit was spent observing the GC in [SiVI] with camera 3 (52″ FOV), the only camera for which the [SiVI] line and continuum filters F196N and F200N are available. No clear detection of [SiVI] is evident at Sgr A*; the 3 σ upper limit for the line flux is 1.0×10^{-18} W m^{-2}, where σ is determined by the local neighborhood of Sgr A*; in areas devoid of stars, σ is about a factor of 12 less. The residuals are larger near the bright stars such as the IRS16 cluster, with large positive and negative deviations much exceeding the residual flux level at Sgr A*. This poor continuum subtraction is due to the large pixels (0.203″), which undersample the PSF and which are very sensitive to which part of the pixel is being illuminated. The non-detection of [SiVI] does not definitively rule out locally high excitation; a suite of observations at even higher excitations, (e.g. [SiVII] at 2.48 μm and [SiIX] at 3.94 μm) would yield a more meaningful interpretation.

4. Paschen α Observations

The interstellar gas in inner parsec of the GC is characterized by a system of ionized filaments as well as winds from an unusual cluster of Wolf-Rayet stars. The filaments are possibly drawn into the GC via loss of angular momentum in clump-clump collisions in the molecular ring $\sim$ 2 pc from the center. The interaction of the streaming filaments with the stellar winds are of interest in understanding the shock physics, stellar evolutionary processes, and mechanisms for accretion of interstellar material onto the black hole. Yusef-Zadeh et al. (1998) and Zhao and Goss (1998) measured proper motions in the radio continuum. Some of the gas appears to follow unbound trajectories; at least one compact

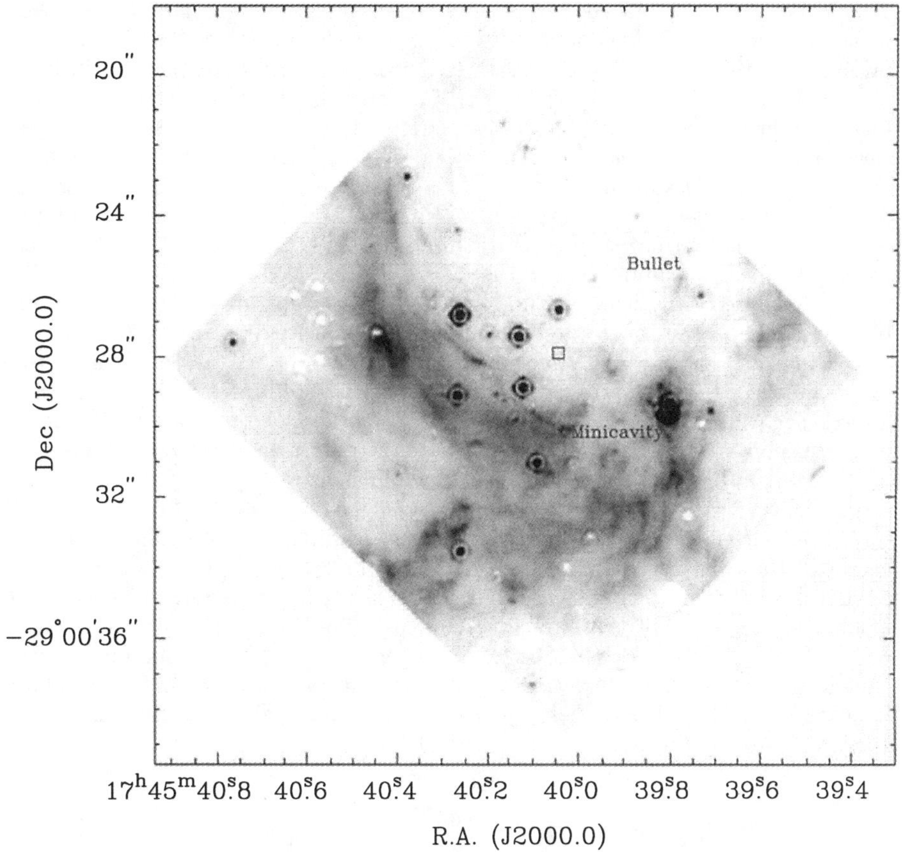

Figure 3. Pa α mosaic of the central parsec (logarithmic inverse grayscale with a dynamic range of 100) with $0.2''$ resolution. Note the wispy, filamentary structure in the ionized gas and the sharp-edged bubble to the lower right known as the "minicavity". The fast-moving "bullet" is also shown (the label is northeast of the feature). A box centered on Sgr A* is shown for re ference.

feature ("the Bullet") is moving faster than 1000 km/s. Tracking the gas proper motions combined with radial velocities from H92 α (Roberts, Yusef-Zadeh and Goss 1996) will reveal the effects of stellar winds and will test whether or not the gas is bound to the gravitational potential.

The central parsec was imaged in 1% filters F187N (Pa α+continuum) and F190N (continuum) in order to produce a map of the ionized gas distribution. The goals are to map out the proper motions of the ionized gas, which is subject to non-gravitational forces, and to image the Sgr A* region in Pa α. The map presented here in Fig. 3 was constructed from 12, 192-sec exposures in each filter obtained on March 5, 1998. The proper motions will be discussed in a future paper, as the second epoch data were obtained in Oct. 1998. The images were aligned with a cross-correlation technique, and the continuum was scaled appro priately and subtracted to produce the Pa α image. The area near Sgr A* will also be observed for any evidence of material flowing into or out of Sgr A*; these faint thermal structures within $0.5''$ of Sgr A* would be obliterated by Sgr A* itself at radio wavelengths.

The complex structure in the NICMOS Pa α map reveals filaments, bubbles, and tails in the inner parsec with improved clarity as compared to images of the ionized gas at radio wavelengths. Pa α emission is also evident in the atmospheres of the hot stars, as seen in the positive stellar residuals. Stars that appear negative may do so from Pa α absorption and/or because of differential extinction between F187N and F190N (note the grayscale is reversed in Fig. 3).

A bubble-like feature known as the "minicavity" is present a few arcsec southwest of Sgr A*; the sharpness of the edge is shown for the first time in Pa α. This image reveals a star (a Pa α source) located very near the geometric center of the mini-cavity, implying that the mini-cavity is a bubble created by a wind from this star. Proper motions and possibly spectroscopy are needed to confirm whether or not the minicavity is expanding from this central star. If it can be shown that there is no outflow from the central object in the minicavity, that would imply that the wind is temporarily inactive or that the minicavity is formed by some other process, such as gravitationally focused winds from the IRS 16 cluster by the black hole.

5. [FeII] Observations

The [FeII] camera 1 observations were taken on Oct. 27-28, 1997 and July 8, 1998. In the first epoch, a total of 8 320-sec exposures were taken in each of the 1% filters F164N ([FeII]+ continuum filter) and F166N (continuum). The July observations were pointed further southeast to better cover the "mini-cavity" and were made of six 192-sec exposures in each filter. Although there are small proper motions of the stars and gas in the two epochs, they are nevertheless combined here in Figure 4 to improve the signal/noise and to generally show the distribution and morphology of [FeII] emission. Comparisons of the [FeII] and Pa α distributions will be very useful in identifying regions of shock-destruction by grains in the ISM and/or regions of relatively low ionization.

The southern edge of the mini-cavity is shown for the first time to be outlined in [FeII]. Some correspondence is seen between the [FeII] emission and the Pa α distributions, but there are notable exceptions. In particular, the bright

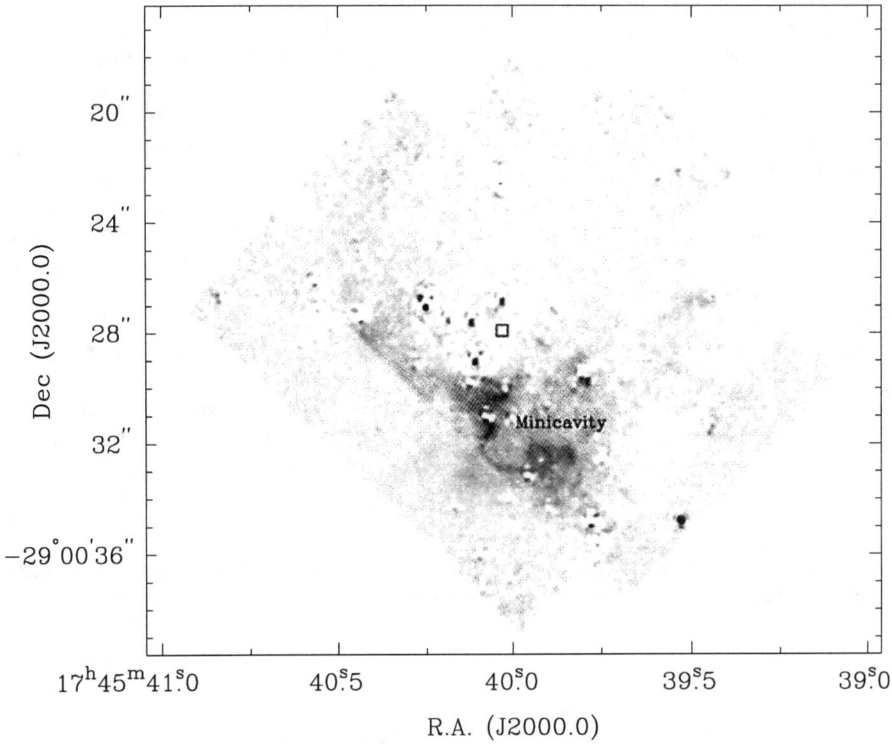

Figure 4. [FeII] mosaic of the inner parsec, combining data from Oct. 1997 and July 1998, displayed in inverse linear grayscale. The map is smoothed slightly with a Gaussian with σ=2 pixels. Note the [FeII] concentration to the south and east of the minicavity. The strong peak at the star in the SW corner coincides with the "AF" star. A box centered on Sgr A* is shown for reference.

peak in Pa α at IRS13 (northwestern edge of the minicavity) is nearly absent in the [FeII]. Since ground-based spectroscopy with a 2.2″ wide slit (Lutz, Krabbe and Genzel, 1993) show a peak at IRS13 in [Fe III], the dominant species in this region is likely to be Fe^{++} (which requires 16.2 eV) rather than Fe^{+} (which requires only 7.9 eV). The harder ionization radiation could originate from the IRS16 cluster as well as IRS13(E) itself. This also implies that the southern part of the mini-cavity is at least partly shielded from this harder radiation. The peaks of the [FeII] distribution are found at the eastern edge of the mini-cavity, in contrast to the Pa α, which has the strongest peaks near IRS1 and IRS 13. Also, the [FeII] appears to delineate a region to the southeast of the minicavity opening to another cavity, possibly indicating a "blowout" (also visible in the Pa α). The [FeII] peaks indicate regions of lower ionization and may also signify where the Fe gas phase abundance is enhanced by grain destruction.

The 1.644 μm [FeII] line is blended with HI (12-4) (Br 12) at 1.6407 μm in the 1% filter, but clearly it is not dominated by the HI line in the diffuse

ISM, since the [FeII] peaks do not correspond to the Pa α peaks. The blending of [FeII] and Br12 may manifest itself most strongly in the stellar atmospheres of some of the GC sources; the Br 12 line may be in emission or in absorption against the stellar continuum. The bright "AF star" in the southwest corner of the mosaic is notable in its line emission. This star has strong HeI and HI lines, and the blended Br12/[FeII] line has been previously detected at a spectral resolution of R$\sim$800 (Libonate et al. 1995).

A curious striated, wave-like pattern is clearly seen in Pa α as well as in the [FeII] image on the southern edge of the mini-cavity, near $\alpha = 17^d45^h39.9^s$, δ=-$29^o00'32''$. The 3 strongest [FeII] peaks match the 3 strongest Pa α peaks in this region and have a characteristic "wavelength" (in an east-west cut) of $\sim 0.74''$ (about 18 pixels), corresponding to 8.9×10^{16} cm. Higher harmonics may also be present (although not always in both lines). This may correspond to a cooling length pattern formed by a driving bulk gas flow, possibly related to the flow from east to west as observed by Yusef-Zadeh, Roberts and Biretta (1998). High spatial resolution proper motion measurements will determine the dynamics of this region, and will be critical to understanding the processes forming these structures.

Acknowledgments. We would like to acknowledge Heino Falcke for clarifying discussions on accretion and jet phenomena.

References

Alexander, T. and Sternberg, A., 1998, submitted to ApJ

Close, L., McCarthy, D. W. and Melia, F., ApJ, 439, 682

Eckart, A. and Genzel, R. 1997 MNRAS 284, 576

Falcke, H. 1998, this volume, p. 148

Genzel, R. Eckart., A., Ott, T. and Eisenhauer, F. 1997 MNRAS 291, 219

Ghez, A. M., Klein, B. L., Morris, M. and Becklin, E., ApJ (in press)

Hollywood, J.M, Melia, F., Close, L. M., McCarthy, D. W., DeKeyser, T. 1995 ApJL 448, 21

Libonate, S., Pipher, J. L., Forrest, W. J., and Ashby, M. 1995, ApJ, 439, 202

Lutz, D. Krabbe, A. and Genzel, R. 1993, ApJ, 418, 244

Lynden-Bell, D., and Rees, M. J. 1971 MNRAS 152, 461

Menten, K. M., Reid, M. J., Eckart, A. and Genzel, R. 1997 ApJ, 475, 111

Marconi, A., Van Der Werf, P.P., Moorwood, A. and Oliva, E. 1996, å, 315, 335

Najarro, F., Krabbe, A., Genzel, R., Lutz, D., Kudritzki, R. P., Hillier, D. J. 1997 A&A, 325, 700

Narayan, R. Mahadevan, R., Grindlay, J. E., Popham, R. G., Gammie, C. 1998 ApJ 492, 554

Roberts, D. A., Yusef-Zadeh, F. and Goss, W. M. 1996 ApJ 459, 627

Yusef-Zadeh, F., Roberts, D. A. and Biretta, J. 1998, ApJ, 499, 159

Zhao, J.-H. and Goss, W. M. 1998 ApJ, 499, 163

The Central Parsecs of the Galaxy
ASP Conference Series, Vol. 186, 1999
H. Falcke, A. Cotera, W.J. Duschl, F. Melia, M.J. Rieke, eds.

Discussion Session I: What is the Central Matter Distribution?

E. E. Becklin[1]

Division of Astronomy, Department of Physics & Astronomy, University of California, Los Angeles, 405 Hilgard Avenue, Los Angeles, CA 90095-1562

A. Cotera[2]

Steward Observatory, The University of Arizona, Tucson, AZ 85721

Abstract. This is an edited transcript of a discussion session about the mass concentration and the evidence for a central black hole in the Galactic Center that took place during the Galactic Center conference "The Central Parsecs" in Tucson, Arizona, Sep. 7-11, 1998. The session was chaired by the moderator as indicated in the author list. The discussions were taped, transcribed, and finally edited by an editor who is listed as co-author. The contributions of the conference participants are preceded by their names and thus references to specific aspects and ideas of this discussion session must also include a reference to the respective speaker(s), e.g., in a footnote.

Discussion

Becklin: We now have a discussion session that I've been asked to organize. We've heard talks which cover several subtopics and I have a few others that I've added that I'd like to bring up, but we mainly want to address the mass distribution in the central parsec in this discussion session. The best place to start is with what do we know, then move on to questions to be asked, and end with future observations and anticipated results. In particular, I want to make sure that we don't miss things we should be doing with the unique NICMOS data set. You've now seen the beautiful data we have gotten with NICMOS which we have yet to analyze and we want to make sure that our analysis correctly addresses the outstanding questions. As a member of the NICMOS instrument team, part of my contribution is to assist Marcia Rieke, the PI on this project, to insure we answer the questions of how the NICMOS data can best be used to compliment the other data, as well as discuss the relationship of the NICMOS

[1]Moderator of the discussion session

[2]Editor of the discussion session

data with the other excellent NIR data that have been presented in this session.

What do we know? We know there is a mass concentration of at least 10^{12} $M_\odot$ pc^{-3}, concentrated in a central source which is about 2.6 $\times 10^6$ $M_\odot$. The mass concentration is at or near Sgr A*, within $\lesssim 0''.1$, which is within the systematic errors. However, I think this is still an open question, because it is a 1σ result at this point. Are they the same? The mass concentration being at exactly the location of Sgr A* is a very important question, and we should not just assume it is until we actually measure it. The mass of Sgr A*, as we have heard from Mark Reid, has to be greater than 3×10^3 $M_\odot$. In addition, we have seen that the sources near Sgr A* may be variable. Certainly, I know from the Keck data we feel that they are variable, and they are located near the mass concentration/Sgr A*. During this discussion I hope that Andrea Ghez will be able to say a little bit about some of the analysis she has done on the variability, because I know she did not have a chance to cover that in her talk. In the central parsec, little is known about angular momentum. Reinhard Genzel alluded to it, but did not make any strong comments regarding the angular momentum of the stars. We do not really see a lot of angular momentum right in the central region, isn't that right, Reinhard?

Genzel: Well, the He stars definitely show angular momentum on a scale of half a parsec, for the innermost stars that is something which we can now do with the velocity determinations we have for the stars nearest to Sgr A*. I think that, as Andrea Ghez showed, and as I presented in the comparison of θ and R, we may be seeing the first hint of evidence that in the innermost regions we have elongated bound orbits.

Cotera: I had a question on that; both groups mentioned stars in bound orbits. Why should the stars whose velocities have been measured be in bound orbits rather than simply passing through the plane? In the viewgraphs of both groups, the positions of the measured stars seemed to be along straight lines, they didn't seem to have any curvature, although you also said you were expecting to start to see curvature around Sgr A*. Why do you think that there has to be curvature, that the stars must be in bound orbits, rather than just coming from outside the central regions and just zipping past?

Genzel: If it is not gravitationally bound to Sgr A* or the innermost star cluster, then it could not be bound to the Galaxy and we would have a serious problem.

Becklin: If it is not bound, then it would have to be a runaway star, and that would be weird. Also, the stars here are very blue, with the same brightness as the other stars in the region, which, as Marcia [Rieke] convincingly showed, indicates they are main sequence stars right at the tip of the main sequence. For them to be runaway star at this evolutionary stage would be really weird.

Gezari: Along exactly the same lines, neither group sees curvature in the motions of these stars. In looking at Andrea Ghez's data, all of the stars did show orderly motion. It seemed that most of the stars had motions which were radial to the position of Sgr A*. This would seem to indicate that either these are extremely eccentric orbits, or we are fortuitously seeing all of these orbits edge-on. I believe there is only one star on the plot with a measured velocity that was approximately circumferential, and I am curious what that tells us.

Becklin: That is an interesting point, but you have to be careful when extrapolating as there are very few orbits measured at this point.

Gezari: There is one showing curvature but it is not radial to Sgr A*.

Ghez: Within the errors there is no significant curvature for any of the orbits.

Gezari: So are we seeing all of these stars orbiting Sgr A* edge-on? It seems unlikely.

Becklin: No, they certainly are not all edge-on.

G. Rieke: It seems to me that at this point there is the possibility that we are guilty of a selection effect. We only look for orbital shapes on the stars with the largest apparent proper motions (as you can see in Andrea's [Ghez] plot). This might be because we are picking out the stars which have the most extreme edge-on orbits at some level.

Genzel: However, that is not what Andreas Eckart and I did. We took all the proper motions, (and future analysis will follow the same method as Andrea Ghez has just to make sure), and just decomposed their trajectories in the V_X, V_Y into projected V_R,V_θ. Doing this we see that in the central parsec we get the same information you have on here, right? In the central arcsecond or so there seems to be a preponderance of radial orbits.

Ghez: The difference is not statistically significant.

Genzel: In the central arc second, it is.

Ghez: Within that radius it's not. There is not a 3σ difference in the measurements

Genzel: We will have to look at this.

G. Rieke: There was actually a question I wanted to ask. It would be very interesting to find out if these velocities align in any interesting way, like the CND, or are they counter-rotating. At one point, Reinhard, I thought you felt that maybe the massive stars were counter-rotating. Certainly one of the things that has been going on within the central parsecs is some degree of dynamic evolution. You would very much like to know if some of the structures we see are causally connected with star formation. We did not hear anything about this in detail in these talks, I wonder where these issues stand.

Genzel: On the He stars where we have radial motions on a scale of 10″, it looks like a pretty convincing case.

Becklin: What was the statistical significance?

Genzel: There are 25 stars. If you put them in a diagram of the coordinates versus the velocity, they seem to occupy largely two quadrants. That looks like a significant result.

Becklin: And it was counter-rotating to the Galaxy?

Genzel: It was counter-rotating to the Galaxy.

Becklin: Even though the M stars, even those within 30″ of Sgr A*, are rotating in the direction of the Galaxy?

Genzel: Yes, that is correct. This is not unexpected, because we have got to get these objects into the central regions somehow. They have to form locally, because they could not have drifted to their present location in a dynamical time scale. So they may have had a very unusual angular momentum when they formed, their relaxing time scale would not be the same as the dynamical time scale. These innermost stars are not that short-lived, so they may be more relaxed than the massive stars, but not fully relaxed in this dense environment as the oldest stars. Still, you would expect dynamical effects to show up.

Actually, I think with the proper motion data we now have we can begin to look for such effects. One problem will be that the proper motion data sets still do not distinguish between late-type stars and early-type stars. You must be able to separate these stars in order to throw out the old stars and then look for effects that can be attributed to the younger, lower mass stars. I think the data sets are good enough that you can begin to ask questions whether from the proper motions you are able to see coherent motion.

Ghez: I think we are also entering into an era, very quickly with the advent of adaptive optics, where the third dimension, the line-of-sight velocity, will be accessible; which, in conjunction with the data we currently have, will be very powerful in determining these orbits as well as their masses.

Yusef-Zadeh: You really can't say whether there is isotropic motion within the central arcsec or so.

Ghez: No.

Genzel: You need the spectra. If they have spectra ...

Becklin: They'll have spectra. They will also be able to examine any spectral features.

Gezari: The fact of the matter is, that what we have seen today is a real picture of the real sky, and those are supposed to be real positions.

Becklin: Those are real positions.

Gezari: Yes, taken a year apart, and so we're seeing a movie right there of the motion of these stars, and they're big motions, and so they have to correspond to some path through the sky, and I'm trying to imagine how those paths can be connected to Sgr A*, and I'm having a hard time accepting that they're edge-on.

Becklin: Well, I don't think they all have to be edge-on.

Gezari: Nearly.

Genzel: Why?

Gezari: Or extremely eccentric orbits.

Becklin: Okay, well, it's a good question, and a theorist, somebody that knows orbits ...

Melia: I hate to belabor the point, but I think this argument that these stars are unbound to the black hole, they're being bound to the Galaxy, is probably not right in this context. The reason is that if you have a star which has a velocity of 200 km s^{-1} or so, a sizable distance away from the black hole, so it is not bound by its gravity, and then it has a trajectory which takes it more or less radially in, it will have a velocity when it gets that close, which is about a factor of $\sqrt{2}$ higher than it would be if it was bound. So I'm not sure that we can rule out the fact that these stars may not be bound, but they're just forming and coming in radially. In this case, we would be overestimating the mass of the black hole, probably by a factor of $\sqrt{2}$, depending on how many of the stars are involved.

Genzel: But these are $\lesssim$ 1000km s^{-1} stars, not 200 km s^{-1}.

Melia: But the thing is that if the star itself is travelling at 200 km s^{-1}, when it falls in that close, it will have a velocity close to 1000 km s^{-1}. So some of these may be unbound.

Becklin: I would say that is fine, except for the fact that they are at the tip of the main sequence, assuming that Ghez's analysis is correct. I have a hard time believing these stars are just randomly wandering in, although that might be the situation.

Townes: The fact that these are all straight lines is rather puzzling at the least. It seems to me the stars which are moving across the line of sight, should be the ones that are orbiting the black hole. At least the largest fraction of them should be in bound orbits, rather than simply falling into the plane. I wonder if anyone has made a calculation to see what the probability is of seeing something like a random infalling star producing the motions we see here.

Becklin: That would be a good calculation. What would the probability be of seeing this kind of motion, to first order?

Townes: The higher velocity ought to be the ones where you're looking perpendicular to the plane of motion.

Becklin: I may add that if you add on the Max-Planck data, which isn't here, the trajectory continues to go straight, right up to 1992. So it's even more puzzling than what we show here.

Yusef-Zadeh: I have a quick question. Did both of you use IRS 21 as a reference star?

Ghez: No, I think the different groups used different algorithms for establishing the reference frame. As I think Andreas pointed out, registration is our limiting problem. The work we presented just minimizes the net velocity between epochs to register two epochs. So, if anything, we are underestimating the velocities. Every year we take all our stars, and match them, and then minimize the net dispersion and use that as our registration: allowing for only translation and rotation. The rotation is tiny, and of course the translation is significant, because we don't center it the same from year to year.

Becklin: So it was a conservative way of doing it.

Ghez: True. No one star sets our registration frame. In addition, we can take just the brightest stars, or the whole set, and basically it gives you the same answer. This is the registration that produces the minimum motion.

Eckart: The reference stars we use are basically stars of the outer cluster, because when we started out we had too few stars, so we use like 10 or 20 stars. The criterion for them being reference stars is that they can be reasonably well isolated, are reasonably bright, and we can determine the positions of those to within 5 or 10 mas. These sets of stars are used as reference stars.

Eckart: One more point before we leave the question of orbits. I think what would shed more light on the subject, is to search for stars which are equally fast or at least very fast in the outer regions, because if these are all straight trajectories, then one should expect some of the stars to be in the outer regions as well.

Ghez: I'll respond to that. In our sample there is no bias towards finding fast stars just near Sgr A*. However, we only see the fast movers; in our sample, we don't see these fast movers at the outer edges of the data.

Becklin: Within 5″! We haven't gone beyond that.

Eckart: With the same flux density level as the stars near Sgr A*?

Ghez: Yes. The criterion for being included is position independent.

Becklin: I'm glad you brought that up, because Andrea has tried very hard to make this very unbiased. She has used every star in the central 5″ that can be detected, measured its velocity and included it in the velocity diagram.

Melia: I'd like to go back to one of the previous points. There is a caveat to the mass limit of Sgr A* which Mark Reid presented. If it is not a stellar-like object or compact object, for example if it is gas, trapped in the potential well of the cluster, then its mass could actually be very small. It could be $10^{-3}M_{\odot}$, then it would trace the position of the centroid of the cluster rather than the black hole candidate. I don't believe that at all, but it needs to be mentioned.

Reid: You would have put the object right in the center of the stellar cluster.

Townes: At any one time, like right now, it's velocity might just happen to be zero.

Becklin: That's always possible.

Melia: I'm not saying I believe it or I don't, I'm just saying that if we are going to consider all scenarios, this needs to be included.

Genzel: Actually, Fulvio and I, and others, have performed the calculation and showed that the results are actually inconsistent with the radiation characteristics of Sgr A*.

Becklin: Okay. Next. Is the mass concentration a black hole? Well, from an astronomical point of view, Andrea Ghez, Andreas Eckart and Reinhard Genzel have made a strong case that it is. I think from the physics standpoint, we still don't know for certain that it's a black hole at this point. We just know that there's a large mass concentration within the central 0.01 pc.

Townes: Do we have really any decent model for that mass concentration?

Becklin: No, there is no decent model, except for a black hole. However, the black hole models also predict that there should be an accretion disk, infalling mass, and there should be some luminosity at other wavelengths than currently measured. There is no smoking gun.

Townes: I think we'll hear some arguments why this is the case in tomorrow's session.

Becklin: Okay, but I would at least like to bring up the question, how can we show it observationally? For instance, can "we" push further in towards Sgr A*. We aren't anywhere near, in terms of the resolution or getting in close with the Keck. We can get 3 times closer in than currently, we would obviously have to go to fainter magnitudes.

Falcke: I'll make this point tomorrow. I think there is in fact, a way to prove it. With mm-VLBI we could have sufficient resolution to see the photon horizon of Sgr A*—literally a black hole. If a black hole is there, with that mass, we have to go to much higher resolution to see it, but we could see it.

Townes: What would you say is the proof of a black hole?

Ghez: Relativistic orbits

Becklin: That and actually seeing the accretion disk.

Townes: Relativistic velocity, again, you still ultimately have to rely on some physical model. Is there anything else which could give this?

Becklin: It's a good question, and I don't have the answer, obviously.

Genzel: I guess if we could prove that the theory of accretion is good enough, that it doesn't radiate, and it doesn't have a surface, that would be pretty good.

Narayan: You need some such argument, some argument that shows that this object doesn't have a surface. Because just having relativistic velocities is not enough, since we know that such objects exist at lower masses. We would then be depending on theory which tells us that it cannot happen with $10^6 M_\odot$. Can we be so confident of the physics?

Bower: I have a question along a different line: what about the variability of certain IR components around Sgr A*? How could that be used to enhance our understanding of the black hole candidate?

Genzel: Okay. Taking these things together, if we could get a bona fide lensing event within a parsec or so, then that Einstein ring would be ~10 mas. If we can be even luckier to have a lensing event which is even faster, and has a smaller Einstein ring, then you could get information at scales which would be really impressive.

Becklin: So lensing would be one way to prove there is a black hole?

Genzel: Lensing could potentially be a very interesting way of doing it. This of course is very difficult, because once we don't have NICMOS, the sensitivity for such an observation will be lost. I would say there is now a fairly good theoretical calculation which indicates that with a sensitivity of about 15th to 16th magnitude at K-band, the probability of seeing a lensing event in which a previously undetected star becomes visible is about a few percent. Then, with a lensing time scale of such an object being a few years, or perhaps even ten years, the probability to see a lensing event at any given time is about ~ 10–12%. So, it's not hopeless, and it will become even more feasible if you could push the limiting magnitude down to 17th, 18th, or even 19th.

Becklin: Well, one way I can think of, Reinhard, would be to have a dedicated system on one of the VLTs with adaptive optics.

Genzel: I'm going to have trouble getting even limited time on the VLT for this project!

Becklin: Kidding aside, we are talking about fundamental physics here—looking for a black hole. And it seems that if we can get our act together, we could dedicate a major portion of one of the four VLTs to this project.

Wardle: Is that really going to tell us if there is a black hole there, rather than just some very compact object that can produce gravitational lensing?

Genzel: No. I didn't intend to represent this as sufficient proof, I'm just saying that you could push inward towards Sgr A* in a fairly dramatic way, down to a resolution of a few mas.

Stolovy: A comment on variability in general; stars in the central stellar cluster, within a few tens of arcseconds, do show some variability over a year, and I think we may have some evidence that a few may be variable on a time scale of a few minutes. Although as I said, variation within the group of stars nearest to Sgr A* is almost nil: except the source S3 near Sgr A*. The bright stars that are farther from Sgr A* do vary from epoch to epoch, but I think that may be a result of pointing errors. We have to be careful because variability in this crowded region is going to be tricky. We have to first determine what is the intrinsic variability, before we look at the ones perhaps closest to Sgr A* in the hopes of catching a lensing event. In addition, as Andrea has shown, the stars near Sgr A* are moving, so you need to know that you're tracking the same objects from year to year.

Reid: Isn't the key to gravitational lensing that the event is achromatic? Can one, in the IR, do the sort of switching between filters which is required to observe such an event.

Becklin: Presently we cannot, but it's possible in the future we will be able to, maybe even build a customized system: once we dedicate one of the VLTs ...!

Ghez: I'm afraid I may have left you with some ambiguity about what we were seeing in terms of variability or what sort of variability we believe we see in our map. Taking our very strict criterion for what we believe is a star, we need to remember we made two assumptions. First, the proper motion sample contains only those stars which are seen in each data set (four epochs), some stars are lost because the data from the 1996 map is poor compared to the other three. Second, many of the stars are rejected from the proper motion sample because they are near the edge of our field of view However, there is one, and only one, star which consistently shows variability that is not missed for these two reasons. And that's S3, and that's the source in the sample which is closest to Sgr A*. So there is something special about S3.

Eckart: Yes, the unusual nature of S3 was very clear from the beginning when we first started to look at the time tracks and S3 continues to be strange.

Ghez: Especially when compared to the entire field of view.

Becklin: Also, it is close to Sgr A*, within a few sigma or ...

Gezari: Are you saying that S3 could be an example of a lensing event?

Ghez: That's one interpretation.

Becklin: It's possible. It could be almost anything at this point. First of all, we don't know exactly where the mass concentration is. We're assuming it's on Sgr A*. But it still may not be.

Narayan: On the very variability of Sgr A* in the K band: I would like to get the actual numbers, if you don't mind. What is the flux of Sgr A*?

Eckart: First of all we have to agree on whether it was detected or not. In the two epochs where we think we detected it properly, it has a flux at K of 15 magnitudes.

Narayan: Which is how many mJy?

Eckart: It's about 13 mJy corrected for extinction.

Becklin: And this now is not S3, this is S12.

Narayan: This is S12, right on Sgr A*?

Ghez: Right.

Eckart: As a conservative upper limit, I always use $m_K \sim 16.3$.

Ghez: Just to add a little fuel to the fire, I will make the statement that I have never seen something with $m_K \sim 15$ at the position of Sgr A*; and our limit, at that *exact* location, is $m_K = 16$.

Eckart: Sixteenth?

Ghez: Sixteenth. So if you allow that position to move, you can fall on something that's 15th magnitude, if, for instance, you put it on S3.

Bower: How accurately can you know that?

Ghez: We only have the two masers, IRS 10EE and IRS 7, to give us the position. Of course with these two you can determine your orientation and your pixel scale, so you don't have any redundancy. But a formal 1σ error is 10 mas. But what you really want is another maser to make sure you understand your uncertainties.

Stolovy: Are you going to say anything about the distortions in the camera?

Ghez: In fact I have not mentioned that yet. Some of the pictures I showed have a very odd boundary, and that's because we have taken advantage of the fact that we're on an alt-azimuth telescope and we turn the rotator off. So we let the image rotate on the array, which averages over the camera distortion. We also have solved for the distortion of the array, because of course we measure the Galactic Center at many different places on the array. So I think we have correctly addressed the distortion. I don't think distortion is an issue for us.

Stolovy: What is the scale of the total correction?

Ghez: I think it's one pixel edge to edge.

Becklin: Around 20 mas.

G. Rieke: I think I can give a little help to Ramesh. Susan, you derived, I thought, a limit of 50 μJy.

Stolovy: That was a source which was within a pixel of where I thought Sgr A* was, but my error box was larger than Andrea's [Ghez]

G. Rieke: But the point is, that the value you get is very consistent with $m_K \sim 16$ and normal reddening. So I think there's kind of a convergence at a plausible limit, right now, $m_K \sim 16$, $m_H \sim 18$, which can be used in theoretical modelling, which is what Ramesh [Narayan] is after.

Becklin: That's a very good point, and I'm glad you brought it up George. George is correct in that everyone is converging on an upper limit of $m_K \sim 16$. However, keep in mind it is a limit.

Narayan: If this detection is correct, even when taking into account the possibility of slight variability, it's a tremendous constraint on models. There are a whole range of models which one could make where the flux is quite a bit below this limit. When it's actually detected, and you can have a definite flux, even some variable flux, there is some value to pin your models on. Even an upper limit immediately constrains the models. That said, it is very important that we figure out whether this detection is right or not.

Becklin: These are all important arguments that we have to use as we go for more telescope time, especially with the upcoming availability of adaptive optics.

Backer: Can anyone in the audience state something about the prospects of finding more masers in this region to help constrain the location of Sgr A*?

Reid: We have found some more. They are variable, they change by factors of a few. I would say there are about 8 or so known right now, and probably with current observing techniques we can push that up to a dozen or so.

Backer: Is this within the region where you can do the cross-registration with the NIR data?

Reid: Not within the central 5″. In the inner 5″ field there are 2, and if we're lucky we'll get to 3 or 4 more.

Townes: What we need is a good maser to spring up in material circulating around the black hole!

Reid: One thing that will help considerably is the VLA upgrades. In the future we'll be able to go a lot deeper.

Becklin: I should also say for the future, which Marcia [Rieke] didn't mention, but NICMOS may get a cryocooler adaptation in the next year or so, and NICMOS could come back on in 1999?

M. Rieke: It's slipped to May of 2000.

Becklin: That would really open up the determination of the proper motions if NICMOS could take another set of observations. However, there is still uncertainty as the cryocooler has to be approved and it has to work properly.

Getting back to the outline for this discussion session. I feel angular momentum is something that ought to be looked at, because how well you can actually get some matter into the black hole depends both on orbital mechanics and angular momentum. You cannot get an accretion disc when you have a lot of stars randomly going around, at least not very easily. So I think the theorists need to look at that. Exactly how much feeding can we expect and how much angular momentum is there?

I have a few other random questions. Why are there so few red giants? This came up in Genzel's talk, but there are still uncertainties. Why are there so few red giants in the center, unless there is some sort of segregation that's gone on, especially if the present models are correct and you can't get rid of the atmosphere? There are missing red giants in the center, there's no doubt about that.

Roberts: Before we move on, I have one more question. Is there some way to tell if the S12 burst is a lensing event or just an intrinsic flare or something similar?

Becklin: Not at this present time. In fact, the different groups still do not agree on S12. The Max-Planck group has seen variable activity. We still only see something very faint at that position. We have to really search for it in our data. Now we just may have missed it, it may have just happened to have come up at the time of the Max-Planck observations.

Roberts: How close in time were the different observations?

Ghez: Ours was June '96.

Becklin: When do you see it the best?

Eckart: '97 and '96. In '97 our observations were in June or July, I think.

Ghez: We have May '97 data and S12 is not there.

Genzel: I don't want to be too technical, but that's not correct according to what you presented.

Ghez: What we see is at the correct position, if you claim that S12 is really on Sgr A*.

Genzel: No, I'm not saying that, because that involves where you place Sgr A*. I'm just saying, relative to the other sources, there is a third source which we call S12. That may be at the same place you are calling Sgr A*. What you see is several things coming together. There is the issue of placing things on or around Sgr A* and a possible variable source. Perhaps what you should all take away with you is that it is a very confusing region.

Ghez: There are too many stars.

Genzel: But in terms of sources showing up and changing and so forth, I'm actually encouraged. I think this particular source, whatever you call it, S12, showed up in the vicinity of Sgr A*. You have to remember we have very different resolutions, and for us there is the deconvolution involved. I think there is a source there, it's probably $m_K \sim 15 - 15.5$, right? It's not 16?

Becklin: We see it at about 16. But we also are not asserting it is at the location of Sgr A*. For comparison, we saw S3 clearly, then it went away. You saw S3 and it changed. We don't see it as consistently as you do, and I don't quite understand that. But it was there, very bright, in 1995, just like you saw it, very bright.

Bower: Do you think there should be a monitoring program at the VLA looking for something like this. Looking for exactly the fluctuations you see in the IR, in the radio.

Becklin: It would be good if we could find a correlation. Of course there would be a different time scale, and it would have to be a different mechanism.

Zhao: What I found was a large variation in Sgr A* between 1991 and 1993. For example at 1 cm there is currently a flux density of 1 Jy, whereas before 1993, it was 1.5 Jy. It was 50% higher than now.

G. Rieke: Just to play devil's advocate on variability, what is the probability of very close binaries? When you're looking at a very small percentage of the population, it strikes me that you could have tight binaries and that would give you a mechanism for variability that would be totally boring.

Becklin: It is a possibility. I think from what we see of S3, it was very bright and then has gone away, basically. So that is not likely to be a binary. Now I have to add that we haven't checked our measurements with those of the Max-Planck group.

G. Rieke: It would be amazing if none of these stars were close binaries.

Becklin: In fact, Mike Werner and I had an experiment to look for these binaries because they were predicted by some models.

Narayan: But would you expect such binaries to show such large proper motions? Just some double-star that happens to be very close to the Galactic Center? It could be moving like crazy.

G. Rieke: I would argue we're only seeing the tip of the iceberg on proper motion. Stars that have the largest radial motions we haven't picked up yet. Maybe in 10 years we can answer your question.

Narayan: I know there's a very strong limit on the proper motion of S12.

Eckart: Again, if you believe the identification.

Ghez: At this point, we're basically working in the noise. Since the position of Sgr A* is very special to us, we are willing to call something a source that elsewhere, with very different criteria may be missed. That was the whole purpose of showing you that messy plot, which is, if you put Sgr A* someplace else, you would still find a source there, at the level that we're starting to say something is a source. I think that to improve the certainty, we will have to wait until we go to AO and get much deeper images at that kind of resolution. Then we can have a more meaningful discussion about variability. Right now I think it is a little premature.

Becklin: Before we go to lunch, I'd like to go through some of the special things that can be done with the NICMOS data, since I'm a team member. I've just put down a couple of the unique opportunities we have with NICMOS, and I'd like to encourage more from the NICMOS team members here, or if anybody from the audience would like to contribute to the discussion.

Genzel: Will you get deep images, say 19th or 20th magnitude?

M. Rieke: We should be able to by co-adding all the images we are going to get. What we have presented here is, basically, as deep as we're going to get at one epoch; which is limited primarily because of what we can get in one orbit.

Becklin: You're looking at the question I have up here of what is the radial dependence of the star concentration?

Genzel: Not only that, but the star formation history, you could go a little off the center, right? For instance, if you are 10″ off, then you do not have the same crowding issues. Could you then get to a limiting H magnitude of 21 magnitudes?

M. Rieke: We have some data that probably should get very close to somewhere near that limiting magnitude, we just haven't reduced it yet. Also whenever you take an image with one camera you get data from the other cameras. Of course Camera 3 is usually useless because of the focus problems. However, with Camera 2 you can get very useful parallels. We just haven't looked at those yet. I'm not certain that we are going to be able to reach the limiting magnitudes that NICMOS is capable of because we just were not able to set up incredibly deep exposures. If it ever gets revived maybe we can get GO time to do that. During the initial planning stages, there were difficulties in allocating GTO time and, as you can imagine, there was a conflict between how much should be spent on the Galactic Center, how much on the Hubble Deep Field.

Sanders: I'm curious about the other component, the mass distribution in the central parsec, that is to say the extended component. You know it's there.

Becklin: The stars?

Sanders: Specifically the old stars, not the bright young new stars. One sees evidence for a mass distribution, the point mass dominates the central parsec, and then does something else take over?

Becklin: Oh yes, you see the red giants that are part of the older population.

Sanders: Is there any hope of seeing the diffuse component?

M. Rieke: In fact, I see a very big difference in the DC level between the central images and just $2'$ away. When we get a little bit further into the data reduction, I think we'll be able to tell you the colors of this diffuse component between the stars.

Becklin: You're not resolving out the extended component?

M. Rieke: Right.

Sanders: Do you know enough to say whether or not there's a core or is there a cusp?

M. Rieke: We did some old work on that.

Becklin: Okay, I've been told by the organizer, we have to go to lunch. I've completely overrun the allotted time. I just want to thank you for the NICMOS ideas, if you've got others, come to Marcia and myself or any others NICMOS team members. I just want to say a few things about the future. Adaptive optics is working and is on its way to a 6m, 8m, or 10m telescope near you: it will revolutionize these types of GC observations. Also, something that has come out of this meeting, I'd like to consider dedicating a major portion of one VLT to this program. I don't know if that's realistic, but with all of us working together under the leadership of Reinhard Genzel, anything is possible!

Session II

Can VLBI Constrain the Size and Structure of Sgr A*?

The Central Parsecs of the Galaxy
ASP Conference Series, Vol. 186, 1999
H. Falcke, A. Cotera, W.J. Duschl, F. Melia, M.J. Rieke, eds.

Sgr A* – Propagation Effects, Variability and VLA Proper Motion

D. C. Backer

Astronomy Department and Radio Astronomy Laboratory, University of California, Berkeley, CA 94720

R. S. Sramek

National Radio Astronomy Observatory, Socorro, NM 87801

Abstract. We provide an introduction to the propagation effects that alter the intrinsic radio emission of Sgr A*. This serves as background for a review of the temporal variability of the flux of Sgr A*. The proper motions from a 17-y experiment at the Very Large Array at 5 GHz and a shorter experiment at 8.4 GHz are presented and compared to that from the Very Long Baseline Array at 43 GHz. The possibility of real differences between experiments from refractive effects in the intervening medium is discussed.

1. Radio Wave Propagation

A plane wave passing through a thin slab of turbulent plasma is modulated with an rms phase of

$$\sigma_\phi = \sqrt{L/a}\delta\phi_a,$$

where L is the slab thickness and a is the characteristic size of the plasma perturbations that produce phase perturbations of $\delta\phi_a = \lambda a r_e \delta n_e$ and r_e is the radius of the electron, about 10^{-13} cm. A plasma pertubation of $\delta n_e \sim 0.001$ cm^{-3} over a scale of $a = 10^{11}$ cm leads to rms phase of 10 radians for a slab of $L = 100$ pc at a wavelength of $\lambda = 10$ cm (3 GHz). If the rms phase exceeds 1 radian over a transverse scale smaller than the first Fresnel zone radius ($a_F = \sqrt{\lambda D}$), then diffractive effects will occur – both diffractive scintillation if the source size is less than $\sim a_F/D$, where D is the distance of the screen, and measureable angular broadening if the source size is less than the scattering angle of the slab, $\theta_s = \lambda\sigma_\phi/a$. This thin screen, discrete plasma perturbation analysis follows that originally stated by Scheuer (1968) and does not include effects of a distribution of irregularities. The review by Rickett (1990) and the paper by Cordes, Weisberg & Boriakoff (1985) formulate the effects of radio wave propagation more thoroughly. Note that in spite of the severe diffractive effects which are possible, the emission measure resulting from these plasma fluctuations is miniscule, 10^{-4} cm^{-6} pc.

The phase fluctuations in the outgoing wavefront are best summarized by a structure function

$$D_\phi(b) =< [\phi(x+b) - \phi(x)]^2 >,$$

which allows definition of the coherence scale $l_\circ$ it via $D_\phi(l_\circ) = 1$; $\theta_s = \lambda/l_\circ$. The observed intensity distribution $I_{\rm obs}(\hat{\mathbf{k}})$ is given by the Fourier transform of the transverse coherence (visibility) function of the source $I_{\rm int}(\mathbf{b})$ filtered by diffractive scattering:

$$I_{\rm obs}(\hat{\mathbf{k}}) = I_{\rm int} * I_s \longleftrightarrow \Gamma_{\rm int}\Gamma_{\rm s} = \Gamma_{\rm int} \exp[-D_\phi(\mathbf{b})/2],$$

where $\mathbf{b}$ is a vector to allow for asymmetric scattering/turbulence. The quantity on the right hand side of the above equation is what is sampled by an interferometer with baseline $\mathbf{b}$ projected on a plane perpendicular to the source direction. Frail *et al.* (1994) report asymmetric scattering for paths to OH masers near Sgr A* that is similar in amplitude and axial ratio to that found for Sgr A* (Lo *et al.* 1999, at this workshop).

When the source is at a finite distance behind the screen, then appropriate corrections for magnification effects are required. The filtering effects of the medium are $\Gamma_s = \exp[-D_\phi(f\mathbf{b})/2]$, where f is the ratio of the slab distance from the source to the observer distance D. The observed size is then $\theta_\circ = f\theta_s$ and the corresponding coherence scale in the slab is $fb_\circ$ where $b_\circ$ is the baseline at which the visibility drops to 0.5.

For Sgr A* we resolve the source brightness distribution on baselines of $\sim$ 500 km at 4 cm and, based on the OH maser results quoted, believe that the scattering occurs within 100 pc of the center. If so, then $f \sim 0.01$ and $l_\circ \sim 5$ km. While one needs to consider the details of plasma dynamics in the slab, this scale is likely to be smaller than the inner scale of the turbulence, the scale at which the cascade of energy from larger scales is dissipated. If so, then the structure function becomes quadratic in baseline as it is dependent only on smooth gradients. A quadratic structure function means a point source ($\Gamma_{\rm int} = 1$) will have a bivariate Gaussian profile, and the apparent size will depend strictly on $\lambda^{2.0}$. This is the dependence reported in the literature for Sgr A* and by Lo *et al.* (1999) in this workshop. Further, we expect a small degree of refraction with respect to $\theta_\circ$, a point to which we will return in the final section.

2. Variability

The blurring effects of scattering in an intervening slab limit the time scale of any refractive modulation of the source flux, scintillation, if indeed there is any. Refractive scintillation depends on the transverse second derivative of the phase perturbations averaged over the transverse averaging scale set by diffractive scattering, $l_d = \theta_\circ(1-f)D$. At 4 cm for Sgr A* this is 2×10^{15} cm. For a transverse velocity of 100 km s^{-1} characteristic of the molecular ring at $\sim$ 100 pc, this is 6 y. The time scale will increase with λ^2, and the modulation is expected to be $\sim$ 25%.

The observed variations (Zhao *et al.* 1991; Backer 1994; Backer & Wright 1993) behave differently. The modulation increases with frequency and the time scales are not strongly dependent on frequency. The report of a flare by Tsuboi et al. (1999) at this workshop further emphasizes the likely intrinsic nature of the variability of Sgr A*. Nevertheless, the environment through which the signals from Sgr A* traverse in passing from the 'atmosphere' of the 2.6×10^6 $M_\odot$ black hole is extreme, and one needs to consider the state of the overlying plasma carefully to exclude its affect on the signals in spite of the very small factor of f that is involved. The severe scattering seen in the pulsar-Be star binary B1257-67xx when the pulsar passes behind, but still very near, the disk of the Be star provides an important lesson (Johnston *et al.* 1996; McClure-Griffiths *et al.* 1998).

3. Proper Motion

We have conducted a proper motion experiment on Sgr A* at the Very Large Array with 8 epochs of observations at 4.8 GHz between 1981 March and 1998 April and 3 epochs at 8.4 GHz between 1989 January and 1998 April. The observations consist of rapid switching between Sgr A* and three reference sources spread in a triangle around Sgr A* with typical separation from Sgr A* of 0.5°. Each epoch consists of 3 days of observations with approximately 4 hour-long blocks on each day. The proper motion we obtain is

$$\mu_\alpha = -2.70 \pm 0.15 \text{ mas y}^{-1}$$

$$\mu_\delta = -5.60 \pm 0.20 \text{ mas y}^{-1}$$

The errors are determined by dividing the data into three subsets in two different ways, by day of epoch and by block. The χ-squares of the fits with errors determined by the internal scatter of measurements each epoch are excellent apart from the right ascension measurement in epoch 7. A complete account of the experiment is nearly ready for submission.

This proper motion can be converted to a peculiar velocity for Sgr A* relative to the galacto-centric inertial reference frame. We transform (μ_α, μ_δ) into galactic coordinates, and then remove first the effects of the peculiar velocity of the solar system relative to the Local Standard of Rest and second the effects of galactic rotation, the net being a secular parallax. There is considerable uncertainty in the latter correction. When translated to velocity at a distance of 8.5 kpc the uncertainty in the longitude motion correction is 23 km s^{-1}. The corresponding uncertainty in latitude motion is an order of magnitude smaller. We estimate the residual (peculiar) motion of Sgr A* in latitude to be

$$v_b = -15 \pm 5 \text{ km s}^{-1}.$$

This is to be compared to the value reported by Reid *et al.*(1999) at this workshop which has a similar amplitude and uncertainty but has the *opposite sign*. We conclude that 0 km s^{-1} is then the best estimate for the peculiar motion in latitude, but must raise the uncertainty to 15 km s^{-1} to include the two measurements. Reid *et al.* summarize the implications of the small peculiar velocity on the mass of Sgr A*. The large inferred mass is a strong argument for associating Sgr A* with an 'atmosphere', using a general term, rigidly attached to the 2.6×10^6 $M_\odot$ black hole discussed by Genzel & Eckart (1999) and Ghez *et al.* (1999) at this workshop and in recent publications. Mass asymmetries in the galactic center on scales from 0.1 pc (*i.e.*, the nearest stars) to 100 pc are at most capable of moving the massive black hole by a few km s^{-1}. Questions about the origin of the black hole and the possibility of an $m = 0$ instability of galactic cores (Miller & Smith 1992) may allow significant peculiar motion of this large mass and its radio-emitting atmosphere. Miller (1996) reports that a speed as large as 70 km s^{-1} with an orbital radius of 0.6 pc (period 10^4 y) is possible.

This report started with discussion of propagation effects, and there is the possibility that refraction in the intervening turbulent plasma and even in the putative mass accretion flow that fuels Sgr A* will lead to time and frequency dependent position shifts. In our best data we find a 5 mas position offset between the differential position of Sgr A* with respect to the same reference sources at the two frequencies. This can result in part from shifts in the centroids of the reference sources. Shifts on the order only 0.5 mas are seen in the location of radio jets relative to the core, and are attributed to opacity differences in the core. The differential position shift may also result from refraction in the intervening ionized material. The discussion in the first section about the regime of diffractive scattering can be used to argue that refractive effects are small. Nevertheless, we can continue to consider time variable refraction. The large scattering diameter requires long time scales for observation of an independent realization of this putative refraction. Therefore there may be a contribution of 'refractive wander' in the apparent proper motion. Continuation of the proper motion measurements at multi-frequencies is required to settle this possible contamination of this fundamental result in galactic astronomy.

Acknowledgments. We are grateful to the NRAO for maintenance of the high quality and long term integrity of the Very Large Array and G. C. Bower for comments on writeup.

References

Backer, D. C. 1994 in "The Nuclei of Normal Galaxies", ed. R. Genzel & A. I. Harris, [Kluwer : Dordrecht], p. 403.

Backer, D. C., & Wright, M. C. H. 1993 ApJ, 417, 560.

Codes, J. M., Weisberg, J. M., & Boriakoff, V. 1985 ApJ, 288, 221.

Frail, D. A., *et al.* 1994 ApJ, 427, L43.

Ghez, A. et al. 1999, this volume, p. 18

Genzel, R., & Eckart, A., 1999, this volume, p. 3

Johnston, S., *et al.* 1996 MNRAS, 279, 1026.

Lo, K.Y., et al., 1999, this volume, p. 72

McClure-Griffiths, N. M., *et al.* 1998 ApJ, 492, L49.

Miller, R. H., & Smith, B. F. 1992 ApJ, 393, 508.

Miller, R. H. 1996 ASP-CS, 102, 327.

Reid, M., et al. 1999, this volume, p. 24
Rickett, B. J. 1990 ARAA, 28, 561.
Scheuer, P. 1968 Nature, 218, 920.
Tsuboi, M., et al. 1999, this volume, p. 105
Zhao *et al.* 1991 ASP-CS, 31, 295.

The Central Parsecs of the Galaxy
ASP Conference Series, Vol. 186, 1999
H. Falcke, A. Cotera, W.J. Duschl, F. Melia, M.J. Rieke, eds.

Intrinsic Size and Shape of Sgr A*: 3.6 AU by <1 AU

K. Y. Lo and Zhi-Qiang Shen[1]

Academia Sinica Institute of Astronomy & Astrophysics, PO Box 1-87, Nankang, Taipei 115

Jun-Hui Zhao and Paul T. P. Ho

Harvard-Smithsonian Center for Astrophysics, 60 Garden Street, Cambridge, MA 02138

Abstract. By means of near-simultaneous multi-wavelength VLBA measurements, we determine for the first time the intrinsic size of Sgr A* to be 3.6 AU by <1 AU or 72 $R_{sc}(\star)$ by $< 20R_{sc}(\star)$, with the major axis oriented essentially north-south, where $R_{sc}(\star) \equiv 7.5 \times 10^{11}$ cm is the Schwarzschild radius for a $2.5 \times 10^6 M_\odot$ black hole. Contrary to previous expectation that the intrinsic structure of Sgr A* is observable only at $\lambda \leq 1$ mm, we can discern the intrinsic source size at λ 7 mm because (1) the scattering size along the minor axis is half that along the major axis, and (2) the near simultaneous multi-wavelength mapping of Sgr A* makes it possible to extrapolate precisely the minor axis scattering angle at λ 7 mm. The intrinsic size and shape place direct constraints on the various theoretical models for Sgr A*.

1. Introduction

Recent proper motion studies of the stars in the vicinity of Sgr A* have provided compelling evidence for the existence of a compact mass of $2.5 \times 10^6 M_\odot$ at Sgr A* (Eckart etal 1997; and Ghez etal 1998). Based on theoretical modeling, Narayan etal (1998) have also shown that the spectral energy distribution (SED) of Sgr A* from radio up to γ-ray can be explained by an advection dominated accretion flow (ADAF), in which the presence of an event horizon would indicate that Sgr A* is in fact a massive black hole.

Over the past two decades since the discovery of this compact nonthermal radio source at the Galactic center (Balick& Brown 1974), VLBI observations of Sgr A* have revealed that the observed sizes follow a λ^2 dependence, and that the apparent source structure can be described by an elliptical Gaussian brightness distribution (e.g. Davies etal 1976; Lo et al. 1981, 1985, 1993; Backer et al. 1993; Rogers et al. 1994; Krichbaum et al. 1998; Bower & Backer 1998). The corruption of the visibility function of Sgr A* due to the scattering by

[1]Present address: National Astronomical Observatory, Mitaka, Tokyo 181

the interstellar electrons has been a persistent problem in resolving its intrinsic structure.

In this paper, we review the results from our recent work (Lo etal 1998) in which we imaged Sgr A* with the Very Long Baseline Array (VLBA) nearly simultaneously at five wavelengths (λ = 6.0, 3.6 2.0 1.35 cm and 7mm). The multi-wavelength imaging, with the same interferometer array, is crucial for our differentiating interstellar scattering effects from the intrinsic source structure of Sgr A*. After taking out the scattering effect extrapolated from the long-wavelength data, we have for the first time determined that the intrinsic size of Sgr A* at 7 mm is asymmetric, being 3.6 AU by <1 AU, assuming a distance of 8.0 kpc to the Galactic center. The elongation is nearly N-S.

In addition, we also discuss the constraints of the radio observations on the existing theoretical models.

2. Observations and Data Reductions

The observations were made with 10 VLBA antennas and 1 VLA antenna at wavelengths 6.0, 3.6 2.0 and 1.35 cm and 7 mm in Feburary 1997. The hour angle coverage was UT12^h-20^h. A standard VLBA mode with BW of 32 MHz for both RCP and LCP was used. The visibility data were produced using the VLBA correlator at the Array Operation Center of the National Radio Astronomy Observatory (NRAO). The data reduction and calibration were discussed in Lo et al. (1998). We emphasize that the calibration is critical to the VLBA measurements, in particular, at the short wavelengths. With global fringe fitting, Sgr A* was detected on the short and intermediate baselines depending on the observed wavelengths while both calibrators NRAO 530 and PKS 1921-293 were detected on all the baselines. The amplitude calibrations of the visibility were done using the system temperature measurement at each site. The elevation dependent opacity corrections were done at the short wavelengths.

3. Data Analysis and Results

3.1. Measurements of The Angular Size and Structure of Sgr A*

An elliptical Gaussian model was fitted by the least-square method to both amplitudes and the phases of the calibrated visibility data to yield a quantitative description of the source structure. Table 1 summarizes the measurements at all five wavelengths. Fig. 1 shows the visibility data as a function of time and the fitted model in both amplitude and closure phase. The robustness of the fit is ensured by the good quality of the data and the availability of many baselines where the structure can be determined. The steadily improved performance of the VLBA played a pivotal role in the quality of this data set.

Fig. 2 shows both the major and minor axis diameters determined from our near-simultaneous multi-wavelength VLBA observations of Sgr A*. Fig. 3 shows the images of Sgr A* at the five wavelengths. The apparent major axis diameters can be fit by

$$\theta_{\rm maj} = (1.43 \pm 0.02)\lambda_{\rm cm}^{1.99\pm0.03} \quad \text{mas,} \tag{1}$$

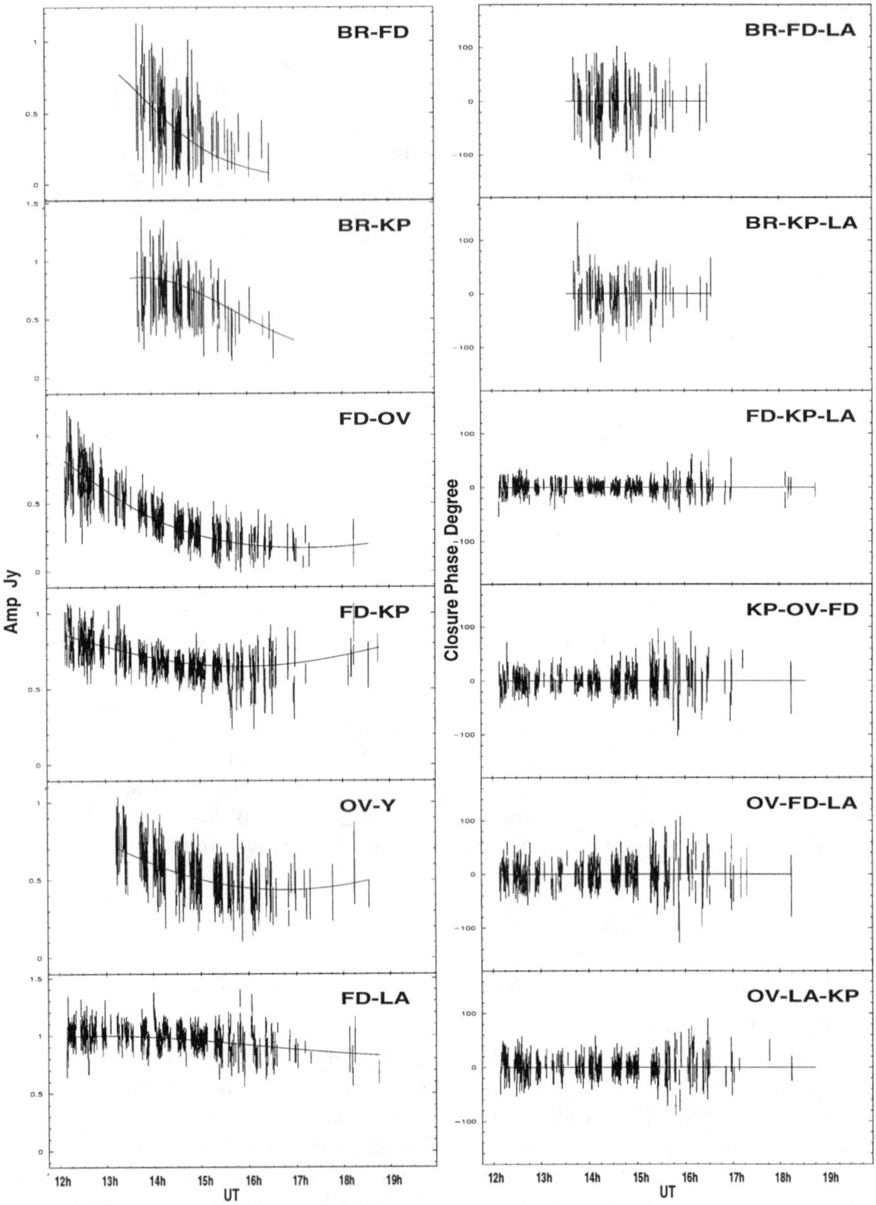

Figure 1. A plot shows the Gaussian elliptical model fitting (solid curves) to the visibility data at 7 mm (vertical bars). Left panels: amplitude vs. baseline pairs (BR-FD, BR-KP, FD-OV, FD-KP, OV-Y, and FD-LA). The maximum baseline lengths of the pairs are 2346, 1914, 1508, 744, 1025 and 607 kilometers, respectively. BR-FD is the longest baseline in NS. Right panels: closure phase triangles.

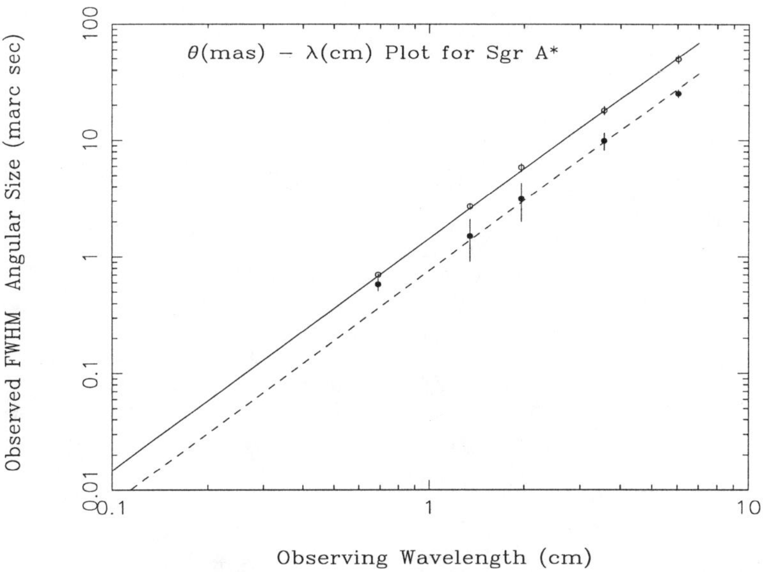

Figure 2. A log-log plot of measured (FWHM) source size versus observing wavelength for Sgr A* (7- 14 February 1997). The solid line represents a 1.43 $\lambda^{1.99}$ fit to the major axis sizes (open circles), while the dashed line a 0.76 $\lambda^{2.0}$ fit to the minor axis sizes (filled circles).

which is in good agreement with the previous results of $1.42\lambda_{\rm cm}^{2.0}$ mas (e.g. Alberdi et al. 1993). This result is also consistent with the value $\beta = 4$ for the power spectrum of the density fluctuations in the intersterllar electrons ($\propto$ k$^{-\beta}$, where k is the wavenumber of the irregularities).

Along the minor axis, a fit to the measurements at all five wavelengths yields $\theta_{\rm min} = (1.06 \pm 0.10)\lambda_{\rm cm}^{1.76\pm0.07}$ mas, which is inconsistent with the interstellar scattering. For a fit to the four points at $\lambda \geq 1.35$ cm, the apparent minor axis size $\theta_{\rm min} = (0.87 \pm 0.23)\lambda_{\rm cm}^{1.87\pm0.16}$ mas, which *is* consistent, within the errors, with the λ^2–dependence expected from the interstellar scattering. If we assume that the λ^2–dependence derived for the major axis also applies to the minor axis, we obtain

$$\theta_{\rm min} = (0.76 \pm 0.05)\lambda_{\rm cm}^{2.0} \quad \text{mas,} \tag{2}$$

and a constant axial ratio of 0.53.

This is the first time that the λ–dependence of the minor axis diameter is determined directly by observations. Our results strongly suggest that interstellar scattering dominates the observed minor axis image at $\lambda \geq$1.35 cm. The elongation of the scatter-broadened image can be caused by an anisotropic scattering medium in the vicinity of the Galactic center.

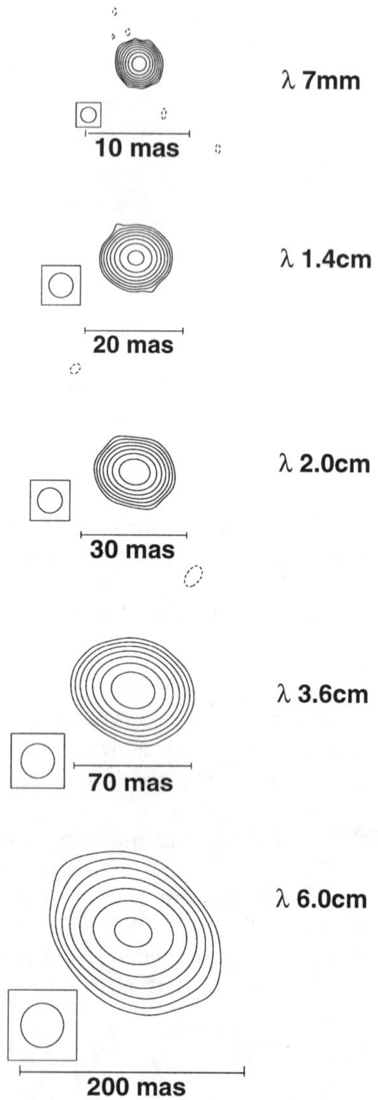

Figure 3. VLBA images of Sgr A* at wavelengths 6.0, 3.6, 2.0, 1.35 cm and 7 mm made with DIFMAP. These images are smoothed to a circular beam of FWHM = 2.62 $\lambda_{cm}^{1.5}$ mas as shown on the left-bottom corner on each image. At 7 mm, FWHM beam = 1.5 mas $\sim$ mean synthesis beam size; and at 6 cm FWHM beam = 38 mas that is close to the mean scattering size at this wavelength. The contours are 2 mJy beam$^{-1}\times$ (-2, 2, 4, 8, 16, 32, 64, 128, 256).

Table 1. Parameters of Elliptical Gaussian Model Fit

λ (cm)	ν (GHz)	S_ν (Jy)	θ_{major} (marc s)	θ_{minor} (marc s)	Axial Ratio	P.A. (°)
6.03	4.97	0.60±0.09	49.6±4.50	25.1±2.00	0.51±0.09	81±3
3.56	8.41	0.73±0.10	18.0±1.53	9.88±1.68	0.55±0.14	78±6
1.96	15.3	0.68±0.06	5.84±0.48	3.13±1.14	0.54±0.21	73±14
1.35	22.2	0.74±0.04	2.70±0.15	1.50±0.59	0.56±0.25	81±11
0.69	43.2	1.03±0.01	0.70±0.01	0.58±0.07	0.83±0.11	87±8

3.2. The Intrinsic Structure of Sgr A*

Comparing the measurement at λ 7 mm with Eq. (2), we find a significant deviation of the observed minor axis diameter ($\theta_{\rm min}$) from the scattering angle ($\theta_{\rm sc}$) which is extrapolated from Eq. (2)

$$\Delta\theta_{\rm min} \equiv \theta_{\rm min} - \theta_{\rm sc} = 0.21 \pm 0.07 \quad \text{mas.} \tag{3}$$

The deviation of the apparent angular diameter (θ_{obs}) from the λ^2–dependence is naturally expected when the intrinsic source diameter ($\theta_{\rm int}$) becomes comparable to the scattering angle, since the observed apparent angular diameter $\theta_{\rm obs} = \sqrt{\theta_{\rm int}^2 + \theta_{\rm sc}^2}$ (Narayan & Hubbard 1988). The deviation described by Eq. (3) implies that the intrinsic angular diameter at 7mm would be $\theta_{\rm int} = 0.45 \pm 0.11$ mas for Sgr A* along the minor axis (PA = –10°, nearly N-S). From the measurements conducted by Bower and Backer (1998), we would have $\theta_{\rm min} = 0.55 \pm 0.11$ mas, $\Delta\theta_{\rm min} = 0.18 \pm 0.11$ mas, and $\theta_{\rm int} = 0.41 \pm 0.17$ mas. Combining the two sets of measurements from both this paper and Bower and Backer, we obtain

$$\begin{aligned} \theta_{\rm min} &= 0.57 \pm 0.06 \quad \text{mas,} \\ \Delta\theta_{\rm min} &= 0.20 \pm 0.06 \quad \text{mas,} \\ \theta_{\rm int} &= 0.44 \pm 0.09 \quad \text{mas,} \end{aligned} \tag{4}$$

The reasons that we can discern the intrinsic source size at λ 7 mm, contrary to previous expectations that intrinsic source size is observable only at $\lambda \leq$ 1mm, can be highlighted by the following two facts. First, the scattering size along the minor axis is half that along the major axis. In addition, the near simultaneous multiwavelength imaging of Sgr A*, with the same instrument, over the same hour angle coverage, and calibrated in a uniform manner, makes it possible to extrapolate precisely the minor axis scattering angle at λ 7 mm.

Along the major axis (PA = 80°, essentially E-W), the observed angular diameter ($\theta_{\rm maj} = 0.70 \pm 0.01$) mas and the extrapolated scattering diameter ($\theta_{\rm sc} = 0.69 \pm 0.01$) mas imply

$$\theta_{\rm int} \leq 0.13 \quad \text{mas.} \tag{5}$$

Combined with the intrinsic diameter along the minor axis (Eq. (4)), this result suggests that the intrinsic source structure of Sgr A* could be elongated along an essentially N-S direction with an axial ratio of <0.3. The derived intrinsic angular size also implies an intrinsic brightness temperature exceeding 1.3×10^{10} K at λ 7 mm.

Finally, the intrinsic size along the minor axis derived at 7mm appears to be larger than the value inferred at λ 1.4 mm by Krichbaum et al. (1998). At λ 1.4 mm, using the marginal detection of Sgr A* with an interferometer with a single baseline, they derived $\theta_{\rm int} = 0.11 \pm 0.06$ mas. Taking our λ 7 mm result and the 1.4 mm measurement together, a preliminary wavelength dependence of the intrinsic source size can be inferred:

$$\theta_{\rm int} = 0.08\lambda_{\rm mm}^{0.9} \quad \text{mas} \tag{6}$$

4. Discussion

At a distance of 8 kpc, the intrinsic angular size of Sgr A* corresponds to a linear size of 3.6 AU by <1 AU. The angular size is also equivalent to 72 $R_{\rm sc}(*)$ by <20 $R_{\rm sc}(*)$, where $R_{\rm sc}(*)$ (= 7.5 $\times10^{11}$ cm) is the Shcwarzschild radius of a 2.5×10^6 $M_\odot$ black hole. The elongation is nearly north-south.

Numereous models, typically involving synchrotron emission, have been proposed for the structure and mechanism of radio emission from Sgr A*. A model proposed by Reynolds & Mckee (1980) based on a pulsar wind that is confined by the ram presure in the Galactic center region can be ruled out due to the fact that Sgr A* is apparently associated with a massive black hole (2.5×10^6 $M_\odot$).

The rest of the models do involve a massive black hole. Melia (1994) modeled Sgr A* as synchrotron radiation from thermal electrons, heated by the dissipation of magnetic energy, as a result from the Bondi-Hoyle accretion process of the mass loss in the winds from the stars in the vicinity of Sgr A*. A jet in a coupled jet-disk system was also proposed to account for Sgr A* (Falcke, Mannheim & Biermann 1993). A dynamically self-consistent model involving a rotating advection-dominated accretion flow (ADAF) suggests that Sgr A* arises from thermal electrons of a two-temperature plasma near the massive black hole (Narayan et al. 1998). In fact, none of the models has predicted both the intrinsic shape and size of Sgr A* as observed at 7 mm. Among these models, ADAF has naturally explained the spectral energy distribution as observed from radio to γ-ray, and predicts a very sensible X-ray limit. The ADAF model may have to incorporate a radio jet or a wind in order to account for the structure of Sgr A*.

Further observations are clearly indicated, especially at wavelengths shorter than 7 mm. Repeated observations are also important to monitor any variation in the intrinsic structure of Sgr A*.

Acknowledgments. The National Radio Astronomy Observatory is a facility of the National Science Foundation operated under cooperative agreement by Associated Universities, Inc.

References

Alberdi, A. et al. 1993, A&A 277, L1

Backer, D. C., Zensus, J. A., Kellermann, K. I., Reid, M., Moran, J. M., and Lo, K. Y. 1993, Science 262, 1414

Balick, B., and Brown, R. 1974, ApJ 194, 265

Bower, G. C., and Backer, D. C. 1998, ApJ 496, L97

Davies, R. D., Walsh, D., Booth, R. S. 1976, MNRAS 177, 319

Eckart, A., and Genzel, R. 1997, MNRAS 284, 576

Falcke, H., Mannheim, K., and Biermann, P. 1993, A&A 278, L1

Ghez, A. M., Klein, B. L., Morris, M., and Becklin, E. E. 1998, ApJ, in press

Krichbaum, T. P. et al. 1998 A&A 335, L106

Lo, K. Y. et al. 1981, ApJ 249, 504

Lo, K. Y. et al. 1985, Nature 315, 124

Lo, K. Y. et al. 1993, Nature 362, 38

Lo, K. Y., Shen, Z., Zhao, J.-H., & Ho, P. T. P. 1998, ApJ Lett. November issue

Melia, F. 1994, ApJ 426, 577

Narayan, R., and Hubbard, W. B. 1988, ApJ 325, 503

Narayan, R., Mahadevan, R., Grindlay, J., Popham, R., and Gammie, C. 1998, ApJ 492, 554

Reynolds, S. P., and McKee, C. F. 1980, ApJ 239, 893

Rogers, A. E. E. et al. 1994, ApJ 434, L59

The Central Parsecs of the Galaxy
ASP Conference Series, Vol. 186, 1999
H. Falcke, A. Cotera, W.J. Duschl, F. Melia, M.J. Rieke, eds.

VLBA Imaging at 7 mm and Linear Polarimetric Observations at 6 cm and 3 mm of Sgr A*

Geoffrey C. Bower and Heino Falcke

Max Planck Institut für Radioastronomie, Bonn, Germany

Don C. Backer and Melvyn Wright

Radio Astronomy Laboratory, UC Berkeley, Berkeley, CA

Abstract. We summarize the results of 7 mm VLBA imaging of Sgr A* and discuss some of the difficulties of accurately constraining the size of Sgr A* with VLBI observations. Our imaging results are fully consistent with the hypothesis that the VLBA image of Sgr A* is a resolved elliptical Gaussian caused by the scattering of an intervening thermal plasma. We show that determination of the minor axis size at 7 mm with the VLBA is very unreliable.

We also present new polarimetric observations from the VLA and from the BIMA array of Sgr A*. At 4.8 GHz, we find an upper limit to the polarization of 0.1%. At 86 GHz, we report a marginal detection of $1\pm1\%$ linear polarization. We discuss the effects of interstellar propagation on the linear polarization and consider the significance of very low intrinsic linear polarization in Sgr A*.

1. 7 mm VLBA Imaging of Sgr A*

We summarize previously-published results of 7 mm VLBA imaging of Sgr A* (Bower & Backer 1998). Observations of Sgr A* at 43 GHz were performed on 29 September 1994 with the VLBA. Fringes on Sgr A* were detected to a maximum (u, v)-distance of 250 $M\lambda$. Fringes to the compact calibrator NRAO 530 were detected on all baselines throughout the experiment. Imaging and modeling revealed a single elliptical Gaussian component with a major axis FWHM of 0.762 ± 0.038 mas in a position angle of $77\overset{\circ}{.}0\pm7\overset{\circ}{.}4$. The axial ratio was measured to be 0.73 ± 0.10. We show in Figure 1 the visibility amplitudes as a function of (u, v)-distance along with model fits. No secondary structure was detected in the vicinity of Sgr A* to a limit of 35 mJy.

The measured size is fully consistent with the expected major and minor axes predicted by lower frequency observation (Yusef-Zadeh et al. 1994, Lo et al. 1998). However, as Lo et al. note, this result combined with their new result implies a significant deviation in the axial ratio from that expected due to scattering. While this is true, poor North-South resolution and systematic errors in calibration make this claim uncertain and difficult to prove.

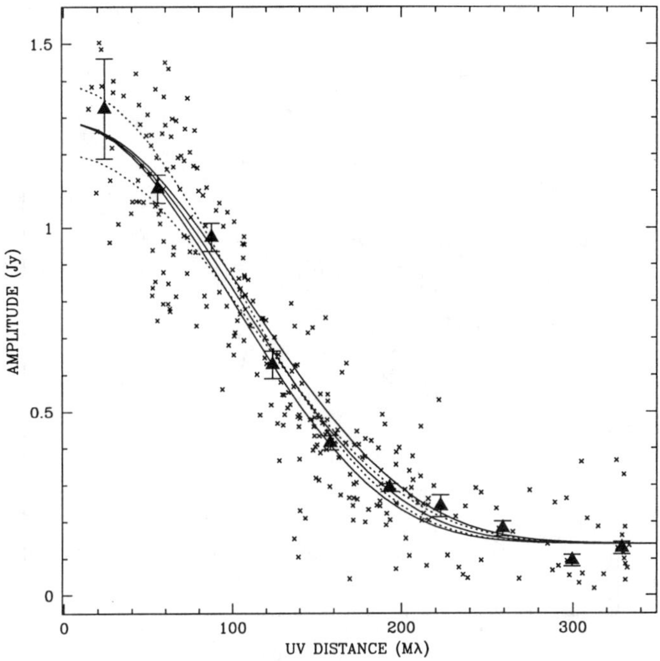

Figure 1. Visibility amplitude as a function of *uv* distance for Sgr A*. The solid lines indicate the expectation for a circular Gaussian with a zero baseline flux of 1.28 Jy and FWHM of 0.694, 0.727 and 0.760 mas. The dashed lines indicate the expectation for a circular Gaussian with zero baseline fluxes of 1.38 Jy and 1.18 Jy with FWHM of 0.760 and 0.694 mas, respectively. The noise bias is added in quadrature to the model. The large triangles indicate the median flux in a 35 Mλ bin. The errors indicate the scatter in the data.

1) The North-South resolution of the VLBA is inadequate to resolve the scattering size at 7mm. The extreme southern declination of Sgr A* combined with the northern latitudes of the VLBA produces very poor North-South resolution. In Figure 2 we show the baselines on which we made detections to Sgr A*. The visibilities that contribute predominantly to the minor axis size are limited to three baselines, two of which involve the Brewster station, one of the VLBA's worst millimeter sites. The maximum North-South resolution of $\sim 100M\lambda$ on a single baseline is considerably smaller than the FWHM of the minor axis scattering size $\sim 190M\lambda$. In fact, the size scale of interest — the difference between the scattered and non-scattered minor axis sizes — is more than 10 times smaller than the beam size in that direction.

2) Observations of Sgr A* with the VLBA are performed at very low elevation and are difficult to calibrate. Variable atmospheric opacity

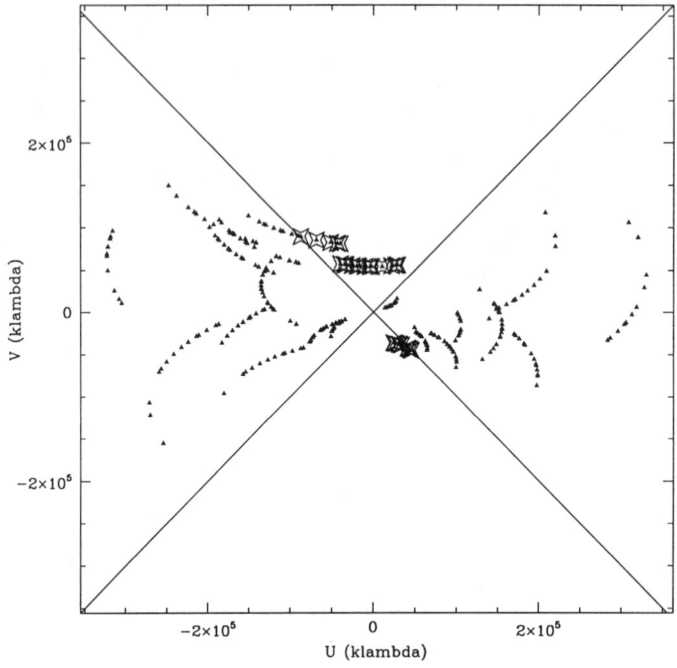

Figure 2. Visibility coverage of detections of Sgr A* at 7 mm by Bower & Backer (1998). The starred visibilities are those in which the minor axis dominates the major axis in determination of the visibility amplitude.

due to water vapor in the troposphere makes amplitude calibration particularly difficult at 7 mm. Atmospheric coherence times can be very short and variable, as well. These effects increase as $\sec z$ with increasing zenith angle, z. Further, *a priori* gain curves at low elevation with VLBA antennas are suspect. The Brewster antenna, which provides the two longest North-South baselines on Sgr A*, never gets above 13° elevation. Hence, it is always looking through at least 4.5 atmospheres. Transferring gains from a compact calibrator, J1730-13 or J1924-29, for example, can improve amplitude calibration but a substantial uncertainty remains due to the large difference in contemporaneous elevation. Amplitude self-calibration is not accurate, either, since the density of visibilities in the (u, v) plane is low.

One can consider the combined effect of the lack of resolution and difficulty in amplitude calibration by plotting the amplitudes as a function of the North-South spatial frequency v, after the amplitudes have been corrected for the East-West structure (Figure 3). If the North-South axis were unresolved, the residual amplitude would be unity. The considerable scatter in the data indicates

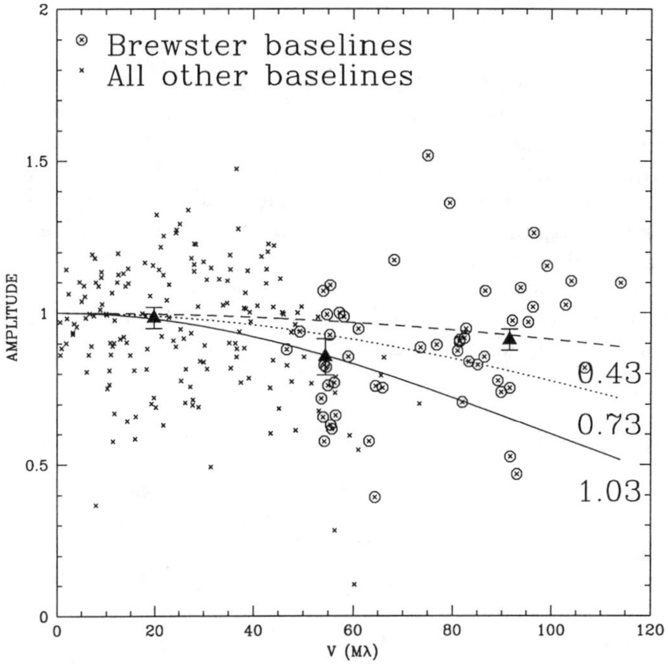

Figure 3. Visibility amplitude corrected for East-West structure against the North-South spatial frequency v. The lines indicate expected amplitudes for the mean axial ratio and plus and minus three sigma deviations in the axial ratio. Circled visibilities indicate those on a baseline with the Brewster station. These dominate for $v > 60M\lambda$. The triangles indicate median amplitudes in three bins of 40 $M\lambda$ in v.

the difficulty in determining the axial ratio accurately. And, as noted above, baselines involving Brewster dominate at large v.

What observations can accurately measure the size of Sgr A* at 7 mm? The use of the Green Bank Telescope with the VLBA will add crucial medium-length North-South baselines to Hancock, Saint Croix, North Liberty and Fort Davis. The Large Millimeter Telescope under construction in Mexico will also add interesting baselines to the Southwestern VLBA stations. Finally, successful adoption of spectral-line calibration methods using the nearby SiO maser VX Sgr will significantly improve calibration.

Higher frequency observations are plagued by the same problems of difficult calibration and poor North-South resolution. These are discussed in more detail by Doeleman et al. (1999) and Krichbaum et al. (1999).

2. Linear Polarimetric Observations of Sgr A* at 4.8 and 86 GHz

Linear polarization stands as one of the few observables of Sgr A* not extensively investigated observationally or theoretically. However, we expect that linear polarization should arise from the cyclo-synchrotron radiation that is responsible for the radio and millimeter wavelength spectrum. In active galactic nuclei, the degree of linear polarization is a measure of the order in the magnetic field. Comparing the evolution of linear polarization to the evolution of total intensity provides a strong argument for the existence of shocks in the relativistic jets of AGN (e.g., Hughes, Aller and Aller 1985). Detection of correlated polarized and total intensity variations in Sgr A* would be convincing evidence for a jet model. Uncorrelated variability would indicate that a different mechanism must be responsible for the radio to millimeter spectrum.

2.1. VLA Observations at 4.8 GHz

The VLA observed Sgr A* on 10 and 18 April 1998 in the A array at 4.8 GHz with a bandwidth of 50 MHz. Instrumental calibration was performed with the compact sources 1741-038 and 1748-253. The right-left phase difference was set with observations of 3C 286. Baselines longer than 100 $k\lambda$ were used for Sgr A*.

We summarize the measured polarizations of Sgr A* and several compact calibrators in Table 1. The calibrators GC 441, W56 and W109 belong to the proper motion survey of Backer and Sramek (1999). Consistency between the results on the two days indicates the accuracy of the results. Polarization was reliably detected from all sources but Sgr A* and GC 441. The measured polarization at the position of Sgr A* is 0.1%. This value is equal to the average off source fractional polarization in the map and is, therefore, an upper limit.

Table 1. Polarized and Total Flux at 4.8 GHz

Source	Date	I (Jy)	P (mJy)	p (%)	χ (deg)
Sgr A*	10 Apr	0.49	< 0.63	< 0.13	...
	18 Apr	0.47	< 0.34	< 0.08	...
1741-038	10 Apr	4.90	10.6	0.22	27
	18 Apr	4.94	15.3	0.31	24
1748-253	10 Apr	0.48	8.4	1.8	-15
	18 Apr	0.48	11.4	2.4	-16
GC441	10 Apr	0.044	< 0.09	< 0.20	...
	18 Apr	0.043	< 0.12	< 0.28	...
W56	10 Apr	0.104	2.2	2.1	-66
	18 Apr	0.104	2.2	2.1	-66
W109	10 Apr	0.098	0.59	0.6	-26
	18 Apr	0.098	0.51	0.5	-17

2.2. BIMA Observations at 86 GHz

Observations of Sgr A* with the BIMA array were conducted on 10 and 14 March 1998 at 86 GHz with a bandwidth of 800 MHz. The BIMA array was configured in its A array with baselines from 20 to 540 $k\lambda$. Nine antennas were switched systematically between right and left circular polarization throughout the 5 hour track. A single antenna observed linear polarization constantly. This permitted phase self-calibration for each 10 sec averaging interval. Amplitude calibration was performed with *a priori* gain information. Phase calibration was done with interleaved scans of NRAO 530.

Instrumental polarization calibration was performed in several ways, producing similar results. D-term solutions were obtained from the calibrator NRAO 530 on 14 March, from Sgr A* itself on both days and from observations of the SiO maser in Orion on 28 January 1998 and 25 February 1998. A comparison of these D-terms shows that the instrumental polarization is stable over two months to a level of 0.3% (Figure 4).

We tabulate the measured polarization on Sgr A* and other sources in Table 2. The results for Sgr A* are marginally consistent with each other and a fractional polarization of $1 \pm 1\%(3\sigma)$.

We note that our observations occurred within days of the peak of the millimeter flare reported by Tsuboi, Miyazaki and Tsutsumi (1990). Our mean flux $S_{86} = 1.5 \pm 0.2$ Jy is lower than but marginally consistent with the peak flux of 2.4 ± 0.4 Jy reported by these authors.

Table 2. Polarized and Total Flux at 86 GHz

Source	Date	I (Jy)	P (mJy)	p (%)	χ (deg)
Sgr A*	10 Mar	1.72 ± 0.075	22 ± 6	1.3 ± 0.3	-41 ± 8
	14 Mar	1.37 ± 0.097	7 ± 10	0.5 ± 0.7	16 ± 41
3C273	10 Mar	24.7 ± 1.1	756 ± 36	3.1 ± 0.1	-1 ± 1
	14 Mar	23.9 ± 1.7	802 ± 96	3.3 ± 0.3	0 ± 3
3C454.3	10 Mar	5.33 ± 0.37	34 ± 11	0.6 ± 0.2	43 ± 10
	14 Mar	4.93 ± 0.13	21 ± 12	0.4 ± 0.2	-41 ± 16
1733-130	14 Mar	2.82 ± 0.010	66 ± 3	2.3 ± 0.1	-42 ± 1

2.3. Interstellar Propagation Effects

The interstellar medium may depolarize a polarized radio wave in two ways: differential Faraday rotation along the many paths that contribute to the scatter-broadened image of Sgr A*; and, significant rotation of the polarization position angle through the observing bandwidth. We consider each of these separately.

The Faraday rotation is

$$\chi_F = \text{RM}\ \lambda^2,$$

where RM is the rotation measure in rad m^{-2} and λ is the wavelength in meters. The scattering region will depolarize the signal if $\delta\chi_F \approx \pi$. Such variations

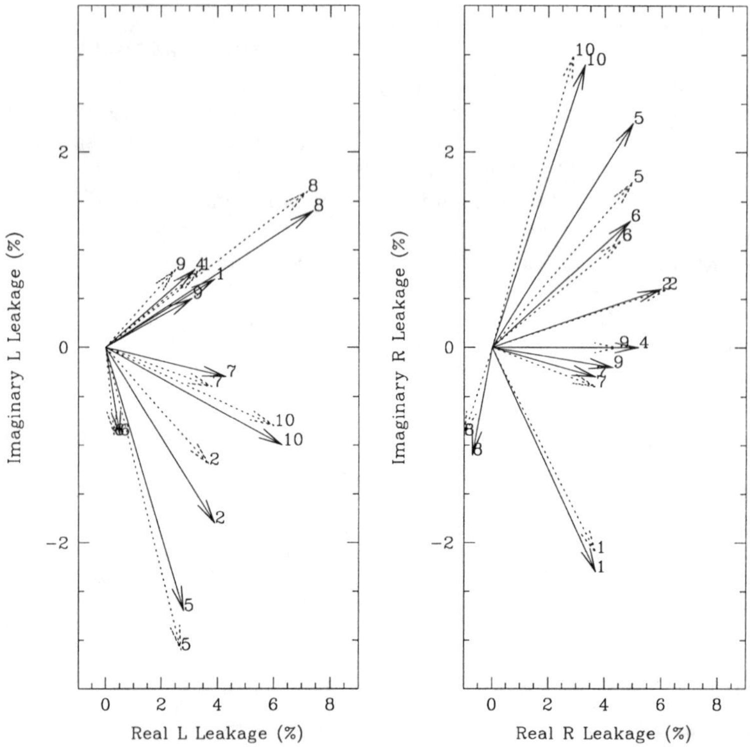

Figure 4. D-terms for all antennas determined from SiO maser observations on 25 Feb 98 (solid lines) and 28 Jan 98 (dashed lines). The terms differ by 0.3% on average.

may occur over multiple turbulent cells in a scattering screen each of which will have a different RM. For $\lambda = 6$ cm, $\delta\mathrm{RM} = 900$ rad m^{-2} (at $\lambda = 3$ mm, $\delta\mathrm{RM} = 3 \times 10^5$ rad m^{-2}). Over the scattering size of 50 mas at 6 cm, this corresponds to $\delta\mathrm{RM}/\delta\theta = 18000$ rad m^{-2}arcsec^{-1}. Observations on the arcsecond to arcminute scale of a nonthermal filament within 1 degree of Sgr A* find a maximum $\delta\mathrm{RM}/\delta\theta = 250$ rad m^{-2}arcsec^{-1} (Yusef-Zadeh, Wardle & Parastaran 1997). Extrapolating their RM structure function to the scattering size implies $\delta\mathrm{RM} \approx 50$ rad m^{-2} These values are well below those necessary to depolarize Sgr A*.

Do the conditions exist to depolarize Sgr A* in the scattering screen? The RM is

$$\mathrm{RM} = 0.8 n_e B L \text{ rad m}^{-2},$$

where n_e is the electron number density in cm^{-3}, B is the magnetic field in μG and L is the size scale in pc. In the thin screen approximation L must be a fraction of the scattering diameter θ_{SgrA*}. We find $L \sim 0.1\theta_{SgrA*}D_{SgrA*} \sim$

10^{-4} pc. This implies $n_e B \sim 10^7$ μG cm^{-3}. This condition is matched by that found by Yusef-Zadeh et al. (1994) for the ionized skins of molecular clouds in the Galactic Center. However, the high temperature and low density model of Lazio & Cordes (1998) predicts no depolarization of Sgr A* for $B < 1$ G.

Finally, depolarization may occur in the accretion region of Sgr A* where the electron density and magnetic field strength are large but the length scale is smaller. In the model of Melia (1994) inside a radius of 0.01 pc, $B \gtrsim 100$ mG and $n_e \gtrsim 10^4$ cm^{-3} for an accretion rate of $10^{-4} M_{\odot}\ y^{-1}$. The ADAF model requires similar conditions (Narayan *et al.* 1998). Whether the extreme conditions in the accretion region can cause a significant δRM depends critically on the scale of turbulence.

We now consider the effect of Faraday rotation on a polarized signal measured over a specific bandwidth. Differentiating χ_F permits us to write the maximum observable rotation measure as a function of bandwidth $\Delta\nu$ and observing frequency

$$\mathrm{RM}_{\mathrm{max}} \approx \frac{1}{2}\frac{1}{\lambda^2}\frac{\nu}{\Delta\nu}.$$

For the observing bandwidths of 50 MHz and 800 MHz at 4.8 GHz and 86 GHz, we find maximum rotation measures of 10^4 rad m^{-2} and 10^7 rad m^{-2}, respectively. The detection of rotation measures in the Galactic Center as large as a few thousand suggests that the low frequency result may in fact be limited by a very high rotation measure (e.g., Yusef-Zadeh, Wardle & Parastaran, 1997). We are currently studying spectro-polarimetric observations of the continuum radiation from Sgr A* at 4.8 and 8.4 GHz in search of linear polarization with extremely high rotation measures (up to 10^7 rad m^{-2}). A preliminary result from this study is that an extreme RM does not depolarize Sgr A*.

It is not yet certain whether interstellar propagation effects can depolarize Sgr A*. Known RM variations are insufficient to depolarize Sgr A*. However, at least one model for the scattering medium of Sgr A* does include the necessary conditions. We now consider the hypothesis of an intrinsically weakly polarized Sgr A*.

2.4. Discussion

The centimeter and millimeter wavelength polarization studies probe different regions of Sgr A*. Spectral studies of Sgr A* (e.g., Falcke et al. 1998) reveal that there may be significantly different sources at centimeter and millimeter wavelengths which may have different polarization structures. A higher fractional polarization at 86 GHz may be the result of observing at lower opacity and over a smaller, more homogeneous region. The polarization of AGN are greater and more variable at millimeter than at centimeter wavelengths (e.g., Nartallo *et al.* 1998).

A polarization fraction less than 1% is uncommon in compact radio sources at centimeter wavelengths (Aller, Aller & Hughes 1992). However, optically thick quasar cores observed with VLBI are frequently weakly polarized (Cawthorne et al. 1993). Such cores may be analogous to the radio source in Sgr A*, which, due to its low power, may not produce the strong shocks in the jet that are the source for higher polarization regions in quasars. Weak polarization is more common in galaxies than quasars or blazars and it is also more common in compact-double sources or sources with irregular morphologies (Aller, Aller & Hughes 1992). A notable source with a very low polarization fraction is the radio galaxy 3C 84, which has a very irregular morphology. If the radio to millimeter spectrum of Sgr A* does arise in a jet, the low power of this jet or environmental effects in the Galactic Center region may limit the magnetic field order. Alternatively, low energy electrons in the synchrotron environment of Sgr A* may depolarize the source.

References

Aller, M.F., Aller, H.D. & Hughes, P.A., 1992, ApJ, 399, 16.

Backer, D.C & Sramek, R.S., 1999, this volume, p. 67

Bower, G.C. & Backer, D.C., 1998, ApJ, 496, L97.

Doeleman, S. et al., 1999, this volume, p. 98

Cawthorne, T.V., Wardle, J.F.C.; Roberts, D.H., Gabuzda, D.C., 1993, ApJ 416, 519

Falcke H., Goss W.M., Matsuo H., Teuben P., Zhao J.-H. & Zylka, R., 1998, ApJ, 499, 731.

Krichbaum, T.P. et al., 1999, this volume, p. 89

Hughes, P.A., Aller, H.D. & Aller, M.F., 1985, ApJ, 298, 301.

Lazio, T.J.W. & Cordes, J.M., 1998, ApJ, 505, 715.

Lo, K.Y., Shen, Z.-Q., Zhao, J.-H. & Ho, P.T.P., 1998, ApJ, 508, L61.

Melia, F., 1994, ApJ, 426, 577.

Narayan, R., Mahadevan, R., Grindlay, J.E., Popham, R.G. & Gammie, C., 1998, ApJ, 492, 554.

Nartallo, R., Gear, W.K., Murray, A.G., Robson, E.I. & Hough, J.H., 1998, MNRAS, 297, 667.

Tsuboi, M., Miyazaki, A., Tsutsumi, T., 1999, this volume, p. 105

Yusef-Zadeh, F., Cotton, W., Wardle, M., Melia, F. & Roberts, D., 1994, ApJ, 434, L63.

Yusef-Zadeh, F., Wardle, M. & Parastaran, P., 1997, ApJ, 475, L119.

The Central Parsecs of the Galaxy
ASP Conference Series, Vol. 186, 1999
H. Falcke, A. Cotera, W.J. Duschl, F. Melia, M.J. Rieke, eds.

VLBI Observations of the Galactic Center Source Sgr A* at 86 GHz and 215 GHz

T.P. Krichbaum, A. Witzel, and J.A. Zensus

Max-Planck-Institut für Radioastronomie, Auf dem Hügel 69, D-53121 Bonn, Germany

Abstract. We[1] summarize previous VLBI observations of Sgr A* at millimeter wavelengths and present new results from VLBI observations at 86 GHz and at 215 GHz. At 86 GHz the measured closure phase is close to zero, consistent with a point-like or symmetric structure of $190 \pm 30\,\mu$as size. At 215 GHz we have (for the first time) detected Sgr A* with a signal-to-noise ratio of 6. This yields a tentative size estimate of $50 \leq \theta \leq 190\,\mu$as, which is larger than the scattering size of 20μas at this frequency. The intrinsic size of Sgr A* thus appears to be a few up to a few ten Schwarzschild radii (for a $2.6 \cdot 10^6$ $M_\odot$ black hole).

1. Introduction

At cm-wavelength the apparent size of the compact VLBI component of the Galactic Center Radio Source Sgr A* decreases approximately quadratically towards shorter wavelengths. This is usually interpreted as angular broadening caused by diffractive scattering by the interstellar electrons.

The location at the dynamical center of the Galaxy, the compactness and high brightness temperature, and the kinematics of the moving gas and the stars around Sgr A* all strongly support the idea that Sgr A* is associated with a supermassive black hole with a mass of $2.6 \cdot 10^6$ $M_\odot$ (eg. Genzel and Eckart 1999). At $D = 8$ kpc distance, the size of Sgr A* must at least be of order of two Schwarzschild radii, which translates into $\theta \geq 2\,R_s/D = 13\mu$as. This size becomes equal to the scattering size near $\lambda = 1$ mm.

Future mm-VLBI observations at $\lambda \leq 3$ mm ($\nu \geq 86$ GHz) will therefore allow to image the direct environment of the black hole (and its event horizon) with an angular resolution of only a few times the Schwarzschild radius at frequencies, where interstellar scattering is not affecting the observed brightness distribution.

The investigation of the compact radio source Sgr A* with mm-VLBI thus is mainly motivated by these two facts: high angular resolution and clear view. In the following we summarize recent results obtained from VLBI observations at 43 GHz and 86 GHz and report on the first VLBI-detection of Sgr A* at 215 GHz (single baseline). VLBI observations at these high frequencies are diffi-

[1] for collaborators see acknowledgement

Table 1. Summary of size measurements from mm-VLBI.

λ [cm]	Axis [mas]	Ratio	*p.a.* [deg]	S_{VLBI} [Jy]	Epoch [yrs]	Reference
0.69	0.70±0.01	0.83±0.1	87± 8	1.03±0.01	1997.10	Lo et al. 1998
0.69	0.76±0.04	0.73±0.1	77± 7	1.28±0.10	1994.75	Bower et al. 1998
0.69	0.74±0.03	0.54±0.2	90±10	2.10±0.10	1992.62	Backer et al. 1993
0.69	0.70±0.10	0.60±0.2	115±20	1.42±0.10	1992.40	Krichbaum et al. 1993[a]
0.69	0.75±0.08	1.0		1.42±0.10	1992.40	Krichbaum et al. 1993[b]
0.35	0.19±0.03	1.0		1.80±0.30	1995.18	Krichbaum et al. 1997
0.35	0.15±0.05	1.0		1.40±0.20	1994.25	Rogers et al. 1994
0.35	0.33±0.14	1.0		1.47±0.75	1993.27	Krichbaum et al. 1994
0.35	0.22±0.19	1.0		1.25±0.35	1993.27	Krichbaum et al. 1994[c]
0.14	0.15±0.04	1.0		3.1±0.1	1995.17	Krichbaum et al. 1998[d]
0.14	0.11±0.06	1.0		2.0±0.2	1995.17	Krichbaum et al. 1998[e]

Notes:
[a]: size of the main component B1 in a 2 component model fit
[b]: size from a single circular Gaussian component model fit
[c]: reanalysis, using the assumption of NRAO 530 being completely unresolved
[d]: case 1, see text and Krichbaum et al. 1998
[e]: case 2, see text and Krichbaum et al. 1998

cult and at present are still affected by the limited number of available antennas, a sparsely sampled uv-coverage and residual calibration uncertainties caused by atmospheric fluctuations (which increase towards shorter wavelengths), and impurities of the receivers (see also the remarks by Bower et al. 1999 and Doeleman et al. 1999). Therefore some of the results presented here still require a careful interpretation and at least for the highest observing frequencies need confirmation by better and more extended experiments.

2. 43 GHz Observations

In 1991–1992 the first VLBA antennas became operational at 43 GHz. Subsequently the Galactic Center source Sgr A* was detected for the first time with VLBI at this frequency (Krichbaum et al. 1993). The data revealed a compact core component B1 of 0.7 mas size and some evidence for an elongation of the overall source structure to the north, along p.a.$= -25°$. With only 3 VLBA antennas (a 4th antenna yielded only few detections) and one closure phase available, this experiment did not allow to unambiguously determine a possible elongation, which seemed to be present nearly parallel to the main axis of the elliptical observing beam. Follow-up 43 GHz-experiments performed in 1992.6 by Backer et al. (1993) and in 1994.8 by Bower et al. (1998) with more VLBA antennas (N= 5) revealed consistently a point source and a refined and more accurate size estimate. From these data, the brightness distribution of Sgr A* is best described by a single and elliptically shaped Gaussian component (major axis: 0.75 mas, axial ratio: ~ 0.6, cf. Tab. 1), which is oriented in east-west direction. This orientation is also seen at the longer cm-wavelengths (eg. Lo et al. 1985, Alberdi et al. 1993) and seems to be wavelength independent (Fig. 2). New multifrequency data from an experiment with 6 telescopes (5 x VLBA + VLA_1) were presented during this conference (Lo et al. 1998, 1999). They reveal – besides the aforementioned east-west orientation – a minor source axis, possibly larger than the extrapolated scattering size: at 43 GHz the axial ratio seems to be ~ 0.8, 40-50 % greater than at lower frequencies. After subtracting the image broadening due to scattering, the authors interpret this as being caused by an intrinsic elongation of Sgr A* towards the north, not far off the orientation seen in the first 43 GHz experiment in 1992.

In Table 1 we summarize the results of all 43 GHz experiments on Sgr A* published so far. The inspection of the individual measurements reveals a very good agreement of the sizes within their uncertainties. We note an indication of flux density variability of the VLBI component, particularly if the data of 1992.6 and 1997.1 are taken into account. We also note that already in the earliest 43 GHz experiments evidence for an east-west elongation of the core component was present (component B1 in Krichbaum et al., 1993). More detailed VLBI monitoring will be necessary to investigate whether the structure of Sgr A* is variable and if the flux density changes are real and related to this.

3. 86 GHz Observations

The 3 mm (86 GHz) VLBI observations of Sgr A* are even more difficult to perform and to analyze than at 43 GHz. The stronger influence of the atmosphere (short coherence times of 10 – 20 sec) and the low aperture efficiency or small size of some of the antennas, still limit the detection sensitivity to 0.2 – 0.4 Jy (7σ). Up to now, all of the 3 mm VLBI observations of Sgr A* revealed visibility data on it only for 1–3 interferometer baselines. This restricts the determination of the brightness distribution to a circular component, precluding the determination of the axial ratio of an eventually elliptical structure.

In 1993.27, Sgr A* was detected with VLBI at 86 GHz on the 350 Mλ baseline between the 30 m antenna at Pico Veleta (Spain) and the 100 m telescope at Effelsberg (Krichbaum et al. 1994). Although the signal-to-noise ratio of the detection was reasonably high (SNR $\simeq 20$), calibration uncertainties made the size estimate difficult. Depending of the details of the calibration, we obtain a size of 0.33 ± 0.14 mas, if the 'a priori' calibration and atmospheric opacity corrections are used, and a size of 0.22 ± 0.19 mas with the additional assumption that a secondary calibrator (NRAO 530) is an unresolved point source[2] (see Table 1).

In 1994.25 a VLBI experiment with three stations yielded data for Sgr A* on the 250 Mλ baseline Kitt Peak to Owens Valley. This and the lack of a detection on the longer baselines to the East Coast (Haystack Observatory) led to a more accurate size measurement of 0.15 ± 0.05 mas (Rogers et al. 1994).

[2]new VLBI images of NRAO 530 show a more complex core-jet structure (Bower et al. 1997).

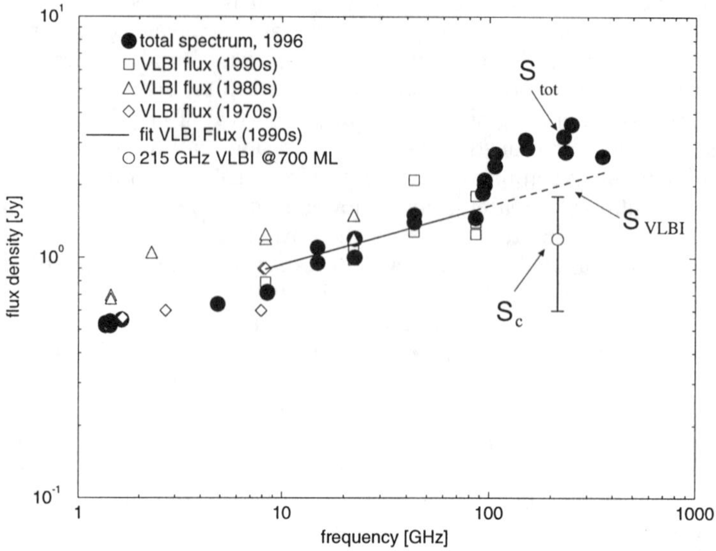

Figure 1. The cm- to mm-spectrum of Sgr A*. Filled circles show the total flux $S_{\rm tot}$. Open symbols denote fluxes of the VLBI component ($S_{\rm VLBI}$) at 3 arbitrarily binned time intervals: open squares for the 1990s, open triangles for the 1980s, open diamonds for the 1970s. The solid line shows a power-law fit to the VLBI spectrum of the 1990s ($S_{\rm VLBI}[{\rm Jy}] = 0.535 \cdot \nu_{\rm GHz}^{+0.24}$), which compares well with the total spectrum (filled circles). Above $\nu > 100$ GHz the difference between total flux and extrapolated spectrum (dashed line) indicates a flux density excess. At 215 GHz a correlated flux of $S_c = 1.2 \pm 0.6$ Jy ($B = 700\,{\rm M}\lambda$) has been measured with VLBI (open circle).

With measurements of the visibility on at least three baselines, information on the source orientation and possible deviations from circular symmetry can be obtained. In March 1995 (1995.18) Sgr A* was observed on the triangle formed by the antennas at Effelsberg, Pico Veleta and a single antenna of the IRAM interferometer on Plateau de Bure (Krichbaum et al. 1997 & 1998). This yielded for the first time a direct measurement of the closure phase, however only for a short time interval (5 scans, GST interval: 15h45 - 17h00). The scan to scan variations of the closure phase are $\leq 15^\circ$ consistent with a closure phase of zero in this hour angle range. This indicates a simple point-like or at least symmetric source structure. The size derived from these observations (0.19 ± 0.03 mas) is consistent with the size estimate of Rogers et al. (1994) and Doeleman et al. (1999). In Table 1 we summarize the size measurements at 86 GHz.

4. 215 GHz Observations

In 1994 it was demonstrated that successful VLBI observations at a wavelength of $\lambda = 1.4$ mm are possible, using the two IRAM instruments at Pico Veleta and Plateau de Bure (Greve et al. 1995). In a follow-up experiment performed

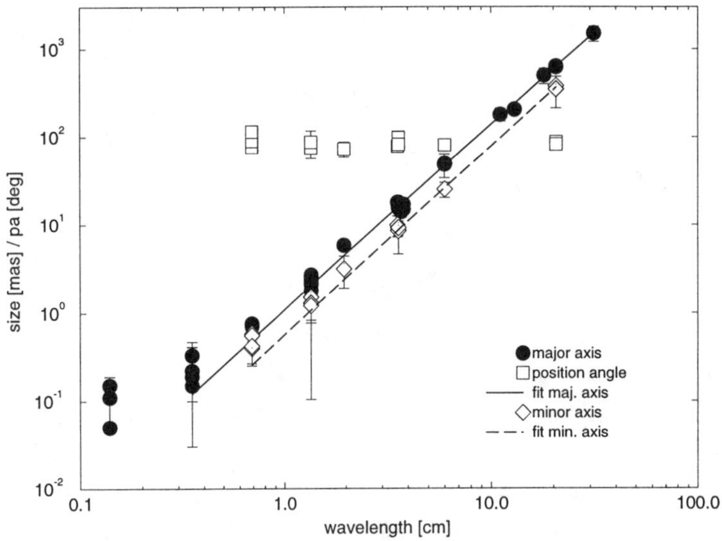

Figure 2. The major source axis (filled circles) of the Gaussian VLBI component of Sgr A*, the minor source axis (open diamonds) and the position angle of the major axis (open squares) plotted versus wavelength. For the least squares fit to the major axis only data for $\lambda \geq 3$ mm were used (solid line), for the fit to the minor axis only data for $\lambda \geq 7$ mm were taken into account (dashed line).

in 1995, we clearly detected 6 out of 8 bright AGN with scan averaged signal-to-noise ratios in the range of SNR$= 7 - 35$ (Krichbaum et al. 1997). In this experiment also Sgr A* was observed. A detailed analysis of the three VLBI scans (each of 6.5 min duration) showed that Sgr A* was detected in 2 out of the 3 scans with a signal-to-noise ratio of SNR$=$ 4.6 and 6.0, respectively (Krichbaum et al. 1998). With a formal detection threshold of SNR$=$ 7 for MK III VLBI data, this detection at present should be regarded as marginal (for a detailed discussion see Krichbaum et al. 1998, and references therein). It is obvious that a confirmation by future 1 mm-VLBI experiments is required.

If the detection is real, the correlated flux density for Sgr A* falls in the range of $S_c = 0.6 - 0.8$ Jy if the 'a priori' calibration is used and $S_c = 1.2 \pm 0.6$ Jy, taking into account also residual systematic calibration errors. Thus the correlated flux appears to be lower than the zero spacing flux extrapolated from VLBI measurements at longer wavelengths ($S_{\rm VLBI} = 2.0 \pm 0.2$ Jy, see solid line in Fig. 1), and also lower than the total flux $S_{\rm tot} = 3.1 \pm 0.1$ Jy determined from flux density measurements in 1996 (Falcke et al. 1998, filled circles in Fig. 1). We note that $S_{\rm tot}$ results from observations with observing beams of one to a few arcseconds, four orders of magnitude larger than the resolution achieved with VLBI at 215 GHz. It is therefore possible that on intermediate angular scales structure components might exist, which are presently not seen. We note that

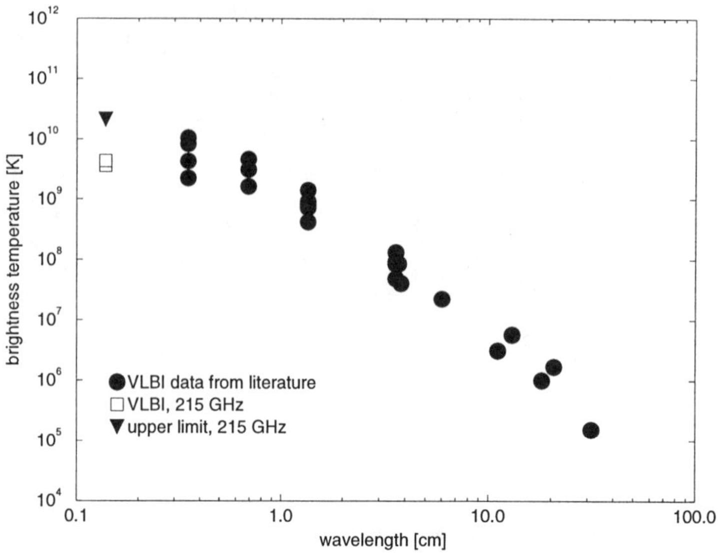

Figure 3. The apparent brightness temperature of the VLBI component of Sgr A* plotted versus wavelength. Circles denote measurements at $\lambda \geq 3$ mm. The filled triangle denotes an upper limit, adopting $\theta \geq 50\,\mu$as at 215 GHz. The open squares represent the best estimate at 215 GHz for 'case 1' and 'case 2' (see text and Table 1).

below 100 GHz, $S_{\rm tot}$ and $S_{\rm VLBI}$ are equal, excluding extended intermediate scale structures at longer wavelengths.

If the correlated flux of Sgr A* indeed is lower than the total flux, the source would be resolved and a size could be estimated. As mentioned above, above 100 GHz there is a discrepancy between the total flux resulting from direct measurements with single antennas or local interferometers ($S_{\rm tot}$, see Fig. 1) and the flux density extrapolation resulting from VLBI observations ($S_{\rm VLBI}$, see Fig. 1). Using as total flux $S_{\rm tot}$, one obtains a size of $\theta_1 = 150 \pm 40\,\mu$as (case 1). If instead for the total flux $S_{\rm VLBI}$ is used, a size of $\theta_2 = 110 \pm 60\,\mu$as is derived (case 2). Both sizes are close to the source size seen at 86 GHz (s. Table 1). In summary it is therefore likely that at 215 GHz the size of Sgr A* lies in the range of $50 \leq \theta \leq 190\,\mu$as.

5. Discussion

In Figure 2 the decrease of source size towards shorter wavelengths is plotted, including all published VLBI measurements of the size of Sgr A* and the new measurements presented in this paper. It is obvious that the major and minor source axis obey almost the same wavelength dependence: fitting a power law ($\theta \propto \lambda^n$) to the data, we obtain for the power law index a value of $n = 2.04 \pm 0.01$ for the major source axis and $n = 1.96 \pm 0.04$ for the minor axis. Down to a

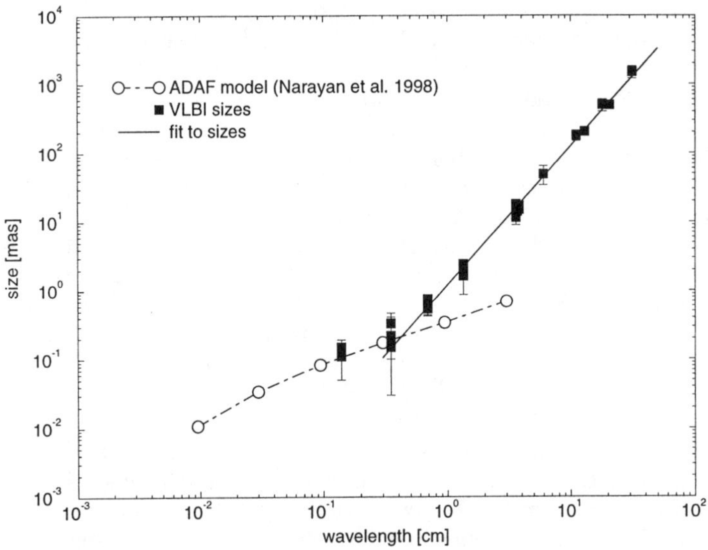

Figure 4. The geometric mean of major and minor axis of the VLBI component of Sgr A* plotted versus wavelength. Open symbols show the size predictions from the advection-dominated accretion flow (ADAF) model of Narayan et al. (1998). At 1.4 mm and 3 mm the predicted and observed sizes agree remarkably well.

wavelength of 1.3 cm, the orientation of the major axis seems to be independent of wavelength, being close to 80° (open squares in Fig. 2). For shorter wavelengths, no data for the position angle are yet available.

From the fit to the sizes, the scattering size at millimeter wavelengths can be estimated empirically: for the major axis we obtain 0.5 mas at 43 GHz, 0.12 mas at 86 GHz and 0.02 mas at 215 GHz. These numbers do not change significantly, if instead of the major axis the geometric mean of the minor and major axis is used (Fig. 4). The comparison with the sizes given in Table 1 clearly shows that, starting from 43 GHz, the discrepancy between measured source size and expected scattering size increases with frequency. It is therefore likely that the scattering screen becomes more transparent towards shorter wavelengths and that we are close to see the underlying intrinsic source structure of Sgr A*. Further evidence for this comes from (i) the recently observed increase of the axial ratio, which at 43 GHz is probably larger than at 22 GHz (see Lo et al. 1998, 1999), and (ii) from an increase of the flux density variability amplitudes towards higher frequencies (Zhao et al. 1991, Tsuboi et al. 1999). Particularly the latter cannot be explained by the scattering effect.

Despite remaining measurement uncertainties at 215 GHz, the size of Sgr A* is at least a factor of 2−3 larger than the scattering size (20 μas) at this frequency. Further evidence for the correctness of this size measurement comes from the the brightness temperature T_B. In Figure 3 we plot T_B versus wavelength (for the calculation of T_B we used all published size and flux density measurements from

VLBI). If we extrapolate the brightness temperature from the longer wavelengths to $\lambda = 1.4\,\text{mm}$ (215 GHz), we obtain $T_B \leq 4.5 \cdot 10^{10}\,\text{K}$. With a total flux of Sgr A* in the range $S_{\text{tot}} = 2 - 3\,\text{Jy}$ and using T_B from above, this yields a size of $\theta \geq 30 - 40\,\mu\text{as}$, again larger than the scattering size.

A lower limit to the size of Sgr A* is given by the Schwarzschild diameter ($2R_s/D = 13\,\mu\text{as}$) of the central black hole. The observable size of Sgr A*, however, should be much larger, if the radiating volume is related to the accretion disk, which for the inner stable orbits should have a diameter of at least a few up to a few ten R_s. The broad band spectral properties and the low (sub-Eddington) luminosity of Sgr A* are well explained by the advection-dominated accretion flow (ADAF) model of Narayan et al. (1998). This model predicts a size of 0.1 mas ($\sim 15R_s$) at 215 GHz. In Figure 4 we display the sizes from the ADAF model together with the measurements. At 3 mm and 1.4 mm wavelength, the agreement is striking.

We note that the existing models predict slightly different variations of the intrinsic source size with the wavelength (eg. Melia et al., 1992, Falcke & Biermann, 1998). This can be used to descriminate between the models, as soon as a deviation of the size from the λ^2-law (scattering) will be confirmed.

Future 1 mm VLBI experiments will be extremely important to improve the accuracy of the size measurement and will help to decide whether Sgr A* harbors a supermassive black hole or not. A break-through probably can be achieved, if Sgr A* can be imaged with 1 mm-VLBI also on the longer (continental, or even transatlantic) baselines. With an angular resolution of up to $20-30\,\mu\text{as}$ reached in this configuration, relativistic effects near the event horizon of a supermassive black hole could be *directly* studied. To provide the necessary sensitivity for this, the existing (eg. BIMA, OVRO, Plateau de Bure) and future (eg. LSA, MMA) interferometers need to participate as phased arrays in such VLBI experiments.

Acknowledgments. The results summarized here are obtained in a collaboration between various observatories including in addition to the authors the following scientists: D. Graham (MPIfR), A. Greve, M. Grewing, J. Wink (IRAM), F. Colomer, P. de Vicente, J. Gómez–González (Yebes), and A. Baudry (Bordeaux). We also like to appreciate very much the support of P.G. Mezger.

References

Alberdi A., Lara L., Marcaide J.M., et al., 1993 A&A, 277, L1

Backer D.C., Zensus J.A., Kellermann K.I., et al., 1993, Sci, 262, 1414

Bower G.C., & Backer D.C., 1998, ApJ, 496, 97

Bower G.C., Backer D.C, Wright M.C.H., et al., 1997, ApJ, 484, 118

Bower G.C., et al., 1999, this volume, p. 80

Doeleman, S., et al., 1999, this volume, p. 98

Falcke H., & Biermann P.L., 1998, A&A, 342, 49

Falcke H., Goss W.M., Matsuo H., et al., 1998, ApJ, 499, 731

Genzel, R., & Eckart, A., 1997, this volume, p. 3

Greve A., Torres M., Wink J.E., et al., 1995, A&A, 299, L33

Krichbaum T.P., Graham D.A., Greve A., et al., 1997, A&A, 232, L17

Krichbaum T.P., Graham D.A., Witzel A., et al., 1998, A&A, 335, L106

Krichbaum T.P., Schalinski C.J., Witzel A., et al., 1994, *in: The Nuclei of Normal Galaxies: Lessons from the Galactic Center*, ed. R. Genzel & A.I. Harris, ASI Conference Series C, Vol. 445, (Kluwer, Dordrecht), p. 411.

Krichbaum T.P., Zensus J.A, Witzel A., et al., 1993, A&A, 274, L37

Lo K.Y., Backer D.C., Ekers R.D., et al., 1985, Nat, 315, 124

Lo K.Y., Shen Z.Q., Zhao J.H., & Ho P.T.P., 1998, ApJ, in press

Lo K.Y., Shen Z.Q., Zhao J.H., & Ho P.T.P., 1999, this volume, p. 72

Melia F., Jokipii J.R., & Narayanan A., 1992, ApJ, 395, 87

Narayan R., Mahadevan R., Grindlay J.E., et al., 1998, ApJ, 492, 554

Rogers A.E.E., Doeleman S.S., Wright M.C.H., et al., 1994, ApJ, 434, 59

Tsuboi, M., et al., 1999, this volume, p. 105

Zhao J.-H., Goss W.M., Lo K.Y., & Ekers R.D., 1991, *in: Relationships between Active Galactic Nuclei and Starburst Galaxies*, ed. A. Filippenko, ASP Conference Series, Vol. 31, p. 295.

The Central Parsecs of the Galaxy
ASP Conference Series, Vol. 186, 1999
H. Falcke, A. Cotera, W.J. Duschl, F. Melia, M.J. Rieke, eds.

Structure of Sgr A* from 3mmλ-VLBI

S. Doeleman and A.E.E. Rogers

Haystack Observatory, Off Route 40, Westford, Mass. 01886

D. C. Backer and M. Wright

Astronomy Department, University of California, Berkeley, CA 94720

G. C. Bower

Max Planck Institut für Radioastronomie, Bonn, Germany

Abstract. We report on 3mmλ-VLBI observations of Sgr A* with an array of three antennas. A brief discussion of calibration precedes our results as it is the most important problem to overcome in high frequency continuum VLBI measurements. Our size estimates of Sgr A* are consistent with past results but cannot improve on them. However, calculation of the closure phase (sum of interferometric phases around a triplet of antennas) does yield direct information on the possible asymmetry of Sgr A*. In our observations we find the closure phase to be consistent with zero which restricts to some extent the type and degree of asymmetry that Sgr A* can exhibit.

1. Introduction

Estimates of the intrinsic size and structure of Sgr A* are difficult to make due to the angular broadening that occurs in the ionized material which lies between us and the Galactic Center. VLBI observations at 86 GHz, and even 215 GHz, provide an opportunity to peer through this material as the broadening decreases quadratically with observing frequency. At these high radio frequencies, the scattering size approaches and is in some cases less than the intrinsic size predicted by radiative emission models of Sgr A*.

Rogers et al. (1994) set lower limits on the size of Sgr A* using 86-GHz VLBI which failed to detect the source on long (~5000 km) baselines. Corresponding upper limits were obtained on the shorter KittPeak-OVRO baseline thus bracketing the apparent size between 0.1 mas and 0.2 mas. These values differed from those of Krichbaum et al. (1994) who proposed a size of ~ 0.33 mas at this frequency which far exceeded the scattering size at 86 GHz of ~ 0.16 mas. More recently (Krichbaum et al. 1998 & 1999), 86-GHz VLBI results using three European antennas have yielded an estimated size near ~ 0.19 mas. Here, we report on observations made in April 1997 in which an array of three antennas were used. First we comment on the importance of proper calibration and why it is essential in deriving size estimates of Sgr A*. A discussion of closure phase follows including what recent measurements can say about the structure of Sgr A*.

2. Calibration

It is now well accepted that interferometric detection of Sgr A* on 1000 km baselines and at high (~86 GHz) frequencies is possible and can even be considered routine. However, translating these detections into credible size estimates still depends on carefully calibrating the measured interferometric visibilities. Sgr A* is at such low declination (−29°) that arrays in the Northern hemisphere can observe for relatively short periods and typically at low telescope elevations. For this reason, coverage in the (u, v) plane (number of fourier components of the sky brightness distribution) is sparse and standard techniques of deconvolution and imaging are seldom used. Rather, one normally assumes that Sgr A* is represented as a circular Gaussian brightness distribution and a best fit to the data is found. The relation between correlated flux density, baseline length and source size can be expressed as:

$$S_c = S_t \exp\left[-\frac{\theta^2\pi^2B^2}{4\ln(2)}\right] \tag{1}$$

where S_c and S_t are the respective correlated and total flux densities, θ is the FWHM source size in radians, B is baseline length in wavelengths. "Zero-baseline" measurements taken with a single dish or connected-element interferometer are often added to the fit and are essential when only a few VLBI data points are available. Fig. 1 shows examples of this relation and the effects on derived source size due to measurement uncertainty in both zero-baseline and moderate baseline correlated flux density.

The 50% uncertainties in Fig. 1 are, unfortunately, quite common at high frequencies and underscore the sensitivity of estimated sizes on calibration which often introduces the largest source of error. Sources of error include atmospheric turbulence which introduces phase fluctuations in the VLBI signal, high frequency phase noise in receiver electronics, errors in telescope gain determination, telescope pointing errors, variable opacity and variable anomolous refraction.

3. Observations

In April 1997 we observed Sgr A* with an array of five telescopes: the phased BIMA array, the KittPeak 12m, Haystack Observatory, the VLBA PieTown site and the phased OVRO array. Due to a phasing problem at BIMA, no detections were obtained on Sgr A* on baselines to that site, and bad weather precluded detections on baselines to Haystack. Projected baseline lengths for the remaining three stations ranged from 80 $M\lambda$ to 400 $M\lambda$. The continuum source NRAO530 and the spectral line source VXSgr were observed as calibrators. The data were correlated at Haystack Observatory.

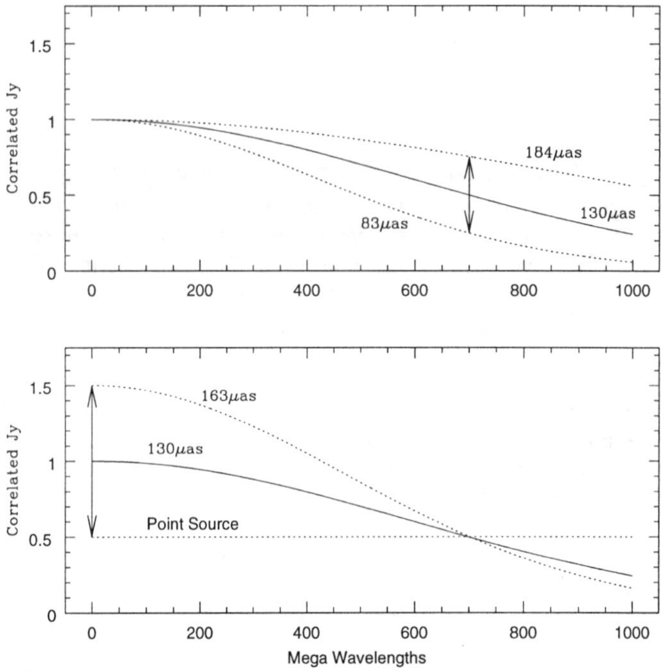

Figure 1. Two different types of miscalibration and their effects on size estimates assuming a circular Gaussian brightness distribution. In the top panel we suppose that the zero-spacing flux density is well known to be 1Jy but at 700 Mλ, there is a 50% uncertainty in the measured correlated flux of 0.5Jy. In the lower panel we take as fixed a 0.5Jy measurement of correlated flux density at 700 Mλ with a 50% uncertainty in the zero-spacing flux density measurement of 1Jy. In both cases the dotted curves show the range of sizes one can obtain.

4. Size Estimates

After correlation the visibility amplitudes were corrected for atmospheric coherence effects by use of incoherent averaging techniques (Rogers et al. 1995). Conversion to Janskys was performed by assuming that all the single dish flux density measured on NRAO530 was compact which allowed us to determine antenna gain as a function of elevation directly from the VLBI data. The resulting scatter in the visibility as a function of baseline length does not produce a unique fit to Eq. 1. The data do suggest, though, that the size of Sgr A* must be less than 0.3 mas FWHM. A separate attempt at calibration using system temperature measurements at each site and previously determined gain curves produced the same results. Thus, our amplitude data, whose errors are dominated by

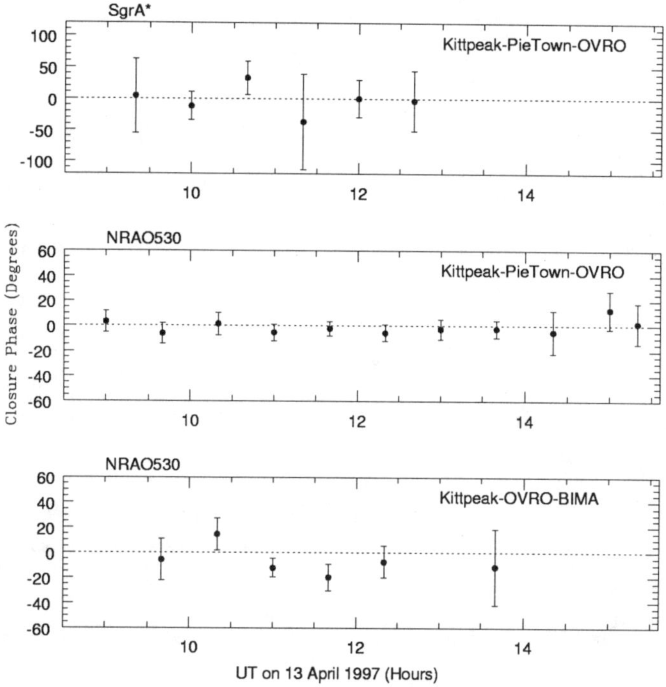

Figure 2. Closure phases for Sgr A* and calibrator NRAO530. On the KittPeak-PieTown-OVRO triangle, both sources exhibit closure phase consistent with zero. A clear non-zero closure phase signature appears for NRAO530 on the KittPeak-OVRO-BIMA triangle of stations. Note the difference in scale between plots of the two sources. Error bars are 3σ.

calibration uncertainty do not establish a firm size estimate but do produce a large upper limit.

5. Closure Phase and Structure

Uncertainties in the closure phase, by contrast, depend only on the Signal to Noise ratio of the VLBI data and are largely insensitive to atmosphere induced phase fluctuations and antenna-based calibration uncertainties. This is because the closure phase is formed by summing the interferometric phase around a triangle of baselines thus canceling all station-based effects. Closure phase is a direct measure of asymmetric structure : a source whose structure is symmetric about the origin yields zero closure phase for any array geometry. Conversely, an asymmetric source will have a non-zero closure phase signature for *some* arrangement of telescopes. Detection of asymmetric structure in Sgr A* could

DUNELM UNIV

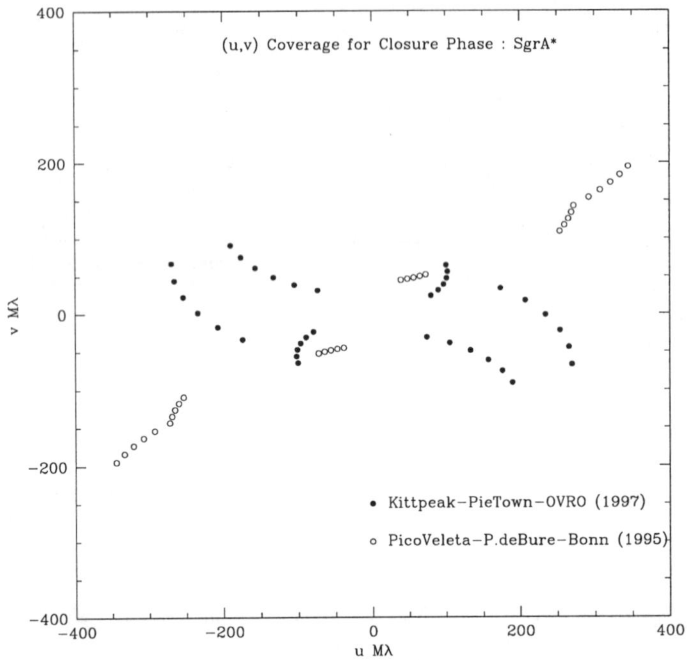

Figure 3. Total set of (u, v) points corresponding to observations producing closure phases on Sgr A*. The open circles are points from 1995 observations with the Bonn, PicoVeleta, Plateau de Bure antennas. The solid dots are taken from the observations herein described with the KittPeak, PieTown, OVRO stations.

help resolve whether core-jet models (Falcke et al. 1993) or symmetric models (Melia 1994; Narayan et al 1998) apply.

Fig. 2 shows the closure phases we measured for Sgr A* and NRAO530 over a six hour period on the KittPeak-PieTown-OVRO triangle. Though there is one point at 10:20 UT which could be non-zero, all closure phases for Sgr A* are consistent with a zero value. This same triangle measures zero closure phase for NRAO530. An example of a significant non-zero closure phase signature is obtained on the KittPeak-OVRO-BIMA triangle for NRAO530. This deviation from zero is corroborated by data on other triangles as well.

With so few data points, one can only say that the structure of Sgr A* is consistent with a symmetric distribution. But to ascertain the strength of that statement we can ask which types of asymmetry are concretely ruled out by the results. To do this we have collected the closure phase data from two separate observations : the results described here and the zero closure phase results described in Krichbaum et al. (1998) on a three station European mm-VLBI array. We assume both of these observations to have measured zero closure

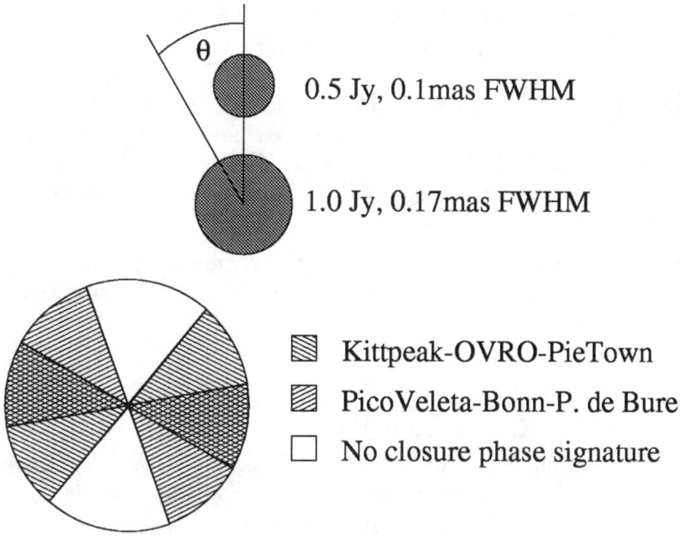

Figure 4. "Straw" model which underscores the need for more closure phase data in determining possible asymmetries in the structure of Sgr A*. Hatched areas of the pie chart show angles for which one would see a non-zero closure phase signal on one of the two arrays. The fact that both arrays see zero closure phase limits the model angle to the unfilled regions.

phase and plot in Fig. 3 the combined (u, v) coverage of both data sets. We can now ask : given a unique structural model of Sgr A*, is it ruled out by this combined data set?

Fig. 4 shows a "straw" model exhibiting a gross asymmetry in the form of two unequal components. The model consists of two circular Gaussian components whose centers are separated by 0.4 mas. If we parameterize the model by the position angle (E of N) of the smaller component then we can determine which orientations would yield non-zero closure phase signatures using the two different arrays. The filled regions of the pie chart show orientations disallowed by measurements while the open regions, which span over 120°, denote angles for which we have no constraints and regions in which such gross asymmetry could exist despite our measurements. Fully one third of the model parameter space remains unconstrained by measurement which emphasizes the need to increase the number of antennas and, hence, (u, v) coverage on the Galactic Center.

6. Summary

We have attempted to point out the importance of calibration in current mm-VLBI efforts to constrain the size of Sgr A*. It should be noted that progress is being made in this area and new methods such as calibration via observation of SiO maser sources and extension of integration times using Water Vapor Radiometers will help alleviate some current difficulties.

Despite the incomplete (u, v) coverage of past observations, some asymmetric models of Sgr A* can be ruled out. But it is clear that more antennas are needed to make definitive observations. We note that if one includes the FortDavis (VLBA) and BIMA baselines to the KittPeak-OVRO-PieTown array, many holes in the (u, v) plot of Fig. 3 fill in making the array more sensitive to asymmetric structure.

Acknowledgments. mm-VLBI necessarily requires the assistance and dedication of many individuals. We thank them and all participating telescopes.

References

Falcke, H., Mannheim, K., Biermann, P.L. 1993, A&A, 278, L1

Krichbaum, T., Witzel, A., Standke, K. J., Graham, D. A., Schalinski, C. J., & Zensus, J. A. 1994, in Proceedings of "Compact Extragalactic Radio Sources", eds. J. A. Zensus & K. I. Kellerman.

Krichbaum, T., et al. 1998, A&A, 335, L106

Krichbaum, T., et al. 1999, this volume, p. 89

Melia, F. 1994, ApJ, 426, 577

Narayan, R., Mahadevan, R., Grindlay, J. E., Popham, R. G., & Gammie, C. 1998, ApJ, 492, 554

Rogers, A. E. E., et al. 1994, ApJ, 434, L59

Rogers, A. E. E., Doeleman, S., & Moran, J. 1995, AJ, 109, 1391

The Central Parsecs of the Galaxy
ASP Conference Series, Vol. 186, 1999
H. Falcke, A. Cotera, W.J. Duschl, F. Melia, M.J. Rieke, eds.

Flare of Sgr A* at Short Millimeter Wavelengths

Masato Tsuboi and Atsushi Miyazaki

Institute of Astrophysics and Planetary Science, Ibaraki University, Bunkyo, Mito, Ibaraki, 310-8512, Japan

Takahiro Tsutsumi

National Astronomical Observatory, Mitaka, Tokyo, 181-8588, Japan

Abstract.
We have performed 3 sets of one-month intensity monitor observations toward Sgr A*, at $\nu = 100$ and 140 GHz from 1996 to 1998 using the Nobeyama Millimeter Array. We got following results;
1) We detected a flare of Sgr A* in March 1998. The flux density at $\nu =$ 100 GHz was flared over $(\Delta S/S =)$ 100% in a week and decreased to the mean flux density within two weeks.@The flux density in the flare at $\nu =$ 140 GHz was also increased over $(\Delta S/S =)$ 100%.
2) The flux densities in the flare are consistent with the power-law spectrum of $\alpha = 0.6 (S \propto \nu^{\alpha})$. On the other hand, the mean flux densities in quiescent phase in 1996–1998 are consistent with the power-law spectrum of $\alpha = 0.3$.

1. Introduction

Time variability of the flux density of Sgr A* at cm wavelengths has been studied in these two decades (e.g. Zhao et al. 1992). The time variability at cm wavelengths has already been established. The variability below $\nu \leq 10$ GHz is, however, mainly dominated by refractive interstellar scintillation. On the other hand, the variability at millimeter wavelengths is believed to show the intrinsic activity of Sgr A*. Wright and Backer reported that significant flux variations at $\nu = 100$ GHz in 1990 (1993). This appears to correspond to the flare observd at cm wavelengths (Zhao et al. 1992). The flare at mm wavelengths has the time scale of one month and the amplitude of the valiability is larger than one at cm wavelengths. However, the short-term variability at mm wavelengths is not well established because of the shortage of systematic monitor observations (Zylka et al. 1995, Serabyn et al. 1997). Then we made the intensity monitor observations at mm wavelengths to probe the intrinsic variability of Sgr A*. It is also important to observe the simultaneous spectrum at mm wavelengths because the non-simultaneous spectrum suffers from the time valiavility.

As well known, Sgr A* at mm wavelengths is a weak (~ 1 Jy) compact component embedded in the extended and strong HII region, Sgr A West. Thus it is difficult to observe Sgr A* itself at mm wavelengths. The emission from the HII region is several times larger than that of Sgr A* even using the most sharpe beam (17") of the Nobeyama 45-m telescope. Therefore the high angular resolution (a few arcsecond or more) observations by interferometers are required to discriminate the compact component from the surrounding extended components. Then, in order to make intensity monitor experiments toward Sgr A*, we observed at $\nu = 100$ and 140 GHz using the Nobeyama Millimeter Array (NMA).

2. Observations

We have performed 3 sets of one-month intensity monitor observations toward Sgr A*, at $\nu = 100$ and 140 GHz from 1996 to 1998 using NMA. As mentioned in the previous section, interferometer observations have ability to discriminate a compact source embedded in extended structures. However, the long-time intensity monitor by the interferometers has some problems. First, the different array configuration of the interferometer could introduce a systematic error in the result caused by sampling different structures in the beams. Second, the array configuration with the longest baseline gives us the highest angular resolution. But this configuration requires a best weather condition. This configuration does not always benefit the long-time intensity monitor observations. Then, we used an intermediate-span array configuration, "C-configuration" of NMA in all observations. The configuration contains the $U - V$ distances of 7–55 kλ at the 100 GHz band and 10–77 kλ at the 140 GHz band, respectively.

The time sequence of our one-month monitor observations is as followings. A couple of observations was separated by 1 days. The pair observations were proceeded every 5 days for 1 month. The 1 day separation is for day-scale variability. The 5 days separation is for week-scale variability. We should observe up to month-scale variability because the total duration of the observation is 1 month.

We used double side band (DSB) SIS receivers at the 100 GHz and the 140 GHz bands as frontends. Data in 1996 were acquired using FX correlator of 320 MHz bandwidth. The observed frequencies in 1996 were $\nu = 102$ and 146 GHz. Data in 1997–98 were acquired using the Ultra Wide Band Correlator (UWBC) of 1 GHz bandwidth. UWBC has been able to observe simultaneously both in the upper side band (USB) and the lower side band (LSB) separated by 12 GHz since 1997 . The observed frequencies in 1997–98 were $\nu = 90$ and 102 GHz for the100 GHz band and $\nu = 134$ and 146 GHz for the140 GHz band, respectively.

Intensity calibration is most important for intensity monitor observations especially at mm wavelenghs. The intensity calibration procedure of our observations is as followings. The instrumental gain and phase stability of NMA were calibrated by alternating observations of Sgr A* and NRAO 530 at about 20 min. intervals. To improve the accuracy of the flux calibration, we also observed Uranus and Sgr B2(M). Uranus was observed at 10 min. intervals, and Sgr B2(M) was observed at 15 min. intervals. The absolute flux density scale was established from the observations of Uranus. The flux calibration strongly depends on weather conditions through phase stability. The validity of the flux calibration was checked by the observations of Sgr B2(M) assuming the flux den-

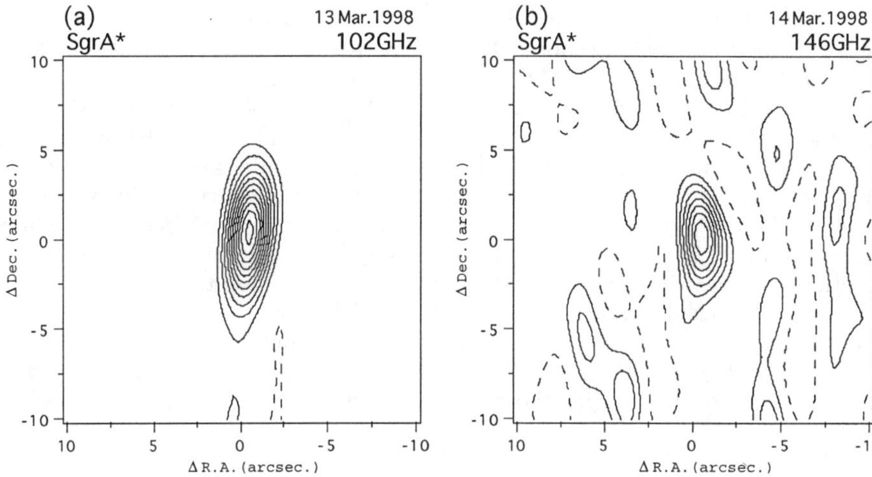

Figure 1. The spatial filtered images of Sgr A* at the 100 GHz and the 140 GHz bands made from the $U-V$ data restricted larger than the 25 kλ in order to suppress the contaminations from the extended components around Sgr A*.
a)The spatial filtered images of Sgr A* at the 100 GHz band. The sysnthesized beam in uniform weight is approximately $\alpha \times \delta = 3" \times 6"$ at the 100 GHz band.
b)The spatial filtered images of Sgr A* at the 140 GHz band. The sysnthesized beam in uniform weight is approximately $\alpha \times \delta = 2" \times 4"$ at the 140 GHz band.

sity was constant. As the result, the absolute flux density scales are accurate to about 15% at the 100 GHz band and 20% at the 140 GHz band, respectively.

We made spatially filtered maps from the $U-V$ data restricted larger than 25 kλ in order to suppress the contaminations from the extended components around Sgr A*. Figure 1a and 1b show spatially filtered images of Sgr A* at 100 and 140 GHz bands, respectively. We used the peak flux densities on the maps. The resultant beam size of the synthesized beam in uniform weight are $\alpha \times \delta = 3" \times 6"$ at the 100 GHz band and $\alpha \times \delta = 2" \times 4"$ at the 140 GHz band, respectively.@

3. Results

3.1. 100 GHz Band

Figure 2 shows all data of the monitor observations in 1996–1998 and figure 3-a shows time variation of the flux density of Sgr A* at $\nu = 102$ GHz in March 1996. The flux density of Sgr A* in March 1996 varies by $\Delta S/S = 30$ % in time scale of 10 days. However, the mean flux density at $\nu = 102$ GHz in March 1996 is similar to that in March 1997 up to $\Delta S/S = 10$ %. The mean flux density is $S = 1.0 \pm 0.2$ Jy at $\nu = 90$ GHz (1997) and $S = 1.2 \pm 0.2$ Jy at $\nu = 102$

GHz (1996–1997), respectively. These values are consistent with the previous data with NMA in 1990 (Tsuboi, Okumura and Ishiguro 1992). Sgr A* in these observation epochs is supposed to have been in the quiescent phase.

In March 1998, we detected a flare of Sgr A* at the100 GHz band (see figure 2). Figure 3-a shows the time profile of the detected flare. The flux density of Sgr A* peaked at March 13, 1998. The maximum flux density is $S = 2.4 \pm 0.4$ Jy at $\nu = 90$ GHz and $S = 2.4 \pm 0.4$ Jy at $\nu = 102$ GHz, respectively. These values are about twice times larger than the mean flux density. Because the flux densities of Sgr A* just before the flare are slighly smaller than the mean value mentioned above, the flux density substantially increased up to three times. The flux density reached to the maximum at most within one week. The flux density decreased to $\Delta S/S = 80\%$ in 3 days after the peak and decreased to the mean flux density at most within 2 weeks. The peak flux density and the time scale of the flare is consistent with those of the flare reported by Wright and Backer (1993).

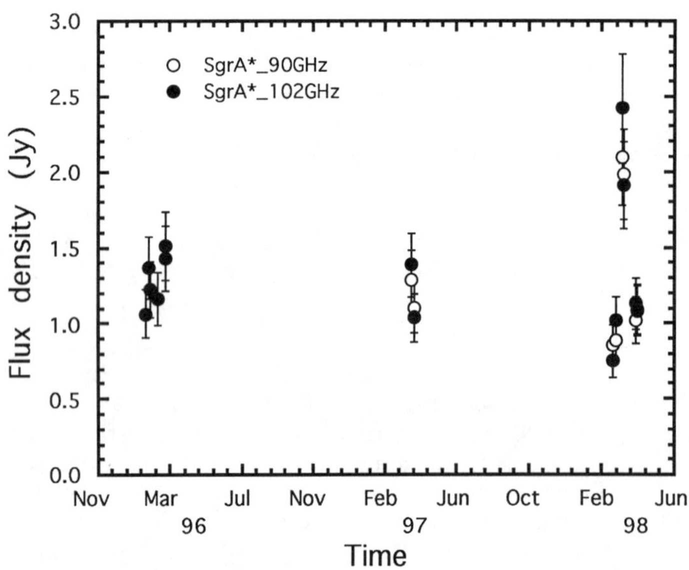

Figure 2. Time variation of the flux density at $\nu = 90$ (open circles) and 102 (filled circles) GHz of Sgr A* in 1996-1998. The mean flux density is $S = 1.0 \pm 0.2$ Jy at $\nu = 90$ GHz in March 1997 and $S = 1.2 \pm 0.2$ Jy at $\nu = 102$ GHz in March 1996 and March 1997, respectively.

We also observed the peak position of Sgr A* in the spatially filtered maps. We detected no difference between the peak positions in the flare and the quiescent phases. The upper limit of the difference is $\Delta\theta \leq 1$". This suggests that the observed intensity variation is caused by the intrisic valiability of Sgr A* rather than the variabilities in the surrounding objects, for example radio novae (Zhao et al. 1992b)

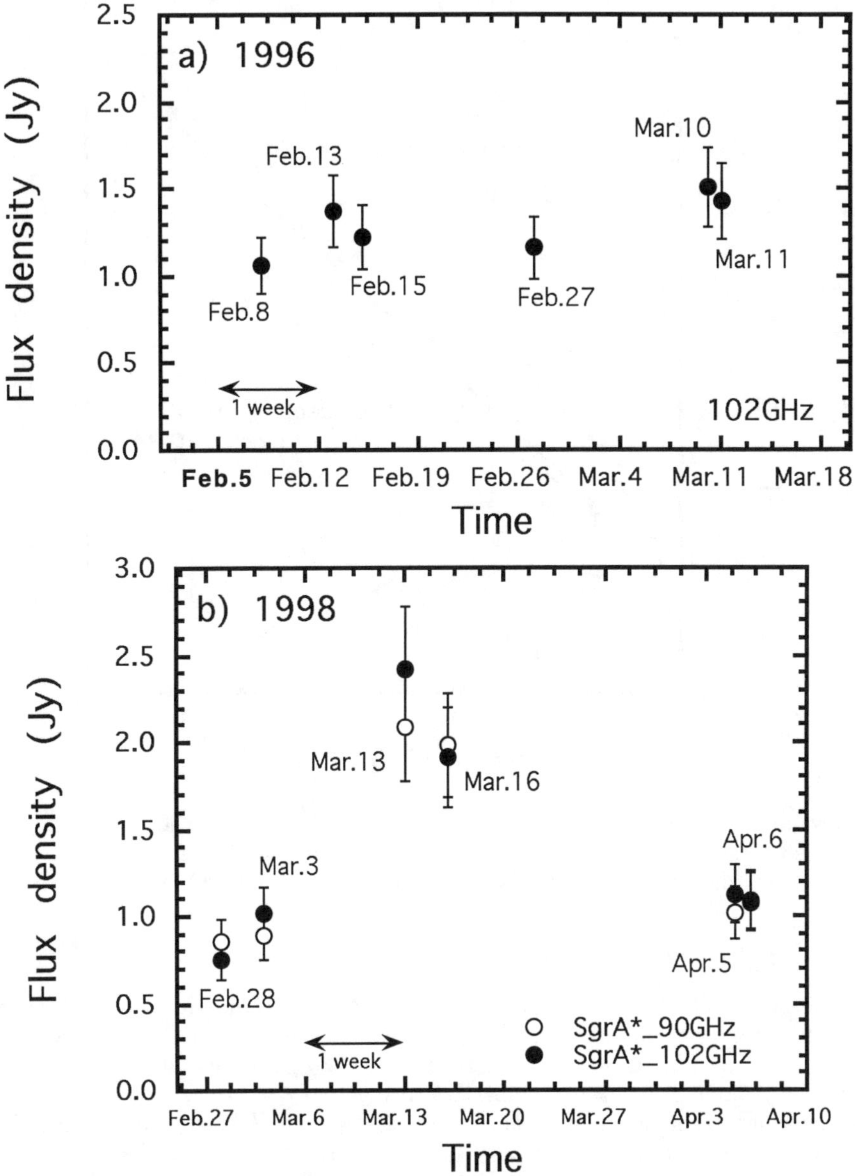

Figure 3. a) Time variation of the flux density at $\nu = 102$ GHz of Sgr A* in March 1996. b) Time profile of the flare of Sgr A* at 90 (open circles) and 102(filled circles) GHz in March 1998. The flux density of Sgr A* peaked at March 13, 1998. The maximum flux density is $S = 2.4 \pm 0.4$ Jy at $\nu = 90$ GHz and $S = 2.4 \pm 0.4$ Jy at $\nu =$102 GHz, respectively. The flux density reached to the maximum at most in a week and decreased to the mean flux density at most within 2-3 weeks.

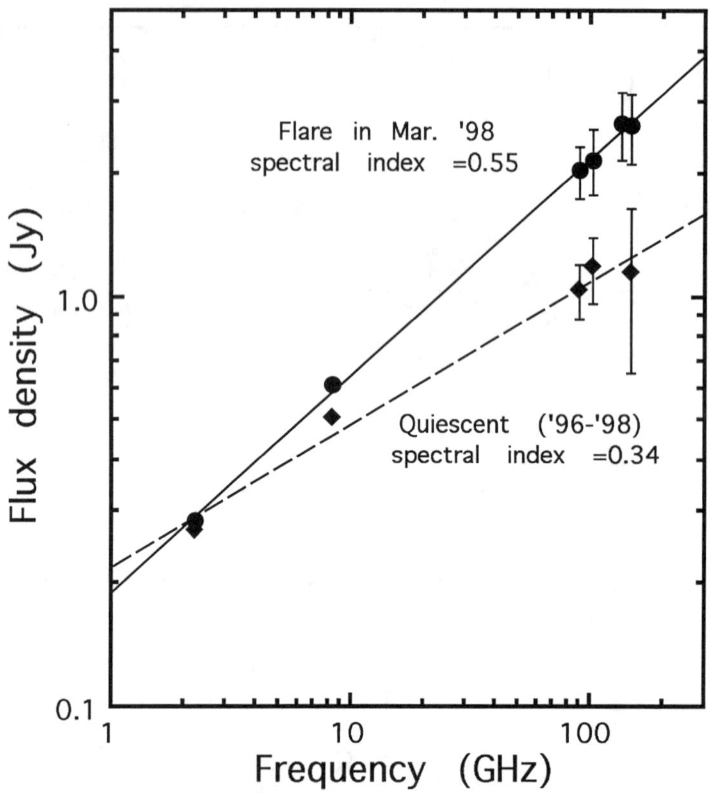

Figure 4. Spectra of Sgr A* from 2.3 to 140 GHz in the flare and the quiescent phases. The plots at ν = 2.3 and 8.3 GHz are from the archival data of the Green Bank Interferometer of NRAO.

3.2. 140 GHz Band

We measured the flux density of Sgr A* at 140 GHz band in the flare. The observation date is the day next after the peak at the 100 GHz band, or March 14, 1998. The flux densities were about $S = 2.7 \pm 0.5$ Jy at ν =134 GHz and $S = 2.6 \pm 0.5$ Jy at ν = 146 GHz, respectively. On the other hand, we also measured the flux density of Sgr A* at ν = 146 GHz at several times in 1996–1998. The mean flux density in the quiescent phase in 1996–1998 are $S = 1.2 \pm 0.4$ Jy at ν = 146 GHz. In the flare, the flux density of Sgr A* at the 140 GHz band increased over $\Delta S/S \geq$100% for the mean flux density.

4. Discussion

Figure 4 shows the simultaneous spectra of Sgr A* from ν = 2 GHz to ν = 140 GHz in the flare observed in March 1998 and the quiescent phases in 1996–1998. The plots at ν = 2 and 8 GHz are obtained from the archival data of the Green Bank Interferometer. The flux densities at ν = 2 and 8 GHz in the flare appear to have no significant difference from those in the quiescent phase. In the flare, the spectrum is explained by the power law spectrum with spectral index of $\alpha = 0.55 (S \propto \nu^{\alpha})$. Our observed flux densities in the flare at the 100 and 140 GHz bands are consistent with the near millimeter excess observed by single-dish telescopes (e.g. Falcke et al. 1998). On the other hand, the excess was not observed in the quiescent phase. The spectrum in the quiescent phase up to ν = 140 GHz is explained by another power law spectrum with the spectral index of $\alpha = 0.34$. This spectrum is consistent with the spectrum extrapolated from lower frequency observations (e.g. Falcke et al.1998, Krichbaum et al. 1997). Thus the near millimeter excess may be expalined by flares of Sgr A*. The excess is presumably caused by the variable component in millimeter wavelengths. Assumed that the duration of the flare is a few weeks, the emission is isotropic, the spectral turn-over is at ν =150 GHz, and the Galactic center distance is 8.5 kpc, then total energy of the flare is an order of 10^{39} erg.

5. Conclusions

Some remarks can be about Sgr A*;
1) We detected a flare of Sgr A* in March of 1998. The flux density at ν = 100 GHz was flared over ($\Delta S/S =$) 100% in a week and decreased to the mean flux density with two weeks.
2) The flux density at ν =140 GHz was also increased over ($\Delta S/S =$) 100 % in the flare.
3) The flux densities in the flare are consistent with the power-law spectrum with $\alpha = 0.6 (S \propto \nu^{\alpha})$.
4) The mean flux densities in the quiescent phase of Sgr A* in 1996–1998 are also consistent with the power-law spectrum with $\alpha = 0.3$.

Acknowledgments. We thank K. I. Morita at Nobeyama Radio Observatory for useful discussion and the staff of NMA group for their assistance during observations. We also thank NRAO for using the archival data of the Green Bank Interferometer.

References

Falcke, H., Goss, W.M., Matsuo, H., Teuben,P., Zhao, J.-H., Zylka, R. 1998, ApJ, 499, 731

Krichbaum, T.P., Graham, D.A., Greve, A., Wink, J.E., Alcolea, J., Colomer, F., de Vicente, P., Baudry, A., Gomez-Gonzalez, J., Grewing, M., Witzel, A. 1997, A&A, 323, L17

Serabyn, E., Carlsrom, J., Lay, O., Lis, D. C., Hunter, T. R., Lacy, J. H. 1997, ApJ, 490, 77

Tsuboi, M., Okumura, S.K., Ishiguro, M. 1992, Nobeyama Radio Observatory Report, 285

Wright, M.C.H., Backer, D.C. 1993, ApJ, 417, 560

Zhao, J.-H., Goss, W.M., Lo, K.Y., Ekers, R.D. 1992, Reletionships Between Active Galactic Nuclei and Starburst Galaxies, (ASP:San Francisco) 31, 295

Zhao, J.-H., Roberts, D.A., Goss, W.M., Frail, D.A., Lo, K.Y., Subrahmanyan, R., Kesteven, M.J., Ekers, R.D., Allen, D.A.,Burton, M.G. 1992b, Sci., 255, 1538

Zylka, R., Mezger, P. G., Ward-Thomson, D., Duschl, W.J., Lesch H. 1995, A&A, 297, 83

The Central Parsecs of the Galaxy
ASP Conference Series, Vol. 186, 1999
H. Falcke, A. Cotera, W.J. Duschl, F. Melia, M.J. Rieke, eds.

Radio Variability of Sgr A* at Centimeter Wavelengths

H. Falcke

Max-Planck-Institut für Radioastronomie, Auf dem Hügel 69, D-53121 Bonn, Germany

Steward Observatory, The University of Arizona, Tucson, AZ 85721

Abstract. Results of two years of continuous monitoring of flux density variations at 8.3 and 2.3 GHz of the Galactic Center super-massive black hole candidate Sgr A* are reported. The average RMS modulation indices are 6% and 2.5% at 8.3 & 2.3 GHz respectively. There is a certain degree of correlation between both frequencies. The timescale of variability at 8.3 & 2.3 GHz is between 50 and 200 days. We cannot confirm a λ^2 dependence of the timescale. At 2.3 GHz a quasi-periodic behaviour with a period of 57 days was discovered which is reminiscent to, though longer than, those found in some compact extragalactic radio sources.

1. Introduction

Sgr A* is believed to be the radio source associated with the $2.6 \cdot 10^6 M_\odot$ (Haller et al. 1996; Ghez et al. 1998 & 1999; Eckart & Genzel 1996; Genzel & Eckart 1999; Zhao & Goss 1999) dark mass concentration in the center of the Galaxy. Since we know very little about this source from other wavelengths, where it is extremely faint (see Falcke 1996 for a review), a detailed study of its radio properties is an important prerequisite for its interpretation. The overall shape of the Sgr A* radio spectrum has been discussed in many papers (e.g., Serabyn et al. 1997; Falcke et al. 1998) and the variability has been investigated by Zhao et al. (1989 & 1992). The spectral index ($S_\nu \propto \nu^\alpha$) of the source tends to be in the range $\alpha \simeq 0.2 - 0.3$ with an increasing value of α at mm-wavelength and a possible cut-off at lower frequencies. At high frequencies the spectrum cuts off in the infrared. A major problem with the investigation of its radio variability is that Sgr A* is at relatively low elevation for most interferometers, that it is embedded in a large confusing structure, and that it becomes scatter-broadened at low frequencies. The confusion especially is a major problem for single-baseline interferometers with short baselines like the Green Bank Interferometer (GBI) that is often used for variability studies. For this reason the exact nature of the variability of Sgr A* has remained inconclusive. Flux density variations are clearly seen between different epochs, but the timescale of the variability at various frequencies is not well determined and it is not clear whether some of the more extreme claims of variability are real or instrumental artifacts. So far, Zhao et al. (1989,1992) probably have presented the largest database of Sgr A* flux-density measurements. They found a number of outbursts at higher frequencies and tentatively concluded that the small-amplitude variability at longer wavelengths is caused by scattering effects in the ISM while the variability at higher frequencies is intrinsic. In this paper new results of a continuous monitoring program of Sgr A* at cm-wavelengths performed with the GBI are presented and evaluated.

2. GBI Observations and Data Reduction

Sgr A* has been part of the NASA/NRAO Green Bank Interferometer (GBI) monitoring program for the past two years. The GBI is a two-element interferometer (26m dishes) with a separation of 2400 meters, operating simultaneously at X- and S-band (8.3 & 2.3 GHz) with 35 MHz bandwidth. The resolution of the pencil beam is 3 and 11 arcseconds and 1 σ noise levels are typically 30 and 6 mJy at X and S-band respectively. The data are publically available but need further processing, since the baseline gains depend on hourangle. In addition observations of Sgr A* will also suffer from severe confusion due to the small baseline and the extended structure of Sgr A West as mentioned in the introduction.

The data were post-processed in the following way: an hourangle dependent gain correction was fitted to 1622-297 which serves as a calibrator to Sgr A*. Absolute gains were obtained using 3C286 as the primary flux density calibrator. This gain corrections were then applied to all sources and outliers were clipped when flux density measurements deviated by more than 3 σ from the median flux density within a 20 day interval. For some calculations the data were further averaged and gridded in three-day intervals. Only data after July 1997 were considered due to initial calibration problems with the GBI. All subsequent observations were made at almost the same hour angle.

Sgr A* was also corrected for confusion. Comparison of the GBI data with contemporaneous observations of Sgr A* at 5 and 8 GHz with the VLA and VLBA (Bower et al. 1999a; Lo et al. 1998; Goss 1998, p.c.) were used to calculate the difference between the GBI-single baseline flux density and the total flux density of Sgr A*, where the 2.3 GHz total flux density was obtained by extrapolation. Thus for an hourangle of ~ 0.88 hrs a flux of 70 and 177 mJy was added to the X and S-band data respectively.

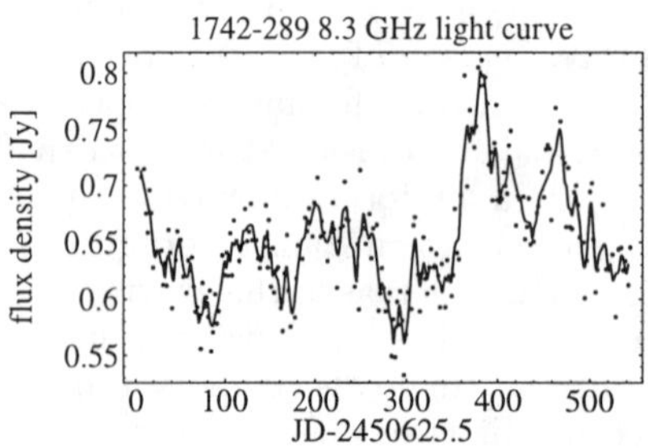

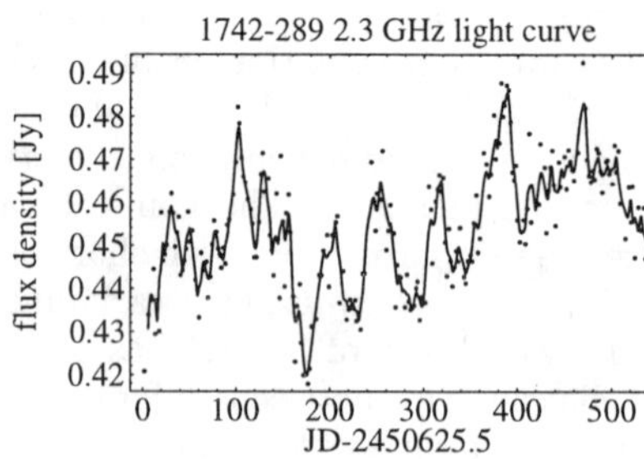

Figure 1. Radio light curves of Sgr A* at 8.3 GHz (left panel) and 2.3 GHz (right panel) measured with the GBI (dots). The solid line is the interpolated light curve for three-day averages.

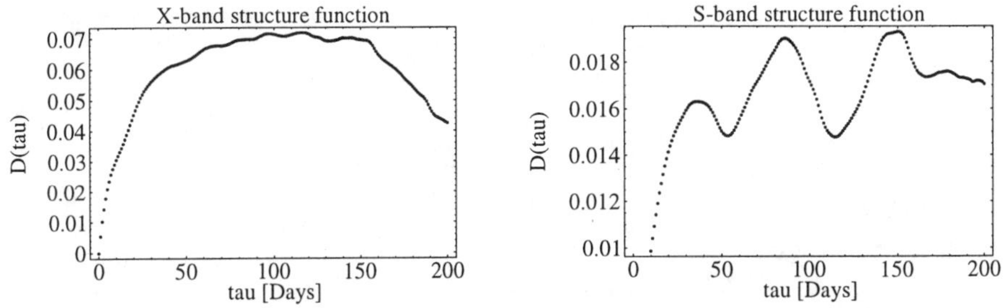

Figure 2. Structure function of the radio light curves of Sgr A* at 8.3 GHz (left panel) and 2.3 GHz (right panel). Maxima indicate a characteristic times scale, minima indicate a characteristic period

The final light curves are shown in Figure 1. One can see a peak-to-peak variability of 250 mJy and 60 mJy with an RMS of 6% and 2.5% at 8.3 & 2.3 GHz, respectively (i.e., modulation index). The median spectral index between the two frequencies for the whole period is $\alpha = 0.27$ ($S_\nu \propto \nu^\alpha$), varying between 0.2 and 0.4. There is a trend for the spectral index to become larger when the flux density in both bands increases.

3. Results

To characterize the variability pattern better, Fig. 2 shows the structure function $D(\tau)$ of the two lightcurves, where

$$D(\tau) = \sqrt{\left\langle (S_\nu(t) - S_\nu(t \pm \tau))^2 \right\rangle} \quad . \tag{1}$$

A maximum in the structure function indicates a characteristic timescale, a minimum indicates a characteristic period. A characteristic period in radio-lightcurves usually does not persist for a long time, and hence, similar to X-ray astronomy, is commonly called a quasi-periodicity, even though the underlying physical processes are probably very different from those seen in X-ray binaries.

Interestingly, the structure functions at both frequencies look very differently. While at both frequencies the characteristic time scale is somewhere between 50 and 200 days, we find a clear signature of quasi-periodic variability at 2.3 GHz, which is not obvious at 8.3 GHz. All the three maxima and the two minima in the structure function are consistent with a period of 57 days.

A cross correlation of the two light curves gives a strong peak near zero time-lag which indicates a certain degree of correlation between the emission at 8.5 GHz and 2.3 GHz (Fig. 3). A slight offset of the peak by 2-3 days is visible (Fig. 3, right panel). Usually such an offset would indicate that the 8.5 GHz light curve precedes the one at 2.3 GHz. This would be qualitatively expected by a model where outbursts travel outwards, from high to low frequencies as for example in a jet model (Falcke et al. 1993), however, the time lag one obtains is also close to the sampling rate and it is not clear how significant this offset really

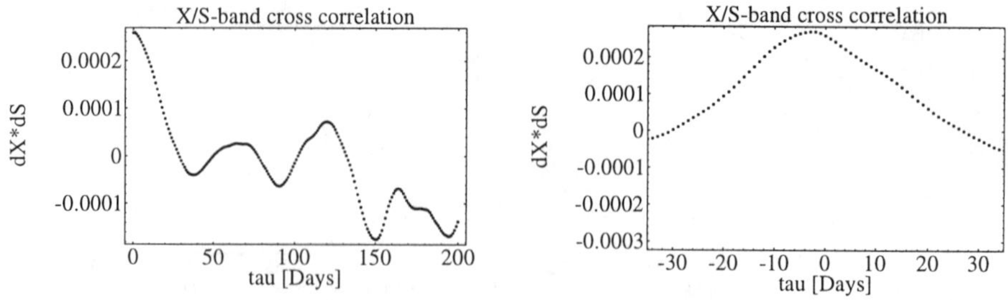

Figure 3. Cross correlation of 8.3 GHz and 2.3 GHz radio light curves of Sgr A*. In the left panel positive and negative time lags have been co-added. The right panel shows a blow-up of the cross correlation (without co-adding positive and negative lags) around zero lag.

is. Another noteworthy feature of the cross correlation is that the 2.3 GHz quasi-periodicity can still be seen. This could indicate that the quasi-periodicity is also present at 8.3 GHz, but is swamped by another, more erratic type of variability.

4. Discussion and Summary

To summarize the results one can say that there is clear evidence for variability of a few percent at cm wavelengths in Sgr A*. The variability does not seem to be consistent with a simple model of refractive interstellar scintillation (RISS) as suggested by Zhao et al. (1989&1992). The timescales at 2.3 GHz and 8.3 GHz both seem to be comparable to the one found at 5 GHz by Zhao et al. (1989&1992) and does not follow a $t \propto \lambda^2$ law. Moreover, the modulation index apparently decreases towards lower frequencies.

The quasi-periodicity is reminiscent to those in some quasar cores. For example the QSO 0917+624 is know to show episodes of quasi-periodicity (Kraus et al. 1999). Unfortunately the frequency of these quasi-periodicities in quasar cores and perhaps also Sgr A*, may not be related to a well defined and constant (e.g., precession) frequency like the QPOs in x-ray binaries, but could simply be due to intermittent periodic phenomena in the accretion disk (e.g., waves) or the jet (e.g., helical motion). In the case of Sgr A* all characteristic timescales associated with a black hole or a relativistic outflow at these frequencies are less than a day and hence one might consider global accretion flow instabilities for such a behaviour. On the other hand the possibility whether the quasi-periodicity could be produced by interstellar scattering needs to be explored as well.

Acknowledgments. Helpful discussions with A. Kraus are gratefully acknowledged. W.M. Goss provided VLA data for calibration purposes. This work was supported by the Deutsche Forschungsgemeinschaft, grants Fa 358/1-1&2. The Green Bank Interferometer is a facility of the National Science foundation operated by NRAO with support from the NASA High Energy Astrophysics program.

References

Bower, G.C, Falcke, H., Backer, D., Wright, M. 1999, this volume, p. 80

Serabyn, E., Carlstrom, J., Lay, O., Lis, D.C., Hunter, T.R., Lacy, J.H. 1997, ApJ 490, L77

Eckart A., Genzel R. 1996, Nature 383, 415

Falcke, H. 1996, in "Unsolved Problems of the Milky Way", IAU Symp. 169, L. Blitz & P.J. Teuben (eds.), Kluwer, Dordrecht, p. 169-180

Falcke H., Goss W.M., Matsuo H., Teuben P., Zhao J.-H., Zylka R. 1998, ApJ 499, 731

Falcke H., Mannheim K., Biermann P. L. 1993, A&A 278, L1

Genzel, R., Eckart, A. 1999, this volume, p. 3

Ghez, A. et al. 1999, this volume, p. 18

Ghez, A. M., Klein, B. L., Morris, M., and Becklin, E. E. 1998, ApJ 509, 678

Haller J., Rieke, M., Rieke, G., Tamblyn, P. Close, L., Melia, F. 1996, Ap.J. 468, 955

Kraus, A. et al., 1999, New Astronomy Reviews, submitted

Lo, K.Y., et al. 1998, ApJ 508, L61

Zhao, J.-H., Goss W.M. 1999, this volume, p. 224

Zhao J.-H., Ekers R.D., Goss W.M., Lo K.Y., Narayan R. 1989, in: "The Center of the Galaxy", IAU Symp. 135, Morris M. (ed.), Kluwer, p. 535

Zhao J.-H., Goss W.M., Lo K.Y., Ekers R.D. 1992, in: "Relationships Between Active Galactic Nuclei and Starburst Galaxies", ASP Conf. Ser. 31, A. Filippenko (ed.), p. 295

The Central Parsecs of the Galaxy
ASP Conference Series, Vol. 186, 1999
H. Falcke, A. Cotera, W.J. Duschl, F. Melia, M.J. Rieke, eds.

Discussion Session II: Can VLBI Constrain the Size and Structure of Sgr A*?

J. A. Zensus[1], H. Falcke[2]

Max-Planck-Institut für Radioastronomie, Auf dem Hügel 69, D-53121 Bonn, Germany

Abstract. This is an edited transcript of a discussion session about radio and VLBI observations of Sgr A*, the compact radio core and suspected black hole candidate at the Galactic Center. The session took place during the Galactic Center conference "The Central Parsecs" in Tucson, Arizona, Sep. 7-11, 1998. It was chaired by the moderator as indicated in the author list. The discussions were taped, transcribed, and finally edited by an editor who is listed as co-author. The contributions of the conference participants are preceded by their names and thus references to specific aspects and ideas of this discussion session must also include a reference to the respective speaker(s), e.g., in a footnote.

Discussion

Zensus: Let us talk about the structure of Sgr A*, especially at millimeter wavelengths: there's some disagreement and there's some conservatism, and there's some progressive interpretation, I might say. There are axial ratios which have been measured and one thing to take home, maybe, is that above 1.3 centimeters, the axial ratio seems to be consistently measured as 0.5, while at higher frequencies, below 1.3 centimeters, there is an indication for a higher axial ratio. Lo et al. have quite impressively concluded that they have measured the intrinsic size to be 72 Schwarzschild radii by less than 20 Schwarzschild radii. I actually think that the other groups have been more or less consistent with their conclusions. I think rightly so, because they all have limited data, and without going back into the detail of the presented evidence, it's too premature to say that any of these other three groups can really rule out their results. Bower summarized his results to be consistent with the Sgr A* size. Krichbaum concludes the same from comparing all of them, and I think Doleman basically says the same. So that's the size. The question is: is there an asymmetry. First, I concluded that I would say, no, now I conclude that I say, yes, both in parentheses. I think there is some hint that there is asymmetry. Experience shows that these things tend to get confirmed, but

[1]Moderator of the discussion session

[2]Editor of the discussion session

I'll be happy to let anybody else say what they feel we should conclude at the end of this session with regard to asymmetry. It seems to me the asymmetry is particularly important because that's one of the keys that allows probing the models. Now, we'll hear more about the models tomorrow. But the asymmetry and the jet structure, those are questions I certainly find interesting.

Bower: I'm curious about your switching your vote from no to yes here. Because as far as I could tell, everyone who presented information on the size had somewhat different answers.

Zensus: I said yes in parentheses: to be confirmed. Because it's not a final conclusion, it's just a hunch.

Bower: In the technical language, noone presented non-zero closure phases on Sgr A*, and in that sense there is no evidence for asymmetry. In the sense of an intrinsic axial ratio — if you want to call that an asymmetry — perhaps we could say yes in parentheses.

Krichbaum: I would like to point out one more item, whether the source is scatter-broadened or not at 43 GHz probably depends on the data quality. However, you can do the extrapolation of the scattering size, if you plot the size versus wavelength, you find that the scattering size equals the Schwarzschild radius of a 10^6 million solar mass black hole near a frequency of 270 GHz. That means future VLBI experiments at about 250, 300 GHz, would provide you with a resolution which is of the size of the Schwarzschild radius. That means in principle in a few years time it should be possible to image an area around the black hole, or as close to the black hole as possible. And we are now approaching with millimeter-VLBI a region where we can probe scales of order 10 Schwarzschild radii, so we are pretty close to a point where we really can rule out several models.

Zensus: We have a tantalizing possibility, but we have ways to go, because we still haven't really said what we find at those lower frequencies, let alone the very highest. It is very encouraging that you do detect Sgr A* at 215 GHz, but these tremendous calibration problems are only going to be much worse at higher frequencies. Following the more conservative path, Bower and Doleman have been advocating, we have to be careful to actually not say that at those—yet higher—frequencies we will be able to answer these questions.

Doleman: I'd also add to that, that there is some light at the end of the tunnel. The VLBA is coming on-line at 86 GHz, we are getting almost one antenna per year at some very good sites. Now that we have Los Alamos and soon Fort Davis, we can significantly improve our data. Especially Fort Davis is very important, because that bridges the gap across this country between the East coast and the Southwest U.S. stations. That gives us an intermediate baseline. Over the next few years we can get a greatly improved UV coverage.

Goss: We're upgrading Fort Davis and Mauna Kea to 86 GHz in the next month.

Zensus: Geoff, do you want to say your sentence again about the bleak outlook and that maybe at the higher frequencies you don't see the same source?

Bower: Well, I think this is a consequence of the spectral data, Falcke et al. and Serabyn et al. have shown. It appears that there are two synchrotron spectral components. This is apparent from the change in the spectral index from the centimeter to the millimeter regime, and it may be that, when we start doing higher frequencies—VLBI at 215 GHz and higher—we will be seeing the sub-millimeter component, which could be an entirely different thing than the component at centimeter wavelengths that people are pursuing for years now. So this object may really remain obscured until all those clouds in the Galactic Center go.

Zensus: Don, you had a nice sentence about the fly-by in your abstract.

Backer: Yes, we will have to wait until we have sent somebody there.

Unknown: I want to volunteer for that flight!

Falcke: I just want to corroborate [Geoff Bower's statement], because we are approaching—really—the black hole at these high frequencies. Whatever model you consider, it will be getting more and more spherical. Even in the jet model this will be the base of the jet—the nozzle—and that may be roughly spherical (i.e. as wide as it is high), so in the worst case it may not tell us much about the model. However, the presence of a black hole itself could produce quite interesting structures.

Townes: Are you saying the structure will be more spherical? Is there no hole in the middle of this thing?

Falcke: That's right, at higher resolution there literally will be a 'black hole' in the emission region that we could observe. The different models probably become more similar with respect to each other, but at least the presence of the black hole could produce a characteristic sub-mm structure. It may not tell us what the nature of the radio emission in Sgr A* is, but "just" tell us that there is a black hole.

Anantharamaiah: I want to bring up a more general question about the structure of Sgr A*. If you look at any of the collapsed objects that we observe, stellar remnants, black hole, binaries, and so on, a jet-like feature is almost always evident. And of course many spectacular jet-like features in galaxies seem to harbor a black hole candidate in the center. How come, this most spectacular black hole candidate in our Galaxy does not have anything that looks like a jet? Is there anything in the way that accretion is happening that is preventing a jet-like feature in Sgr A*. We have a compact source, and we are struggling to see whether there is a jet-like feature at all. Is there a problem, or is it just okay?

Falcke: Wait until tomorrow morning [the theory session] ...

Narayan: According to our story, there are ADAFs in other galaxies as well. People have found a black hole in M87, for instance, probably accreting at a similar rate as in our Galaxy—when you normalize by the mass of the black hole. In Eddington units it is not a bright system, but M87 does have a jet. It's a mystery to me, why several systems which all probably have ADAFs, some have very prominent jets and some have, as you say, nothing at all, as in the case of our Galaxy. Heino [Falcke] claims it is a jet, but if so, it is really tiny, unlike what we are used to seeing as jets.

Of course some people are speculating that it might say something about the spin of the black hole. This is of course pure speculation. It could be that you need a certain amount of spin before you get a really powerful jet. There is a recent paper by Blandford and Begelman which has picked up one idea that we had earlier, namely that all ADAFs will have enormous outflows. A large part of the gas that is flowing in is going out, so that all ADAFs have outflows, and then the question is why should some of them become ripe and ready and others not? That is how nature is, and I don't think anyone has an idea why.

Bower: Can any of the low-frequency observers comment on whether there's anything that looks like a double- or single-lobed radio source that you could possibly associate with a jet.

Kassim: Our map itself has a very suggestive impression of an outflow and possible large scale lobe, but of course that's not symmetrical around Sgr A*.

Doleman: Would that be evidence for past activity of Sgr A*? Aren't there some indications, Heino, that perhaps this jet was on for a certain amount of time.

Falcke: There is the Galactic Center spur, discovered by Sofue, Reich, & Reich, which is pointing out of the Galaxy, starting just above the Galactic Center. It's a very steep spectrum feature and might be a remnant of past activity. But of course it could also be a supernova remnant.

Yusef-Zadeh: But there is also another feature that is at a scale of about a few arcseconds or so, just outside Sgr A*, and you see it's extended. The structure is horizontally coming out of Sgr A*. It's not clear, really, if it is associated with it, but we thought that it may be just the outflow from IRS 16 being focused by Sgr A*. You do clearly see an extended emission ending at Sgr A* and sort of moving to the right. It is connected to the Minicavity. So on a scale of a few arcseconds, there are some features that seem to be associated with Sgr A*.

Sofue: Is this free-free emission?

Yusef-Zadeh: Free-free emission, correct. And they have a high velocity. They have very high proper motion. But the interesting thing is that it's exactly perpendicular to the N-S direction Jun-Hui Zhao was talking about.

Sellgren: I don't know the answer to this question, so I'm farming it out. Does anyone here happen to know whether the spin axis of the jet in other active galaxies is always along the minor axis of the disk?

Falcke: There is the example of NGC 4258. The axis of the jet is perpendicular to what you would expect from the orientation of the disk of the galaxy.

Sellgren: So there's no reason to expect that a jet — if there is one — would come out at a right angle to the plane of the Galaxy.

Falcke: In Seyfert galaxies it's the same, jets come out in all directions.

Stolovy: I think it was very interesting, in the last talk, that the spectral index changed during the flare, and it increased—was more positive. With a jet—wouldn't you expect a more negative spectral index?

Falcke: In a jet you will see self-absorbed synchrotron components coming out of the nozzle. At a given frequency the shape of the spectrum will first become more inverted (i.e. positive) and steepen later.

Stolovy: Is that what is seen in GRS 1915+105?

Falcke: Yes, exactly. Basically, the outburst travels from high frequencies to low frequencies. So, first we're expecting the spectral index to increase and then to decrease.

Melia: We will of course be talking a lot more about this tomorrow morning, but I just want to say one thing in answer to Anantha's question, and that is that the systems that he pointed out have a preferred plane which define the axis. A binary system has a binary plane, an accretion disk also defines a plane, which then defines an axis perpendicular to the disk. I would suggest that the simplest explanation why Sgr A* wouldn't have a jet—if it doesn't have one—is because it is not a disk. It has not a well-defined plane and no axis perpendicular to it.

Narayan: M87 doesn't have a disk, close to the black hole, for certain. That's the same problem as in Sgr A*, that there is no disk-like luminosity from the center.

Melia: I'm not talking about the disk, I'm talking about something like an ADAF which extends very close to the black hole, very close to an event horizon.

Zensus: Okay, we just have a few more minutes. All areas need more data, few as much as this. What about the discrepancy between the VLA and the VLBA measurements of the proper motion of Sgr A*? Do we need to worry about that? Is it a problem of the data, or of understanding the refractive properties.

Backer: It's still possible that this can be reconciled.

Zensus: So that's something to be watched. The detection at yet higher frequencies, sizes, asymmetries, and also variability of the total spectrum, all remain to be studied ...

Doleman: There's this unknown excess contribution to the high radio frequencies. I'm still a little unsure as to what that's all about. Thomas [Krichbaum] did a careful extrapolation from the lower frequencies, and he was suggesting that perhaps the upper part of the spectrum is not part of the deal. It could be interpreted as not part of Sgr A*. But what is that? Someone was telling me beforehand that it can't be free-free or dust because it is too low temperature. Why do you see this excess? What is that excess?

Beckert: In our model this excess corresponds to the core, but this cannot explain the centimeter measurements, which is probably an extended structure. So we will not see the whole thing. VLBI is seeing just the core.

Krichbaum: The problem can be solved if if you have an interferometer like the MMA which can give us the zero-spacing flux.

Townes: Are there some plans for millimeter astronomy down at the South Pole? The water vapor is very, very low there. Calibration problems will be much lower. Are the Australians thinking about doing that?

Becklin: There's a 10-meter millimeter dish that goes by Tony Stark.

Townes: If there is one, why shouldn't it be used down there to look at this? The seeing is not remarkable in the visible regions, but seeing in the millimeter range should be very good.

Becklin: I believe there's only one dish, so there's no interferometer.

Townes: Well, the Australian VLBI Array: what is the highest frequency they can do?

Goss: They're going to one centimeter now.

Lazio: I have a question going back to the refractive shifts of the position of Sgr A* . Could that still be a possibility.

Backer: It's still a possibility.

Lazio: What shift are we talking about?

Backer: Three milli-arcseconds [between 8 and 5 GHz].

Lazio: The point with this is: in one session we heard about determining the IR position of Sgr A*. Since the registration depends on masers located around, and if there are refractive position shifts, I'm wondering on how this is going to affect our ability to register Sgr A* in the IR.

Backer: If you look, as you no doubt have, at how refractive effects occur, it's when there's a very steep spectrum of irregularities. And the steeper the spectrum of irregularities is, the larger the shift can be. It can shift by as much as the scattering diameter. That's if you want to just invent a refractive medium: it can be all refraction, there would be no diffraction, and it could just be dithering the image around. So I don't think there's a way to say what it would be. If we find this $\lambda^{2.0}$ index for the size, then with some confidence, we're probably beyond the inner scale of the turbulence in terms of what we're sampling. At that point I think the scattering theory says there are no refractive shifts. I'm raising a little specter just to air all the possible avenues of thinking about this, while being somewhat convinced that there's probably not a refractive term. And there's some limits on it, when we deal with dual frequency measurements. I didn't mention it, but we had dual frequency measurements with our old Green Bank Interferometer (GBI) of the Sgr A* shift. And we put a very strong limit on a refractive component to the Sgr A* proper motion.

Bower: But for the case of registering the infrared frame, those are measurements at 43 GHz and I think you wouldn't expect any refractive shifts.

Backer: Not by more than about the scattering size. That's a fair statement.

Yusef-Zadeh: I should say also that we just detected radio continuum emission from the photosphere of IRS 7. And the difference between the position of the continuum emission from the photosphere of IRS 7 and Sgr A* is quite consistent with the shift that Menten et al. found. It really matches beautifully. This is now at the wavelength of 1.2 centimeters.

Zensus: Okay. I would like to come to an end of the discussion session, so I'll leave a few more comments open.

Becklin: This is along Charlie's question. Are there any plans for a Merlin type array in Chile? Do they have the millimeter array going in? Or the SEST telescope?

Zensus: Well, I think there are going to be possibilities, because you have the MMA there, you have any number of ideas to put other antennas in that vicinity. There are plans for the Japanese to put an array in the next valley, so I see longer baselines than the MMA baselines.

Becklin: But is somebody thinking about it in terms of optimizing UV coverage?

Zensus: The VLBI community is a small part, but it's part of that MMA community.

Becklin: But that's critical, isn't it?

Backer: It's probably about as likely as you getting a VLT ...

Zensus: Okay. Let me answer the question of the workshop afternoon with a qualified yes: VLBI, can constrain the structure of Sgr A*, but a lot has still to be done, a lot will remain to be learned, and we're looking forward to tomorrow's and the rest of the week's session. Thank you very much, everybody.

Session III

How to Reconcile Models and Observations of Sgr A*?

The Central Parsecs of the Galaxy
ASP Conference Series, Vol. 186, 1999
H. Falcke, A. Cotera, W.J. Duschl, F. Melia, M.J. Rieke, eds.

The Nature of Sgr A* – An Introduction

Fulvio Melia
Physics Department and Steward Observatory, The University of Arizona, Tucson AZ 85721

Abstract. Our understanding of the nature of Sgr A* has evolved rather rapidly since the early 90's, in part due to the growing interest among theorists and to a growing body of observational constraints. We will briefly review what now appears to be widely accepted, and highlight the areas where the observational uncertainties allow for a variance of assumptions that may lead to different interpretations.

1. Introduction

The Galactic center (GC) has long been suspected of harboring a central mass concentration, which appears to be coincident with the unique, non-thermal radio source, Sgr A*. Haller et al. (1996) used the velocity dispersions of stars at $\gtrsim 0.1$ pc from Sgr A* to derive a compact mass of $\sim 2 \times 10^6\ M_\odot$. This is consistent with the value of $\sim 2.5 - 3.2 \times 10^6\ M_\odot$ derived more recently by Genzel et al. (1996), using the radial velocities and velocity dispersions of ~ 25 early-type stars and of ~ 200 red giants and supergiants within the central 2 pc. A third technique for tracing the central gravitational potential is based on the acquisition of proper motions for the ~ 50 brightest stars within the radial range $\sim 0.004 - 0.4$ pc (Eckart & Genzel 1996). These stellar motions seem to require a central dark mass of $(2.61 \pm 0.35) \times 10^6\ M_\odot$ (Genzel et al. 1997), in good agreement with earlier ionized gas kinematics and the velocity dispersion measurements.

Of course, showing that the GC must contain a centralized mass concentration does not necessarily imply that this dark matter is in the form of a compact object with a few million solar masses. It does not even imply that the unusual radio source Sgr A* must be associated with it. VLBA images of Sgr A* with milliarcsecond resolution (Bower & Backer 1998) show that at $\lambda 7$ mm, its size is 0.76 ± 0.04 mas, or roughly 6.2×10^{13} cm, much smaller than the present limiting region within which the $2 - 3 \times 10^6\ M_\odot$ are contained. So the dark matter may be distributed, perhaps in the form of white dwarfs, neutron stars, or $\sim 10\ M_\odot$ black holes (e.g., Haller et al. 1996), though the latest stellar kinematic results appear to rule out the first two possible constituents (Genzel et al. 1996).

Whatever the composition of a distributed mass concentration is, one is left with the task of accounting for the nature of Sgr A* itself. It is likely that many of Sgr A*'s characteristics are associated with the liberation of gravitational energy as gas from the ambient medium falls into a central potential well (Melia 1994; see also Ozernoy 1989 for an alternative conclusion regarding wind accretion). There is ample observational evidence in this region for the existence of rather strong winds in and around Sgr A* itself (from which the latter is accreting), e.g., the cluster of mass-losing, blue, luminous stars comprising the IRS 16 assemblage located within several arcseconds from the nucleus. Measurements of high outflow velocities associated with IR sources in Sgr A West (Krabbe et al. 1991) and in IRS 16 (Geballe et al. 1991), the H_2 emission in the circumnuclear disk (CND) from molecular gas being shocked by a nuclear mass outflow (Genzel et al. 1996; but see Jackson et al. 1993 for the potential importance of UV photodissociation in promoting this H_2 emission), broad Brα, Brγ and He I emission lines from the vicinity of IRS 16 (Hall et al. 1982; Allen et al. 1990; Geballe et al. 1991), and radio continuum observations of IRS 7 (Yusef-Zadeh & Melia 1991), provide clear evidence of a hypersonic wind, with a velocity $v_w \sim 500 - 1000$ km s^{-1}, a number density $n_w \sim 10^{3-4}$ cm^{-3}, and a total mass loss rate $\dot{M}_w \sim 3 - 4 \times 10^{-3}\ \dot{M}_\odot$, pervading the inner parsec of the Galaxy.

Even so, the observations do not yet provide sufficient information for us to identify the physics of accretion when the infalling gas penetrates to within about 10^3 or 10^4 Schwarzschild radii of the central object. The captured plasma is magnetized and highly ionized, but it is not clear how much specific angular momentum it carries, what the intensity of the magnetic field is, what the relative importance of non-thermal to thermal particles is, and whether the plasma separates into a two-temperature fluid. In this review, we discuss the emission characteristics associated with the trapped gas and we will summarize how the variance of assumptions (consistent with the observations) results in a range of different interpretations. We will show, however, that it is very unlikely that Sgr A* is associated with a distributed dark matter, which cannot provide the steep gravitational potential required to account for Sgr A*'s spectrum.

2. Size Constraints and the Brightness Temperature

One benefit of our relative proximity to Sgr A* is that we can resolve it, or at least resolve its scatter-broadened image, and place meaningful constraints on the *global* properties of the emitting region. VLBI observations (e.g., at 86 GHz) indicate that Sgr A*'s size is no larger than ~ 1 A.U., which requires that its brightness temperature be greater than $\sim 10^{10}$ K (e.g., Rogers et al. 1994; Krichbaum et al. 1998). A lower limit on its size is provided by the observed absence of refractive scintillation at $\lambda\lambda$ 1.3 and 0.8 mm, which translates into a minimum size of ~ 0.1 A.U. Correspondingly, the upper limit on the brightness temperature (at 0.8 mm) is $\sim 0.5 \times 10^{12}$ K (Gwinn et al. 1991).

Sgr A* therefore shines below the Compton limit (at approximately 10^{12} K; Begelman et al. 1984). This is the maximum brightness temperature of incoherent synchrotron emission from an electron plasma; above this temperature, the radiation is heavily Comptonized to a frequency well beyond the GHz range. However, if the emitting particles are Maxwellian, they must reach relativistic energies, i.e., their temperature must exceed $\sim 5 \times 10^9$ K, in order for them to be efficient synchrotron emitters (see, e.g., Melia 1994; Mahadevan et al. 1997).

3. Mass Condensation Onto the Central Nucleus

These characteristics are consistent with the idea that Sgr A*'s spectrum results from the energy liberated by a compressed hot plasma bound to the central gravitational potential. Here, we will first consider the possibility that the potential well is associated with a distributed cluster of dark objects, and then discuss the issues pertaining to a black hole potential.

3.1. Condensation Into a Dark Cluster Potential

The Galactic center wind appears to be produced by the early-type stars enclosed (in projection) within the Western Arc, the Northern Arm, and the Bar. Thus far, 25 such stars have been identified (Genzel et al. 1996), and all appear to be located within the central parsec surrounding Sgr A*. Figure 1 shows the positions (relative to Sgr A*) of these winds sources, in which the size of the circle marking each position corresponds to the relative mass loss rate (on a linear scale) for that star. Note that due to clustering some of the stars are combined into a single wind source in this figure, and some are outside the field of view.

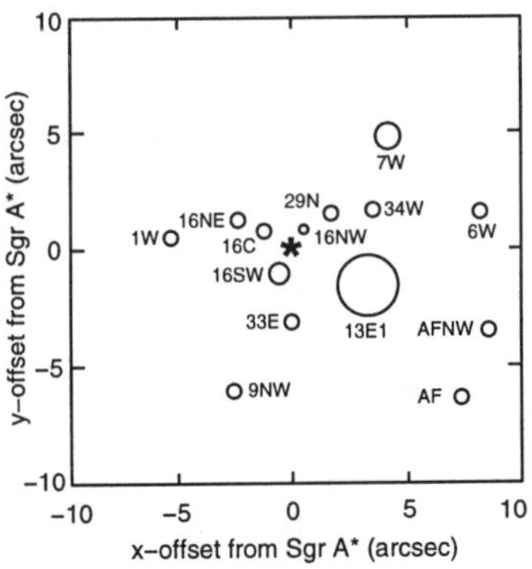

Figure 1. Location of some of the wind-producing stars relative to the position of Sgr A* indicated by the * symbol. The radius of each circle corresponds (on a linear scale) to that star's mass loss rate. Setting the scale is 13E1, with $\dot{M} = 7.9 \times 10^{-4}\ M_\odot\ \mathrm{yr}^{-1}$.

The gravitational potential of a dark cluster can be represented by an "η-model" (Haller & Melia 1996). This function mimics an isotropic mass distribution with a single parameter, and it is scaled so that the total dark cluster mass is $2.7 \times 10^6\ M_\odot$ (e.g., Ghez, et al. 1998). Melia & Coker (1998) used the 3D hydrodynamics code ZEUS to simulate the flow of the Galactic center wind through this distributed dark matter using the η-potential, and one of the key results of this calculation is summarized in Figure 2. A more complete description of this simulation and the results is given by Coker & Melia (1999a). Figure 2 shows the angle- and volume-averaged density and temperature for the whole central $0.''7$ region, using a bin size of $0.''0025$. (The results are only very weakly dependent on the bin size, so we shall not dwell on this here.) In this figure, the density rises gradually to the middle, and reaches an average value of roughly 10^8 cm^{-3}. In contrast, the central density for a gas falling freely into a black hole potential with the same central mass approaches $\sim 10^{13}$ cm^{-3} (Melia 1994). The temperature similarly rises to the middle, but it levels off within about 0.004 pc, and the average is never greater than about 10^7 K. This is to be compared with the temperature profile of the gas falling into the black hole, where T attains values as high as 10^{10} K or more. This is critical because the electrons begin to emit significantly via the synchrotron process when they become relativistic above a few times 10^9 K. This gas can at best therefore only emit cyclotron radiation, but the emissivity is a strong function of T and is here insignificant compared to bremsstrahlung. The flattening of the density and temperature profiles shown in Figure 2 is a direct consequence of the shallowness of the cluster potential compared to the steep potential gradients encountered by the gas falling into the black hole (e.g., Melia 1994). The magnetic field, which is coupled to the physical state of the gas, behaves in a similar fashion, though it is clumpier due to the uneven dissipation in regions of gas compression and rarefaction.

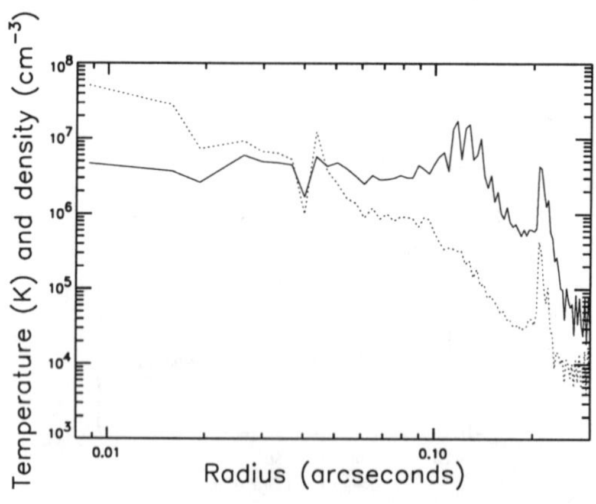

Figure 2. Angle- and volume-averaged density (dotted) and temperature (solid) as a function of radius (in arcseconds) from the center. The flattening of these distributions at small radii (i.e., $< 0.''1 \approx 0.004$ pc) is clearly evident.

Gas flowing through a dark cluster therefore does not produce the type of condensation and high temperature required to account for Sgr A*'s spectrum (see § 2 above, and Melia 1994).

3.2. Condensation Onto a Massive Black Hole

Let us instead consider the physical state of the gas when the gravitational potential well deepens as the plasma approaches the event horizon. In the classical Bondi-Hoyle (BH) scenario (Bondi & Hoyle 1944), the mass accretion rate for a uniform hypersonic flow is $\dot{M}_{BH} = \pi {R_A}^2 m_H n_w v_w$, in terms of the accretion radius $R_A \equiv 2GM/{v_w}^2$. With the conditions at the Galactic center (see above), we would therefore expect an accretion rate $\dot{M}_{BH} \sim 10^{22}\,\mathrm{g\,s^{-1}}$ onto the black hole, with a capture radius $R_A \sim .02\,\mathrm{pc}$.

In reality the flow past Sgr A* is not likely to be uniform, so this value of $\dot{M}_{BH}$ may be greatly underestimated. For example, one might expect many shocks to form as a result of wind-wind collisions within the IRS 16 complex, even before the plasma reaches R_A. With this consequent loss of bulk kinetic energy, it would not be surprising to see the black hole accrete at an even larger rate than in the uniform case. The implications for the spectral characteristics of Sgr A*, and thus its nature, are significant. Coker & Melia (1997) have therefore undertaken the task of simulating the BH accretion from the spherical winds of a distribution of 10 individual point sources located at an average distance of a few R_A from the central object. The results of these simulations show that the accretion rate depends not only on the distance of the mass-losing star cluster from the accretor but also on the relative spatial distribution of the sources.

Figure 3 shows a logarithmic negative grey scale image of the density profile for a slice running through the center of the accretor, for one of these simulations taken 2000 years after the winds are "turned on". The image is 6 R_A on a side with the $0.1R_A$ radius accretor located at the center. One stellar source can be clearly seen to the upper right of the accretor; the other 9 sources are either not in the plane of the slice or are off to the right. The image spans 4 magnitudes of density with white being $\lesssim 0.01n_w$ and black being $\gtrsim 10n_w$. Once the stellar winds have cleared the region of the original low density gas, all such simulations point to an overall average density ($\sim 10^3\,\mathrm{cm^{-3}}$) in agreement with observations.

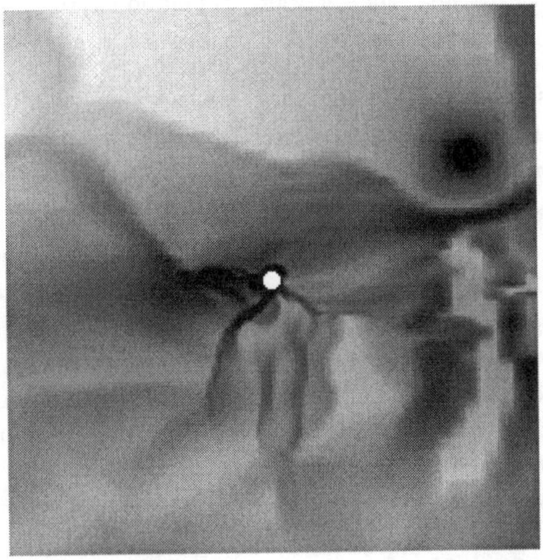

Figure 3. A logarithmic negative grey scale snapshot, 6 R_A on a side, of the gas density for a simulation taken 2000 years after the winds are "turned on". It is a slice in the $\hat{x}-\hat{z}$ plane (with $\hat{z}$ to the right) through the accretor (marked with a filled white circle). White corresponds to a number density $\lesssim 0.01n_w$ and black corresponds to a number density $\gtrsim 10n_w$.

Wind-wind collisions from the multiple sources reduce the effective v_w, thereby increasing the effective R_A and therefore $\dot{M}$. In Figure 4 we present the mass accretion rate, $\dot{M}$, and the accreted specific angular momentum, λ (in units of cr_g, where r_g is the Schwarzschild radius), versus time, starting two crossing times ($\sim$ 800 years) after the winds are "turned on". The average value for the mass accretion rate once the system has reached equilibrium is $\dot{M} = 2.1 \pm 0.3\dot{M}_{BH}$.

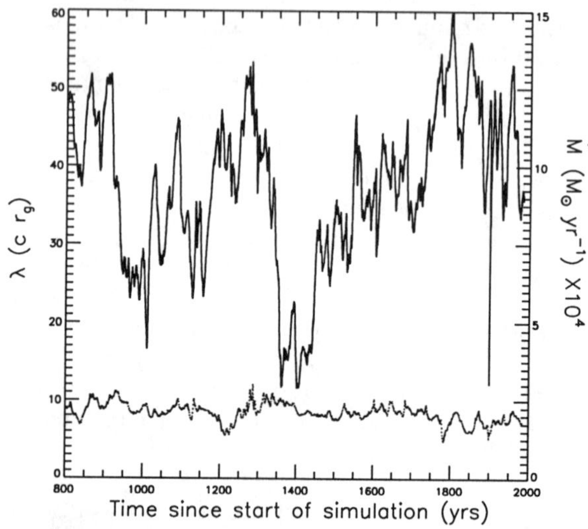

Figure 4. The upper solid curve is accreted specific angular momentum λ (in units of cr_g). The scale for λ is on the left side. The lower dotted curve is the mass accretion rate $\dot{M}$ (10^{-4} $M_\odot\,yr^{-1}$) versus time. The scale for $\dot{M}$ is shown on the right side.

The mass accretion rate shows high frequency temporal fluctuations (with a period of $\lesssim 0.25\,$yr) due to the finite numerical resolution of the simulations. The low frequency aperiodic variations (on the order of 20% in amplitude) reflect the time dependent nature of the flow. Thus, the mass accretion rate onto the central object, and consequently the emission arising from within the accretor boundary, is expected to vary by $\lesssim 20-40\%$ (since in some models the luminosity may vary by as much as $\propto \dot{M}^2$) over the corresponding time scale of < 100 years, even though the mass flux from the stellar sources remains constant.

Similarly, the accreted λ can vary by 50% over $\lesssim$ 200 years with an average equilibrium value of 37 ± 10. Since previous uniform simulations resulted in $\langle\lambda\rangle \sim 3 - 20$, it appears that even with a large amount of angular momentum present in the wind, relatively little specific angular momentum is accreted. This is understandable since clumps of gas with a high specific angular momentum do not penetrate to within 1 R_A. The variability in the sign of the components of λ suggests that if an accretion disk forms at all, it dissolves, and reforms (perhaps) with an opposite sense of spin on a time scale of $\sim$ 100 years.

The captured gas is highly ionized and magnetized, so it radiates via bremsstrahlung, cyclo-synchrotron and inverse Compton processes. However, the efficiency of converting gravitational energy into radiation is quite small (as little as 10^{-4} in some cases), so most of the dissipated energy is carried inwards (Shapiro 1973; Ipser & Price 1982; Melia 1992). In fact, if the magnetic field is a negligible fraction of its equipartition value (see below), Sgr A* would be undetectable at any frequency, except perhaps at soft X-ray energies. But as the plasma continues to compress and fall toward smaller radii, one or more additional things can happen, each of which corresponds to a different theoretical assumption, and therefore a potentially different interpretation, which we explore next.

4. The Physics of Accretion Toward Small Radii

The questions one may ask include the following: (1) Does the flow carry a large specific angular momentum so that it forms a disk with lots of additional dissipation? (2) Does the flow produce a radiatively dominant non-thermal particle distribution at small radii (e.g., from shock acceleration), or does thermal emission continue to dominate the spectrum? (3) Does the flow lead to an expulsion of plasma at small radii that forms a non-thermal jet, which itself may then dominate the spectrum? These, either individually or in combination, have led to a variance of assumptions about the nature of the inflowing gas that then form the basis for the development of different interpretations.

Observationally, the key issue is why the infalling gas maintains a low radiative efficiency. Beckert & Duschl (1997) suggest that shocks in the accreting plasma produce a power-law electron distribution, which is truncated by strong cooling. This forms a "quasi" mono-energetic distribution. The overall emission, which is strictly non-thermal, is suppressed by constraining the number density of relativistic particles and the intensity of the magnetic field (at about 5 to 10 Gauss).

Falcke, Mannheim and Biermann (1993), on the other hand, assume that the infalling plasma eventually produces a jet of power-law electrons whose number density varies with radius in the expulsion. The overall emission, which is a sum of non-thermal components, is also suppressed by constraining the particle number density and hence the equipartition magnetic field, both of which are assumed to be scaled by a slowly accreting fossilized disk.

In the picture developed by Narayan et al. (1998), the infalling gas is assumed to carry a very large angular momentum, so that a disk forms with an outer radius at more than 10^5 Schwarzschild radii. To suppress the overall emission, which now includes the additional dissipation of this large angular momentum, it is also assumed that the electron temperature is much lower than that of the protons ($T_e \ll T_p$). In fact, $T_e < 10^{10}$ K. Since the electrons do the radiating, the efficiency remains small even though the protons are very hot.

A significantly more detailed discussion of these ideas is presented elsewhere in this volume.

5. Is the Magnetic Field Sub-Equipartition?

The idea that Sgr A*'s low radiating efficiency is due to a sub-equipartition magnetic field B deserves further attention, especially in view of the fact that the actual value of B depends strongly on the mechanism of field line annihilation, which is poorly understood. Two processes that have been proposed are (i) the Petschek (1964) mechanism, in which dissipation of the sheared magnetic field occurs in the form of shock waves surrounding special neutral points in the current sheets and thus, nearly all the dissipated magnetic energy is converted into the magnetic energy carried by the emergent shocks; and (ii) Van Hoven's (1979) tearing mode instability, which relies on resistive diffusion of the magnetic field and is very sensitive to the physical state of the gas. In either case, the magnetic field dissipation rate is a strong function of the gas temperature and density, so that assuming a fixed ratio of the magnetic field to its equipartition value may not be appropriate.

Kowalenko & Melia (1999) have used the Van Hoven prescription to calculate the magnetic field annihilation rate in a cube of ionized gas being compressed at a rate commensurate with that expected for free-fall velocity onto the nucleus at the Galactic center. An example of these simulations is shown in Figure 5, for parameter values like those pertaining to the Galactic center. Whereas the rate of increase $\partial B/\partial t|_f$ in B due to flux conservation depends only on the rate $\dot{r}$ of the gas, the dissipation rate $\partial B/\partial t|_d$ is a function of the state variables and it is therefore not necessarily correlated with $\dot{r}$. Although these attempts at developing a physical model for magnetic field dissipation in converging flows is still rather simplistic, it is apparent from the test simulations that the equipartition assumption is not always a good approximation to the actual state of a magnetohydrodynamic flow, and very importantly, that the violation of equipartition can vary in degree from large to small radii, in either direction. As such, calculations in which the predicted spectrum relies critically on the contribution from magnetic bremsstrahlung, and yet assume equipartition of the magnetic field with the gas throughout the domain of solution, may be greatly underestimating the importance of the deviations of B from its B_{eq} value.

5.1. The Calculated Spectrum Using a Sub-Equipartition B Field and Comparison with Observations

The best fit spectra reported in Melia (1994) were consistent with a black hole mass of $\approx 10^6\ M_\odot$, but they are not consistent with the now more accurately known value of $\sim 2.6 \times 10^6\ M_\odot$, which accounts for a factor ~ 7 increase in $\dot{M}$. In addition, the earlier calculations integrated the cyclo-synchrotron emissivity out to the lowest 20 harmonics only, which can miss some of the contribution to the flux by the highest temperature gas at the smallest radii (Mahadevan, Narayan & Yi 1996). Coker & Melia (1999b) have therefore recalculated the spectrum produced by a quasi-spherical infall in Sgr A* using its updated mass value and a complete set of harmonics in the emissivity, and an empirical fit to the magnetic field based on the simulations of magnetic dissipation discussed above. Some of the results from this effort are shown in Figure 6.

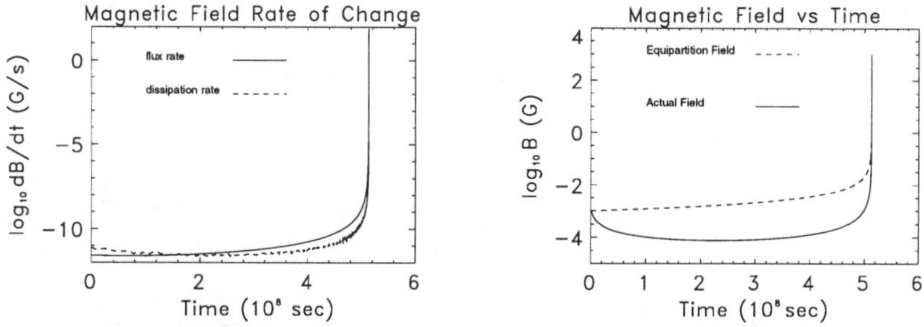

Figure 5. The *left* panel shows the magnetic field rates of change $\partial B/\partial t|_f$ and $\partial B/\partial t|_d$ due, respectively, to flux conservation and resistive dissipation as functions of time in units of 10^8 seconds. The compression rate is here assumed to be the free-fall velocity at the accretion radius. Solid curve: the rate of increase due to flux conservation; dashed curve: the rate of decrease due to resistive dissipation. The *right* panel shows the magnetic field (solid curve) calculated as a function of time from the rates displayed in left panel. By comparison, the equipartition field B_{eq} is shown here as a dashed curve. The rapid increase in B toward the end of the simulation is associated with the accelerated rate of change in the physical parameters as the gas flows inwards toward a zero radius.

Without invoking additional suppression mechanisms for the radiative efficiency, such as a two-temperature flow, we conclude that the observations delimit the magnetic field intensity in Sgr A* to a value $\sim 5 - 10$ Gauss. With the magnetic field shown in Figure 6, the fit to the data is rather promising, including the apparent excess at sub-mm wavelengths (e.g., Falcke et al. 1998). It should be noted that this calculation does *not* include the possible impact of a two-temperature flow, nor does it include the possible contribution to the emissivity from non-thermal particles (see above).

6. Closing Thoughts

The emission in Sgr A* requires a very deep potential well, so the case for a massive black hole rather than a distributed dark matter has grown stronger. Whether the radiation mechanism is thermal or non-thermal, the radiative efficiency of the infalling gas appears to be very low ($< 10^{-5}$). All things considered, this low efficiency is probably due to either a sub-equipartition magnetic field (for either thermal or non-thermal models), or to the separation of the gas into a two-temperature plasma with $T_e \ll T_p$. Clearly what is required in the future is a much higher-resolution hydrodynamic simulation that can follow the gas down to tens of Schwarzschild radii. The current limit on the accreted specific angular momentum appears to be inconsistent with the formation of a large disk (fossil or otherwise), but no one can really be sure yet.

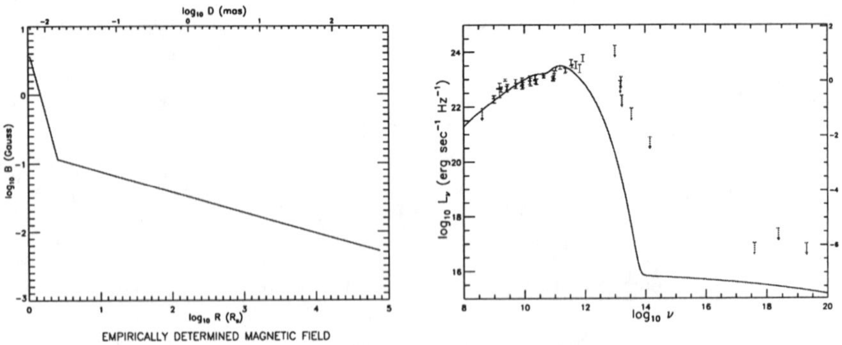

Figure 6. The *left* panel shows an empirical fit to B (as a function of r in units of the Schwarzschild radius r_g) required in order to produce the spectrum shown on the right. The shape for B was motivated in part by the simulation shown in Fig. 5, including the relatively flat distribution over most of the infalling region, with a steepening toward small r. The *right* panel shows the spectrum produced by a quasi-spherical accretion flow (Melia 1994), but here using the empirical fit for B shown on the left, rather than assuming an equipartition magnetic field. See Coker & Melia (1999b) for data sources.

References

Allen, D., Hyland, A., & Hillier, D. 1990, MNRAS, 244, 706.

Beckert, T. & Duschl, W.J., 1997, AA, 328, 95.

Begelman, M.C., Blandford, R.D. & Rees, M.J. 1984, Rev. Mod. Phys., 56, 255.

Bondi, H, & Hoyle, F. 1944, MNRAS, 104, 273.

Bower, G.C. & Backer, D.C. 1998, ApJ Letters, 496, 97L.

Coker, R.F. & Melia, F. 1997, ApJ Letters, 488, L149.

Coker, R.F. & Melia, F. 1999a, this volume, p. 214.

Coker, R.F. & Melia, F. 1999b, ApJ, submitted.

Eckart, A. & Genzel, R. 1996, Nature, 383, 415.

Falcke, H., Mannheim, K., & Biermann, P.L. 1993, AA, 278, L1.

Falcke, H. et al. 1998, ApJ, 499, 731.

Geballe, T., Krisciunas, K., Bailey, J., & Wade, R. 1991, ApJ Letters, 370, L73.

Genzel, R., et al. 1996, ApJ, 472, 153.

Genzel, R., Eckart, A., Ott, T. & Eisenhauer, F. 1997, MNRAS, 291, 219.

Ghez, A., et al. 1998, ApJ, in press.

Gwinn, C.R., et al. 1991, ApJ, 381, L43.

Hall, D., Kleinmann, S., & Scoville, N. 1982, ApJ Letters, 260, L53.

Haller, J.M., & Melia, F. 1996, ApJ, 464, 774.

Haller, J.M., et al. 1996, ApJ, 456, 194.

Ipser, J.R. & Price, R.H. 1982, ApJ, 255, 654.

Jackson, J.M., et al. 1993, ApJ, 402, 173.

Kowalenko, V. & Melia, F. 1999, MNRAS, in press

Krabbe, A., Genzel, R., Drapatz, S., & Rotaciuc, V. 1991, ApJ Letters, 382, L19.

Krichbaum, T.P., Graham, D.A., Witzel, A. et al. 1998, AA, 335, L106.

Mahadevan, R., Narayan, R. & Yi, I. 1996, ApJ, 465, 327.

Mahadevan, R., Narayan, R. & Krolik, J. 1997, ApJ, 486, 268.

Melia, F. 1992, ApJ Letters, 387, L25.

Melia, F. 1994, ApJ, 426, 577.

Melia, F. & Coker, R. 1998, ApJ, in press.

Narayan, R., Mahadevan, R. et al. 1998, ApJ, 492, 554.

Ozernoy, L. 1989, in The Center of the Galaxy, ed. M. Morris (Kluwer: Dordrecht), p. 555.

Petschek, H.E. 1964, Proc. of the Symposium on Physics of Solar Flares (NASA SP-50), 425.

Rogers, A.E.E., Doeleman, S., Wright, M.C.H., et al. 1994, ApJ Letters, 434, L59.

Shapiro, S.L. 1973, ApJ, 185, 69.

Van Hoven, G. 1979, ApJ, 232, 572.

Yusef-Zadeh, F. & Melia, F. 1991, ApJ Letters, 385, L41.

The Central Parsecs of the Galaxy
ASP Conference Series, Vol. 186, 1999
H. Falcke, A. Cotera, W.J. Duschl, F. Melia, M.J. Rieke, eds.

Advection-Dominated Accretion in the Galactic Center and the Appearance of Sgr A*

Thomas Beckert and Wolfgang J. Duschl[1]

Institut für Theoretische Astrophysik, Universität Heidelberg, Tiergartenstr. 15, 69121 Heidelberg, Germany

Abstract. We present a modified description of advection-dominated accretion flows with a geometrical viscosity. A simplified treatment of the dynamics of the flow in the inner relativistic part is suggested and results are compared with the Galactic Center source Sgr A*.

1. Introduction

The existence of a massive black hole in the Galactic Center is a keystone in the understanding of stellar dynamics in the inner parsec of the galaxy (Ghez et al. 1999; Genzel & Eckart 1999) as it is a test case for the theory of accretion flows in active and non-active galactic nuclei (Mezger, Duschl & Zylka, 1996). Standard thin accretion disks fail to explain the luminosity and spectrum of the radio source Sgr A* and several models for the accretion and emission processes have been suggested. Optically thick synchrotron emission from a jet-disk system (Falcke et al., 1996) or from an advection-dominated accretion flow (ADAF) were investigated (Narayan et al., 1998), while optically thin emission by highly relativistic particles with a quasi-monoenergetic distribution (e.g., Beckert & Duschl 1997) is also a possible mechanism. Mahadevan (1998) propose secondary electrons from pion decay in ADAFs as the synchrotron emitting particles and showed a way to connect the accretion flow as an energy source with the relativistic electrons. We present a simplified relativistic treatment of advection-dominated accretion in the Galactic Center and propose a switch between standard and ADAF accretion to distinguish AGNs and sub-luminous galactic centers.

2. Dynamics of Advection-Dominated Accretion Onto Black Holes

Advection dominated accretion flows are geometrically thick and a description in spherical coordinates is appropriate. In order to avoid complications due to frame dragging by a rotating black hole we use the Schwarzschild metric and hope to see also the characteristics of accretion into a slowly rotating hole. The effects of strong gravity and the existence of a horizon are correctly described in this way.

[1]also at: MPI für Radioastronomie, Auf dem Hügel 69, 53121 Bonn, Germany

2.1. Energy Advection and Projectors

The dynamics of an accretion flow into a black hole is governed by energy and momentum conservation. Energy losses due to radiation have to be calculated and applied separately. Angular momentum is transported into the black hole and to larger radii than those under consideration so a momentum flux through the boundaries of the disk occurs. We will neglect magnetic torques in our treatment and consider only a contribution of the magnetic pressure in the flow to the total pressure by scaling it as $\beta = p_{\rm Gas}/p_{\rm Mag}$. The dynamics can then be derived from the divergence of the stress-energy tensor

$$T^{ij}_{\ ;j} = 0 \quad \text{with} \quad T^{ij} = \tilde{\rho}\frac{V^iV^j}{c^2} + pg^{ij} - 2\mu D^{ij} + c^{-2}\left(Q^iV^j + V^iQ^j\right) \quad . \tag{1}$$

The density $\tilde{\rho}$ contains a contribution from the specific enthalpy of the gas like $\tilde{\rho} = \rho\left(1 + \frac{h}{c^2}\right)$ with the rest mass density ρ. V^i are the components of the 4-velocity of the flow, p the isotropic pressure and g^{ij} the metric tensor. The viscous part of the stress-energy tensor contains the kinematical viscosity $\mu = \nu\rho$ and the shear-stresses D^{ij}. We have added a contribution by a possible heat flux Q^i which has to be orthogonal to the velocity. The 0-component of the divergence in (1) is an energy equation for the flow, but does not resemble the second law of thermodynamics. This is formulated for a system at rest so we have to project the divergence in (1) onto the 4-velocity of the flow. The tool for this is a symmetric projector

$$P^{ij} = g^{ij} + c^{-2}V^iV^j \qquad P^i_j\left(T^{jk}_{\ ;k}\right) = 0 \quad . \tag{2}$$

After the projection we get an energy equation like

$$\rho T\frac{\mathrm{D}s}{\mathrm{D}\tau} = 2\mu D_{ij}D^{ij} - Q^i_{\ ;i} - c^{-2}Q_iA^i \quad . \tag{3}$$

Here we have used the substantial derivative $\frac{\mathrm{D}s}{\mathrm{D}\tau} = V^js_{,j}$ and the acceleration $A^i = \frac{\mathrm{D}V^i}{\mathrm{D}\tau}$. In the following we treat stationary and axisymmetric flows with radial infall.

2.2. Isothermal Poloidal Stratification

In optically thin plasma the heat produced locally by viscous friction will be either maintained as internal energy or radiated away. Diffusive transport by radiation will not occur. Instead of a radiative transport we expect turbulence driven by the magneto-rotational instability (e.g., Balbus & Hawley, 1998) to transport heat effectively perpendicular to the flow. So we approximate the poloidal structure by an isothermal plasma in hydrostatic equilibrium. The poloidal stratification is given by

$$\frac{\partial p}{\partial \theta} = \rho(1 + h/c^2)\sin\theta\ \cos\theta\ r^2\Omega^2 \quad . \tag{4}$$

With the assumed isothermal equation of state we can integrate and obtain

$$\rho(\theta) = \rho_0 \exp\left[-\frac{\cos^2\theta}{2}\left(\frac{r\Omega}{c_s}\right)^2\left(1 + \frac{\epsilon}{c^2} + \left(\frac{c_s}{c}\right)^2\right)\right] \quad . \tag{5}$$

One obtains the criterion for the flow to be a thin disk as

$$\frac{c_s^2}{1+\frac{h}{c^2}} \ll (r\Omega)^2 \quad . \tag{6}$$

We take $c_s^2 = \frac{\partial p}{\partial \rho} \propto T$ at constant temperature for an ideal gas. The internal energy and so the enthalpy rises strongly with temperature when the particles become relativistic and the disk thickness will shrink in the relativistic inner part close to the black hole.

2.3. Momentum Transport and Viscosity

In a stationary state as we are interested in the angular momentum distribution is determined by diffusive transport due to an effective turbulent viscosity

$$-\frac{\dot{M}}{2\pi}\left(\left(1+\frac{h}{c^2}\right)\frac{\partial}{\partial r}(r^2\Omega)+\frac{r^2\Omega}{\Sigma c^2}\frac{\partial P}{\partial r}\right)=\frac{\partial}{\partial r}\left(\nu\Sigma r^3\frac{\partial}{\partial r}\Omega\right) \quad . \tag{7}$$

For the viscosity we take a parametrization

$$\nu = \alpha(r-r_s)v_\phi \qquad \alpha \approx \mathcal{R}_c^{-1} \approx 10^{-1\ldots-3} \tag{8}$$

proposed by Duschl, Strittmatter & Biermann which is discussed by R. Auer (1999). For the self-similar solution of Narayan & Yi (1994) this viscosity has the same radial dependence as the standard form by Shakura & Sunyaev (1973) . Whenever the ADAFs deviate from the power-law behavior the viscosity is not coupled to the thermal state of the flow but rather to the dynamical structure[1]. The parameter α is proportional to the inverse of the critical Reynolds number $\mathcal{R}_c$ in the flow. MHD-instabilities may produce turbulence at rather small $\mathcal{R}$, so we take $\alpha = 0.1$. The radial momentum balance

$$\left(\frac{Mc^2}{r^2}-\left(1-\frac{3M}{r}\right)r\Omega^2+v_r\frac{\partial v_r}{\partial r}\right)\left(1+\frac{h}{c^2}\right)+\frac{\left(1-\frac{2M}{r}\right)+\frac{v_r^2}{c^2}}{\Sigma}\frac{\partial P}{\partial r}$$
$$=\frac{4}{3\Sigma}\left[\left(\frac{\partial}{\partial r}\frac{\nu\Sigma}{r}\left(\frac{\partial}{\partial r}rv_r\right)\right)-\frac{2v_r}{r}\left(\frac{\partial}{\partial r}\nu\Sigma\right)\right] \tag{9}$$

includes the viscous braking in a converging flow and the inertia and pressure terms. The centrifugal force reverses sign at $r = 3M = 3/2R_S$. The length scale $M = G\mathcal{M}/c^2$ is the gravitational radius with $\mathcal{M}$ the mass of the black hole.

2.4. Heating and Cooling of the Flow

The thermal balance in a stationary state is given by

$$Q_{\rm adv} = Q_{\rm vis}^{+} - Q_{\rm heat} - Q^{-} \quad . \tag{10}$$

---

[1]The length scale over which angular momentum is transported by eddy-exchange is limited by the distance to the horizon at r_s.

The flow cools via bremsstrahlung, synchrotron- and inverse Compton losses (Q^-), while the disk is heated by viscous friction

$$Q^+_{\rm vis} = \nu \Sigma r^2 \left[\frac{4}{3} \left(\frac{\partial}{\partial r} \left(\frac{v_r}{r} \right) \right)^2 + (\partial_r \Omega)^2 \right] \tag{11}$$

and the divergence of a possible heat flux $Q_{\rm heat}$. The heat flux in collisionless plasma is approximated by $Q_{\rm heat} = \Phi \Sigma c_s^3$. In a magnetized plasma the heat flux is expected to be strongly depressed so we take $\Phi = 10^{-3}$. The heat generated in the flow is not completely radiated but is advected with the flow according to

$$Q_{\rm adv} = \Sigma v_r c_V \left[\frac{\partial T}{\partial r} - (\Gamma_3 - 1) \frac{T}{\rho} \frac{\partial \rho}{\partial r} \right] \quad . \tag{12}$$

The thermodynamic quantities c_V, Γ_3 are taken from the relativistic equation of state for an ideal gas. The thermal balance must be treated for electrons and ions separately with an exchange of heat between the two species. This is assumed to occur only by Coulomb collisions but is subject to debate, as energy transfer due to resonant scattering of electrons by whistler waves can not be excluded.

3. Numerical Treatment

The vertically averaged dynamical equations for stationary flows presented above allow for a self-similar solution in the Newtonian limit and a constant advection ratio (Narayan & Yi, 1994), so that $Q_{\rm adv} \propto Q^+_{\rm vis}$ replaces the energy equation (10). The self-similarity implies constant logarithmic gradients for all variables

$$\frac{{\rm d}\Omega}{{\rm d}r} = \left(\frac{{\rm d}\ln\Omega}{{\rm d}\ln r} \right) \frac{\Omega}{r} \tag{13}$$

as shown here for the angular velocity. The differential equations are decoupled and one obtains a set of algebraic equations, which can be easily solved. We use a logarithmic radial grid and solve the algebraic equations with the full energy balance at all points. From the local solutions we recalculate the actual logarithmic gradients for a next iteration step. In doing so we avoid inner regularity conditions at critical (sonic) points. The steep gradients present in the solutions at 3 Schwarzschild radii (Fig. 1) lead to a slow convergence and the solutions shown are not completely relaxed to a stationary state. The cooling of the flow by radiation is unimportant at very small accretion rates and the radiation processes have no back-reaction on the dynamics of the flow. We include bremsstrahlung, synchrotron and inverse Compton-scattered synchrotron radiation in a rough approximation in the energy equation and derive the spectra from the ADAF-solutions afterwards.

4. Mass Accretion in the Galactic Center

The accretion rate necessary to power the Galactic Center source Sgr A* is $10^{-9...-10}$ $M_\odot$/yr for a thin accretion disk. But upper limits from IR-measurements allow only a mass flux of 10^{-11} $M_\odot$/yr or lower. The inefficiency of

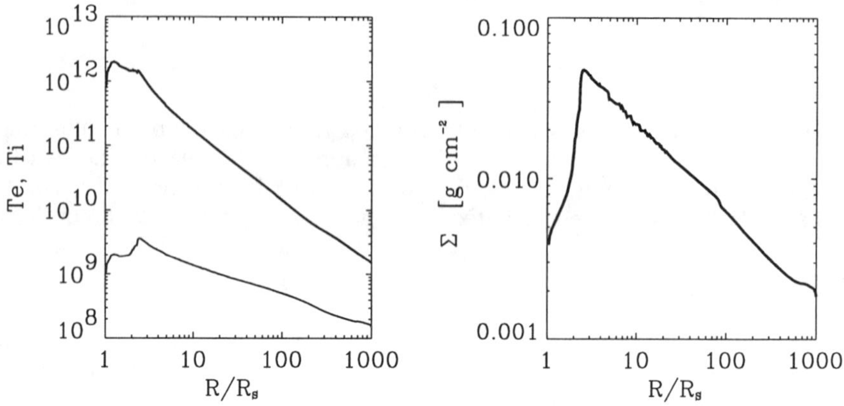

Figure 1. Electron (lower curve) and ion temperatures (upper curve) as a function of radius and the surface density of an ADAF with a mass flow rate of 10^{-5} $M_\odot$/yr. Density and temperatures deviate significantly from the self-similar form at the last stable circular orbit at $3R_S$ where angular momentum does not provide a potential barrier anymore.

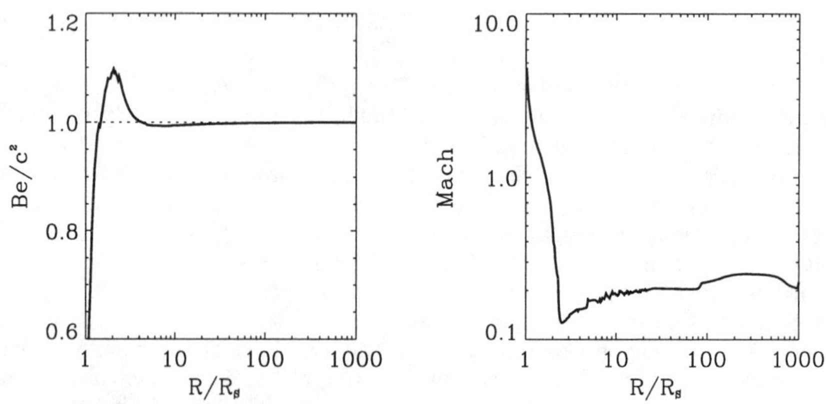

Figure 2. Specific Bernoulli number and radial Mach number of the flow. The Bernoulli number includes the rest mass of the particles, so a zero-energy flow has a Bernoulli number of 1. The Mach number shows a transition from sub-sonic to supersonic behavior inside the last stable circular particle orbit.

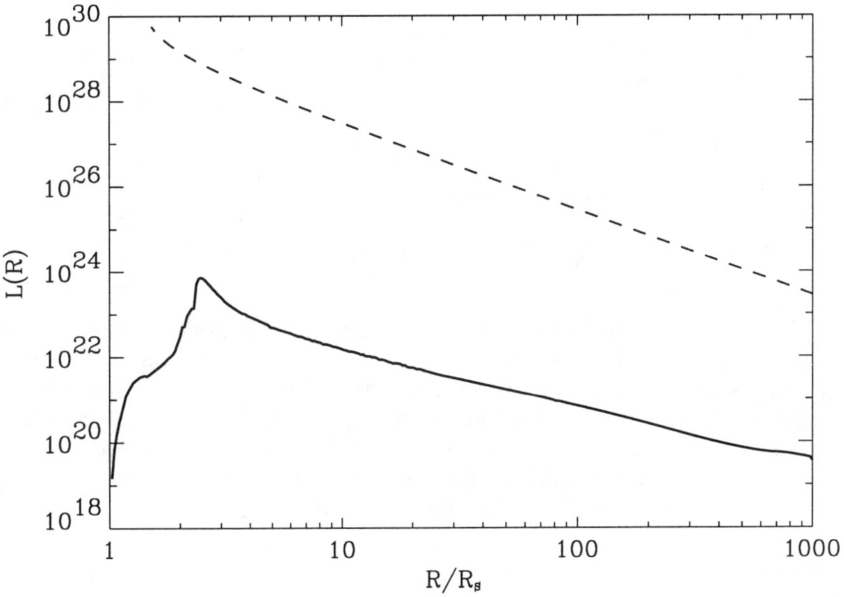

Figure 3. Local luminosity of ADAF and standard thin accretion disk for a mass accretion rate of 10^{-5} $M_\odot$/yr. The thin disk luminosity (dashed line) is considerably larger than the ADAF emission (solid line) and ends at 1.5 black hole radii, where the centrifugal force reverses its direction and is unable to support a stationary thin flow.

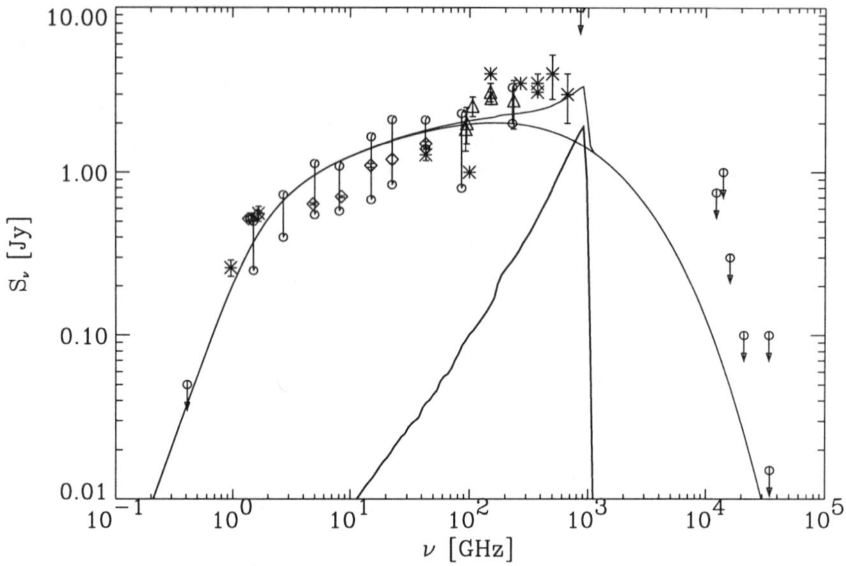

Figure 4. A comparison between minimal ADAF model and radio-IR observations. We have combined single measurements with variability ranges marked by symbols with circles at both ends. The emission by the ADAF itself peaks at 10^3 GHz. Optically thin synchrotron emission from relativistic electrons provide the flux at almost all other frequencies. The combined spectrum is also shown and falls a little short for the measured flux at 100-1000 GHz.

ADAF-emission processes point to much larger accretion rates, which are determined from a simultaneous fit of sub-mm and X-ray measurements. The critical parameters are the mass of the black hole, here $\mathcal{M} = 2.6\,10^6$ $M_\odot$, the distance to the Galactic Center $D = 8$ kpc and the hydrogen column density for soft X-ray absorption $N_H = 6\,10^{21}$ cm^{-2}. We assume that most radio observations are dominated by optically thin synchrotron emission from relativistic electrons and the ADAF contributes significantly in the sub-mm regime only. This minimal ADAF-model is calculated with a plasma-$\beta = 10$ and is correspondingly gas pressure dominated. The derived mass accretion rate is $1 \cdot 10^{-5}$ $M_\odot$/yr. Temperatures and surface density of this model is shown in Fig. 1. In contradiction to the analytical solutions and Narayan & Yi (1995) we find Bernoulli numbers less than 1, corresponding to the rest mass at infinity as seen in Fig. 2. The radiation inefficiency is shown in Fig. 3, where the local luminosity is compared to the luminosity of a thin disk at the same accretion rate.

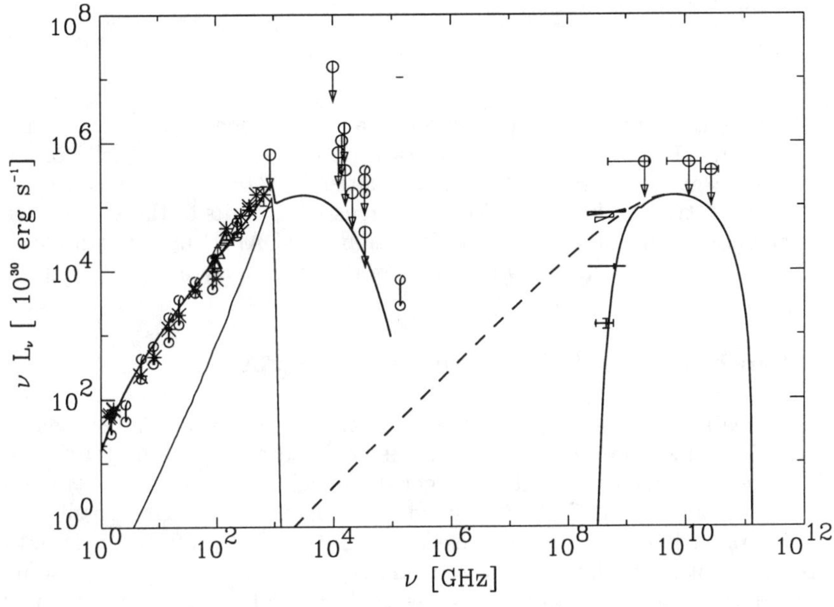

Figure 5. The radio to X-ray emission of Sgr A* and the minimal ADAF-model. ROSAT, *Einstein* and GINGA observations are included. Both the uncorrected and the corrected fluxes corresponding to a column density of $N_{\rm H} = 6\,10^{21}$ cm^{-2} are shown. The intrinsic spectrum of free-free emission is given by the dashed line. According to our model most radio observations see optically thin synchrotron emission, but the energy output is dominated by the ADAF, both in the sub-mm and X-ray regime. The Comptonized synchrotron emission emerges at IR-optical frequencies and is not included.

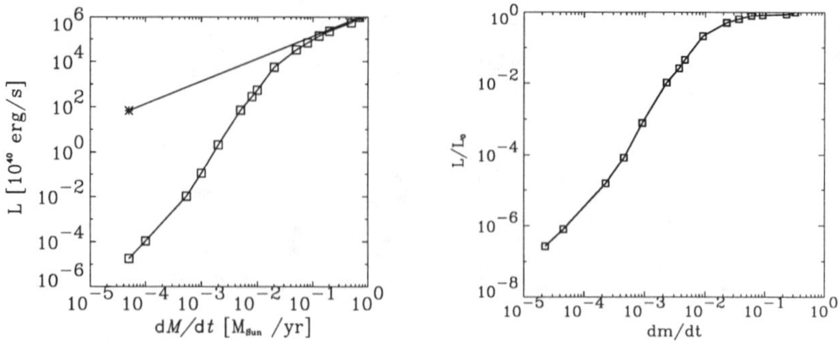

Figure 6. The total luminosity of standard (upper line) and ADAFs as a function of mass accretion rate (left plot) for a 10^8 $M_\odot$ black hole. Squares mark calculated luminosities of ADAFs in our treatment. The relative luminosity of ADAFs versus $\dot{M}$ in units of the Eddington accretion rate shows the rapid deviation of ADAFs from the standard disk luminosity within two decades in the mass flow rate.

5. Synchrotron and X-ray Spectrum of Sgr A*

The optically thick self-absorbed synchrotron radiation of the ADAF emerges in the radio and sub-mm regime and rises steeply to a maximum at 10^3 GHz. It shows an even steeper cut-off for larger frequencies seen in Fig. 5. Most of the radio observations can not be explained by the ADAF itself. One possible source for this radiation are relativistic electrons of $E = 203$ MeV emitting optically thin synchrotron radiation in a weak magnetic field $B = 2.8$ G. Assuming a homogeneous source its radius is at least $R = 4.5\,10^{13}$ cm, much larger than the synchrotron emitting region of the ADAF and thus providing an extended halo. These relativistic electrons can be accelerated in magnetic reconnection sheets (Schopper, Lesch & Birk, 1998) and maintained in a quasi-monoenergetic distribution (Jauch 1999) or they may be secondary electrons from charged pion decays (Mahadevan, 1998; Markoff et al. 1997).

6. How to Switch On an AGN and How to Switch it Off Again

For mass accretion rates below 0.1 $M_{\rm Edd}$ there are two possible states of accretion into black holes. The standard thin disk (Frank, King & Raine, 1992), which loses all the energy produced by viscous friction locally, and the advection-dominated accretion flow (ADAF) described above. When we compare the radially integrated luminosity of the two kinds of accretion flows at the same accretion rate, we find roughly the same luminosity for a critical mass flow rate of $\alpha^2 \dot{M}_{\rm Edd}$. At smaller rates the luminosity of ADAFs drop dramatically, while the luminosity of efficiently radiating flows is proportional to the mass accretion rate $L_0 = \eta \dot{M} c^2$. The ratio of the luminosities for ADAFs to L_0 gives the amount of specific energy advected into the black hole and the inefficiency of the heat transfer between electrons and ions.

As both states of accretion are available at the same time in a stationary flow, there must be a physical reason, which enforces the flow to be advection-dominated or not. If a mechanism is at work deciding in which state the flow will be, it provides a switch to turn a luminous accreting black hole into a quiet sink of matter or the other way round.

Let us, for the moment being, assume that, whenever available, an accretion disk chooses to operate in the advection dominated regime, which is indicated by the extremely low luminosity of the Galactic Center. Then comparatively small changes in the mass flow rate through the disk may result in dramatic changes of the efficiency of the accretion process, i.e., the luminosity of the AGN. This highly non-linear behavior of the luminosity as a function of the accretion rate has the potential of acting almost as a switch between an AGN phase of a galaxy and a non-active phase without having to change the mass flow rate by a similar amount as the luminosity. In our example of Fig. 6, at a mass flow rate of $\sim 10^{-4}\,M_{\odot}/yr$ an increase of the rate by two orders of magnitude results in almost a million times larger a luminosity.

Acknowledgments. This work was supported by the DFG through SFB 328 *Evolution of Galaxies.*

References

Auer, R., & Duschl, W.J., 1999, this volume, p. 157

Balbus S. A., Hawley J. F., 1998, Reviews of Modern Physics 70, 1

Beckert T., Duschl W. J., 1997, A&A 328, 95

Falcke H., Biermann P. L., 1996, A&A 308, 321

Genzel, R., & Eckart, A., 1999, this volume, p. 3

Ghez, A., et al., 1999, this volume, p. 18

Frank J., King A., Raine D. 1992, Accretion Power in Astrophysics, 2. ed., Cambridge: University Press

Jauch, Z., & Duschl, W.J., 1999, this volume, p. 169

Mahadevan R., 1998, Nature 394, 651

Markoff, S., Melia, F., Saresevic, I. 1997, ApJ, 489, L47

Mezger P. G., Duschl W. J., Zylka R., 1996, A&AR 7, 289

Narayan R., & Yi I., 1994, ApJ 428, L13

Narayan R., & Yi I., 1995, ApJ 452, 711

Narayan R., Mahadevan R., Grindlay J. E., Popham R. G., Gammie C., 1998, ApJ 492, 554

Schopper R., Lesch H., Birk G. T., 1998, A&A 335, 26

Shakura, N.I., Sunyaev, R.A. 1973, A&A, 24, 337

The Central Parsecs of the Galaxy
ASP Conference Series, Vol. 186, 1999
H. Falcke, A. Cotera, W.J. Duschl, F. Melia, M.J. Rieke, eds.

The Jet Model for Sgr A*

H. Falcke

Max-Planck-Institut für Radioastronomie, Auf dem Hügel 69, D-53121 Bonn, Germany

Steward Observatory, The University of Arizona, Tucson, AZ 85721

Abstract. In this paper the jet model for the supermassive black hole candidate Sgr A* in the Center of the Galaxy is reviewed. The most recent model, with a reduced set of parameters, is able to account for all major radio properties of the source: size, structure, flux density, and spectrum. The model requires a minimum jet power of $\sim 10^{39}$ erg/sec and in a symbiotic jet/disk system implies a minimum accretion rate of a few times $10^{-8} M_{\odot}$/yr for a radio loud jet or $\sim 10^{-5} M_{\odot}$/yr for a radio quiet jet. Low near-infrared limits on the Sgr A* flux then imply that the accretion flow onto the central black hole must be radiatively deficient, but most likely has a high viscosity. Within the jet model the high-frequency part of the Sgr A* spectrum is self-consistently explained as the nozzle of the outflow. In a symbiotic model this innermost region of the jet could possibly be identified with the innermost region of an advection dominated accretion disk, a Bondi-Hoyle accretion flow, or any other type of under-luminous accretion process. The compact nozzle region is of particular importance since it can be used as a background photon source against which the central black hole could be directly imaged with future mm-VLBI experiments.

1. Introduction

Sgr A* has been at the center of an extensive discussion at this workshop. An overwhelming number of observations now demonstrate that this compact radio source in the center of the Galaxy is associated with a dark mass of $2.6 \cdot 10^6 M_{\odot}$ (Haller et al. 1996; Ghez et al. 1998 & 1999; Eckart & Genzel 1999; Zhao et al. 1999). In the past Sgr A* was only detected in the radio and many properties were associated with it only because of it being the most unusual radio object in the entire region (see Falcke 1996a for a review). Now we have much more direct information about its various properties. For example Ghez et al. (1999) show that the central dark mass concentration does indeed peak at the position of Sgr A* without any a priori assumptions about its nature. Stolovy et al. (1999) find that the only short-term variable source in their NICMOS near-infrared images of the very center also coincides with Sgr A*. Eckart & Genzel (1999) also claim the detection of a variable NIR source as a counterpart to Sgr A*. Cotera et al. (1999) report a detection of an unusual and narrow mid-infrared ridge at the position of Sgr A* but it is unclear how and if that is related to the central dark mass concentration.

While the amount of dark mass itself and its association with Sgr A* now seems well established, the nature of Sgr A* is not. Any model needs to explain the compact size, the possibly elongated structure (Lo et al. 1998), the inverted radio spectrum, the cut-off of the spectrum towards the infrared, and the faintness of emission at other wavelengths. At present it is still completely unclear which of the proposed emission models, if any, would be a fair description of the actual processes in the Galactic Center. Models proposed so far include advection dominated two-temperature accretion flows (ADAFs, Narayan et al. 1995), Bondi-Hoyle accretion (Melia 1994), emission from mono-energetic electrons (Duschl & Lesch 1994), or a jet (Falcke, Mannheim, & Biermann 1993). Even though they involve different physical assumptions, all models usually assume the presence of a supermassive black hole in Sgr A*; a comparison of the various approaches is given in Falcke (1996a) and Melia (1999). This paper will now mainly focus on the jet model and its recent improvements.

2. The Jet Model

The jet model originally proposed by Falcke et al. (1993) assumes that the radio emission from Sgr A* can be explained as emission from a radio jet, similar to, but with much lower power than, those seen in other nearby active galaxies or even quasars. Since the original publication the model has been refined further in Falcke (1996b) and Falcke & Biermann (1999). 'Refined' in this context means that the model was further simplified to have fewer free parameters: equipartition factors were set to unity, the electron distribution was assumed to be mono-energetic[1], and the velocity field of the jet was calculated self-consistently.

The basic idea is to start with a certain power $Q_{\rm jet}$ which is used to accelerate plasma through a nozzle of characteristic scale height $Z_{\rm nozz}$. The plasma is treated as an almost relativistic proton/electron gas with a sound speed of $\simeq 0.4c$ and an adiabatic index $\Gamma = 4/3$. The magnetic field is assumed to be tangled and in equipartition with relativistic particles. Moreover, the combined energy of relativistic particles and magnetic field is assumed to be in equipartition with the kinetic energy of the outflow. With these equipartition assumptions the particle density and magnetic field energy densities then follow directly from the input parameter $Q_{\rm jet}$ at the sonic point inside the nozzle.

For the purpose of calculating the radio emission one can treat the acceleration region as a black box which pushes the plasma through the sonic point. In a one-dimensional treatment the flow can then be described as a pressure-driven wind with the jet proper velocity $\gamma_{\rm j}\beta_{\rm j}$ given by the Euler equation to be

$$\left(\frac{\left(\frac{\Gamma+\xi}{\Gamma-1}\right)(\gamma_{\rm j}\beta_{\rm j})^2-\Gamma}{\gamma_{\rm j}\beta_{\rm j}}\right)\frac{\partial\gamma_{\rm j}\beta_{\rm j}}{\partial z}=\frac{2}{z} \tag{1}$$

[1]In contrast to Duschl (1999) the mono-energetic distribution here is not an essential ingredient of the model since the final spectrum will result from a superposition of many self-absorbed components.

and with $\xi = (\gamma_j\beta_j/(\Gamma(\Gamma-1)/(\Gamma+1)))^{1-\Gamma}$ (Falcke 1996b).

Assuming a vacuum, the jet will expand transversally with roughly sound speed after leaving the nozzle. The resulting pressure gradient along the jet (Fig. 1, left) will accelerate the flow to bulk Lorentz factors of a few (Fig. 1, right). This is of course an extremely simplified treatment and if there are additional acceleration processes (centrifugal, magnetic) the velocity would be higher still. The resulting shape of the outflow is roughly conical and depicted in Figure 2 (left). The axial ratio is of the order 3 to 4:1.

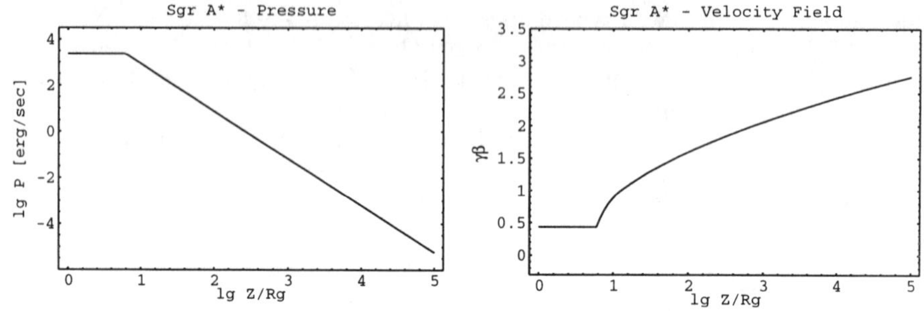

Figure 1. Run of pressure (left) and proper velocity (right) in the Sgr A* jet as a function of radius (in gravitational radii $GM_\bullet c^{-2}$) calculated for a simple pressure driven outflow of relativistic plasma.

Based on this velocity field, density distribution, and geometry it is then possible to calculate the expected radio spectrum of Sgr A*. Here, another free parameter is the energy distribution of the radiating particles (electrons or pairs). The simplest approach is to use a quasi mono-energetic electron distribution with a characteristic energy γ_e (Duschl & Lesch 1994). However, any other distribution with a similar characteristic energy (e.g., power-law with low-energy break or even thermal distributions) would lead to similar results.

As shown in Falcke & Biermann (1999) all these simplifications yield the following expressions for the observed flux density and angular size of Sgr A* observed at a frequency ν as a function of jet power. For a source at a distance D, with black hole mass $M_\bullet$, size of nozzle region $Z_{\rm nozz}$ (in $R_{\rm g} = GM_\bullet/c^2$), jet power $Q_{\rm jet}$, inclination angle i, and characteristic electron Lorentz factor γ_e, the observed flux density spectrum is given as

$$\begin{aligned} S_\nu &= 10^{3.03\cdot\xi_0}\ {\rm m}Jy\left(\frac{Q_{\rm jet}}{10^{39}{\rm erg/sec}}\right)^{1.27\cdot\xi_1} \\ &\cdot\left(\frac{D}{10{\rm kpc}}\right)^{-2}\left(\frac{\nu}{8.5{\rm GHz}}\right)^{0.20\cdot\xi_2}\left(\frac{M_\bullet}{2.5\cdot 10^6 M_\odot}\frac{Z_{\rm nozz}}{10R_{\rm g}}\right)^{0.20\cdot\xi_2} \\ &\cdot\left(3.9\cdot\xi_3\left(\frac{\gamma_{\rm e,0}}{200}\right)^{-1.4\cdot\xi_4} - 2.9\cdot\xi_5\left(\frac{\gamma_{\rm e,0}}{200}\right)^{-1.89\cdot\xi_6}\right), \end{aligned} \quad (2)$$

with the correction factors ξ_{0-6} depending on the inclination angle i (in radians):

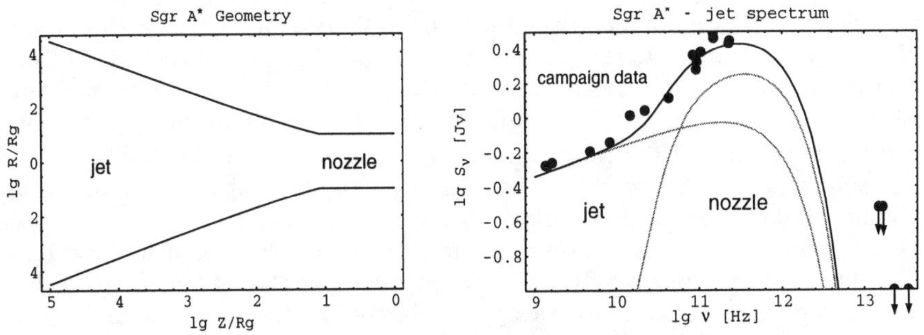

Figure 2. Left: geometry of the jet model for Sgr A*. Note the logarithmic scale on the vertical axis ending at and not going through zero. Right: radio spectrum of Sgr A*. The radio data are from Falcke et al. (1998) and references therein. The solid line is the prediction of the jet model, where the gray lines indicate which parts are contributed by the jet and the nozzle respectively.

$$\begin{aligned}
\xi_0 &= 2.38 - 1.90\,i + 0.520\,i^2 && (3)\\
\xi_1 &= 1.12 - 0.19\,i + 0.067\,i^2 && (4)\\
\xi_2 &= -0.155 + 1.79\,i - 0.634\,i^2 && (5)\\
\xi_3 &= 0.33 + 0.60\,i + 0.045\,i^2 && (6)\\
\xi_4 &= 0.68 + 0.50\,i - 0.177\,i^2 && (7)\\
\xi_5 &= 0.09 + 0.80\,i + 0.103\,i^2 && (8)\\
\xi_6 &= 1.19 - 0.29\,i + 0.101\,i^2. && (9)
\end{aligned}$$

Likewise, the characteristic angular size scale of the emission region is given by

$$\begin{aligned}
\Phi_{\rm jet} &= 4.7 \cdot \chi_0 \,\text{mas}\, \sin i \\
&\cdot \left(\frac{\gamma_{\rm e,0}}{200}\right)^{1.77\cdot\chi_1} \left(\frac{D}{10\text{kpc}}\right)^{-1} \left(\frac{\nu}{8.5\text{GHz}}\right)^{-0.89\cdot\chi_1} \\
&\cdot \left(\frac{Q_{\rm jet}}{10^{39}\text{erg/sec}}\right)^{0.60\cdot\chi_1} \left(\frac{M_\bullet}{2.5\cdot 10^6 M_\odot}\,\frac{Z_{\rm nozz}}{10R_{\rm g}}\right)^{0.11\cdot\chi_2}, \qquad (10)
\end{aligned}$$

with the angle dependent correction factors

$$\begin{aligned}
\chi_0 &= 4.01 - 5.65\,i + 3.40\,i^2 - 0.76\,i^3 && (11)\\
\chi_1 &= 1.16 - 0.34\,i + 0.24\,i^2 - 0.059\,i^3 && (12)\\
\chi_2 &= -0.238 + 2.63\,i - 1.85\,i^2 + 0.459\,i^3, && (13)
\end{aligned}$$

where again the inclination angle i is in radians. All correction factors are normalized to yield unity at $i = 1$ rad ($\sim 57°$).

For Sgr A* we are left with four basic unknown quantities in this equation: the inclination angle i, the jet power $Q_{\rm jet}$, the characteristic electron Lorentz factor $\gamma_{\rm e}$, and the relative nozzle size $Z_{\rm nozz}$. The expectation from fitting of other radio cores is that $Z_{\rm nozz}$ will be close to the jet origin of order 10 $R_{\rm g}$ and $\gamma_{\rm e}$ is of order 10^2. Moreover, given our position in the Galaxy it seems unlikely that Sgr A* would be pointing right at us and the inclination angle should probably be larger than say 30°.

If one assumes that the mm-bump in the spectrum is in fact intrinsic to Sgr A* and is produced by the innermost region of the jet/accretion disk (i.e., the nozzle) one can use this feature to effectively constrain $\gamma_{\rm e}$ and $Z_{\rm nozz}$ for a given $Q_{\rm jet}$. In addition, for a given flux density and size of the core at frequencies below the bump, one can also solve for the jet power and the inclination angle.

Unfortunately, the exact size of Sgr A* is still uncertain. Lo et al. (1998) claim that the intrinsic size of Sgr A* is ~ 0.45 mas at 43 GHz for a flux density of 1.1 Jy. Even though these numbers should be considered to be very tentative at present (see Bower et al. 1999a for a discussion of the uncertainties involved), we will use these number to fit the jet model. Figure 2 shows the predictions of the jet model for the Sgr A* radio spectrum in comparison to the data. The overall spectrum including the submm-bump is well reproduced for a nozzle size of $Z_{\rm nozz} = 12 R_{\rm g} = 35\mu$as (needs to be multiplied by two if we see both sides), a characteristic electron Lorentz factor of 125, an inclination angle of $\sim 45°$, and a jet power of $Q_{\rm jet} \gtrsim 10^{38.7}$ erg/sec. The predicted spectral index of the low-frequency part of the spectrum is $\alpha \simeq 0.17$ ($S_\nu \propto \nu^\alpha$) which agrees well with the observed one. The predicted size as a function of frequency roughly scales as $\nu^{-0.9}$ and is shown in Fig. 3 together with results from recent mm-VLBI experiments. As far as the observations are concerned, one has to point out that the measurements at λ3 and λ1.3 mm do not probe the NS direction, and hence for a jet geometry along the axis seen by Lo et al. (1998) could underestimate the true size of Sgr A* (see Doleman et al. 1999 for a discussion of this important detail).

3. Discussion

3.1. Results of the Jet Model

In general it can be said that the jet model explains all known observational details of Sgr A* reasonably well. This shows that the claim by Lo et al. (1998) that none of the existing model could reproduce their data is unfounded. In fact, all the observational results Lo et al. claim (i.e., elongation, intrinsic size, intrinsic size scaling with frequency) had already been well predicted earlier by the jet model within reasonable margins. Nevertheless, one needs to point out that the core size within the current version of the jet model cannot be very accurately predicted since it depends quite sensitively on the electron Lorentz factor, leaving an inherent uncertainty of a few.

Another important benefit of the jet model is that it not only fits a single source, Sgr A*, but the same relatively simple equations presented here can be used to explain in detail well-constrained radio cores in other galaxies like

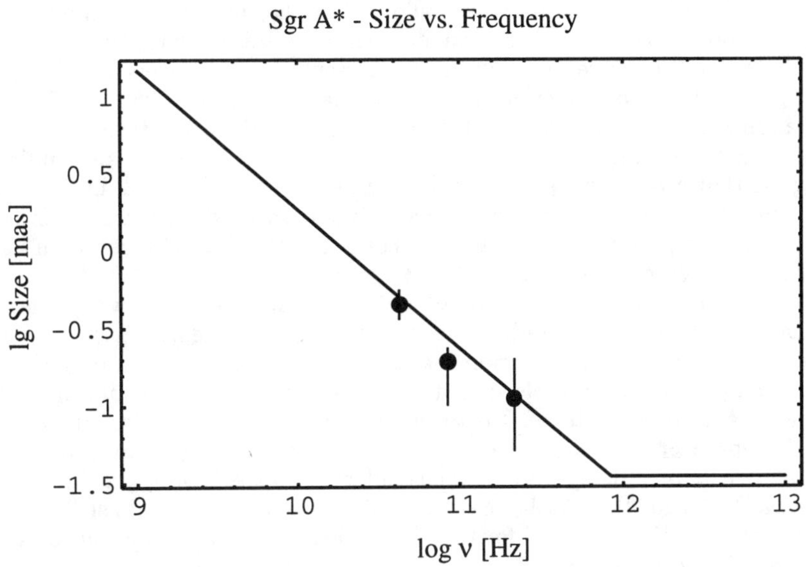

Figure 3. Intrinsic size of Sgr A* (major axis, probably north-south) versus frequency as predicted in the jet model (solid line). The parameters (i, $\gamma_{\rm e}$, $Q_{\rm jet}$) have been chosen to match the intrinsic NS size of 0.44 mas at λ7mm given by Lo et al.(1998). The data points at λ3 and λ1.3 mm were taken from Krichbaum et al. (1999) and Doleman et al. (1999). The error bars indicate the reported range of values. Note that the sizes at λ3 and λ1.3 mm are not for the NS direction.

M81 or NGC4258 and even the galactic superluminal jet source GRS1915+105 (Falcke & Biermann 1996, 1999).

3.2. The Accretion Flow in Sgr A*

Moreover, the jet model allows one to make interesting inferences for the nature of the accretion process in Sgr A*. Our current understanding is that jets are formed in the innermost region of an accretion disk near a compact object. This seems to be true for young stars as well as for neutron stars and black holes. Jet and disk therefore cannot be considered in isolation but form a symbiotic system (Falcke & Biermann 1995). In many cases one finds that the jet power is a significant fraction of the total accretion power and is comparable to the accretion luminosity.

Sgr A* adds another twist to this story. Since the jet model used here basically assumed the most efficient type of radio jet, we can place a firm lower limit on the jet power. Any modifications to the model, other than assuming unreasonably small inclination angles, will only lead to an increased energy demand and a larger $Q_{\rm jet}$. If we assume that half of the energy released in an accretion process onto a rotating black hole is used to drive the jet, we need a *minimum* accretion rate of $5{\cdot}10^{-8} M_{\odot}$. Interestingly, any standard accretion disk model with such an accretion rate would violate the current upper limits on the NIR emission of Sgr A* as presented for example in this volume. Hence, we must conclude that the accretion disk in Sgr A* must be radiatively deficient, e.g., is optically thin and advection dominated (Narayan et al. 1995). As discussed in Falcke & Heinrich (1994) the disk in Sgr A* would turn optically thin when the viscosity parameter is $\alpha \gtrsim 10^{-3}$. The jet model therefore also indirectly suggests that a relatively efficient viscosity mechanism is at work in Sgr A*.

Unfortunately, it is not easy to give a similarly robust estimate for an upper limit of the accretion rate. From studies of other AGN (see discussion in Falcke et al. 1995; Falcke & Biermann 1995) we know that there is a dichotomy of the radio-luminosity of jets, the nature of which is completely unknown. At least in quasars the radio cores of radio-quiet jets seem to be roughly a factor hundred less luminous for a given accretion rate than radio-loud jets. Assuming that Sgr A* would physically resemble such a radio-quiet quasar core, would imply that in fact the accretion rate needed to power Sgr A* is of the order $10^{-5} M_{\odot}$/yr. This is close to the value needed in the ADAF model, though not quite yet the value predicted by 3D hydrodynamic simulations of the GC region (Melia 1999; Coker & Melia 1999). In either case a radio-quiet jet model would still require a very high 'inefficiency factor' for the production of thermal radiation to explain the low NIR limits for Sgr A*.

The idea of the jet/disk-symbiosis could be pushed even further for Sgr A*: an interesting part of the jet model is the nozzle, which is supposed to be the innermost part of the jet and in more luminous AGN is usually invisible due to energy-loss arguments. In a symbiotic system this region would be the interface between the accretion flow (inflow) and the jet (outflow) and it cannot really be claimed to belong to either one—the nozzle could be as much part of the disk as it is part of the jet. This would allow one to easily combine existing models for the accretion flow (e.g., Melia 1999, Coker & Melia 1999; Narayan et al. 1995) with the jet model, by dropping the assumption that these models explain the cm-spectrum and assuming that their mm-emission components are in fact the black box labeled 'nozzle' in the jet picture. An ADAF or Bondi-Hoyle accretion flow with outflow in its innermost region might do just that and explain what the nozzle really is.

3.3. Predictions

A number of predictions from the jet model can be made that can be tested in the near future. Sgr A* should become resolved at 3 and 1 mm in the NS direction once a suitable mm-VLBI array is available. From analogy to other radio cores one would expect a polarization at the percent level at mm-wavelengths where interstellar propagation effects become negligible (Bower et al. 1999a&b). The most likely direction of the magnetic field is probably along the jet axis (NS?). Because the outflow travels from small to large scales and from small to large wavelengths one would expect that radio outbursts appear first at high frequencies and then propagate to longer wavelengths. The time scale for this delay could be relatively short. The model also predicts a certain level of x-ray emission, since the relativistic electrons in the nozzle will inverse-Compton scatter their own synchrotron radiation into the soft x-ray regime. The luminosity, however, will be relatively low, of the order $\lesssim 10^{34}$ erg/sec with a relatively unusual, curved spectrum. A more detailed calculation was presented already in Beckert & Duschl (1997).

4. Outlook

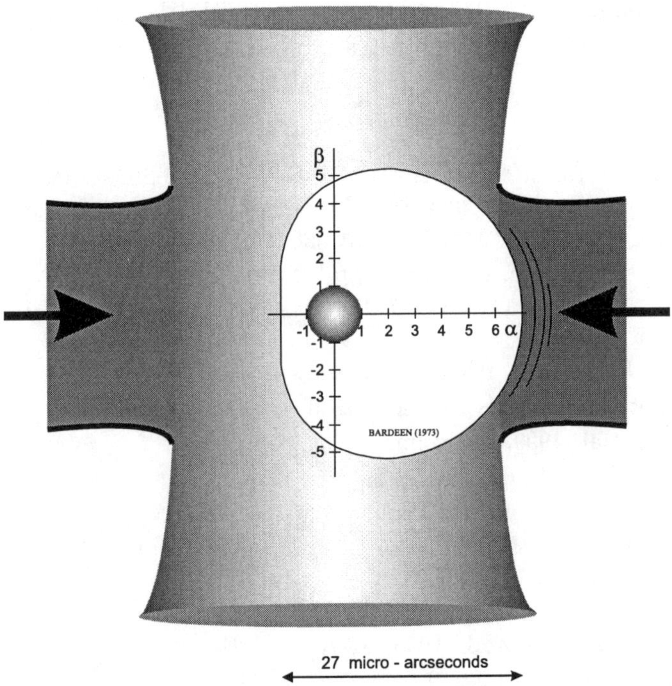

Figure 4. Sketch of the inner region of Sgr A* with accretion flow and nozzle surrounding the black hole. Overlaid is an appropriately scaled reproduction of a Figure from Bardeen (1973), showing the 'hole' of photons absorbed by the black hole if observed against a background source. A similar process could apply to Sgr A* and its compact, high-frequency emission component.

The most intriguing consequence of the radio spectrum and its modeling is the suggestion that the mm-part of the spectrum corresponds to an ultra-compact region in Sgr A*. This follows from a few simple arguments (e.g., Falcke et al. 1998) as well as from basically all models proposed. Within the jet model, the size of the component used here to fit the spectrum corresponds to 35μas. This is interestingly close to the suspected black hole — six Schwarzschild radii. Bardeen (1973) made an interesting calculation, where he showed that photons from a background source behind a rotating black hole would be absorbed within an asymmetric disk of 4.5 Schwarzschild radii diameter. Even though the mm-emission region in Sgr A* is not at infinity behind the black hole one would expect a similar effect here. Figure 4 shows the Bardeen absorption disk overlayed on a sketch of Sgr A* with accretion disk, jet and black hole. Future 220 GHz VLBI experiments—first results of which have already been reported (Krichbaum et al. 1999)—will have enough resolution to start imaging such effects directly. At this frequency scattering will also not be major problem any more. This experiment could provide the final proof, not only for the validity of a specific model, but also for the existence of black holes in general.

Acknowledgments. This work was supported by the Deutsche Forschungsgemeinschaft, grants Fa 358/1-1&2.

References

Bardeen, J.M. 1973, in "Black Holes", C. DeWitt & B.S.DeWitt, Gordon and Breach Science Publisher, New York, p. 215

Beckert, T.,& Duschl, W. J. 1997, A&A 328, 95

Bower, G.C. et al. 1999a, this volume, p. 80

Bower, G.C., Backer, D.C., Zhao, J.-H., Goss, W.M., & Falcke, H. 1999b, ApJ, submitted

Coker, R.C., & Melia, F. 1999, this volume, p. 214

Cotera, A., et al. 1999, this volume, p. 240

Doleman, S., et al. 1999, this volume, p. 98

Duschl, W.J., & Lesch, H. 1994, A&A 286, 431

Falcke, H. 1996a, in "Unsolved Problems of the Milky Way", IAU Symp. 169, L. Blitz & P.J. Teuben (eds.), Kluwer, Dordrecht, p. 169-180

Falcke, H. 1996b, ApJ 464, L67

Falcke, H., & Biermann, P. L. 1995, A&A 293, 665

Falcke, H., & Biermann, P. L. 1996, A&A 308, 321

Falcke, H., & Biermann, P.L. 1999, A&A 342, 49

Falcke H., Goss W.M., Matsuo H., Teuben P., Zhao J.-H., & Zylka R. 1998, ApJ 499, 731

Falcke H., & Heinrich O. 1994, A&A 292, 430

Falcke H., Malkan M., & Biermann P.L. 1995,A&A 298, 375

Falcke H., Mannheim K., & Biermann P. L. 1993, A&A 278, L1

Genzel, R., & Eckart, A., 1999, this volume, p. 3

Ghez, A. M., Klein, B. L., Morris, M., & Becklin, E. E. 1998, ApJ 509, 678

Ghez, A. et al. 1999, this volume, p. 18

Haller J., Rieke, M., Rieke, G., et al. 1996, Ap.J. 468, 955

Krichbaum, T., et al. 1999, this volume, p. 89

Lo, K.Y., et al. 1998, ApJ 508, L61

Melia, F. 1999, this volume, p. 127

Melia F. 1994, ApJ 426, 677

Narayan R., Yi I., & Mahadevan R. 1995, Nature 374, 623

Stolovy, S., et al. 1999, this volume, p. 39

Zhao, J.-H., et al. 1999, this volume, p. 224

The Central Parsecs of the Galaxy
ASP Conference Series, Vol. 186, 1999
H. Falcke, A. Cotera, W.J. Duschl, F. Melia, M.J. Rieke, eds.

Accretion Disk Instabilities in the Vicinity of Black Holes

R. Auer

Institut für Theoretische Astrophysik, Heidelberg; Sonderforschungsbereich 328: Evolution of Galaxies, Universität Heidelberg

W.J. Duschl

Institut für Theoretische Astrophysik, Heidelberg; Sonderforschungsbereich 328: Evolution of Galaxies, Universität Heidelberg; Max-Planck Institut für Radioastronomie, Bonn

Abstract.
Accretion disks around massive black holes will themselves become massive. The α-viscosity can no longer be valid in the case where the accretion disk mass is comparable to the mass of the accretor, as shown by Duschl *et al.* (1998). Using a modified viscosity prescription, one that takes into consideration the massive disk, leads to noticeably different disk structure. We present preliminary results of black holes with masses ranging from 10^6 to 10^{10} $M_\odot$ and mass accretion rates ranging from 10^{-9} to 10^1 $M_\odot$/a.

1. Introduction

The α-parameterization of the viscosity in accretion disks as introduced by Shakura & Sunyaev (1973) has been extraordinarily successful. It has also been extraordinary in that it has remained so little understood. This problem has certainly been rationalized away, so that even in the paper of Shakura & Sunyaev (1973) they realize that "This result allows us to compute the external appearance of the disk, ... without deciding upon the mechanism of angular momentum transfer", yet also realize that the answer lies with the observations and deductions of Taylor (1936 a,b). In this paper we attempt to reconcile what works (α-parameterization) with what is known (Taylor), thus providing another link in the chain towards a complete understanding of the mechanism for non-magnetically induced viscosity in accretion disks.

2. A New Viscosity

Fundamentally, we note that a viscosity is a combination of a scale velocity and a scale length; $\nu = \Delta s \cdot \Delta v$. If in an accretion disk we can determine a radial scale length Δs, we can clearly associate a scale velocity with that scale length: $\Delta v \approx \left|\frac{\mathrm{d}v}{\mathrm{d}s}\right| \cdot \Delta s$ where $\left|\frac{\mathrm{d}v}{\mathrm{d}s}\right| \approx \frac{v_\phi}{s}$. Arguments by Duschl *et al.* (1998)

and a re-analysis of the experiments performed by Taylor (1936 a,b) suggest that the viscosity can be well-described by $\beta s v_\phi$, implying that an increase in angular momentum results in an increase in the viscosity and hence the stability of the disk. In this prescription, $\beta = 1/\mathcal{R}_{\rm crit}$ where $\mathcal{R}_{\rm crit}$ is the critical Reynolds Number with a value of $\sim 10^2$ to 10^3. Using $\Delta v \approx \xi \cdot v_\phi$, we find $\Delta s \approx \xi \cdot s$ and $v = \Delta s \cdot \Delta v = \xi^2 s v_\phi$, implying that $\xi = \sqrt{\beta}$.

Now we must make one more restriction. The scale velocity $\Delta v = \sqrt{\beta}\, v_\phi$ should not exceed the speed of sound in the medium. We will cap the scale velocity at a maximum of $\sqrt{\alpha}\, c_{\rm s}$ where $\alpha \approx 1$. We then find that $\Delta s = \sqrt{\alpha}\, c_{\rm s} \frac{s}{v_\phi}$ and $\nu = \alpha c_{\rm s}^2 \frac{s}{v_\phi}$ which, in purely Keplerian disks, reduces to $\nu = \alpha h c_{\rm s}$. In disks with high sound speeds or low angular momentum, the β- viscosity will appear as $\nu = \gamma c_{\rm s} \sqrt{s\, h}$.

3. Discussion and Conclusions

The new prescription for the viscosity was, for the purpose of this paper, used in a simple one-zone model of an accretion disk; specifically for accretion disks around massive black holes such as is believed to lie at the center of our Galaxy. And, since we would like to show that in the limit of non-massive disks, this viscosity provides the same result as a one-zone model of an α-disk, we provide comparisons between these two models in all of our figures. It is important to note here that we are not implying that magnetically-induced viscosity is unimportant. We are not saying anything about magnetic viscosity at all, rather, we are concentrating on two things: *1)* gaining a better understanding of the physical processes which makes the "mechanical" viscosity the way it is and *2)* what effect does this change in understanding, based on observable physics, have on those models which already exist.

In Figure 1, for example, we compare solutions to the α- and β-models by showing the central temperature $T_{\rm c}$ as well as the effective Temperature $T_{\rm eff}$ as a function of radius. Those calculations that were performed with a pure α-disk are labeled as such throughout this paper. The scale on the right hand side of the figure shows the relative mass of the disk in comparison to the mass of the black hole. Differences begin to appear as soon as the disk mass is $\sim$ one hundredth of the central mass, the mass of the black hole. This simple figure already shows large differences due to the viscosity prescription. The most marked of these is the temperature at high radii s. Namely, it can be shown that massive disks with an α-prescription must have a *constant* effective temperature (Duschl et al. 1998)—all the way out to infinity: an intuitively impossible result. We see this constant temperature at high s in Fig. 1 for the α-viscosity, but not for the viscosity described in Section 2 of this paper (which, for simplicity, we will call the β-viscosity throughout the rest of the paper, even if that nomenclature is mildly misleading).

Similar comparisons, though less spectacular, can be made when looking at the surface density Σ as a function of the mass accretion rate. In Figure 2 we see both the surface density and the viscosity ν as a function of $\dot{M}$. In this figure, the viscosity is shown primarily in the bottom half of the graph, though in both plots the one-zone model using a traditional α-viscosity is shown as long dashed lines.

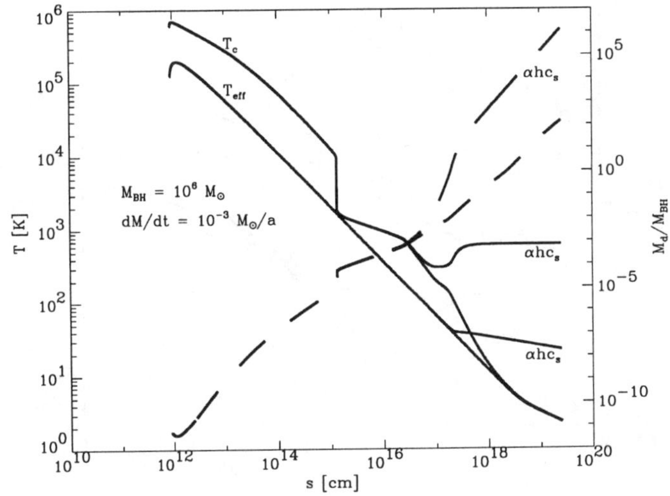

Figure 1. Comparison: Pure standard α-disk vs. new prescription. With the standard prescription, we expect the temperature of the disk to be independent of the radial distance as soon as the disk becomes massive. This is clearly seen here for the α-viscosity. The new prescription leads to much more acceptable results.

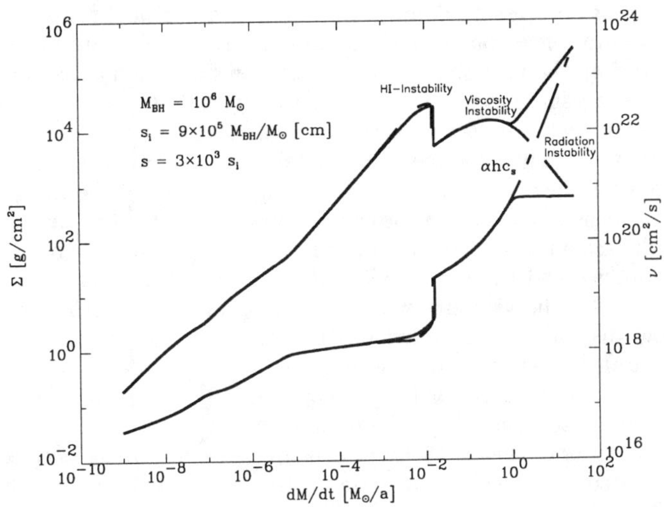

Figure 2. Solid lines: β-viscosity. Dotted lines: α-viscosity. In the top half, the column density Σ is shown as a function of accretion rate $\dot{M}$. Instabilities are possible whenever multiple accretion rates exist for a certain Σ. In the bottom half, the viscosity for the models are shown.

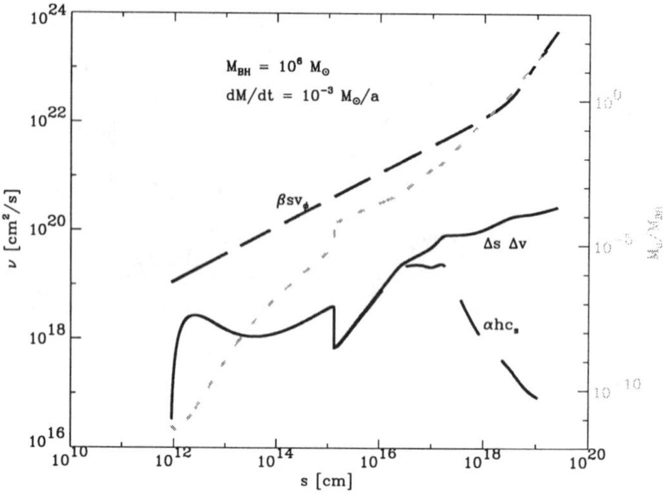

Figure 3. Model of a single accretion disk. Here, the angular velocity is always supersonic, so that Δv is always $\sqrt{\alpha}\, c_{\rm s}$. The dotted line is the ratio of disk mass to stellar mass and the dashed lines indicate viscosities of pure β- or pure α-viscosities.

The point made here is very subtle: the "radiation"-instability which appears when the radiation pressure dominates over the mechanical pressure and which has caused quite a few concerns, does not arise in the new prescription. In the new prescription, the viscosity becomes "sub-sonic" at high accretion rates, and the radiation pressure never gets a chance to dominate the structure of the disk.

In order to get a better picture of the behavior of the viscosity, we have included Figure 3. This is a one-zone model of an accretion disk with a typical central mass for the objects we are examining (10^6 $M_\odot$). The ratio of the disk mass to the mass of the central object is shown with a short-dashed gray line, for which the scale is shown on the right side of the graph. The pure α-viscosity falls off at high s and the pure β-viscosity remains at a "supersonic" and thus unphysical level. The viscosity which was used in the model is labeled $\Delta s\, \Delta v$ and, at low disk masses is equivalent to the pure α-viscosity, from which it deviates at a disk mass of $\sim 10^{-3}$ $M_{\rm BH}$.

When the central accretor is very massive (10^{10} $M_\odot$), the mass of the disk does not contribute as much to the dynamics of the system (see Figure 4). It is interesting to note that we seldom end up using $\nu = \beta s v_\phi$ because we often find that the sound speed is low. Interestingly enough, we have found disks in which the disk is not yet massive, but $c_{\rm s}$ is so large (and accordingly, in the 1-zone model, the scale height h) that the β-viscosity becomes important. However, around massive black holes with high accretion rates we find that the β-viscosity is the dominant factor at the high end of the accretion rates, where the β-viscosity works against the radiation pressure instability.

All of these models were calculated as one-zone models and show the typical uncertainties of such models such as that the HI-instabilities appear as jumps in

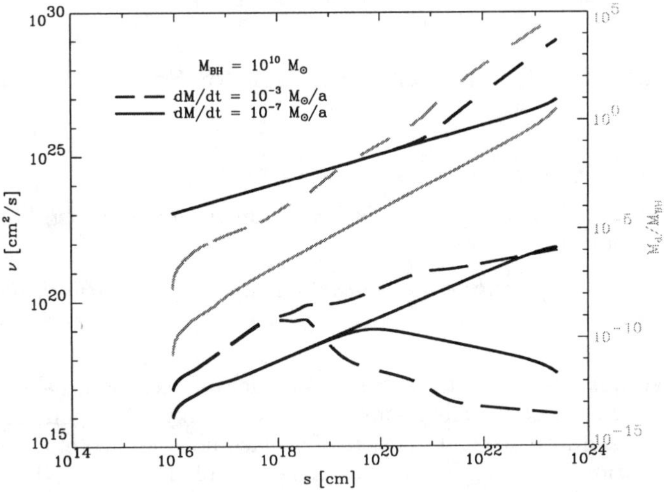

Figure 4. In accretion disks with very massive central objects, one finds no instabilities. The mass of the disk, however, still grows comparable to the central mass, as can be seen in the grey lines, which denote the ratio of disk mass to central mass for the two models. As in the previous figure, we only show the viscosities as a function of radius, the top depiction being a pure β-disk, the bottom a pure α-disk.

the parameter space (these jumps can clearly be seen in three of the four figures shown). For such cases, a more detailed numerical model would be advisable.

References

Duschl, W.J., Strittmatter, P.A., Biermann, P.L. 1998, AAS Meeting, 192, # 66.17; ApJ(submitted)

Shakura, N.I., Sunyaev, R.A. 1973, A&A, 24, 337.

Taylor, G.I. 1936a, Proc. Roy. Soc. Lond., 157, 546

Taylor, G.I. 1936b, Proc. Roy. Soc. Lond., 157, 565

The Central Parsecs of the Galaxy
ASP Conference Series, Vol. 186, 1999
H. Falcke, A. Cotera, W.J. Duschl, F. Melia, M.J. Rieke, eds.

Disks with Jet, ADAF, or EDAF for Sgr A*

A. C. Donea[1], H. Falcke[2] and P.L. Biermann[2]

[1] *Astronomical Institute of the Romanian Academy, Cutitul de Argint, 5, Bucharest, Romania*

[2] *Max Planck Institute of Radioastronomy, Auf dem Hügel, 69, Bonn, Germany*

Abstract. We investigate various models of accretion disks for Sgr A*, one of the most puzzling sources in the Galaxy. The generic image we have taken into account consists of a black hole, an accretion disk, and a jet. Various accretion models are able to explain the low NIR flux of Sgr A*: a standard accretion disk with a jet, an ADAF, or an EDAF (Ejection Dominated Accretion Flow) model. We find that all of these models are conceptually similar. The accretion model which allows the formation of the jet at the innermost edge of the disk requires a sub-keplerian gas motion and a very large base of the jet. The large base of the jet may be unrealistic for Sgr A*, since the jet model and the observations suggest that the jet is collimated and anchored in the disk in a very narrow region of the disk close to the black hole. Alternatively, one can think of a jet plus wind model (EDAF), where most of the energy goes out without being dissipated in the disk. The model resembles the ADAF model at small radii. At large radii the energy is ejected by a wind.

1. Introduction

As it is well known the Galactic Center harbors a black hole candidate of $(2.5 \pm 0.4) \cdot 10^6 M_\odot$ mass (Eckart & Genzel 1997). The accretion rate inferred from Bondi-Hoyle accretion of stellar winds should be small and should be of order $10^{-4} \dot{M}_\odot$ (Genzel et al. 1994, Melia 1992), but the exact number will depend critically on the distribution of stellar wind sources in the Galactic Center (Coker & Melia 1999). What is peculiar about the central part of our Galaxy is that, for a radiative efficiency of 10% of the accretion flow, the accretion luminosity should be higher than 10^{40} erg/s. This directly contradicts the bolometric luminosity of $\sim 10^{37}$ erg/s inferred from observations. This means that a standard accretion disk does not work. To have such a low luminosity the accretion disk in Sgr A* has to be radiatively deficient.

There are several accretion models which in principle can produce such low luminosities: ADAF (advection dominated accretion flow) models (Narayan, Yi & Mahadevan 1995, Narayan & Yi 1994), disks driving jets from the innermost regions (Donea & Biermann, 1996) or an EDAF (ejection dominated accretion flow) model (see below). All models provide a low luminosity which can explain the data of Sgr A*.

As it has been emphasized by Falcke & Biermann (1995, 1999) a symbiosis between the accretion disk, jet, and the central black hole has become a widely accepted model for active galactic nuclei and recently also for some of the micro-quasars. The strong radio emission in these systems comes from relativistic jets emitting synchrotron radiation. Based on the jet–disk symbiosis, Falcke (1999) has explained the radio emission in Sgr A* as being due to a radio jet. The model fits well all current radio observations. In the following we will make use of this model and the jet-disk symbiosis idea to calculate a number of accretion disk models that would be consistent with the observed properties and upper limits of Sgr A*.

2. Accretion Disk with Jet

Donea & Biermann (1996) have put forward an accretion disk model with a jet starting at the inner regions of a disk. The jet extracts energy, mass and angular momentum from small radii of the disk. They have written the conservation laws of mass and angular momentum taking into account the extraction of mass, angular momentum and energy by the jet (Newtonian treatment). The new conservation mass law is:

$$\dot{M} = -2\pi R \Sigma u^R + \dot{M}_{jet} \tag{1}$$

where $\dot{M}$ is the rate of mass accretion rate in the disk. The mass flow rate into the jet is $\dot{M}_{jet}$. The angular momentum conservation law is:

$$2\pi R \Sigma v_R(R)\ R^2\Omega = 2\pi R \nu \Sigma R^2 \Omega' + \dot{M}_{jet} v_\phi(R_{jet}) R_{jet} + C \tag{2}$$

ν is the kinematic viscosity of the disk. R_{jet} is the radius where the base of the jet starts. The constant C is related to the boundary conditions of the accretion flow (Frank et al. 1985).

We assume that the jet starts between $R_{\rm ms}$, the last marginally stable orbit in the absence of a jet and the radius $R_{\rm jet}$ with $R_{\rm jet} > R_{\rm ms}$. (Donea & Biermann 1996). The presence of the jet will modify both the behaviour of the infalling matter across the radius $R_{\rm jet}$ and the structure of the relativistic disk.

The local energy dissipation in the disk becomes:

$$D^*(R) = \frac{3GM\dot{M}}{8\pi R^3}\left[(1-q_m) - (1-q_m)\left(\frac{R_{jet}}{R}\right)^{1/2}\right] \tag{3}$$

where

$$q_m = \frac{\dot{M}_{jet}}{\dot{M}}, \tag{4}$$

As we mentioned before, the gravitational potential energy released between $R_{\rm ms}$ and the outer radius of the jet, $R_{\rm jet}$ goes into the jet. Therefore, the total energy carried outwards by the jet is strongly dependent on the mass and angular momentum of the black hole. Q_{jet} is the total power of the jet – including the rest mass energy of the expelled matter – and is expressed as:

$$Q_{jet} = L_{disk} - L_{disk}^{jet}$$

L_{disk}^{jet} is the total luminosity of a disk modified by the presence of the jet and L_{disk} is the total luminosity of the disk if there are no physical conditions to drive the jet.

The power of the jet depends on the the way in which the jet gets energy from the black hole/disk system. Total power of the jet and the radio activity of the AGN are probably directly connected to the inner geometry of the system near the event horizon. Adopting the point of view that formation of jets is possible under all circumstances (Falcke & Biermann 1995), be it for an AGN or a stellar size system, we investigate the effects of the presence of the jet on the structure of the accretion disk, and implicitly on its emission spectrum.

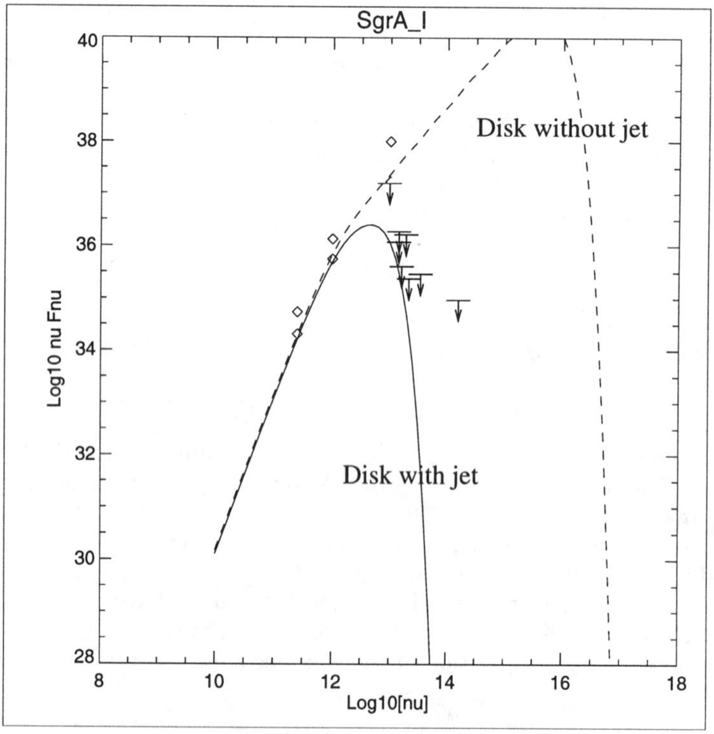

Figure 1. The emission from an accretion disk with a jet starting at large radii in the disk. $M = 2.5 \cdot 10^6 M_\odot$; Kerr black hole: $a_\star = 0.9982$; $\dot{M} = 2 \cdot 10^{-6} M_\odot/yr$; fitting the data below the NIR requires an extremelysh large base of the jet.

We applied the above disk–jet model to the case of Sgr A*. The spectrum from the disk is below the NIR flux if one considers the emission from a thin disk surrounding a Kerr black hole of $2.5 \pm 0.4 \cdot 10^6 M_\odot$. In Figure 1 we show a spectrum from a disk with jet (the dashed line) and without the jet (continuous line). As it is expected from the standard models (Shakura & Sunyaev (1973), Novikov & Thorne 1973) the standard model does not fit data at all. If one takes into account the jet, then the model is completely different: the high energy part of the spectrum is cut off due to the extraction of energy by the jet from small radii.

The parameters of the model (see Figure) are :

- low accretion rate: $\dot{M} = 2 \cdot 10^{-6} M_{\odot}/yr$.
- black hole mass: $2.5 \cdot 10^{6} M_{\odot}$.
- Kerr black hole: $a_{\star} = 0.9982$.

Interestingly, the NIR limits are now so stringent that fitting the data with such a model would require an extremely large base of the jet, from $1.23 R_{\rm g}$ to $7000 R_{\rm g}$. The outer radius of the disk is $2 \cdot 10^{5} R_{\rm g}$. The luminosity of the disk is $L_{\rm disk} = 3.7 \cdot 10^{36}$ erg/s and Q_{jet} takes out 99.5% of the total gravitational energy.

This is not a realistic situation for Sgr A*, since we know that the radio emission is much more concentrated. Nevertheless, these results point into two very interesting directions for modifying accretion disk theory. For one, the disk model with large fraction of energy extraction resembles an ADAF model (Narayan & Yi, 1994) in various points:

a) the region between the radius of the last marginally stable circular orbit R_{ms} and the radius where we consider to be the outer radius of the jet R_{jet} no longer shows a standard disk-like structure;

b) the flow becomes subkeplerian in the region where the jet is formed;

c) at these radii there is no radiative cooling any more of the flow—the energy is advected in a jet instead of being radiated away.

3. The Ejection Dominated Accretion Flow

A second way to think of our results, is that for such a large region where energy extraction is required one can hardly speak of a 'jet' any more. Rather, one could think of a continuous process, where a magnetically driven wind across the accretion disk would extract a large fraction of the dissipated energy. The jet would then only be the most highly collimated and fastest feature in such a wind.

Like in the ADAF model, the energy of the disk would no be radiated away, but advected away. Unlike the ADAF model, however, the channel through which the energy is advected would no longer be the disk, but rather the wind. In light of these connections with the existing ADAF theory we have labeled our approach EDAF—ejection dominated accretion flow.

Indeed, the fact that accretion disks are able to expel a large fraction by, most likely magnetically driven, jets opens up the possibility that a related process could also drive large scale winds from an accretion disk. Since it is often suggested that the viscosity and dissipation in accretion disks is also a result of magnetic fields, a deeper connection between viscosity, dissipation, and wind ejection may exist. In cases where wind ejection dominates the energy dissipation, a substantial part of the accretion energy could be extracted by the wind from the accretion disk. In this case the large radius of $7000R_{\rm g}$ we obtained earlier may be the radius where a significant wind coexists with the disk (EDAF models). As pointed out in Falcke (1999) application of the jet model indeed supports a non-standard, radiatively deficient accretion disk in Sgr A*. The EDAF model would be a situation where the wind goes off from the accretion disk at every radius and where at each radius a fraction of the available energy is put into a wind.

Falcke & Melia (1997) solved the accretion disk equation for the case of deposition of mass and angular momentum onto the disk by an infalling wind. The reverse case, when the wind takes out mass and angular momentum, would be an EDAF case. The equations remain the same, only the sign of the mass loss/gain rate $\dot{\Sigma}_{\rm w}$ at each radius changes.

We have used the code described in Falcke & Melia (1997) to simulate such a situation. The important parameter for these calculations is the energy extraction efficiency: we assume that at each radius a fraction $f_{\rm EDAF}$ of the locally dissipated energy is put into kinetic and internal energy of the wind. The wind is assumed to leave the disk with escape speed and with the specific angular momentum and internal energy corresponding to the local conditions in the disk. The local mass loss rate $\dot{\Sigma}_{\rm w}(r)$ is then completely determined by $f_{\rm EDAF}$.

To facilitate the connection to the jet model we also assumed that the wind extraction efficiency approaches unity at the inner $12R_{\rm g}$, i.e. all the energy is used to drive the jet modeled in Falcke (1999). Because of the continuous mass loss throughout the disk, the accretion rate at small radii and the available energy is significantly reduced with respect to a standard disk model. Moreover, the emitted radiation spectrum will also look markedly different (i.e. flatter) from a standard α−disk.

The F_ν emission spectrum from an EDAF we have calculated for the situation in Sgr A* is shown in Figure 2. The bump at low frequencies is due to heating of the disk by the surrounding stars, which we had to take into account as well. The spectrum lies below the NIR limits. In Figure 2 the radio spectrum from the jet assumed to be produced at the innermost radii by the remaining accretion flow is also shown. The great disadvantage of such a model is that $f_{\rm EDAF} = 99.995\%$ of the locally produced energy has to go into a wind rather than into heat for an initial accretion rate of $\dot{M} = 10^{-6}M_\odot/\text{yr}$. Even though we do not have any arguments for how large the value of $f_{\rm EDAF}$ really has to be, this seems uncomfortably high. On the other hand this high 'inefficiency factor' is very similar to those required for ADAF models. In both cases one wonders whether such values are physically meaningful. Perhaps various combinations of the models explored here (e.g. ADAF+EDAF) could lead to a more convincing solution to the inefficiency problem, or the mass accretion rate on Sgr A* is significantly lower than the already low value used in our calculations.

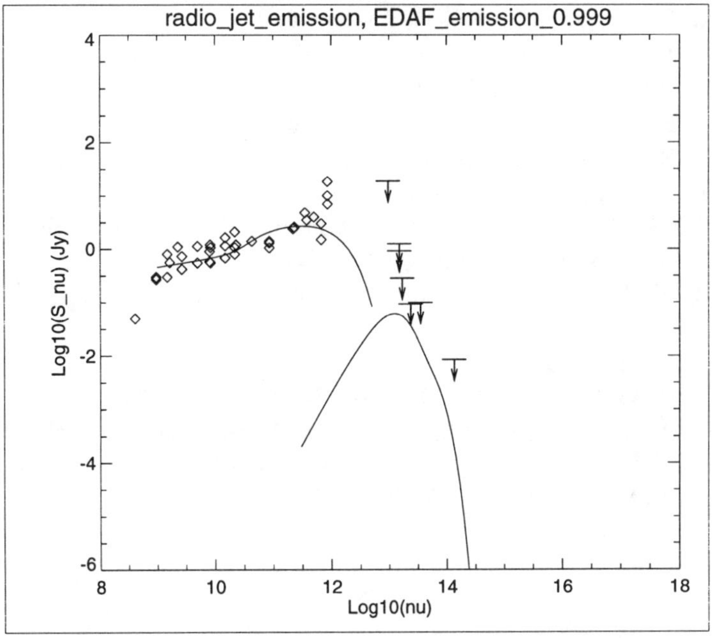

Figure 2. The radio emission and the spectrum from an EDAF (Ejection Dominated Accretion Flow) model of Sgr A*.

4. Conclusions

We conclude that there are in principle various ways to explain the low NIR flux and that all the models (jet, EDAF, ADAF) are conceptually very similar. However, each of these models seems to require rather extreme parameters to explain the inefficiency of the accretion flows.

The basic picture for the very central part of our Galaxy, explored here, consists of an accretion disk, a black hole and a jet. The accretion disk may be an EDAF or a "standard" disk at large radii. The angular momentum and part of the dissipated energy goes out with the wind. Perhaps at small radii, the disk becomes an ADAF with a jet. The jet–disk model allows the formation of an jet at the innermost radii and this would automatically produce a subkeplerian, disipationless flow similar to an ADAF. In any case a large fraction of the energy produced in the accretion disk in Sgr A* has to be either advected into the black hole or ejected by a wind or jet.

Acknowledgments. A. C. Donea is grateful to Dr. Angela Cotera and Sera Markoff for making her attendance to this conference possible. HF is supported by DFG grant Fa 358/1-1&2.

References

Coker, R., Melia, F. 1999, this volume, p. 214

Donea A. C., Biermann P.L. ,1996, A&A, 316, 43

Eckart A., Genzel R., 1997, MNRAS, 284, 576

Falcke, H., 1999, this volume, p. 148

Falcke H., Biermann P.L., 1995, A&A 293, 665

Falcke H., Biermann P.L., 1999, A&A 342, 49

Falcke H., Melia F., 1997, ApJ 479, 740

Frank J., King A.R., Raine D.J., 1985, Accretion Power in Astrophysics, Cambridge University Press, Cambridge, 1985

Genzel R., Hollenbach D., Townes C, 1994, Rep. Prog. Phys., 57, 417

Melia F. 1992, ApJ, 387, L25

Narayan, R., Yi, I., 1994, ApJ, 428, L13

Narayan, R., Yi, I., Mahadevan, R., 1995, Nature, 374, 623

Novikov I.D., Thorne K.S., 1973, Black Hole Astrophysics, in: DeWitt C., DeWitt B. (eds.), Les astres occlus, Gordon & Breach, New York, p.343-450,

Shakura N.I., Sunyaev R.A., 1973, A&A 24, 337

The Central Parsecs of the Galaxy
ASP Conference Series, Vol. 186, 1999
H. Falcke, A. Cotera, W.J. Duschl, F. Melia, M.J. Rieke, eds.

Acceleration Processes Towards a Quasi-Monoenergetic Electron Distribution

Zeno Jauch[1], Wolfgang J. Duschl[1,2]

Institut für Theoretische Astrophysik, Universität Heidelberg, Tiergartenstraße 15, 69121 Heidelberg, Germany

Abstract. We present the results of an investigation of acceleration processes in galactic centers and show that a quasi-monoenergetic electron distribution (pile-up) forms in quite a natural way under the combined action of diffusive shock wave acceleration, stochastic acceleration on turbulent MHD waves and radiation losses. We follow the temporal evolution of both, the electron and the photon distribution. This requires the numerical solution of two kinetic equations, i.e., a fully coupled set of nonlinear partial integro-differential equations. We apply the results to the non-thermal radio emission from the Galactic Center source Sgr A*, which we interpret as being optically thin synchrotron radiation from these quasi-monoenergetic electrons.

1. Introduction

The radio spectrum of the Galactic Center source Sgr A* has three characteristic features: a flux $S_\nu \propto \nu^{1/3}$, a high frequency cut off at $\nu \approx 1000$ GHz, and a low frequency turn over at $\nu \approx 1$ GHz. It has been shown by Duschl & Lesch (1994) and Beckert et al. (1997) that this spectrum can be interpreted as being optically thin synchrotron radiation from a quasi-monoenergetic distribution of electrons.

Consequently, the next question that has to be asked is: How can a quasi-monoenergetic electron distribution be achieved? The answer is, that this can be done using a combination of acceleration processes and radiation losses (Schlickeiser 1984).

Consider a spherically symmetric source, where a magnetized plasma is accreted into a black hole. In most accretion scenarios (Bondi-Hoyle, accretion disk) shock waves and turbulent MHD waves (e.g. Alfvén waves) will be present, so that electrons gain energy through diffusive shock acceleration (first order Fermi process) and diffusion in momentum space (second order Fermi process). The energy gain is limited because of the unavoidable synchrotron losses. .

The crucial point for the formation of a quasi-monoenergetic distribution of electrons, a so called pile-up, is that the time scale for energy gain through shock

[1]Sonderforschungsbereich 328 Evolution of Galaxies, Universität Heidelberg, Germany

[2]Max-Planck-Institut für Radioastronomie, Auf dem Hügel 69 53121 Bonn, Germany

waves $t_{\rm acc}$ must be small compared to the mean residence time of the electrons inside the source $t_{\rm esc}$.

2. The Model

The consistent treatment of electron acceleration and the subsequent radiation requires the solution of two coupled kinetic equations. We use spatially averaged kinetic equations, which implies that all the physical processes occur at any place inside the source. The electron and photon distributions are then described by the spectral number densities $n_{\rm e}(\gamma, t)$ and $n_\gamma(x, t)$, where γ is the electron Lorentz factor, x the photon frequency normalized to $m_{\rm e}c^2/h$ and t the time. The kinetic equations for the two distributions including electron injection, acceleration, escape from the source, synchrotron radiation and synchrotron self-absorption read:

$$\begin{aligned}
\frac{\partial n_{\rm e}(\gamma,t)}{\partial t} &= \frac{\partial^2}{\partial \gamma^2}\left\{\frac{\gamma^2}{t_{\rm diff}} n_{\rm e}(\gamma,t)\right\} - \frac{\partial}{\partial \gamma}\left\{\left(\frac{4}{t_{\rm diff}} + \frac{1}{t_{\rm acc}}\right)\gamma n_{\rm e}(\gamma,t)\right\} - \\
&\quad - \frac{n_{\rm e}(\gamma,t)}{t_{\rm e}^{\rm esc}} + Q_{\rm e}^{\rm inj} - L_{\rm e}^{\rm syn} \qquad (1)\\
\frac{\partial n_\gamma(x,t)}{\partial t} &= -\frac{n_\gamma(x,t)}{t_\gamma^{\rm cross}} + Q_\gamma^{\rm syn} - L_\gamma^{\rm ssa} \quad . \qquad (2)
\end{aligned}$$

In these equations, Q describes source and L loss terms, the diffusion of the electrons in momentum space due to the second order Fermi process is described through the time scale $t_{\rm diff}$. The mean residence time of the photons is given by the light crossing time $t_{\rm cross} = R/c$, where R is the source radius. The numerical solution of the above system of equations is done using a so called method of lines (Schiesser 1991), which means that the partial differential equations are semi-discretized in x and γ and the resulting system of stiff ordinary differential equations is integrated forward in time using a semi-implicit Euler method with extrapolation (Deuflhard 1984).

3. Results

The figures 1 - 5 show the results of a run of our code. The following input parameters have been chosen:

source radius	$R = 3 \cdot 10^{13}$ cm
magnetic field strength	$B = 4.5$ G
acceleration time scale	$t_{\rm acc} = 131\, t_{\rm cross}$
diffusion time scale	$t_{\rm diff}/t_{\rm acc} = 12$
mean residence time	$t_{\rm esc} = 10^6\, t_{\rm cross}$

At $t = 5\, t_{\rm esc}$ the steady state is reached and a pile-up has formed. Figures 1 and 2 show the evolution of the electron and photon distribution towards

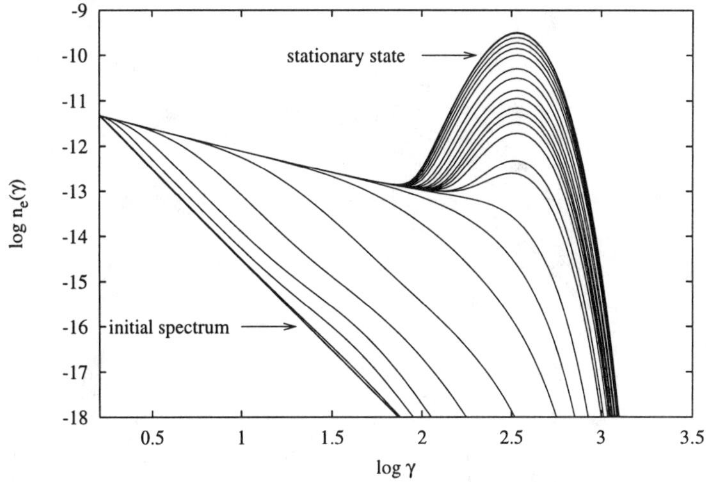

Figure 1. Evolution of the electron distribution $n_e(\gamma, t)$ towards steady state. To start the integration with a non-empty grid, the initial distribution was chosen to be a power law with spectral index -4.

steady state. The evolution of the photon distribution at four specific frequencies ($x = 8 \cdot 10^{-12,-11,-10,-9}$, labeled (1), (2), (3) and (4), respectively) as a function of time is given in figure 3, while figure 4 shows the spectral index of $xn_\gamma(x,t)$ at $x = 3 \cdot 10^{-10}$. Last, but not least, figure 5 shows a fit of the steady state spectrum to the radio data of Sgr A*, where the numerical solution is plotted together with an analytical solution, that can be obtained from the radiative transfer equation in the case of a homogeneous and isotropic source which contains a strictly monoenergetic distribution of electrons ($n_e \propto (\gamma - \gamma_0)$). The differences in the high and low frequency part of the spectrum are due to the finite width of the pile-up in the numerical solution. In the steady state, the electrons have a density $N_e = 3.5 \cdot 10^3$ cm^{-3}, the pile-up lies at $\gamma_{max} = 365$ and the energy density of the relativistic electrons u_e is roughly in equipartition with the energy density of the magnetic field u_B: $u_e/u_B = 1.3$.

4. Discussion

We have developed a code that is capable of following the temporal evolution of a distribution of relativistic electrons under the combined influence of energy gains (diffusive shock acceleration and stochastic acceleration on turbulent Alfvén waves) and losses (synchrotron radiation and escape from the source). If the electrons are confined long enough ($t_{esc} >> t_{acc}$) inside the acceleration region, a pile-up of electrons at a certain Lorentz factor γ_{max} can form and the resulting spectrum of synchrotron photons has the characteristic flux $S_\nu \propto \nu^{1/3}$. Applied to the Galactic Center source Sgr A*, we obtain a good fit to the radio data and source parameters in agreement with observations and previous

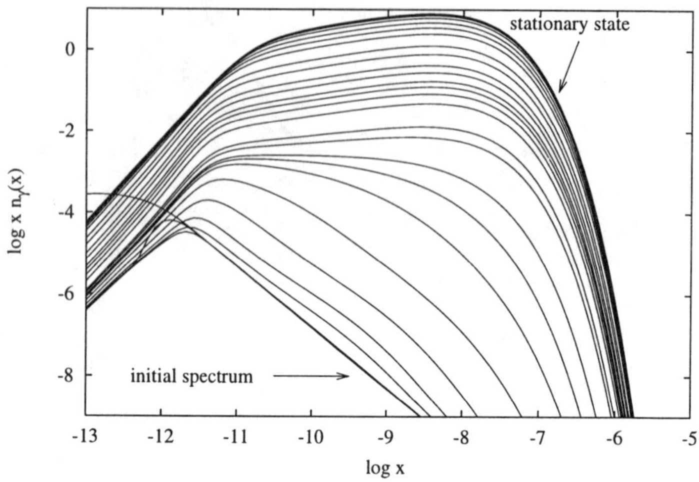

Figure 2. The temporal evolution of the corresponding synchrotron photon distribution, plotted as $xn_\gamma(x,t)$. The initial values $n_\gamma(x,0)$ are just the synchrotron photons emitted by $n_e(\gamma,0)$.

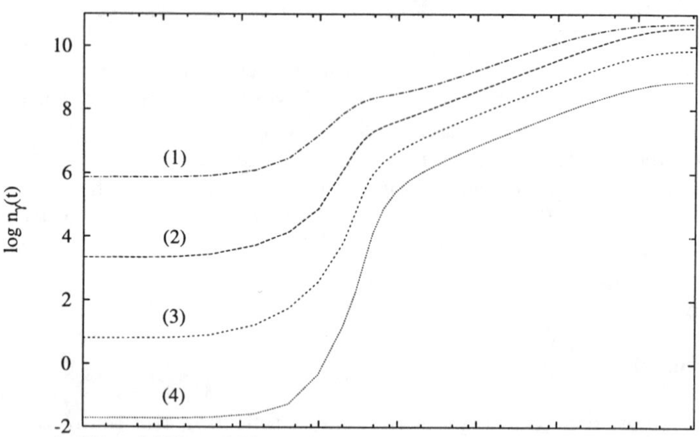

Figure 3. The photon distribution $n_\gamma(x,t)$ as a function of time at four specific frequencies, namely $x = 8 \cdot 10^{-12,-11,-10,-9}$ (curves (1), (2), (3) and (4) respectively).

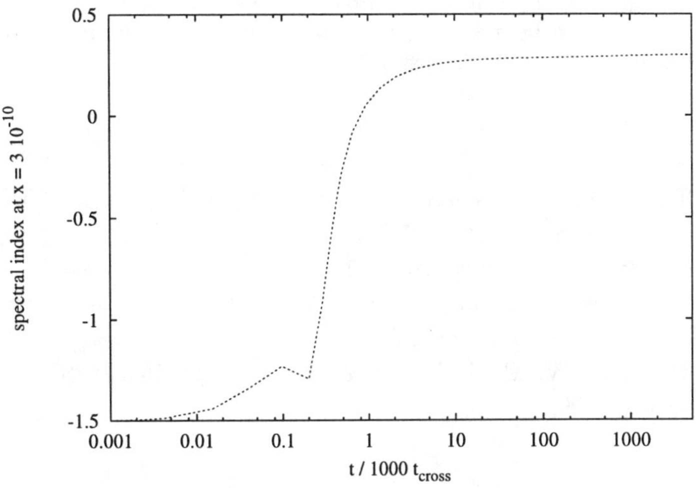

Figure 4. The spectral index of $xn_\gamma(x,t)$ at $x = 3\cdot10^{-10}$ as a function of time.

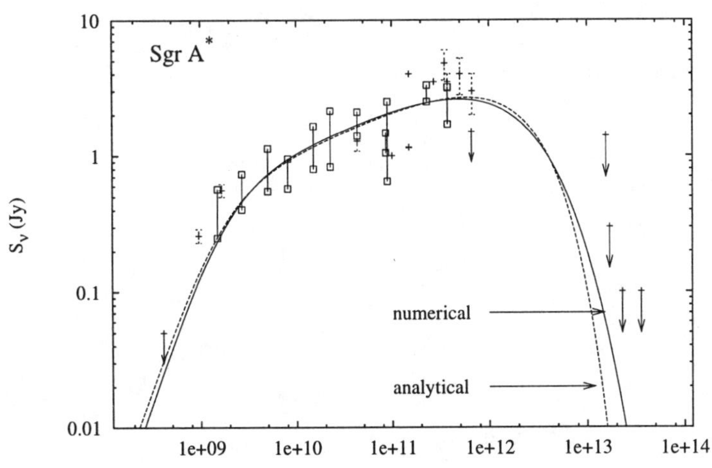

Figure 5. The observed radio spectrum of Sgr A* together with the best fit spectra resulting from a quasi-monoenergetic (numerical) and a strictly monoenergetic (analytical) electron distribution, with $\gamma_0 = \gamma_{\max}$.

theoretical work. Our code is well suited for the investigation of time-dependent phenomena like the radio outbursts observed by Zhao et al. (1992) and can easily be extended to include further radiation processes like inverse Compton scattering and non-thermal Bremsstrahlung.

References

Beckert, T., Duschl, W.J., Mezger, P. G. & Zylka, R. 1996, A&A, 307, 450

Deuflhard, P. 1984, SIAM Rev., 27, 505

Duschl, W.J. & Lesch, H. 1994, A&A, 286, 431

Schiesser, W.E. 1991 The Numerical Method of Lines, Academic Press Inc.

Schlickeiser, R. 1984, A&A, 136, 227

Zhao, J.-H., Goss, W. M., Lo, K.-Y. & Ekers, R. D. 1992, ASP Conference Series, 31, 295

The Central Parsecs of the Galaxy
ASP Conference Series, Vol. 186, 1999
H. Falcke, A. Cotera, W.J. Duschl, F. Melia, M.J. Rieke, eds.

Discussion Session III: How to Reconcile Models and Observations of Sgr A*?

R. Genzel[1]

Max-Planck-Institut für extraterrestrische Physik, Postfach 1603, 85740 Garching, Germany

H. Falcke[2]

Max-Planck-Institut für Radioastronomie, Auf dem Hügel 69, D-53121 Bonn, Germany

Abstract. This is an edited transcript of a discussion session on theoretical models for Sgr A*, the compact radio core and suspected black hole candidate at the Galactic Center. The session took place during the Galactic Center conference "The Central Parsecs" in Tucson, Arizona, Sep. 7-11, 1998. It was chaired by the moderator as indicated in the author list. The discussions were taped, transcribed, and finally edited by an editor who is listed as co-author. The contributions of the conference participants are preceded by their names and thus references to specific aspects and ideas of this discussion session must also include a reference to the respective speaker(s), e.g., in a footnote.

Discussion

Narayan: I was not going to give any talks, but a couple of issues were discussed the last two days, and I thought I'd just very briefly try and expand on two points. One is shown on my viewgraph. The top band shows our standard model, the ADAF model of Sgr *, with all the radio data, and upper limits in infrared and X-rays. As Rohan [Mahadevan] emphasized, one of the critical elements in this model is that all the heat energy in the gas—the energy that is not being radiated—disappears through the black hole when it reaches the center. Now, if you didn't have a black hole, but you had some other object with a surface that energy would then get re-radiated from the surface and come out somewhere in the spectrum. Then you could ask, where would it appear? Now you all can decide what size you want, and I decided to put a star with a radius of 3 Schwarzschild radii in there—something like a big neutron star. This would be like a $2 \cdot 10^6 M_\odot$ neutron star, with a radius of a few times the Schwarzschild radius. Then you take all that directed energy, thermalize it on the surface, and assume

[1]Moderator of the discussion session

[2]Editor of the discussion session

that it comes out as a black body. That's another assumption. We don't know what spectrum we could have. Let's assume it's black body. You all know what's going to happen. The efficiency is suddenly going to jump up four orders of magnitude. So if you take the same accretion rate that you have in our model, you get a spectrum that's shown by the dark line in this viewgraph. It's got a peak way up in the optical, and it violates quite severely the infrared limits. Incidentally, I think Susan Stolovy's upper limit would come a lot lower than what's drawn here. You do have to reduce the accretion rate by more or less five orders of magnitude, before you can satisfy the infrared limits. I think it just emphasizes how important these infrared limits are. They constrain any surface that is not a black hole. It also constrains, as Falcke and Melia have shown, any thin-disk component. The moment you put a thin disk in this system, it gives a whopping component, right here in the infrared. The fact that the limit is so low constrains any disk component quite severely—that is if the disk is actually radiating. If, like Heino [Falcke] is saying now, you have a completely cold disk, and all the energy goes into a jet/wind, then you can do anything you want with the disk.

The second topic I wanted to talk about, is this business of having winds from the accretion flow. In the case of an ADAF the wind is fairly natural, and the reason is straightforward: the gas is not radiating. It's got all this thermal energy, so in fact, it's similar to having zero binding energy. And as we showed in our very first paper, it's not even zero, it's actually positive binding energy, due to the viscosity. The gas has actually got a positive energy as it's flowing in, which means it can very easily escape from the system. So you would expect quite massive winds, not just from the center, but all the way down the ADAF. Starting from the outside, matter will be going out everywhere and less and less of the mass might accrete to the center. This is something we have never put into any of our models, but Blandford and Begelman have now said, we'd better start doing this. They say that winds could be so severe that the mass accretion rate may be a very serious function of radius: going as the radius to the power of p, and p may be as large as unity. That's what they speculate. Of course we don't really know what the value of p may be.

We also heard from Heino [Falcke] about the EDAF [ejection dominated accretion flow]. More or less the same idea: a big amount of energy going out in the jet or wind, and not much going into the black hole.

Now what I wanted to show was how this affects our model. We can make the accretion rate a function of radius: a function of r^p for different choices of p. Let's make an assumption. Let's assume that whatever flows out doesn't produce any radiation, so we only see the accretion rate going down—that we know how to do. As far as the jet is concerned, we don't have good models for how the jet might radiate, and we just ignore it completely. Then we can calculate spectra, shown in the diagram on the viewgraph. The dotted line is a no-wind model, you can see, as you turn on the wind—fairly modest winds with $p = 0.2, 0.4, 0.6$, the radio completely disappears. So the one nice thing that the ADAF model is doing, explaining this millimeter excess and its peak in the right place, is gone completely. Basically you'll be left with having to explain the radio by some other model: a jet or whatever else.

I would say that this rules out winds, but there may be a way out. There is one other parameter in these models which we normally don't touch at all, and which Rohan [Mahadevan] didn't even mention. It's built into the model, where we say that most of the viscous energy goes into the protons, and hardly anything goes into electrons. So, there's a little parameter δ, which we set equal to zero, or 10^{-2}, or some small number: δ is what fraction of energy goes into the electrons. What if it is not small? Maybe half the energy goes into the electrons, half goes into the protons, or some other number in between. Would that change the results? The answer is yes. You take one of these strong wind models, e.g. with $p = 0.4$, and if you make δ small, it will give you a spectrum where the radio is completely in disagreement with the data. But as you tune up this δ, you can build up the radio emission. Basically you're making the electrons hotter. You make the electrons hotter because you're putting more energy into the electrons, and suddenly you find that you can get a model with a wind, with a modest δ, that fits the data as well as our old models that had no wind, and had no δ.

What does this mean? I think it means that with the available data, there is a degeneracy in these ADAF models. So far we were sitting in a region with no winds and very weak heating of electrons. But now we find that we can make a whole range of models. You can have more and more wind, and you just put more and more energy into the electrons and you can get pretty decent models all along a track between those extremes. We can say for sure that models down here don't work, they don't explain the observations. I think this gives us new opportunities for understanding winds and ADAFs, and maybe more observations can help us to discriminate along this track. But for now it makes the business a little dirty. There's a degeneracy in the models. You can explain the data with a whole family of models, not just with one.

Townes: Can you tell us how much inflow you can allow and not get too much radiation?

Narayan: These models start from a radius of 10^4 Schwarzschild radii and end at the black hole. The mass that's actually going into the black hole is a lot less when you put these winds in. What you might have thought was that the mass on the outside could go up, because we are always concerned that our accretion rate is too low. Can we increase it by a factor of 10, 50, or whatever? In the models with $p = 0.3 - 0.6$ you can ask what is the accretion rate at 10^4 Schwarzschild radii? And even in those strong wind models the accretion rate is only going up by about a factor of two or three—not much more than that. So the accretion rate is still limited to the same $2 \cdot 10^{-5}$ solar mass per year. Of course you might speculate to have an ADAF extending much further out, let us say going off to 10^6 Schwarzschild radii, with a wind all the way out from there. However, we haven't computed this yet, because I think the X-rays will give you a problem. Perhaps we can push it up a bit more. But it doesn't really allow you to increase the accretion rate as much as you might have hoped

Krichbaum: Could you state something about the geometry you would expect? How much would an orientation of the disk with respect to the line of sight affect such a model? For example, is there a completely circular geometry, or is it more a flattened geometry?

Narayan: It is flattened. But how much it's flattened depends on some of the details, depends on the adiabatic index of the gas, depends on the wind. I think, the more wind you put in, the more flattened it gets, because it becomes closer to Keplerian rotation. The reason ADAFs are normally not very flattened is that they're highly sub-Keplerian. Winds can change this. But I assume axial-symmetry, of course, so it also matters at what angle you view it. If you're viewing it edge-on, you might expect more or less flattening, maybe two-to-one. Three-to-one was mentioned [by Zhao], but this is going to be hard.

Krichbaum: So, VLBI observations with very elliptical shapes would cause a problem.

Narayan: It would fail the property checks at some level, yes. But I won't say it's the final word yet. With the winds, I don't know the answer.

Genzel: I guess it's time to put things together and try to see whether we can have a general discussion. In listening to all of this, I thought we could structure it into three basic steps. One is the input situation—basic inputs into these models. One input of course is the black hole mass, which we think we have established. Then we have the radio spectrum, we discussed in detail yesterday, and we have the X-ray flux.

Now the critical thing we heard about today is the accretion rate. Obviously, that is where you clearly can see the differences in these models. In Fulvio's [Melia] Bondi-Hoyle models, the accretion rates are 5 to 10 by 10^{-4} solar masses per year. In the ADAFs we heard that there are a few by 10^{-6} to less or equal 10^{-5} solar masses per year. And then in Heino's [Falcke] starved, accretion disk—let's call it the jet/disk model—you had even 10^{-7} solar masses per year. That's a clear, big difference, and that's something we should discuss, i.e., the question whether we do or do not agree from the basis of the observations, that any of these numbers seem plausible or not.

Of course there are other questions one might have. One is on the amount of variability one can allow, given that we think we know on average, how much accretion you should have from nearby stars. And I think in this respect we perhaps should review again Coker & Melia's detailed numerical models, because they do say something about the variability. There's the point, which was raised to me by Jerry Ostriker some time ago, which is, how do we know that there is not a hot bubble in the center? Can we exclude that there is a hot bubble which basically does not allow a lot of accretion? You basically make the gas much hotter in the vicinity of the black hole, and get the accretion down in that way. And similarly, the wind could also decrease the amount of accretion. That's the first set of things which I suggest we perhaps could discuss: What do we know about the accretion?

The next set which I would list is actually the very basic question: Do we need an event horizon? Independent of the details of these seemingly disparate models, do we need an event horizon? Do we need a black hole in all of these models in order to make it consistent with the observations? And there, we've heard today two inputs. One was by Fulvio [Melia], who basically showed us that if you make a distribution of dark objects, never mind whether it is stable or not, of the size roughly now constrained by the observations you get into trouble. That size would be a hundredth of a parsec or something like this. If you have a surface of a hundredth of a parsec, then we probably do not need an event horizon. On the other hand such a large surface would not produce enough radiation to explain Sgr A*. I could use this argument now the other way around, and say, there actually is a hypothetical configuration, which doesn't have much radiation coming from it—not even enough to explain the radio emission from Sgr A*.

Narayan: But Reinhard, where do you get 0.01 pc from?

Genzel: I'm just going to the one extreme, and saying, we might allow a dark cluster of objects, make the potential as shallow as possible. How much radiation is coming out of this? That's what I believe your model shows. And I think in this limit [of a large radius] we do not need an event horizon. And the other limit, that was Ramesh's [Narayan] point, suppose we take the radius to be about three Schwarzschild radii and there we have way too much radiation. So clearly, such a compact star, like you've just described, is not possible. Obviously, there also must be a continuum in between. And I think we might want to think about where do we go from one regime into another.

Then we come to the detailed physics of the models. And now let me try taking three subsections here. One is the Bondi-Hoyle type models, plus the question of low B-fields. Is that a realistic assumption? Do we know basic physics which speaks against this? Here we also have the issue of cancellation of angular momentum. At this point I want to ask, is everyone convinced that these models show that enough angular momentum is cancelled that you actually can create a Bondi-Hoyle type flow? Here, we have the situation that the efficiency which one is getting is marginal.

In the ADAF models, the key issue is the existence or non-existence of the two-temperature solution. Is it established that such a two-temperature solution actually can exist? Some theoreticians still tell me they don't necessarily believe this.

And then of course we had the issue of alpha versus beta accretion disk, the high-viscosity situation in the alpha disk versus the lower one used by Duschl. Moreover, I would raise the issue of disagreement of ADAF models, with observations of external galaxies, in particular NGC 4258 and some nearby ellipticals where dark masses have been observed. Like in the Galactic Center, sub-millimeter observations now exist, and differ from the Sgr A* situation very much in that there is no sub-millimeter peak, but rather show a decline, like in a non-thermal spectrum.

Finally, we have the low-efficiency, the low-accretion jet/disk models. Here my basic question would be, on the variability of $\dot{M}$, since this is the most extreme of these cases, independent of the issue of the radio spectrum and how the radio spectrum might be explained. Is it at all realistic that such a low $\dot{M}$ may occur for a long enough time that we can expect it to observe?

These are the points I wrote down while I was listening to the talks. You're welcome to add points to this, and then I would suggest we should go maybe through these basic three steps. Are there other points you would like to suggest we should add to this list?

Yusef-Zadeh: I have another point. As an observer I'd like to know if there's a frequency dependence of the centroid of Sgr A*, based on different models that exist.

Genzel: I actually forgot that, I would like to ask our theoretical colleagues to give us predictions. The prediction, you will give me in a sealed envelope, please: your best estimate and your limits for the X-ray flux to be observed by AXAF. We'll seal this, keep them, and open them when AXAF has made its first measurements. Seriously, I think the X-ray flux is a key measurement, which will become available very soon. Then, Farhad, you wish to get ...

Yusef-Zadeh: ... the centroid of the emission as a function of frequency. I think we lose it by scattering at cm-wavelengths now. With higher frequencies, we do expect to see emission coming from different places.

Genzel: Right, and then we add the other point which was discussed here, the size of Sgr A*.

Biermann: I think you may want to add the VLBI observations and the spectrum. They constrain some models—not yet, but soon.

Genzel: Yes, spectrum and size as a function of λ.

Goss: How about time variations?

Genzel: Yes. That's a good one.

Genzel: Okay, let's perhaps first discuss the accretion rate.

Falcke: I just want to make the point, that our $10^{-7} M_{\odot}/\mathrm{yr}$ is just a lower limit needed to power the radio jet.

Genzel: Okay, we're starting the discussion already, but let me just say, Fulvio [Melia] made the point that basically within a factor of 2 or so, the accretion rate is given. That's how I understood his talk. If I take this seriously, if I believe him, you guys are out.

Falcke: As I said, first of all, we need at least $10^{-7} M_\odot/\text{yr}$ to power the jet. This has to be at the inner edge of the accretion disk. So anything lower than that wouldn't have enough power to make a jet. If we would forget our egos and say, okay, look what nature can do: It is absolutely possible that you have an ADAF disk and a jet outflow. It's very well possible. And that would mean we'd have an accretion rate of $10^{-5} M_\odot/\text{yr}$ and a jet which—we don't know why—takes just a tiny fraction of the total accretion power.

Genzel: So if I understand, what Heino does, he, when pushed with this argument, immediately deserts his model. Sorry. Not in terms of the jet ...

Falcke: Not in terms of the jet!

Genzel: .. not in terms of the jet, in terms of the disk, and joins Ramesh [Narayan] et al. on the ADAF.

Falcke: I think that the Galactic Center is actually the only case, where the jet model now, given all the IR limits, gives a good argument for an ADAF or say an ADAF-like solution, though perhaps not as extreme as Ramesh [Narayan] wants it.

Genzel: Okay, good, I understood. If I were so bold and say, look, I find the issue of the jet perhaps presently not so threatening as the issue of the basic accretion rate, I would say, we don't have a third case [of extremely low accretion rate]. You join the forces if you're forced to accept a mass accretion rate greater than 10^{-7}. But I'm sorry, gentlemen, you still are out.

Narayan: So maybe I can say, whether that can be pushed up. 10^{-5} is obtained for a particular value of this viscosity, the α-parameter. This is already very large, around 0.3.

Genzel: That is large.

Narayan: Some people would like 0.1, but it's 0.3. The model is not very sensitive to α, but of course it is at some level. And the limit for the accretion rate really comes from the X-rays where we have the ROSAT limit and the ASCA limits, and in fact it turns out, also the GRANAT limits in hard X-rays. These are all fairly tight limits that you have to satisfy. You can increase α and make the accretion rate larger. That's one way to do it, but nobody would like to go beyond $\alpha = 1$, and $\dot{M}$ just goes like α. So you're not really gaining much from pushing α.

The other way is to do it through the wind. But you have to start the wind at a very large radius, maybe at a fraction of a parsec, and have the wind operate all the way down to the center. I suspect X-rays will still kill the model, but I haven't done the experiment. So I will say that we have to wait and see, whether one could start a wind much farther out, so that [the accretion rate in the center] comes down to 10^{-5}, but starts [further out] at 10^{-4}. I don't know whether this is possible, but this is the one other escape that I can see. If neither of these works, Fulvio [Melia] should go and ask whether he can reduce that number.

Genzel: If you introduce the wind here, can I go over to the hot bubble and ask Fulvio? Suppose we create the wind, which would seem to create more shocks, is it possible in your calculations to maintain a hot bubble, or is it all going to cool so rapidly that no lowering of the accretion rate can be obtained that way?

Melia: Well, the issue here is, how does the Bondi-Hoyle accretion rate depend on the velocity, whether it's the sound speed or the kinetic velocity. The dependence for both cases is the same; it's always c_s^{-2}. So the point is that if the wind has a high kinetic energy but a very low temperature, then the dominant velocity scale is the flow velocity, whereas if the wind collides and thermalizes its energy, the velocity scale will be dominated by the sound speed. As long as energy is conserved, the accretion rate is probably not going to change very much, regardless of whether most of the velocity comes from the bulk flow or whether it's the sound velocity. Another way to think of it is that, if the black hole is surrounded by this hot gas, it will capture anything out up to where the energy per unit mass—as given by the sound speed component—is less than the gravitational specific energy density. So it doesn't really matter whether it's a hot gas with a low kinetic velocity or whether it's a cold gas with a high velocity, the scale is going to be the same.

G. Rieke: I haven't checked this recently, but when I last worked it out, the X-ray flux was already embarrassingly small for stellar processes. So while I'm not qualified to put something in your sealed envelope, I wonder what would happen to these models if the X-ray flux becomes an order of magnitude lower than what's being assumed now, since that seems like that's quite possible.

Genzel: Yes. In fact, I once made a calculation just on the basis of winds creating shocks out there, which could basically explain the ROSAT flux in the central 20 arcseconds. So, I also would say, more likely than not, the AXAF flux will turn out to be a tenth of what we have here.

Coker: Actually, my first comment is that our Bondi-Hoyle accretion rate is almost a lower limit. There's a lot of work being done on the He-I stars. It's actually likely that there's more gas flow in the region than what we assume now. So, it's quite distinctly possible that the Galactic Center has actually a much larger mass flux. And no matter what model you have, you have to figure out where that gas is going. You don't see it in the ISM; it's going somewhere. So, the natural conclusion is that it's going down into the black hole.

Genzel: Okay. Fulvio, there was a comment "you better check your numbers". Can't you make this [accretion rate] ten times smaller?

Melia: We've taken your advice . . .

Genzel: Don't blame it on me!

Melia: We've used a thousand km/s now for the wind velocity, which decreased the mass influx. What you did was to also increase the mass of the black hole to $2.5 \cdot 10^6$ solar masses, and the accretion rate scales as the mass squared. So, those two effects have basically cancelled out. That's why you still end up with an accretion rate on the order of 10^{-4} solar masses per year.

But there's a comment about Heino's [Falcke] jet feature. It's true that quantities like the magnetic field are in equipartition, but that is usually between the magnetic field and the relativistic particles. When one really talks about equipartition, the way I mean it, it is equipartition with all the available energy that's there. So, if the jet power corresponds to an accretion rate that's very low, then that means that the properties of the jet are not coupled to the global properties of accretion. It's really the global properties of the accretion that correspond to a much higher accretion rate, even if it's 10^{-5}.

Falcke: For my jet model [and your accretion rate] your terminology would mean sub-equipartition.

Melia: That's right. This would be greatly sub-equipartition.

Falcke: I just want to mention, that we actually observe this situation in other cases. There are quasar jets and Seyfert jets—radio-loud and radio-quiet jets. Quasar jets are in equipartition with the accretion disk, while Seyfert jets are a hundred times below equipartition. If Sgr A* were a Seyfert jet, it would require an accretion rate of $10^{-5} M_\odot/\text{yr}$.

Genzel: Well, Ramesh [Narayan], Fulvio [Melia] is not moving much, and in addition Robert [Coker] told us that, the accretion rate may even be greater because you're not really seeing at the present time all the O and B stars.

Beckert: Isn't there a problem with the brightness temperature and the observed size of the source? If you take this brightness temperature seriously, do we not get problems with the X-rays? This is really an optically thick synchrotron source, and the temperatures are kind of unique. We must have Bremsstrahlung and we must have a certain amount of X-rays.

Yusef-Zadeh: I have a point to make about the X-ray data. I think a lot of people use ROSAT data for the X-ray emission from Sgr A*, but that might not be really associated with Sgr A*. You have to look at higher frequencies, i.e., hard X-rays. If you get hard X-rays, ASCA data actually shows a lot of emission in this area, quite near Sgr A*. In fact, the temperatures are about 10^8 Kelvin.

Genzel: Can I point out that the basic problem we have is $L/(\dot{M}c^2)$. This is around 10^{-7}. The ROSAT luminosity is only 10^{34} ergs/sec, while ASCA is somewhat higher. So, in fact the problem is already bad enough with the bigger number from ASCA. It would get worse if you take ROSAT only. It's really the $\dot{M}$, as you see, which makes the problem so bad.

Townes: I want to ask about the hot bubble model, which is an attractive model in a certain sense, but is there really a realistic hot bubble which could be sustained and would not cool quickly through radiation and so on?

Genzel: Well, Ostriker's point comes from a calculation he's made for cooling flows in ellipticals. There he basically has shown that there's a limit cycle behavior of a cooling flowing in ellipticals with a black hole in the center, where initially you get a lot of accretion and the black hole radiation goes up. But then it produces winds—also from star formation—which brings down the accretion and the black hole luminosity goes down. Thus there's a sort of oscillating type behavior.

Townes: Oscillation? OK, then that's not a permanent thing.

Genzel: No, no, it's not a permanent thing.

Townes: You don't have a problem with the X-rays in the past?

Genzel: Not ROSAT, but the hard X-rays, yes. Fulvio [Melia], can you say something about variability. How much variability can you tolerate in your numerical simulations.

Coker: We can expect a large variability at high frequencies. Like the flare discussed yesterday, for example, that matches fairly nicely.

Genzel: What is the frequency? What's the time scale?

Coker: 200 GHz observations—we would expect flares and outbursts out there, no problem.

Genzel: Let's concentrate on $\dot{M}$. The question is, how much variability amplitude do you get, on what time scale?

Melia: It's about 20 to 30 percent.

Genzel: So, it's not a factor of 10.

Coker: No, it's not a factor of 10 at the moment, no. I can imagine it easily being a factor of 2 or 3.

Becklin: Does the $\dot{M}$ come from the He-I stars?

Melia: Yes.

Becklin: Is it possible that they are losing mass anisotropically? Could we just be very unlucky and there are ten stars going off in the wrong direction?

Najarro: It could be. It seems that we don't have so many good emission lines. When we first modelled the stars we had only two, three good lines, and two other lines, and were happy that we could get along with spherical models. It could well be that there's clumpiness. But to just infer the presence of clumpiness we would need very high signal-to-noise spectra, for instance of the hydrogen lines. And then we could infer if there's something like clumpiness or any asymmetries of the winds. But there I come to another point. Where did you get these thousand km/s from? The accretion rate scales with the velocity cubed, isn't it? So a factor of 2 makes almost an order of magnitude

Genzel: Fulvio [Melia] already said this: It's only going to get worse.

Najarro: Not necessarily. If you take the main contributors, which are the IRS 16 stars, maybe IRS 13, they have almost 500 km/sec [wind velocity]. So, the accretion rate goes up a factor of 8, but if by some count these winds are a little bit accelerated when they get there and it's 2000 instead of thousand, then you have 10^{-5} for $\dot{M}$.

G. Rieke: You have to remember, these stars are in a pretty strong magnetic field. The Lamor radius is less than a parsec, and the particles aren't going to go very far, before they would get very isotropic. And furthermore, I wonder if people have included the effects of the magnetic field in bringing stuff around. Would that affect how you feed a black hole or don't feed it?

Melia: They haven't yet, but the expectation would be that it would make the accretion rate go up even higher.

G. Rieke: I would think so too, because it would trap the particles, slow them down, and bring them back around the black hole.

Melia: So this re-iterates Rob's point, that this accretion rate in many ways is a lower limit.

Reid: If anybody asked me how well one could estimate the accretion rate based on the observations, I would have thought if you got within one or two orders of magnitude you're doing very well. There must be a million parameters effecting this, including velocities, densities, and distances you can infer throughout the center. In addition, it looked to me from your plot that there were some stars that contributed most of the wind. And we saw yesterday from the lack of accelerations of a few stars, that it didn't look like things were at the same distance radially as they are on the sky. How sensitive is it all to these parameters?

Melia: This is an excellent point. If you remember the image that I showed, there's one particular star, 13E1, which contributes a large fraction of the total accretion rate, and hence the value is sensitive to these parameters.

Genzel: I'm very pleased to see so much discussion is going, but you realize, we only have 15 minutes left...

Zhao: I want to make a point. From free-free emission we see in the radio, we probably can show you some estimated upper limits for the ionized gas that fills the Sgr A* region, which I will talk about this afternoon. And then there is the issue about the low-temperature 10^4 K gas and the high-temperature gas which probably goes up to 10^5 K and is produced by the He I stars. The interaction between the two probably reduces the accretion rate.

Melia: The higher temperature definitely lowers the accretion rate. The interaction has the opposite effect, because what it does is to dissipate the bulk flow and also takes away any momentum, so it makes the gas sit there longer—a little closer, with less kinetic energy. So, these interactions can actually increase the accretion rate, but the increase in temperature would lower it, because it would increase the sound speed and make the gas slightly less bound. It's not clear in which way it will go in the end.

Genzel: Okay, I would like to go on. I personally would conclude from this discussion that probably this number for the accretion rate here is actually a pretty serious number. A factor pushing it up is adding more stars, a factor pushing it down is that maybe the most important stars are a little farther away than we thought. This may have some effect and could push this down by a factor of ten, but it won't make that problem go away. It would also make things a little bit more comfortable for the ADAF model. Let's go to the next issue, do we need an event horizon? That seems to me such a basic conclusion, that we should really explore this issue, because if we have this, then that's a fantastic conclusion. Do we need an event horizon?

Melia: I think the case is very strong. Right now we can't think of any alternative using the spectrum of Sgr A* without having a very deep potential well, which means we have to have a point-like object. If we do have a point-like object, then we have Ramesh's [Narayan] very good argument that we just don't see the energy being dissipated at the surface. So those two factors go together—I think—and they make the case quite strong.

Genzel: Of course one could converge this argument into something which one should present to a larger audience, which is how about making that object here just large enough that Fulvio [Melia] and his model can just explain the model of Sgr A*. He couldn't explain Sgr A* with [a very large] size, but [if you shrink the size] at some point you'll be able to get high enough temperatures of the electrons and explain Sgr A*, and then suppose we take that surface and make your calculation [of how much energy would be dissipated on the surface] and see how bad the factor is.

Townes: Take a really crazy model of a central black hole which isn't very big and then bring back the neutrinos. I don't think anybody has really worked through the neutrino collapse theory.

Genzel: Yes, but they are relativistic and I don't think they are going back in there.

Townes: They don't have to be relativistic, now that they have mass. That's a crazy model, but on the other hand I don't think anybody's going to disprove it so easily.

While we are talking of this inflow rate again, it seems to me one can always find a reason why, temporarily, the source is not radiating very much. Therefore, I think the Russian's X-ray work is a critical thing in trying to say over the last 500,000 years what has happened. For a decade or so, there are lots of ways you can say, here's one of these fluctuations, something has blown the material away temporarily. That's easy. So, the long term is much more critical.

Back to the Neutrinos: I wish somebody could examine some of these neutrino models and see whether a small black hole would be a possibility after all.

Sellgren: I think there have been a paper or two in ApJ [by Viollier & Tsiklauri] which I haven't read but apparently someone is working on it. And the other thing that occurs to me is that, as you know, 99 percent of the universe's mass is dark matter and we don't know what that is. So perhaps there is something in the Galactic Center we don't know yet.

Genzel: Well, we can do simple things. For example, we can exclude fermions. The Chandrasekhar mass is a million solar masses for electron-mass particles. To have overall zero net charge the object would have to consist of electrons and positrons which would annihilate and radiate very strongly. You could, however, put in some heavy bosons.

Okay. So you all think we need an event horizon? I don't know whether this is possible, but it would be good to get you people together to look at that question, because that seems to me like something which people outside of the Galactic Center community would be very, very interested in: seeing whether or not one can make such a case?

Narayan: I'll just make one comment on that. A critical part of our claim is that it's a two-temperature plasma with a huge amount of thermal energy in the protons. Electrons are much cooler. So, the two-temperature is really important, the fact that not all the energy is being stored, but is disappearing.

Genzel: Not necessarily, Ramesh. I would look at this more generally—more simply as an observer. I would say, okay, the details of these models are very important, but both have in the end the goal to produce a low efficiency. So, if you look just at this point, how do you get a low efficiency in the case where you do not have a black hole? That is something which is independent now of the detailed physics. And it seems to me that also the black body limit is something which is very general in terms of the size of the object where you convert the energy.

Okay, well, people really seem to be convinced of this, then we have a wonderful case. Should we now go to the details of the physics? Ramesh, do you believe that there are low B-fields? Is that possible—lower than equipartition?

Narayan: With angular momentum and differential rotation, etc., I think B-fields will build up because there are instabilities—Balbus-Hawley for instance. Those things you can't stop. The B-field will build up to some fraction of equipartition. Whether exactly to a strict equipartition or 10 or 3 percent of it, we cannot tell. But it will build up to some reasonable level, that would be my guess. And it's a fairly generic argument, because these instabilities exists.

Melia: I would agree with that. I would say that this effect would be stronger at smaller radii which would tend to make the B-field gradient steeper still. But at larger radii where the differential rotation is not very significant, if there is not much angular momentum, it's possible that the dissipation rate dominates over the generation rate. Those two might converge at small radii. So at small radii with a lot of angular momentum I definitely agree that B-fields will be produced. Maybe that's what we see in the sub-millimeter excess. But at larger radii it's still not clear. And, it's obvious that the dissipation rate is going to be very sensitive on what mechanism one actually uses for calculating it, that's uncertain. One can't really be sure whether the value of the B-field at larger radii is accurate or not. The best we can do here is to use the calculations as a guide, and then maybe infer the magnetic field empirically and see if it makes any sense. But it's pretty clear that with this type of accretion rate, if we don't have a sub-equipartition field, the picture is not going to work. So I think an equipartition field would be ruled out.

Roberts: Are there any attempts to tie the magnetic field at larger radii to whatever values might be needed in the accretion region.

Melia: Yes. We started with typically a milligauss field at the capture radius—that's a standard value—and then evolved that inwards. Actually, the calculations do depend on what the magnetic field at the large radius is. For example, starting with the milligauss field, produces a very different profile at smaller radii than, let's say, starting with a tenth of a milligauss.

Roberts: So, the better observers constrain that number, the better can you do your calculation?

Melia: Yes, we typically assume one or two milligauss, if we change it by a factor of 10, this will drastically change the profile.

Biermann: I just want to note that massive stars have magnetic fields already. If you inject mass from the winds of these stars [you will have a highly magnetized gas]. Massive stars are pretty strong non-thermal radio emitters, so you know that the magnetic fields must be very strong. This could be self-consistently put in this scenario. Then you could ask the question, already raised before, why does ASCA find 10^8 K gas? Does this relate to these stars? Because, this X-ray emission is actually extended, not only in our Galaxy, but also in other galaxies even without an AGN, and this defies explanation—at least in the published literature. So, there indeed could be a hot bubble: 10^8 K, that's beyond escape speed. In terms of the Bondi-Hoyle accretion modelling, if this 10^8 K gas is really there, and the ASCA people make a very strong point of it, then you would just lose everything out of the Galaxy in some sort of chimney or so.

Melia: We'll actually hear a little bit more about that on Friday, during the morning session. But you raised some good points, and I think the next generation calculation has to be made more realistic like that. It actually has to start with the properties of the winds at the sources and follow those, rather than starting with the wind itself at larger radius because of poor resolution.

Genzel: I think it is very important that Peter and George have raised the issue that the magnetic field has to be included.

Stolovy: There was an observation, by the OSSE group that there was a fountain perpendicular to the Galactic plane. This was seen in annihilation radiation at 511 keV, on a very large distance scale, i.e. kiloparsecs. Can anyone give any more details? Is that thought to originate from a single object or from the Galactic Center, or what?

Melia: The resolution is so poor that it's more likely to be something like Sgr A East, rather than Sgr A West. It's much more extended, but again, on Friday morning we'll hear about the high-energy process, which we can probably connect to this, because we know what the non-thermal particle distribution is at Sgr A East, and that could be what produces this fountain. But because the resolution is so poor, it's likely that that's associated with a diffuse source, rather than Sgr A*.

Wardle: Back to the dissipation and the flow. I find it hard to understand why you manage to keep the magnetic field so far below equipartition, because the dissipation is very sensitive to the length scale that we have in the magnetic field—you need very small length scales to dissipate efficiently. And yet the process that's building up the field is a large-scale process, it's essentially the compression in the flow, which operates on much larger scales, and I don't see how that injects small scale structures in the magnetic field.

Melia: But that is on much smaller time scales. What you say is right, but it's an interplay between the flow time scale as the gas gets compressed, and the time scale that we get from the dissipation process, the resistive diffusion.

Wardle: But the dissipation wipes out the small-scale structure of the field, and you're just building up the large-scale structure.

Melia: Right. So one has to assume what the turbulent eddies are, whether they are on a large or small scale. There's a lot of uncertainty here, there's no question about it. It's just a fact that we don't even know what the right scale is.

Genzel: Let me say, I think it seems to me that since this is becoming a reasonably important conclusion for several of the models, it would be perhaps good, like in the case of observers, if another theory group makes a calculation, and sees whether they agree or not agree on this important issue.

Two-temperature solutions: who of the non-ADAF people would like to address this. I don't want to always talk and then turn to Fulvio. So, any one of the others? How serious do we, who don't understand this, have to take the comments by prominent theoreticians when they say: "Oh yes, two-temperature solutions, that's an interesting issue, which has not been solved since 20 years"?

Biermann: I think that these two-temperature solutions depend on the argument that Coulomb collisions are too inefficient. On the other hand, we do know from the solar wind that in many cases, magnetic field instabilities govern the exchange of momentum, and that needs to be worked out. And I think as long as we don't know to what degree magnetic field instabilities and micro-instabilities, which are very efficient, are ruled out in this context we can't make a firm conclusion. We know from the solar wind that the Coulomb collisions are totally irrelevant. Yet it behaves to a very good approximation like a fluid, but it should not do so, since the mean free path is bigger than the solar system. [The energy distribution] is always Maxwellian, even though an anisotropic Maxwellian is expected. And so, from the solar wind we can gain a clue whether we don't have these micro-instabilities, which make these two temperatures approach much more closely than it is used in these [ADAF] models. It may not kill the model, but it would be good to check this in some other environment where we know that these micro-instabilities are very, very strong.

Genzel: Peter, that's a very good suggestion, maybe, Ramesh [Narayan] might want to contact the theorists in the space plasma field. But often the problem is that their parameters of space are so far away from ours that they just shrug their shoulders and say "Well, you know, we don't even understand the solar wind".

Narayan: I talked to some plasma physicists and I told them, don't just say that there are processes that would equilibrate. Tell us what process that is. Give us a process and we'll calculate whether or not it's very important for our conditions, because we know the density, we know the temperature, we know the magnetic field, we know the whole thing. As far as I can tell, this is an issue where people basically say, what Peter has said, "Oh, there must be micro-instabilities, we see it happening here, we see it happening there". I don't think there's a particular mechanism that somebody can point out and say "Here is this particular instant which happens when the field is like this or the density is like that. This isn't something we can put our finger on. Therefore we are not able to compare it to our conditions.

Mahadevan: Don't we also have observations of the sun where we see oxygen at higher temperatures than other elements?

Biermann: Yah, in the solar wind of course we have all kinds of interesting complications, because of the shocks on the solar surface, which do all kinds of things that nobody understands. When I teach students I always say "We are trying to understand quasars, with all these beautiful models, and we don't know what drives the solar wind". We don't know what produces these funny element abundances in the solar wind and the energetic particles. We don't know any of these things.

Reid: In Rohan's model, protons basically are responsible for producing [pairs which produce] the radio emission at centimeter wavelengths. Is there any clear prediction that comes out of that model which can be tested observationally? Because that would then effectively prove that there are protons.

Mahadevan: As far as variability is concerned, a change of the accretion rate could be a factor. In that case a definite correlation between the gamma-ray flux, the amount of optically thin synchrotron emission from the [pairs produced by the] protons, and the Bremsstrahlung. You should observe basically a variability on the same order of time scale. That would be one thing that could be observed. One definite thing of these models, is that it produces a turnover in the near infrared from the $\pi^{\pm}$ production of pairs with a definite energy. That turnover can be observed. Of course, that would just say that you know there are pions.

Genzel: Okay, let's go on. I'm looking at the time a little bit. Let's go to the next point, maybe skip over the alpha versus beta viscosity, because that's another can of worms. Let's discuss this challenge made by some of your Cambridge colleagues, that the lack of sub-millimeter/millimeter emission in some of the ellipticals with dark masses, really excludes ADAFs.

Falcke: Let me just comment on the other galaxies. For example, we have our model applied to the radio core in NGC 4258. Here we know that the radio emission is resolved into a jet and so the only thing the ADAF model fits now in NGC 4258 is one X-ray data point. The point is that if you go with the jet model, you can explain the sub-millimeter emission as well as the spectrum down to centimeter wavelengths—then there is no sub-millimeter problem as in the ADAF model. We can also look at some of the other galaxies with similar cores and resolve them with VLBI into jets as well. I would be very, very careful to explain the radio emission in any of the galaxies with the ADAF model. There is ample evidence that those radio cores in AGN and low-luminosity AGN are in fact jets. And the reason why we have an almost unique sub-millimeter bump in Sgr A* is because of it low power and the long cooling time scales. In many other AGN the sub-millimeter-bump would just cool away.

Genzel: If I may transform this into my own words, the emission seen in the radio is from the jet, so that's not relevant. The ADAF contribution in the radio is much lower than that, but since in those other galaxies presumably the problem is not as severe as in the Galactic Center, we'd better solve the Galactic Center problem first, and then we may look at these other galaxies. They are not as threatening. Is that how I can say this in one sentence.

Falcke: Well ... I probably would object to calling the radio emission not relevant, be it in the Galactic Center or somewhere else.

Genzel: We have already talked about the variability of $\dot{M}$; now to the predictions! It seems to me, the X-ray flux is a key thing, and you've put down your numbers. Fulvio, have you put down your number?

Melia: Yes, in a sealed envelope.

Genzel: That's good. I don't want to challenge you, but I think it would be very useful. I think the believability of a theory is much stronger if you can make a good prediction.

Narayan: But with our model the issue is the other way round. We use the X-rays to normalize $\dot{M}$. We don't predict the X-rays, we need them.

Genzel: I understand this, Ramesh, but on the other hand, we've established up here that, if anything, you should produce more X-rays, because that would allow a larger accretion rate.

How predictive are these models in terms of size and centroid versus λ and time variations?

Falcke: I just want to make a number of predictions. The X-ray emission, has a lower limit, and that is synchrotron-self-Compton (SSC) emission from the jet nozzle. This will be somewhat lower, say a factor 3 to 10 then the ROSAT flux. AXAF with its high resolution can check that. This could mean that the X-ray flux detection from ROSAT and ASCA of Sgr A* is some other process. But at some point we should see SSC from Sgr A*. The radio centroid of Sgr A*, should shift with frequency, i.e. of the order milli-arcseconds at cm wavelengths. At very high frequencies, where the nozzle is, it should remain roughly constant. The size and the shift should scale with $\nu^{-0.9}$. It's a very clear prediction to that order. It might level off towards higher frequencies near the black hole. The time variations are also very crucial. For outbursts the high frequencies should lead the low (radio) frequencies. Like in AGN-jets, the outburst would first be visible at high frequencies and then travel towards lower frequencies, if you sample it rapidly enough. Of course, at some point it would be nice to see indications of a jet structure. So, there's a number of clear predictions.

Genzel: Okay, are there predictions in the other models?

Melia: I think as far as the centroid is concerned, it might depend on frequency. Probably most of these pictures would share the same basic dependence. The point is that we're looking at self-absorbed corners at each frequency, and so depending on what the structure of that surface is, the centroid could actually shift. So I'm not sure that that's a very good prescription.

Narayan: We have size vs. lambda for some of the models. It's not a simple powerlaw, but it's close enough. The centroid shouldn't change, because I think the velocities are pretty axisymmetric. The flow may be slightly flattened, and so there may be a little elongation, but the center itself should not change as you go to higher frequencies, until you come right close to a black hole. We have a rotating flow, which will have some Doppler effects, and one side might look a little brighter than the other side. Probably this would happen at around 300 GHz. It is probing the event horizon. The only other thing is what Rohan said, there should be coupled variations, at various parts of the spectrum. The things that come from the inner parts of the flow. In his model, for instance, all of the radio, sub-millimeter bump, and infrared—if there is some—should go up and down together, since they all come from the same part of the ADAF. It is just variations in $\dot{M}$. X-rays come from farther out, they will do their own thing.

Genzel: Allright. I hope this was an interesting discussion. Let's thank our speakers.

Session IV

How are the Various Components of Sgr A West Related?

The Central Parsecs of the Galaxy
ASP Conference Series, Vol. 186, 1999
H. Falcke, A. Cotera, W.J. Duschl, F. Melia, M.J. Rieke, eds.

The Interaction of Two Prominent Galactic Center Sources: Sgr A East and the Molecular Ring

F. Yusef-Zadeh

Northwestern University, Evanston, IL. 60208

S. R. Stolovy

University of Arizona, Tucson, AZ 85721

M. Burton

University of New South Wales, Sydney, NSW, Australia

M. Wardle

University of Sydney, Sydney, NSW, Australia

F. Melia

University of Arizona, Tucson, AZ 85721

T. J. W. Lazio and N. E. Kassim

Naval Research Lab., Washington, DC 20375-5351

D. A. Roberts

University of Illinois, Urbana, IL 61801

Abstract.
We present a synthesis of a number of recent observations in the near–IR H_2 and [Fe II] lines, OH (1720 MHz) maser line and various radio continuum measurements using the NICMOS of the HST, UNSWIRF on the AAT and the VLA. These observations suggest that the outer edge of the CND is collisionally excited whereas the inner edge is likely to be heated predominantly by the IRS 16 cluster. The velocity and spatial correlation of H_2 and OH (1720 MHz) as well as the spatial distribution of radio continuum emission at 90cm suggest that Sgr A East is responsible for shocking the gas and interacting with the circumnuclear ring at the Galactic center.

1. Introduction

The picture that has emerged from a suite of multi-wavelength observations of the Galactic center is that it contains a rotating, clumpy molecular ring (also known as the Circumnuclear Disk- CND) which is heated by a centrally concentrated source (the IRS 16 cluster) of UV radiation. Within the ring's central cavity, three "Arms" of ionized gas (Sgr A West) are generally in orbital motion around the center, which is believed to contain a $\sim 2.5\times10^6\ M_\odot$ black hole (Genzel and Eckart 1999; Ghez et al. 1999). The coincidence in the geometry and kinematics of the southwestern edge of the molecular ring and ionized gas suggests that the ionized gas is dynamically coupled to the inner edge of the circumnuclear ring. The western edge of the ring (sometimes called the Western Arc) is out of the plane of the sky and obscuring the background stars and absorbing Br γ emission from the western arm causing the ratio of H92α to Br γ to increase (see recent reviews by Genzel, Hollenbach and Townes 1994; Morris and Serabyn 1997). Although the CND is often described as having a circular geometry and kinematics, (e.g. Morris and Serabyn 1997 and the references cited therein), the ring structure shows deviations from circular geometry. A prominent gap in the distribution of molecular gas in the CND shows non-circular moving gas (e.g. Jackson et al. 1993). The CND gap is located where an ionized "streamer" appears to run along the extension of the Northern Arm beyond the CND. The thermal ionized gas in the streamers show that the N. Arm of Sgr A West is not coupled dynamically and spatially to the molecular gas in the inner edge of the CND.

On a larger scale the non-thermal structure known as Sgr A East is thought to be a shell-type explosive event, possibly a supernova remnant (SNR), surrounding the thermal source Sgr A West. The recent discovery of OH(1720MHz) masers at the interface of the 50 $\mathrm{km\,s^{-1}}$ molecular cloud and Sgr A East provides strong evidence that these two are physically interacting with each other (e.g. Mezger et al. 1996; Zylka et al. 1992; Serabyn et al. 1992; Yusef-Zadeh et al. 1996). The masers are detected to the southeast of Sgr A East (sources A, D-G) as well as to the northwest of the CND. The maser spots surrounding the Sgr A East shell have radial velocities near 50 $\mathrm{km\,s^{-1}}$ close to the systemic velocity of the SNR. The expansion of a supernova into the Sgr A East molecular cloud (i.e. the 50 $\mathrm{km\,s^{-1}}$ cloud) is considered to be responsible for the production of the shocked OH(1720 MHz) maser emission.

The nature of the association between the two systems, Sgr A East and Sgr A West and their corresponding molecular clouds, is not well established. Radio continuum studies have pointed out the curious alignment of the western edge of Sgr A East and the Western Arc of Sgr A West (Ekers et al. 1983; Yusef-Zadeh and Morris 1987; Pedlar et al. 1989). Sgr A East is behind Sgr A West, a conclusion based on low-frequency observations showing a hole in the distribution of ionized gas toward the Arms of Sgr A West (Yusef-Zadeh and Morris 1987; Pedlar et al. 1989). These observations imply that a segment of the Sgr A East shell must lie behind the Galactic center by a distance that is estimated to range between a few to hundreds of parsecs.

The relative location of Sgr A East with respect to the Galactic center is of interest for a number of reasons. First, the separation between the two systems is relevant to the role that the relativistic particles of Sgr A East may play in upscattering the CND's IR and UV photons to high energies. The strength of this interaction depends on the distance between Sgr A East and the CND. This process was recently incorporated into the modeling the high-energy emission from the Galactic center based on EGRET measurements (Melia et al. 1998a,b; Fatuzzo et al. 1999). In this model, Sgr A East is considered to be within 5pc of the Galactic center. Second, the disturbance of the molecular ring by Sgr A East also depends upon the separation between the two objects. Recent studies suggest that the N. Arm of Sgr A West may delineate the inner edge of an infalling "tongue" of neutral gas toward the Galactic center. The obvious question that needs to be addressed is whether Sgr A East is responsible for the observed deviation from circular motion at the location of the gap in the CND and for the infalling cloud. Lastly, the closeness of Sgr A East to the Galactic center and its unusual energetics brings up the possibility that the sources responsible for the explosion are the family of young, massive stars in the IRS 16 cluster. This hypothesis may provide an insight into the nature of massive star formation within the same parent cloud in the inner 5pc of the Galactic center.

A signature of an interaction between Sgr A East and the CND should be shocked gas as the blast wave of Sgr A East compresses the gas in the CND. The unambiguous detection of shocked (as opposed to UV-heated) gas associated with the Galactic center has been difficult. The intensity ratios of the v=2–1 and 1–0 S(1) lines of H_2 (Gatley et al. 1984; Burton & Allen 1992; Pak et al. 1996) are often consistent with collisional excitation rather than fluorescence, but the high density of the molecular gas at the Galactic center and the intense UV radiation field in the region allow the line ratios from UV-irradiated gas to resemble those of shock-heated gas (Sternberg & Dalgarno 1989). However, the presence of the OH 1720 MHz line, and the absence of the 1665/1667 MHz lines provides a clear diagnostic of shocked molecular gas, as the FIR radiation field from warm dust in UV-heated clouds would pump the latter transitions. Here we report preliminary results of our new H_2 and [FeII] and 1.2 and 90cm radio continuum observations and examine the interaction picture of Sgr A East with the CND.

2. Observations

We observed the Galactic center in the 1–0 S(1) transition of H_2 line using both the NICMOS on the HST as well UNSWIRF on the AAT. The UNSWIRF observations were carried out using four overlapping frames with a size of 90″, a pixel size of 0.77″ and channel maps separated by $\approx$39 km s^{-1} (Figs. 1,2 and 3). Using camera 3 of NICMOS, a total of 12 overlapping frames each with a 52″ size and a pixel size of 0.203″ was observed in both the line and the continuum. Each frame was dithered by 16″ in each principal directions of the detector, and the final image was constructed by mosaicing all 12 frames, a portion of which is shown in Figure 2. The 90cm continuum observations were carried out in the BnA (Fig. 4) and A-array configurations (Figs. 5 and 6) of the Very Large Array of the National Radio Astronomy Observatory[1]. Camera 3 of NICMOS was also used to observe the [FeII] line emission from the central 110″ × 160″, a subsection of which is shown in Figure 8b. The NICMOS images were produced

[1]The National Radio Astronomy Observatory is a facility of the National Science Foundation, operated under a cooperative agreement by Associated Universities, Inc.

by subtracting images in the 1% filter centered on the 1.64μm [FeII] line. The continuum was adjusted to minimize the stellar residuals, but both positive and negative residuals remain. Some of the residuals are due to differences in intrinsic colors and locally patchy extinction. These observations have provided wealth of structural details with subarcsecond resolution capability of the NICMOS, the velocity information from UNSWIRF, and the ability to discriminate line from continuum emission that UNSWIRF gives.

The high resolution 1.2cm image were based on combined A and B array configurations whereas the low-resolution image was based on C configuration. The astrometry of the radio and IR images was done by using the position of IRS 7 with respect to Sgr A* (Menten et al. 1995). A detailed account of these observations will be given elsewhere.

3. Results

3.1. H_2 Filament and H_2 Gas Inside the CND

Figure 1 shows contours of 1–0 S(1) H_2 line emission measured by UNSWIRF superposed on the greyscale 6cm continuum emission from Sgr A West and East (Yusef-Zadeh and Morris 1987). A segment of the Sgr A East nonthermal shell is seen to the NE near $\alpha = 17^h42^m32^s, \delta = -28^058'00''$ whereas the spiral-shaped structure of Sgr A West is evident near the center of the Figure. The peaks located in the NE (N. lobe) and SW (S. lobe) parts of the ring are consistent with limb-brightening of the inner edges along the principal axis. An extended H_2 feature with a plume-like appearance is noted within the ring. This new feature appears to terminate at the position of IRS 7 (a more detailed account of this feature will be given elsewhere). The distribution of H_2 line emission appears like a ring surrounding Sgr A West except in the region near $\alpha = 17^h42^m30^s, \delta = -28^058'45''$, where the HCN distribution indicates non-circular motion. at the location of the inner gap (Jackson et al. 1993). A long filamentary structure with an extent of $\approx 1'$ is discovered to the NW of the CND. A gap is noticed along the extent of the filament near $\alpha = 17^h42^m29^s, \delta = -28^058'30''$ which shows a similarity in its spatial distribution to the inner gap of the CND. The crosses in all the figures show the positions of OH (1720MHz) masers and the star symbol is coincident with the position of Sgr A*. The C and B masers with velocities of 43 and 134 kms^{-1} are seen near the outer and inner gaps of H_2 distribution.

Figure 2 shows the same contours as in Figure 1, but they are now superposed on the greyscale H_2 image based on the NICMOS observations. Because of the complex nature of Galactic center sources, the stellar continuum in NICMOS images has not been removed completely. There is an overall agreement between the NICMOS and UNSWIRF data in spite of the contamination from stellar continuum in NICMOS images. The new H_2 filament appears strikingly sharp with a width of about 0.5″ in the NICMOS image.

Figure 3 shows the velocity distribution of H_2 emission based on one 90″ field centered near the N. lobe as observed with UNSWIRF. The crosses indicate the position of the masers. The N. lobe of the CND appears prominantly at higher velocities near the 120 and 159 kms^{-1} channel maps. The new H_2 filament

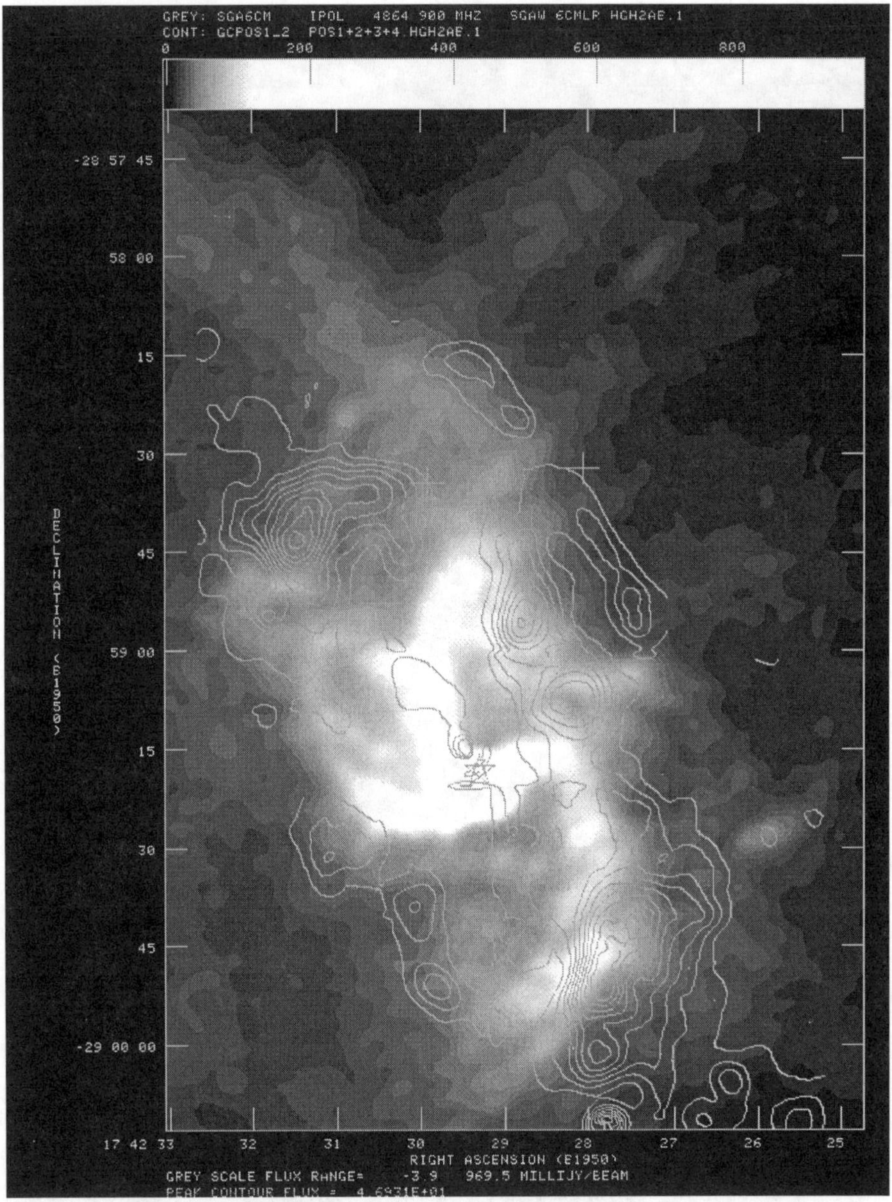

Figure 1. Contours of H_2 1–0 S(1) emission from the CND with a pixel size of 0.77″ is superposed on a 6cm radio continuum image of thermal emission from Sgr A West with a spatial resolution of 0.67″ × 0.4″. A segment of the nonthermal Sgr A East shell is seen to the NE. The star symbol represents Sgr A* and the crosses show the position of OH (1720 MHz) masers.

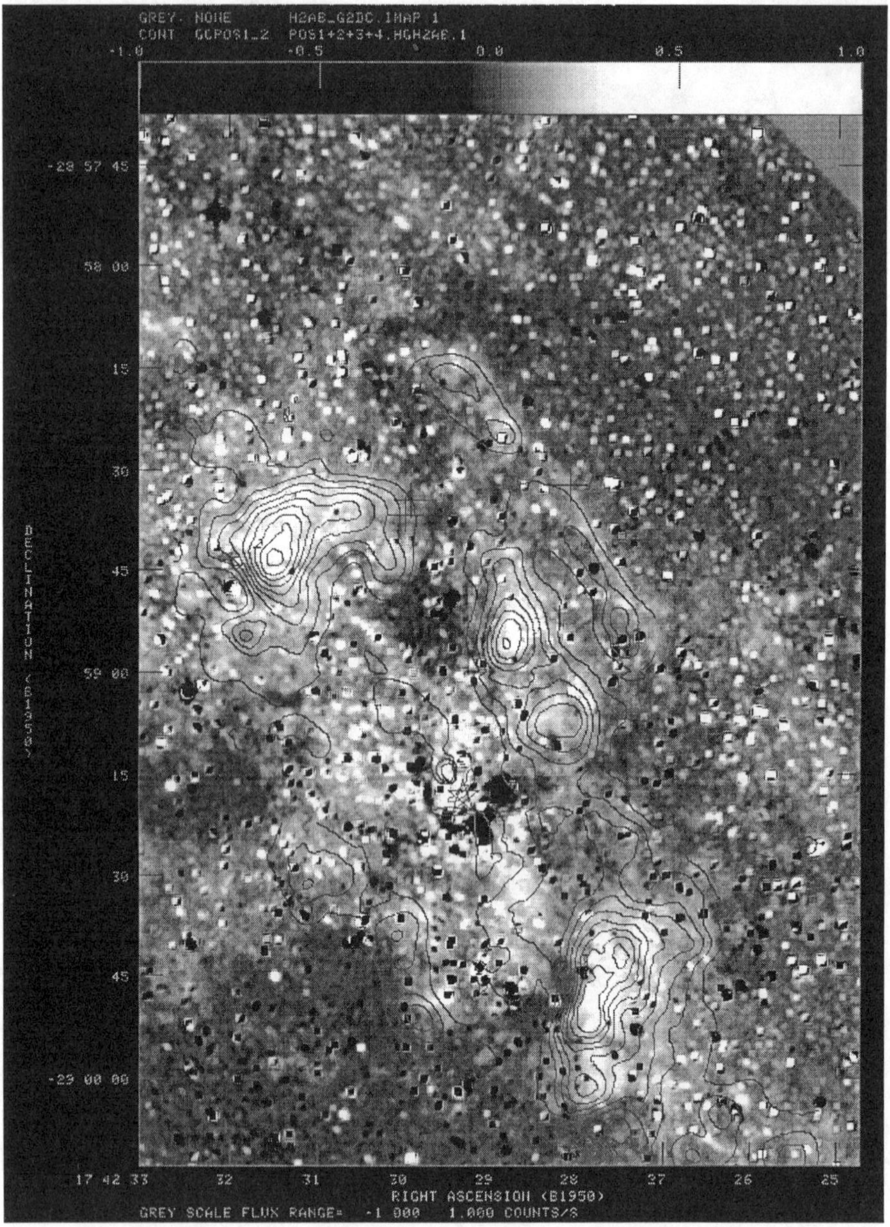

Figure 2. The same as Figure 1 except that the greyscale image is replaced by the HST/NICMOS H_2 image. The greyscale image is convolved by a Gaussian having a size twice the pixel size of 0.203″.

appears as a coherent velocity feature and is best seen in the 41.2 and 80.4 kms^{-1}channel maps. The C maser with a velocity of 43 kms^{-1} is probably related to the new H_2 filament, whereas the B masers, with velocities of about 134 kms^{-1} are associated with the N. lobe. The overall velocity structure of the new filament is also similar to that of the inner edge of the CND near the Western Arc, suggesting that the new filament is part of the CND but outlining the outer edge of the CND. A ridge of weak H_2 emission is also seen in the 41 and 80 kms^{-1} channel maps. This feature appears to connect the two gaps, as seen in the new filament and in the inner edge of the CND, suggesting that the H_2 features are indeed associated with each other.

3.2. Radio Continuum Features at 90 and 1.2cm

Figure 4 shows the greyscale distribution of 90cm emission superposed on contours of H_2 emission. The hole at the center is due to free-free absorption of the arms of Sgr A West against Sgr A East. The 90cm image also shows the ionized streamers in absorption at the position of the inner and outer gaps. The elongated structure of Sgr A East, which is saturated in this image, delineates the bright shell of Sgr A East. The western edge of the shell is weaker in its surface brightness due to optical depth effect but what is clear is a deviation from a shell-like geometry to the west of the CND. We also note that the CND is clearly surrounded by nonthermal emission (Pedlar et al. 1989) not only to its western edge but also to its eastern edge. The prominent eastern shell of Sgr A East lies near $\alpha = 17^h42^m38^s$ which is about $1.5'$ east of the eastern edge of the CND. Figure 5 presents a more close up view of the 90cm emission with a higher resolution where contours of HCN emission (Wright et al. 1987) appear to be surrounded by Sgr A East. In addition, the 90cm emission shows an elongated weak nonthermal feature within the hole. The striking feature in this image is the continuum emission crossing the bar of ionized gas associated with Sgr A West. This shows clearly a lack of free-free absorption of a segment of the bar against Sgr A East.

Figure 6 shows contours of 90cm emission superposed on a greyscale 1.2cm continuum image of Sgr A West showing the extent of the elongated low-frequency feature with respect to Sgr A West and Sgr A*. The presence of low-frequency 90cm emission from the direction of the bar and Sgr A* is surprising because the bar of ionized gas is known to be the brightest feature with the highest ionized density in Sgr A West. If Sgr A East lies behind Sgr A West, the bar should have the largest optical depth at 90cm. This suggests that the Sgr A East shell is not totally behind Sgr A West as the elongated feature which is probably a nonthermal feature associated with Sgr A East crosses the hole created by free-free absorption. In fact, the comparison of the molecular and ionized material with the 90cm image, as presented in Figures 5 and 6, gives the impression of an anti-correlation between the molecular material and the 90cm emission. The lack of nonthermal emission from Sgr A East versus the free-free absorption of thermal Sgr A West against Sgr A East will be considered in more detail elsewhere.

Radio continuum emission at λ1.2cm traces the distribution of thermal gas in Sgr A West and is unlikely to be contaminated by the nonthermal Sgr A East given its steep radio spectrum in radio wavelengths (Pedlar et al. 1989). Figure 7

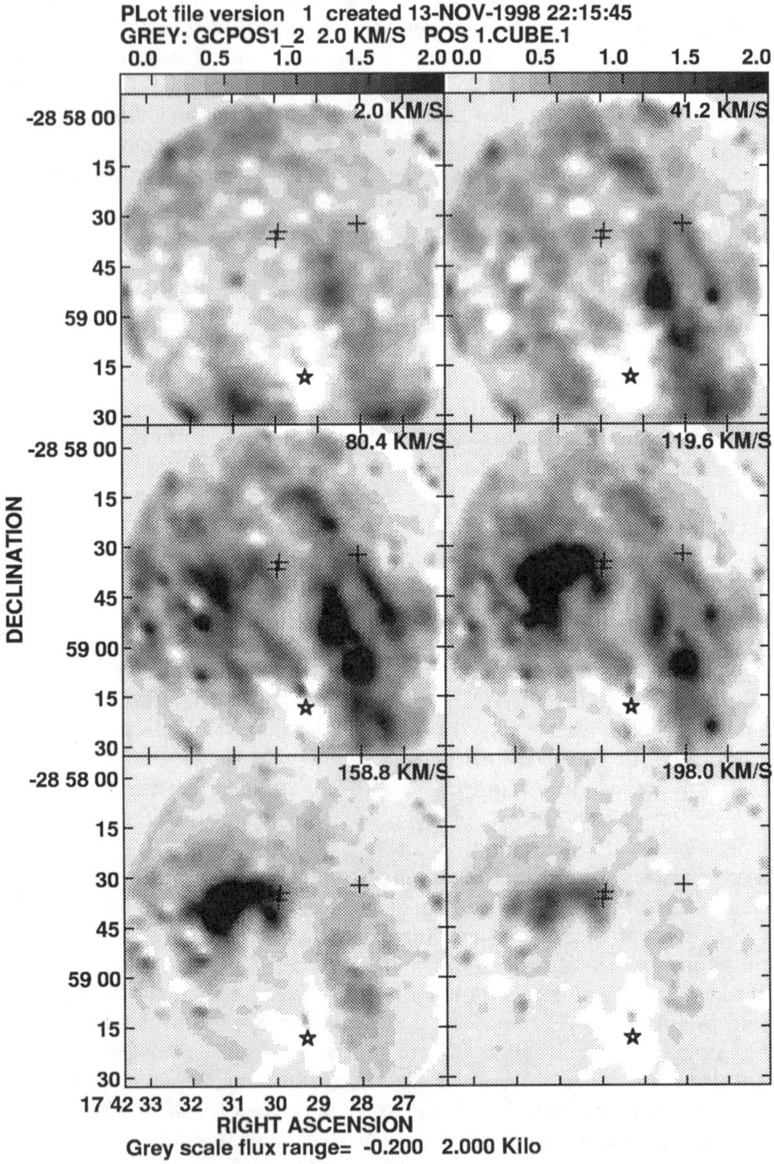

Figure 3. The H_2 velocity distribution of a field centered on the N. lobe of the CND is shown in reverse greyscale. The LSR channel velocities are shown on top right corners and the crosses represent OH maser sources, C to the right and B to the left. The high and low values are shown in black and white, respectively. The white features are artefacts from imperfect continuum subtraction.

shows contours of HCN emission superposed against the 1.2cm greyscale of Sgr A West. This figure shows the prominent three arms of Sgr A West and a faint Arm of ionized gas which appears to delineate the outer western edge of the CND. This new and extended ionized feature runs between $\alpha = 17^h42^m27^s, \delta = -28^059'50''$ and $\alpha = 17^h42^m28^s, \delta = -28^059'15''$ and has a typical brightness between 0.5 to 1.5 mJy/beam with a beam size of $1.58'' \times 0.76''$. There is also a peak of 90cm emission coincident with the new high-frequency 1.2cm feature near $\alpha = 17^h42^m26.5^s, \delta = -28^059'30''$ with a flux density of 80 mJy in a beam of $9.5'' \times 5.1''$. It appears that the western edge of the CND is sandwiched by two ionized layers from east and west along the inner and outer edges of the CND, respectively. The eastern or the inner layer of the CND is traced by the S. arm (i.e. the Western Arc) and is ionized by the strong UV radiation field originated by the IRS 16 cluster. The western or the outer layer of ionized gas is identified as a new and weakly emitting arm of ionized gas tracing the outer edge of the CND. The latter can either be due to photoionization by the radiation field from nearby stars lying outside the CND or by shocked gas as a result of an interaction with Sgr A East as a J-type shock is driven into the CND.

3.3. Dark Patches

We note prominent dark patches distributed throughout the H_2 images. These dark features which are best seen in NICMOS images in Figure 2 are extended and appear to be concentrated prominantly along the inner and outer gaps of the CND and along the remarkably extended structure NW of the H_2 filament. Although the lack of proper removal of stellar continuum contributes to the absorption features in NICMOS images, all the features noted here are extended, have been detected in continuum images and show atomic or molecular couterparts. Estimates of the extiction toward these patches will be given elsewhere.

An example of the dark feature in Figure 2 is located near the N. lobe but to the south of the inner gap of the CND near $\alpha = 17^h42^m30^s, \delta = -28^058'50''$. This feature coincides with the well-known OI and dust feature within the CND (Davidson et al 1992; Jackson et al. 1993; Telesco et al. 1996). The OI feature is thought to lie in the CND with H_2 density of 3×10^5 cm^{-3} and a mass of 300 $M_\odot$ suggesting that the dark feature in this region is located at the Galactic Center. The strongest support for this suggestion comes from the distribution of radio continuum emission at 1.2cm and the distribution of [FeII] feature based on the VLA and HST/NICMOS measurements. Figure 8a shows the N. Arm of Sgr A West and a faint semi-circular shell structure surrounding the northern tip of the N. Arm. This weak shell-like ionized feature with a typical brightness of 0.2 mJy within a beam of $0.3'' \times 0.2''$ coincides with the outer rim of the OI/dust feature and the dark H_2 patch. This shell of ionized feature suggests that the the dark patch is photoionized externally.

Figure 8b shows the distribution of [FeII] in the same region as shown in Figure 8a. A dearth of [FeII] emission corresponding to the northern half of the N. Arm is seen at the position of H_2 dark patch. Both the [FeII] and H_2 features are surrounded by a rim of ionized gas in the 1.2cm image. We note a lack of correlation along the Northern arm in 1.2cm and [FeII] images, as shown in figures 8a and b. The northern half of the N. Arm also shows a sudden drop by a factor of 2 in its brightness to the north of $\alpha = 17^h42^m29.7^s, \delta = -28^059'00''$

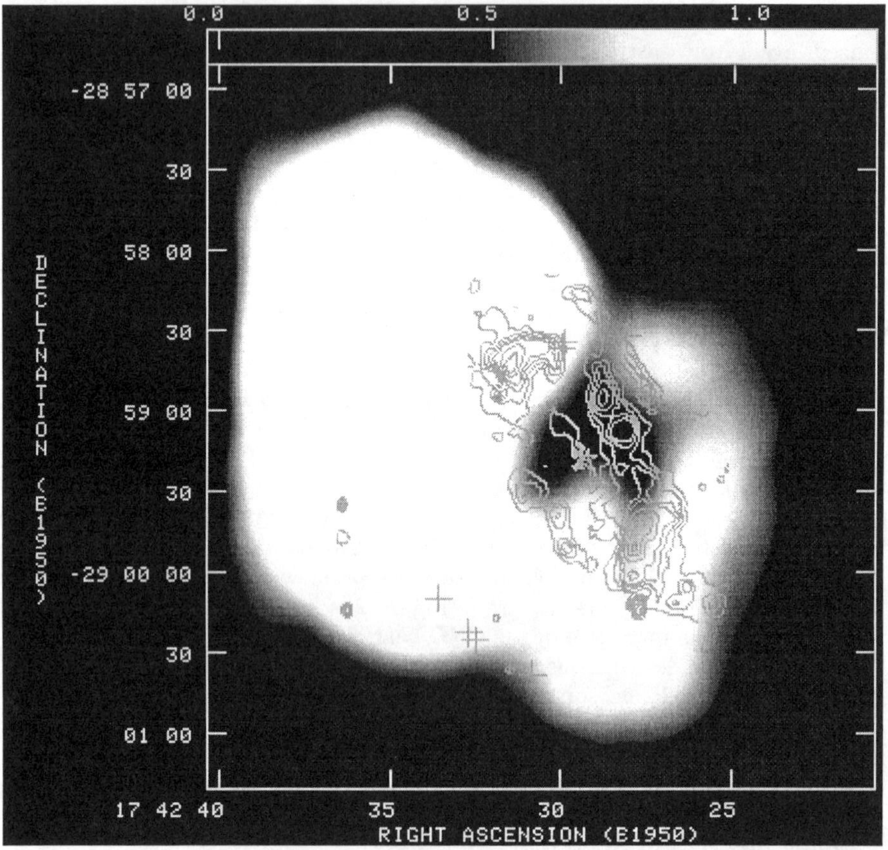

Figure 4. Contours of H_2 distribution superimposed on the 90cm emission from Sgr A East with a spatial resolution of 16.5″ × 10″. Sgr A West is absorbed against Sgr A East at this low frequency.

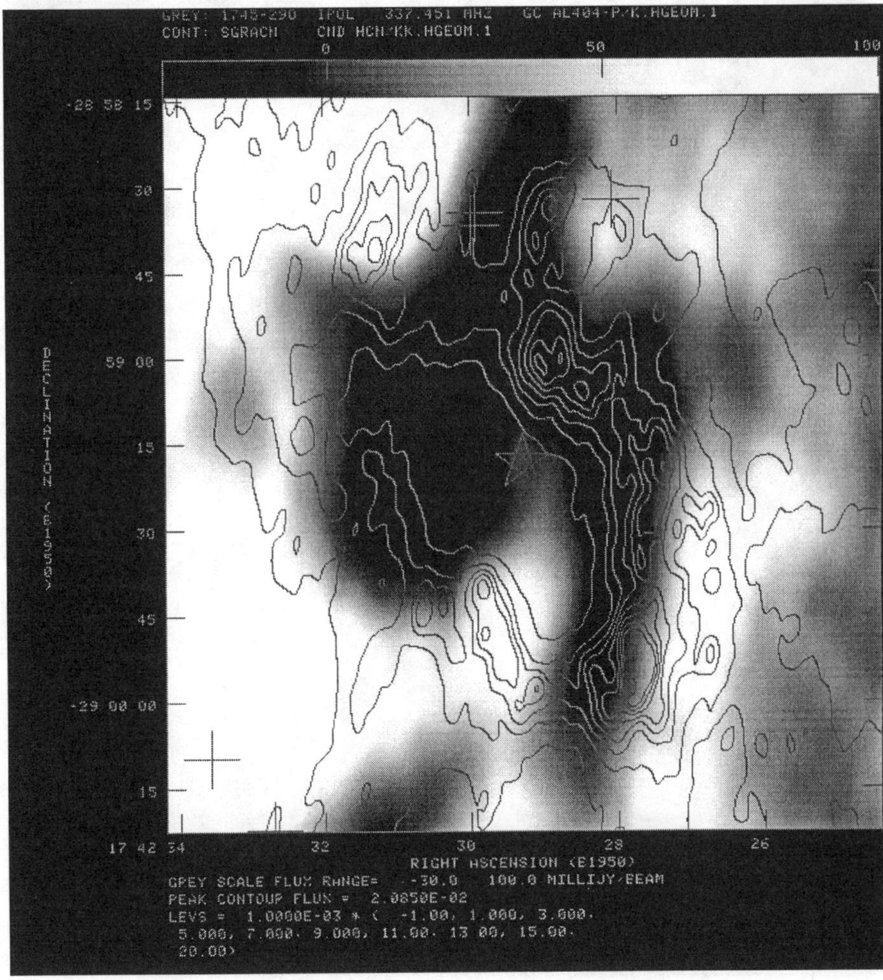

Figure 5. Contours of HCN emission (Güsten et al. 1987; Wright et al. 1987) with a spatial resolution of 5-7″ superposed on the central region of the 90cm distribution with a spatial resolution of 9.5″ × 5.1″.

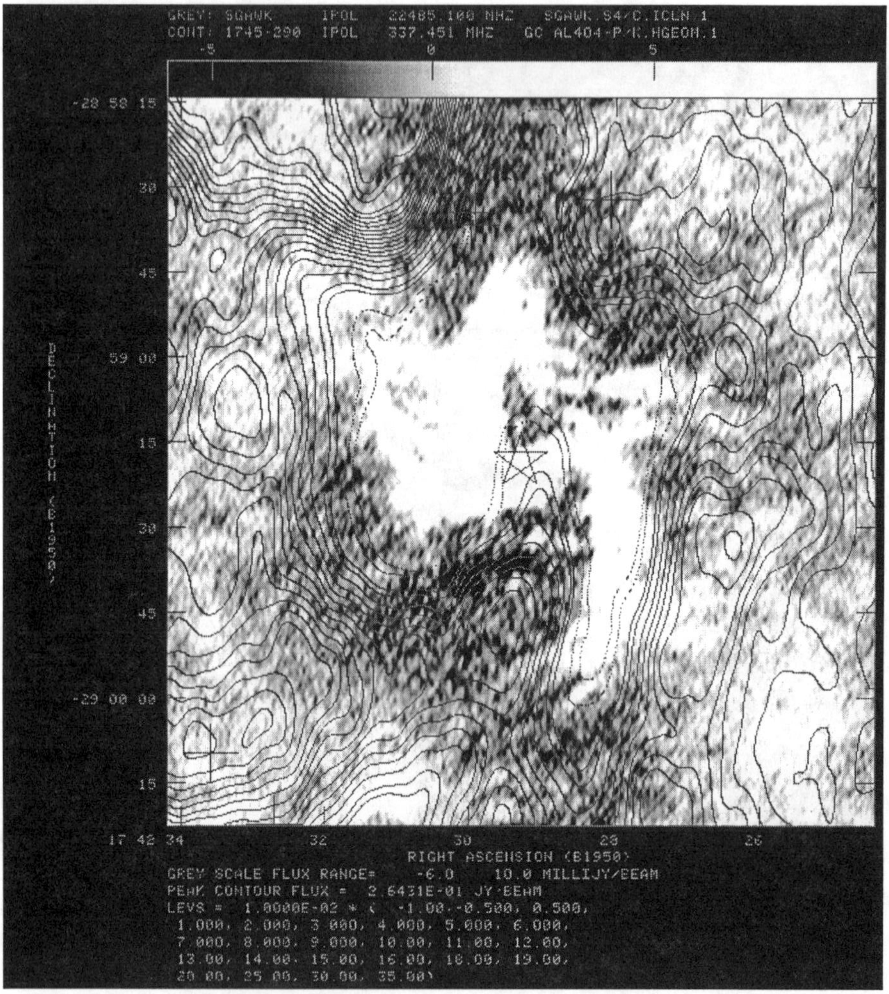

Figure 6. Contours of 90cm emission as seen in figure 5 are superposed on the greyscale distribution of 1.2cm emission with a spatial resolution of $1.6'' \times 0.8''$. The star marks the position of Sgr A*.

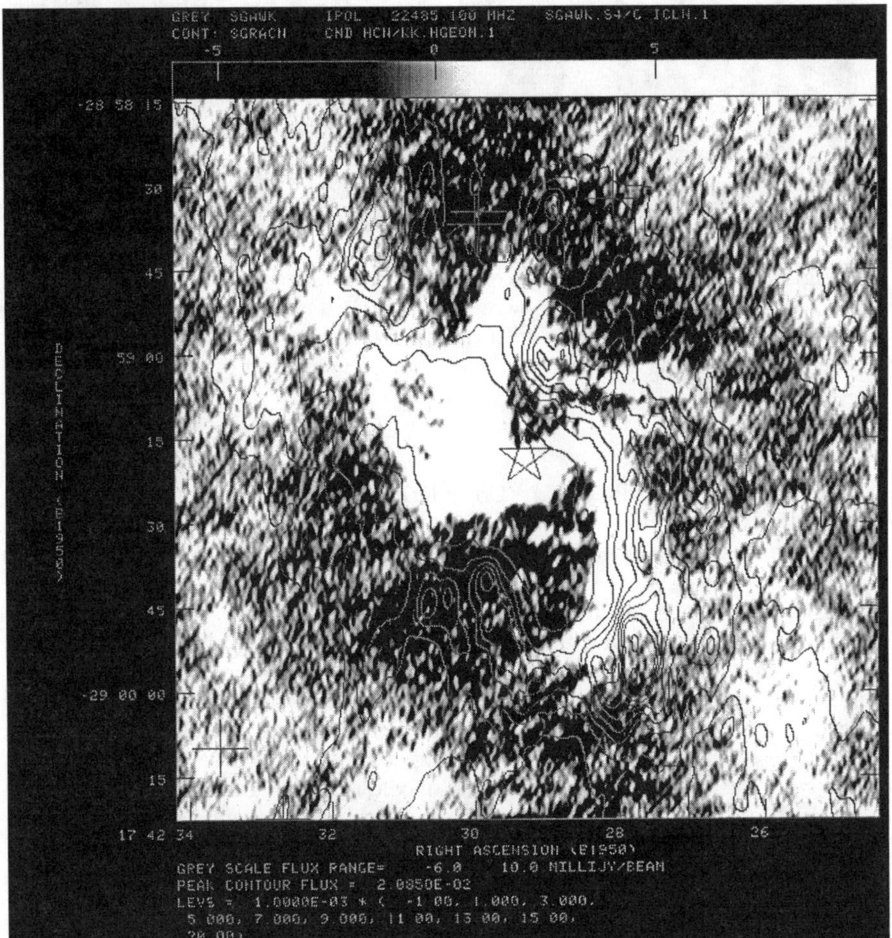

Figure 7. A greyscale distribution of ionized gas at 1.2cm as shown in figure 6 superimposed on contours of HCN distribution

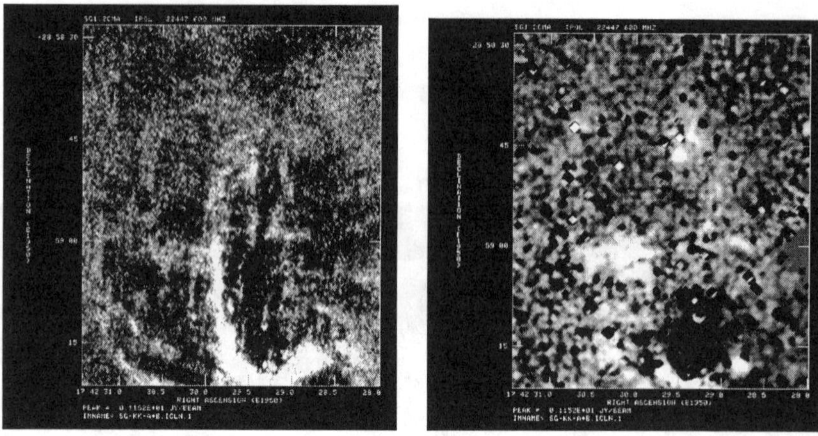

Figure 8. (left) A greyscale distribution of ionized gas at 1.2cm with a resolution of $0.3'' \times 0.2''$ showing the Northern arm and a shell of weak ionized gas around its northern tip. (right) A greyscale distribution of [FeII] ionized gas showing a lack of emission centered on the northern half of the N. arm.

(Yusef-Zadeh and Wardle 1993). Considering that free-free emission at 1.2cm is optically thin, the drop in the emission measure as the Northern arm crosses the shell is a strong support for the association of the OI/dust feature with the N. arm.

The coincidence of OI/dust emitting feature with the dark features and an ionized rim surrounding them suggests that they are all located within the CND. Further support comes from a drop in the emission measure of optically thin free-free emission from the N. Arm at 1.2cm. A detailed examination of the ionized gas in Sgr A West shows also a number of dramatic dark features which can not be explained by a lack of short-spacing data. These dark continuum features are likely to delineate the region of cold dust and gas where the emission measure drops significantly. Recent proper motion results show the direction of the ionized flow supporting the infall picture of the N. Arm toward the Galactic center (Yusef-Zadeh, Roberts and Biretta 1998; Zhao and Goss 1998). The large-scale distribution of dark features in the H_2 images suggests that the 50 $\mathrm{km\,s^{-1}}$ could be responsible for the infall of neutral material toward the Galactic center as evidenced by the disturbed molecular material in the CND at the location of the inner and outer gaps. The location of the dark patches along the N. Arm, near Sgr A*, suggests that the atomic and molecular gas survives well within the cavity of ionized gas.

We also note the darkest feature lying close to the IRS 16 cluster with very bright continuum emission. Because of the strong stellar emission from the IRS 16 cluster, this dark feature may be due to artefacts from imperfect continuum subtraction. However, it is interesting to note that the dark feature coincides with the highly blue shifted H_2CO, OH, HI and HCO^+ absorption features, with velocities of about −190 $\mathrm{km\,s^{-1}}$(Marr et al. 1993; Pauls et al. 1993; Yusef-Zadeh et al.1993; Yusef-Zadeh 1994; Zhao et al. 1995). The kinematic and spatial distribution of the absorbing gas suggest that the gas is located at the Galactic center (see Liszt and Burton 1995 for an alternative interpretation).

4. Discussion

The new H_2 filament running parallel to the nonthermal shell of Sgr A East supports the picture that the impact of the nonthermal shell of Sgr A East with a molecular cloud is responsible for shocked H_2 emission from the filament. Additional support for the shock hypothesis comes from the sheet-like morphology and the location of maser source C along the filament. The radial velocity of the C maser is about 43 $\mathrm{km\,s^{-1}}$ which is within the H_2 velocity range of velocities of 40 and 80 $\mathrm{km\,s^{-1}}$ as shown in Figure 3. These so-called supernova masers are thought to be excellent tracers of shocks and are seen primarily in regions where SNR's are interacting with molecular clouds (Wardle et al. 1998; Green et al. 1998; Frail et al. 1996). The lack of any thermal continuum or Br γ emission at the location of the new filament is consistent with the C-type shocks being responsible for exciting the OH gas (Wardle et al. 1998).

The compelling evidence that the new filament traces the shocked molecular gas by the expansion of the Sgr A East shell into the cloud raises the question of the location of the filamentary shocked gas with respect to the CND. It is possible that either the shocked gas is associated with the 50 $\mathrm{km\,s^{-1}}$ cloud and it happens to be aligned fortuitously along the outer edge of the CND or that the outer edge of the CND is shocked externally as a result of its interaction with Sgr A East. What follows is a series of arguments in support of the outer boundary of the CND being collisionally excited by the expansion of the nonthermal Sgr A East into the CND.

First, the anti-correlation of the streamers of ionized gas and molecular gas at the positions of the inner and outer H_2 gaps show compelling morphological evidence for the association of the streamers of Sgr A West and the new H_2 filament which traces the shocked gas associated with the 50 $\mathrm{km\,s^{-1}}$ molecular cloud. Second, the inner and outer boundary where the gaps are observed show similar velocity distributions suggesting that they both must be part of the molecular ring. A ridge of H_2 emission linking the new filament to the inner edge of the CND furthers the support for the physical association between them.

Third, the evidence that Sgr A East and the CND are physically interacting with each other comes from detailed 90cm observations of the Galactic center. In these images, the western and eastern edges of the CND appear to be engulfed by a shell of radio continuum emission at 90cm. The emission at these low-frequencies are due to nonthermal processes and is expected to be associated with Sgr A East. In addition, the evidence for low-frequency emission from the region of the bar and absorption of the arms of Sgr A West add further evidence that Sgr A East surrounds the CND. The high frequency radio emission delineating the outer edge of the Western Arc of the CND could also be used as an argument for the interaction hypothesis if the radiation field outside the CND is not strong enough to ionize the outer edge of the CND. A lack of [FeII] emission from the northern half of the N. arm and its correlation with OI/dust emission and H_2 dark patches as well as a number of kinematically disturbed features in the CND all could be used as implications of a scenario in which a cold dust cloud plunges into the CND by an energetic phenomena.

Lastly, the recombination line emission toward Sgr A East and Sgr A* show stimulated thermal gas at velocities near 50 $km s^{-1}$ (Anantharamaiah et al. 1999; Roberts and Goss 1993). It is possible that the thermal ionized gas is associated with the 50 $km s^{-1}$ molecular cloud. If so, a part of the 50 $km s^{-1}$ molecular cloud with which Sgr A East is interacting must be in front the of Galactic center.

5. Summary

The conclusion that Sgr A East and the CND are interacting is inescapable based on various lines of evidence and arguments given above. A further test of this model could be done by measuring the intensity ratio of v=2–1 and 1–0 S(1) line emission from the region already observed presented in v=1–0 S(1) line. The implication of such an interaction is that the inverse Compton scattering of IR and UV photons should not be ignored in modeling the high-energy emission from the Galactic center. Furthermore, the disturbance of the gas in the inner edge of the CND is caused by the impact of Sgr A East. These suggest that Sgr A East is responsible for pushing gas cloud into the cavity. In this scenario, the N. Arm traces the trajectory of the infalling cloud as it is ionized by the strong UV radiation field of the IRS 16 cluster. The three-dimensional motion of ionized gas based on proper motion and radial velocity measurements (Yusef-Zadeh, Roberts and Biretta 1998; Zhao and Goss 1998) is consistent with this picture. The dark patches seen within the inner 30″ of Sgr A*, in the CND coincident with OI feature, in the inner and outer gaps and the extended dark feature to the NW of the filament in NICMOS image of H_2 distribution gives the impression that the 50 $km s^{-1}$ molecular cloud may in fact be responsible for feeding the Galactic center (see Coil and Ho 1998). Lastly, this interaction picture of Sgr A East and the CND may shed some new light on the possibility of the common origin of the IRS 16 cluster and the progenitor of stars responsible for the formation of Sgr A East.

Acknowledgments. We thank Mel Wright for providing the HCN image and M. Ashley, S. Ryder and J. Storey for assistance with using UNSWIRF. Basic research in radio astronomy at the Naval Research Laboratory is supported by the Office of Naval Research. TJWL is supported by an NRL-National Research Council Associateship.

References

Anantharamaiah, K.R. et al. 1999, this volume, p. 422

Burton, M. & Allen, D.A. 1992, in Proc. Astr. Soc. Australia, 10, 55

Coil, A.L. and Ho, P.T.P. 1998, ApJ, in press.

Davidson, J., et al. 1992, ApJ, 387, 189

Genzel, R. & Eckart, A. 1999, this volume, p. 3

Ekers, R.D., van Gorkom, J.H., Schwarz, U.J. & Goss, W.M. 1983, A.A. 122, 143

Fatuzzo, M. et al. 1999, this volume, p. 560

Frail, D. A., Goss, W. M., Reynoso, E. M., Green, A. J. & Otrupcek, R. 1996, AJ, 111, 1651

Gatley, I., Jones, T. J., Hyland, A. R., Beattie, D. H., & Lee, T. J. 1984, MNRAS, 210, 565

Genzel, R., Hollenbach, D.J. & Townes, C.H. 1994, Rep.Prog.Phys., 57, 417

Ghez, A. et al. 1999, this volume, p. 18

Green, A.J., Frail, D.A., Goss, W.M. & Otrupcek, R. 1997, AJ, 114, 2058

Güsten et al. 1987, ApJ, 318, 124

Jackson, J.M., Geis, N., Genzel, R., Harris, A.I., Madden, S., Poglitsch, A., Stacey, G.J., Townes, C.H. 1993, ApJ 402, 173

Liszt, H. & Burton, W.B. 1995, ApJS, 98, 679L

Marr, J.M., Rudolph, A.L., Pauls, T.A., Wright, M.C., Backer, D.C. 1993, ApJ, 400, L29

Melia, F., Fatuzzo, M., Yusef-Zadeh, F. & Markoff, S. 1998a, ApJ, 508, L65

Melia, F., Yusef-Zadeh, F. & Fatuzzo, M. 1998b, ApJ, in press.

Menten, K.M., Goss, W.M., Ho, P.T.P. 1995, ApJ, 450, 1227

Mezger, P.G., Duschl, W.J. & Zylka, R. 1996, A.A.Rev., 7, 289

Morris, M. & Serabyn, G. 1997, ARA&A 34, 645

Pak, S., Jaffe, D.T. & Keller, L.D. 1996, ApJ, 457, L43

Pauls, T., Johnston, K.J., Wilson, T.L., Marr, J.M. & Rudolph, A. 1993, ApJ, 403, L13

Pedlar, A., Anantharamaiah, K.R., Ekers, R.D., Goss, W.M., van Gorkom, J.H., Schwarz, U.J., & Zhao, J.-H. 1989, ApJ 342, 769

Roberts, D. & Goss, W.M. 1993, ApJS, 86, 133.

Serabyn, E., Lacy, J.H., & Achtermann, J.M. 1992, ApJ, 395, 166

Sternberg, A. & Dalgarno, A. 1989, ApJ, 338, 197

Telesco, C.M., Davidson, J.A. & Werner, M.W. 1996, ApJ, 456, 541.

Wardle, M., Yusef-Zadeh, F., & Geballe 1998, submitted to ApJ.

Wright, MCH, et al. 1987, in *The Galactic Center*, ed., DC Backer, 133, New York: AIP

Yusef-Zadeh, F., Lasenby, A. & Marshall, J. 1993, ApJ, 410, L27

Yusef-Zadeh, F. & Morris, M. 1987, ApJ, 320, 545

Yusef-Zadeh, F., Roberts, D.A., Goss, W.M., Frail, D. & Green, A. 1996, ApJ 466, L25

Yusef-Zadeh, F., Roberts, D.A. & Biretta, J. 1998, ApJ 499, L159

Yusef-Zadeh, F. & Wardle, M. 1993, ApJ, 405, 584

Zhao, J.H. & Goss, W.M. 1998, ApJ, 499, L163

Zhao, J.H., Goss, W.M., Ho, T.P.T. 1995 ApJ, 450, 1227

Zylka, R., Mezger, P.G. & Lesch, J. 1992, A.A. 261, 1197

The Central Parsecs of the Galaxy
ASP Conference Series, Vol. 186, 1999
H. Falcke, A. Cotera, W.J. Duschl, F. Melia, M.J. Rieke, eds.

3D Simulation of the Gas Dynamics in the Central Parsec of the Galaxy

R.F. Coker[1]

Physics Department, The University of Arizona, Tucson, AZ 85721, USA

Fulvio Melia[2]

Physics Department and Steward Observatory, The University of Arizona, Tucson, AZ 85721, USA

Abstract. It is thought that many characteristics of the gaseous features within the central parsec of our Galaxy, are associated with the accretion of ambient plasma by a central concentration of mass. Using a 3D hydrodynamical code, we have been simulating this process in order to realistically model the gaseous flows in the center of our Galaxy. In the most recent simulation, we have taken into account the multi-point-like distribution of stellar wind sources, as well as the magnetic heating and radiative cooling of these stellar winds. As expected, we find that the structure of the flow is significantly different from that due to a uniform medium. We also investigate the possibility that Sgr A* is due to a distributed mass concentration instead of the canonical point mass of a black hole. We discuss the physical state of the accreting gas and how our results suggest that Sgr A* is unlikely to be associated with a "dark cluster".

1. Introduction

The hydrodynamics of the interstellar medium (ISM) within the central parsec of the Galaxy has long been thought to be dominated by the gravitational potential due to a central mass concentration (for a review, see Mezger, Duschl, & Zylka 1996). For example, the Western Arc, one of the dominant kinematic features of the region, appears to be in circular rotation about Sgr A*, a unique compact radio source whose lack of proper motion suggests it lies at the dynamical heart of the Galaxy (Roberts & Goss 1993; Backer 1994). Stellar motions (Genzel et al. 1997), gas kinematics (Herbst et al. 1993), and velocity dispersion measurements (Eckart & Genzel 1999) together suggest the presence of $\sim 2.6 \times 10^6 M_\odot$ of dark mass located within $\sim .01$ parsecs of Sgr A*. Note that the Galactic center (GC) is at a distance of ~ 8 kpc so that 1 arcsecond ~ 0.04 parsecs.

---

[1]NASA GSRP Fellow.

[2]Presidential Young Investigator.

However, showing that the GC must contain a centralized dark mass concentration does not necessarily imply that it is in the form of a single compact object nor does it imply that Sgr A* must be associated with it. Stellar kinematic arguments (Genzel et al. 1996) rule out a distribution of neutron stars or white dwarfs. One possibility is that the dark mass distribution consists of $\sim 10M_\odot$ black holes; cluster evolution calculations (Lee 1995) have shown that this is at least feasible although stability arguments (Maoz 1998) suggest the cluster lifetime would be $\sim 10^8$ years, considerably less than the age of the Galaxy. In this paper, we wish to determine if, stability arguments aside, the spectrum of Sgr A* could be due to a plasma trapped within the potential well of a dark cluster of arbitrary objects.

In addition to large scale gaseous features, there is ample evidence for the existence of rather strong stellar winds in and around Sgr A* itself. The key wind sources appear to be the cluster of mass-losing, blue, luminous stars comprising the IRS 16 assemblage, located within several arcseconds of Sgr A*. A variety of observations over the years (for a review see Morris & Serabyn 1996) provide clear evidence of a hypersonic wind, with a velocity of $v_w \sim 500 - 1000$ km s^{-1}, a number density $n_w \sim 10^{3-4}$ cm^{-3}, and a total mass loss rate $\dot{M}_w \sim 3 - 4 \times 10^{-3} M_\odot$ yr^{-1} pervading the inner parsec of the Galaxy. If the dark matter is distributed, it is likely that a portion of this wind is captured by the dark compact cluster and that it settles within the cluster's potential well. Although the potential well of a cluster does not include a cusp such as that due to a black hole, the trapped plasma might conceivably still account for at least some of Sgr A*'s radiative characteristics.

In §2 we discuss the simulation of the gas flow through a distributed dark mass cluster. In §3 we describe the resulting spectrum and present some preliminary semi-analytical spectral calculations based on Sgr A* being a point mass while in §4 we summarize our analysis.

2. The 3D Hydrodynamical Model

In the classical Bondi-Hoyle scenario (Bondi & Hoyle 1944), the mass accretion rate for a uniform hypersonic adiabatic gas flowing past a centralized mass is

$$\dot{M}_{BH} = \pi {R_A}^2 m_H n_w v_w \ , \tag{1}$$

where $R_A \equiv 2GM/{v_w}^2$ is the accretion radius and M is the mass of the centralized object(s). At the GC, for the conditions described in the Introduction, we would therefore expect an accretion rate $\dot{M}_{BH} \sim 10^{21-22}$g sec^{-1}, with a capture radius $R_A \sim 0.01 - 0.02$ pc. Since this accretion rate is sub-Eddington for a $\sim$ one million solar mass concentration, the accreting gas is unimpeded by the escaping radiation field and is thus essentially in hydrodynamic free-fall starting at R_A. Our initial, simplistic, numerical simulations of this process, where we assume a point object and uniform flow (Ruffert & Melia 1994; Coker & Melia 1996) have verified these expectations.

2.1. The Stellar Wind Sources

The GC wind, however, is unlikely to be uniform since the winds from many stars contribute to the mass ejection. We assume that the early-type stars enclosed (in projection) within the Western Arc, the Northern Arm, and the Bar produce the observed wind. Thus far, 25 such stars have been identified (Genzel et al. 1996), though the stellar wind characteristics of only 8 have been determined from their He I line emission (Najarro et al. 1997); the relevant characteristics of these 25 stars are summarized in Table 1. Two sources, IRS 13E1 and IRS 7W, seem to dominate the mass outflow with their high wind velocity ($\sim$ 1000 km sec^{-1}) and a mass loss rate of more than $2 \times 10^{-4}\, M_\odot$ yr^{-1} each. Unfortunately, the temperature of the stellar winds is not well known, and so for simplicity we have assumed that all the winds are Mach 30; this corresponds to a temperature of 10^{4-5}K. In addition, for the sources that are used in these calculations, their location in z (i.e., along the line of sight) is determined randomly with the condition that the overall distribution in this direction approximately matches that in x and y.

Table 1. Parameters for Galactic Center Wind Sources

Star	x[a](arcsec)	y[a](arcsec)	z[a](arcsec)	v (km sec^{-1})	$\dot{M}(10^{-5}\, M_\odot$ yr$^{-1})$
IRS 16NE	-2.6	0.8	5.5	550	9.5
IRS 16NW	0.2	1.0	7.3	750	5.3
IRS 16C	-1.0	0.2	-7.1	650	10.5
IRS 16SW	-0.6	-1.3	4.9	650	15.5
IRS 13E1	3.4	-1.7	-1.5	1000	79.1
IRS 7W	4.1	4.8	-5.1	1000	20.7
AF	7.3	-6.7	8.5	700	8.7
IRS 15SW[b]	1.5	10.1		700	16.5
IRS 15NE[b]	-1.6	11.4		750	18.0
IRS 29N[c]	1.6	1.4	3.5	750	12.9
IRS 33E[c]	0.0	-3.0	1.5	750	12.9
IRS 34W[c]	3.9	1.6	-6.4	750	12.9
IRS 1W[c]	-5.3	0.3	7.8	750	12.9
IRS 9NW[cd]	-2.5	-6.2	-3.8	750	12.9
IRS 6W[c]	8.1	1.6	3.6	750	12.9
AF NW[cd]	8.3	-3.1	-2.1	750	12.9
BLUM[b]	9.2	-5.0			
IRS 9S[b]	-5.5	-9.2			
Unnamed 1[b]	1.3	-0.6			
IRS 16SE[b]	-1.4	-1.4			
IRS 29NE[b]	1.1	1.8			
IRS 7SE[b]	-2.7	3.0			
Unnamed 2[b]	3.8	-4.2			
IRS 7E[b]	-4.2	4.9			
AF NWW[b]	10.2	-2.7			

[a]Relative to Sgr A* in l-b coordinates (negative x is east and negative y is south of Sgr A*)
[b]Star not used in these calculations
[c]Wind velocity and mass loss rate fixed (see text)
[d]Star position changed slightly due to finite physical resolution

The sources are assumed to be stationary over the duration of the simulation. The stars without any observed He I line emission have been assigned a wind velocity of 750 km sec^{-1} and an equal mass loss rate chosen such that the total mass ejected by the 14 stars used here is equal to $3 \times 10^{-3}\, M_\odot$ yr^{-1}. Note that although we have matched the overall mass outflow rate to the observations, we have only used 14 of the 25 stars in the sample. There are two principal reasons for this: (1) stars further away than 10 arcsec (in projection) from Sgr A* are outside of our volume of solution and therefore could not be included, and (2) due to our computational resolution limits, we needed to avoid excessively large local stellar densities.

In a complex flow, generated by many wind sources, the wind velocity and density are not uniform, so the accretion radius is not independent of angle. To set the length scale for the simulation, we shall therefore adopt the value $R_A = .018\text{pc}$ (for which $1'' = 2.3\ R_A$) as a reasonable mean representation of this quantity.

2.2. The Dark Cluster Potential

We wish to study the emission characteristics of a hot, magnetized plasma "trapped" within the dark cluster's gravitational potential well. Following Haller & Melia (1996), we will represent the gravitational potential of the dark cluster with an "η-model" (Tremaine et al. 1994). This function represents an isotropic mass distribution with a single parameter. We here restrict our examination to the case $\eta = 2.5$ since this provides the closest approximation to a King model that is physically realizable (i.e., a nonnegative distribution function). We scale the mass so that $2 \times 10^6\ M_\odot$ are enclosed within 0.01 pc and the total integrated mass of the dark cluster is $2.7 \times 10^6\ M_\odot$. Thus, writing r in units of R_A we get for the enclosed mass as a function of r,

$$M_\eta(r) = 2.7 \times 10^6 \left(\frac{13.84r}{1 + 13.84r}\right)^{5/2} M_\odot\ . \tag{2}$$

A more recent assessment of the enclosed mass (Genzel et al. 1997) places a yet more rigorous constraint on the possibility of a distributed dark matter component. These newer observations may indeed invalidate the idea that *any* realistic stable distribution of objects can account for the observed gravitational potential.

2.3. The Hydrodynamics Code

We use a modified version of the numerical finite difference algorithm ZEUS, a general purpose code for MHD fluids developed at NCSA (Stone & Norman 1992; Norman 1994). The code was run on the massively parallel Cray T3E at NASA's Goddard Space Flight Center under the High Performance Computing Challenge program. Zone sizes are geometrically scaled by a factor of 1.02 so that the central zones are ~ 20 times smaller than the outermost zones, mimicking the "multiply nested grids" arrangement used by other researchers (e.g., Ruffert & Melia 1994). This allows for maximal resolution of the central region (within the computer memory limits available) while sufficiently resolving the wind sources and minimizing zone-to-zone boundary effects. The total volume is $(40R_A)^3$ or $\sim (0.7\text{pc})^3$ with the center of the spherically symmetric dark cluster distribution being located at the origin.

The density of the gas initially filling the volume of solution is set to a small value and the velocity is set to zero, while the internal energy density is chosen such that the initial gas temperature is $\sim 10^2$ K. Free outflow conditions are imposed on the outermost zones and each time step is determined by the Courant condition with a Courant number of 0.5. The 14 stellar wind sources are modeled by forcing the velocity in 14 subregions of 125 zones each to be constant with time while the densities in these sub-volumes are set so that the total mass flow into the volume of solution from each source is given by Table 1. Also, the magnetic field of the winds is assumed to always be at equipartition with the thermal energy density. The angular momentum and mass accretion rates are calculated by summing the relevant quantity in zones located within $0.1R_A$ of the origin.

We assume the magnetic field is perfectly tangled and thus ignore the effects of the magnetic field on the large scale kinematics. We take the medium to be an adiabatic polytropic gas, with $\gamma = 5/3$. Building on previous work (Melia 1994; Coker & Melia 1997), we have included a first order approximation to magnetic dissipative heating as well as an accurate expression for the cooling due to magnetic bremsstrahlung, thermal bremsstrahlung, line emission, radiative recombination, and two-photon continuum emission for a gas with cosmic abundance. For magnetic heating, we assume that the magnetic field never rises above equipartition. If compression and flux conservation would otherwise dictate a magnetic field larger than the equipartition value, the field lines are assumed to reconnect rapidly, converting the magnetic field energy into thermal energy, thereby re-establishing equipartition conditions. The cooling function includes a multiple-Gaussian fit to the relevant cooling emissivities provided by N. Gehrels (see Gehrels & Williams 1993, and references cited therein), though with the thermal bremsstrahlung portion supplanted with more accurate expressions that are valid over a broader range of physical conditions and with the inclusion of magnetic bremsstrahlung. For details on the cooling expressions used in the hydrodynamics code as well as the spectrum calculations below, see Melia & Coker (1999). Note that cooling due to Comptonization and any pair production have not yet been included since they are not thought to be significant in the vicinity of Sgr A*. Also, it is assumed that the optical depth is small throughout the volume of solution.

2.4. Results of the 3D Simulation

The integrated emissivity along the line-of-sight, at ~ 1450 years after the start of the simulation, is shown in Figure 1. The grey scale is logarithmic with solid white corresponding to a frequency-integrated intensity of $\sim 1.1 \times 10^5$ erg cm^{-2} s^{-1} steradian^{-1}, and black corresponding to $\sim 1 \times 10^{-2}$ erg cm^{-2} s^{-1} steradian^{-1}. Sgr A* is located at the center of the image. Of particular interest in this image is the appearance of streaks of high-velocity, high-density gas ("streamers") that are very reminiscent of features, such as the so-called "Bullet" seen near the Galactic center (Yusef-Zadeh et al. 1996). In our simulation, these structures are produced predominantly within the wind-wind collision regions, and in the future, we shall consider in greater detail the possibility that the observed high-velocity gas components near Sgr A* are produced in this fashion. Note the dominant role played by IRS 13E1 to the lower right of the compact radio source.

After reaching equilibrium several sound crossing times (~ 1000 years) after the start of the simulation, the enclosed mass and energy begin to fluctuate

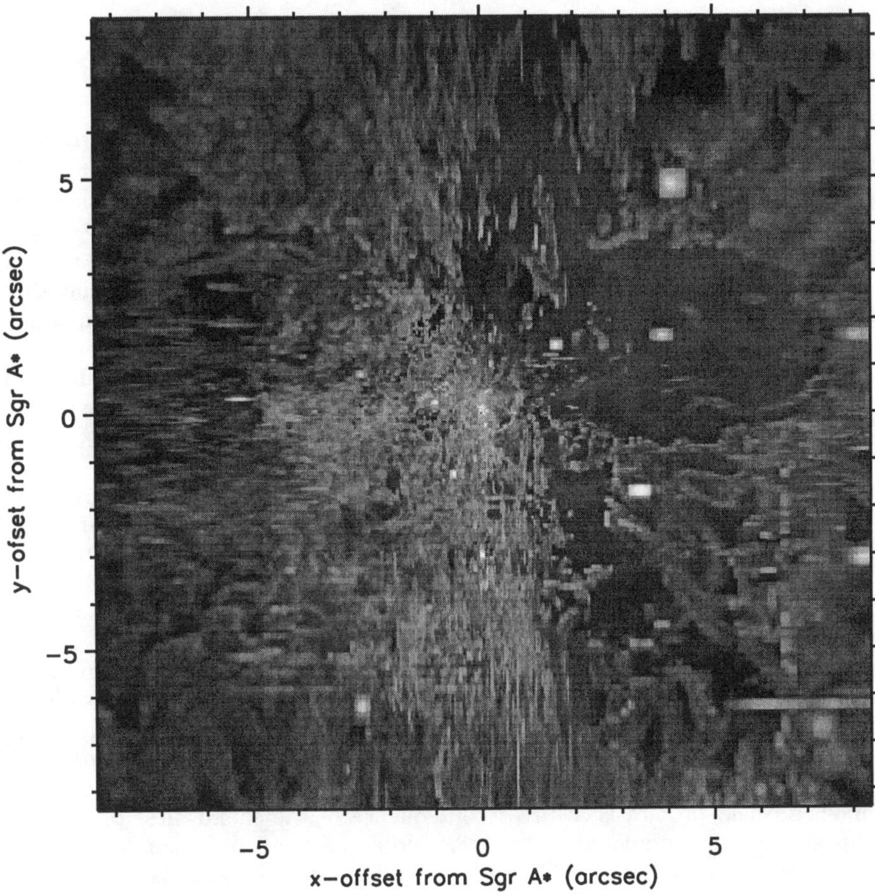

Figure 1. The line-of-sight integrated emissivity along the z-axis.

aperiodically on time scales of less than a few decades and with an amplitude of up to 50%, reflecting the turbulent cell nature of the flow in and out of the central region. Typically 2.7×10^{-3} $M_{\odot}$ of gas is trapped within the cluster at any given time. Although the gas is generally supersonic, with most of the energy in kinetic form, the thermal energy can be boosted rather suddenly when the enclosed magnetic field energy is dissipated. This occurs when strong shocks pass through the central region; the shocks compress the field sufficiently to the point where it reaches, or even surpasses, equipartition and dissipation ensues.

3. The Spectrum of Sgr A*

3.1. The Spectrum Due to a Dark Cluster

In order to calculate the observed continuum spectrum, we assume that the observer is positioned along the negative z-axis at infinity and we sum the emission from all zones that are located at a projected distance, R_{xy}, of less than $0.1\,R_A$. Since the size of Sgr A* (at λ7 mm) is $\approx 10^{12-13}$ cm (Bower & Backer 1998) and the smallest cell size in our simulation is 7×10^{14} cm, in order to minimize the inaccuracy due to numerical fluctuations, we have calculated the spectrum from a central region roughly 10 times this size ($0.1R_A$), to include at least 100 zones. Thus clearly our predicted spectrum constitutes an upper limit to the actual emission expected from Sgr A*. At the temperature and density that we encounter here, the dominant components of the continuum emissivity are electron-ion and electron-electron bremsstrahlung; we ignore line emission in these spectral calculations.

As discussed in Melia (1998), the density in the central region reaches a peak value of roughly 10^8 cm^{-3} and the temperature is never greater than about 10^8 K. Thus, since electrons begin to emit significant synchrotron radiation only above a few times 10^9 K, the gas can only emit cyclotron radiation. However, the cyclotron emissivity is insignificant compared to bremsstrahlung so that the final spectrum, as shown in Figure 2, is a bremsstrahlung spectrum that,in the radio, falls more than 4 orders of magnitude short of the observed luminosity of Sgr A* (compare with Figure 6b in Melia 1998). Figure 2 shows spectra for 3 points in time during the 3D hydrodynamical simulation. The solid, dotted, and dashed curves are at approximately 1200, 1300, and 1400 years, respectively, after the start of the simulation. The flattening of the density, temperature, and magnetic field profiles is a direct consequence of the shallowness of the dark cluster potential compared to the steep potential gradient encountered by gas falling into a black hole. The high-energy shoulder of the emission shown in Figure 3 is characteristic of the highest temperature T_{max} ($\sim 10^8$ K) attained by the gas and is consistent with the brightness temperature limit associated with Sgr A*. Similarly, the X-ray and γ-ray emission is significantly below the observed upper limits (Predehl & Trümper 1994; Goldwurm et al. 1994).

3.2. The Spectrum Due to a Black Hole

Although we have not yet run a 3D hydrodynamical simulation using the gravitational potential due to a point mass, we have undertaken semi-analytical calculations that improve upon earlier work (Melia 1994). Although details will be presented elsewhere (Coker & Melia 1999), we here present some preliminary results. We now explicitly solve for the velocity of the spherical flow using the relativistic Euler equation and integrate over μ, the cosine of the angle between the line-of-sight and the flow. We also include effects due to inverse Compton scattering and the index of refraction.

In recent work (Lo et al. 1998) it has been suggested that, at mm wavelengths at least, the intrinsic size of Sgr A* varies as ν^α with $-1.9 < \alpha < -0.7$. It is difficult to compare the absolute observed size, which is usually the FWHM of a best-fit Gaussian (e.g. Krichbaum et al. 1999), to a theoretical size, which is usually given as the surface of last scattering where $\tau(\nu) \equiv 1$; the observational

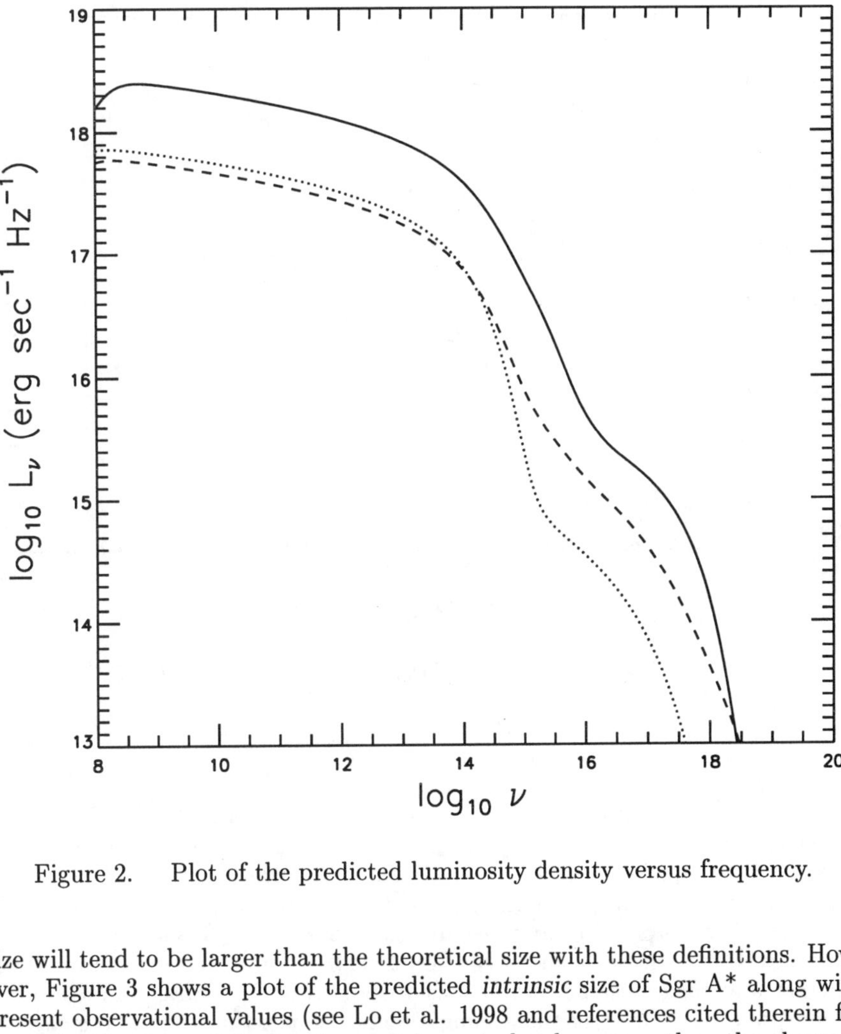

Figure 2. Plot of the predicted luminosity density versus frequency.

size will tend to be larger than the theoretical size with these definitions. However, Figure 3 shows a plot of the predicted *intrinsic* size of Sgr A* along with present observational values (see Lo et al. 1998 and references cited therein for details). The slope of the line is ~ -0.6, somewhat less steep than the observed value. This is not surprising since this particular model does not produce enough low frequency magnetic bremsstrahlung at large radii. The slope should steepen when a more optimal fit to the spectrum is found. Note that the lower size bound on the plot corresponds to the size of the black hole itself ($1R_s \sim 6\mu$as).

4. Summary

It does not appear, based on these calculations, that the gravitational potential of a distributed dark mass, be it due to a compact cluster of stellar remnants or an even more exotic collection of objects (e.g. Tsiklauri & Viollier 1998), can compress the gas from stellar winds to the point where the tem-

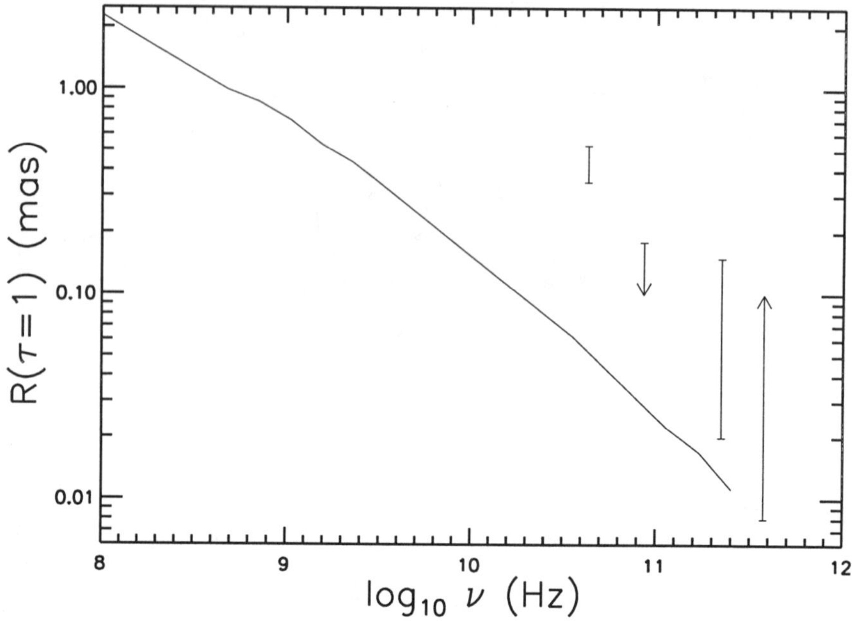

Figure 3. Plot of the predicted size of Sgr A* versus frequency (Coker & Melia 1999).

perature, density, and magnetic field can produce an observationally significant cyclotron/synchrotron emissivity at GHz frequencies. Although the spatial resolution of the simulation near the origin can be improved over that used here, it is unlikely that it will alter the results; the lack of compression is due to the inherently flat potential of any distributed dark cluster. Thus, aside from the issue of whether any dark cluster can account for the observed GC potential and whether any such cluster is stable over a significant fraction of the age of the Galaxy, our 3D hydrodynamical simulation has shown that the radio emissivity from the gas trapped within any such cluster cannot reproduce the spectrum of Sgr A*. We are left to conclude that either Sgr A* is unrelated to the accreting or trapped GC gas or that Sgr A* is the signature of an accreting, massive black hole. However, models of the structure and mechanism of radio emission from such a black hole are now faced with further constraints due to recent observations of the intrinsic source structure of Sgr A*. Coupled with the observed spectral energy distribution of Sgr A*, we may be close to differentiating between the various accretion models.

Acknowledgments. This work was partially supported by NASA grant NGT-51637 and has made use of NASA's Astrophysics Data System Abstract Service.

References

Backer, D.C. 1994, in "The nuclei of normal galaxies", ed. R. Genzel & A.I. Harris (Dordrecht: Kluwer), 403.

Bondi, H, & Hoyle, F. 1944, MNRAS, 104, 273.

Bower, G.C. & Backer, D.C. 1998, ApJ, 496, L97.

Coker, R., & Melia, F. 1996, in ASP Conf. Vol. 102, The Galactic Center: 4th ESO/CTIO Workshop, ed. R. Gredel (San Francisco: ASP), 403.

Coker, R. & Melia, F., 1997, ApJ, 488, L149.

Coker, R. & Melia, F., 1999, in preperation.

Gehrels, N. & Williams, E. D. 1993, ApJ, 418, L25.

Genzel, R., & Eckart, A. 1999, this volume, p. 3

Genzel, R., Thatte, N., Krabbe, A., Kroker, H., & Tacconi-Garman, L.E., 1996, ApJ, 472, 153.

Genzel, R., Eckart, A., Ott, T. & Eisenhauer, F. 1997, MNRAS, 291, 219.

Goldwurm, A., et al. 1994, Nature, 371, 589.

Haller, J.M., & Melia, F. 1996, ApJ, 464, 774.

Herbst, T.M., Beckwith, S.V.W., Forrest, W.J., Pipher, J.L. 1993, AJ, 105, 956.

Krichbaum, T. et al. 1999, this volume, p. 89

Lee, H.M. 1995, MNRAS, 272, 605.

Lo, K.Y., Shen, Z.-Q., Zhao, J.-H., & Ho, P. 1998, ApJ, 508, L61.

Maoz, E. 1998, ApJ, 494, L181.

Melia, F. 1994, ApJ, 426, 577.

Melia, F. 1998, this volume, p. 127

Melia, F. & Coker, R. 1999, ApJ, in press.

Mezger, P.G., Duschl, W.J. & Zylka, R. 1996, A&A Rev., 7, 289.

Morris, M. & Serabyn, E. 1996, ARA&A, 34, 645.

Najarro, F., Krabbe, A., Genzel, R., Lutz, D., Kudritzki, R., & Hillier, D. 1997, A&A, 325, 700.

Norman, M. 1994, BAAS, 184, 50.01.

Predehl, P. & Truemper, J. 1994, å, 290, 29.

Roberts, D.A. & Goss, W.M. 1993, ApJS, 86, 133.

Ruffert, M., & Melia, F. 1994, A&A, 288, L29.

Stone, J.M., & Norman, M.L. 1992, ApJS, 80, 753.

Tremaine, S., Richstone, D. O., Byun, Y., Dressler, A., Faber, S. M., Grillmair, C., Kormendy, J., & Lauer, T. R. 1994, AJ, 107, 634.

Tsiklauri, D., & Viollier, R.D. 1998, ApJ 500, 591

Yusef-Zadeh, F., Roberts, D.A., Goss, W.M., Frail, D., & Green, A. 1996, ApJ, 466, L25.

The Central Parsecs of the Galaxy
ASP Conference Series, Vol. 186, 1999
H. Falcke, A. Cotera, W.J. Duschl, F. Melia, M.J. Rieke, eds.

Proper Motions and Structure of the Radio Components Near Sgr A*: Evidence for Ionized Gas Falling Into the Massive Black Hole at the Galactic Center

Jun-Hui Zhao

Harvard-Smithsonian CfA, MS 78, Cambridge, MA 02138

W. M. Goss

NRAO-AOC, P. O. Box 0, Socorro, NM 87801

Abstract.
Proper motions and radio structures of the HII components near Sgr A* have been determined based on high resolution (0.1″) observations at 13 and 7 mm during the past 7 years (1991-1998). Proper motions up to 30 mas yr^{-1} (or 1200 km s^{-1}) have been observed for a number of prominent features. These proper motions, measured from compact HII components, can be explained by motions of a clumpy HII gas accelerated by either strong winds and outflows or explosive events at the Galactic Center. Large pattern speeds due to wave or shock motions in the circumnuclear medium at the Galactic Center can also be responsible for some of the observed proper motions. In addition, the HII components near Sgr A* (within the central 2.5″or 0.1 pc) show filamentary structures suggesting that the HII components are tidally disrupted by tides due to the concentrated mass at the Galactic Center. Based on the proper motion measurements, we also infer that the Minicavity is a transient phenomena (< 500 years old), created by the interactions between the orbiting ionized streamer and the fast moving materials at the Galactic Center. The proper motion of the Tip, NE of the Minicavity, suggests that a portion of the orbiting ionized gas components will lose angular momenta and move towards Sgr A* as a result of these interactions. High precision measurements of the proper motions for the emitting HII components can be used to track the motions of the accreting gas in proximity to the super massive black hole associated with Sgr A*.

1. Introduction

The stellar proper motion measurements at the Galactic Center have shown that the IR stars in the vicinity of Sgr A* are moving with large velocities, up to 1400 km sec^{-1} (Eckart et al. 1997; and Ghez 1998). The inferred central mass density of 10^{12} $M_\odot$ pc^{-3},which surpasses the mass densities found at the center of any other galaxy, provides strong evidence for the existence of a massive black hole (2.6×10^6 $M_\odot$) at the center of the Galaxy. Also, proper motion of Sgr A* with respect to background quasars indicates that the intrinsic motion of Sgr A* is less than 20 km s^{-1} (Backer 1996; Reid et al. 1998); this compact radio source is the signpost of the massive black hole. An important problem is to understand how the gaseous material accretes onto the black hole. Because of its proximity, the Galactic Center provides a unique laboratory to study the accretion process.

The proper motions of the ionized gas measured with respect to Sgr A* suggest that the bulk motion of the ionized gas is consistent with the kinematics dominated by the central black hole (Yusef-Zadeh, Roberts, & Biretta 1998; and Zhao & Goss 1998, hereafter ZG) while peculiar motions of some components with large transverse velocities may be explained by either local dynamics due to stellar winds and explosive events or large pattern speeds due to wave motions (ZG).

In this paper, we present a continuation of proper motion measurements of the HII components at the Galactic Center, comparing the early epoch data (1991) with new VLA observations made in 1998. We will also highlight some new results based on these new VLA observations and proper motion measurements of the HII components close to Sgr A*. Both the structures and proper motions of the ionized components in the proximity of Sgr A* are discussed. We also discuss a model for the Minicavity and a possible accretion process suggested by the proper motion measurements.

The differential motions of the radio components presented in this paper are with respect to Sgr A*. The conversion factor of proper motion to transverse velocity is 1 mas yr^{-1} = 38 km s^{-1} at a distance of 8.0 kpc (Reid 1993).

2. Observations

The new observations were made with the Very Large Array (VLA) at wavelengths 7 and 13 mm in the A array on 1998 May 8 and B array on 1998 September 9. A standard continuum mode with total bandwidth of 100 MHz was used. We recalculated u, v, w coordinates and the correct orientation of the synthesized aperture for the equinox J2000 using the AIPS task UVFIX. The flux density scale of the observations was determined using the calibrator 3C 286, assuming flux densities of 2.50 Jy at 13 mm and 1.45 Jy at 7mm, respectively. The quasar NRAO 530 was used to monitor the gains for both amplitude and phase. Further self-calibration was also applied.

The observations during 1991/1992 and 1996/1997 have been discussed by ZG (1998).

3. Results

Images of the radio emission in the vicinity of Sgr A* have been made from the 1998 observations. The angular resolutions at 7 and 13 mm are $0.06''$and $0.1''$, respectively. The r.m.s. noises of the 7 and 13 mm images are $\sim$ 0.1 and 0.2 mJy $beam^{-1}$. In Fig. 1, the new 1998 image at 13mm is compared with the previous image observed in 1991 at the same wavelength (ZG).

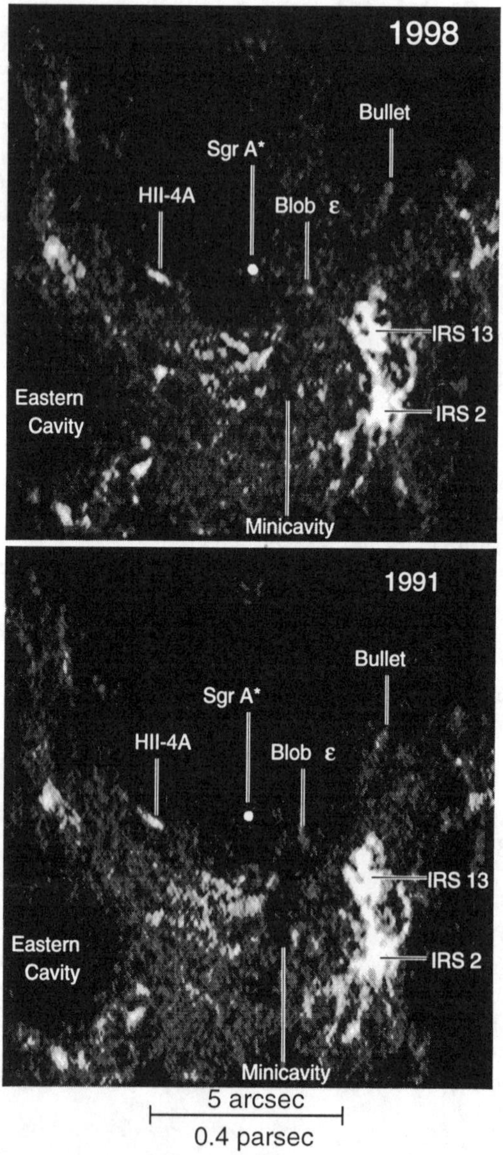

Figure 1. *(See Plate II for color version)* Radio images of the ionized gas in the vicinity of Sgr A* observed with the VLA in its A and B configurations at 13 mm in 1998 (top) and 1991 (bottom). The FWHM beam is $0.10''$. The images were made using maximum entropy method (MEM). The flux density of Sgr A* varies. The flux densities of Sgr A* on the observing dates of 1991JUN22, 1992JAN19, 98MAY08 and 1998SEP08 were 0.97 ± 0.03, 1.08 ± 0.03, 1.06 ± 0.03, and 1.37 ± 0.05 Jy, respectively. The peak intensity in both images was unified.

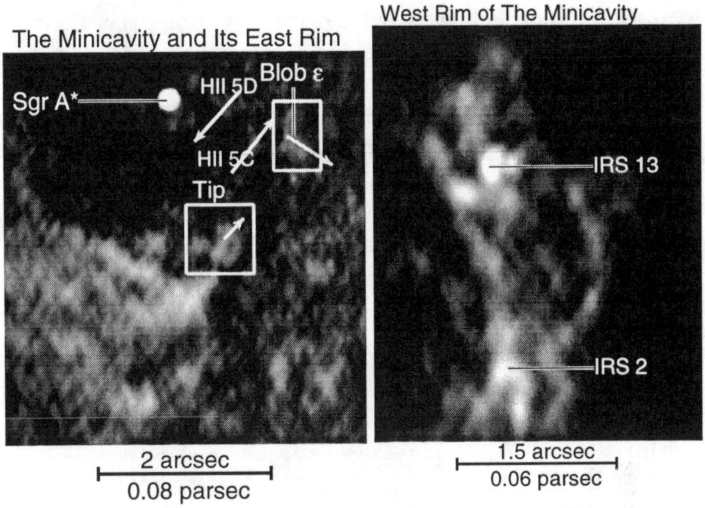

Figure 2. *(See Plate III for color version)* Images (at 13 mm) at the two epochs are overlaid to illustrate proper motions of the ionized components. Purple is 1991 and green 1998. Left panel (Fig. 2a) is for the region east of the Minicavity including Blob ϵ and the Tip, and right panel (Fig. 2b) is for the region west of the Minicavity, IRS 13 and IRS 2. The FWHM beam is $0.10''$. The proper motion vectors in the left panel are on a relative scale with the value for HII 5c = 1160 km s^{-1}.

3.1. Structure and Proper Motions of HII Components

As shown in Fig. 1, the ionized gas near Sgr A* is quite complex. Many small scale filamentary structures ($\sim 0.1''$ in width and up to a few arcsec in length) are observed in the center 10 arcsec (0.4 pc) region. Various HII features in the vicinity of Sgr A* are labeled. The emission at 13 mm (except for Sgr A*) is dominated by the free-free radiation arising from thermally ionized gas components. In the optically thin case, the observed emission intensity (I_ν) is proportional to the emission measure (EM = $\int n_e n_i \, dl$, where n_e and n_i are the volume densities of electrons and ions, and l is the path length along the line of sight) and is a weak function of electron temperature ($I_\nu \propto T_e^{-0.35}$), *i.e.* the intensity drops only by a factor of two if the electron temperature increases by an order of magnitude.

Table 1 summarizes the proper motion measurements of the HII components near Sgr A* from both an early paper (ZG) and this paper. These results are confirmed with the new epoch data obtained in 1998. Fig. 2 shows a comparison of the two epochs (1991 & 1998). The left panel (Fig. 2a) shows various features within the Minicavity and along the NE rim of the Minicavity. Among these features, Blob ϵ and the feature at the tip of the SE rim of the Minicavity (hereafter the Tip) are relatively strong and closer to Sgr A*. The proper motion of Blob ϵ is $\mu_\alpha = -19.8 \pm 2.0$, $\mu_\delta = -12.6 \pm 2.3$ mas yr^{-1}, while the Tip has a proper motion of $\mu_\alpha = -9.2 \pm 0.4$, $\mu_\delta = -8.8 \pm 0.6$ mas yr^{-1}. Both components have a filamentary morphology, stretched roughly along the direction of their own proper motion.

Table 1. Proper Motions of HII Components

Source	μ_α mas y^{-1}	μ_δ mas y^{-1}	v_t km s^{-1}	PA degree
From ZG (1998)				
HII 2-B (Bullet)	-12 ± 3	28 ± 3	1200 ± 150	-26
HII 4-A	13 ± 3	8 ± 3	590 ± 180	65
HII 7 (IRS 13)	-3.8 ± 0.7	0.9 ± 1.0	150 ± 35	−85
HII 8 (IRS 2)	-4.6 ± 0.6	7.9 ± 0.8	350 ± 38	−34
HII 9-C (Loop)	5.6 ± 2.1	-17 ± 3.3	680 ± 140	-160
HII 9-F (Loop)	-9.6 ± 3.6	17.2 ± 4.4	750 ± 85	-30
HII 12 (IRS 9)	-17 ± 5	-8 ± 5	710 ± 250	-110
From this paper				
HII 5-AB (Blob ϵ)	-19.8 ± 2.0	-12.6 ± 2.3	892 ± 79	79
HII 5-C	-19 ± 3	24 ± 4	1160 ± 140	-38
HII 5-D	20 ± 3	-22 ± 3	1130 ± 110	138
HII 5-D	20 ± 3	-22 ± 3	1130 ± 110	138
HII 11 (Tip)	-9.2 ± 0.4	-8.8 ± 0.6	484 ± 19	−46

The right panel (Fig. 2b) clearly shows that the ionized gas in the IRS 13 and IRS 2 region (or western rim of the Minicavity) moves to the west with proper motions of $\mu_\alpha = -3.8 \pm 0.7$, $\mu_\delta = 0.9 \pm 1$ mas yr^{-1} for HII 7(IRS 13) and $\mu_\alpha = -4.6 \pm 0.6$, $\mu_\delta = 7.9 \pm 0.8$ mas yr^{-1} for HII 8(IRS 2).

In addition, Fig. 3 shows low level (3 - 6σ) emission features distributed between Sgr A* and Blob ϵ. Two extended components (HII 5C and HII 5D) which can be identified in both 1998 and 1991 images show noticeable position offsets between the two epochs. The proper motions are $\mu_\alpha = -19 \pm 3$, $\mu_\delta = 24 \pm 4$ mas yr^{-1} for HII 5C and $\mu_\alpha = 20 \pm 3$, $\mu_\delta = -22 \pm 3$ mas yr^{-1} for HII 5D, corresponding to transverse velocities of 1160±140 km s^{-1} (PA=–46°) and 1130±110 km s^{-1} (PA= 138°) respectively. The proper motion vectors are illustrated in Fig. 2a.

3.2. Detailed Structure and Proper Motions of the IRS 13 Complex

Fig. 4 shows images of the IRS 13 complex, west of the Minicavity, observed with the VLA at wavelengths 7 and 13 mm at three epochs during the past seven years. In early VLA A-array observations (1996) of IRS 13 at wavelengths of 7 and 13 mm, a double feature was revealed with a compact, optically thick component to the east (IRS 13E) and an optically thin component to the west (IRS 13W). In the 1998 observations, the structure of this complex has varied significantly, the position of the optically thick component has shifted westward ($\Delta\alpha(1998 - 1996) = -44.5 \pm 8.6$, $\Delta\delta(1998 - 1996) = 13.1 \pm 13.0$ mas). The time separation of the two epochs for the 7mm observations is about 1.4 yr. A proper motion of $\mu_\alpha = -31 \pm 6$, $\mu_\delta = -9 \pm 9$ mas yr^{-1} is associated with the optically thick component.

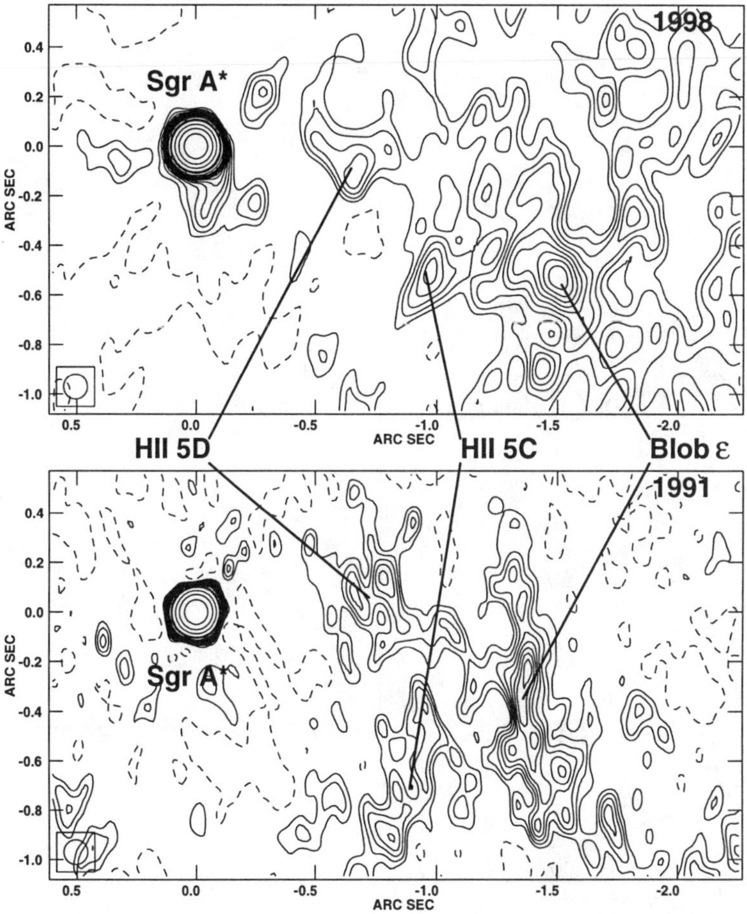

Figure 3. Contour images (at 13 mm) of the region containing both Sgr A* and Blob ϵ. Top (Fig. 3a) is 1998 and bottom (Fig. 3b) is 1991. The contours are 0.1 mJy/beam $\times$(–2, 2, 3, 4, 5, 6, 7, 8, 9, 10, 12, 16, 24, 32, 64, 128, 256). The FWHM is 0.10″.

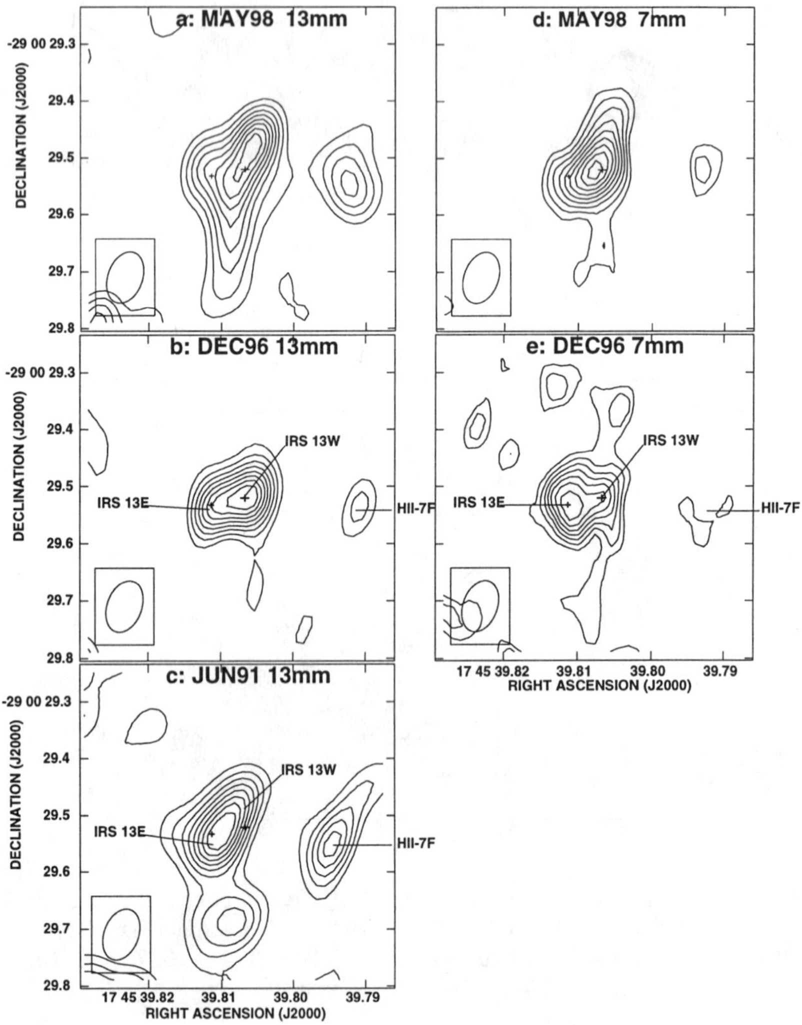

Figure 4. Images of the IRS 13 complex observed at 7 and 13 mm. The images at both wavelengths are smoothed to a FWMH beam ($0.09'' \times 0.062''$, P.A. = –21.2°). Contours are 0.4 mJy beam^{-1} × (–4, 3, 4, 5,...,10). These images were observed at three epochs as labeled. The crosses mark the positions for the double components derived from the 7 mm data from 1996.

4. Discussion

4.1. The IRS 13 Complex: A Stellar Wind from A Hot Star

The position of the compact radio source IRS 13E is close to a HeI emission line star (IRS 13E) (Krabbe et al. 1995; ZG). Based on its IR properties, this star can be classified as a hot WN9 star (Krabbe et al. 1995). The large proper motion measurement of the compact component suggests that the transverse velocity of the stellar wind is 1230 $\pm$ 240 km s^{-1} (PA=–106°) based on the observations from 1996 and 1998. The direction of the stellar outflow appears to be in the same direction as that of the expansion of the Minicavity on the western side, suggesting that the stellar wind is influenced by the gas dynamics in the Minicavity region. The mean angular size and brightness temperature of the compact component are 0.09″ (~680 AU) and 700 K, respectively. For an assumed $T_e = 1 \times 10^4$ K, the inferred emission measure (EM) and electron density are 6×10^8 cm^{-6} pc and 4×10^5 cm^{-3}, respectively.

The total flux densities of the IRS 13 complex (including both the E and W components) are 13.2±1.0 and 11.0±0.7 mJy at 7 and 13 mm. The mass loss rate can be estimated (Panagia 1991):

$$\dot{M}_{\rm wind} = 9.3 \times 10^{-5} M_\odot \ {\rm yr}^{-1} \left[\frac{S_\nu}{1 \ {\rm mJy}}\right]^{3/4} \left[\frac{d}{10 \ {\rm kpc}}\right]^{3/2} \left[\frac{\nu}{10 \ {\rm GHz}}\right]^{-0.45} \times$$

$$\left[\frac{T_e}{10^4 \ {\rm K}}\right]^{-0.075} \left[\frac{v_t}{1000 \ {\rm km \ s^{-1}}}\right],$$

where S_ν is the radio flux density of the free-free emission from the wind, d is the distance, T_e is the electron temperature, and v_t is the terminal velocity. Using the values derived from the radio observations, we infer a mass loss rate of 3 $\times10^{-4}$ $M_\odot$ yr^{-1}. This value is consistent with that derived from HeI emission line observations (Najarro et al. 1996).

4.2. Structure and Proper Motion of HII Components Within 0.1 Parsec: Tidal Disruption and Gas Falling Towards Sgr A*

The proper motion of the HII components in the vicinity of Sgr A* is consistent with the radial velocities observed in spectral lines (e.g. the H92α line by Roberts, Yusef-Zadeh, & Goss 1996), suggesting that the proper motions also trace the kinematics of ionized gas which are influenced by the central gravitational field (ZG; Yusef-Zadeh et al. 1998). There are also some components which show peculiar motions with either larger velocities in comparison with the radial velocity or a direction of motion which is inconsistent with the global proper motion patterns. These peculiar motions have been observed from the Bullet, the Loop, HII 4A and the optically thick component in IRS 13 complex (ZG; Yusef-Zadeh et al. 1998). The peculiar motions of these components can be explained either by large pattern speeds of wave motions at the Galactic Center or by acceleration due to both the local stellar winds and possible explosive events (ZG).

The small scale filaments shown in Fig. 1 may be the results of tidal instabilities. The critical density against tidal disruption at the Galactic Center is

(Sanders 1992):

$$n_{\rm HII} > 4 \times 10^{10} \left[\frac{\rm M_{BH}}{10^6\ \rm M_\odot}\right] \left[\frac{\rm 0.1pc}{\rm r}\right]^3 \qquad {\rm cm}^{-3},$$

where M_{BH} is the mass of the central black hole and r is the distance between the HII component and the black hole. Within a radius of 1 pc from the black hole, any HII components with a density below the critical value (10^8 cm^{-3}) are subject to tidal instability and will form elongated filaments. As these HII filaments move closer to the black hole (at Sgr A*), the filaments will eventually break into substructures. Within r= 0.1 pc, the critical value of the HII density exceeds 10^{11} atoms cm^{-3}. Nearly all the observed HII components in Sgr A West are subject to these tidal disruptions as they approach Sgr A*. Blob ϵ, HII 5C, HII 5D and the Tip are the closest HII components to Sgr A* . The projected distance between these components and Sgr A* is < 0.1 pc (2.5″). The observed morphology of these features is consistent with this picture. The proper motion measurements indicate that the transverse velocities of HII 5C, HII 5D and the Tip are 1160±140 km s^{-1} (PA = –38°), 1130±110 km s^{-1} (PA = 138°), and 480±20 km s^{-1} (PA = –46°), respectively. The proper motion patterns determined in the vicinity of Sgr A* also provide evidence that the debris of the HII components are moving towards Sgr A*.

There is one exception to this trend; Blob ϵ appears to move away from Sgr A* with a transverse velocity of 892±79 km s^{-1} (PA = – 122°). The nature of this source is still unknown. One explanation for this component is that the gaseous outflow winds which originate from the IRS 16 complex are focused by the gravity of the massive black hole at Sgr A*. Both Blob ϵ and the Minicavity would then result from this Galactic Center wind process (Melia et al. 1996). However, the detailed proper motions of the HII components in the region between Sgr A* and the Minicavity do not fit into this picture. Alternatively, the peculiar proper motion suggests that Blob ϵ may be associated with an ionizing star that is moving in an orbit with large eccentricity. Identification of this HII component with a stellar object will be necessary to test this hypothesis.

4.3. The Minicavity: A "Fueling Tank" for Sgr A* and the Accretion Rate

The Minicavity is a distinct hole in the distribution of the radio continuum emission with an angular diameter of 2″, located in one of the ionized streamers (the Bar) (Yusef-Zadeh, Morris, & Ekers 1989). Near-infrared spectroscopy showed that a [FeIII] emission feature coincides with the Minicavity (Lutz, Krabbe, & Genzel 1993). Based on both [FeIII] and Brγ observations, Lutz et al. suggest that the Minicavity corresponds to an expanding bubble created by a fast (> 1000 km s^{-1}) wind from one or several sources within the Minicavity. In addition, a neutral molecular streamer at v_r = –200 km s^{-1} was observed towards Sgr A*. The OH line observations suggest that the negative velocity molecular streamer impacts on the ionized Bar (Zhao, Goss & Ho 1995). The Minicavity might result from these interactions.

In addition, as a result of the interactions, the ionized gas components interior and on the east side of the Minicavity tend to lose their orbital momenta and then shift to lower angular momentum orbits, *i.e.* these gas components will fall onto Sgr A*. The interactions can provide a mechanism for Sgr A* to capture the ionized gas from the orbiting streamer in the Bar. The observed proper motion of the Tip is consistent with this picture. The accretion rate can be estimated based on that the infalling gas moves towards Sgr A* via the northern opening of the Minicavity. Therefore, the accretion rate is proportional to the mass flux ($m_H \cdot \sigma \cdot n_e \cdot v_{HII}$, where m_H is the mass of atomic hydrogen) across the northern opening with a cross section of σ:

$$\dot{M}_{acc} = 7.9 \times 10^{-4}\ \eta\ M_\odot\ yr^{-1} \left[\frac{r_{op}}{0.1\ pc}\right]^2 \left[\frac{n_e}{10^3\ cm^{-3}}\right] \left[\frac{v_{HII}}{10^3\ km\ s^{-1}}\right],$$

where r_{op} is the radius of the northern opening of the Minicavity, n_e is the mean electron density within the cavity, v_{HII} is the mean velocity across the northern opening and η is the accretion coefficient. For $r_{op} = 0.04$ pc, $n_e = 5 \times 10^3$ cm^{-3}, $v_{HII} = 500$ km s^{-1} (estimated from the current radio observations), an accretion rate of $\dot{M} = 3 \times 10^{-4}\eta\ M_\odot\ yr^{-1}$ at the Galactic Center is inferred.

Acknowledgments. The National Radio Astronomy Observatory is a facility of the National Science Foundation operated under cooperative agreement by Associated Universities, Inc.

References

Backer, D. C. 1996, in: "Unsolved Problems of the Milky Way", eds. L. Blitz & P. Teuben, (IAU Symp. 169), p. 193

Eckart, A., and Genzel, R. 1997, MNRAS, 284, 576

Ghez, A. M., Klein, B. L., Morris, M., and Becklin, E. E. 1998, ApJ, 509, 678

Krabbe, A., Genzel, R., Eckart, A. et al. 1995, ApJ, 447, L95

Lutz, D., Krabbe, A., & Genzel, R. 1993 ApJ, 418, 244

Melia, F. Coker, R. F., & Yusef-Zadeh, F. 1998, ApJ, 460, L33

Najarro, F., Kudritzki, R. P., Krabbe, A., Genzel, R., Lutz, D., Hiller, D. J. 1996 in The Galactic Center: 4th ESO.CTIO Workshop, ed. Roland Grebel, (ASP) p. 203

Panagia, N. 1991 in The Physics of Star Formation and Early Stellar Evolution, eds. C. J. Lada, N. D. Kylafis (Kluwer Academic Publishers, Dordrecht, The Netherlands), p. 565

Reid, M. J. 1993 ARA&A, 31, 345

Reid, M. J., Readhead, A. C. S., Vermeulen, R.C., & Treuhaft, R. N. 1998, this volume, p. 24

Roberts, D. A., Yusef-Zadeh, F., & Goss, W. M. 1996, ApJ, 459, 627

Sanders, R. H. 1992, Nature, 359, 131

Yusef-Zadeh, F., Roberts, D. A., Biretta. J. 1998, ApJ, 499, L159

Yusef-Zadeh, F., Morris, M., & Ekers, R. D., 1989, in The Center of The Galaxy, ed. M. Morris (Dordrecht: Kluwer), p. 443

Zhao, J.-H., & Goss, W. M. 1998, ApJ, 499, L163 (ZG)

Zhao, J.-H., Goss, W. M., & Ho, P. T. P. 1995, ApJ, 450, 122

The Central Parsecs of the Galaxy
ASP Conference Series, Vol. 186, 1999
H. Falcke, A. Cotera, W.J. Duschl, F. Melia, M.J. Rieke, eds.

Sgr A* in the Mid-Infrared Reference Frame: The Relationship Between the Radio and Infrared Sources in the Central Parsec

Dan Gezari

NASA/Goddard Space Flight Center, Greenbelt, MD 20771

Abstract. Comparison of the ionized gas and warm dust in the Sgr A West complex shows shifts in the peak positions between the infrared and radio data, as well as shifts in the positions of the IRS source peaks within the infrared image data, which could reveal the influence of the luminous sources which heat the dust and ionize the gas. No clear trend can be found in the comparison of radio/infrared or near-infrared/mid-infrared source positions which can be attributed to the effects of an exceptional source at any specific location (such as near Sgr A*), or to the results of any global process (such as ionization by a central radiation field). Thus the observations suggest that the observed position shifts are due to local effects and conditions at the individual sources in each case, and lend no additional weight to the argument that a dominant luminous or massive object is present near the position of Sgr A*.

1. Introduction

There is growing enthusiasm for the idea that a massive black hole is located at the Galactic Center. Three important questions related to this idea can be addressed with mid-infrared imaging observations: 1) Can a counterpart to the Sgr A* radio source be identified at any other wavelength? 2) Is there any observational evidence that Sgr A* interacts with other sources in the central parsec? 3) Do the radio, infrared and stellar source distributions suggest the influence of central dominant source, or a $\sim 10^6$ $M_\odot$ mass concentration, near Sgr A*?

This paper is concerned with the relationship between the mid-infrared source distribution, the radio emission and stellar sources in the central parsec. Detailed comparison of the ionized gas and warm dust in the Sgr A West complex could reveal the influence of the luminous sources which heat the dust and ionize the gas. Shifts in the peak positions between the infrared and radio data have been observed, as well as shifts in the positions of the IRS source peaks within the infrared image data. Shifts should not be unexpected in many cases, since near-infrared, mid-infrared and radio emission generally arise from entirely different physical source components (2.2 μm from stellar sources, 12.4 μm from extended warm dust, and 2-cm from gas ionized by the stars). However, any systematic trends or global effects could be the results of large-scale ionization, heating or tidal processes at work in the central parsec.

In this study we have used images of the Sgr A West complex at eight wavelengths between 4.8 and 20.0 μm which we made at the 3-m IRTF at Mauna Kea with our mid-infrared array camera system (Gezari and Dwek 1991, Gezari 1992, Gezari, Dwek and Varosi 1999). The camera uses a 58 x 62 pixel gallium doped silicon (Si:Ga) photoconductor detector array which is sensitive from 5 - 17 microns. A complete description of the array camera system is given by Gezari et al. (1992).

2. Discrepancies Between Radio and Mid-Infrared Source Positions

At first glance, the large-scale mid-infrared (our 12.4 μm image) and radio continuum emission (2-cm VLA data; Yusef-Zadeh 1996; private communication) distributions in Sgr A West appear nearly the same (Figure 1). This kind of similarity is the exception rather than the rule among Galactic mid-infrared sources with radio emission. In most familiar cases the radio and mid-infrared brightness distributions are not at all alike. However, close inspection reveals that most of the compact radio and infrared IRS source peaks in the central parsec do not coincide exactly. The differences are real, and not due to plate scale discrepancies, field rotation, distortion, etc.

At five of the nine IRS sources (IRS 2, 5, 7, 10W and 21) the 12.4 μm and 2-cm peaks are displaced 3σ or more. In the cases of IRS5 and IRS10W the radio sources are double-peaked, the weaker of which coincides with the infrared peak. In the cases of IRS2 and 21, the infrared and radio peaks are separated along the axis of the east-west bar, tangent to the direction toward Sgr A*. None of the remaining IRS sources coincide exactly, although those shifts are either too small ($\sim 1\sigma$) or the sources are too extended to be measured accurately. The 12.4 μm peak of IRS7 is located at the tip of the radio bow-shock and wake (Gezari 1997), resulting in an apparent $\sim$0.5 arcsec displacement between centroids of the 12.4 μm and 2-cm peaks when observed with the same (1 arcsec) spatial resolution.

The only IRS sources for which their 12.4 μm peak coincides in position exactly with their peak at 2-cm (to within the $\pm$0.1 arcsec infrared astrometric errors) are IRS1 and IRS10W (remember that IRS3 is not detected as a 2-cm source). However IRS1 is not a simple case, complicated by the fact that the 2.2 μm and 12.4 μm peaks do not coincide exactly (the 12.4 μm the peak coincides with the elongated 2-cm peak, but the 2.2 μm source is displaced toward the north-eastern end of the 2-cm source).

The various displacements, taken together, show no global trend or sense in which they could be evidence for the influence of a central luminous source, radiation field, or massive object located near the position of Sgr A*. Local conditions and embedded stellar sources at the 2-cm peak positions seem to be the most reasonable explanations for the observed differences in the radio and mid-infrared distributions.

3. Infrared Source Position Shifts

Several infrared sources in the central parsec show small but unambiguous changes in peak position at various wavelengths from 2.2 μm to 20.0 μm in our image data. Shifts in peak position with wavelength can be seen in the cases of IRS1, 2, 5, 7, 10W and 21. There are several possible natural (rather than instrumental) causes for the small shifts observed in near- and mid-infrared positions of some IRS peaks. Observed source brightness distributions can change with wavelength due to source temperature gradients, wavelength-dependent opacity effects, and the presence of multiple sources in only partially resolved clusters. Or, systematic shifts could indicate the effects of a large-scale radiative or tidal process. These phenomena are complicated further by optical resolution effects near the seeing/diffraction resolution limit.

First, IRS3 is the only compact infrared source whose peak does not show any shift at wavelengths between 2.2 μm and 20.0 μm, so it can be used with confidence as a fiducial mark for the registration of image data at all infrared wavelengths. However, since there is apparently no other unambiguous point source in the field, the plate scale of all images must be known and matched with high accuracy (not a trivial requirement). The peaks at IRS 1, 3, 7 and 10W do coincide at the shorter infrared wavelengths (2.2 - 8.7 μm), removing some of the ambiguity in the alignment process.

In the case of IRS1 the 12.4 μm peak position shifts $\sim$0.3 arcsec west compared to the shorter wavelengths (2.2 - 8.7 μm) because the near-infrared source is a star which is offset toward one end of the elliptical 2-cm peak, while the 12.4 μm source is centered on the 2-cm source (as discussed above). This configuration is likely to be caused by diffraction effects, and the fact that the 12.4 μm peak is asymmetrical and elongated slightly in the same sense as the 2-cm peak due to increased dust density away from the 2.2 μm stellar source. The same conditions would also account for the $\sim$0.2W arcsec shift seen in the 12.4 μm peak of IRS10W. IRS5 shows a similar monotonic change in position with wavelength. However, IRS5 shifts 0.3S, 0.1E at 12.4 μm (the opposite direction). The sense of these shifts argues against any central or global process effecting the compact sources in the central parsec.

Temperature gradients in asymmetrically distributed circumstellar material are the most likely cause of color-dependent shifts in the peaks of IRS 2 and IRS21. IRS2 is the most dramatic example of this effect, being a strong mid-infrared source which has relatively weak 2.2 μm emission; the peak in the mid-infrared emission at IRS2 apparently occurs because the density increases faster than the temperature decreases moving south and east away from neighboring IRS13. IRS21 is interesting in that the 12.4 μm peak is located between several 2.2 μm point sources and shifts $\sim$0.5 arcsec from 2.2 to 12.4 μm. The 4.8 μm peak is roughly midway between the 12.4 μm source and one of the 2.2 μm sources, again suggestive of a temperature gradient across the ridge of the northern arm.

As with the radio-infrared shifts the important point here is that, while the infrared shifts are real and anomalous, the relative positions of compact infrared sources in the central parsec do not suggest the influence of a central field or luminous source near the position of Sgr A*.

4. The Position of Sgr A* Among the Mid-Infrared Sources

Considerable effort has been made to identify an infrared counterpart to the radio source Sgr A*. Detections of Sgr A* have been claimed in the near-infrared (Close et al. 1995, Eckart et al. 1995, Rosa 1996) as well as in the mid-infrared (Stolovy et al. 1996), however, all of these results remain problematic. Upper limits for mid-infrared emission from Sgr A* have been set by Gezari et al. (1994) and Cotera (1997). Menten et al. (1997) succeeded in registering the radio continuum and near-infrared (2.2 μm) positional reference frames with $\pm$0.03 arcsec, and find the position of Sgr A* relative to the SiO maser source at IRS7 to be (α, δ) = -0.040, = -5.583 arcsec. No unambiguous infrared counterpart to Sgr A* has yet been identified, although a very weak. peculiar looking cluster of 2 μm sources is seen at that position (Close et al 1995, Menten et al. 1997).

We independently determined the position of Sgr A* relative to the mid-infrared sources in the central parsec by a direct correlation of our 12.4 μm image and the 2-cm continuum VLA data of Yusef-Zadeh (1996, private communication), comparing structure in the extended emission as well as the compact peaks, and relaxing all preconceptions about which sources must coincide. Our registration of the mid-infrared and radio source distributions (Gezari 1997) places Sgr A*(2-cm) at an offset of (α, δ) = -0.1,-5.4 ($\pm$0.1) arcsec relative to the IRS7 source peak in the 12.4 μm image. Our result for the position of Sgr A* relative to IRS7 in the 12.4 μm image differs from the result obtained by Menten et al. (1997) by 0.2 arcsec, a discrepancy which exceeds our astrometric uncertainty.

In fact, both results are apparently correct. The difference is due to small but real differences at 2.2 and 12.4 μm in the shape of IRS7 and position of the source centroid. For reasons described above, neither IRS1 nor IRS7 can be considered an unambiguous astrometric reference at all infrared wavelengths. Following the same alignment procedure as above using the 2.2 μm data of Herbst, Beckwith and Shure (1993), we obtained a position for Sgr A* relative to the IRS7 2.2 μm continuum star of (α, δ) = -0.1,-5.6 ($\pm$0.1) arcsec, essentially the same result as that of Menten et al. (1997). We also compared newly obtained NICMOS Brα images (Stolovy 1998), NICMOS 3-color near-infrared continuum images (Rieke 1998) and new 2.2 μm Keck images (Ghez 1998), and our mid-infrared image set. Initial inspection of the preview images confirms suggests that there may be a slight difference between the positions of the peak of IRS7 at 2.2 and at 12.4 μm. However, a more rigorous analysis using final, calibrated NICMOS and Keck images will be needed to determine the credibility of this result.

The 0.2 arcsec displacement between the 2.2 and 12.4 μm peaks could easily be accounted for by a mechanism which is observed in the Orion Ney-Allen Nebula, for example. In our 12.4 μm image (Gezari & Backman 1994) the extended Ney-Allen Nebula, a clearly resolved dust shell ~3 arcsec SW of Θ^1C Ori, one of the four luminous Trapezium O stars. Despite their high luminosity, the O stars are not detected at 12.4 μm with groundbased instrumentation because their photospheric emission is negligible compared to the environmental background noise level. Only the extended Ney-Allen dust cloud is detected at 12.4 μm. This cluster is about 20 times closer to us than the Galactic Center. At the distance of the Galactic Center, the ~3 arcsec separation between the star near-infrared star Θ^1C Ori and mid-infrared emission peak would measure only 0.15 arcsec, small enough to be misinterpreted as a single source. This is the magnitude of the shifts we have identified between the some of the near-infrared and mid-infrared IRS sources in the central parsec.

In any event, the small discrepancy between our result and that of Menten et al. (1997) may be moot since no conspicuous candidate for an infrared counterpart to Sgr A* lies within the error box of the positional discrepancy.

5. Conclusion

There is little evidence for perturbation of the mid-infrared brightness contours by the stellar sources, and no significant correlation between the positions of HeI line stars (Krabbe et al. 1991, 1995) and mid-infrared brightness features. There is also no obvious correlation with dust color temperature features or our modeled source opacity (dust mass) distribution (Gezari and Dwek 1991, Gezari, Dwek and Varosi 1999). This is consistent with the conclusion (Gezari 1997) that the HeI stars are located between the dense dust clouds in the Sgr A West complex and heat the dust clouds externally from regions of reduced dust density surrounding the HeI stars, between the IRS sources. But the IRS sources must be heated internally by other kinds of luminous objects. Unfortunately our previous results (Gezari 1997) suggesting higher color temperatures along the inner edge of the central cavity (which we had interpreted as allowing for the possibility of heating by a dominant luminous source near Sgr A*) were incorrect, and have proven to be artifacts of processing.

No clear trend can be found in the comparison of radio/infrared or near-infrared/mid-infrared source positions which can be attributed to the effects of an exceptional source at any specific location (such as near Sgr A*), or to the results of any global process (such as ionization by a central radiation field). Thus the observations suggest that the observed position shifts are due to local effects and conditions at the individual sources in each case. While they do exclude the presence of an exotic object, the observations lend no additional weight to the argument that a dominant luminous or massive source is present near the position of Sgr A*.

References

Close, L. M., McCarthy, D. W. J., and Melia, F. 1995, Ap J, 439, 682

Cotera, A. 1997, I. A. U. Symposium No. 184: "The Central Regions of the Galaxy and Galaxies," Kyoto, Japan

Eckart, A., Genzel, R., Hofmann, R., Sams, B. J., and Tacconi-Garman, L. E. 1995, ApJ, 445, L26

Gezari, D. Y. 1992, "The Center, Disk, and Bulge of the Galaxy", ed. L. Blitz, Kluwer Academic Publishers (Dordrecht), 23

Gezari D. Y. 1997, IAU Symposium No. 184: "The Central Regions of the Galaxy and Galaxies," Kyoto, Japan

Gezari, D. Y. and Backman, D. E. 1994, "Astrophysics with Infrared Arrays", ed. I. McLean, Kluwer Acad. Pub. Co., 195

Gezari, D. Y. and Dwek, E. 1991, IAU General Assembly Comm. 33, Buenos Aires, abstract

Gezari, D. Y., Dwek, E., and Varosi, F. 1999, Submitted to ApJ

Gezari, D. Y., W. Folz, L. Woods and Varosi, F. 1992, P. A. S. P., 104, 191

Gezari, D. Y., Ozernoy, L., Varosi, F., McCreight, C. and Joyce, R. R. 1994, "Nuclei of Normal Galaxies," ed. Genzel, R. and Harris, A., Kluwer Pub. Co., 445, 427

Herbst, T. M., Beckwith, S. V. W., Shure, M. 1993, ApJ, 411, L21

Krabbe, A., Genzel, R., Drapatz, S. and Rotaciuc, V. 1991, ApJ, 382, 119

Krabbe, A., Genzel, R., Eckart, A., Najarro, F., Lutz, D., Cameron, M., Kroker, H., Tacconi-Garman, L. E., Thatte, N., Weitzel, L., Drapaz, S., Geballe, T., Sternberg, A., and Kudritzki, R. 1995, ApJ, 447, L95

Menten, K., Reid, M. J., Eckart, A., and Genzel, R. 1997, ApJ, 475, L119

Rieke, M. J. 1998, private communication.

Rosa, M. R. 1996, in "The Galactic Center", ASP Conf. Series, 102, 189

Stolovy, S. R. 1998, private communication.

Stolovy, S. R., Hayward, T. L. and Herter, T. 1996, ApJ, 470, L45

Yusef-Zadeh, F. 1996, private communication

The Central Parsecs of the Galaxy
ASP Conference Series, Vol. 186, 1999
H. Falcke, A. Cotera, W.J. Duschl, F. Melia, M.J. Rieke, eds.

Mid-Infrared Imaging of the Central Parsec with Keck

Angela Cotera[1]

Steward Observatory, University of Arizona, Tucson, AZ, 85721

Mark Morris, Andrea M. Ghez, Eric E. Becklin, Angelle M. Tanner

Department of Physics and Astronomy, University of California Los Angeles, Los Angeles, CA, 90024

Michael W. Werner

California Institute of Technology, Jet Propulsion Laboratory, Pasadena, CA, 91109

Susan R. Stolovy

Steward Observatory, University of Arizona, Tucson, AZ, 85721

Abstract. We have obtained high resolution images of the central ~1.3 parsecs of the Galactic Center in the Mid-Infrared (MIR), using the camera MIRLIN on Keck II. The near-diffraction-limited narrow-band images, taken at 8.7μm, 12.5μm, 20.3μm, and 24.5μm, give unprecedented detail of the MIR emission in the region. Although there is no obvious *point source* at Sgr A*, there is MIR emission at the position of the black hole candidate. *Preliminary* upper limits on the possible contribution from a point source are derived. In addition, a color-temperature map for 12.5μm/20.3μm is presented, showing a strong gradient in dust temperature with distance from the IRS 16 stellar cluster and embedded sources.

1. Introduction

At this workshop we have seen that measurements of arguably the best black hole candidate, Sgr A*, have improved dramatically at all accessible wavelengths in the last five years. At 2.2μm, no definitive point source has been identified at the position of Sgr A*, although an upper limit of ~16 magnitude seems to be appropriate at this time (Genzel & Eckart 1999). As models (e.g., Donea et al. 1999, Narayan et al. 1998, Melia 1994) are tailored to reconcile the virtually non-existent NIR flux with the strong radio flux, one spectral region in which they differ is in the MIR. Therefore, in order to discriminate between these models, detection and measurement in the MIR is crucial.

---

[1] SETI Institute

In addition to measuring the MIR flux at the position of Sgr A*, high resolution MIR imaging in the central parsecs reveals significant morphological detail for both the diffuse dust and the numerous compact MIR peaks within the central parsecs (Figure 1). The diffuse dust emission is resolved into numerous streamers, ridges, and holes, many of which correspond to features seen in radio images. In particular, the IRS 7 tail (see Yusef-Zadeh & Morris 1991) is well resolved in the images at all four wavelengths, allowing us to determine for the first time the dust temperature of this feature. In addition, there are numerous similarities between the diffuse dust morphology and the NICMOS Paα images (Stolovy et al. 1999) opening the door to an enhanced understanding of the relationship between the dust and gas.

Capitalizing on the high resolution detail available at all four observed wavelengths, we are able to derive detailed color-temperature maps. With these maps, the debate over whether the heating of the dust in the region is the result of external or internal sources (e.g. Gezari et al. 1996, Smith et al. 1990, Gezari, et al. 1985, Rieke & Lebofsky 1982) may finally be resolved. The color-temperature maps clearly show that compact sources such as IRS 1 and IRS 3 must in fact be the heat source of the dust in their immediate vicinity, verifying they are most likely dust enshrouded stars; however, the diffuse emission shows a strong radial gradient, indicating a heat source in the direction of the IRS 16 cluster of hot stars.

In future work we will address whether the diffuse dust can be heated entirely by the IRS 16 stellar cluster or whether some of the diffuse dust must be heated by sources closer to the various bars, streamers and cavities. If the IRS 16 cluster is the only heat source for the diffuse dust, several features, for example the streamer seen southeast of IRS 1 (see Figure 1), will be required to have unencumbered access to the cluster: from this we can investigate whether these features are in front of, or possibly behind, the Northern Arm. Thus the temperature distribution of the dust may help us to decipher the three-dimensional structure of the region.

In this conference proceedings we present the observations, give preliminary estimates of the maximum possible MIR flux at Sgr A*, and briefly discuss the dust temperature map. Future work will address the data in greater depth including a well constrained limit on the flux at Sgr A* (Morris et al. 1999), the dust temperature and opacity distribution (Cotera et al. 1999), and the nature of the compact MIR sources and their relationship to the NIR stars and the radio sources.

2. Observations and Data Reduction

All observations were made at the Keck II Observatory using MIRLIN. MIRLIN was developed at the Jet Propulsion Laboratory and is a mid-infrared large-well imager utilizing a Rockwell Si:As BIB 128×128 pixel array. The pixel size for these observations is $0\farcs138 \pm 0\farcs001$. On 13 March 1998 observations were made at 8.7 μm, with an on-source integration time of 196s. On 15 March 1998 observations at 24.5 μm were obtained, with a total on-source time of 261s. Sky subtraction was done by chopping north-south 20–30″ and nodding 60–90″ east-west. When objects were found in the nod positions, a Gaussian was fit

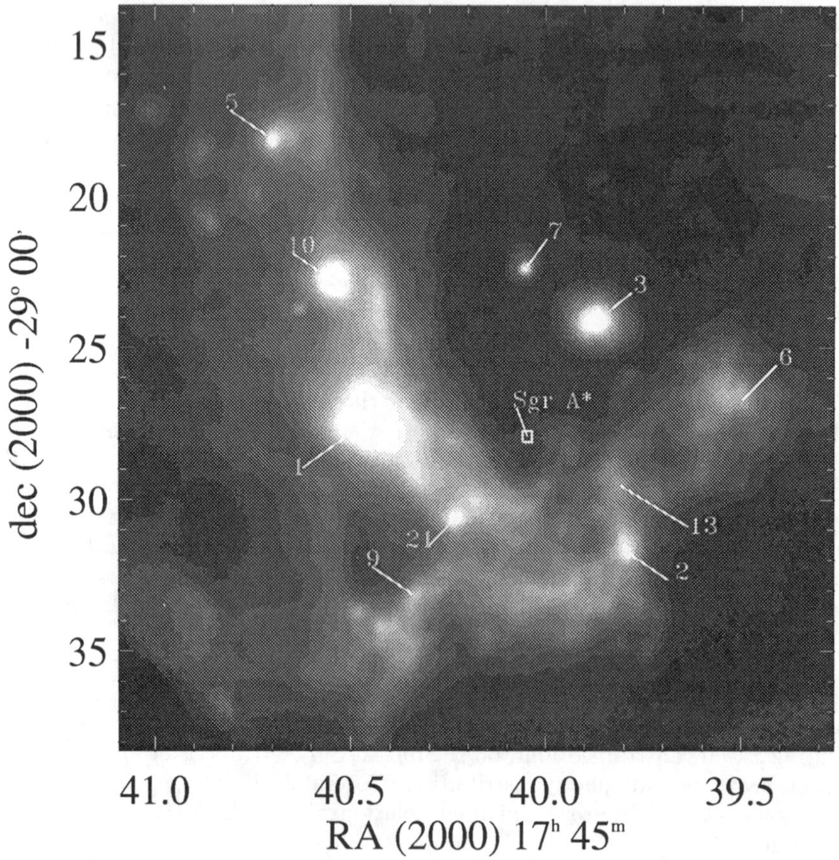

Figure 1. 12.5 μm linear grey-scale image of the central $\sim$ 1.3 pc. The position of Sgr A* is marked (the box indicates the 3σ positional error). Positions of previously identified sources are labelled although some of the sources have been resolved into several components.

to the object and the object removed. On 15 June 1998 both 12.5 μm and 20.3 μm observations were made with on-source integration times of 568s and 456s respectively. Seeing was very good for all three nights with the measured FWHM for both the 8.7 μm and 12.5 μm images $\sim0''.32$. For 20.3 μm the measured FWHM is $\sim0''.46$, and at 24.5 μm, $\sim0''.60$.

One question that arises when using a 10m class telescope at MIR wavelengths is whether or not the images are diffraction limited. Between the March observations and June observations, the size of the Lyot stop within the dewar was changed to both eliminate stray background and better match the aperture at Keck II. Therefore, in March the effective aperture was 10.5m whereas for the June observations the aperture was 9.2m. Since the theoretical definition of the diffraction limit is the diameter to the first diffraction minima, the best observational comparison which can be reliably measured is the value of the Full Width Zero Intensity (FWZI). Using apertures from above, we find for our observations the diffraction limit is $\sim0''.42$ at 8.7 μm, $\sim0''.68$ at 12.5 μm, $1''.14$ at 20.3 μm, and $1''.18$ at 24.5 μm. The measured FWZI for the images are $\sim0''.69$ for the 8.7 and 12.5 μm, $\sim1''.01$ at 20.3 μm, and $\sim1''.24$ at 24.5 μm. Therefore the images at 20.3 μm and 24.5 μm are diffraction limited and the images at 12.5 μm are at the diffraction limit.

All data were reduced using IDL routines developed specifically for use with MIRLIN data. After removing the sky and instrument effects, the images were subpixellated to a pixel scale of $0''.046$/pixel prior to being shifted and added. Throughout the night the telescope rotated; a record of the rotations was kept and the standards which are used for both calibration and point-spread-function (PSF) subtraction were smeared to simulate the effects of the source observation rotations. The images were then shifted and added by maximizing the correlation function for each successive frame in the region in which it overlapped the previously co-added frames. During the June run, two different orientations were obtained. Each orientation was reduced separately, then the subsequent images were co-added by again maximizing the correlation function for the two images.

For the June observations three standards were taken throughout the night which allows for a good determination of both airmass correction and flux calibration. By first determining the airmass correction, then flux calibrating the stars corrected to 1 airmass, the flux calibration from the three standards was consistent to within an estimated error of $\sim3\%$. For the March observations only two standards per night were available and therefore the airmass corrections and subsequent flux calibrations are not as consistent, resulting in an estimated error of $\sim8\%$.

3. Preliminary Estimate of the MIR Flux at Sgr A*

The primary source of error in determining the position of Sgr A* in our images is the need to offset from IRS 7 (Stolovy et al. 1999). The four final images were aligned by centroiding on IRS 3, however the resulting centroid of IRS 7 is slightly different at each of the different wavelengths with a total positional error of $\pm0''.054$ in RA and $\pm0''.058$ in Dec. When combined with the errors in the offsets of Menten et al. (1997) and the pixel size error, the formal positional

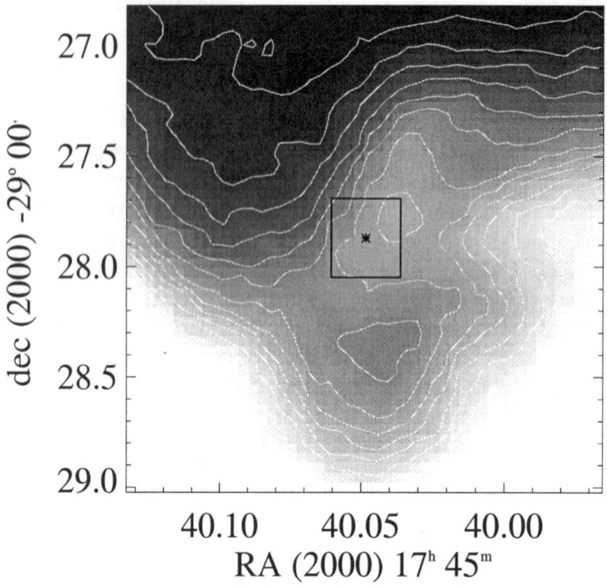

Figure 2. Grey scale 12.5 μm image of the central $\sim 2.3'' \times 2.3''$. The derived position of Sgr A* is marked with an asterisk and the box indicates the 3σ positional error. Contours are every 0.5 mJy/pixel starting at 2.5 mJy/pixel.

error in the determined position of Sgr A* in our images is $\pm0\overset{''}{.}055$ in RA and $\pm0\overset{''}{.}060$ in Dec. The position is shown in Figure 2, with a 3σ error box.

As can be seen in Figure 2, there is no distinct point source at the location of Sgr A*, although there are two peaks within the 3σ error box. It is anticipated that future work (Morris et al. 1999) will improve the astrometry and positional errors and thereby eliminate these peaks from within the 3σ error box. These images do, however, confirm a ridge of emission passing through the position of Sgr A*, and the MIR peak just to the northwest first detected at 8.7 μm by Stolovy et al. (1996) and Cotera et al. (1999). The ridge of emission is apparent at all of the wavelengths observed, although due to the higher signal-to-noise in the 12.5 μm image, it is most easily seen at that wavelength.

To date, we have used two methods to make a first, plausible estimate of the upper limit on the MIR flux from a point source at the position of Sgr A*: aperture photometry and point spread function (PSF) subtraction. There are problems with both methods which are primarily the result of the significant background MIR flux in the region. The question that needs to be answered for both methods is how to properly subtract the sky background at Sgr A* when the sky varies greatly immediately adjacent to Sgr A*?

For a first photometric estimate, we have used a position $0\overset{''}{.}5$ north and $0\overset{''}{.}5$ east of Sgr A* to estimate the background. This position represents the minimum background MIR emission at Sgr A* thereby giving a true upper limit on the flux: the background is very likely much higher. Another problem with trying to perform photometry is that due to the crowded conditions, an aperture of diameter 0.″55−1.″10 (depending on wavelength) has to be used to derive a measured flux. If a larger aperture is used, other distinguishable objects in the region are included within the aperture. Therefore, the measured fluxes have to be scaled by a factor derived using the same aperture on the standard stars in order to estimate the total flux from a point source at the location of Sgr A*. The measured apertures represent only ~50–60% of the PSF flux, introducing an additional error of ±5–10% in the final estimate. Using this method we estimate the upper limit on the flux at Sgr A*, uncorrected for extinction, to be 70±20 mJy at 8.7μm, 631±50 mJy at 12.5 μm, 1230±110 mJy at 20.3 μm, and 880±75 mJy at 24.5 μm. Before using these numbers, however, one should note that if a position south of Sgr A* is used for background subtraction, the fluxes are significantly lower, and in some cases negative. In addition, we do not ascribe this flux to Sgr A* since there is no point source at the correct location.

The other method used to determine the fluxes relies on a visual determination of the best subtraction of a normalized PSF derived from the appropriate standard star. A line cut through the position of Sgr A* is used to determine when a subtracted point source has removed the observed flux, such that a hole is created in the background emission. In order to determine an upper limit on the flux which might be contributed at the position of Sgr A* from a point source, a line cut was taken going from the northeast to the southwest across the position of Sgr A*: the reader should note that this cuts across the ridge and therefore includes all the flux from the ridge. As can be seen in Figure 3, the sensitivity of this methods varies with wavelength and the final determination is somewhat subjective. The numbers and errors quoted here are a best estimate of when the subtracted flux at the position of Sgr A* has removed most of the flux observed without creating an obvious hole. As can be seen in the figure, the residual flux on either side of the position of Sgr A* indicates the emission comes from a diffuse source, rather than a true point source: reinforcing the idea that the emission may not be from Sgr A*. Using this method, we estimate upper limits on the flux at Sgr A* to be 20±5 mJy at 8.7μm, 125±25 mJy at 12.5 μm, 250±50 mJy at 20.3 μm, and 300±100 mJy at 24.5 μm. When a line cut along the ridge of emission from the southeast to the northwest through both the position of Sgr A* and the MIR peak northwest of Sgr A* was made, the derived fluxes were significantly lower.

4. Dust Temperatures

One of the ongoing debates among Galactic Center MIR observers in the past 10–15 years has centered on whether the dust is heated by internal (e.g. Gezari 1996; Smith et al. 1990; Rieke & Lebofsky 1982) or external (e.g. Gezari et al. 1985; Becklin, Gatley & Werner 1982) sources. With our high resolution and signal-to-noise images, the situation is significantly clarified using the observed color temperature maps. A thorough examination of the data is in progress (Cotera et al. 1999), but even from preliminary work we can address the location of the heating sources. Assuming blackbody radiation and using linear interpolation of the extinction curve of Rieke & Lebofsky (1985) for $\lambda < 15\mu$m, and Mathis (1990) for $\lambda > 15\mu$m, we plot contours of constant temperature on the 12.5μm/20.3μm

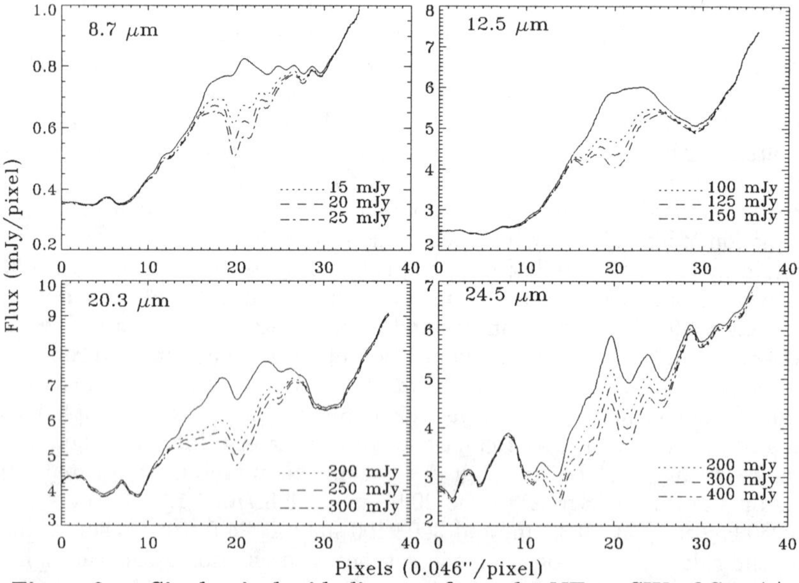

Figure 3. Single pixel wide line cut from the NE to SW of Sgr A*. The solid line is the observed flux along the cut, the other are the result of PSF subtraction of the given fluxes at the position of Sgr A* (pixel 20±1).

image (Figure 4). In Figure 4, all negative data has been set to zero, and the image has been slightly smoothed to enhance the visual clarity.

Examination of Figure 4 quickly answers the question of whether the heat sources are internal or external with a resounding "both are responsible." The compact sources long associated with MIR temperature peaks, e.g. IRS 1, 3, 7, 10, etc. (Becklin et al. 1978), are temperature peaks in our dust temperature map. The hot dust surrounding these objects is most likely the result of stellar heating of dust produced during a previous or ongoing mass loss phase of the associated star. Whether all of the Lyα emission is being re-radiated at MIR wavelengths in these compact sources, or whether some of the emission escapes and is thus available to heat the diffuse dust will be addressed in future work. A new feature seen in this high resolution map is the fact that the compact sources often associated with the Northern Arm (IRS 1, 21, 20 and 5) are offset from the temperature peaks of the diffuse emission. In addition the steepest gradient in the diffuse component is located on the western side of the Norther Arm, in the direction of the known cluster of hot stars at the position of Sgr A*/IRS 16. The temperature peak of the diffuse emission ($\sim$210 K) is actually just south of the position of Sgr A*/IRS 16 cluster at approximately $17^h\ 45^m\ 40.1^s\ -29^o\ 00'\ 30''$. Therefore, although the individual sources are also temperature peaks, they are most likely not heating the diffuse dust of the Northern Arm; preliminary indications are that the heat source is the IRS 16 cluster.

The temperature map reveals more detail, however, than just whether the heating sources are internal or external. Among the more notable morphological details is the warm ridge of emission running north-south through the position of IRS 2 and 13, which in this data set is resolved into several discrete sources within the diffuse emission. Excluding IRS 3 and 7, some of hottest dust in the region is along this ridge with peak temperatures $\gtrsim$220 K. In addition, seen for the first time is a narrow ridge of warm dust to the southeast of IRS 1. The uniformity of the dust temperature along this north-south ridge implies that it is heated externally, possibly by the Lyα radiation emitted from the IRS 16 cluster (in which case it would have to be either in front of, or behind, the Northern Arm), by IRS 21, or by an unknown nearby hot star. Finally, also seen for the first time in a dust temperature map, we can discern the tail of IRS 7 with a temperature peak of $\lesssim$200 K. The temperature peak of the IRS 7 tail is well within the tail, not along the edge closest to IRS 7 and a model for the heating needs to be addressed in the future. One possibility may be a star near the temperature peak seen in HST/NICMOS images of the region (Stolovy private communication).

5. Some Final Thoughts

These observations contain a wealth of new information on the nature of the central few parsecs of the Galaxy in the MIR. The data will permit a well constrained determination of the flux from the position of Sgr A*, although at present there does not appear to be a distinct point source at the position of Sgr A*. Future work on the distribution of the diffuse dust, its temperature, opacity and the dust/gas ratio in the region will help to clarify the nature of the Northern Arm, the Bar, and the other diffuse features in the area. The

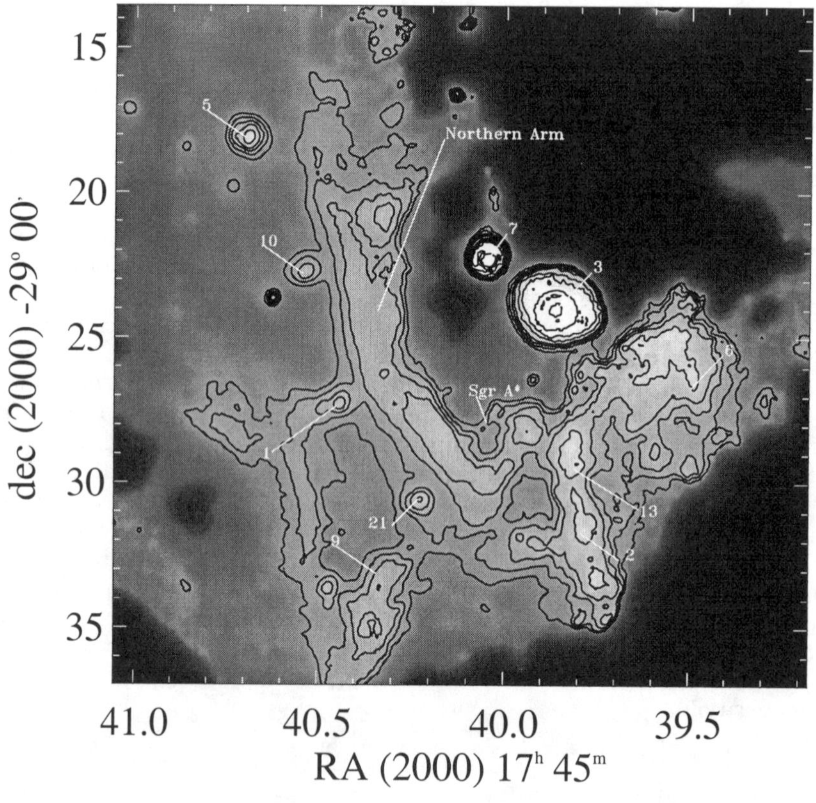

Figure 4. 12.5μm/20.3μm color-temperature map. In order to see the contours more easily, a smoothing routine has been applied to the image. Contours are 190, 195, 200, 205, 215, 225, 240, 265K.

detailed images will also allow comparison of the MIR emission and the stellar component to determine the frequency of dust enshrouded stars in the region and whether the diffuse component can be heated by hot stars located outside the IRS 16 cluster. Although the data in and of itself is very exciting, much work remains to reconcile the MIR images with the plethora of objects seen within the central few parsecs of the GC at other wavelengths.

Acknowledgments. We would like to thank Dr. David Koerner and Dr. Michael Ressler for obtaining the 8.7 μm and 24.5 μm images, and Ken Marsh for his development of the software used in the data reduction. This research was carried out in part at the Jet Propulsion Laboratory, California Institute of Technology, under a contract with the National Aeronautics and Space Administration.

References

Cotera, A.S., Morris, M., Werner, M. W, Ghez, A., Tanner, A., & Becklin, E. 1999, ApJ, in preparation

Becklin, E. E., Matthews, K., Neugebauer, G., & Willner, S. P. 1978, ApJ, 219, 121

Becklin, E. E., Gatley, I., & Werner, M. W. 1982, ApJ, 258, 135

Donea, A. C., Falcke, H., & Biermann, P. I. 1999, this volume, p. 162

Genzel, R., & Eckart, A. 1999, this volume, p. 3

Gezari, D. Y. et al. 1985, ApJ, 299, 1007

Gezari, D. Y. 1996, in *The Calactic Center: 4th ESO/CTIO Workshop*, ed. R. Grendal (A.S.P. Conf. Series, Vol. 102), p. 491

Melia, F. 1994, ApJ, 426, 577

Mathis, J. S. 1990, ARA&A, 28, 37

Menten, K. M., Reid, M. J., Eckart, A., & Genzel, R. 1997, ApJL, 475, 111

Morris, M., Ghez, A., Cotera, A., Werner, M. W., Becklin, E., & Tanner, A. 1999, ApJ, in preparation

Narayan, R., Mahadevan, R., Grindlay, J. E., Popham, R. G., & Gammie, C. 1998, ApJ, 492, 554

Rieke G. H., & Lebofsky, M. J. 1982, in *The Calactic Center*, ed. G. R. Riegler, & R. D. Blandford (AIP Conf. Proc., No. 83), p. 194

Rieke G. H., & Lebofsky, M. J. 1985, ApJ, 288, 618

Smith, C. H., Aitken, D. K., & Roche, P. F. 1990, MNRAS, 246, 1

Stolovy, S. R., Hayward, T. L., & Herter, T. 1996, ApJ, 470, 45

Stolovy, S. R., et al. 1999, this volume, p. 39

Yusef-Zadeh, F., & Morris, M. 1991, ApJ, 371, 59

The Central Parsecs of the Galaxy
ASP Conference Series, Vol. 186, 1999
H. Falcke, A. Cotera, W.J. Duschl, F. Melia, M.J. Rieke, eds.

The Circumnuclear Disk and Ionized Gas Filaments as Remnants of Tidally Disrupted Clouds

R.H. Sanders

Kapteyn Astronomical Institute, Groningen, The Netherlands

Abstract. Sticky particle calculations indicate that a coherent structure, a dispersion ring, forms when a cloud on a low angular momentum orbit passes close to the dynamical center of a potential containing a point mass. The cloud is tidally stretched and differentially wrapped, and dissipation in shocks organizes the gas into a precessing off-set elliptical ring which can persist for many rotation periods. The morphology and kinematics of the circumnuclear disk (CND) between 2 and 5 pc and the Northern arm in the inner 1 pc are well-represented by such structures. In the case of the Northern Arm, strong shocks which arise during the formation of the dispersion ring can lead to star formation even in the near tidal field of a massive black hole.

1. The Relation Between Gas Flow and Periodic Orbits

To what extent can the gas structures in the inner few parsecs of the Galaxy be understood in terms of gas flow in a gravitational field? I am referring to the well-studied Circumnuclear Disk (CND) and the Ionized Gas Filaments within the central cavity of the CND (Morris & Serabyn 1996, Mezger et al. 1996). The morphology and kinematics of the ionized gas filaments have been modeled by motion along Keplerian orbits appropriately projected on the plane of the sky (Serabyn et al. 1988, Herbst et al. 1993, Roberts et al. 1996); the success of such models strongly suggests that the motion is primarily orbital and that other possible mechanisms (e.g., stellar winds, magnetic fields, supernova explosions) do not significantly influence the overall, systematic pattern of flow. But the material is, afterall, gaseous, and how well can gas motion be approximated by orbital motion?

It is obvious that the hydrodynamic equation of motion written in Lagrangian form is the equation of motion of a particle when the pressure gradient and viscous stress terms are omitted. This means that, in a gravitational field, if pressure forces are negligible (including thermal, turbulent, viscous, and magnetic), then gas elements move on orbits. Moreover, for steady state inviscid gas flow in a gravitational field the gas streamlines are non-intersecting periodic orbits. Of course, not all orbits can be streamlines because orbits may loop and there cannot be two values of the fluid velocity at one point.

Fig. 1a shows a typical orbit in a gravitational field resulting from a point mass embedded in an approximate isothermal sphere. This mass distribution is chosen to mimic that implied by stellar kinematics in the inner few parsecs of the Galaxy, as described by Genzel & Eckart (1999) and Ghez et al. (1999) in this volume (the details are given by Sanders 1998, Paper 1). The orbit is typical of that in a general axisymmetric potential, but the hard core, the point mass, causes an abrupt bending of the orbit at closest approach (strong scattering). Now one might say that such an orbit could not possibly be a gas stream line because of the loops. However, it is possible to find a rotating frame in which the orbit appears as in Fig. 1b; this is a possible gas streamline. It was the proposal of Lindblad (1956, in connection with the problem of spiral structure) that a gas cloud on such an orbit would disperse to form such a structure which would precess at a fixed angular velocity, hence the term "dispersion ring". In this case, the precession would be counter to the sense of particle motion with an angular velocity of -26 km $s^{-1}pc^{-1}$.

One could find a set of such orbits covering a range of energy, all of which precess at about the same rate. Such a configuration is shown in Fig. 1c. For these orbits the maximum distance to the center varies between 1.7 and 2.2 pc and the tangential velocity at this point varies between 0.18 and 0.27 times the circular velocity. It is evident that these "streamlines" crowd at the point of closest approach to the point mass and this would cause a density enhancement there (in spite of the higher velocities). It is actually impossible to maintain this structure forever because the outermost orbits precess (or rather, regress) a bit faster than the inner most orbits (remember, nearer the center the potential is closer to Keplerian). So after several orbital periods we might expect the configuration to appear as it does in Fig. 1d. Here we see that the orbits have crowded on the leading edge which would give the appearance of a one-arm spiral. In fact, the orbits have begun to cross which means that dissipative forces (shocks) will intervene and the orbits can no longer represent streamlines. This will cause a loss of energy and the gas will settle, on some timescale, to a circular ring at a radius appropriate to the specific angular momentum.

2. The Fate of Clouds on Low Angular Momentum Orbits

Of course I can create a structure like this artificially by loading gas on such streamlines, and it would persist for much longer than a characteristic orbital period. But how might such a structure arise naturally? In particular, how might it arise in the inner few parsecs of the Galaxy? Let us consider the fate of a gas cloud launched on a low angular momentum orbit – an orbit which will carry the cloud near the central point mass. We will look at two cases: 1) A clumpy cloud initially between five and ten parsecs from the center on an orbit which will carry it to within one or two parsec of the center (initial tangential velocity, V_t, is 0.4 of the circular velocity, V_c). 2) A small cloud in a very low angular momentum orbit; i.e., initially between two and three parsecs on an orbit which takes it to within 0.1 to 0.2 pc of the center ($V_t = 0.2V_c$). The first case may be relevant to the circumnuclear disk (CND) and the second to the Northern Arm.

I have done these calculations by means of a sticky particle code (Paper 1) in which the motion of 4000 particles is computed in the gravitational potential described above. The effects of dissipation are included by allowing particles to interact: two particles exert a force on one another proportional to their velocity

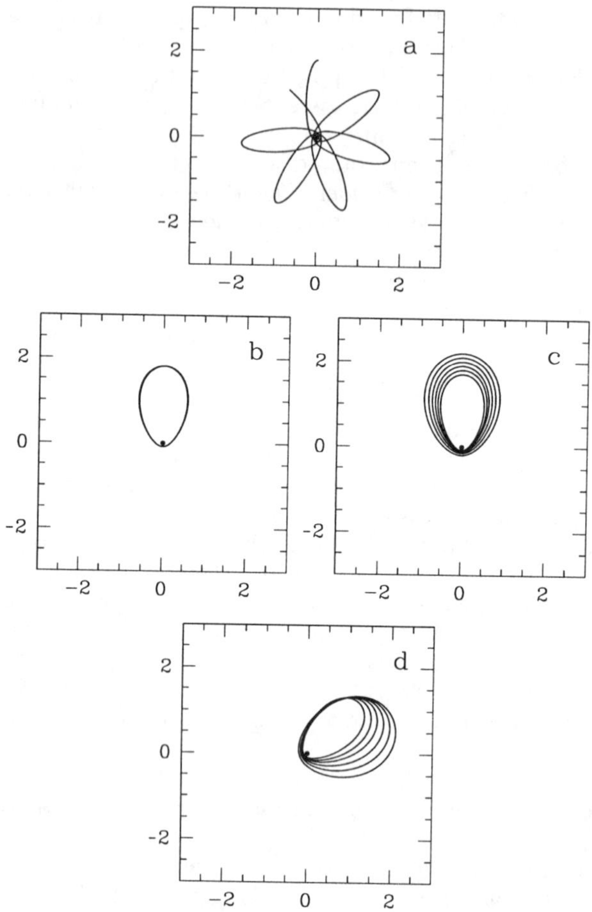

Figure 1. a) An orbit in the gravitational field of an isothermal sphere containing a point mass. The initial distance from the center is 1.8 pc and the tangential velocity is 20% of that required for a circular orbit. Distance is in parsecs.
b) The same orbit in a frame rotating counter to the sense of particle motion with $\Omega_p = -26$ km s^{-1}pc^{-1}.
c) A collection of such orbits in the rotating frame which could serve as gas streamlines for a persisting oval structure in the Galactic Center.
d. The same structure after several orbital periods; the appearance has changed due to differential precession. In all cases here the sense of particle motion is counter-clockwise and the precession is clockwise.

difference, but only if they are approaching. The procedure is similar to SPH codes but ordinary gas pressure forces are neglected; this mimics a bulk viscosity which is only effective if the flow is converging. Similar schemes have been used in simulations of accretion disks (e.g., Syer & Clark 1992); the rationale is that dissipation in strong shocks is the only important gas dynamical effect if the flow is highly supersonic (as it is).

In the first case (the "CND" cloud), the time evolution is described in Paper 1. After several rotation periods, the material from the disrupted cloud has formed an offset asymmetric ring with an extended arm. The entire structure is precessing at a mean rate of about -12 km $s^{-1}pc^{-1}$ (counter to the sense of rotation). Gradually, on a timescale of two to three million years the structure circularizes and the cavity shrinks. The reason for this circularization over a timescale of about 10 rotation periods, is differential precession. Over the rather large range in radius covered by the cloud, the outer parts precess more slowly than the inner parts leading to crossing of streamlines and strong dissipation in shocks. Non-the-less, the structure, as it appears 0.85 to 1 million years after the initial passage, does bear a strong resemblance to the CND, both in morphology and kinematics. This is shown in Fig. 2 as a particle plot with proper motion vectors projected onto the plane of the sky. Note the asymmetric central cavity; because the streamlines crowd on the side nearest the point mass, the highest gas density is observed there (north-western side). On the opposite side the ring is quite tenuous just as observed in the actual CND (Güsten et al. 1987)

The dispersion ring model is not entirely relevant to this model for the CND; there is just too much differential precession over the width of the ring. However, the structure does persist considerably longer than a characteristic rotation period (2.5×10^5 years at r = 3.5 pc). The important point is that the tidal stretching and differential wrapping of an initially clumpy cloud is consistent with the observed kinematics and morphology of the CND.

The dispersion ring model maybe more relevant to the ionized filaments within the central cavity of the CND. In Fig. 3 we see the results of the second calculation: that of a small cloud, initially 2.0 to 2.8 pc from the center, on a very low angular momentum orbit ($V_t = 0.2V_c$) which carries it within 0.2 pc of the center. The cloud is initially tidally stretched into a very long filament which repeatedly collides with itself. By an epoch of 1.5×10^5 years, a clear dispersion ring is formed. This coherent and relatively long-lived structure precesses at an angular speed of -26 km $s^{-1}pc^{-1}$. It is longer-lived than the CND model in terms of orbit times because there is less differential precession due to the smaller initial size of the cloud. In the eighth frame of Fig. 3 (t = 1.75×10^5 years), the cloud is well-described by the dispersion orbits shown in Fig. 1d. Note in particular the much higher density of gas on the side where the material is approaching the center due to the differential precession of streamlines, just as in Fig. 1d. This could explain why the Northern Arm is not observed to be a complete ring.

When this structure as it appears in the eighth frame is appropriately projected onto the plane of the sky (for this projection the orbital plane almost coincides with that of the CND cloud), the morphology and distribution of radial velocity again bears a strong resemblance to that observed for the Northern Arm (Gezari et al. 1996, Herbst et al. 1993, Roberts & Goss 1993). In Fig. 4

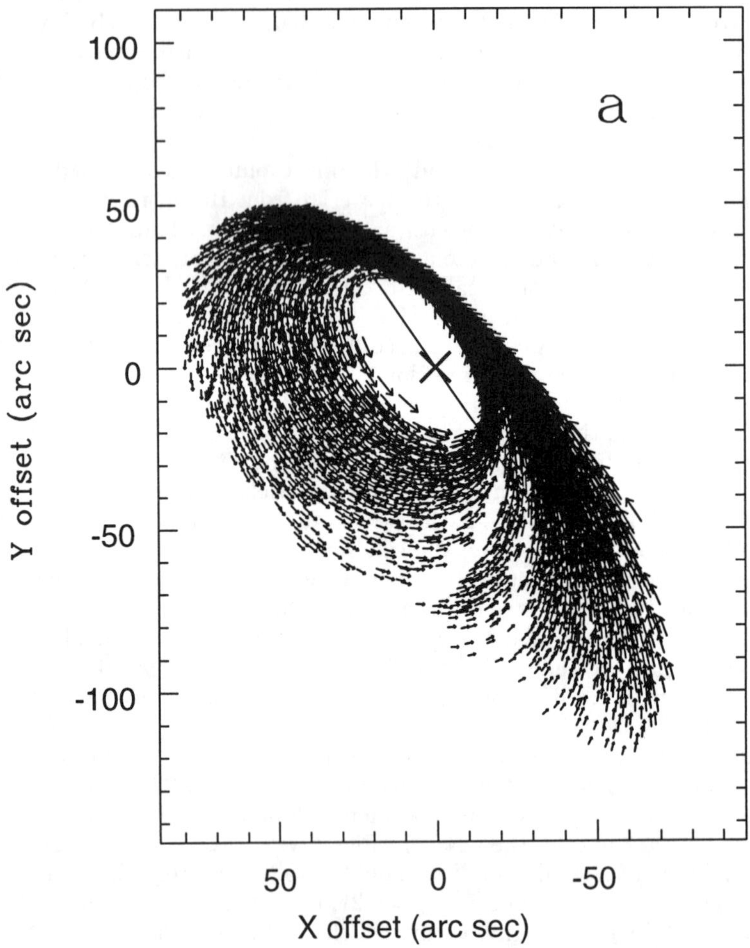

Figure 2. The projected CND model (from Paper 1) 8.5×10^5 years (three orbital periods) after launching the clumpy cloud on its low angular momentum orbit. The orbital plane is inclined 60 degrees with respect to the plane of the sky and the position angle of the line-of-nodes (solid line) is 35 degrees. The north-western side (upper right) is the near side in this projection. Units are in arc seconds offset from the dynamical center, presumably Sgr A*. This is a particle plot where the the arrows at the location of the particles indicate the sense and magnitude of proper motion. The asymmetric central cavity and the higher gas density on the side nearest the dynamical center, observed characteristics of the CND, are model-independent properties of a dispersion ring.

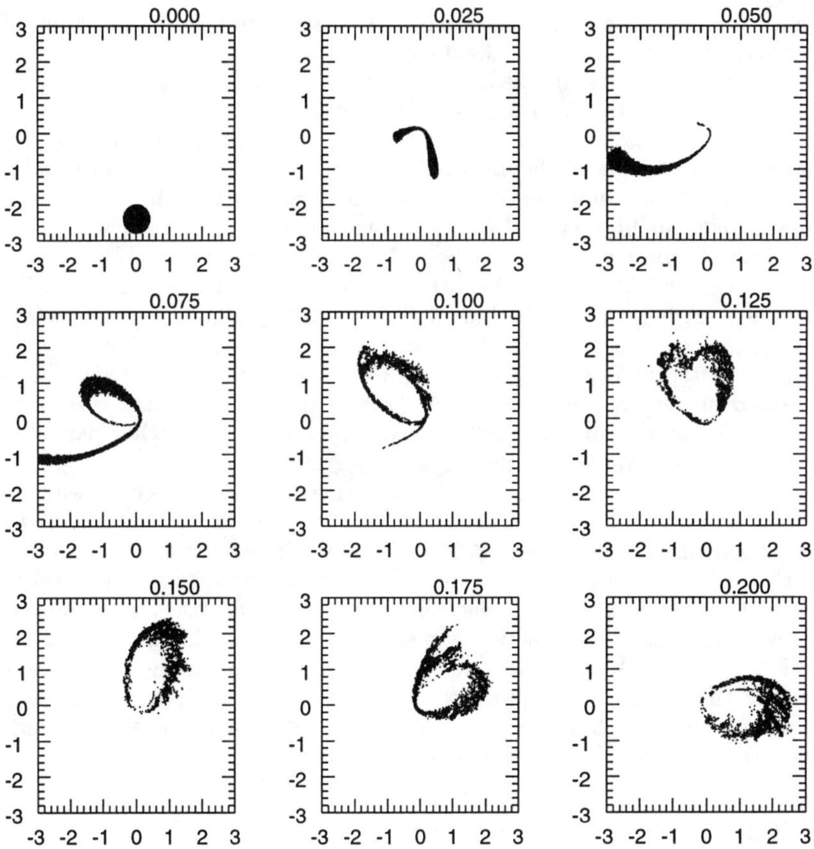

Figure 3. Time evolution of a small cloud (from Paper 1), initially at a mean distance of 2.4 pc from the center, on a very low angular momentum orbit ($V_t = 0.2V_c$). The distance is in units of parsecs and the time since beginning of the infall is given above each frame in units of one-million years

we see the morphology and the projection of the velocity vectors of on the plane of the sky. The sense of proper motion agrees with that recently observed by Yusef-Sadeh et al. (1998) and by Zhao & Goss (1998).

3. Star Formation in Strong Shocks

While the dispersion ring is forming, the tidally stretched cloud collides with itself near the point mass (Fig. 3, fourth and fifth frames). The collision velocities exceed 100 km/s. Assuming highly efficient radiation of the thermal energy, which is likely in the molecular gas, the resulting strong shocks will lead to significant compression of the gas and high gas densities; indeed the gas densities might well exceed the Roche limit (10^{13} to 10^{14} particles cm^{-3}) in the tidal field of a 2.5×10^6 $M_\odot$ point mass at a distance of 0.1 pc (Phinney 1989). Such a mechanism might very well be the explanation for the young stars observed within a few tenths of a parsec of Sgr A*, the putative massive black hole at the Galactic Center (Allen et al. 1990, Eckart et al. 1993, Krabbe et al. 1995).

The sticky particle algorithm applied here allows one to determine the local value of the compression at the location of a particle ($-\nabla \cdot V$). This may then be used as a criterion for star formation. If the compression exceeds some arbitrary threshold (in this case 2000 km $s^{-1}pc^{-1}$), that particular particle is re-tagged as a star; thereafter, there are no "viscous" interactions with other particles, and the particle's motion is determined only by gravity.

The results of such a calculation are shown in Fig. 5 where we see both the gas and star distribution after roughly one and two precession periods of the dispersion ring. We see that the gaseous dispersion ring resembling the Northern Arm can easily persist for two precession periods or roughly 10 to 20 characteristic orbital periods; there is no need for this structure to be a highly transient manifestation of a tidally disrupting cloud on the first passage by the black hole. Basically, the dispersion ring appears as an "attractor" in the phase space of the system – a well-known phenomenon in dissipative systems. The stars, on the other hand, without any such mechanism for self-organization, slowly diffuse throughout the available phase space.

4. Comments

We may conclude that the tidal disruption of a small cloud in a potential containing a point mass will lead to a long-lived gas structure – a dispersion ring – which precesses counter to the sense of gas rotation. Such objects resemble, morphologically and kinematically, the gaseous structures seen in the inner few parsecs of the Galaxy – the CND and, within the central cavity of the CND, the Northern Arm. While forming a dispersion ring, a low angular momentum cloud (passing near the central point mass), is stretched into a long filament which collides with itself several times. The resulting strong shocks lead to high compression and can be the sites of star-formation, even in the near tidal field of the black hole. This may explain the presence of massive stars less than several million years old within 0.1 pc of Sgr A*.

Of course, I have not dealt with the problem of initial conditions. How are clouds launched on such orbits to begin with? We know from a number of obser-

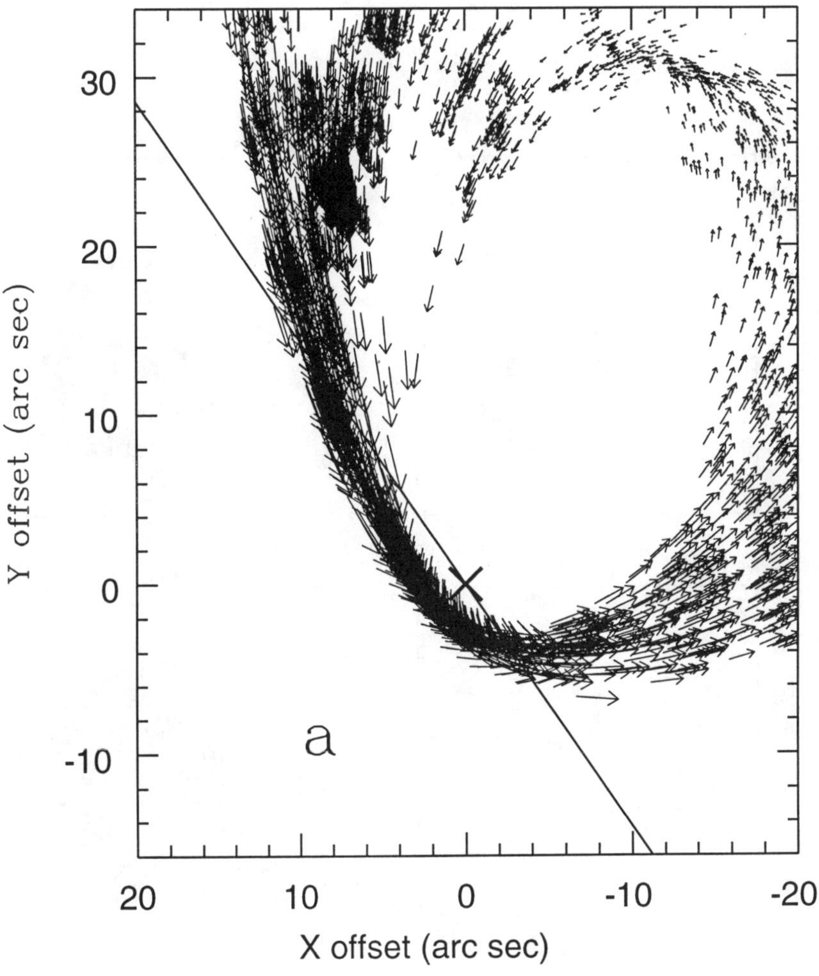

Figure 4. The Northern Arm model as seen in the eighth frame of Fig. 3 (1.75×10^5 years) projected onto the plane of the sky (from Paper 1). This is a particle plot where the arrows indicate the sense and magnitude of the proper motion. Coordinates are offsets in arc seconds from the dynamical center. Here again the north-western side (upper right) is the near side; the plane of the Northern Arm model coincides with the plane of the CND model to within 10 degrees. The model matches the observed morphology and kinematics of the Northern Arm. Note in particular the higher density on the western side (due to orbit crowding as in Fig. 1d); this gives the appearance of an incomplete ring.

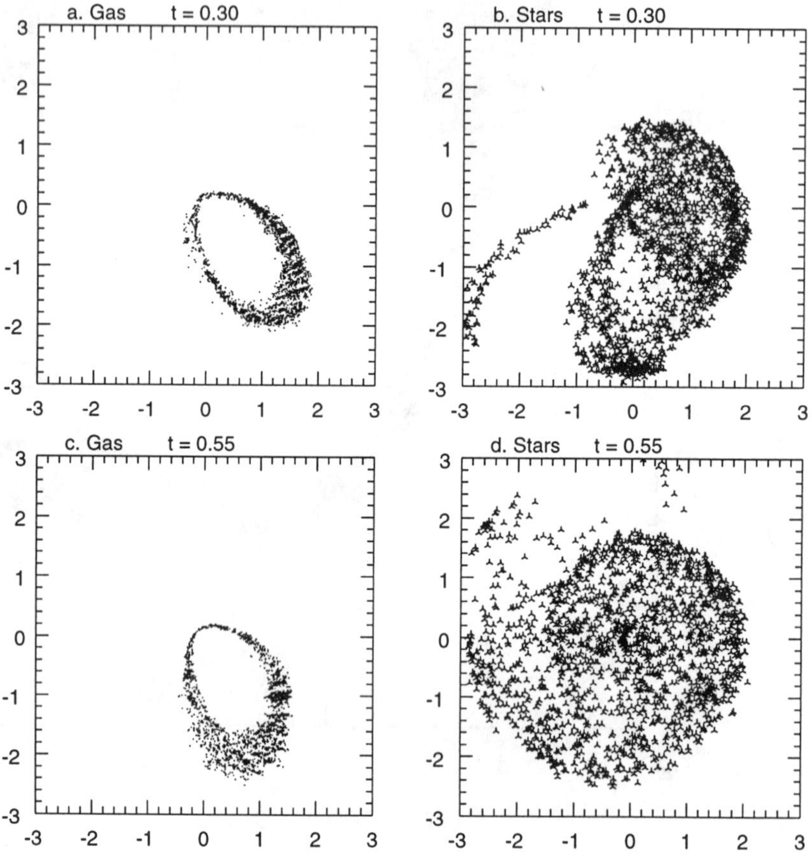

Figure 5. a,b) Gas and star distributions at t = 3×10^5 years corresponding to more than one complete precession of the dispersion ring. c,d) Gas and star distributions at t = 5.5×10^5 years or almost one precession time later. Distance is in units of pc. This figure is from Paper 1.

vations (Morris & Serabyn 1996) that the interstellar medium in the inner 200 pc of the Galaxy is highly inhomogeneous and very turbulent – most of the gas is found in massive molecular clouds and the random velocities of these clouds are a considerable fraction of the circular velocity. The decay of supersonic turbulence occurs through cloud-cloud collisions so occasionally, through this process, a low angular momentum cloud will be created. Bursts of star formation or the occasional flaring of the black hole are possible mechanisms for maintaining the turbulence.

Such a scenario might have general relevance to accretion onto massive black holes in active and normal galactic nuclei. Accretion may proceed via a series of such tidal disruptions of clouds on low angular momentum orbits. If this were true, we would expect accretion to be highly episodic but also highly inefficient with most gas disappearing into star formation.

References

Allen, D.A., Hyland, A.R., Hillier, D.J. 1990, MNRAS, 244, 706

Eckart, A., Genzel, R., Hormann, R., Sams, B.J., Tacconi-Garman 1993, ApJ, L77

Güsten, R., Genzel, R., Wright, M.C.H., Jaffe, D.T., Stutzki, J., Harris, A.I. 1987, ApJ, 318, 124

Genzel, R., & Eckart, A. 1999, this voume, p. 3

Gezari, D., Dwek, E., Varosi, F. 1996, in IAU Symp.169, Unsolved Problems of the Milky Way (eds. L. Blitz & P. Teuben), Dordrecht: Kluwer, 231

Ghez, A., et al. 1999, this volume, p. 18

Herbst, T.M., Beckwith, S.V.W., Forrest, W.J., Pipher, J.L. 1993, AJ, 105, 956

Krabbe, A., Genzel, R., Eckart, A., Najarro, F., Lutz, D., Cameron, M., Kroker, H., Tacconi-Garman, L.E., Thatte, N., Weitzel, L., Drapatz, S., Geballe, T., Sternberg, A., Kudritzki, R. 1995, ApJ, 447, L95

Lindblad, B. 1956, Stockholms Obs.Ann., 19, No. 7

Mezger, P.G., Duchl, W.J., Zylka, R. 1996, A&A Rev. 7, 289

Morris, M., Serabyn, E. 1996, Ann.Rev.A.Ap., 34, 645

Phinney, E.S. 1988, in IAU Symp.136, The Center of the Galaxy (ed. M. Morris), Dordrecht: Kluwer, 543

Roberts, D.A., Goss, W.M. 1993, ApJS, 86, 113

Roberts, D.A., Yusef-Zadeh, F., Goss, W.M. 1996, ApJ, 459, 627

Sanders, R.H. 1998, MNRAS, 294, 35 (Paper 1)

Syer, D., Clarke, C.J. 1992, MNRAS, 255, 92

Serabyn, E., Lacy, J.H., Townes, C.H., Bharat, R. 1988, ApJ, 326, 171

Yusef-Zadeh, F., Roberts, D.A., Biretta, J. 1998, ApJ, 503, L191

Zhao, J.-H., Goss, W.M. 1998, ApJ, 499, L163

The Central Parsecs of the Galaxy
ASP Conference Series, Vol. 186, 1999
H. Falcke, A. Cotera, W.J. Duschl, F. Melia, M.J. Rieke, eds.

A Gas Streamer Feeding the Circumnuclear Disk

Alison L. Coil

Astronomy Department, University of California, Berkeley, CA 94720

Paul T. P. Ho

Harvard-Smithsonian Center for Astrophysics, Cambridge, MA 02138

Abstract.
We present VLA maps of NH_3(1,1) and NH_3(2,2) emission in the inner 10 pc of the Galaxy, covering Sgr A*, the circumnuclear disk (CND), and the 20 km sec^{-1} cloud. In the central 3 pc surrounding Sgr A* we map a portion of the CND which is connected to a long, narrow 2 pc by 10 pc streamer of clumpy molecular gas located towards the south. This southern streamer appears to be carrying gas from the nearby 20 km sec^{-1} cloud to the CND. We find a velocity gradient along the streamer, with progressively higher velocities as the gas approaches Sgr A*. The gas is the northern tip of the streamer is heated and the line width increases dramatically, which may be a kinematic signature of accretion onto the CND.

1. Introduction

Immediately surrounding Sgr A West is the circumnuclear disk (CND), located 1.5 pc to 7 pc from Sgr A*. The CND is dense (10^5 cm$^-3$), clumpy, and turbulent with large line widths ($\geq$40 km sec^{-1}). The gas is predominately rotating around the nucleus at 110 km sec^{-1}, but non-circular motions are also present. The CND appears to be only semi-complete and possibly not in equilibrium. It may be part of an accretion process which funnels gas towards the nuclear region. The question remains as to how the disk itself is fed.

To the south of the CND lies a giant molecular cloud (GMC) named the 20 km sec^{-1} cloud (M-0.13-0.08 in Galactic coordinates). Ho et al. (1991) map in NH_3(3,3) a long gas streamer which connects the 20 km sec^{-1} cloud to the location of the CND. They propose that this gas streamer feeds the circumnuclear region with gas from the 20 km sec^{-1} cloud. No velocity gradient along the streamer is detected within their spectral resolution of 10 km sec^{-1}, and no increase in line width is seen, prompting an interpretation that the gas is moving towards Sgr A* across the line of sight. Hints of this streamer are also seen in HCN (Marr et al. 1993) and submillimeter continuum emission (Dent et al. 1993). However, these studies also find no clear velocity gradient in the streamer along the line of sight.

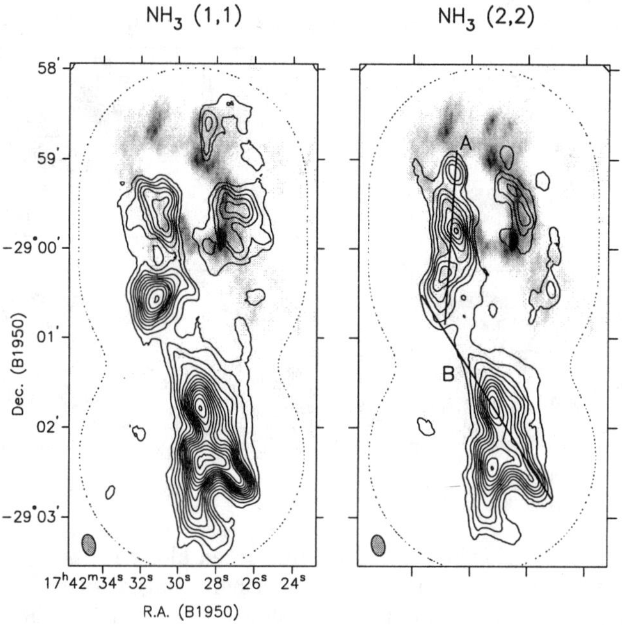

Figure 1. Velocity-integrated maps of NH_3(1,1) and (2,2) emission in the central 10 by 15 pc of the Galaxy are shown in contours over a false color image of the circumnuclear disk (CND) in HCN(1-0) emission (Marr et al. 1993). A 2 pc by 10 pc streamer connects the 20 km sec^{-1} GMC to the south with the CND to the north. NH_3 emission also traces portions of the eastern and western sides of the inner edge of the CND. The contours are at integer levels of 0.4 Jy $beam^{-1}$ km sec^{-1}, and the 9" by 14" synthesized beam is shown in the lower left corner. The straight lines labeled A and B correspond to position-velocity cuts shown in Figure 2.

In order to further probe the morphology and kinematics of this streamer, we observed NH_3(1,1) and (2,2) emission surrounding Sgr A* and the region directly south with the VLA in C/D configuration. With a 12 MHz bandwidth and a spectral resolution of 4.9 km sec^{-1}, we made three-pointing mosaics in MIRIAD with a $2'$ primary beam per pointing.

2. Results

2.1. Morphology

Figure 1 presents the velocity-integrated NH_3(1,1) and (2,2) emission in contours overlaid on a false color HCN map of the inner region of the CND. A long, narrow, 2 pc by 10 pc north-south streamer can be seen in both maps. The southern half of this streamer extends into the 20 km sec^{-1} cloud, and the streamer ends to the north at the location of the CND where little HCN emission

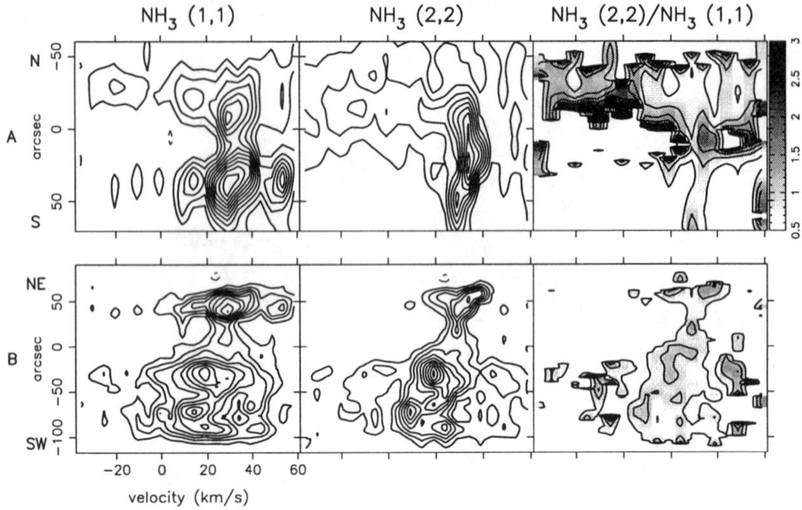

Figure 2. Position-velocity diagrams taken along A and B as shown in Figure 1 along the northern and southern lengths of the streamer. NH_3(1,1) and (2,2) emission are shown, as well as the divided NH_3(2,2)/(1,1) ratio which indicates heating, where the darker regions trace hotter gas. A clear redshift is seen along the southern part of the streamer, where no significant heating is apparent. A dramatic increase in line width is seen at the northern top of the streamer, where the gas is heated. This may be a result of accretion of the streamer onto the CND.

is seen. A semi-complete ring is mapped in NH_3 around the CND. The greatest morphological differences between the NH_3(1,1) and (2,2) emission are around the nuclear region, where the NH_3(1,1) emission traces the CND at slightly larger radii. The northern tip of the streamer extends further into the interior of the disk in the NH_3(2,2) map. The northeastern quadrant of the CND is missing in the NH_3 maps due to the limited spectral window of this experiment, which does not cover the range of HCN velocities observed in the disk.

2.2. Kinematics

Position-velocity diagrams of the streamer are shown in Figure 2 for both the NH_3(1,1) and (2,2) transitions, as well as the divided line ratio of the two, which gives an indication of temperature. The black regions, where the NH_3(2,2)/(1,1) ratio is the highest, trace the hotter gas. The lower row of diagrams from the southern half of the streamer (cut B) show an overall redshifting of the gas of about 5 to 8 km sec^{-1} $arcmin^{-1}$, with the velocities increasing towards the nuclear region. The NH_3(2,2)/(1,1) diagram indicates that there are no significant temperature effects in this region of the streamer.

In the top row of diagrams from cut A along the northern section of the streamer, the most obvious feature is the dramatic increase in line width of the gas where the streamer meets the CND as seen in the NH_3(2,2) diagram.

The (2,2) line traces the intrinsic line width of the gas better than the (1,1) line, which has unresolved hyperfine structures that artificially broaden the line widths. The (2,2) diagram shows that the gas just south of the CND has narrow line widths, 10 km/s, while at the northern tip of the streamer, located inside the CND in projection, the line widths increase significantly and emission is seen across our entire sampled velocity range, from 60 km sec^{-1} to -25 km sec^{-1}. The NH_3(2,2)/(1,1) diagram of the northern part of the streamer shows that the gas at the northern tip is heated, where the blue-shifted wing of emission is seen in black.

3. Discussion

Whether this gas streamer is seen only in projection against the Galactic Center or whether it is actually approaching the central mass concentration can only be determined if we see the direct effects of the deepening gravitational potential well on the purported infalling feature. Our new data support this accretion theory with morphological, thermal, and kinematic evidences.

In Figure 1 it is of special significance that the streamer stops at the CND. This can also be seen in Figure 2 where in the top row the gas moving into the region at 30-35 km sec^{-1} with narrow line widths clearly stops in the center of the diagram, where the gas just above to the north has much larger line widths. The comparison of the different NH_3 transitions suggests that the termination of the streamer is not a temperature effect.

We find that the northern tip of the streamer, located nearest to Sgr A* in projection, is stronger in NH_3(2,2) emission than in (1,1) emission, indicating that the gas is heated (Figure 2, top row, diagram on the right). This would seem to indicate that the gas is located at the distance of the Galactic Center and is being heated by the nuclear processes.

Velocity gradients are detected along the streamer. The expected gradient at a radius of 5 pc for a central mass of 10^7 $M_\odot$ is 9 km sec^{-1} pc^{-1}, roughly 3 or 4 times more than we observe. Thus, our detected gradient suggests that if this motion is infall, the bulk of the motion is across the line of sight, with the geometry such that the streamer is more to the side of Sgr A* rather than along the line of sight.

Finally, the velocity dispersion of the observed motions indicate that the line width of the gas increases substantially at the northern tip of the streamer, where the CND is located. Other studies of the southern streamer have coarser spectral resolutions than our 5 km sec^{-1}, and this increase in line width has not been detected before. It is expected that if the gas is moving from the 20 km sec^{-1} cloud to the nuclear region, it would interact with the CND and become disrupted, resulting in an increase in the line widths of the gas, regardless of whether the motion is predominantly along or across the line of sight. The disk is known to have inherently large line widths, and our detection and measurements of a component with increased line widths concretely place the northern tip of the streamer at the location of the CND and are consistent with the gas in the southern streamer accreting onto the CND. This kinematic evidence does not reflect chemistry or abundance effects and is the strongest argument for the streamer transporting gas to the nuclear region along the southern streamer.

4. Summary

We map a streamer which appears to be feeding the nuclear region with molecular gas from the nearby 20 km sec^{-1} GMC. The streamer originates from the northeastern edge of the 20 km sec^{-1} cloud, south of Sgr A West, and ends to the north at the location of the CND, just southeast of Sgr A*. We detect a clear velocity gradient along its length of 5 to 8 km sec^{-1} $arcmin^{-1}$. The line widths of the gas increase substantially in the northern portion of the streamer, with FWHMs of $\geq$50 km sec^{-1}, possibly indicating accretion onto the CND. The gas at the location of the CND also appears to be heated, as indicated by the ratio of the NH^3(2,2)/(1,1) emission. Our morphological, kinematical, and thermal data strongly support the thesis that gas is falling in towards the circumnuclear region from the 20 km sec^{-1} cloud along the southern streamer and accreting onto the CND.

A more detailed report of the data will be published in Coil & Ho (1999). We would like to thank Mel Wright for providing the HCN maps of the CND.

References

Coil, A. L. & Ho, P. T. P. 1999, to appear in ApJ, 513.

Dent, W. R. F., Matthews, H. E., Wade, R., & Duncan, W. D. 1993, ApJ, 410, 650

Ho, P. T. P., Ho, L. C., Szczepanski, J. C., Jackson, J. M., Armstrong, J. T., & Barrett, A. H. 1991, Nature, 350, 309

Marr, J. M., Wright, M. C. H., & Backer, D. C. 1993, ApJ, 411, 667

The Central Parsecs of the Galaxy
ASP Conference Series, Vol. 186, 1999
H. Falcke, A. Cotera, W.J. Duschl, F. Melia, M.J. Rieke, eds.

The Minispiral in the Galactic Center – Exploring a Data Cube With a Three Dimensional Method

Bernd Vollmer[1] and Wolfgang J. Duschl[2]

Institut für Theoretische Astrophysik der Universität Heidelberg
Tiergartenstr. 15 D-69121 Heidelberg

Abstract. A three dimensional method for the analysis of data cubes is presented. It is applied on Ne[II] line data of the ionized feature in the inner 3×4 pc of the Galactic Center, called the Minispiral. The data cube can be viewed as a whole and different structures can be easily and clearly separated. A simple model of rotating material in different plains is fitted in three dimensions. Thus the location of the emitting material can be determined in the real three dimensional space and can be viewed as such. It is shown that the emission features of the ionized gas can be explained by a minimum number of 3 plains.

1. Introduction

In order to understand the accretion process near a very massive central object (e.g. a black hole) at the parsec-scale, the gas motions around the Galactic Center are of special interest. It is the only place where we can observe the kinematics of the gas within these scales. The knowledge of the dynamics is crucial for the understanding of how the central engine is fueled. In the inner 7pc a disk-like structure of molecular gas is rotating with velocities of about $v \sim 100$ km/s around the Galactic Center (Serabyn et al. 1986, Güsten et al. 1987, Jackson et al. 1993). These velocities of the *circumnuclear disc* (CND) correspond to Keplerian rotation around a central mass of several million solar masses. The CND has a sharp inner edge at ~2pc where the column density of the gas drops by an order of magnitude. Inside this edge a feature of ionized gas is located, called the *Minispiral* (Lo & Claussen 1983, Lacy et a. 1989). The western side of the Minispiral is kinematically connected with the CND and is interpreted as its inner ionized edge. There were attempts to interpret the kinematics of the ionized gas as a one-armed spiral. Roberts & Goss (1993) interpreted the velocity field rather as independent streamers in their high resolution recombination line data.

[1]also at the Observatoire de Paris, Meudon

[2]also at the Max-Planck-Institut für Radioastronomie, Bonn

2. Observations

The Ne[II] line data at 12.8μm were already published in Lacy et al. (1991). The spectral dispersion is about 16.5 km s^{-1} per pixel, the spectral resolution being about 2 pixels FWHM. An area of $75'' \times 90''$ is mapped with a $2''$ resolution.

3. The Visualization

As we are dealing with three dimensional data, the best way to analyze the restricted phase space as a whole is the visualization in all three dimensions: the right ascension, the declination, and the LSR velocity. In this representation brighter points are closer to the observer. At each time one single intensity level is shown. The data cube can now be rotated into the position where a certain structure can be seen most clearly. With the help of the plotted coordinate axis one can now easily analyze the structure of the emission and identify connected features.

4. The Different Emission Features

Lacy et al. (1991) already separated the Sgr A West complex into four different components: the Northern Arm, the Western Arc, the Eastern Arm and the Bar. Figure 1 shows the Minispiral at two different intensity levels. One can easily rediscover them as entities in the restricted phase space. As already pointed out by Aitken et al. (1991), presenting polarization measurements in the central parsec, the Bar is the most complicated structure. We can confirm this and are able to show, that it consists of two spatially and kinematically separated features which we call Bar1 and Bar2.

5. The Model

In the inner 10pc of the Galactic Center the gravitational potential is due to the central star cluster and a very high density central mass of about 3 10^6 $M_\odot$. The enclosed mass is given by

$$M(R) \;=\; \left(3\;10^6 \;+\; 1.6\cdot 10^6\; R_{pc}^{1.25}\right)\; M_\odot\;.$$

This is in good agreement with the values given by Genzel et al. (1996).
The main principle of this model is the fact, that *the path of every mass rotating in a central force potential lies in a plain.* Consequently, the gas is assumed to be on Keplerian orbits around the center which is Sgr A*. The discrete model is constructed in the following way:

- point masses are distributed uniformly within the plain
- the Keplerian velocity is assigned to each point mass corresponding to the gravitational potential mentioned above
- a random velocity component of up to 40% of the Keplerian velocity is added in three dimensions

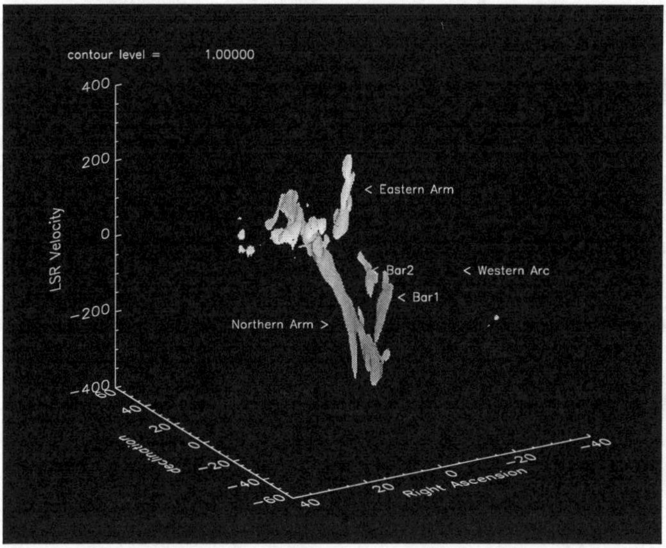

Figure 1. The Sgr A West complex shown in an intensity level of 10^{-2} erg $s^{-1}cm^{-2}sr^{-1}$. The two different components of the Bar can clearly be seen.

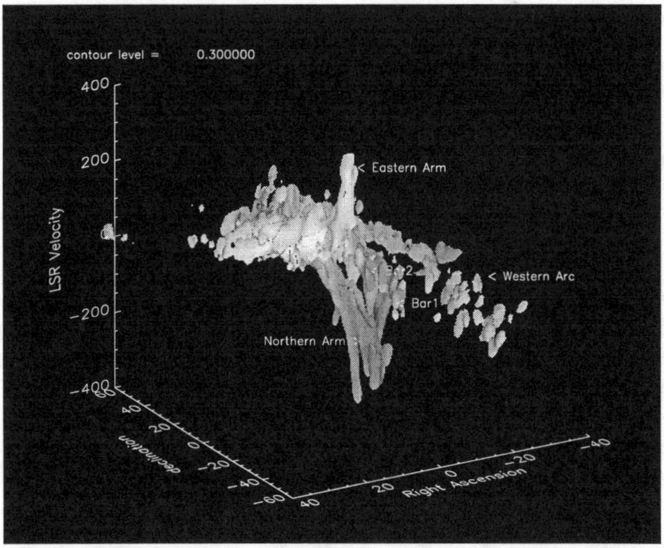

Figure 2. The Sgr A West complex shown in an intensity level of $6{\cdot}10^{-2}$ erg $s^{-1}cm^{-2}sr^{-1}$.

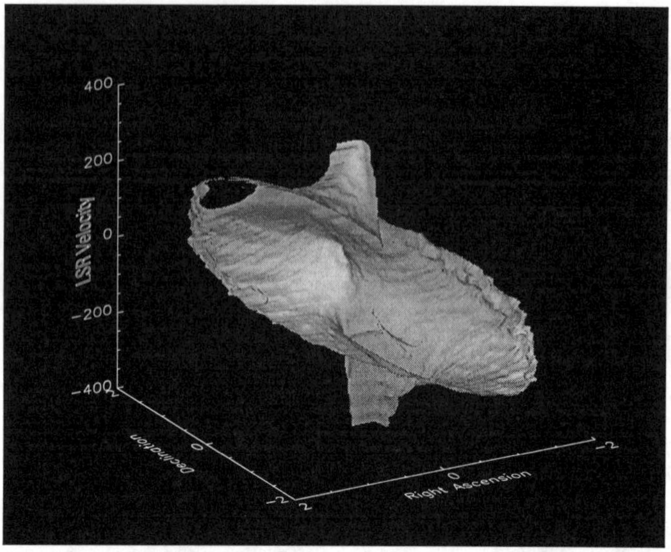

Figure 3. Model data cube for the first plain. The hole on the left-hand side is a smoothing artifact.

- an infall velocity of 5% of the Keplerian velocity is added
- a data cube corresponding to the observed line data cube is constructed
- this cube is smoothed in order to cover all possible random velocities

It should be mentioned that giving the mass points random velocities in all three directions or giving the mass distribution in the plain a non-zero height is equivalent. The model appeared to be sensitive to the absolute value of the accretion velocity. Higher values give less positive fitting results.
With this model distribution we can now construct a cube in the restricted three dimensional phase space similar to the Ne[II] line data cube. This cube for the first plain we fitted is shown in Figure 3. The shape of the feature depends on the underlying gravitational potential and the location of the plain with respect to the observer's LOS. The 'noses' which can be seen in the central part of the cube are due to the increasing velocity for smaller radii in a point potential. They depend very sensitively on the projection angles of the plain and are the best means to find out the correct location with respect to the LOS.
For simplicity we choose a Cartesian coordinate system to describe these rotation angles. In the beginning the plain is located in the x-y-plain and is then rotated successively around the x-, y-, and z-axis. The observer's LOS is parallel to the y-axis.
The minimum number of plains to fit all features in the line data cube is three. Table 1 gives their locations. The main plain which contains the CND is called the plain (i).

As it is not easy to see how the plains are located to each other we have decided to give them a radial profile in the z-direction (non-zero height). We

Table 1. Rotation angles for the different plans in a fixed coordinate system. α_1 is the angle around the x-axis, α_2 around the y-axis, α_3 around the z-axis. in the beginning the plain lies in the x-y-plain, the LOS is along the y-axis.

plain	α_1	α_2	α_3
(i)	103°	20°	30°
(ii)	40°	25°	-5°
(iii)	18°	-28°	17°

then display the plains as disks with a curvature in their height profile which an observer can much more easily locate in space. In order to simplify the picture we only display one surface of a disk (this can be imagined as a dessert plate rotated in space). Figure 4 shows the plain(i) and (iii) treated in this way.

If we consider the sense of rotation of the plains indicated by an arrow, it is quite remarkable that plain(iii) shows a counter rotation with respect to plain(i). The sense of rotation of plain(i) is confirmed by Zhao & Goss (1999).

6. Fitting the Data

The two cubes described above can now be directly compared. Residuals can be built and emission features in the line data cube can be assigned to one of the three plains. This gives:

- plain(i): Western Arc, Northern Arm, Bar1
- plain(ii): Eastern Arm (western part), Bar2, emission feature beyond the rim of the Western Arm to the west
- plain(iii): Eastern Arm (eastern part), IRS 7 complex

Consequently, two or more emission features are assigned to a single plain. This fact makes it very probable that the main features of the whole Minispiral can be explained by material rotating in these plains. Figure 5 shows the Minispiral and the residual of the above mentioned procedure. The resting material could not be assigned to one of the plains.

7. Reconstruction in the Real Three Dimensional Space

As we now know the location of the different plains, we can reconstruct the exact location of emitting material of *one single intensity level* within the plains. We have chosen $3{\cdot}10^{-2}$ erg $s^{-1}cm^{-2}sr^{-1}$, in order to cover all features of interest. For each plain the overlap of the line data cube with the model cube is calculated and can afterwards be easily de-projected. Figure 6 shows this de-projection for plain(i). The plain is seen face-on. At the right is shown how this looks like in the real three dimensional space and no longer in the restricted phase space

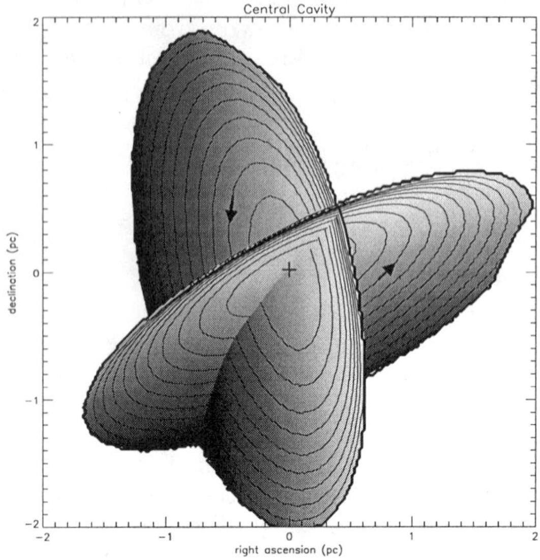

Figure 4. The left surface of plain(i) together with the lower surface plain(iii). The arrows indicate the sense of rotation. The cross is the position of Sgr A*.

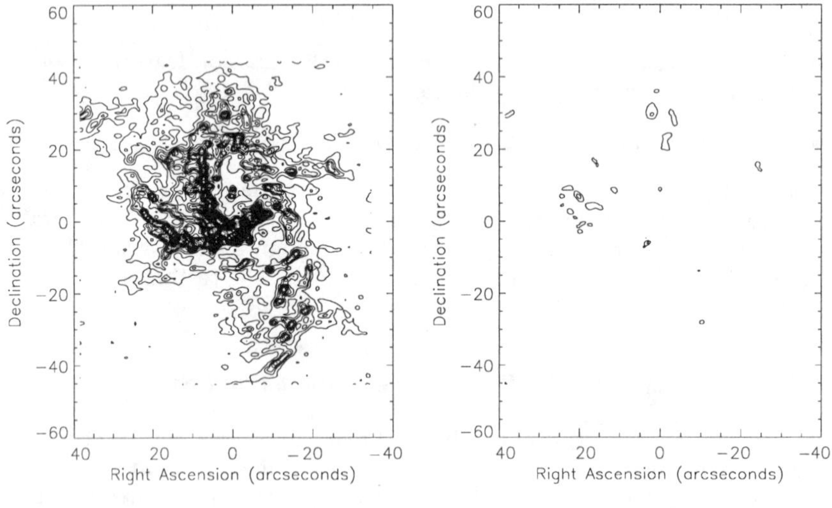

Figure 5. Contour representation of the original data and the final residual with the same contour levels.

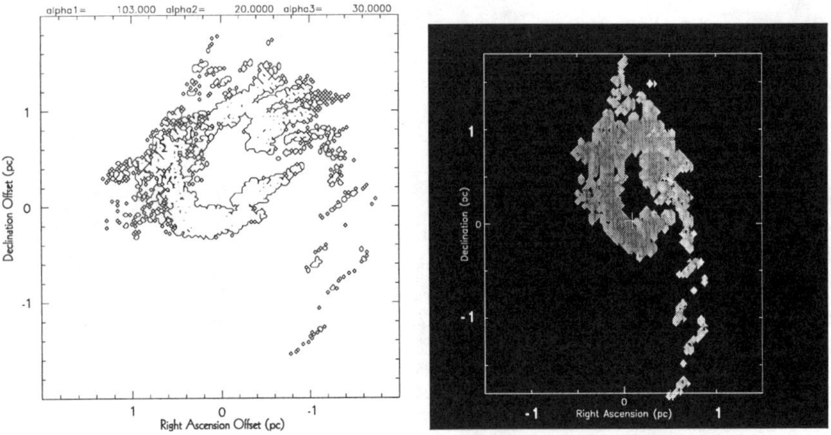

Figure 6. Left: contour plot of the distribution of emitting material in plain(i) seen face-on. The granular structure is an effect of the discrete character of the model. Right: The same distribution shown as the observer on the earth would see it.

(the LSR velocity axis is replaced by the real z-axis). We recall that brighter features are nearer to the observer. The distributions of emitting material in plain(ii) and plain(iii) are shown in Figure 7. After each procedure for one plain the residual was taken to fit the next plain. Thus the unphysical gaps in the emitting material in Figure 7 are due to ambiguities, where we have already assigned the emitting material to another plain.

Taking together the reconstructed emitting material in all three plains gives Figure 8. We recall that this is the Minispiral in the real three dimensional space and that it can now be rotated in each direction in order to have a real three dimensional impression. So we are able to locate easily one feature with respect to another *in all three dimensions.*

As the former reconstruction dealt with an intensity level which had a cut-off to cover the main features, it is now interesting to lower this level in order to see if there is an extended low emission measure gas component located in the plains. Thus we lowered the intensity level by a factor three. Inspecting the data line cube at that intensity showed that this is already in the noise but the features can still be recognized. Figure 9 shows the result of this reconstruction. It turned out that a tenuous gas component exists in plain(i) and it seems that there is another one in plan(ii), whereas the material in the third plain forms separated closed regions without connection. In order to get an impression about the spatial conditions we have rotated the disk seen there by an angle of 155° around the z-axis (Figure 10). Plain(i) and (ii) are seen edge-on.

8. Conclusions

We used [NeII] line emission data (12.8μm) from Lacy (1991), in order to investigate the detailed kinematic structure of the inner 3$\times$4 pc of the Galaxy. This

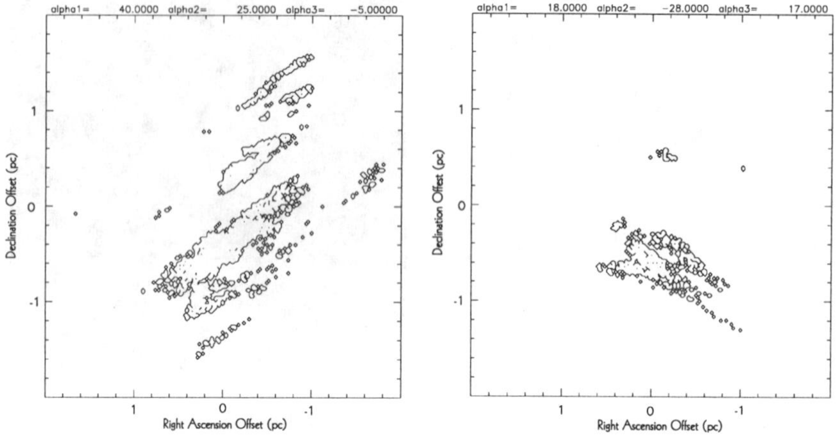

Figure 7. Left: contour plot of the distribution of emitting material in plain(ii) seen face-on. Right: contour plot of the distribution of emitting material in plain(ii) seen face-on.

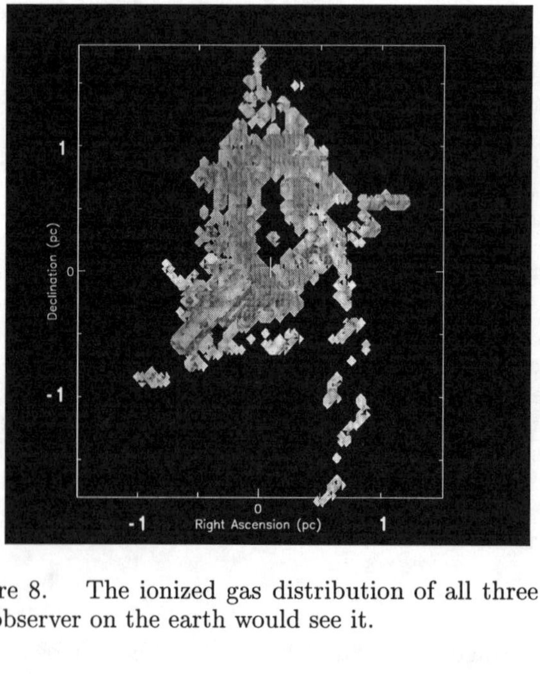

Figure 8. The ionized gas distribution of all three plains shown as the observer on the earth would see it.

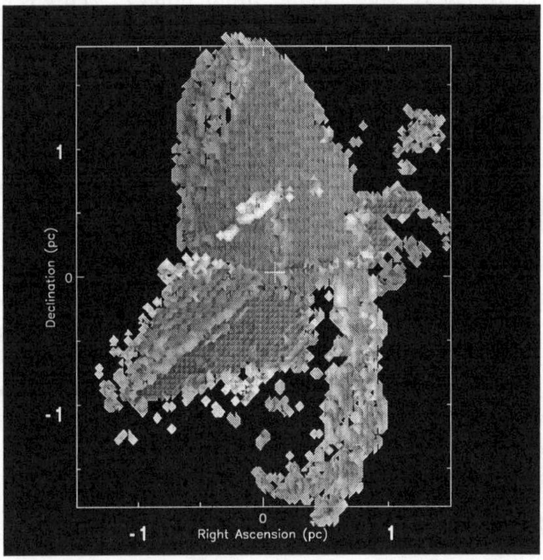

Figure 9. Corresponding representation to Figure 14. For the calculation of the overlap a lower intensity threshold for the data cube was used.

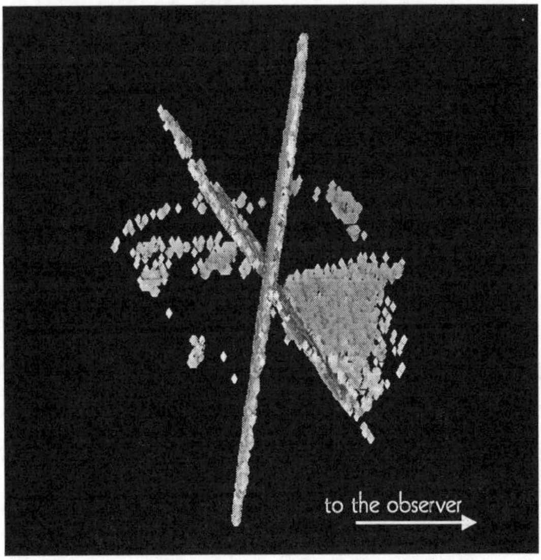

Figure 10. The same cube as in Figure 9 rotated by an angle of 155° around the z-axis. Plain(i) is seen vertically edge-on; plain (ii) is also seen edge-on.

infrared line is a reliable tracer of the ionized inner edge of the circumnuclear disk (CND) as a detailed comparison with H92α line data from Roberts & Goss (1993) showed.
With the help of a three dimensional visualization of the data cube we were able to distinguish between the different components of the ionized gas features called the *Minispiral.*
Within this restricted three dimensional phase space connected emission features and their velocity gradients can be easily and clearly separated. We could thus more clearly identify a substructure in the feature called *Bar* by Lacy (1991).
In order to make a de-projection and to locate the emission in the real three dimensional space, the inner 2pc of the Galactic Center were modeled as accumulations of mass which rotate on circular Keplerian orbits in certain plains around a central mass of 3×10^6 $M_\odot$. In addition the mass contribution of the stellar content was taken into account and a radial accretion velocity component was assumed. Omitting the stellar mass distribution or taking a higher radial velocity than 5% of the Keplerian velocity gave less positive results.
The minimum number of plains to explain the main features of the Minispiral is three.
A tenuous gas component in the plain of the CND seems to form an almost entirely filled disc.
An important conclusion of this work is, that an inflow of disc material towards the Galactic Center is seen.
Furthermore, we can give a possible explanation for the counter-rotation of the early type stars in the Galactic Center which were found by Genzel et al. (1996).

References

Aitken D.K., Gezari D., Smith C.H., et al., 1991, ApJ, 380, 419
Genzel R., Thatte N., Krabbe A. etal., 1996, ApJ, 472, 153
Güsten R., Genzel R., Wright M.C.H., et al., 1987, ApJ, 318, 124
Jackson J.M., Geis N., Genzel R., et al., 1993, ApJ, 402, 173
Lacy J.H., Achtermann J.M., Serabyn E., 1991, ApJ, 380, L71
Lo K.Y., Claussen M.J., 1983, Nature, 306, 647
Roberts D.A., Goss W.M., 1993, ApJS, 86, 133
Serabyn E., Güsten R., Walmsley C.M. et al., 1986, A&A, 169, 85
Zhao, J.-H., Goss, W.M. 1999, this volume, p. 224

The Central Parsecs of the Galaxy
ASP Conference Series, Vol. 186, 1999
H. Falcke, A. Cotera, W.J. Duschl, F. Melia, M.J. Rieke, eds.

Discussion Session IV: Morphology and Gas Dynamics Within Sgr A West: How are the Various Components of Sgr A West Related?

P.G. Mezger[1]

Max-Planck-Institut für Radioastronomie, Auf dem Hügel 69, D-53121 Bonn, Germany

A. Cotera[2]

Steward Observatory, The University of Arizona, Tucson, AZ 85721

Abstract. This is an edited transcript of a discussion session about Sgr A West the thermal 'mini-spiral' in the Galactic Center that took place during the Galactic Center conference "The Central Parsecs" in Tucson, Arizona, Sep. 7-11, 1998. The session was chaired by the moderator as indicated in the author list. The discussions were taped, transcribed, and finally edited by an editor who is listed as co-author. The contributions of the conference participants are preceded by their names and thus references to specific aspects and ideas of this discussion session must also include a reference to the respective speaker(s), e.g., in a footnote.

Discussion

Mezger: To start off our discussion, what I have done is I have written down the talks from this session, starting with the speakers, and have noted down a number of topics which may just reflect my own ignorance, and so you can easily substitute your own questions. This is the time I have allotted to the Yusef-Zadeh talk, 10 minutes, the rest 5 minutes, this would end the discussion at 6:00 and this would probably regain for me the esteem of Angela Cotera, which I may have lost by letting the session run long. Okay. I would like to start with Yusef-Zadeh's beautiful review of the topic for this session. What I have noted down partly reflects the discussion I had with Miller Goss this morning or yesterday. Specifically, where is Sgr A relative to Sgr A West, is it behind or is in front of Sgr A West. Would someone like to comment on this? Miller, you expressed a pretty strong opinion that it is behind Sgr A West.

Goss: I'll let Anantha answer that, we have been working on this together.

[1]Moderator of the discussion session

[2]Editor of the discussion session

Anantharamaiah: Well, if you look at the frequencies where Sgr A West becomes optically thick, at low frequencies of 90 cm, and remember there is also ionized gas surrounding Sgr A West that seems to be associated with it (although this is not the only ionized gas in the region), at the lowest frequency (for instance one of the images that Farhad showed) if you make a cross cut at the position of Sgr A*, you see no evidence of Sgr A* in that image. At 90 cm, you see no evidence of Sgr A*. But if you make a cross cut, taking into account the brightness distribution of Sgr A East, which is well-defined from higher frequencies, you see that Sgr A West is clearly in absorption against Sgr A East. Sgr A West is in absorption! Then at the position of Sgr A*, you see no evidence of any emission. Now clearly we cannot escape the conclusion that if you put Sgr A* in front of Sgr A West, we would have seen a little bump, actually a reasonably big bump, on this cross cut. We don't see any evidence of Sgr A *.

Mezger: Okay, I think that's convincing. There is a second question here, that is, is it the thermal plasma which is also responsible for the distortion of Sgr A*?

Lazio: I'll say right off the bat: Sgr A West could be contributing to the scattering of Sgr A*, but it's not the dominant factor and I'll discuss this more on Thursday. Sgr A West includes little if anything from the plasma distortion that we see in Sgr A*.

Mezger: You convinced me the other day ... was it you who mentioned that the distortion happens within 150 pc?

Lazio: Yeah.

Mezger: Okay, that would coincide with the relatively thin thermal plasma which fills the Nuclear Bulge. And then, Farhad was also talking about the relation between Sgr A West and Sgr A East, and mentioned that part of East is in front of Sgr A West. Would anybody like to ask questions or comments on this?

Lazio: I have a question for Farhad [Yusef-Zadeh]. Have you made some assumptions of how the magnetic field is in that area and then asked how much path length would you need to produce the emission that you see in front of Sgr A West.

Yusef-Zadeh: Well, it's really not in absorption. What I showed was basically a loop-like, coherent structure which basically extends to the ring that surrounds the CND. So it was part of the large scale structure, and I mentioned that it is very likely to be part of Sgr A East, which is dropping the whole Sgr A West in the CND.

Lazio: As I recall, you had a close-up of Sgr A West and there's a finger of material that extends down into it.

Yusef-Zadeh: That's part of the loop that is associated with Sgr A East.

Novak: Yeah, I had a very similar question. I was thinking about the non-thermal emission, that is what you call part of Sgr A East. A non-thermal emission that sticks up in front of Sgr A*. I was wondering if that could be something maybe more interesting, something associated with Sgr A* itself, may be even a jet—a non-thermal jet—, something like that?

Yusef-Zadeh: I think it's more likely to be part of Sgr A East.

Mezger: Please speak up a little bit louder.

Cotera: So what is the answer to number one?

Mezger: OK, the answer to this first question seems to be that Sgr A* is located behind the bar in this mini-spiral. If it is not seen at 90 cm, the only explanation is that it is covered by a thermal plasma which is becoming opaque at 90 centimeters.

Beckert: That's unless the emission is not intrinsic to Sgr A*. If Sgr A* does not emit at that wavelength, you can't absorb it.

Mezger: Yeah!

Beckert: It would be nice to have a low frequency measurement of the region.

Mezger: Okay. It leaves something for the next conference.

Zhao: Actually we know Sgr A West is orbiting gas, and Sgr A* is probably associated with a black hole in the middle of this region, so these two components we know are already related, so this leaves Sgr A* sitting embedded in Sgr A West, right.

Mezger: Embedded or behind it, yeah.

Yusef-Zadeh: I hope you can now see what I was talking about. This feature basically continues on and splits the hole that Anantha and Pedlar saw. So there is some non-thermal emission, which should have been totally absorbed. I mean, we should not see any emission here, because that's where the bar is. If the bar is in front of Sgr A East, this feature should have not been there.

Anantharamaiah: I want to make one comment on this. If you look at the larger picture here, of Sgr A West and Sgr A East and the other emission associated with it, there is a big halo surrounding Sgr A West and Sgr A East; there is plenty of radio emission along with such structures. Now while Sgr A West could be in front of Sgr A East, and there is all of this halo emission which we think could be in front of both of these components, they're at 8 kiloparsecs along the line of sight within the Galactic Center. We need something more to associate the little piece of emission that we see, with Sgr A East, I think.

Yusef-Zadeh: Well, certainly this needs a higher resolution observation.

Zhao: What year was this image taken?

Yusef-Zadeh: This was taken maybe six months ago.

Beckert: I think there is a problem with this. Sgr A East has extended emission over 10 parsecs, let's say. So if you had 99 percent of the synchrotron emission behind, and just one percent in front, it fits perfect. So it's possible.

Anantharamaiah: I suppose it could be in the halo, or it could be Sgr A East.

Beckert: I think what the data really showed was that Sgr A East and West are extremely closely associated. Yet East is not 50 parsecs behind West.

Yusef-Zadeh: That was one of the physical arguments I used to say that these two are related, that's all. But it is important to know for sure that these two are really interacting with each other.

Zylka: I wanted to bring up that some work I did with collaborators, 10 years ago, 8 years ago, showed something very similar. In our data Sgr A West and East, show up in exactly this close relationship.

Yusef-Zadeh: Thanks, I forgot to mention that you've done a lot of work on this picture in the past. So in some sense the idea is not new, but now we just have more evidence for it.

Goss: I think Farhad [Yusef-Zadeh] has very nicely resolved the molecular hydrogen that's northwest of Sgr A West, and it's very consistent with this. And the fact that there is an OH maser in that gas at +50 km s^{-1}.

Stolovy: Farhad, could you put up your plot of the proper motion and the radial velocity color picture you had? Because I think that really does show that, if you just look at the kinematics from the H92α recombination line, it's clear that if the gases are inflowing along the Northern Arm, they are red-shifted to the north, become blue-shifted and then further out are red-shifted again. [The Northern Arm] has to be going behind Sgr A* and then coming towards the observer. The only other explanation for that is that it's outflowing instead. There are only two geometries where you reproduce the observed data. So we all think it's inflowing, if you look at the proper motion, it's pretty clear.

Yusef-Zadeh: I don't think anybody has done modelling of the 3-dimensional velocity structure, but I think Doug [Roberts] has done some work based only on the radio recombination line measurements. And I think all of the ionized gas seems to be in the plane of the CND. So just based on radio velocity measurements, now that we have two more directions, and we can place a nice constraint on the relative relationship between Sgr A* and the ionized gases. One thing I want to stress is if you look at the global flow in this region, it shows the velocity of a big chunk of the ionized gas, and you do clearly see a global, organized motion in the direction towards Sgr A. If you then look at individual spots, you see basically the same velocity structure. So, we have a very good idea what the sense of direction is now.

Mezger: Yeah, I think we have one more question.

Vollmer: I've tried to reconstruct the motion, the real 3-dimensional space out of the restricted phase space we have. It turned out that most of the Northern Arm and the Western Arm are in the same plane. I think that the Western Arm is in front.

Roberts: But there are different models. There's also a model that describes the circumnuclear Western Arc as part of the inner edge of the Circumnuclear Disk which is rotating with the same parameters. The Northern Arm extends much further north as Farhad [Yusef-Zadeh] showed that it extends past the Bridge where the molecular hydrogen is and passes in front of the molecular cloud. If that's falling in, as in Bob models, that could be under two different types of orbital motions that aren't necessarily related. I don't think it's clear that it's all part of one orbital component.

Anantharamaiah: I don't think it's one component. I agree with the fact that the Northern Arm is falling in and the Western Arm is a different component.

Mezger: I would suggest you find a good place to discuss this over a Margarita. Let's just get back to one point, I mean Farhad is always good for a wild idea at the end of his presentation. Could you show once more those clouds passing over the Galactic Center and leave the stars behind?

Yusef-Zadeh: My low-tech simulations?

Mezger: George, you wanted to make a comment on this picture?

G. Rieke: I could take the standard dynamical friction for stars out of any basic source and the most optimistic time scale to settle within one parsec of Sgr A* is 10^7 years. If you're a little more realistic and put in a smaller core radius than one parsec, it takes longer. And so I don't understand the time scales that are involved. That's basically Reinhard's [Genzel] comment also.

Melia: I think the point that Farhad is trying to make is that if star formation occurs in the molecular cloud with a relatively normal IMF, and the cloud is moving with a velocity of 3 km s^{-1} through this region, presumably the stars will also be moving through this region with an average velocity of $\sim$50 km s^{-1}. It's not that we're invoking dynamical friction to bring the stars in, we're seeing the stars already moving through this region. But then the central cluster acts as a filter because gravitational friction will predominately decelerate the heaviest stars relative to the lighter stars. So the net result is that the heaviest stars will be trapped in the middle, whereas the lighter stars will continue to pass on through. So it's not that dynamical friction is bringing them in, dynamical friction is just stopping the heavier ones and allowing the lighter ones to break through.

G. Rieke: But Fulvio, that requires that they be lined up with the black hole in this little cluster that's a fraction of the space.

Melia: No, no, the stars are being formed throughout the cloud. There are many, many stars.

Yusef-Zadeh: In fact, if you look at Sgr A East, the HII regions, that sit on the eastern side of Sgr A East, could be the same type of stars we're talking about but that didn't get filtered and passed through the Galactic Center.

G. Rieke: If they don't get filtered, they go into very eccentric orbits, and those orbits have to be reduced by some frictional mechanism. Yet we see them all concentrated within a couple of tenths of a parsec.

Melia: We see about 16 of them—16 to 20 of them. Within this cloud there will be hundreds, thousands of them.

G. Rieke: In the 60 $M_\odot$ range we see 16 of them.

Cotera: But there is a possible 60 $M_\odot$ star just outside of Sgr A East. It's right where the HII regions are, there is a WR star, actually a WN6 star.

Yusef-Zadeh: But also, if you look at this giant molecular cloud, the 50 km s^{-1} cloud, it extends all the way up to the Arc. It doesn't just end at Sgr A East, it extends over a much larger region. In addition, you see all kinds of hot stellar sources in that molecular cloud.

G. Rieke: Alright, but these stars are really exceptional stars. In the whole Galaxy we'd have trouble finding 16 more.

Mezger: OK, well, we'll talk about the stars more tomorrow. Thank you very much for your contribution. We come now to the next paper, which was by Robert Coker on models of the Gas Dynamics and here I just noted down—after some talks over coffee—the following question: How realistic is the model with the radial infall? That's where I would like to step out and leave it to the theorists to discuss this question. Any volunteers?

Falcke: The stars which produce the wind that is to be accreted could be moving in a coherent circular motion, around Sgr A*, and I wonder how you then would end up with accretion without any angular momentum.

Coker: Well, clearly there's going to be some cohesive aggregation. On the other hand, if you look at the clumps that get to smaller and smaller radii towards Sgr A*, you are going to have smaller and smaller incremental angular momentum before it starts to fall into the center. So as you get into the center of Sgr A*, clearly you're going to have smaller and smaller angular momentum. So, I don't think we have to worry about having the gas that gets to 100 R_g, having a large amount of angular momentum. A large angular momentum clump has never fallen in, in any of the models. So combined with the fact that the shock-shock stellar winds are so strong, and the fact that stellar motions themselves are on the order of 100 km s^{-1}, as compared to 1000 km s^{-1} for the wind, I don't think that the net result, including the actual velocity of the stars, is going to change the angular momentum we see accreted at Sgr A* very significantly.

Falcke: I see what you're saying, however, the gas already within the central few parsecs must already have a large angular momentum and that could reduce the accretion rate. There's just one other question I had: have you taken into account IRS7? Since we so clearly see that there seems to be some kind of tail which is most likely a result of a (stellar) wind from the central cluster?

Coker: We're only accreting about 1/50 of the actual mass flow (or thereabouts) from the central region. Actually, I think it's even smaller than that. In fact the constraints are placed by IRS7, that's where we come up with our arguments as to why there would be that much mass in the region. However, it's still going to be a very small fraction of the amount accreted because the total mass has a mass less than 10^{-3}: and that's relatively conservative still. So you're only getting a fraction of the available mass nearest to Sgr A*.

Falcke: Wait a minute, you just said you get only 2 percent moving towards the center, so there is 98 percent outflow?

Mezger: OK, this appears to be another discussion which can be deferred to later, over a Margarita. We come to the next paper by Jun-Hui Zhao on proper motions of the radio components around Sgr A*.

Cotera: I have a question for Jun-Hui. You showed the Minicavity on one side, and it seemed to be expanding away from the center, and you had a velocity distribution. Do you see that on the other side?

Zhao: Not much. On the other side the intensity is not as large. We have some observations which show a slight structure change but not a velocity.

Cotera: With the star seen by Marcia Rieke and Susan Stolovy in the center of the Minicavity, it would be interesting to see if the gas in the Minicavity is expanding away from that.

Stolovy: Well, I'll have that data in October. We have a second run coming up, where we hope to get the proper motions.

Mezger: I found this a beautiful work, what you did, but I couldn't fully understand what kinematic information you got out of these observations.

Stolovy: Okay, the original idea for the Minicavity was that you had some kind of an exotic plasma where you had winds from the IRS16 cluster and perhaps IRS13, but mostly from IRS16, going by the black hole and getting gravitationally focused and blowing into the Western Arm to create this bubble. But now I think from the NICMOS data, and this is only based on morphology, not any real proper motions yet, it looks like a bubble around the star. There's a star smack in the middle, it's actually a very well detected star. I'm sure other people have seen it, but nobody's labelled it or said anything about it. It just really looks like the Minicavity is a shell around the star, that's expanding. So it doesn't have a lot to do with Sgr A*, or a black hole, probably.

Zhao: I have a comment about that. Actually I think that this is a kind of temporary feature. If you calculate the expansion velocity and the size of the cavity, it only lasts about 500 yrs.

Stolovy: Yeah, about 500 yrs, and that's assuming an expansion velocity of $\sim$ 100 km s^{-1}.

Zhao: This indicates that there are lots of interactions going on from the stellar outflow, with molecular streams that are in that region. In addition, the gas also, as we know, is travelling in the gravitational potential of the black hole, right? So this gas actually has low angular momentum, and is losing angular momentum, going to a lower state close to the black hole and actually is a gradually feeding Sgr A*, which is, to my understanding, what we can learn from the observations of HII components of the proper motions. But I think we need more data.

Stolovy: From your data it's hard to really say for certain though. You took a bulk average of one epoch and a bulk average of another and said that there was a shift, but you really need to look at individual clumps.

Yusef-Zadeh: I should say something. Within the field, there are several of these clumps that are very near Sgr A*, they have a fairly good signal to noise, and you can see them in all prior data. When we look at the proper motion of these, they're not going toward Sgr A*. They're moving away from Sgr A*. Whether it's really associated with Sgr A* or not is a different thing, but at least for the bright blobs, they're moving sort of radially, in some directions, away from Sgr A*.

Mezger: You had some comments.

Roberts: Just that what Farhad showed was a superposition of my H92α recombination line radial velocities on top of the proper motion studies. We have to keep that in mind. Jun-Hui is measuring only the velocity in the plane of the sky. Now, on top of that, in the Minicavity and on top of IRS 13, you have velocities which are the highest negative radial velocities in the whole area, $\pm$300 km s^{-1}. So that's coming towards us, away from Sgr A*, so that has to be a different component. If Jun-Hui is right and there is a sort of infall like this, then it can't be the same gas that we're seeing as a bulk gas.

Zhao: There's a point I want to make, while we're looking at this graphic. If you think of the tidal forces in that region, all of the different gas features are broken into the pieces by the gravitational tides. Then this kind of orbit wraps around with the gas gradually feeding into the Sgr A* region. All this gas is broken into smaller pieces, which are very low density, and it is difficult to observe this kind of gas. This is a problem we have. We need more data of the kind that can see this gas. We need observations, that cover the short baselines and can pick up this kind of low-density gas.

Wardle: You're losing a lot of angular momentum in that sort of scenario.

Zhao: Yeah.

Wardle: How are you doing that?

Zhao: Interactions.

Wardle: With what?

Zhao: With the local stellar wind. From the IRS 16 complex for example, there is a bunch of stellar winds.

Wardle: There are also fairly strong magnetic fields.

Zhao: Yes, we also need to consider the magnetic fields. I think this is something we need to think about. How do these gases fall into Sgr A* in this environment? Clearly from all the maps we only see one pipeline feeding into the center. There is gas that looks like it is close in the Sgr A* region.

Novak: I just think you have to still consider the 3-dimensional velocities since we have a line-of-sight velocity which appears to be consistent with a hyperbolic orbit, or at least a very badly kinked orbit.

Zhao: Another point: to actually determine radial velocities is very difficult, because in the gas that is falling into this kind of gravitational potential well, there are very large velocity dispersions. There is a wide profile in the radial recombination line, probably very wide.

Novak: Perhaps it would be best to look in NeII, or some other line besides a recombination line, that attracts a larger bandwidth.

Anantharamaiah: Can you give some idea about the range of transverse velocities that you can deduce from these observations.

Zhao: Yeah, probably you cover up to ± 500 km s^{-1}. There is probably gas outside that window, which is where we've been searching in the last 20 years. We need to open up the window of available velocities more by searching for this kind of high velocity features, but dynamically, it probably is there.

Yusef-Zadeh: I think NIR observations have a much better chance to look at the radial velocity than radio observations. For the observations we're talking about, we are at only a few percent level in radio recombination lines, and it's a really, really tough region to look for anything near Sgr A* in recombination lines. So I think Brγ lines, for example, or Paα spectroscopy can be done for these features, very close to Sgr A*. That would be really good.

Mezger: OK, let's move on to the next two papers on the MIR observations. I think we can deal with both in one discussion. First the paper by Dan Gezari and then by Angela Cotera on new Keck observations. Of course, I understand the quality of the new MIR data is much higher resolution than the older data, but we haven't yet evaluated the new data? Nevertheless, I encourage you to ask a few questions to either Dan Gezari or Angela Cotera.

Becklin: I'd like to make a comment about your comment that there should not be a correlation between the dust and the gas, because one goes as the density—which is true, infrared goes as the density—and the free-free goes as the density squared. Therefore you wouldn't expect this type of correlation. However, you have to account for the Lyα heating, so that complicates things and makes the correlation tighter than you might think.

Mezger: Yes, but of course between 10 and 20 μm there is a very strong weighting towards high dust temperatures, it's no longer a linear relation. That may also change things. The only remark I will make is that I would not be too surprised.

Becklin: Obviously in Orion, it is not a one-to-one correspondence.

Mezger: Okay, are there further questions.

Anantharamaiah: What is the actual dust temperature that one may expect when it is derived?

Cotera: Actually, Dan probably has better numbers on that at this point, we've not gotten to that yet.

Gezari: The peak temperature is about 300 K, and cooler temperatures are about towards 150 K; at least that's where we start to run out of data at ~20 μm, where we start to run out of temperature resolution.

Mezger: When we analyzed the spectrum of the central parsec, we could separate the spectrum into about three dust components. The highest temperature was 350 K, just what you say. Also, it was a very small amount of dust which was associated with this emission, which is radiating very efficiently.

Gezari: For the isolated, heavily embedded stars like IRS 3 and IRS 7, the temperature was much higher, going up to 1000–1300 K. However, these are pretty much isolated sources, and not part of the Sgr A West/Sgr A* complex.

Anantharamaiah: Since there is such a remarkable similarity between thermal radio emission at Sgr A West and the infrared emission structures that are shown, and we know that the radio emission comes from maybe 10^4 K gas, these temperatures are only 100-1000K, does this present a problem?

Gezari: No.

Mezger: Okay, if there are no more questions on this, then we come to the last paper, by Bob Sanders on a model for how you explain the formation of the nuclear disk. George?

G. Rieke: I'm struck by how clumpy all the gas distributions are that we see in these regions. At this distance, the rotation time is roughly 100,000 years, so things have probably made a couple of orbits, and I thought the theoretically determined distribution indicated a relatively smooth distribution rather than the clumpy way these regions appear in the radio observations. I was curious, why are things so clumpy? It's really important to determine this because we're trying to understand how stars form in this region.

Sanders: I only consider the motion of gas in the gravitational field. That is to say, I'm not considering any other hydrodynamic effects like winds, or supernova shells, or radiation pressure or magnetic fields. By that, I don't mean to say that these effects aren't important, I'm just saying how much progress we can make by just considering gas motion within the gravitational field, towards understanding the global kinematics of these structures, the very global kinematics.

G. Rieke: The question still remains that if the gas has been through a couple of orbits, yet it looks clumpy...

Unknown: That is my question. Why do you assume 10^5 yrs for the inner edge of the CND to form.

G. Rieke: Well, 10^4 yrs makes it even worse, right? I mean, it depends on what radius we're picking for this calculation. Also, it's not like the gas just got to it's present location and was just quietly sitting there and therefore it became clumpy. It should be smoothed out.

Beckert: That's if it has been around at least a few times. What if it is just now sweeping into the center, and we are only seeing, say, 3/4 of one orbit, so the time scale would be then 3/4 of the orbital time scale.

G. Rieke: So you're hypothesizing we're seeing gas that really fell in a fraction of an orbit ago. That's the only thing I can think of, because otherwise it seems it would have to be smooth.

Sanders: My point is, these features can be rather more long-lived than one orbit. As far as the basic morphology and kinematics are concerned, the sort of model I'm presenting here is really very similar to the old orbital models of Serabyn and Lacy and more recently of Herbst et al. and Roberts et al. It's very similar. The only difference is that I'm saying that you can get this structure, which actually can persist for many rotations and that you don't have to be seeing the region at such a particular, special instant of time. That particular structure can persist for much longer than previously thought.

Townes: Let me complicate the problem more, by again pointing out the difference in velocity of the components: the irregularity in velocity of rotation an so on. On the other hand, one should also note that there are batches of gas floating around inside, which can bump into this rotating shell from time to time, creating irregularities. In addition, if we go back 100,000 years, we don't know, but there might have been a star falling into the black hole with some kind of very big energy release. So all those things, it seems to me, are possible. It'd be very interesting if we could nail down what might have done this.

Sanders: I'm just arguing that before we can understand all these details, before we put too much into such detailed calculations, let's understand the global properties. Let's try to have a good understanding of the global morphology, the global kinematics before we put in other possibilities.

Let's not brutalize the problem at this point, so that we can first really understand the basic structure.

Roberts: Some of the different models have something in which there are a few structures, such as the Northern Arm, falling in perhaps with a shorter time frame than the Circumnuclear Disk, which maybe has lasted over several rotations. If these different features are colliding then that's a natural place to start looking at more detailed models. When you start putting together these different features, it seems it would be easier to make it clumpier and less structured.

Reid: It's an important point. You look at the radial mass of the Northern Arm, for example, there are really sharp bends in it, kinks, and looking at ionized material, it's probably now being controlled by magnetic fields that twist around the Arm. So, a lot of the substructure we're seeing has got to be related to the magnetic fields and the ionized gas.

Roberts: But the magnetic field values have to be very large, given the motions we know from the proper motions and the radial velocities from the gas kinematics. I think that the ionized gas is dragging the field lines along, not being controlled by them.

Townes: Magnetic fields as measured so far are too small.

Reid: Then it's hard to explain why there are nice sharp kinks in the ionized gas you see in the good radio maps.

Sanders: Let me say that, after his proper motion talk, Reinhard Genzel made the point that these old models of say, Serabyn and Lacy and so forth, they worked pretty well, explaining the kinematics, the radial velocity of the northern circuit extending northward. He suggested then that somebody should do the calculations for proper motions, for the transverse velocity, now that we have this information. And, I've done it. I mean, this calculation I've showed you is essentially the older models with the proper motion calculations. I think this general pattern of proper motions is what is seen. The fact we have motion along the Northern Arm toward Sgr A*, that sweeps it towards the west. That's the general pattern. I'm very encouraged by the agreement between calculated and observed proper motions.

Unknown: There must be a difference between a model with 4000 particles or if you will, with 500 clumps of $30M_{\odot}$ in the CND. This is a completely different viscosity, if you have 4000 or 500 clouds. This should change the lifetimes significantly which should change a lot of things. One final question maybe a comment. I'd like to come back to magnetic fields. We are talking maybe about completely different scales. In magnetic fields—where it's charted—are the fields large, say on the scale of minutes?

Various: No.

Unknown: The beam dilution effect might be very important.

Roberts: If you see linear structures that are much larger than the beam size, then there have to be ordered magnetic fields over this region.

Unknown: But there are small kinks, right?

Roberts: The kinks are in a linear structure that's thick, so . . .

Mezger: Eric, you wanted to say something

Becklin: I'd just like to make one comment for George to think about. I mentioned this a couple years ago at the IAU. For sources that are knots along the Northern Arm, maybe there are massive stars that happen to be moving just with the Northern Arm at that point, and are therefore accreting matter. So that you have star formation that occurs when they're in the Arm where the density is high and then they go out into the IRS16 sources.

Yusef-Zadeh: What we know, really, is that the stellar system and the ionized system are very decoupled from each other: at least within the inner 10–15″. You really don't see any correspondence at all between radial velocity measurements and proper motion measurements between stars and ionized gas.

Sanders: These sources that Eric was talking about don't have any velocities for the compact portions.

Yusef-Zadeh: Yeah, that would be good to find out.

Mezger: Let me take the chance here, during Angela's talk I noted down there is an MIR source coinciding with Sgr A*, it appears to be extended dust emission. Can I take this home and say that none of the infrared observers contradicts the statement?

Cotera: There's a mid-infrared source coinciding with Sgr A*, but rather than saying it *is* Sgr A*, such that we'll have to write another paper in three years saying it's not Sgr A*, we're saying there is an extended source coincident with Sgr A*. As to what it is, we'll have to do a more careful analysis of the data.

Becklin: It appears to be dust but we haven't proven that.

Cotera: Exactly.

Gezari: What I would say is, that there is resolved mid-infrared emission about at the position of Sgr A*, and it's very intriguing, and I'm afraid it keeps this issue alive for a while. However, I think it's difficult to identify this with a non-thermal point source.

Mezger: OK, we've reached the end of this session. I'd like to thank everyone for a good session.

Session V

How, When, Where and Why are Stars Formed in the Galactic Center?

The Central Parsecs of the Galaxy
ASP Conference Series, Vol. 186, 1999
H. Falcke, A. Cotera, W.J. Duschl, F. Melia, M.J. Rieke, eds.

The Stellar Content of the Galactic Center

R. D. Blum

Cerro Tololo Interamerican Observatory, Casilla 603, La Serena, Chile

S. V. Ramírez

Astronomy Dept., The Ohio State University, 172 W. 18th Ave., Columbus, OH 43210, USA

K. Sellgren

Astronomy Dept., The Ohio State University, 172 W. 18th Ave., Columbus, OH 43210, USA

Abstract.
We briefly review the current stellar census of the bright end of the K−band luminosity function in the central few parsecs of the Milky Way. New H and K−band spectra are presented for a sample of cool stars. The broad H_2O absorption feature near 1.9 μm can be used in combination with the 2.3 μm CO feature to distinguish between AGB and supergiant stars.

1. Introduction

The Galactic Center (GC) is the most complex "star forming region" in the Galaxy. Evolved emission–line stars (with massive progenitors: $40 \lesssim M_\odot \lesssim 100$) from a recent burst of star formation (Allen et al. 1990, Krabbe et al. 1991, Libonate et al. 1995, Krabbe et al. 1995, Blum et al. 1995*a*, *b*, Tamblyn et al. 1996) co–exist with less massive red supergiants ($10 \lesssim M_\odot \lesssim 25$), intermediate age asymptotic giant branch (AGB) stars ($5 \lesssim M_\odot \lesssim 10$), and yet older red giants (Lebofsky et al. 1982, Sellgren et al. 1987, Blum et al. 1996*a*). All of these in a physically small volume with extreme density (Eckart & Genzel 1997, Ghez et al. 1998; see also contributions by Genzel & Eckart 1999 and Ghez et al. 1999).

Our ultimate goal will be to construct a star formation history for the central few parsecs which explains the observed numbers and types of stars found there. In these proceedings, we outline what is currently known about the stellar census in the bright end of the K−band luminosity function, a necessary first step in achieving our final goal. Because of the large and variable extinction over the field (Blum et al. 1996*b*), this requires near infrared spectra.

2. K–band Luminosity Function

Figure 1 reproduces the K–band luminosity function (KLF) for the central ~ 4 parsecs from Blum et al. (1996*b*). Color coding has been added to indicate the relative numbers of stars which have been identified as hot (*blue*) and cool (*red*) from near infrared spectra. The magnitudes in Figure 1 are *dereddened* as described in Blum et al. (1996*b*). It is clear that cool stars dominate the bright end of the KLF in the GC, though the hot stars are more luminous bolometrically.

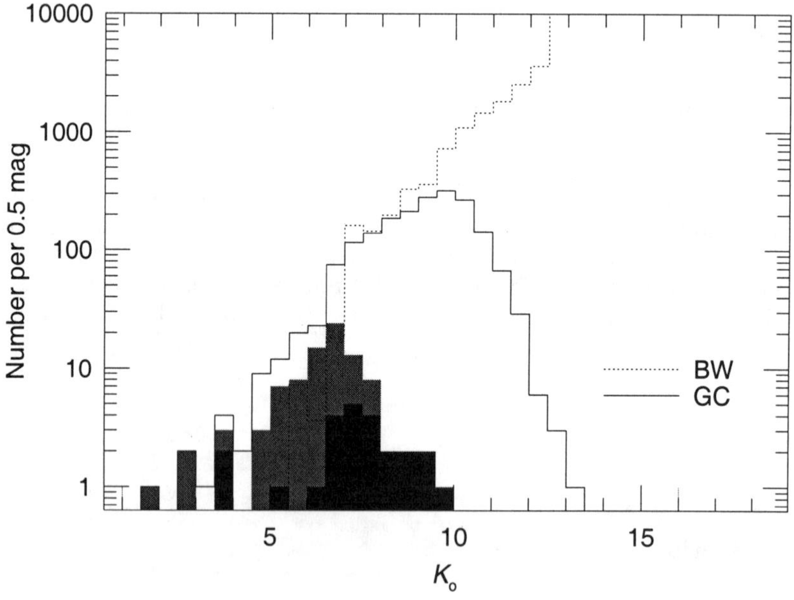

Figure 1. K–band luminosity function for the stellar population in the central ~ 4 pc of the Galaxy. Spectroscopically identified stars have their contributions coded as *gray* for cool and *black* for hot; see text. The hot stars include all the presently known objects, though fainter members may as yet be undetected. There are a large sample of fainter cool stars not represented here but listed by Genzel et al. (1996).

The census of hot stars depicted in Figure 1 includes *all* those identified in the references above and a preliminary indication that IRS 8 may also be a He I emission–line star (from our own unpublished spectra). This brings to 27 the total of young massive stars identified in the GC by 2 μm spectroscopy (to our knowledge, this is a complete list of such stars identified in the literature). Fainter Wolf–Rayet type stars may yet have escaped detection in the crowded GC field.

The *dotted* line in Figure 1 represents the luminosity function in Baade's Window (BW). It is adapted from Tiede et al. (1995). The stellar population in BW is an old one ($\gtrsim$ 10 Gyr), so it is immediately apparent that star formation has been proceeding in the GC from a few Myr ago to roughly as long ago as the age of the universe. It is already clear that star formation has not proceeded at a constant rate over the last 10 Myr or so (Tamblyn & Rieke 1993, Krabbe et al. 1995), because there are too few red supergiants relative to blue. Rather, the most recent burst can be fit as an exponentially decaying event about 3–5 Myr ago (Krabbe et al. 1995). It remains to be seen if a periodic burst scenario can explain the older stars as well. This, in turn, immediately raises questions about the nature of periodic transport of molecular material to and subsequent star formation in the GC. These will be the focus of much future research in the GC.

3. *H* and *K*−band Spectra of Cool Stars

In an effort to completely delineate the younger stellar population in the central parsecs, we have begun obtaining spectra for all the stars brighter than the brightest stars in BW ($K_\circ \leq 5.5$). We currently have completed roughly one–half of these. Tiede et al. (1996) speculate that the brightest few bins in Figure 1 corresponding to BW may be foreground stars. Hence, we plan to "fill in" the complete sample of objects with $K_\circ \leq 7.0$ to continue our investigation of the recent star formation history in the GC.

Due to the strong differential reddening in the field (Blum et al. 1996*b*), near infrared spectra are required in order to obtain accurate individual stellar properties for use in studying the star formation history. Cool, young stars are easily identified by their strong 2.3 μm absorption which increases in strength for stars with lower $T_{\rm eff}$. The CO strength also increases for more luminous stars, so that CO alone can not be used to distinguish between more luminous, warmer supergiants (more massive, younger) and cooler, less luminous AGB stars (lower mass, somewhat older). This degeneracy can be broken by observing the strong H_2O absorption band which lies near 1.9 μm in the region of the near infrared spectrum from the blue end of the *K*−band to the red end of the *H*−band. These effects in the H_2O and CO feature strengths of cool stars are well known (e.g. Kleinmann & Hall 1986).

In the GC it is essential to cover the *H* and *K*−band spectrum. The strong reddening will mask the H_2O absorption in the observed spectrum otherwise seen adequately in the *K*−band spectrum alone. This is demonstrated in Figure 2. Figure 2 shows the observed spectra for several of the GC objects. The dominant effect is the blueward attenuation of light due to interstellar reddening. Figure 3 shows similar disk stars whose spectra have been divided by a low order fit to the continuum. In effect, the H_2O can be measured like an equivalent width, and combined with the CO strength, $T_{\rm eff}$ and $M_{\rm bol}$ estimated. Without the *H*−band spectrum, the H_2O is difficult to measure since its effect is similar to additional extinction.

Figures 3, 4, 5, and 6 demonstrate how the various GC stars can be easily placed into relatively accurate $T_{\rm eff}$ and $M_{\rm bol}$ bins through measurement of CO and H_2O strengths. We observe nearby disk stars with MK spectral types which we then compare to our GC stars. The depression to the blue in all these spectra is due to a combination of intrinsic H_2O and choice of continuum points in the observed spectra. All the spectra presented in Figures 2–6 were obtained in 1997 July or 1998 May on the CTIO 4m using the crossed–dispersed grating of the facility spectrometer (IRS).

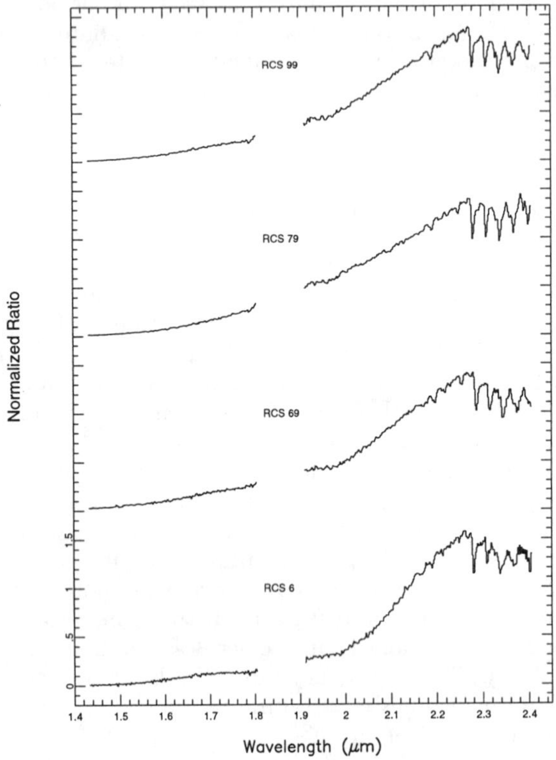

Figure 2. *H* and *K*−band spectra of typical Galactic Center stars. The strong CO absorption at 2.3 μm imediately distinguishes these objects in the bright portion of the *K*−band luminosity function as cool asymptotic giant branch (AGB) stars, or M supergiants. Strong interstellar reddening attenuates the observed spectrum to the blue (here the object is shown as a ratio to a telluric standard star).

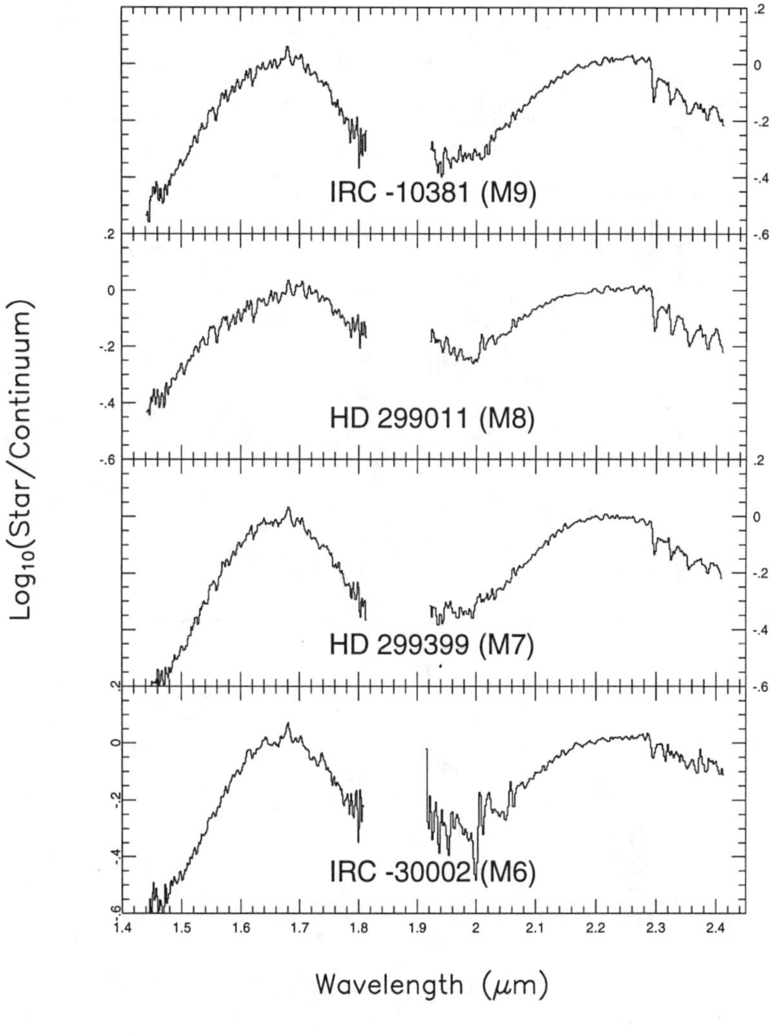

Figure 3. Representative "disk" long period variables (LPVs). These stars (the most luminous asymptotic giant branch stars) are distinguished from more massive red supergiants by their extreme H_2O absorption centered near 1.9 μm.

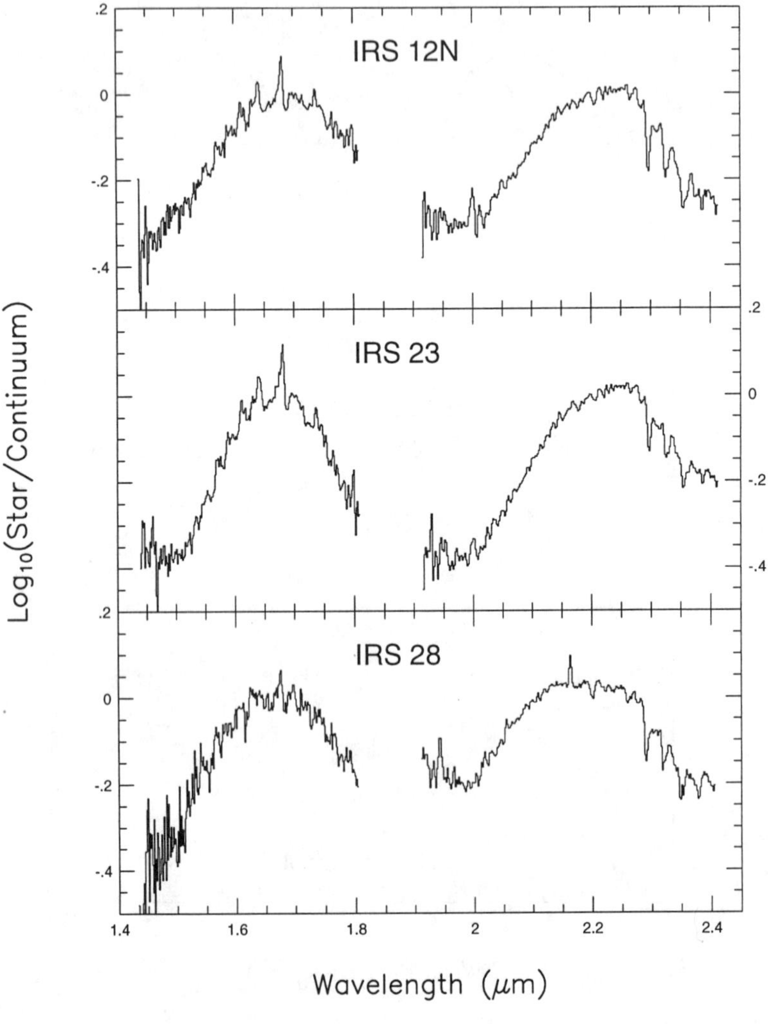

Figure 4. Same as Figure 3, but for Galactic Center stars. These objects are confirmed photometric variables, though they do not have measured periods.

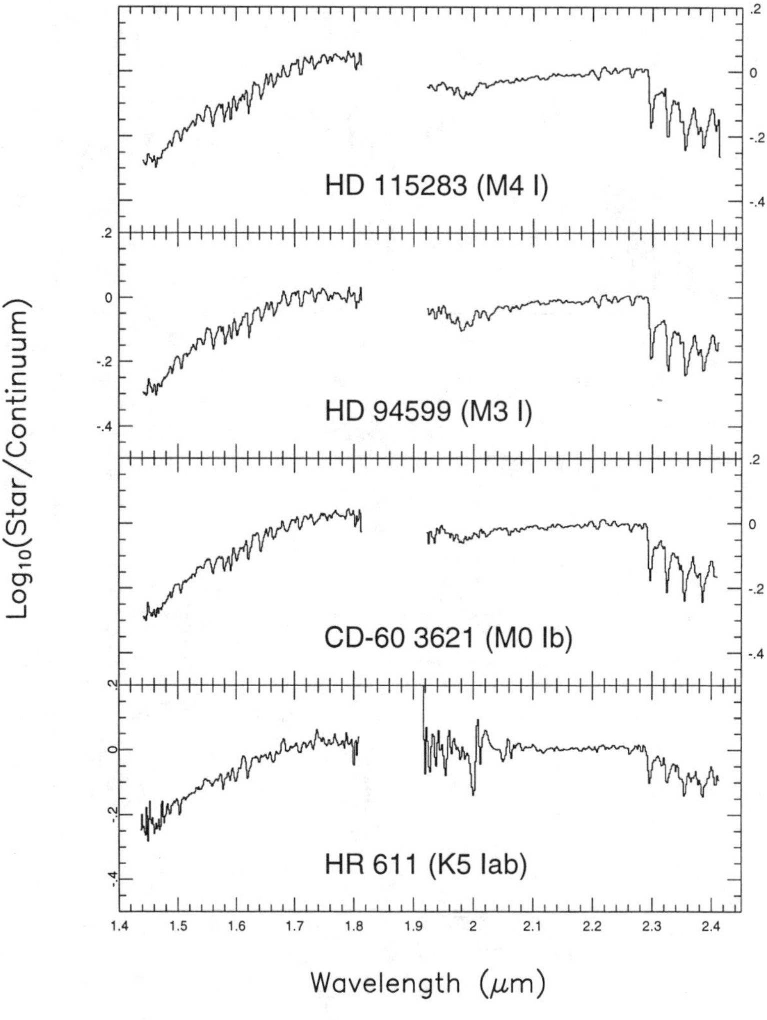

Figure 5. Representative "disk" supergiants. These stars are distinguished from the long period variables by their lack of strong H_2O absorption centered near 1.9 μm and presence of strong CO 2.3 μm absorption. Compare to Figure 3.

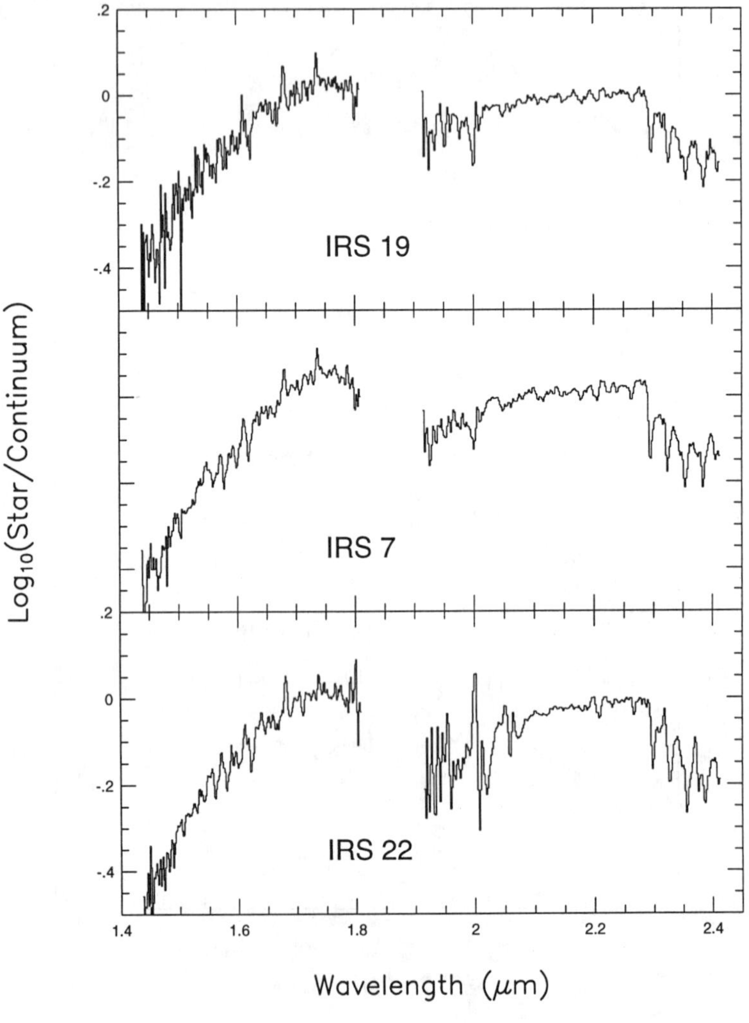

Figure 6. Same as Figure 5, but for Galactic Center stars. Note the easily observed difference in H_2O strength for these objects which can have similar *observed* JHK colors and CO strengths relative to those in Figure 4.

References

Allen, D. A., Hyland, A. R., & Hillier, D. J. 1990, MNRAS, 244, 706

Blum, R. D., DePoy, D. L., & Sellgren, K. 1995*a*, ApJ, 441, 603

Blum, R. D., Sellgren, K., & DePoy, D. L. 1995*b*, ApJ, 440, L17

Blum, R. D., Sellgren, K., & DePoy, D. L. 1996*a*, AJ, 112, 1988

Blum, R. D., DePoy, D. L., & Sellgren, K., 1996*b*, ApJ, 470, 864

Eckart, A., & Genzel, R., 1997, MNRAS, 284, 576

Genzel, R., & Eckart, A., 1999, this volume, p. 3

Genzel, R., Thatte, N., Krabbe, A., Kroker, H., & Tacconi–Garman, L. E. 1996, ApJ, 472, 153

Ghez, A. M., Klein, B. L., Morris, M., & Becklin, E. E. 1998, ApJ, in press

Ghez, A., et al. 1999, this volume, p. 18

Kleinmann, S.G. & Hall, D.N.B. 1986, ApJS, 62, 501

Krabbe, A., Genzel, R., Drapatz, S., & Rotaciuc, V., 1991, ApJL, 382, L19

Krabbe, A., et al. 1995, ApJ, 447, L95

Lebofsky, M. J., Rieke, G. H., & Tokunaga, A. T. 1982, ApJ, 263, 736

Libonate, S., Pipher, J. L., Forrest, W. J., & Ashby, M. L. N. 1995, ApJ, 439, 202

Sellgren, K., Hall, D.N.B., Kleinmann, S.G., & Scoville, N.Z. 1987, ApJ, 317, 881

Tamblyn, P., Rieke, G. H., Hanson, M. M., Close, L. M., McCarthy, D. W., JR., & Rieke, M. J. 1996, ApJ, 456, 206

Tamblyn, P. & Rieke, G. H. 1993, ApJ, 414, 573

Tiede, G. P., Frogel, J. A., & Terndrup, D. M. 1995, AJ, 110, 2788

The Central Parsecs of the Galaxy
ASP Conference Series, Vol. 186, 1999
H. Falcke, A. Cotera, W.J. Duschl, F. Melia, M.J. Rieke, eds.

A Complete K-Band Luminosity Function of the Central 30 pc

Sabine Philipp[1], Peter G. Mezger, and Robert Zylka[1]

Max-Planck-Institut für Radioastronomie, Auf dem Hügel 69, 53121 Bonn, Germany

Abstract. We have obtained a K band mosaic image of an area $\Delta\alpha \times \Delta\delta \sim 650'' \times 710''$ centered approximately on Sgr A*. With a detection limit of $S'_{\rm K} \sim 100\,\mu$Jy and a completeness limit of $\sim 2000\,\mu$Jy our source counts will only be complete for early O-stars, Supergiants and the most luminous red Giants; less luminous stars will merge into an unresolved background continuum. Special care was taken in our observing procedure to recover all of this continuum emission. 6.1×10^4 individual sources were separated from the background continuum by fitting them with modified Lorentzian distributions. For the low-luminosity sources we construct a model K-band luminosity function (KLF) based on a Salpeter IMF, which is compared with the KLF of Baade's Window after readjustment to match the stellar surface density in the mosaic. We obtain the following results:

The emission of both point like sources (interpreted as luminous stars) and an unresolved continuum (interpreted as low-mass main sequence (MS) stars) can be fitted by a KLF $\propto S'^{\alpha}_{\rm K}$ consisting of four power-law approximations with different exponents α: The modeled sectors which relate to MS stars have the functional dependences $\propto S'^{-0.6}_{\rm K}$ and $S'^{-0.8}_{\rm K}$; the observed KLF, which relates to Giants and Supergiants becomes gradually steeper. Compared to the KLF of Baades Window (Tiede et al., 1995) the KLF of the mosaic shows a clear overabundance of stars with high K band flux densities which is strongest within the central 24″ ($\sim$ 1 pc).

1. The Nuclear Bulge: An Introduction

In a recent review by Mezger et al. (1996; referred to as MDZ 96) we defined the central volume of size $2R \sim 600$ pc as Nuclear Bulge (NB). More recent investigations of the distribution of stars and dust in the central region show this to be a very unfortunate choice of wording. The generic relations between Galactic Disk (GD) and NB appear to be much stronger than those of Galactic Bulge (GB; $R \sim 3$ kpc) and NB.

---

[1]Institut für Theoretische Astrophysik, Tiergartenstraße 15, 69121 Heidelberg, Germany

Evaluation of COBE-DIRBE observations in the wavelength range $\lambda\,1.25$ – $240\,\mu$m (Launhardt et al., in prep.) show a box-type shape of the NB of size $2R_l \times 2R_b \sim 200 \times 30\,\mathrm{pc}^2$ in the NIR, where stellar radiation dominates; and a somewhat larger size in longitude of $2R_l \sim 300$ pc at MIR/FIR wavelengths, where dust emission dominates. The physical conditions in the NB are rather extreme. In a volume $V_{\mathrm{NB}} \leq 10^{-4} V_{\mathrm{GD}}$, the NB contains $\sim 3\,\%$ of all molecular gas and $\sim 1\,\%$ of the stellar mass of the Galactic Disk and accounts for $\sim 5\%$ of the dust luminosity and $\sim 10\%$ of the formation rate of massive stars in the Galaxy (with the latter estimate based on the Lyman continuum photon production rate in the NB; for details see MDZ 96).

We have begun an investigation of the physical state of the NB. Paper I (Philipp et al., 1998) reports on an investigation of the stellar K band emission in an area $\Delta\alpha \times \Delta\delta \sim 650'' \times 710''$ (corresponding to $R_{equiv} \sim 15.8$ pc) centered approximately on Sgr A*. This area is synonymously referred to as the mosaic or the central 30 pc. We derive a K band luminosity function (KLF) for the mosaic stars and compare it to the KLF of Stars in the GB and GD. Paper II (Mezger et al., in prep.) gives an interpretation of the KLF of the central 30 pc. Here we summarize the essential results of paper I and give preview of some preliminary results discussed in paper II.

2. Observations and Data Reduction

The K band observations presented and discussed here were made with the ESO-MPG 2.2m telescope on La Silla using the IRAC2B camera. A NICMOS3 chip with 256×256 pixels produces images of size $71.2'' \times 71.2''$. The mean visual extinction towards stars in the central arcseconds is $A_{\mathrm{V}} \sim 31$ mag, corresponding to a K band optical depth of $\tau_{\mathrm{K}} \sim 3.2$. With a detection limit of $S'_{\mathrm{K}} \sim 100\,\mu$Jy for individual sources and a completeness limit for source counts of $S'_{\mathrm{K}} \sim 2000\,\mu$Jy it became soon clear that most of the Main Sequence (MS) stars later than spectral type O9 would not be detected as individual sources but rather merge into a background continuum (see also paper I, Fig. 9). Special care was therefore taken in our observing procedure to recover this continuum and to eliminate propagating errors in both position and flux density level which can occur in concatinating observations. 494 images were joined together for the mosaic shown in Fig. 1. 6.1×10^4 sources could be separated from the extended background by fitting Lorentzian distributions to the observations. The K band surface brightness integrated over the mosaic amounts to a flux density of $S'_{\mathrm{K,integr}} \sim 752$ Jy, of which ≤ 131 Jy ($\sim 20\,\%$) are estimated to be contributed by stars in Galactic Bulge and Disk. High-luminosity stars in the NB observed as individual sources and low-luminosity stars observed as an extended background contribute about equally ($\sim 50\,\%$ each) to the integrated flux density.

3. Observed K-band Luminosity Functions (KLFs)

The KLF normalized to an appropriate area is obtained by counting the number of (reddened) K band sources $N(S'_{\mathrm{K}})$ within fixed logarithmic ($d\log S'$) or linear (dS') intervals, respectively. Linear and logarithmic KLFs are related by

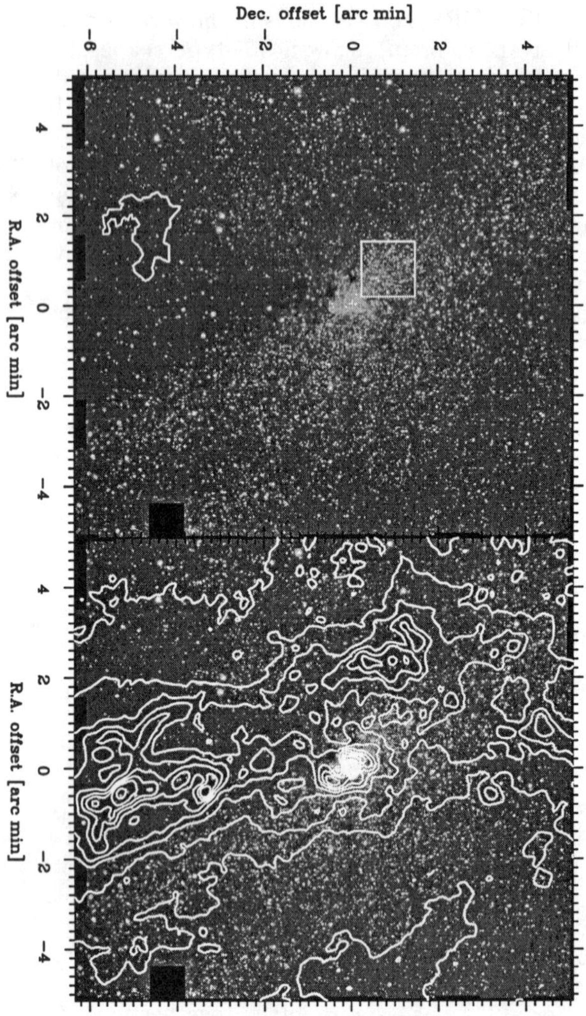

Figure 1. a) K band mosaic of the central $\Delta\alpha \times \Delta\delta \sim 650'' \times 710''$ centered approximately on Sgr A*. The black square indicates "no usable data". White framed contours indicate subareas of the mosaic $5400''$ on a side, which are referred to in the text. The rectangle is centered on the synchrotron source Sgr A East, the irregularly shaped area is centered on the compact cloud core M−0.13 − 0.08 (see text sect. 3.3).
b) Overlaid on the K band mosaic is a contour map representing the dust emission observed at λ 1.2 mm with MPIfR bolometer arrays in the IRAM 30-m MRT ($\Theta_b = 11''$). Contour levels are: 90, 240, 390, 540, 690, 840, 990, 1180, 1680, 2180, 2680, 3180 and 3680 mJy/11″ beam. 100 mJy/11″ beam corresponds to a $\tau_{1.2\text{mm}} = 4 \times 10^{-4}$ for $Z/\text{Z}_\odot \sim 2$ and $T_{\text{dust}} \sim 40\,\text{K}$. The corresponding $\lambda 2.2\,\mu\text{m}$ opacity is $\tau_{2.2\mu\text{m}} = 0.98$.

$$\frac{dN(S')}{dS'} = 0.43 S'^{-1} \frac{dN(S')}{d\log S'} \tag{1}$$

We derived KLFs for three areas indicated in Fig. 1a:

1. The total mosaic.
2. A rectangular area centered on Sgr A East, where a strong explosion has blown off $\sim 5 \times 10^4$ yrs ago most of the gas and dust in the NB facing the sun.
3. An irregular shaped area centered on "M−0.13 − 0.08" which is opaque for K band emission (i.e. $\tau_K > 6$). The area referred to here and in paper I as M−0.13 − 0.08 has been chosen as the area of highest K band extinction within the mosaic. The absorbing dust was believed to be associated with gas which formes part of the 20 km/s cloud. Recent spectroscopy of the gas (Zylka et al. 1999) shows, however, that in this area of highest extinction another cloud with $v_{LSR} \sim -135$ km/s overlaps with the +20 km/s. The high negative velocity gas – which probably contributes most to the K band extinction – actually peaks at M−0.114 − 0.124. For consistency reasons we continue to refer to this high extinction region as M−0.13 − 0.08.

These KLFs together with their cumulative source numbers and flux densities normalized to 1 arcsec^2 are shown in Fig. 2a – c. It is obvious that the highest star density occurs towards Sgr A East, the lowest towards M−0.13 − 0.08. Figure 3 shows the difference KLF between Sgr A East and M−0.13 − 0.08, which is in essence the KLF of the stars in the NB, together with the difference KLF related to stars in the Galactic Bulge and Disk. Although the stars in the NB are both farther away and subject to more extinction, their KLF contains the stronger sources, most of which are probably very early O stars, Giants and Supergiants. There is a clear overabundance of K band luminous stars visible in the KLF of the NB.

4. The Radial Distribution of High- and Low-Luminosity Stars

To facilitate the discussion we consider in the following all resolved high-luminosity stars in the NB as young stars; and we consider as old stars all low-luminosity stars observed as an extended background. Obviously, this is a rather crude separation and there exists a considerable overlap of the two populations, specifically in the range between detection and completeness limit. Combination of the K band flux densities discussed here with H band flux densities (see paper II) should help to clarify the situation.

Figure 4 shows the K band surface brightnesses of different components obtained by integration of the mosaic (Fig. 1) in concentric circles of radius R. The three curves relate to i) the total surface brightness, ii) that of the luminous stars and iii) the extended background formed by low-luminosity stars, respectively. Within $R \leq 10''$ the surface brightness is dominated by young high-luminosity stars; farther out high- and low-luminosity stars contribute about equally. In Fig. 5 we fitted King profiles to the observed radial distribution of high- and low-luminosity stars and obtained core radii of $7''$ and $30''$, respectively. The obvious conclusion is that high-mass star formation during the past $\sim 10^7$ –

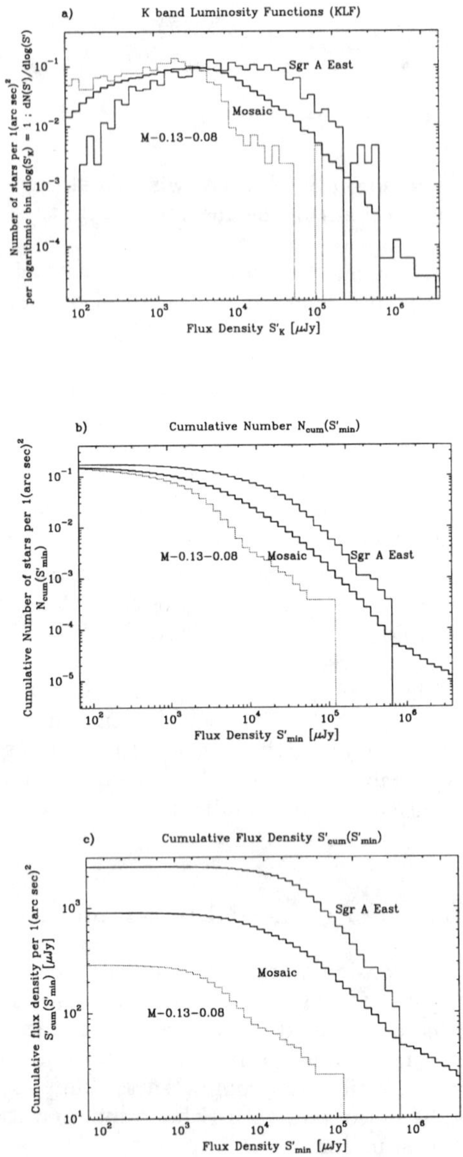

Figure 2. a) KLF of the mosaic image and of the two subareas indicated in Fig. 1a, all normalized to a solid angle of 1″. Sources are counted in logarithmic bins of width 0.08.
b) Cumulative number $N_{\rm cum}(S'_{\rm min})$ of stars with flux densities $S'_{\rm K} \geq S'_{\rm min}$ computed with the KLFs shown in Fig. 2a.
c) Cumulative flux density of stars $S'_{\rm cum}(S'_{\rm min})$ with flux densities $S'_{\rm K} \geq S_{\rm min}$ computed with the KLFs shown in Fig. 2a.
(for the labelling of M−0.13 − 0.08 see text sect. 3.3).

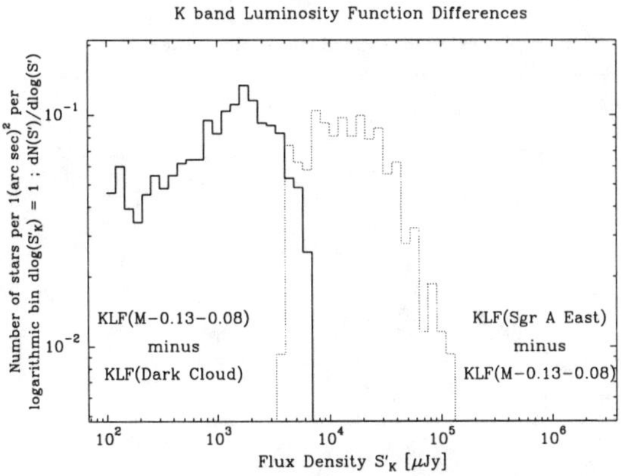

Figure 3. Difference KLFs: (Sgr A East) minus (M−0.13 − 0.08) and (M−0.13 − 0.08) minus (Dark Cloud).

10^8 yrs attained its maximum within the central 14″ ($\sim$ 0.6 pc) and decreased from the center of the NB outward.

5. A Model KLF of the Low-Luminosity Stars

To construct a model KLF of the low-luminosity stars we made two assumptions:

1. The formation rate of stars with $M_* \leq 20\mathrm{M}_\odot$ is described by the Salpeter IMF.
2. The total stellar mass contained within the mosaic ($R \leq 383''$) is $\Sigma M_* \sim 1.2 \times 10^8 \mathrm{M}_\odot$ (as suggested by the kinematics of gas and stars).

Then the number of stars in the mass interval M_*, $M_* + dM_*$ is

$$dN = 1.81 \times 10^7 \left(\frac{M_*}{\mathrm{M}_\odot}\right)^{-2.35} dM_* \tag{2}$$

and the total number of MS stars which account for the mass of $1.2 \times 10^8 \mathrm{M}_\odot$ is obtained by integration over the mass range $0.06 \leq M_*/\mathrm{M}_\odot \leq 20$ (corresponding to stars of spectral type M8 – O9) yielding $N_{\mathrm{cum}} \sim 6 \times 10^8$ MS stars. We computed the K band flux density of MS stars as a function of their mass M_* using the stellar parameters given in Lang (1992) reddened for an optical depth of $\tau_{\mathrm{K}} \sim 3.2$ (corresponding to $A_{\mathrm{V}} \sim 31$ mag and $A_{\mathrm{K}}/A_{\mathrm{V}} \sim 0.112$). The result is shown in paper I, Fig. 9. This relation is well represented for the above star mass interval[1] by the power-law approximation

$$\left(\frac{S'_{\mathrm{K}}}{\mu\mathrm{Jy}}\right) = 1.6 \left(\frac{M_*}{\mathrm{M}_\odot}\right)^{2.3} \tag{3}$$

[1]In paper II we show that this mass-luminosity relation holds for all MS stars.

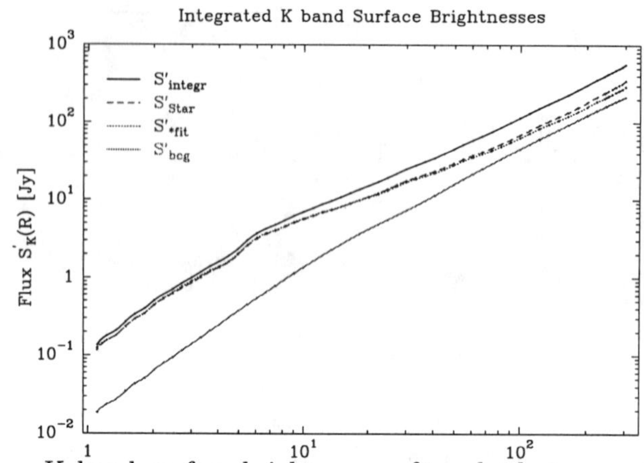

Figure 4. K band surface brightnesses of resolved stars and unresolved continuum based on our observations, integrated in concentric circles of radii R centered on Sgr A* and including the flux density of IRS 7. Heavy solid line: Total integrated flux density $S'_{\rm integr}$; light solid curve: Flux density $S'_{\rm bcg}$ contributed by the unresolved background emission. Dotted and dash-dotted curves: Flux density $S'_{*\rm fit}$ and $S'_{\rm Star} = S'_{\rm integr} - S'_{\rm bcg}$, respectively, which both relate to resolved stars with $S'_{\rm K} > 100\,\mu$Jy or $m_{\rm K} < 17\,$mag.

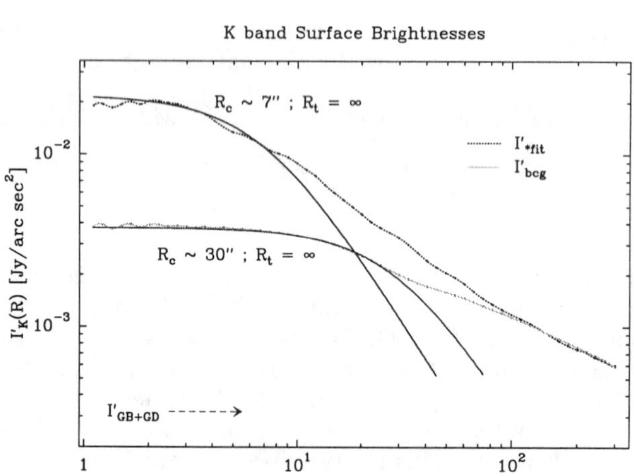

Figure 5. The radial distribution of K band surface brightnesses observed for resolved ($I'_{*\rm fit}$) and unresolved ($I'_{\rm bcg}$) stars together with King profiles and fit parameters $R_{\rm c}$ and $R_{\rm t}$. Also shown is the estimated contribution $I'_{\rm GB+GD}$ of stars located in the Galactic Bulge and Galactic Disk.

Combination of equations (2) and (3) yields the linear model KLF

$$\mathrm{KLF(mod)} = \frac{dN(S'_{\mathrm{K}})}{dS'_{\mathrm{K}}} = 1.03 \times 10^7 S'^{-1.6}_{\mathrm{K}} \tag{4}$$

valid within the limits S'_{min}(M8 star) $\sim 3 \times 10^{-3}\,\mu$Jy and S'_{max}(O4 star) $\sim 4.5 \times 10^4\,\mu$Jy. The corresponding logarithmic KLF, normalized to an area of 1 arcsec^2 is

$$\frac{dN(S'_{\mathrm{K}})}{d\log S'_{\mathrm{K}}} = 58.5 S'^{-0.6}_{\mathrm{K}} \tag{5}$$

(see eq. (1)). This relation is shown in paper I, Fig. 7 as "model KLF" for $S'_{\mathrm{K}} \leq 10^3\,\mu$Jy, i.e. for MS stars with masses $M_* \leq 20\mathrm{M}_\odot$. Also shown in this Fig. is the "observed KLF" of the mosaic (see this paper Fig. 2a).

It should be mentioned here that new investigations (to be published in paper II) which will be briefly discussed in the following section clearly indicate that it is not justified to extend the model KLF eq. (4) beyond MS stars of $\geq 1\mathrm{M}_\odot$ whose MS lifetime is shorter than the lifetime of the Galactic and Nuclear Bulge. With increasing stellar mass an increasingly large fraction of all stars ever formed in the NB will have moved off the MS and become Giants and Supergiants. Therefore, for $M_* > 1\mathrm{M}_\odot$ the IMF – which describes the mass distribution of a newly born generation of stars – has to be replaced by the Present Day Mass function (PDMF), which e.g. decreases very steeply in the solar neighbourhood.

6. Towards a Complete KLF of the Central 30 pc

Here we summarize results dicussed in paper II. For a complete model KLF we have stated the following conditions:

1. It must join smoothly with the observed KLF for flux densities $S'_{\mathrm{K}} \sim$ completeness limit $\sim 2 \times 10^3\,\mu$Jy.
2. The cumulative flux density

$$\int_{S'_{\mathrm{K}}=2\times10^6}^{3\times10^{-3}} S\frac{dN(S)}{dS}dS = S_{\mathrm{integr}} \sim 752\,\mathrm{Jy} \tag{6}$$

   must be equal the integrated flux density S_{integr} of the mosaic.
3. A specific power-law sector of the KLF cannot be selected arbitrarily but must be physically plausible.

In paper I, Fig. 7 we show the model KLF $\propto S'^{-0.6}_{\mathrm{K}}$ given by eq. (5) together with the observed KLF. The model KLF does not join smoothly with the observed KLF. In this Figure we therefore introduce a second model KLF $\propto S'^{-1.0}_{\mathrm{K}}$ which is an extension of the observed KLF in the range $S'_{\mathrm{K}} \sim 5 \times 10^3 - 5 \times 10^4\,\mu$Jy. Although this model KLF combined with the observed KLF fulfills the above conditions 1) and 2) approximately, the extension of the observed KLF $\propto S'^{-1.0}_{\mathrm{K}}$ is physically not justified.

In paper II we assume that the ensemble of stars in the NB consists of Population II stars, to that in the Galactic Bulge, superposed on Population I stars which formed whenever enough gas has assembled in the Nuclear Bulge. We then consider a KLF of stars in the GB, as determined e.g. by Tiede et al. (1995) in Baade's window at $l \sim 1°$, $b \sim -3.9°$. Within the limits $8 \times 10^4 \geq S'_{\rm K,BW}/\mu{\rm Jy} \geq 2 \times 10^2$ this KLF can be approximated by a power law $S'^{-0.8}_{\rm K,BW}$. For $S'_{\rm K,BW} \geq 8 \times 10^4\,\mu$Jy the KLF turns sharply downward meaning that in BW the surface density of K band luminous stars decreases rapidly. This supports our earlier findings (see Figs. 2a and 3) of an overabundance of K band luminous stars in the NB. For $S'_{\rm K,BW} < 2 \times 10^2\,\mu$Jy the KLF again turns sharply downward due to the detection limit of the surveys Tiede et al. used. To compare the BW KLF with the KLF of the mosaic we use the relation $S'_{\rm K,mosaic} = e^{-\tau_{\rm K}} S'_{\rm K,BW}$. For $A_{\rm K,mosaic} \sim 3.5\,$mag, $A_{\rm K,BW} \sim 0.13\,$mag, $\tau_{\rm K} \sim 3.1$ and $e^{-\tau_{\rm K}} \sim 4.5 \times 10^{-2}$; hence the above limits for the BW KLF $\propto S'^{-0.8}_{\rm K,BW}$ translate into $3.6 \times 10^3 \geq S'_{\rm K,mosaic}/\mu{\rm Jy} \geq 9$. It is noteworthy (see paper I, Fig. 9) that $S'_{\rm K,mosaic} \leq 3.6 \times 10^3\,\mu$Jy is also the upper limit of MS stars and Giants.

The BW KLF shifted towards the central 30 pc thus becomes

$$\frac{dN(S)}{d\log S} \sim 1.77 S'^{-0.8}_{\rm K,mosaic} \tag{7}$$

To have it smoothly join with the model KLF $\propto S'^{-1.0}_{\rm K,BW}$ we have to multiply the above KLF by a factor 55.6 which tells us about the increase of the surface density of medium and low-mass MS stars from Baade's Window to the central 30 pc of the NB. This increase compares reasonably well with the differences of surface brightnesses in NB and GB+GD, respectively (see Fig. 5).

Figure 6, taken from paper II, shows what we consider as a physically plausible KLF of all stars in the central 30 pc. It consists of four sectors whose powerlaw approximations are also given in Fig. 6. Its interpretation in terms of different populations of MS and evolved stars is the subject of paper II.

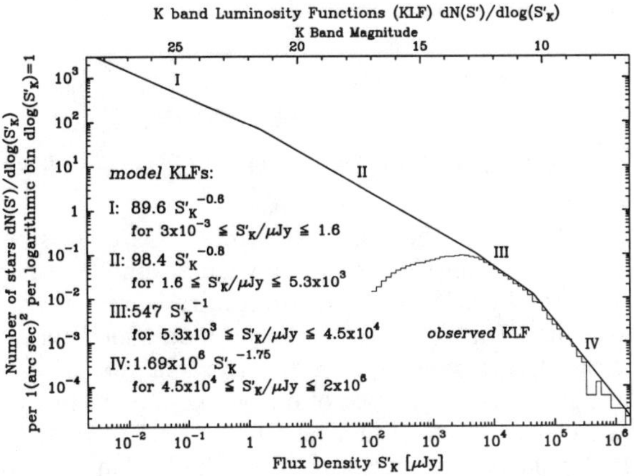

Figure 6. The four-sectors KLF which fits the observation of the mosaic best together with their power-law approximations (see text).

References

Lang K.R., 1992, Astrophysical Data: Planets and Stars, 132ff.

Mezger P.G., Duschl W.J., Zylka R., 1996, AAR 7, 289 (MDZ96)

Philipp S., Zylka R., Mezger P.G. et al. 1998, A&A in press (paper I)

Tiede G.P., Frogel J.A., Whitford A.E, 1995, AJ 110, 2788

Zylka R., Güsten R., Philipp S. et al., 1999, this volume, p. 415

The Central Parsecs of the Galaxy
ASP Conference Series, Vol. 186, 1999
H. Falcke, A. Cotera, W.J. Duschl, F. Melia, M.J. Rieke, eds.

Properties of Stars in the Galactic Center

Thomas Ott, Andreas Eckart, Reinhard Genzel

Max-Planck-Institut für extraterrestrische Physik, Garching

Abstract. We present the results of a time-series analysis of high resolution photometry covering 7 consecutive years. Some long-period variables already discovered by other authors are confirmed and are probably Miras or luminous blue variables. We also report on short period variables in the range of 3-10 days which give a clue to the actual masses and radii of the stars in the innermost parsec of our galaxy. In addition, polarimetric data were recorded on an observing run in 1995. Finally, some objects with unusual polarization vectors are studied in more detail regarding their extendedness compared to point-sources. This search for embedded objects in the galactic center gives a hint on where these objects have formed and how old they are.

1. Introduction

In order to better understand the evolution and dynamics of the central stellar cluster of the Milky Way it is necessary to investigate individual cluster members in detail. The presence and nature of 2 important constituents of the Galactic Center stellar cluster has just recently been revealed by spectroscopic measurements (e.g. Sellgren et al. 1990, Krabbe et al. 1995, Genzel et al. 1996). 80% of these are suspected to be late-type giants on the asymptotic giant branch, as seen by their deep CO absorption bandheads. These stars are situated in a part of the HRD where stellar variability on timescales of few months to several years is to be expected. Recent studies already discovered several late-type variables (Tamura et al. 1996, Blum et al. 1996).

The remaining 20% of the stars are either objects with steep, red and featureless spectra or stars that exhibit strong He-emission (Krabbe et al. 1995). The number density of He-stars peaks towards the few central arcseconds where the bright IRS 16-cluster is located. Some of those could be Luminous Blue Variables (LBV's) which are also supposed to show variations on a broad range of timescales.

Polarization studies of the Galactic Center region in the IR have been carried out by several groups (Knacke and Capps 1977, Lebofsky et al. 1982, Aitken et al. 1998). The angular resolution of these investigations was seeing limited with a FWHM of 1.0" - 1.5" at best. Mostly larger apertures of a few arcseconds diameter were used. The result of these investigations is that in the mid-infrared the polarization is dominated by warm dust grains that are aligned by magnetic fields in the mini-spiral gas streamers. In the near infrared the polarization is dominated by dust scattering in the Galactic plane. The first high angular resolution polarimetric study of the Galactic Center in the near-infrared at the diffraction limit of a 3.5m telescope was conducted by Eckart et al. (1995). This study covered the north-eastern region of the central parsec, in which several sources were found that show large polarization vectors, differing from the mean flux-weighted polarization of all sources in the central 0.25 parsec. This is 4% at a position angle of 25°, within 7° parallel to the Galactic plane. Most of the polarizations therefore can be explained via absorption by interstellar dust grains which are aligned in the large-scale magnetic field in the spiral arms of the galaxy.

2. Observations and Data Reduction

All data were acquired using the MPE speckle-camera SHARP I installed on the New Technology Telescope (NTT) on La Silla which is operated by the European Southern Observatory (ESO). Our data base covers the interval from 1992 up to the year 1998 with 1 to 2 observing runs per year. Each night, we take several thousands of short-exposure frames with integration times of $\sim$0.5 seconds, which is a good compromise of obtaining a high signal-to-noise ratio while still being able to resolve the speckle-pattern. These on-source integrations were interleaved by sky measurement exposures. Since the field of view of SHARP I is 12".8×12".8, the Galactic Center region was imaged as a mosaic from 4 pointings, placing the central IRS 16 complex alternately in the 4 quadrants of the detector array. This way we recorded the central 20"×20" region, which is 1pc×1pc for a Galactic Center distance of 8 kpc.

The polarimetric data were taken on July 1995 using a warm, rotatable wire grid which is mounted in front of the SHARP I entrance window. The data acquisition, reduction and calibration for both the direct speckle imaging and the polarization measurements are described in Eckart et al. (1994) (see also Christou 1991).

2.1. Deconvolution

To correct for the seeing background in the SSA process, the images were deconvolved in order to produce clean images at the diffraction limit of the NTT, which is 0".13 in K-Band (see also Eckart et al. 1994). As a point spread function (PSF) we used the weighted mean of the brightest sources in the SSA images. In order to be able to check our photometric results on the individual sources in this crowded region, we used 3 different deconvolution algorithms, namely LUCY, CLEAN and a Wiener filtering as a linear one-pass method, and compared their results. These 3 deconvolution methods can be summarized as follows:

- The Lucy-Richardson algorithm is an iterative procedure where the solution of pointlike sources, convolved with the PSF is compared with the initial picture
- The Högbom CLEAN algorithm iteratively subtracts the PSF from the individual sources
- The Wiener filtering divides the image by the PSF in Fourier space

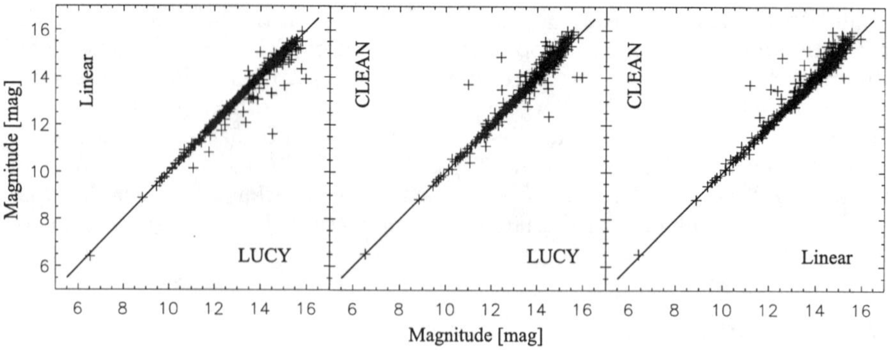

Figure 1. Comparison of aperture photometry results obtained using 3 different deconvolution methods.

In Fig. 1 we show a comparison of the photometry obtained using these 3 deconvolution methods. All 3 methods yield very comparable results up to a magnitude of $m_K > 13$. Therefore, we only regard stars brighter than this value in our variability study. We were able to get light curves of 522 individual stars. In our final plots, we used the mean of those 3 methods with their standard deviation as error bars. The flux calibration was done using 8 different stars which were visible in all parts of our mosaic, using the photometry data published in Blum et al. 1998. We assumed that the mean flux of these 8 stars is constant for all epochs. This assumption is supported by the resulting flat light curves of the individual calibrators.

2.2. Variability

The variability study was performed doing aperture photometry on the deconvolved images. The standard error σ_i of a single measurement i then is a measure of the standard deviation of these 3 measurements. To get a measure for the probability that a given star shows variability, we performed a reduced χ^2-test. This test is defined by:

$$\chi^2 = \frac{1}{\nu} \sum_i \frac{|m_i - \overline{m}|}{\sigma_i^2}$$

Here, ν is the number of degrees of freedom, m_i is the magnitude derived from a single measurement, σ_i its error, and $\overline{m}$ the mean magnitude of this star. Out of the 522 stars with $m_K < 13$, 202 show a value of $\chi^2 > 2.5$. We chose this value to be the cutoff for possible variability because it is just above the highest χ^2 value of 2.44 that we obtained for the flux calibrator IRS 16NW, as we suppose the calibrators to not be variable. This value of $\chi^2 = 2.5$ is well above the peak of the distribution which occurs at $\chi^2 \approx 2.0$. The percentage of stars with possible variability thus is 39%. Approximately 80% of the stars are situated in the asymptotic giant branch where variability is to be expected with periods ranging from several years to a few months. As we can only detect variability with periods up to 3 years, this means that every other star in the Galactic Center on the asymptotic giant branch could show detectable variability with periods less or equal to 3 years.

2.3. IRS 16SW

In the following, we want to emphasize on the source IRS 16SW which was claimed by Tamura et al. (1996) to become fainter by $0^m.1$ in their data set over the course of 3 days. This significant variation over a very short time scale was, however, not especially emphasized by the authors. With our better time coverage, we are able to show that IRS 16SW is indeed variable over these time scales. It even shows periodicity with a period of 9.72 days.

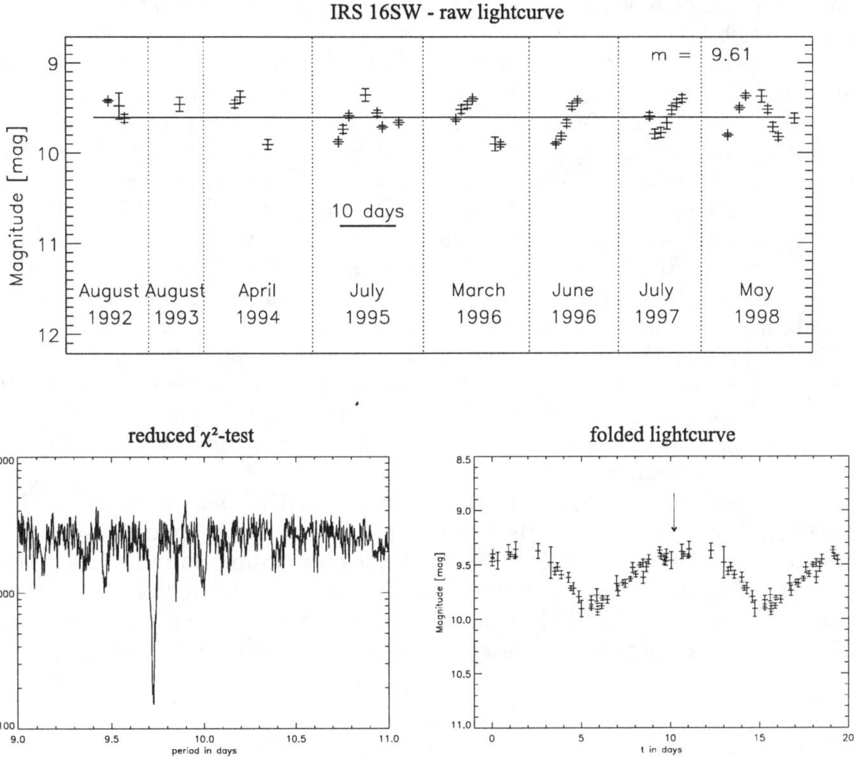

Figure 2. Light curve of IRS 16SW and determination of its period. Top: Raw light curve; Bottom left: Result of the χ^2-test used to determine the period of 9.72 days; Bottom right: Data folded back into a 2-period interval. The arrow indicates the possible secondary minimum.

In Fig. 2, top, we show the raw light curve as deduced from our data. Since the longest observing run in May 1998 was just slightly longer than the period of IRS 16SW, we performed a reduced χ^2-test of 1000 periods in the range between 9 and 11 days in order to get a plot of a single period. In the reduced χ^2-test, the absolute minimum corresponds to the period that best fits to the observed data. There are, of course, other local minima in the distribution which correspond to the harmonics of the period of 9.72 days at 19.44 days, 29.16 days, and so on. There is, however no minimum at a period of 4.86 days, so the period given above should be the correct value, since there is no reason why IRS 16SW should be variable in one of its harmonics.

For the nature of this source, 3 possible answers can be given, looking solely at the length of the period: IRS 16SW could be a Cepheid, it could be an eclipsing binary or IRS 16SW shows brightness variations due to some other effect such as an uneven distribution of sun-spots, non-radial oscillations...

For the third possibility, one would expect variations in amplitude and period over long time scales, which is not supported by our data. Although one can never completely rule out these mechanisms, here we concentrate on the first 2 possibilities.

The period-luminosity function for classical Cepheids of population I in K-Band is $\langle M_K \rangle = -3.80 \cdot [\log(P) - 0.8] - 5.46$ (Hindsley and Bell, 1990). The mean K-Band magnitude for IRS 16SW is $\langle m_K \rangle = 9.61$ (see Fig. 2 top), then taking a Galactic Center distance of $r = 8kpc$ and an extinction $A_K = 3.3^m$, its absolute magnitude is $M_K = -6.2^M$. Taking the period-luminosity function given above and the derived period of $P = 9.72d$ (see Fig. 2 bottom left), the mean absolute K-Band magnitude for such a Cepheid should be $\langle M_K \rangle = -8.2^M$, so IRS 16SW would be 2^M too faint to be a Cepheid. Cepheids of the population II (W-Virginis variables) are 2^m fainter and thus would match the absolute magnitude. However, Cepheids in most cases show a steeper slope while brightening than while becoming fainter, which is not reflected in our light curve. Moreover, Krabbe et al. (1995) have presented a spectrum of this source. It shows strong He I emission lines which would not be expected for Cepheids of either population I or II. For these reasons we rule out the possibility that IRS 16SW is a Cepheid.

The symmetrical shape of its light curve and a hint for a second local minimum, however, suggest that this star is an eclipsing binary. We give an estimation of the mass of IRS 16SW, with the only assumption that it is a semi-detached β Lyrae system as proposed by the shape of the light curve. Kepler's third law relates the period P, the separation r and the sum of the masses $M_1 + M_2$ of a 2-body system independent of inclination and ellipticity:

$$P^2 = \frac{4\pi^2 r^3}{G\,(M_1 + M_2)}$$

In this equation, G denotes the gravitational constant. Solving for the total mass, we can give a relation between the separation of the components and the mass M of the system, which yields $M = (M_1 + M_2) = \frac{4\pi^2 \cdot r^3}{G \cdot P^2}$. We now can give a *lower* limit to the mass of IRS 16SW when considering a minimum separation which can be derived the following way: The radius of IRS 16SW has been estimated by model calculations to be of the order of $\sim 90R_\odot$ (Najarro et al. 1997). Assuming that the total luminosity originates predominantly from the main component, this would be the minimum separation possible for an eclipsing binary. Inserting this value into above equation then gives a lower limit to the total mass of $M = 104M_\odot$.

Finally, we give an estimate of the spectral resolution necessary for confirming the binarity of IRS 16SW using its bright Brγ emission line feature. In the case of the separation of the 2 components of 130 $R_\odot$, an upper limit to the orbital velocity for the system would be 677 km/sec. In order to spectrally resolve the Doppler-shift in the emission line, one then would need a minimum spectral resolution $\lambda/\Delta\lambda$ of ~ 1000.

2.4. Polarimetry

Unfortunately, during the July 1995 observing run, seeing conditions were only medium, so polarization values could only be computed for 48 sources brighter than $m_K \approx 11$. The deconvolution using the LUCY-algorithm was used to produce 0".5 resolution maps in which all the brighter sources could clearly be separated. In Fig. 3 we present the polarization vectors for all sources in 0.5" apertures. The standard errors are $\delta p = 1.8\%$ for the degree of polarization and $\delta q = 19°$ for the position angle.

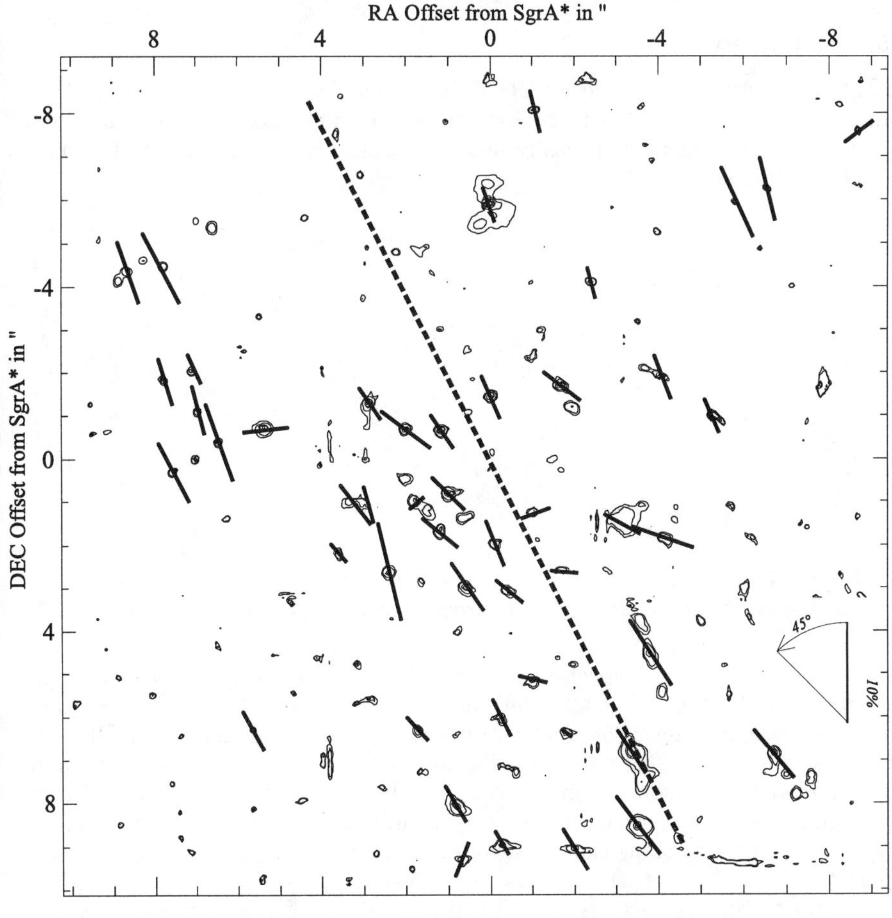

Figure 3. Polarization vectors of the 48 brightest sources using an aperture of 0.5". The orientation of the galactic plane, to which most of the polarization vectors are parallel, is indicated by the dotted line.

Most of the sources show polarization vectors with absolute values of $p \approx 4\%$ and position angles (P.A.) nearly parallel to the Galactic plane. The flux-weighted polarization in the entire region is $4.1 \pm 0.6\%$ at P.A. $30° \pm 10°$ (for comparison, see Eckart et al. 1995). This is, within the errorbars, parallel to the Galactic plane. As there is no reason for local anisotropies to be aligned in any special direction, the overall polarization can best be explained by anisotropic scattering by aligned particles in the spiral arms of the Milky Way along the line of sight from the Earth to the Galactic Center.

We discovered that the sources IRS 21 and IRS 1W are extended and show significant variations from this general alignment (see Fig. 3). Both of them could be young stars which have formed just recently in the northern arm of the "mini-spiral" or have gathered dust from infalling material out of the northern arm.

2.5. Extended Objects

In Fig. 4, we show the profiles of IRS 21 and IRS 1W in comparison to 2 other sources which were selected to be brighter than and of comparable brightness to the extended objects. It is easily seen that both show a broader profile than the comparison stars.

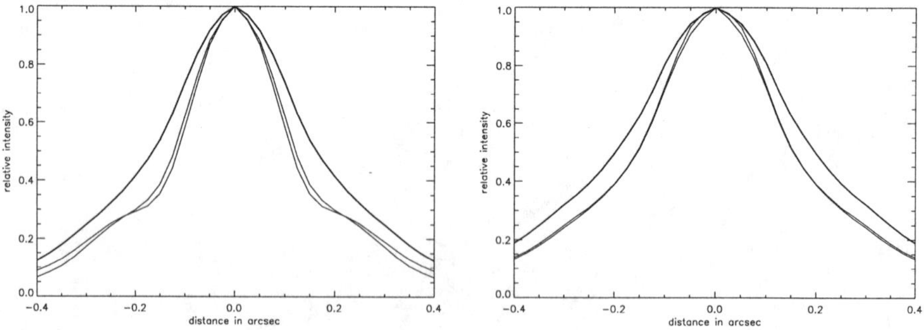

Figure 4. Profiles of the extended sources IRS 21 and IRS 1W (top curves) compared to two other sources (bottom curves).

In Fig. 5, we present spectra for both sources taken with the MPE imaging spectrograph 3D (Weitzel et al. 1996) on La Silla in March 1996. None of these spectra show significant 2.058 μm He I line emission. In the spectrum of IRS 1W, Brγ emission is seen. The nature of this emission line is still unclear, but could be explained by this source being located within the northern mini-spiral arm which consists of hot gas with bright Brγ emission. Due to a strong spatial gradient of this emission, the background subtraction may not be complete. Apart from this feature, the spectra of both sources can best be described to be red and featureless (see also Krabbe et al. 1995). This most probably is the result of IRS 21 and IRS 1W being heavily obscured by a circumstellar dust shell. This is supported by their 10 μm emission (Smith et al. 1990, Gezari 1992) and their unusual properties in polarized light with respect to the other cluster members.

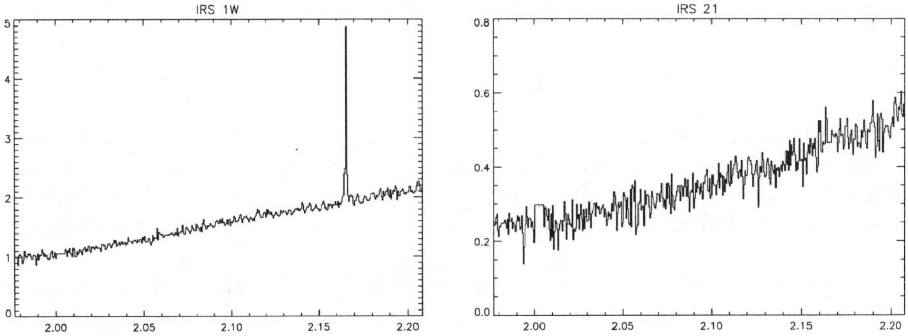

Figure 5. K-short spectra of the extended sources IRS 1W (left) and IRS 21 (right).

The polarization of the source IRS 21 is 9.8%±1.3% at P.A. 14° ± 10° in a 0.5" aperture. Eckart et al. (1995) derived in their diffraction limited image a polarization of 17% at P.A. 16° in a 0.25" aperture. The difference in the degree of polarization is due to the larger aperture we had to use in this crowded area. In any case, we can confirm that the degree of polarization is at least twice the value for the mean polarization in the whole field. The degree of polarization of IRS 1W is 4.6%±2.5%, and thus comparable to the overall polarization in the Galactic Center region. With −84° ± 8°, its P.A. differs significantly from the overall P.A., very consistent with Fig. 4 in Eckart et al. (1995). Aitken et al. (1991) have presented mid-infrared polarization data at a wavelength of 12.4μm. At this wavelength, the dust is in emission. The general orientation of the mid-infrared polarization vectors of these 2 sources agrees to within 20° with the orientation derived from our near-infrared data. The orientation of the position angles is perpendicular to the magnetic field lines which follow the northern arm of the mini-spiral.

A straightforward interpretation of the measured polarization in IRS 21 and IRS 1W would be Mie-scattering in an assymetrical dust shell (Bastien and Ménard 1990). However, the dust grains are most likely aligned by a >2mG (Aitken et al. 1998) magnetic field along the mini-spiral arms due to the Davis-Greenstein effect (Davis and Greenstein 1951). Therefore, combining the results from all wavelengths, we propose the following scenario for both sources: Since the polarization vectors show a similar alignment in the near- and mid-infrared, it seems implausible that the near-infrared polarization results from scattering by aligned dust grains associated with the mini-spiral. Polarization by single scattering would result in polarization vectors parallel to the magnetic field lines, which are thought to be responsible for the alignment of the elongated dust grains. It is therefore likely that the same mechanism which is responsible for the mid-infrared polarization is also applicable to the near-infrared, namely hot dust emission of aligned particles in the dust shells surrounding both objects.

3. Conclusion

Our high angular resolution near-infrared observations revealed polarized and embedded stars in the central stellar cluster. This finding as well as the detection of a short period eclipsing binary with a minimum total mass of about $100M_{\odot}$ can be interpreted in the context of the star formation process in the Galactic Center.

A population of young luminous hot stars as traced by the nuclear radiation field combined with the presence of embedded stars indicates ongoing star formation in the central stellar cluster. Recent subarcsecond angular resolution line imaging (Eckart et al. 1995, Tamblyn et al. 1996) and 3D imaging spectroscopy (Krabbe et al. 1995, Genzel et al. 1996) have shown that several of the brightest members of the central cluster - the so called IRS 16 complex - are He I emission line stars. These objects show in their spectra prominent broad He I and HI recombination lines in emission. From non-LTE, stellar atmosphere modeling of the observed emission characteristics of several of the He I-stars Najarro et al. (1994, 1997) have inferred that these objects are moderately hot (17,000 to 30,000 K), very luminous (1 to $30\times10^5L_{\odot}$) massive stars whose helium rich surface layers are expanding as powerful stellar winds with velocities of 200 to 800 km/s and mass loss rates of 1 to $70\times10^{-5}M_{\odot}$/year. These stars provide the dominant fraction of the total luminosity of the central parsec.

This soft average radiation field with a small harder spectral component from evolved stars indicates that the Galactic Center has recently - about 3 to 9×10^6 years ago - experienced a period of significant star formation. New massive stars which exhibit a much harder spectrum as the average one currently observed must now be formed at a much lower rate. The observation of stars embedded in extended dust clouds, however, also indicates that the formation of stars has not ceased completely. The fact that these embedded stars are located in or very close to the mini spiral arms shows that infalling gas is probably of great importance for the star formation process in the deep potential well of the central stellar cluster.

The detection of a short period binary system in the central stellar cluster allows us to compare this result to binary frequency investigations in other stellar clusters (e.g. Yan and Mateo 1994, Yan and Reid 1996). The detection of a single eclipsing binary under the comparatively ($\leq$ 600 stars in the K-band images) small number of observed stars indicates that star-binary and binary-binary encounters or even captures between single stars in the high density environment of the central cluster must play an important role in the formation of stars in the deep potential well of the central stellar cluster.

Future near-infrared observations with even higher sensitivity and/or resolution as provided by the HST or the new 8 to 10 m class telescopes will allow us to put these investigations on much firmer statistical footing. In addition short time photometric and high resolution spectroscopic monitoring of a larger sample of stars in the central parsec of the Milky Way is required to improve the statistics for the detection of binary systems which can give important clues to the understanding of star formation and the evolution of the central stellar cluster.

4. Acknowledgments

We thank the Director General and staff of ESO for their continuing support of our operation of SHARP at the NTT. N. Ageorges and R. Hofmann have contributed to the construction of the SHARP camera and to a number of observing runs during various phases of this program.

References

Aitken, D.K., Gezari, D., Smith, C.H., McCaughrean, M. and Roche, P.F. 1991, ApJ, 380, 419

Aitken, D.K., Smith, C.H., Moore, T.J.T. and Roche, P.F., 1998, MNRAS, 299, 743

Bastien, P., and Ménard, F., 1990, ApJ, 364, 232

Blum, R.D., Sellgren, K. and DePoy, D.L. 1996, ApJ, 470, 864

Christou, J.C. 1991, Exp. Astron., 2, 27

Davis, L., and Greenstein, J.L., 1951, ApJ, 114, 206

Eckart, A., Genzel, R., Hofmann, R., Sams, B.J., Tacconi-Garman, L.E. and Cruzalebes, P. 1994, in Genzel, R., Harris, A.I., eds, NATO ASI Series, The Nuclei of Normal Galaxies. Kluwer:Dordrecht, p. 305

Eckart, A., Genzel, R., Hofmann, R., Sams, B.J. and Tacconi-Garman, L.E. 1995, ApJ, 445L, 23

Genzel, R., Thatte, N., Krabbe, A., Kroker, H. and Tacconi-Garman, L.E. 1996, ApJ, 472, 153

Gezari, D., 1992, in Blitz L., ed., The Center, Bulge and Disk of the Milky Way. Kluwer, Dordrecht, p.23

Hindsley, R.B. and Bell, R.A. 1990, ApJ, 348, 673

Högbom, J. 1974, A&A (Supp.), 15, 417

Knacke, R.F. and Cabbs, R.W. 1977, ApJ, 216, 271

Krabbe, A. et al. 1995, ApJ, 447L, 95

Lebofsky, M.J., Rieke, G.H. and Tokunaga, A.T. 1982, ApJ, 263, 736

Lucy, L.B. 1974, AJ(Letters), 79L, 745

Najarro, F., Hillier, D.J., Kudritzki, R.P., Krabbe, A., Genzel, R., Lutz, D., Drapatz, S., Geballe, T.R., 1994, A&A, 285, 573

Najarro, F., Krabbe, A., Genzel, R., Lutz, D., Kudritzki, R.P. and Hillier, D.J. 1997, A&A, 325, 700

Sellgren, K., McGinn, M.T., Becklin, E.E., Hall, D.N., 1990, ApJS, 359, 112

Smith, C.H., Aitken, D.K., and Roche, P.F., 1990, MNRAS, 246, 1

Tamblyn, P., Rieke, G.H., Hanson, M.M., Close, L.M., McCarthy, D. W., and Rieke, M.J. 1996, ApJ, 456, 206

Tamura, M. and Werner, M.W. 1996, ApJ, 467, 645

Weitzel, L., Krabbe, A., Kroker, H., Thatte, N., Tacconi-Garman, L.E., Cameron, M., and Genzel, R. 1996, A&A (Supp.), 119, 531

Yan, L. and Mateo, M. 1994, AJ, 108, 1810

Yan, L. and Reid, I.N. 1996, MNRAS, 297, 751

The Central Parsecs of the Galaxy
ASP Conference Series, Vol. 186, 1999
H. Falcke, A. Cotera, W.J. Duschl, F. Melia, M.J. Rieke, eds.

Massive Young Stars in the Arches Cluster

E. Serabyn

California Institute of Technology, 320-47, Pasadena, CA 91125

D. Shupe

IPAC, California Institute of Technology, 100-22, Pasadena, CA 91125

D.F. Figer

Dept. of Physics and Astronomy, UCLA, Los Angeles, CA 90095

Abstract. The star formation rate per volume in the Milky way's nucleus exceeds that in the Galactic disk by several orders of magnitude, but the nuclear stellar initial mass function remains an unknown. Indeed, even the simple identification of stars on the main sequence remains problematic, due both to the presence of numerous bright giant stars in the region, and to the high extinction along the line of sight. The best place to search for main sequence stars is thus in a compact cluster which is suspected of being young. Here we discuss our high resolution near-infrared observations of the "Arches" cluster, which have revealed a remarkable collection of young, massive, main-sequence stars. The Arches cluster contains approximately 10^2 massive O-star candidates in a rather compact volume, dwarfing all other known young Galactic clusters. Its central density evidently exceeds that in the R136a cluster in 30 Doradus. The Arches cluster may in fact represent a low-end analog of the young "super star clusters" found in external galaxies.

1. Introduction

The proximity of the Milky Way's nucleus allows for observation of the workings of a galactic nucleus in unparalleled detail. Of special interest in this regard are the star-formation modes which occur in the turbulent nuclear environment deep within a galaxy's central gravitational well. Studies of star formation in the Milky Way's nucleus can thus potentially provide a template for understanding nuclear star formation in general, and starburst nuclei in particular.

Compared to the star-formation occurring in the large-scale Galactic disk, that proceeding in our Galaxy's nucleus stands out both from the point of view of the more extreme (dense, hot, turbulent) interstellar environment found in the nuclear nursery (Güsten 1989), as well as from the characteristics of the young stars discovered to date (several compact clusters of rare Wolf-Rayet-like stars have been found in the central 30 pc; Morris & Serabyn 1996). Together with the overall ionizing radiation field traced by high-frequency radio emission

(Mezger, Duschl & Zylka 1996), this suggests that massive stars are formed in abundance in the nucleus. However, high extinction and confusion by the numerous giant stars also populating the region have hampered our ability to define the characteristics of the stellar population in our Galaxy's nucleus.

2. Observations

To minimize confusion with other more elderly stars in the nucleus, it is best to target a compact grouping of stars which is suspected of being young. As indicated by the presence of bright WR-like stars (Nagata *et al.* 1995; Cotera *et al.* 1996), the "Arches" cluster (named after the nearby Arched filaments) seemed a prime candidate for the search for an associated young main-sequence population. Herein we briefly review our observations of this cluster, which consist of high-resolution J, H, and K′ imaging, and an HK grism slit scan at a resolving power of 300, carried out with the near-infrared camera (NIRC) at the Keck I telescope. A false-color image of the Arches stars, its color-magnitude diagram, and an analysis of the implications (about 100 to 120 young, massive O/WR stars born in a compact volume of about 0.3 pc radius, implying a stellar density exceeding that of R136a in the Large Magellanic Cloud's 30 Doradus region) have been presented in Serabyn, Shupe & Figer (1998; hereafter S98). To complement that work, we present herein the individual J, H and K′ images, and also several spectra which highlight the different classes of cluster stars: a dozen or so hot WR-like stars with bright emission lines from H and He, a few bright late-type stars showing CO absorption features, and numerous slightly fainter stars which show featureless spectra (at our spectral resolution) consistent with their being upper main-sequence stars. We have thus succeeded in detecting the upper main-sequence in this cluster.

3. Results

Our J, H, and K′ images, acquired on a night of 0.45″ seeing at the Keck Observatory, are shown in Figures 1 through 3. Next, a color-magnitude diagram covering the inner 9″ region (in three annuli of 3″ width) is shown in Figure 4. The total area covered by Fig. 4 is larger than that discussed in S98, and so includes additional cluster and background stars. In particular, a few additional stars show up just to the right (about 0.3 mag) of the slightly curved main vertical track near H-K = 1.8 to 2 (note the three rightmost 6-pointed star symbols at $K < 12$). These three redder stars occupy a location in line with their being on the red giant branch, and CO features are seen in at least two of them (see Fig. 6 for the topmost 6-pointed star in Fig. 4). The main (nearly) vertical track in Figure 4, located just 0.3 mag to the left (blue) of these CO-stars, then occupies the expected location for the upper main sequence, and consists of a slightly sloped upper portion of about a dozen WR-like stars, and just below these, a more nearly vertical track of bright O-type supergiants and fainter main sequence O-stars.

Further evidence that the weaker stars on the main vertical track are indeed main sequence stars can be found in our spectra, a few examples of which are presented in Figures 5 through 7. These grism spectra cover both the K and

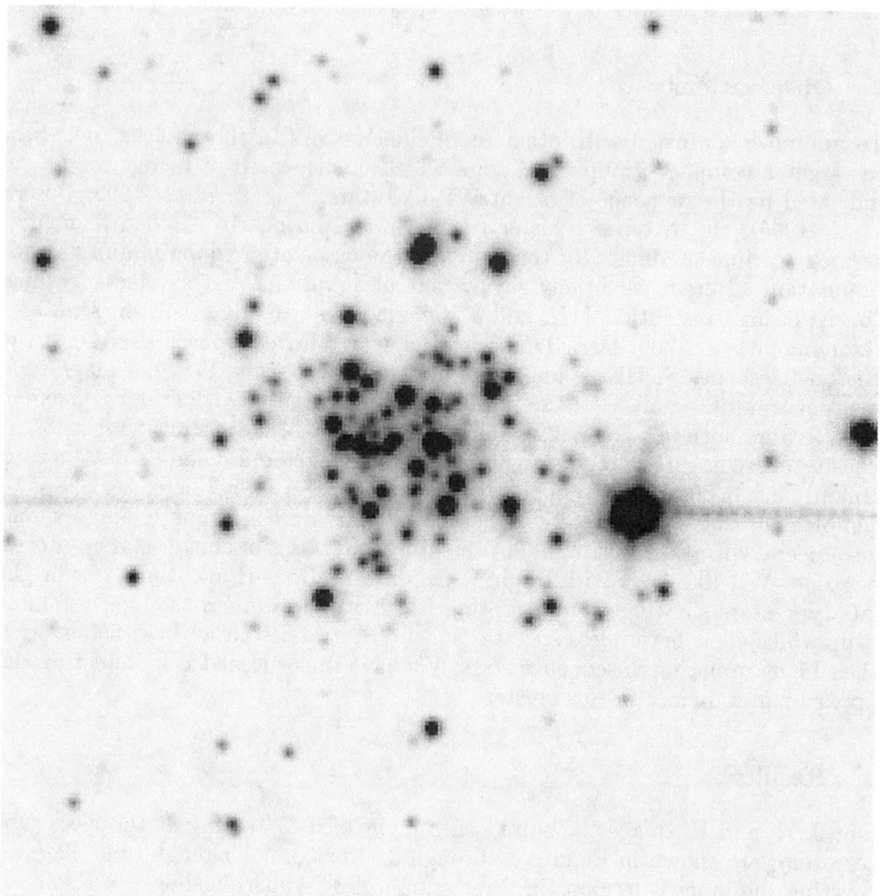

Figure 1. J-band image of the Arches cluster. North is up and east is to the left. The image is 38.4 arc sec on a side, and the FWHM seeing is 0.45 arc sec. 343 stars are seen. The very bright star to the west of center is foreground. A region of high extinction is evident in the southwest corner.

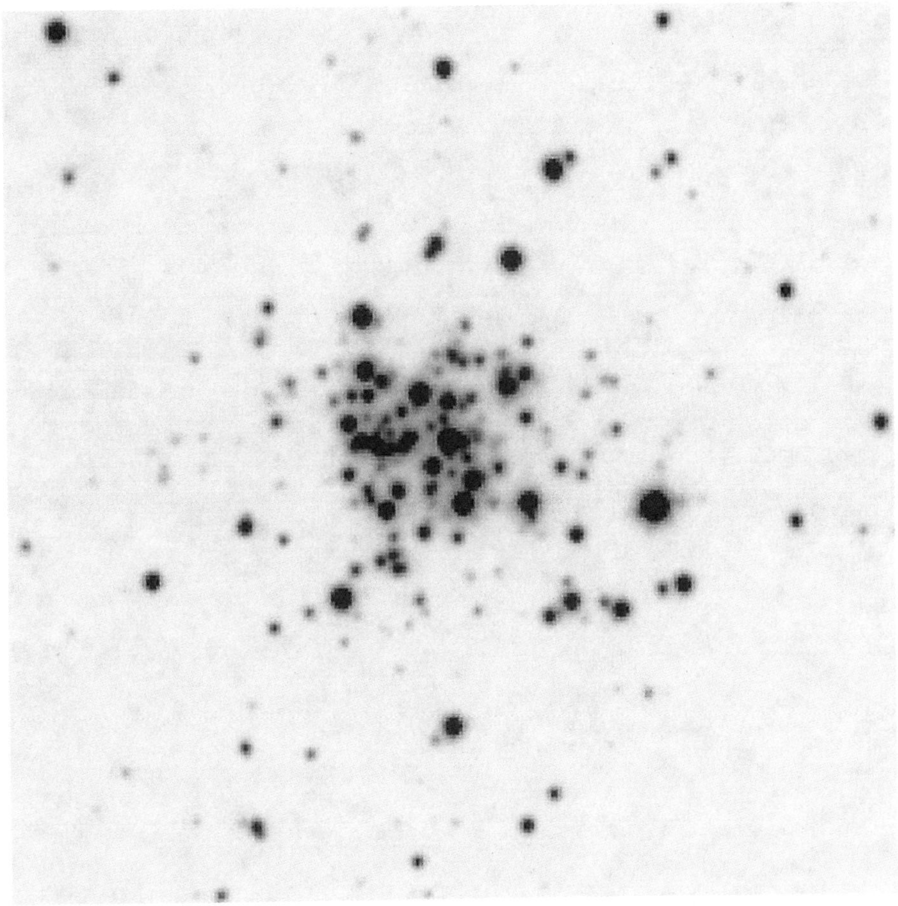

Figure 2. Corresponding H-band image of the Arches cluster. 714 stars are seen.

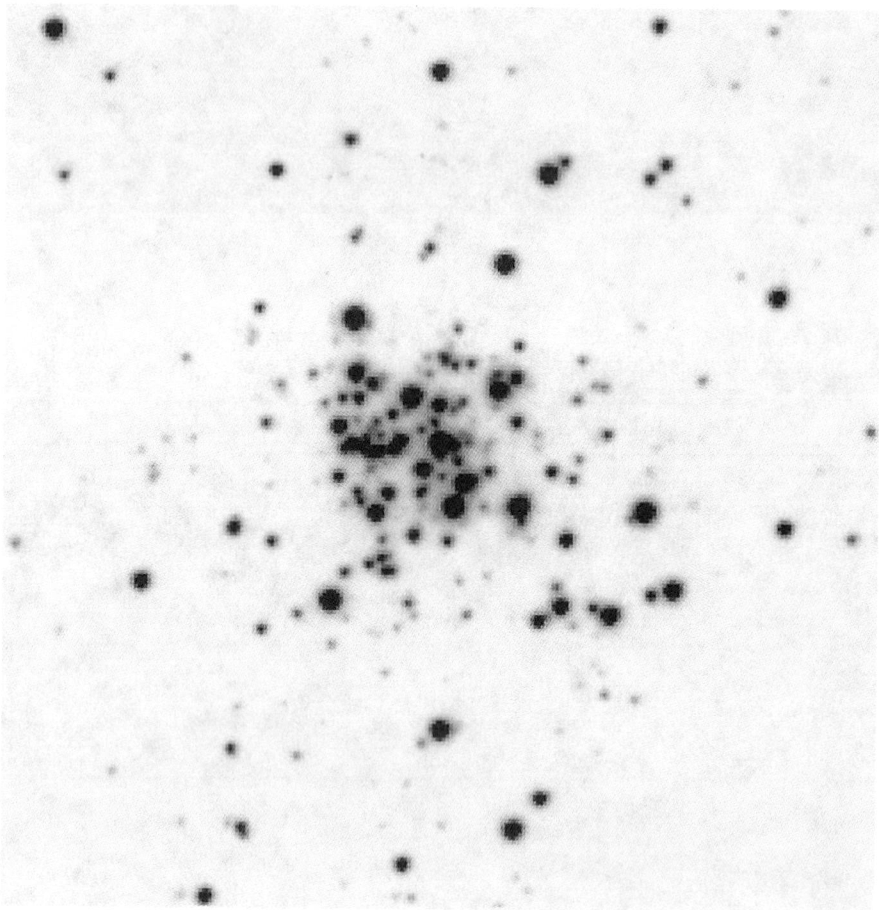

Figure 3. Corresponding K' image of the Arches cluster. 904 stars are seen.

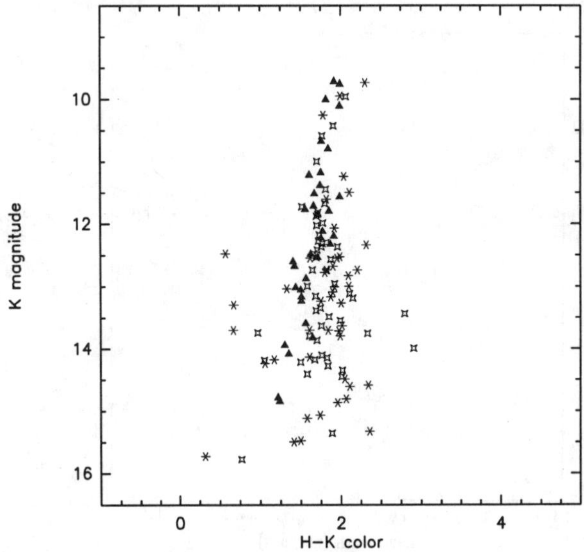

Figure 4. Arches cluster color-magnitude diagram. The stars within 3 arcsec of the cluster center are shown as filled triangles, those between 3 and 6 arcsec are shown as 4-pointed open stars, and those between 6 and 9 arcsec as 6-pointed stars.

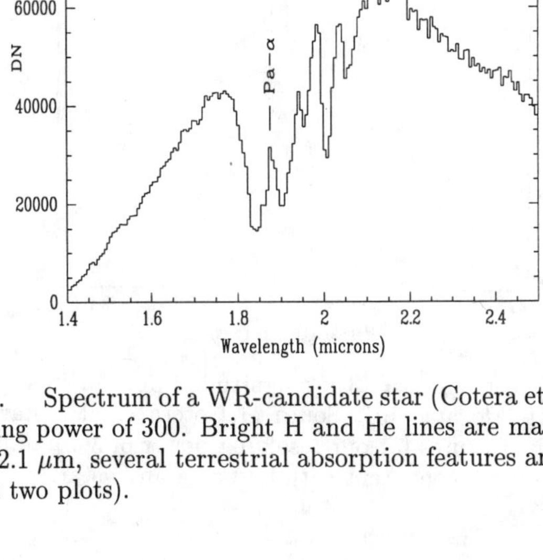

Figure 5. Spectrum of a WR-candidate star (Cotera et al.'s star 8) at a resolving power of 300. Bright H and He lines are marked. Between 1.8 and 2.1 μm, several terrestrial absorption features are seen (also in the next two plots).

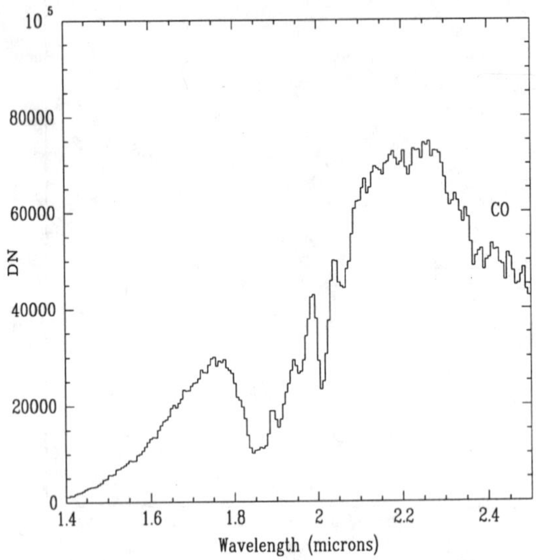

Figure 6. Spectrum of an evolved star (Cotera et al.'s star A) with deep CO absorptions at the long wavelength end of the spectrum.

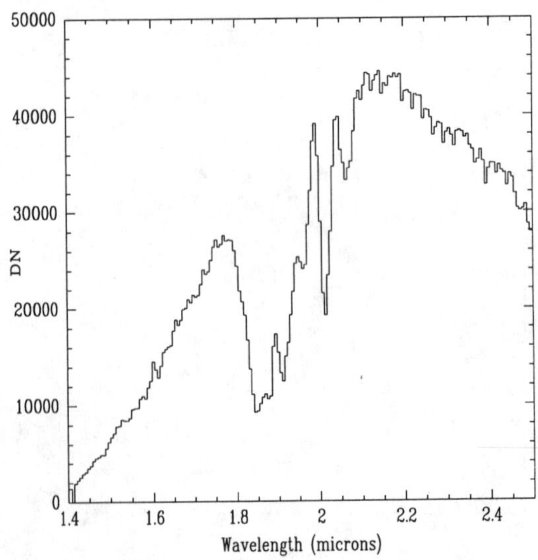

Figure 7. Spectrum of a 13th magnitude upper main-sequence (O-star) candidate (located just South of Cotera et al.'s star 8). The lack of spectral features (at our resolving power of 300) and the star's luminosity are both consistent with its being an O star.

H bands. The three spectra are dominated by a Rayleigh-Jeans rise in the longer wavelength K window, and extinction-induced falloffs in the H band, with several terrestrial spectral features in the intermediate 1.8 to 2.1 micron region. Due to the high interstellar extinction, these spectra have not been divided by a calibrator. In Figures 5 through 7, three quite distinct stellar spectra are evident: the first is from a WR-like star with bright H and He lines, the second from a late-type star with CO absorptions, and the third has a featureless spectrum (at this resolution), consistent with an upper main-sequence O star (Hanson et al. 1996). We thus take the absence of both CO bandhead absorptions and H and He emission lines in the third spectrum as evidence that the somewhat fainter stars are neither late-type nor WR-like stars, but are instead young, upper main-sequence O stars.

4. Discussion

The Arches cluster is perhaps the most massive agglomeration of O stars in the Galaxy, with a spatial density higher even than that of R136a in 30 Doradus (S98). With 100-120 massive stars in the WR/O-range (> 20 $M_\odot$), this cluster is clearly sufficient to provide the ionizing photon flux necessary to ionize the Arched filaments. However, the projected positions of the various ionized cloud surfaces relative to the cluster allow room for substantial doubt as to whether this single cluster can indeed be the sole ionizing source for the entirety of the Arched filaments complex.

One point not addressed in S98 due to space limitations was whether or not this rather compact and massive cluster (few 10^4 $M_\odot$) is bound. This question has bearing not only on the cluster itself, but also has the potential to shed light on the same question in recently discovered super-star-clusters in external galaxies. There are two points which must be considered: is the cluster bound against its own internal kinetic energy, and is it bound against tidal gravitational shear from the external gravitational field? These questions would ideally be addressed using the cluster's velocity dispersion, but a measurement of this quantity is not yet available. Nevertheless, we can make substantial headway based on what we have.

With a mass of a few 10^4 $M_\odot$ (S98), application of the virial theorem to the cluster gives a random velocity of order 10 km/s. With a size of roughly 0.3 pc, and an age of 1-2 Myr (S98), roughly 100 crossing times would already have elapsed. The survival of this tight cluster for such a large number of crossings is thus a good indicator that the cluster is likely bound against its own internal energy. (If the average velocity were lower, fewer crossings would have occurred, but then the lower average energy implies a bound cluster anyway; if the velocities were higher, even more crossings would already have occurred, again suggesting an inability to disperse).

As to the second point, boundedness in the face of the external gravitational shear requires that the average cluster density exceeds the average density of the Galactic mass enclosed interior to its orbital radius of about 25 pc. Taking 10^8 $M_\odot$ as an estimate of the enclosed mass (Lindqvist *et al.* 1992), the cluster density must then exceed about 10^4 $M_\odot$ pc^{-3} to be bound. In fact, S98 estimated a central cluster density of about 3 $\times 10^5$ $M_\odot$ pc^{-3}. Thus, initial indicators suggest a bound state for the cluster. Of course, this does not preclude of few of the more energetic stars having escaped from the cluster, as for example a few of the bright outliers seen around the cluster core.

The Arches cluster may thus differ in this respect from the nearby Quintuplet cluster, which has a similar mass, and similar WR-like stars, but appears to be unbound. If so, this difference may simply reflect a more compact initial state for the Arches cluster, as escape velocity and cluster density both decrease with increasing cluster radius.

What does this cluster then tell us about the initial mass function in the nucleus? Unfortunately, source confusion due to the cluster's high spatial density currently limits the reliability of the available number counts. This important question is thus being pursued further with even higher resolution data from the Hubble Space Telescope (Figer *et al.* 1999).

Acknowledgments. We thank the W.M. Keck Foundation for enabling this research, as well as the conference organizers for providing the opportunity to discuss our results.

References

Cotera, A.S., Erickson, E.F., Colgan, S.W.J., Simpson, J.P., Allen, D.A. & Burton, M.G. 1996, ApJ, 461, 750

Güsten, R. 1989, in IAU Symp. No. 136, *The Center of the Galaxy*, ed. M. Morris, (Kluwer: Dordrecht), 89

Figer, D., et al. 1999, this volume, p. 329

Hanson, M.M., Conti, P.S. & Rieke, M.J. 1996, ApJS, 107, 281

Lindqvist, M., Habing, H.J. & Winnberg, A. 1992, A&A 259, 118

Mezger, P.G., Duschl, W.J. & Zylka, R. 1996, A&A Rev., 7, 289

Morris, M. & Serabyn, E. 1996, ARA&A 34, 645

Nagata, T., Woodward, C.E., Shure, M. & Kobayashi, N. 1995, AJ, 109, 1676

Serabyn, E., Shupe, D. & Figer, D.F. 1998, Nature 394, 448

The Central Parsecs of the Galaxy
ASP Conference Series, Vol. 186, 1999
H. Falcke, A. Cotera, W.J. Duschl, F. Melia, M.J. Rieke, eds.

HST/NICMOS Imaging of the Quintuplet and Arches Clusters

Donald F. Figer, Mark Morris

Division of Astronomy, Department of Physics & Astronomy, University of California, Los Angeles, 405 Hilgard Avenue, Los Angeles, CA 90095-1562

Sungsoo S. Kim

Korea Advanced Institute of Science and Technology, Department of Physics, Space Science Laboratory, Deajon, 305-701, Korea

Eugene Serabyn

JPL, 171-113, 4800 Oak Grove Drive, Pasadena, CA 91109

Abstract. We present images, luminosity functions, and color-magnitude diagrams for the Quintuplet and Arches clusters obtained with *HST*/NICMOS in the F160W and F205W filters. The images allow the detection of stars over 6 magnitudes fainter than ever before and reveal multiple star systems in many cases where previous data revealed single stars; in fact, one of the original Quintuplet stars is seen to be extended. For the first time, we clearly identify the main sequence in the Quintuplet cluster, and we extend earlier detections of the main sequence in the Arches cluster to $< 10\ M_{\odot}$. The fields are corrected for incompleteness using synthetic images based upon the observed number density profiles as a function of cluster radius. A tentative estimate of the initial mass function is given for the Arches cluster, finding that the slope is somewhat shallower than the Salpeter value.

1. Introduction

Our analysis of recently obtained *HST*/NICMOS images continues the trend of advances in the past ten years in studying the three massive clusters in the Galactic Center (see Figer et al. 1998 and references therein). The Central cluster was first to be studied, and is now known to contain over 30 stars having $M_{\rm initial} > 20\ M_{\odot}$ (Genzel et al. 1996). Serabyn, Shupe, & Figer (1998; 1999) published Keck I images revealing over 100 O-stars packed within 0.6 pc in the Arches cluster (G0.121+0.017), suggesting that this cluster is the densest in the Galaxy. Figer, McLean, & Morris (1999) published a list of ≈ 30 stars with $M_{\rm initial} > 20\ M_{\odot}$ in the Quintuplet cluster (G0.167−0.060), including more bona fide Wolf-Rayet stars than any other Galactic cluster. All three clusters are quite similar in most respects, except for age; the Central and Quintuplet clusters are $\approx$ 3-5 Myr old, while the Arches cluster is substantially younger, $\tau_{age} \lesssim 2$ Myr.

The extraordinary sample of young stars in these three clusters provides an excellent opportunity to observe the products of recent star formation in the Galactic Center, a region which is relatively extreme in properties which are thought to affect star formation. Morris (1993; and references therein) suggested that the Galactic Center environment may alter star formation with respect to that observed in the disk of the Galaxy, in particular through a flat initial mass function (IMF) for large masses and by an elevated lower mass cutoff. These, and other suggestions concerning star formation in the Galactic Center, have gone untested by a direct determination of the mass function. With our data set, we may able to test these ideas.

In this paper, we present initial results from an analysis of our F160W and F205W images obtained with *HST*/NICMOS with the final goal of determining the initial mass function of the Quintuplet and Arches cluster.

2. Observations

The Quintuplet and Arches clusters were imaged in a mosaic pattern using *HST*/NICMOS in the NIC2 aperture (19$''$.2 on a side). In addition, four nearby "control" fields were imaged in order to sample the background stellar population; they are arranged in a symmetric cross-pattern and separated by 59$''$ from the centers of the cluster mosaics. All fields were imaged on UT 1997 September 13/14, in F110W ($\lambda_{\rm center} = 1.10\,\mu$m), F160W($\lambda_{\rm center} = 1.60\,\mu$m), andF205W($\lambda_{\rm center} = 2.05\,\mu$m), with exposure times of 256 seconds per image. The plate scale was 0$''$.076 pixel^{-1} (x) by 0$''$.075 pixel^{-1} (y), in detector coordinates. The Quintuplet cluster was imaged in a 4×4 mosaic while a 2×2 was used for the smaller Arches cluster. The +y axis of the detector was oriented 135° east of north in both cases. The images were reduced via the standard NICMOS pipeline (CALNICA, CALNICB; MacKenty et al. 1997).

Figures 1 and 2 show the F205W cluster images. A dense collection of stars can be seen in the Arches cluster, concentrated within a radius of 6$''$ of the cluster center. The Quintuplet cluster is significantly more spread out. It is difficult to distinguish unambiguously cluster members from the local background population in this image; however, note that the average separation of the massive cluster stars from the centroid of the distribution is about 25$''$according to Figer et al. (1999), indicating a region which spans most of the mosaic. The faintest stars which can easily be seen in these figures have $m_{\rm F205W} \approx 19$. Note that two of the original Quintuplet cluster members are multiple (#192 #251, #243) or extended (#231), where the nomenclature in parentheses is from Figer, McLean, & Morris (1999).

3. Data Reduction

We extracted stellar photometry from the images in order to identify main sequences in the clusters by examining the resultant luminosity functions and color-magnitude diagrams. Star-finding, PSF-building, and PSF-fitting proce-

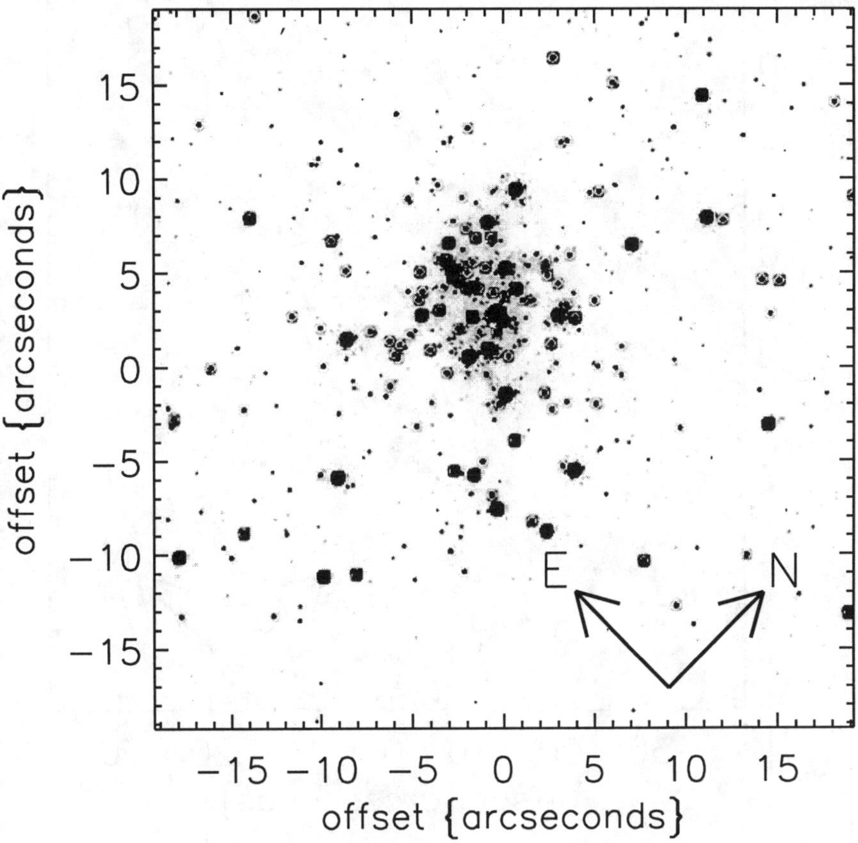

Figure 1. F205W image of the Arches cluster.

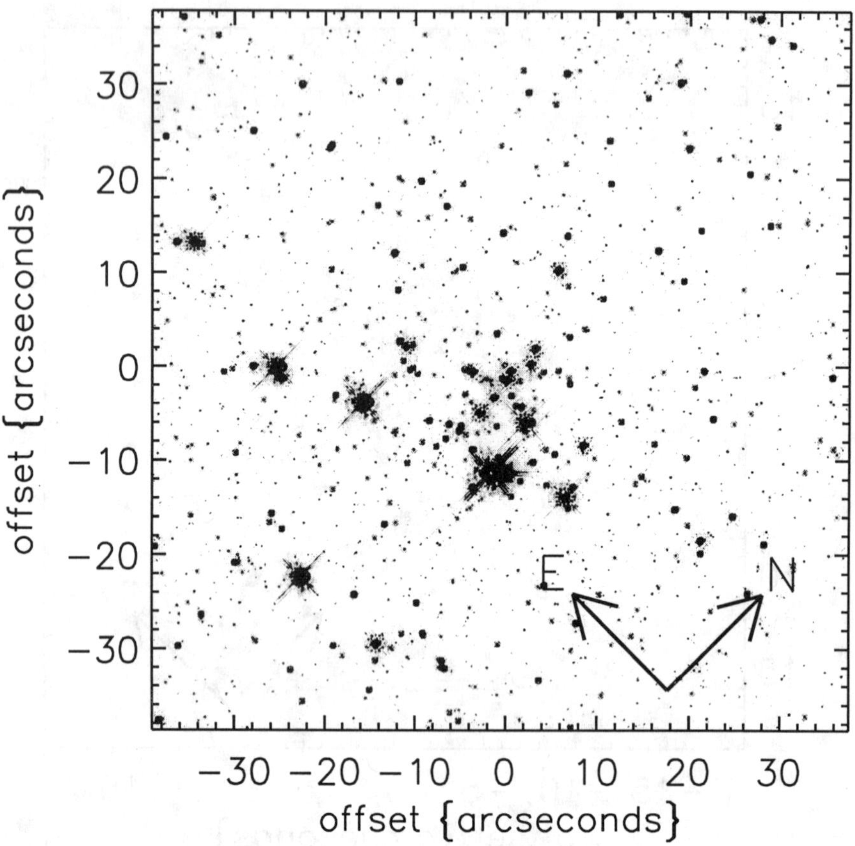

Figure 2. F205W image of the Quintuplet cluster.

dures were performed using the DAOPHOT package (Stetson 1987) within the Image Reduction and Analysis Facility (IRAF).

We combined PSF candidate stars from several different images keeping their locations in the frame to build linearly-varying PSFs (Krist & Hook 1997). We chose the size of the PSF to just cover the second ring of the Airy pattern. The DAOFIND routine in the DAOPHOT package was applied to the data with a detection threshold of 4σ above the background level. Many "bogus" stars (mounds on the secondary rings and noise peaks on the spider diffraction spikes) appear clustered around the brightest stars in the images. In general, we did not manually remove them from the photometry list because it would have required arbitrary decisions in heavily crowded regions; however, we did remove these bogus stars around the brightest few dozen stars in the Quintuplet cluster because they significantly altered the luminosity function near $m_{F205W} \approx 16$. In addition, many false stars were later removed while the PSF was being fitted to the data. The number of false stars which still remain after PSF-fitting is so small compared to the whole sample ($< 0.5\%$) that it does not affect the luminosity function (LF) significantly. We can partially account for the false stars because similar stars will be found in our synthetic images, i.e., false counts in the real images will be offset when we make the completeness corrections based upon the synthetic images.

Stellar fluxes were estimated using DAOPHOT's PSF-fitting routine, ALLSTAR. PSF-fitting for the cluster images was performed on each image, not on the mosaiced images, because the ALLSTAR routine was not able to fit the spatially varying PSF onto the mosaiced image. A $3''$ radius aperture was used to extract photometry, and appropriate filter-dependent aperture corrections were made. Instrumental magnitudes were calibrated and converted into the VEGAMAG photometric system of the *HST*, defined such that the absolute magnitude of Vega would be 0.02 in all bandpasses (Keyes et al. 1997). Typical DAOPHOT magnitude errors (MERRs) of cluster image photometry were 0.03 mag, and we rejected stars with DAOPHOT error greater than 0.2 mag.

4. Completeness Corrections

The amount of crowding in the images differs greatly from image to image. The cluster center regions are extremely crowded such that the detection of stars is limited by diffracted light from neighboring bright stars. The control fields are moderately crowded by background stars. We attempted to correct the star counts by estimating incompleteness due to crowding.

For cluster fields, we performed Monte Carlo simulations to make artificial frames, accounting for the radial number density gradient of the clusters. The input luminosity function (LF) for the synthetic images was obtained by extrapolating the observed LF from $m_{F205W} = 14$ to fainter magnitudes with a power law $L^{-0.3}$, which roughly corresponds to the expected LF of the cluster with a mass spectrum $d(\log N)/d(\log m) \propto \Gamma \approx -0.5$. We did not consider the LF at $m_{F205W} > 21$ because such stars are below our detection limit, and the total contribution of those stars to the background enhancement is negligible. The positions of stars in the synthetic images were chosen to follow the radial number density profile of stars with $m_{F205W} < 15$, whose LF was not significantly

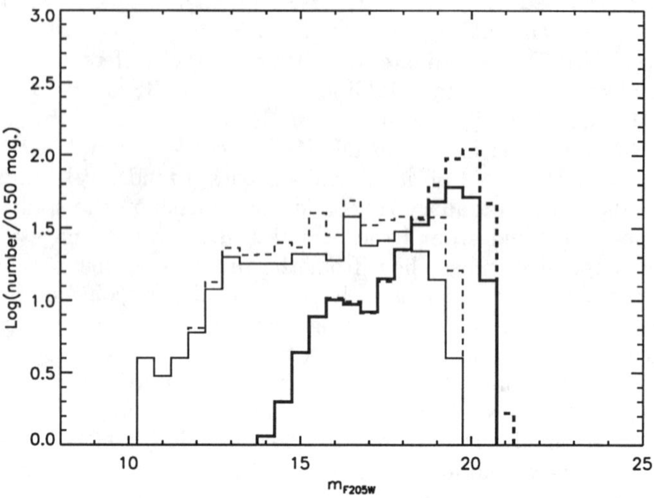

Figure 3. Observed luminosity function for Arches cluster (*thin*) within a radius of 6″ and nearby control fields (*thick*) normalized to the area of the cluster. The solid lines represent the observed luminosity function. The dashed lines represent counts after correcting for incompleteness in all plots.

affected by crowding. The intrinsic LFs were then recovered by dividing the observed LFs by the recovery fractions.

5. Results and Discussion

The extracted luminosity functions are shown in Figures 3 and 4. The plots for both clusters reveal a shallow function with respect to those for the control fields. This is expected for a reasonable initial mass function for young coeval clusters, in the case of the cluster fields, and for an old population, in the case of the control fields. The sharp falloff at dim magnitudes is an expected result of confusion, as opposed to any intrinsic dropoff of faint stars. The peak in counts at $m_{F205W} \approx 16$ in the control field luminosity function is due to very old red giant stars. The increased counts in the Quintuplet cluster field is due to the large size of the mosaic, i.e., the plots are not a function of surface area; however, note that the control field counts have been normalized to the chosen size of the clusters in both plots.

The color-magnitude diagrams are given in Figures 5 and 6. The plots are restricted to the cores of the clusters, radius = 6″ in the case of the Arches, and radius = 12″ in the case of the Quintuplet. Reddened isochrones, covering $M_{\rm initial} = 0.8\ M_\odot$ to $120\ M_\odot$ from the Geneva models are overplotted (Meynet et al. 1994). The leftmost isochrone corresponds to an age of 1 Myr. The

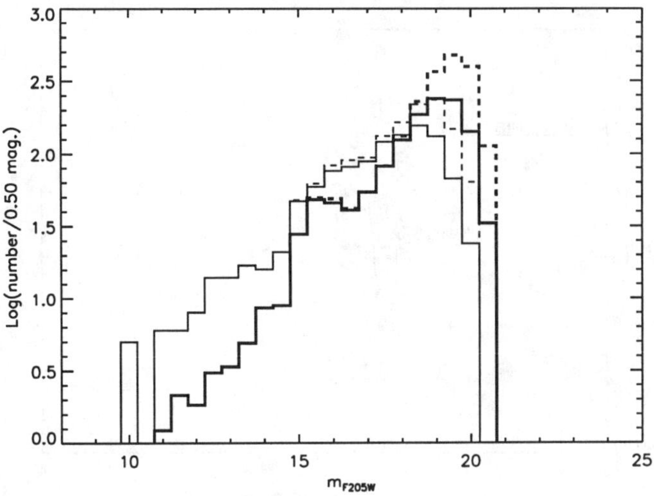

Figure 4. Observed luminosity function for Quintuplet cluster (*thin*) within a radius of 12″ and nearby control fields (*thick*) normalized to the area of the cluster. The solid lines represent the observed luminosity function.

rightmost isochrone corresponds to an age of 10 Gyr. The models have been reddened for $A_K = 3.0$. For both clusters, we see that most of the stars fall near m_{F205W}-$m_{F160W} \approx 1.5$. The distribution of stars widens at the faint end due to increased photometric errors. The extension of the distribution toward redder colors is due to increased extinction suffered by stars embedded within molecular clouds at the Galactic Center.

Figure 7 is a color magnitude diagram for the Quintuplet cluster control fields. A similar diagram can be made for the Arches cluster control fields. The figure shows a striking contrast to those for the cluster fields in that relatively few stars are brighter than $m_{F205W} \approx 15$. This can be explained by the fact that the control fields contain very old stars. Given this fact, we conclude that the cluster fields contain young stars down to $m_{F205W} = 16$. This provides the first clear identification of the main sequence in the Quintuplet cluster, and it confirms the earlier identification of the main sequence in the Arches cluster. We can also see a main sequence turn-off in both clusters starting at $m_{F205W} = 12$ in the Arches cluster and $m_{F205W} = 13$ in the Quintuplet cluster. The difference in the cutoffs reflects the difference in ages of the clusters. Note that stars above the turn-off in the Quintuplet cluster are randomly distributed in color, whereas similar stars in the Arches cluster demonstrate the expected trend of redder colors for brighter stars and a blueward trail for the very brightest few stars. In the case of the Arches cluster, the most massive stars are probably still burning hydrogen in their cores as indicated by the evolutionary models. They correspond to the emission-line stars seen by Nagata et al. (1995) and Cotera et

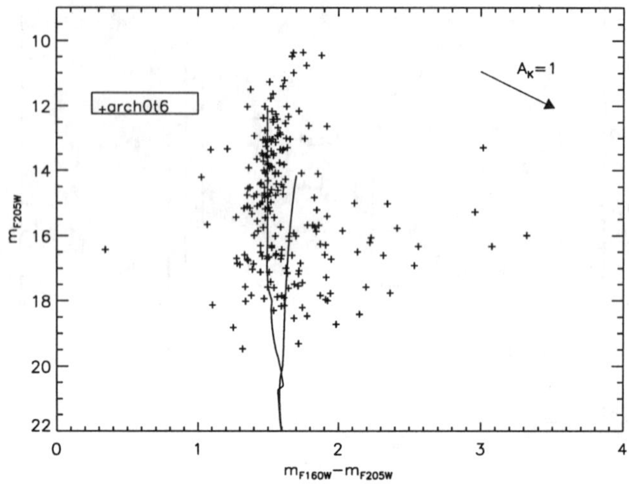

Figure 5. Color-magnitude diagram for the Arches cluster within a radius of 6″ of the cluster center. The color is m_{F160W}-m_{F205W} and is similar to H−K.

al. (1996), and are seen to have strong Paschen-α emission in our own NICMOS images.

The luminosity function can be transformed into an initial mass function assuming a relation between m_{F205W} and $M_{initial}$. Such a relation depends upon age, metallicity, and the mass-loss rates of stars with strong winds, i.e., the most massive stars. We assume an age of 2 Myr, twice solar metallicity, and the "$2\times\dot{M}$" Geneva models for the Arches cluster. The resultant mass function for the Arches cluster is shown in Figure 8. The background sample measured in the control fields was subtracted. We limited the sample to an annulus between 3″ and 9″ because the inner core is hopelessly incomplete for all but the brightest stars.

Linear fits have been made through the completeness-corrected data for intermediate-mass stars and high-mass stars. In both cases, the slopes to the mass functions are somewhat shallower than the Salpeter value (Γ −1.35). More importantly, note that the linear fits are rather poor for the observed distribution. In fact, there appears to be a flattening of the slope between Log(M/$M_\odot$) = 1.1 and 1.7. The number of stars in the highest mass bin is misleading. These stars have been put in this bin because they are brighter than stars with $M_{initial}$ = 120 $M_\odot$, where the Geneva models end. We will further investigate the behavior of the IMF slope in both clusters in Figer et al. (1999)

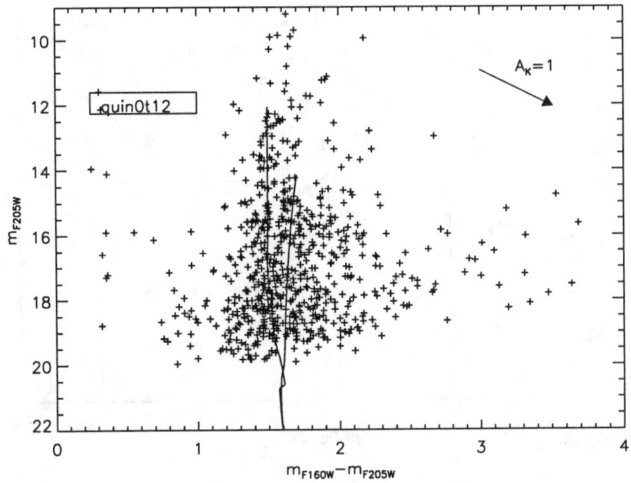

Figure 6. Color-magnitude diagram for the Quintuplet cluster within a radius of 12″ of the cluster center. The color is m_{F160W}-m_{F205W} and is similar to H−K.

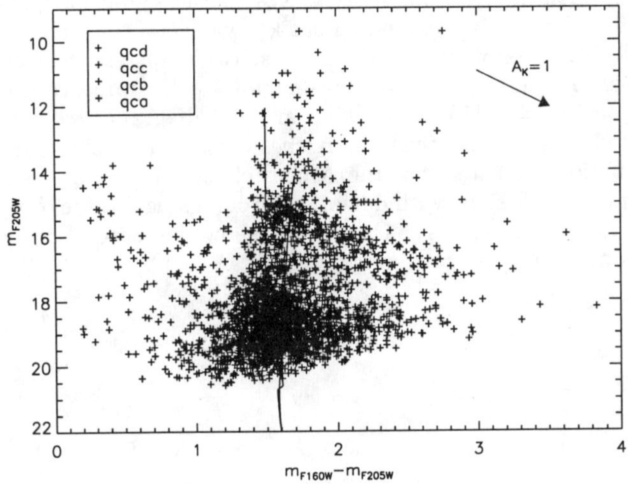

Figure 7. Color-magnitude diagram for the control fields near the Quintuplet cluster. The color is m_{F160W}-m_{F205W} and is similar to H−K.

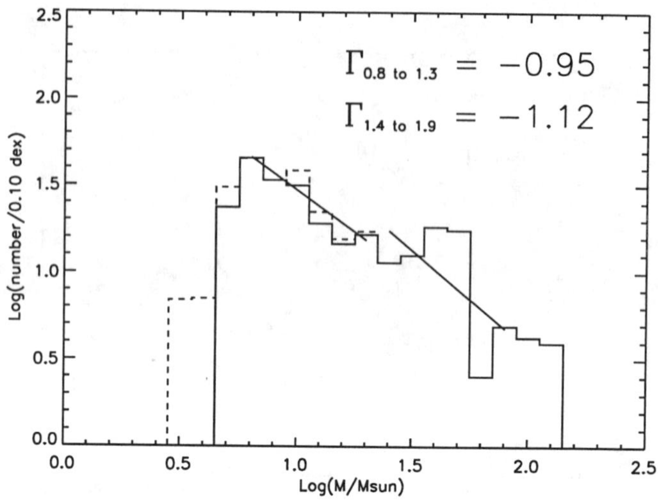

Figure 8. Observed mass function for Arches cluster in the annulus between 3″ and 9″ of the cluster center. The data have been corrected for incompleteness, and background counts have been subtracted.

6. Conclusions

We have clearly identified the main sequences in the Quintuplet and Arches clusters down to below 10 $M_\odot$, extending our knowledge of the stellar content in these clusters by an order of magnitude in mass. Both clusters show evidence of post main sequence evolution in their color-magnitude diagrams, consistent with the spectral types previously observed. From these diagrams, we confirm that the Quintuplet cluster is older than the Arches cluster. A preliminary analysis of the mass function indicates that the Arches cluster has an age near 2 Myr, and a near-Salpeter IMF slope with deviations from a single value over the mass range between 10 $M_\odot$ and 100 $M_\odot$. Our images show that several of the original Quintulet stars are multiple or extended.

Acknowledgments. We thank the late Chris Skinner of STScI for providing assistance in performing the observations. We thank Christine Ritchie of STScI for assisting in calibrating the data. We thank L. E. Bergeron of STScI for "cleaning" the images of bad pixels using his "imclean" IDL script. We also thank the NICMOS team for providing assistance in reducing our data. Support for this work was provided by NASA through grant number GO-07364.01-96A from the Space Telescope Science Institute, which is operated by AURA, Inc., under NASA contract NAS5-26555.

References

Cotera, A. S., Erickson, E. F., Colgan, S. W. J., Simpson, J. P., Allen, D. A., & Burton, M. G. 1996, ApJ, 461, 750

Figer et al. 1999, in preparation

Figer, D. F., McLean, I. S., & Morris, M. 1999, ApJ, in press

Figer, D. F., Morris, M., & McLean, I. S. 1996, in 4th ESO/CTIO Workshop on The Galactic Center, ed. R. Gredel, (San Francisco: ASP), 263

Figer, D. F., Najarro, F., & Langer, N. 1998, in proceedings of "Unsolved Problems in Stellar Evolution," (Baltimore: STScI), in press

Genzel, R., Thatte, N., Krabbe, A., Kroker, H. & Tacconi-Garman 1996, ApJ, 472, 153

Keyes, T., et al. 1997, HST Data Handbook, Ver. 3.0, Vol. I (Baltimore: STScI)

Krist, J. & Hook, R. 1997, The Tiny Tim User's Guide, Ver. 4.4

MacKenty, J. W., et al. 1997, NICMOS Instrument Handbook, Ver. 2.0 (Baltimore: STScI)

Meynet, G., Maeder, A., Schaller, G., Schaerer, D., & Charbonnel, C. 1994, A&A Supp., 103, 97

Morris, M. 1993, ApJ, 408, 496

Nagata, T., Woodward, C. E., Shure, M., & Kobayashi, N. 1995, AJ, 109, 1676

Serabyn, E., Shupe, D., & Figer, D. F. 1998, Nature, 394, 448

Serabyn, E., Shupe, D., & Figer, D. F. 1999, this volume, p. 320

Stetson, P. 1987, PASP, 99, 191

The Central Parsecs of the Galaxy
ASP Conference Series, Vol. 186, 1999
H. Falcke, A. Cotera, W.J. Duschl, F. Melia, M.J. Rieke, eds.

Theoretical Modelling of Hot Stars

F. Najarro

Intituto de Estructura de la Materia, CSIC, Serrano 121, 28006 Madrid, Spain

D. J. Hillier

Department of Physics and Astronomy, University of Pittsburgh, 3941 O'Hara Street, Pittsburgh, PA 15260, USA

D. F. Figer

Department of Physics & Astronomy, University of California, Los Angeles, CA 90095-1562, USA

T. R. Geballe

Gemini N. 8m Telescope,670 N. A'ohoku Place, Hilo, HI 96720, USA

Abstract. Recent progress towards model atmospheres for hot stars is discussed. A new generation of NLTE wind blanketed models, together with high S/N spectra of the hot star population in the central parsec, which are currently being obtained, will allow metal abundance determinations (Fe, Si, Mg, Na, etc). Metallicity studies of hot stars in the IR will provide major constraints not only on the theory of evolution of massive stars but also on our efforts to solve the puzzle of the central parsecs of the Galaxy. Preliminary results suggest that the metallicity of the Pistol Star is 3 times solar, thus indicating strong chemical enrichment of the gas in the Galactic Center.

1. Introduction

At the beginning of this decade, the detection of the He I emission line cluster by Krabbe et al. (1991) triggered a substantial improvement in the atmospheric models for hot stars in the near-infrared. In a significant first step the codes were tuned, and then used as diagnostic tools to extract the astrophysical information present in the H and He I infrared lines (e.g., Najarro et al. 1994, 1997, Crowther & Smith 1996). Reliable values for luminosities, temperatures, mass-loss rates, helium abundances and ionizing photons were obtained for the brightest members of the He I cluster (Najarro et al. 1994, 1997). With these values, strong conclusions could be reached about the nature of these objects and their rôle on the energetics of the central parsec. Though important constraints were placed as well on the theory of the evolution of massive stars, the metal content of these objects could not be determined. Indeed, the metallicity issue in the Galactic Center is still a source of controversy. Based on measurements of the gas-phase, Shields & Ferland (1994) obtained twice solar metallicity from argon and nitrogen emission lines while a solar abundance was derived for neon. For the cool stars, and based on LTE-differential analysis with other cool supergiants, Carr et al. (1996) and Ramirez et al. (1999) have obtained strong indications for a solar Fe abundance. It is therefore crucial to obtain metallicity estimates from direct analysis of the hot stars to see if the values are compatible with those obtained from the cool-star and gas-phase analyses. It is also of importance since the metallicity has a substantial influence on stellar evolution.

To tackle this problem, two conditions were required: observations with sufficiently high S/N and spectral resolution to detect confidently the metal lines, and models for hot stars accounting for metals such as Fe, Mg, Ca, etc and their effects on the photospheric and wind structure (line blanketing).

In this paper, and using the Pistol Star (Figer et al. 1998) as template, we report impressive recent progress in both observations and theoretical models for hot stars. Preliminary results on metallicity determinations will also be presented.

2. Improved Observations

The advances in ground based infrared astronomy in this decade are striking. The introduction of very efficient spectrographs suitable for mid-high resolution observations have opened the near-infrared sky to quantitative spectroscopy. In the field of hot stars, high quality IR-spectra have recently appeared in the literature (Morris et al. 1996, Hanson et al. 1996, Figer et al. 1997, Fullerton & Najarro 1997, Blum et al. 1997). Except for Fullerton & Najarro's sample (R$\sim$10000), all these spectra were obtained with low-mid (R$\sim$500-2000) resolution, which is enough to classify the stars but insufficient in most of the cases to perform accurate quantitative spectroscopic studies.

Examples of mid-high resolution observations of early type stars with strong winds are shown in Fig. 1 where a K-band spectrum of the P Cygni type star HDE 316285 (Fullerton & Najarro 1997) and the H-band of the Pistol Star obtained with UKIRT-CGS4 with a resolution of R$\sim$5000 are displayed. The the spectra shown in Fig. 1 demonstrates the necessity of high resolution and S/N –the same spectral region observed with R$\sim$600 did not show the presence of the important He I 1.700μm diagnostic line. The number of new observational constraints provided by the new spectroscopic data is striking when compared to the previous use of only some H and He lines. The availability of high-quality IR spectroscopic data will be substantially improved in the near future with NIRSPEC on the Keck II and ISAAC on the VLT. Of concern is the presence in Fig. 1 of fairly well isolated lines of Si II, Mg II and Fe II, an invaluable gift for an infrared spectroscopist willing to determine metal abundances and to whom the UV and optical spectra of the star have been swallowed by extinction. Now that the lines are there, "only" the appropriate tool is required: a unified photosphere-wind NLTE-model, accounting for the presence of metals and correctly treating line blanketing. Impressive efforts have been made within

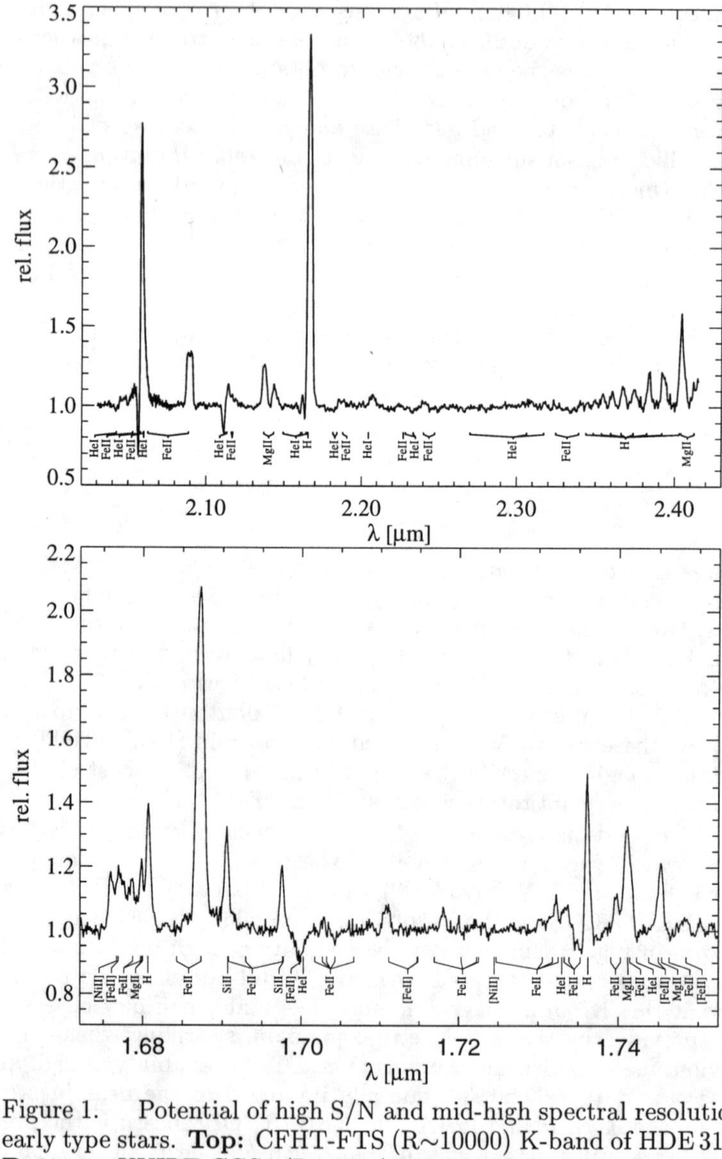

Figure 1. Potential of high S/N and mid-high spectral resolution for early type stars. **Top:** CFHT-FTS (R∼10000) K-band of HDE 316285. **Bottom:** UKIRT-CGS4 (R∼5000) H-band of the Pistol Star

the last years in this field by the Munich-group and Hillier & collaborators (see Pauldrach et al. 1997 and Hillier 1997 for excellent reviews in this topic).

In the next section we will present the new generation of line blanketed models developed by John Hillier and compare them to the formerly used standard model.

3. New Unified Models vs. Standard Model

The standard model which we used to analyze the He I stars in the central cluster (Najarro et al. 1994, 1997) is basically an iterative, non-LTE method presented by Hillier (1987, 1990) to solve the radiative transfer equation for the expanding atmospheres of early-type stars in spherical geometry, subject to the constraints of statistical and radiative equilibrium. Steady state is assumed, and the density structure is set by the mass-loss rate and the velocity field via the equation of continuity. The transfer equation is solved using either the Sobolev approximation, corrected for the diffuse radiation field, or by means of an accurate solution in the comoving-frame. The atmosphere was considered to consist of H, He, N and C, though the latter two species were used basically to account better for cooling processes. The standard model was then prescribed by the stellar radius, R_*, the stellar luminosity, L_*, the mass-loss rate $\dot{M}$, the element abundances, and the velocity field, $v(r)$.

The new model (Hillier & Miller 1998) incorporates new species: O, Mg, Ca, Si, Na, Al and Fe. The new code can still treat lines and continuum separately using the Sobolev approximation or comoving frame solution, as in the standard old model, or can be switched to "blanketing". In this case, no distinction between lines and continuum frequencies is made and the transfer (moment) equations are solved consecutively from the highest to lowest frequency. Hence, the effect of continua on lines and lines on the continua as well as overlapping lines is automatically handled. In addition, the new code accounts for X-ray ionization, clumping, level dissolution, and a smooth transition to the "exact" photospheric stratification based on the method by Santolaya-Rey et al. (1997).

Below we compare a set of models with stellar parameters close to those of the Pistol Star to illustrate the differences between the old standard model and the new code. We consider the old standard model ("Old"), the new model were Fe is included but no blanketing is used ("New NBL") and the new model with Fe and blanketing ("New BL").

The two classical and related effects of line blanketing: blocking and backwarming, are clearly shown in Fig. 2 where the flux distribution and the temperature structure are displayed for the three models under consideration. In Fig. 2 we see that in order to preserve luminosity the temperature gradient (and hence the temperature) has to be increased in the inner photosphere to drive out the photons blocked in the UV. The blocked UV flux emerges in the visual and infrared giving rise to a flux excess over the standard model. We note that these are indeed blanketing effects as in the inner parts, $\log n_e > 12$, the "New NBL" model shows the same behavior as the "Old" one. In the wind, $\log n_e < 12$, the inclusion of metals and Fe (and not the blanketing) are responsible for the different temperature structure.

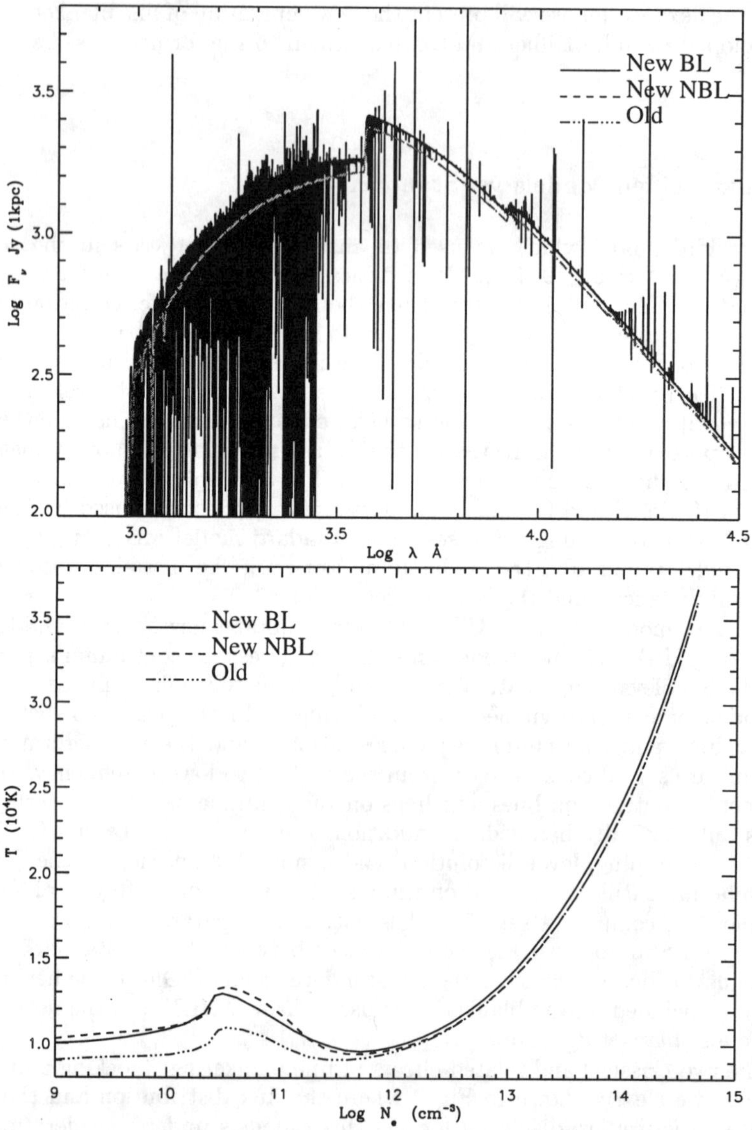

Figure 2. Effects of the inclusion of Fe and line blanketing on the flux distribution (**top**) temperature structure (**bottom**). Flux blocking and redistribution as well as backwarming are produced by blanketing. See discussion in text.

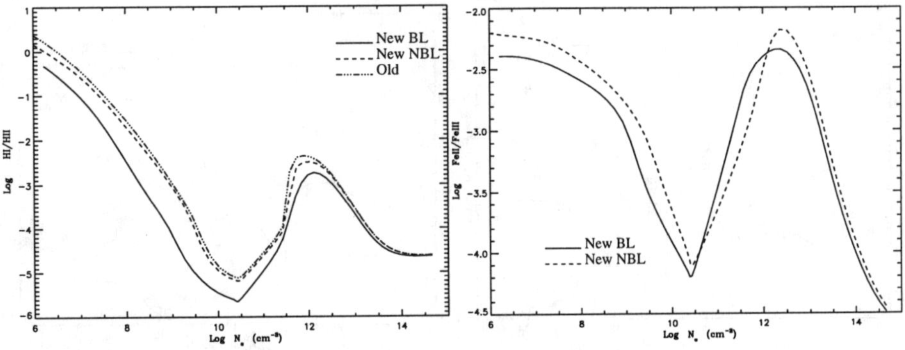

Figure 3. Standard Old vs. New Blanketed models: Ionization structure. **(left:)** For Hydrogen significant changes are only introduced when blanketing is incorporated (New LB). The Fe ionization structure is also altered by blanketing.

The effects on the ionization structure of H and Fe are shown in Fig. 3. For these two reference species, we note again the importance of line blanketing, which can significantly change the ionization structure and hence affect the line profiles. The inclusion of metals without blanketing makes only minor changes to the ionization structure and the line profiles.

Finally, in Fig. 4 we consider both NLTE and blanketing effects on two important diagnostic metal lines present in the K-band spectra of HDE 316285 and the Pistol Star (see Fig. 1). These are the Fe II 2.089μm line and the Mg II 2.13/4μm doublet. The Fe II 2.089μm ($z^4F_{3/2}$-$c^4F_{3/2}$) transition, despite its rather low gf value (gf$\approx 6\times10^{-5}$), can be tremendously amplified due to very efficient UV continuum fluorescence pumping. This is the case when the line is formed in the outer regions of the wind, where the population of the upper level can exceed its LTE value by factors up to 10000 (see Fig. 4). Such conditions are met by stars such as HDE 316285 and the Pistol Star with very dense winds. There, the Fe II 2.089μm line is formed in the region with $10^{6.5} \leq n_e \leq 10^9$. Similar fluorescent mechanism operates on the much more stronger Mg II $5s^2S_{1/2}$-$5p^2P_{1/2}$ / $5s^2S_{1/2}$-$5p^2P_{3/2}$ 2.1432/2.1368μm doublet (gf$\approx$1.2 and 2.4 respectively). These lines, though formed closer to the photosphere ($10^{9.5} \leq n_e \leq 10^{12}$) are already affected by NLTE effects as the departure coefficient of the upper level takes off in this region (see Fig. 4). It is also important to note that, though these lines are driven by a NLTE mechanism, i.e., UV pumping, their absolute strengths (i.e., exact run of the departure coefficients) are intimately coupled to blanketing effects as shown in Fig. 4.

4. Metallicity Determinations: Preliminary Results

In this section we present preliminary results from quantitative spectroscopic studies using line blanketed models for HDE 316285 and the Pistol Star from which, for the first time, abundances for Fe and Mg are obtained. We have chosen the galactic P Cygni type star HDE 316285 as a calibrator for our method.

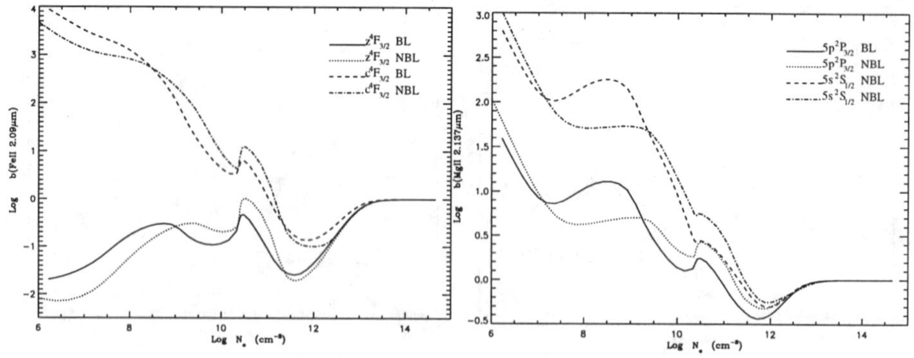

Figure 4. NLTE and blanketing effects on the formation of metal lines in the K-band. Departure coefficients for the upper and lower levels of the Fe II 2.089μm (left) and Mg II 2.13/4μm (right) doublet lines. Note the very efficient pumping to the upper levels thru UV continuum fluorescence. See further discussion in text.

The scope of the investigation on HDE 316285 is to check for consistency between optical and IR analysis before proceeding to investigate GC Center sources for which only IR-spectra are available. We started our analysis of HDE 316285 using the stellar parameters we have derived recently from a spectroscopic investigation using the "Old" model (Hillier et al. 1998).

First results show only small corrections to the stellar parameters derived by Hillier et al. (1998) and will be discussed elsewhere. Figure 5 shows the excellent agreement between the synthetic spectra of the new line blanketed models and the observations, both in the optical and the infrared K-band. This is a very important result as it confirms the consistency of the results between optical and IR studies, and hence provides confidence to study Galactic Center sources. Further consistency of our model with recently obtained ISO-SWS observations of HDE 316285 is also found (Najarro et al. in prep.).

The derived magnesium abundance is twice solar, while for iron an abundance four times solar is obtained. The latter result is somewhat surprising as similar enhancements should be expected for Mg and Fe (Langer priv. comm.) for this kind of star. Looking for possible sources which could lead to this discrepancy we first investigated the quality of the atomic data. While we are confident about the quality of the much simpler Mg II model atom, the Iron Project atomic data for Fe II (Nahar 1995) show discrepancies with Kurucz semiempirical values, in some cases up to factor of 5 or more. Detailed comparison of model results with the observations seem to favor Kurucz gf values and not those from the Iron Project. We're currently combining both data sets in oder to obtain an optimized Fe II atom. However, the uncertainties in some gf values are not enough to remove the factor of two in the derived Fe abundance.

Recent work suggests that the discrepancy may be due to the neglect of the $Fe^{2+} + H \longleftrightarrow Fe^{+} + H^{+}$ charge exchange reaction. This reaction is exothermic in the left to right direction and requires a non negligible amount of H I to work efficiently. This the case in the outer wind of HDE 316285 and the Pistol Star where the recombination of hydrogen (see Fig. 3) provides enough H I. Thus, as

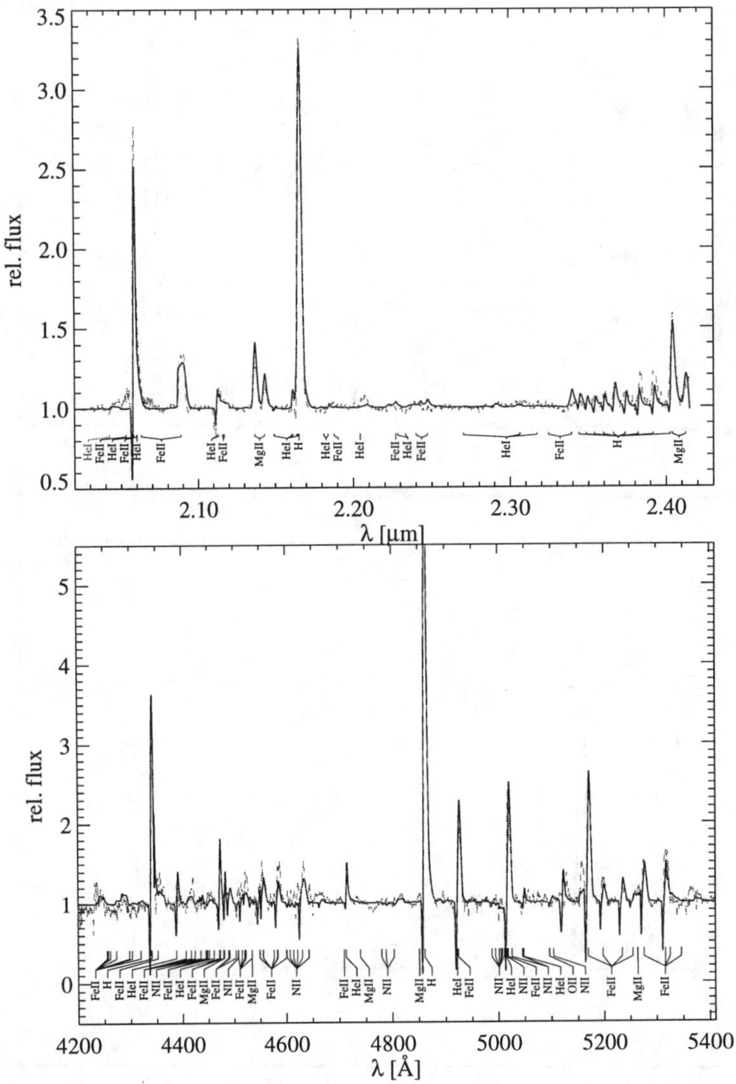

Figure 5. Comparison of the blanketed model (solid) and the observed (grey dashed) spectrum of HDE 316285. **Top:** K-band. **Bottom:** Optical 4200≤ $\AA$ ≤ 5400 region. Excellent agreement is found in both bands not only for H and He I, but also for the Mg II and Fe II lines

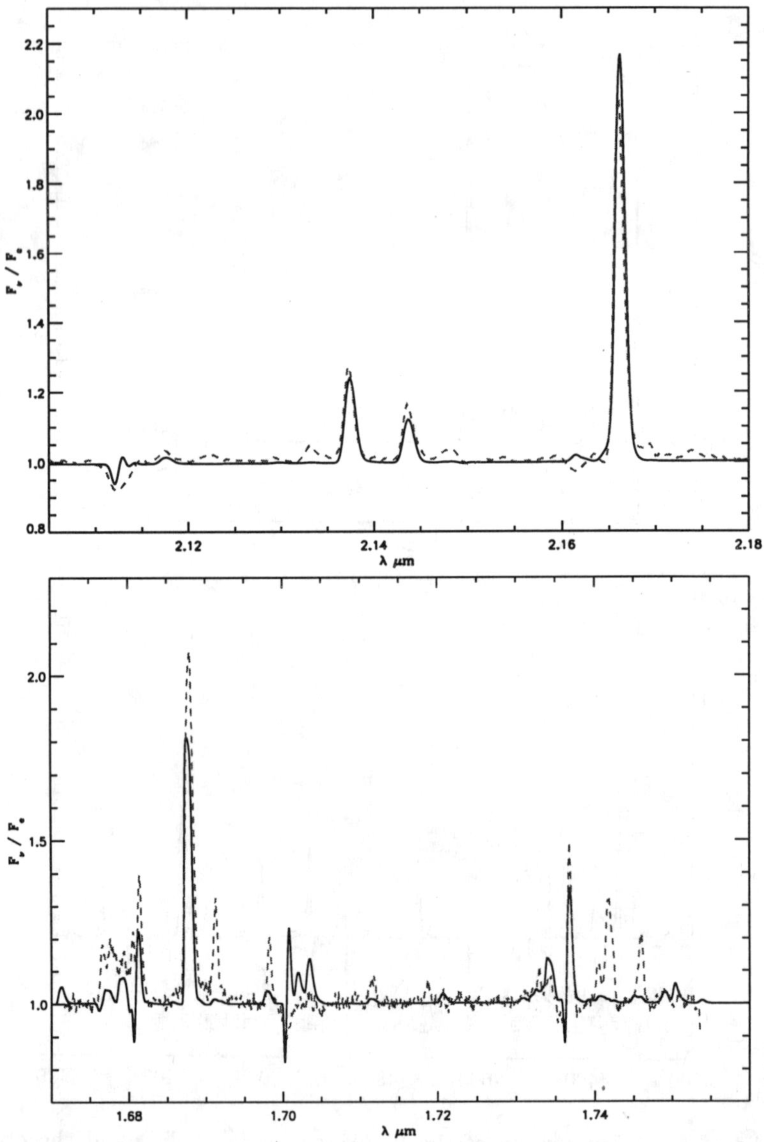

Figure 6. Comparison of the blanketed model (solid) and the observed (grey dashed) spectrum of the Pistol Star. The K-band (**Top:**) is well reproduced. Missing lines in the fit to the H-band (**Bottom:**) are from Si II (neglected in the code) and discrepancies in the Fe II lines are due to uncertainties in the Fe II model atom. See also discussion in text.

Fe III strongly recombines into Fe II in this zone, the strength of the Fe II lines formed there, such as Fe II 2.089μm and the optical Fe II lines (see Fig. 5), are strongly increased and a lower Fe abundance is required. First results show that when charge exchange reactions are included the models can reproduce the observations with a Fe abundance twice solar, consistent with the results obtained for Mg.

Preliminary results for the Pistol Star show this trend as well. While a Mg abundance of roughly 3 to 4 times solar is required to match the observed spectra, we obtain a Fe abundance about 8 times over solar when charge exchange reactions are neglected. Figure 6 shows the best current model fits to the K- and H-band spectra of the Pistol Star.

The new models including Fe and Mg favor the low temperature model vs the high temperature one proposed in Figer et al. (1998), solving the dichotomy on the Pistol Stars luminosity. From Fig. 6 we see, that although the K-band is fairly well reproduced by our models, some discrepancies appear in the H-band. The He I 1.700μm line in our models appears severely contaminated by Fe II lines. The strength of these lines is overestimated by our models while that of the strong Fe II 1.868μm line is underestimated. This discrepancies are removed if the Iron Project oscillator strengths are replaced by Kurucz semiempirical values as discussed above. Also missing are the Si II lines as Si was not included in our models.

We are currently studying the effects of charge exchange reaction on the Pistol Star. While we are confident that this effect will reduce the derived Fe abundance (maybe by a factor of two) and give a similar enhancement to that of Mg, we find it really hard to reconcile the Pistol Star with having solar abundance as derived by Ramirez et al. (1999) from the cool stars.

The Pistol Star is not the only Galactic Center object showing the presence of Fe and other metal lines. Metal lines also appear in the low-mid resolution spectra of IRS16NE (Tamblyn et al. 1996) and in spectra of the recently discovered second LBV candidate in the quintuplet (Figer et al. 1999). The upcoming operation start of ISAAC on the VLT and especially NIRSPEC (with R$\sim$25000) on the Keck II telescope will provide new high quality infrared spectra of hot stars in the Galactic Center, for which reliable abundance determinations will be made. It is evident that a new, fascinating research window has opened – the field of quantitative infrared spectroscopy of early type stars.

Acknowledgments. F. N. acknowledges grants of the DGYCIT under PB96-0883 and ESP98-1351.

References

Blum, R.D, Ramond, T.M., Conti, P.S., Figer, D.F., Sellgren, K., 1997, AJ, 113, 1855

Carr, J.S., Sellgren, K., Balachandran, S.C., 1996, in: The Galactic Center, 4th ESO/CTIO Workshop, ed. R. Gredel, Astron. Soc. Pacific Conf. Series, Vol. 102, p. 212

Crowther, P.A., Smith, L.J., 1996, A&A, 305, 541

Figer, D.F., McLean, I.S., Najarro, F., 1997, ApJ, 486, 420

Figer, D.F., Najarro, F., Morris, M., McLean, I.S., Geballe, T.R., Ghez, A.M., Langer, N., 1998, ApJ, 384, 404

Figer, D.F., , McLean, I.S., Morris, M., 1999, ApJ, (in press)

Fullerton, A.W., Najarro, F., 1997, in: Properties of Hot Luminous Stars, ed. I. Howarth, Astron. Soc. Pacific Conf. Series, Vol. 131, p. 47

Hanson, M.M., Conti, P.S., Rieke, M.J., 1996, ApJS, 107, 281

Hillier, D.J., 1987, ApJS, 63, 947

Hillier, D.J., 1990, A&A, 231, 116

Hillier, D.J. 1997, in IAU Symp. 189, Fundamental Stellar Properties: the Interaction between Observation and Theory, eds. T.R. Bedding, A.J. Booth, J. Davis, Kluwer, p. 209-216

Hillier, D.J., Miller, D.L., 1998, ApJ, 496, 407

Hillier, D.J., Crowther, P.A., Najarro, F., Fullerton, A.W., 1998, A&A, 340, 483

Krabbe, A., Genzel, R., Drapatz, S., Rotaciuc, V., 1991, ApJ, 382, L19

Morris, P.W., Eenens, P.R.J., Hanson, M.M., Conti, P.S., Blum, R.D, 1996, ApJ, 470, 597

Nahar, S.N., 1995, A&A, 293, 967

Najarro, F., Hillier, D.J., Kudritzki, R.P., Krabbe, A., Genzel, R., Lutz, D., Drapatz, S., Geballe, T.R., 1994, A&A, 285, 573

Najarro, F., Krabbe, A., Genzel, R., Lutz, D., Kudritzki, R.P., Hillier, D.J., 1997, A&A, 325, 700

Pauldrach, A.W.A., Lennon, M., Hoffmann, T.L., Sellmaier, F., Kudritzki, R.P., Puls, J., 1997, in: Properties of Hot Luminous Stars, ed. I. Howarth, Astron. Soc. Pacific Conf. Series, Vol. 131, p. 258

Ramirez, S., et al., 1999, this volume, p. 373

Shields, J.C., Ferland, G.J., 1994, ApJ, 430, 236

Santolaya-Rey, A.E., Puls, J., Herrero, A., 1997, A&A, 323, 488

Tamblyn, P., Rieke, G.H., Hanson, M., Close, L.M., McCarthy, D.W., Rieke, M.J., 1996, ApJ, 456, 206

The Central Parsecs of the Galaxy
ASP Conference Series, Vol. 186, 1999
H. Falcke, A. Cotera, W.J. Duschl, F. Melia, M.J. Rieke, eds.

Spatially Resolved Dust Shells in the Central Parsec

A. M. Tanner, A. M. Ghez, M. Morris, E. E. Becklin

UCLA Department of Astronomy and Physics, Los Angeles, CA 90095

Abstract.
Recent diffraction-limited K-band (2.2 micron) imaging observations with the Keck 10m telescope have unambiguously resolved, for the first time at this wavelength, the unusual cool stellar stellar source IRS 21. We have estimated the intrinsic size of this object to be ~1000 AU, placing it among the largest stellar objects known. These measurements, along with modeling of its SED, lead us to conclude that IRS 21 is comprised of an optically thick dust shell surrounded by an optically thin and geometrically thin dust layer which scatters some of the near-infrared light and re-radiates the rest into the far-infrared. We will present both the images and detailed SED modeling of this source.

1. Introduction

IRS 21 is one of the more unusual sources in the central parsec of our Galaxy. Observations of this cool source reveal polarization (Eckart et al. 1995), an IR excess (Blum et al. 1996) and a featureless spectrum (Krabbe et al. 1995). A wide variety of classifications have been proposed for IRS 21 which include a protostar (Zinnecker et al. 1996), an embedded early-type star (Krabbe et al. 1995), a dusty Wolf Rayet (WR) star (Figer et al. 1996) and a high density dust clump (Gezari et al. 1985).

We present new diffraction-limited K (2.2μm) band images of IRS 21 and measure its spatial extent. These images have a resolution of 0.05", which at a distance of 8 kpc corresponds to 400 AU, surpassing all near-IR ground and space-based measurements to date. The 2.2μm size, combined with modeling the spectral energy distribution (SED), constrain the nature of this object.

2. Observations

The spatially resolved structure of IRS 21 was serendipitously discovered in the initial Shift-and-Add (SAA) data collected for a proper motion study of the central parsec by Ghez et al. (1998) in 1995 June. However, in these images IRS 21 is near the edge of the field of view. In subsequent observations presented here, IRS 21 and the nearby point source IRS 33E were centered in the array to facilitate the study of IRS 21's structure (see Fig. 1,2).

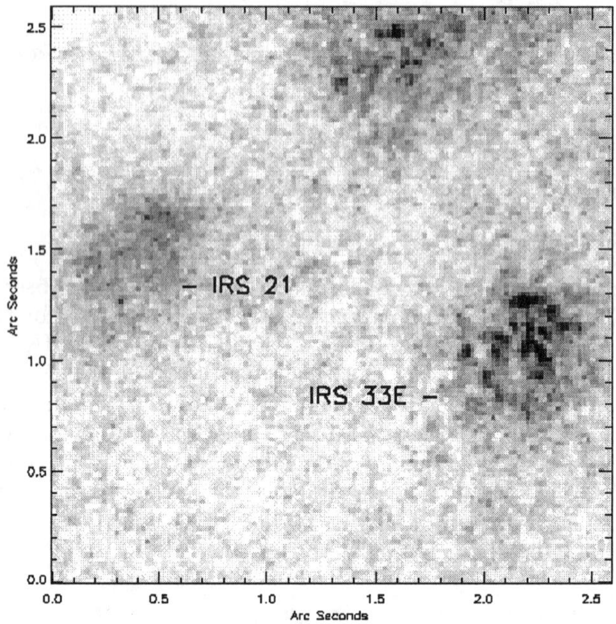

Figure 1. The central ~2.5"×2.5" of a single short exposure (t_{int} = 0.15s) or specklegram is shown. The instantaneous point spread function is composed of a pattern of speckles and is clearly seen at IRS 33E. The speckles are an interference pattern of the wavefronts delayed by the turbulent structure of the Earth's atmosphere. It is clearly evident, even in this short exposure image, that IRS 21 (K=10.2) is extended since it shows no speckles despite being about the same flux as IRS 33E (K=9.9).

3. Intrinsic Size of IRS 21

The Shift-and-Add images of IRS 21 clearly show that it differs from an unresolved point source (see Fig. 3).

To extract information on the intrinsic size and shape of IRS 21, a 1-D maximum likelihood deconvolution was performed on the radially averaged intensity profile of IRS 21 using IRS 33E as the PSF (see Fig. 4). We also minimized the χ^2 of the convolution of the PSF and a Gaussian with consistent results of R = 500±75 AU.

4. Spectral Energy Distribution

We have modeled a dereddended (A_V=31) SED constructed from the literature (see Fig. 5 for references). The shape of the SED suggests there are two com-

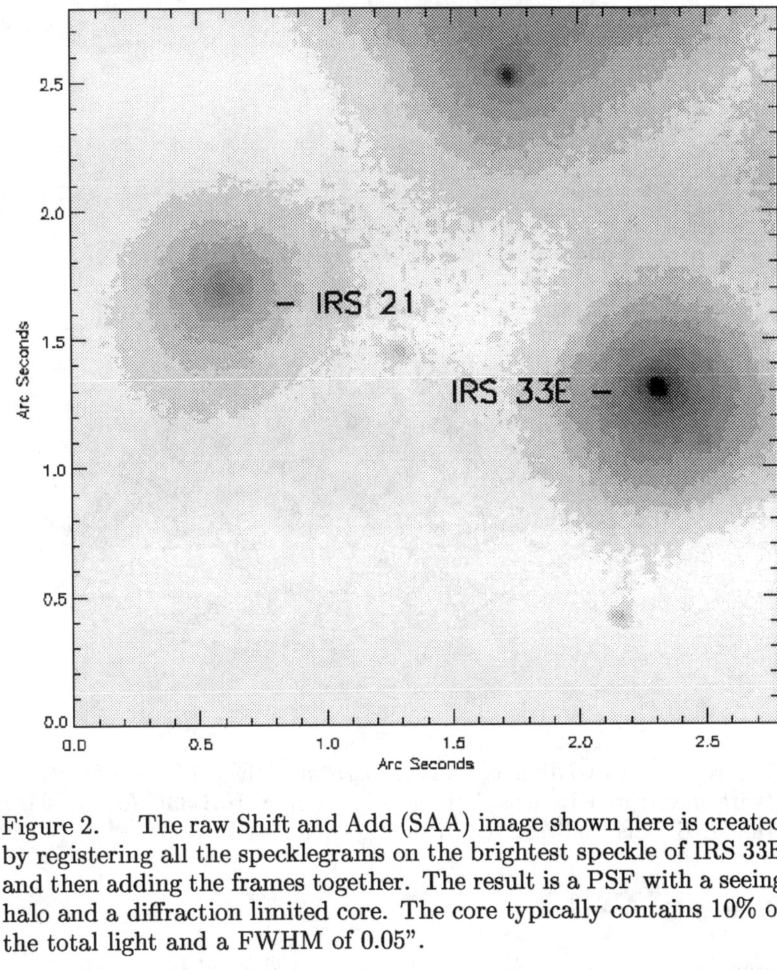

Figure 2. The raw Shift and Add (SAA) image shown here is created by registering all the specklegrams on the brightest speckle of IRS 33E and then adding the frames together. The result is a PSF with a seeing halo and a diffraction limited core. The core typically contains 10% of the total light and a FWHM of 0.05".

ponents: a warm component composed of the K, L and L' points and a cool component composed of the 8.7 and 12.4μm data.

5. Cool Component

We first used the 8.7 and 12.4 μm photometry to estimate a color temperature for the cool component (Gezari 1998). To fit a model to the photometry, we used a blackbody at the resulting color temperature and varied the size of the emitting region to minimize the χ^2. No additional dust extinction was considered in addition to line-of-sight extinction. The best fits resulted in $T_{cool} = 210$ K, $R_{cool} = 480 \pm 100$ AU $L_{cool} = 1.8 \times 10^4$ $L_{\odot}$.

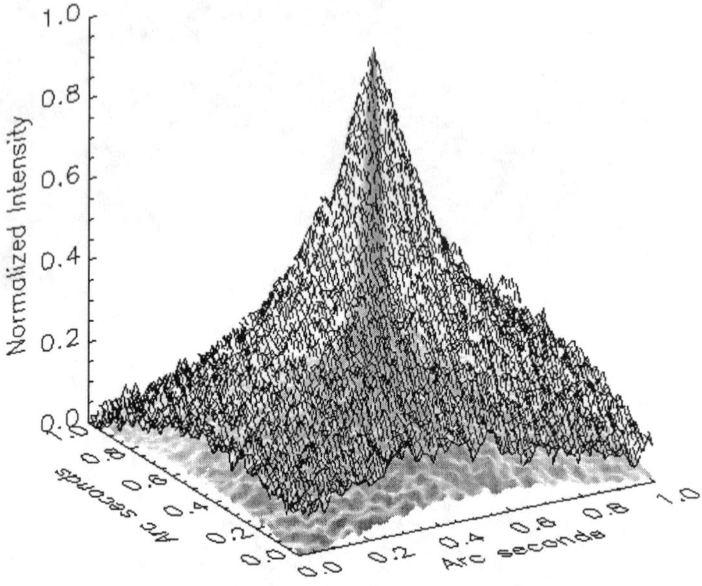

Figure 3. A two dimensional comparison of the SAA data for IRS 21 (wire mesh) and its neighboring point source, IRS 33E (solid) shows that IRS 21 is well resolved in the raw SAA images.

6. Warm Component

The warm component can be fit with a single reddened blackbody. The additional extinction (on top of the line-of-sight extinction) is estimated assuming 0.2μm silicon grains and a uniform extinction through the intervening dust. However, this solution is not unique due to the interplay between the temperature and extinction. We have chosen to assume that the central source is an optically thick inner dust shell. An upper limit on the temperature of the dust grains is set by the condensation temperature (T $\sim$ 1500K). This solution corresponds to a luminosity of 1.6×10^4 $L_\odot$ and a $\tau(2.2\ \mu\mathrm{m}) = 1.2$.

7. Discussion

The most striking result of this analysis is the agreement between the measured K-band extent and the inferred size of the cool component of the SED. This suggests that the cool component is responsible for scattering the 2.2μm emission arising from the warm component. This is consistent with the observed polarization (17%), which most likely arises from single scattering (Eckart et

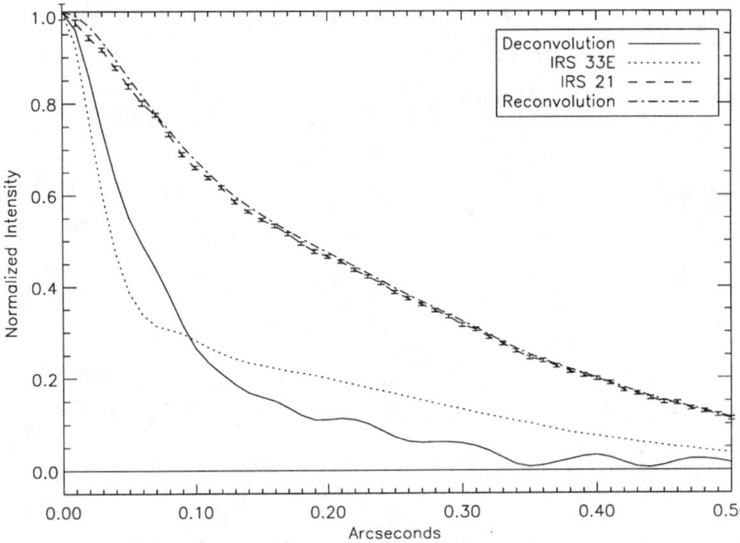

Figure 4. The deconvolution result for IRS 21 using IRS 33E as the PSF is shown. We have plotted the PSF (dotted), IRS 21 (dash), the intrinsic profile of IRS 21 from deconvolution (solid) and the convolution of the intrinsic profile and the PSF (dot-dash). The surprisingly large size of IRS 21's observed radial profile is a result of contributions to the flux from the seeing halo.

al. 1995). The observed polarization also suggests the dust has an asymmetric radial distribution, however whether the dust is a product of mass loss from IRS 21 or is intrinsic to the Northern Arm is still unclear. If the central source is indeed heating the cool component, a simple dust shell model breaks the degeneracy of our earlier fits and requires a temperature and luminosity of 1360 K and 1.1×10^4 $L_\odot$ for the warm component. This is consistent with an optically thick, warm central dust shell.

Our data eliminates the interpretation of IRS 21 as an externally heated dust clump and suggests that the system is composed of cool dust nearby a warm optically thick dust shell. The cool dust scatters light from the warm dust shell and the warm dust shell heats the surrounding dust producing the cool SED component. This model is consistent with any one of the following models: a Wolf Rayet star, a protostar, dust embedded early-type star as well as an embedded late-type star.

Acknowledgments. We thank Mike Jura for useful discussions. Support for this research was provided through an NSF Young Investigator Award to AMG.

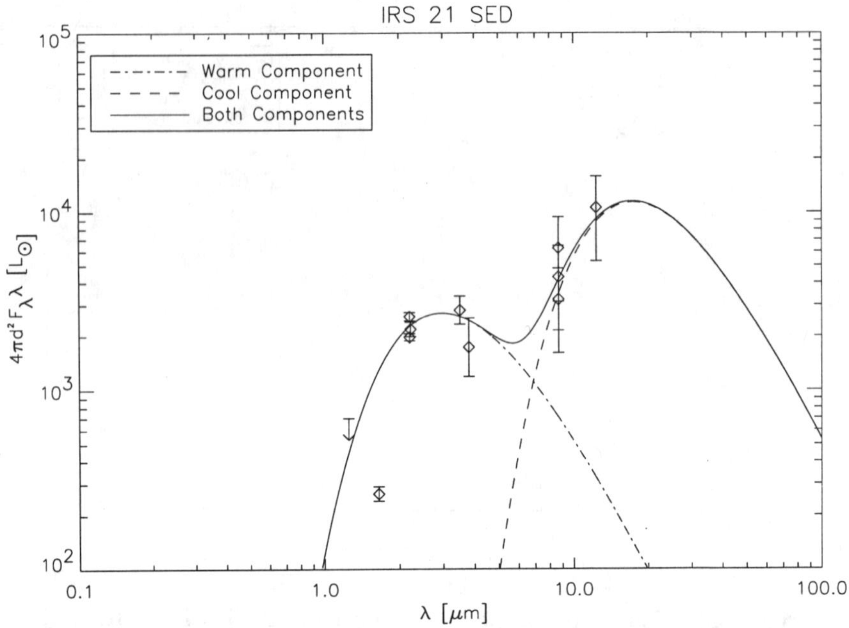

Figure 5. The SED of IRS 21 is shown. The points are dereddened assuming A_V=31, an extinction law from Rieke et al. (1989) and zero points from Tokunaga (1999). J - Blum et al. 1996; H - Blum et al. 1996; K - Blum et al. 1996, DePoy & Sharp 1991, Simon et al. 1990; L - Tollestrup et al. 1989; L' - Simons & Becklin 1996; 8.7μm - Stolovy et al. 1996, Gezari private com., Cotera private com.; 12.4μm - Gezari private com.

References

Blum, R. D., Sellgren, K., & DePoy, D. L. 1996, ApJ, 470, 597

DePoy, D. L., & Sharp, N. A., 1991, AJ, 101, 1324

Eckart, A., Genzel, R. Hofmann, R., Sams, B.J., and Tacconi-Garman, L.,E., 1995, ApJ, 445, L23

Figer, D. F., Morris, M., & McLean, I. S. 1996 in the 4th CTIO/ESO Workshop on the "Galactic Center", ed. R. Gredel, ASP Conference Series, vol 102, 263

Ghez, A., Klein, B. L., Morris, M., & Becklin, E. E., 1998, ApJ, 509, 678

Gezari, D. et al, 1985, ApJ, 299, 1007

Krabbe et al., 1995, ApJ, 447, L95

Reike, G., Rieke, M., & Paul, A. E., 1989, ApJ, 336, 752

Simon, M., et al., 1990, ApJ, 360, 95

Simons, D., and Becklin, E. 1996 AJ, 111, 1908

Stolovy, S., Hayward, T. L., Herter, T., 1996 ApJ, 470, L45

Tokunaga, A.T., 1999, to appear in Astrophysical Quantities, 4th edition, ed. A. Cox, AIP Press

Tollestrup, E. V., Capps, R. W., Becklin, E. E., 1989, AJ, 98, 204

Zinnecker, H., Stanke, T., Kaufl, H., 1996, The Messenger, 84, 18

The Central Parsecs of the Galaxy
ASP Conference Series, Vol. 186, 1999
H. Falcke, A. Cotera, W.J. Duschl, F. Melia, M.J. Rieke, eds.

Discussion Session V: How, when, where and why are stars formed in the Galactic Center?

K. Sellgren[1]

Astronomy Department, Ohio State University, 174 West 18th Av., Columbus, OH 43210

A. Cotera[2]

Steward Observatory, The University of Arizona, Tucson, AZ 85721

Abstract. This is an edited transcript of a discussion session about star formation in the Galactic Center that took place during the Galactic Center conference "The Central Parsecs" in Tucson, Arizona, Sep. 7-11, 1998. The session was chaired by the moderator as indicated in the author list. The discussions were taped, transcribed, and finally edited by an editor who is listed as co-author. The contributions of the conference participants are preceded by their names and thus references to specific aspects and ideas of this discussion session must also include a reference to the respective speaker(s), e.g., in a footnote.

Discussion

Sellgren: I would like to start with the radio image of the Galactic Center and point out the regions where we know star formation is happening. First there is Sgr A West, where we have the Central Cluster with a dozen or so evolved massive stars. There is the Quintuplet Cluster near the Pistol, and the Arches Cluster near the base of the E1 Filament. There are new stars in the process of forming in Sgr B2, and then there are a number of other HII regions scattered about: Sgr D, at least two regions on the other side of Sgr A East, and so on. In addition, there are numerous supernova remnants. So we know that star formation is currently going on not just within the central parsec, but all along the central few hundred parsecs of the Galactic Plane near the Galactic Center.

The star clusters that had detailed studies of their stellar population are these three clusters, the Central Cluster, the Quintuplet Cluster and the Arches Cluster. They have in common that they are on the order of a few million years old, and they contain numerous very short-lived transitional stars such as Ofpe/WN9 stars, Wolf-Rayet (WR) stars, and other hot

[1]Moderator of the discussion session

[2]Editor of the discussion session

emission line stars. One significant difference between the clusters is that we believe there is a massive black hole at the center of the Central Cluster, but for the moment, we don't have any evidence, nor can I imagine there will be massive black holes at the center of the other two clusters. Very early on, people used to say that the strange thing about the stars in the central few parsecs of the Galaxy is that they are somehow affected by the environment of a black hole, and now we know that in fact there are at least two other stellar clusters with very similar number densities of stars, sizes and ages. So, the presence of massive stars, cannot be the result of the black hole creating a unique environment.

One question that people have brought up briefly, but which I think we should discuss in more detail, is: what are the analogues to the three clusters that we know of so far in the central hundred parsecs? Are these clusters going to turn into globular clusters? We don't know of any globular clusters younger than 10 billion years or so, so how can we assess what these clusters will become? Are they an analogue to the giant HII regions, or super star clusters we see in other galaxies like 30 Dor? Are they an analogue to starburst nuclei, e.g. as in M82? Sometimes you see nuclear rings of star formation around Active Galactic Nuclei (AGN), and there is always the question which comes first, the star formation or the AGN? Are we seeing a similar sort of star formation associated with perhaps past AGN activity in our Galaxy? So one thing I want to bring up is, what are the analogues in the rest of the universe?

Another topic to consider is that the environment in the central 100 parsecs is very different from regions of star formation elsewhere in the Milky Way. We know that within the GC the gas pressures are very high, the lines are very broad, the rotation timescale is very small such that there are large shears, the magnetic fields are strong so you see many nonthermal filaments which you don't see in solar neighborhood star formation regions. There is the question of metallicity in the Galactic Center which Paco Najarro has argued is high, although that is still an open question. In addition, at least within the Central Cluster, you have star formation happening in what is essentially an old globular cluster. The Central Cluster of stars has a higher density of old stars than a dense globular cluster, and in the middle of all these old stars we have a large number of massive stars either forming there or moving there from somewhere very nearby where they did form. So again, this is very different from the star formation environment we see within the solar neighborhood.

There are also questions about the IMF [initial mass function]. Are there any low mass stars in the Galactic Center? How would a lack of low mass stars affect the mass-to-light ratio, since we estimate the stellar mass from the K light of the more massive and luminous stars, but all the mass in our assumed IMF is in the low mass stars we currently do not detect in the K band? What triggers star formation in the Galactic Center? Is it just supernovae or is it from material perhaps driven inward by a barred potential? Are we interacting with satellites of the Milky Way, such as the LMC or SMC or the Sagittarius dwarf, all of which are being ripped apart and may be falling into the Milky Way? There is also the question of whether or not our black hole has, in the past, been an AGN such that we are currently either at the beginning or ending of a period of high activity which somehow relates to star formation in our Galaxy. These are a few of the questions that I thought would be interesting to explore.

As to things for observers and theorists to think about: for the observers, we have found three clusters so far which have stars which are only a few million years ago. Where are the clusters of protostars? Where are the stars which are only 10^5 years old? Where are objects which are 10 million years old or 100 million years old? Do those clusters exist? Have they been ripped apart? Can we find any low-mass star formation, similar to the Taurus group, where you have a big molecular cloud and lots of T-Tauri stars forming, but no O stars. From a theoretical perspective it would be helpful to be able to quantify what star formation is like in the Milky Way; Gene Serabyn addressed this in his talk. What we need is a quantitative basis for comparison with other places in our Galaxy and other galaxies. So these are the questions which I thought would be good to address in this discussion. If people want to start making comments, that would be good.

Cotera: Just a quick comment. We see these $10^6 - 10^7$ year old stars not just in clusters. There are isolated stars spread throughout the region which appear to be similar in many respects to the cluster stars, so we can't just look at formation in clusters. They are forming throughout the region. That seems to me a problem. Wouldn't it be easier to form clusters in these really dense molecular clouds, where, once you begin core collapse, you get a bunch of it, all at the same time—forming the cluster. What I don't understand is how you can see that and these isolated randomly spread out stars here and there. I don't have the answer, I'm just adding another question.

Sellgren: Have you looked to see if there are other stars nearby?

Cotera: There are obviously other stars nearby, deciding whether they are in the Galactic Center or are line of sight stars is not always straight forward. In addition to the clusters, there are 6 or 7 more we've found that are isolated which have the strong emission lines, that have the color of the Galactic Center, and nothing else around them. In the H1-H8 region there are 3, that's between the Arches Cluster and the Center, then another two or so near the Sgr A East HII regions.

Figer: Based upon their embeddedness, you would estimate younger ages. For instance, the age of the sources you're talking about, Sgr A East regions A–D, still look like they have a lot of molecular material around them. Region G, south of Sgr A West, by itself is still fully embedded, unless you found a star there.

Cotera: Inside of D or G?

Figer: G.

Cotera: I haven't been able to get data on G.

Figer: I have it and it's highly embedded. So you do perhaps see things forming? You see things perhaps a few hundred thousand years old, but they haven't peeled away their envelopes in the A through D regions. Then you see the clusters, a few megayears old. So when you see these clusters and think it represents a large burst, there is still other activity at a shorter time scale.

Yusef-Zadeh: Then you also have to worry about the extinction, because in the clusters which have been observed, the extinction is not very high. However, in the Sgr A East cluster, for example, the column density is much higher, and the visual extinction is more than 50 magnitudes. So you're really picking up only the ones that have low extinction.

Figer: Well, if you look at the Lyman continuum flux, inferred from the free-free emission, you're still going to detect the star even with the extinction. If you look at the inferred Lyman continuum flux from region A through D, for example, you don't get very big numbers. It looks like they each contain a single B star or a late type O star.

Yusef-Zadeh: There is quite a bit of the Galactic Center which isn't included in what has been presented, for example, Sgr C is another HII region which might contain stars. Of course, the extinction must be very large there.

Cotera: I've looked in Sgr C and haven't seen any clusters.

Figer: I've looked there, too. We can't find a cluster but perhaps that might be a candidate region where there are clusters that you just can't see due to extinction.

Unknown: We know that the brighter IRS 16 stars are very young. Are there comparably young stars in the inner one or two parsecs?

Sellgren: Do you mean between the Central Cluster and the molecular ring?

Unknown: Yes. In the mini-spiral.

Sellgren: I think so, they are concentrated within the central parsec, but I think there are a few stars further away. Bob [Blum], you had one that was $\sim 25''$ away?

Blum: Yes. Yesterday we saw the images of the IRS 16 cluster and there are other identified hot stars along the bar region.

G. Rieke: That's not quite right. There are not stars like IRS 16NE, IRS 16SW, etc., outside of the central 0.4-0.5 parsec. At least if there are, the numbers are insignificant.

Sellgren: I think the question is if there were any, and there certainly is at least one.

G. Rieke: There may be some, but not in comparable numbers.

Unknown: What about the MIR sources extending up the Northern Arm, would you say they are young?

Figer: No one knows what they are.

G. Rieke: We don't have any spectra of them. Actually, we do have spectra, but they are featureless and don't show anything.

Unknown: But could there be comparably young stars up along the Northern Arm.

Figer: I don't think that's been ruled out. In fact, it has been suggested before that stars are forming along the Northern Arm and those are the MIR sources.

Sellgren: I hope someone will be able to figure out what these thermal sources are on the Northern Arm, because they're so dusty, you don't even notice their luminosity much at wavelengths other than the MIR, all the radiation is being absorbed by the circumstellar material. We don't know if they're variable old stars on the AGB, OH/IR stars, WC, evolved massive stars, protostars in the process of forming, you just don't know, and we can't get any information from the spectra, except for the luminosity. So if anybody could think of a solution, that would be good.

Anantharamaiah: We studied pieces of ionized gas in the center of Sgr A West, and one has a good estimate of how much ionizing flux is needed in order to just maintain Sgr A West. Does the stellar content in the Central Cluster or in the surrounding region account for this needed ionizing radiation?

Figer: Yes. I think so, I think it's $N_{Lyc} \sim 10^{50.5}$ from the stars, maybe a little bit more.

Najarro: I think something like 80–90% of the needed flux was accounted for by the stars.

Anantharamaiah: What about the Arches Cluster and the Arched Filaments? Is all the ionizing radiation accounted for?

Figer: The same is true for the other two clusters and their associated HII regions. For the Arches Cluster, we estimate $N_{Lyc} \sim 10^{51}$, the Arched Filaments require $N_{Lyc} \sim 10^{51}$.

Simpson: We have studied the ionization structure, of the Arched Filaments, actually all over the region, and the ionization excitation never falls off with distance from the clusters as fast as you would expect if the clusters were the sole source of ionization.

Figer: That's really a question of 3D morphology, though, isn't it? It's a question of where the cluster truly is with respect to the surface of that molecular cloud.

Simpson: You mean, in other words, you really want the cluster to be very much in the foreground.

Figer: If the Arches are a good distance behind the cluster, and you happen to see them very close to each other in projection, then you won't expect it to drop off.

Cotera: But if you do that, you've just decreased your total effective luminosity, as seen by the Arched Filaments, and then you're not going to get sufficient ionizing flux to account for the Arches right?

Yusef-Zadeh: The point is that the total luminosity that you get from the radio continuum can be, I think, explained by the cluster.

Figer: The luminosity of the cluster is 10^8, the required luminosity of the associated molecular cloud is 10^8. The ionization requirement is 10^{51} the cluster produces 10^{51}. The question is, how do we account for the morphology?

Cotera: What I think Janet Simpson is saying is that from the FIR line observations, if the cluster is the ionization source, you don't get a sharp enough drop-off, assuming we have the correct distance which is needed to capture all the ionizing radiation. In order for the Arches Cluster to ionize the Arched Filaments, it has to be able to capture all of the emergent radiation. If it is too far away, in order to account for the fact that we don't see a gradient in the FIR, then there is a dilution which means the ionization balance isn't there.

Figer: You lose the covering factor. My guess is the problem is trying to figure out how to make all of this happen.

Cotera: I personally would like it if the Arches Cluster ionized the Arched Filaments. But, I'm not sure that you can say yet that other sources aren't required.

Goss: We started out on Sgr A West and then jumped to the Arched Filaments. I'd like to go back to the initial question Anantharamaiah raised. I take it that you can explain the total luminosity in Sgr A West in terms of the known stars, but what about the color of the radiation field? Can you explain that in terms of the He radio recombination line limits and the [Ne II] lines, etc? Is that well understood?

Figer: Well, this is very complex, because you have a few hot stars, and then you have bigger, colder stars. So there isn't just one temperature.

G. Rieke: I think the problem is, a lot of the ionizing flux comes from the stars that are on or near the main sequence, and those aren't the ones that have received our attention. Peter Tamblyn tried to answer this question in his thesis, and I think Krabbe et al. have done a set of population synthesis models, and what they showed is that there is a delicate balance. The answer is—maybe. However, one of the problems is, you're dealing with statistics of small numbers, and so your models get torqued around by what star hadn't formed, rather than just smooth, non-statistical models.

Goss: You mean by matching the total number with the color.

G. Rieke: Yeah. Also, you have to keep in mind that the formation of a single massive star can dominate the radiation field for a little while, and then what do we compute? I think the answer is it is sort of consistent.

Mezger: To answer your question about the temperature of the radiation field, you can get the total luminosity from the dust luminosity and infrared heating, then you can get the Lyman continuum photon production rate from the free-fee emission, and compare them. If you do this simple thing, you get an effective temperature of between 25,000–30,000K.

Goss: Which is then nicely consistent with the He temperatures.

Mezger: Yes.

Sellgren: Okay, any other comments.

Falcke: I have a question, aimed perhaps at George Rieke, about the connection between AGN's and star formation. If we could see the Galactic Center from far away, would we expect to see any AGN-like features? For instance, should we get strong Hα from a narrow line emitting region that we perhaps have seen in the Paα image, Marcia [Rieke] has shown? Could we mistake the hot stars we see for some AGN-like activity? Or would we be able to completely rule out AGN-activity in our own Galactic Center?

Sellgren: I think Joe Shields tried to tackle this question a few years back—Shields & Ferland. They actually took all the MIR observations and the radio observations and tried to synthesize them into what you would see in a large beam. Because of course when you look at our Galactic Center, we have arcsecond resolution or subparsec resolution, whereas when we look in another galaxy we see on a scale of hundreds of parsecs. His conclusion is that if we were to observe the Milky Way from far away, we would have a weak starburst galaxy. At least that's what I recall.

G. Rieke: It depends on how you observe it, doesn't it? Logically, if you see a point like nucleus, you tend to get very excited; that was sort of the point I was making in M33. There is another galaxy where you see a point light like nucleus. But as soon as you go and take a full spectrum, you say "Oops, no high excitation lines."And this creates a problem you have to straighten out. This is happening with HST, looking at LINERs, where you have $0\farcs5$ nuclei, but the spectrum turns out to indicate it's stellar, you have star formation.

Sellgren: I think the question was whether or not it was like a starburst and obviously it's not a starburst in the sense of M82, which has about 10 times the star formation activity. Robert [Sanders] do you have a comment?

Sanders: Yes, in my remark after George's talk, I didn't mean to trivialize the problems with star formation in the near tidal field of a black hole. I think it's an important process we need to seriously consider. I think it's fair to say that star formation is not well understood in any environment, even a normal environment. One thing that does seem to assist in star formation is compression of the gas in shocks. We see that in the classic theory of spiral structure, which seems to work very well. We see star formation behind the spiral shock. In the Galactic Center, in the Sgr A region (I say this apart from any model, even my own model) what we see is very high velocities along these filaments, such as the Minispiral. We see velocities of 100–200 km s^{-1}, and we see very strong gradients in those velocities as well. That is to say, the velocities change direction and magnitude, over short distances. So there have to be shocks in this region—strong shocks. As Binney showed 10 years ago, in strong shocks you can exceed the Roche limit even at a distance of 0.1 pc. There must be strong shocks, because the motion is highly supersonic and changes over short distances. These strong shocks lead to compression. Probably a kind that's unique in star formation.

Sellgren: Well, there are clouds running into each other all over the place in the Galactic Center, and there are shocks, someone should go look for emission lines which are the spectroscopic signature of shocks.

G. Rieke: In defense of the statement, we don't know that shocks are happening now. Binney's prediction was there were strong shocks a million years ago, and that may be harder to test.

Sanders: We know that IRS 16 stars are a few million years old. That gives us an upper limit, but is there a limit on how young the stars can be?

Najarro: Some of the IRS 16 stars have a lot of He at their surface, so they must be at least 2 Myr old. But, modeling has become very difficult recently. Five years ago we all believed that tracks with different masses will always run parallel to the temperature axis. But now when one includes rotation, and other complications, it really becomes a mess. So it's very difficult. The best constraint would be to get the log g value from the spectroscopic investigations of those stars: but you cannot see the photospheres. The winds are so dense that you cannot get a clue of what the actual mass is, or any log g effect. Maybe when NIRSPEC comes on line and we can go for the O and B stars, we will be able to get log g out of that. Then we will be able to get masses and put more constraints on the other stellar parameters. Give us one year.

Sellgren: If you followup the spectroscopic binary, and get the radial velocity curve you can get masses.

Najarro: Well, you need to take detailed spectra to know how the masses are distributed.

Sellgren: Don't you just have to know how the radial velocity varies, and that gives you the mass of the individual components and everything else just falls out from there?

Wardle: One of the things that resists our fit is magnetic fields. It actually makes it quite hard to start with clouds that are supported by magnetic fields and get them to collapse or partially collapse through compression.

It's very difficult when magnetic fields are around, which we know they are.

Stolovy: We have to consider shocks and there is considerable evidence that we see high excitation silicon and iron in the central parsecs which is usually evidence that there have been shocks. It's sort of circumstantial evidence ...

G. Rieke: The shocks which we are seeing now are not shocks in the gases but stellar shocks. These are a different kind of shock waves.

Stolovy: Yes.

Narayan: Have people looked for young stellar objects in the Galactic Center area? Because one of the signatures of these objects would be bipolar outflows and although the region is very obscured, they might show up as bipolar outflows. I don't know whether the right observations exist.

Sellgren: What about planetary nebulae? Did anyone find a planetary nebula in their observations? I'm still wondering if we're ever going to find any low-mass stars.

Lazio: I know that there could certainly be searches in the radio for planetary nebulae in the Galactic Center. I don't know if anybody has looked with the VLA. Have they?

Goss: Yes, one group has been at the VLA. But their search was over a much larger region. They were looking over a region with clouds, perhaps 5 degrees on the sky.

Lazio: I'm looking at the inner degree for pulsars, I suspect my data set would be useful for planetary nebulae, if you're interested in talking about it later.

Unknown: Did you find any pulsars?

Lazio: Uh, wait till tomorrow.

Najarro: Because with so many massive stars, you could see some.

Lazio: We really expect to find some neutron stars.

Najarro: X-ray binaries should be detectable with AXAF.

M. Rieke: If we go away from the very central $5''$ or so to the outer parts of our HST images on the central parsecs, we move down to about 22 magnitudes at H and 20 at K. We will not be able to get to low mass main sequence stars, but that is plenty deep enough to find K giants, which are less than 20 $M_\odot$ stars and into a very interesting region. I'm encouraged to press my data to the limits and to get there. I think the prospects for seeing stars on the main sequence in the Galactic Center are rather exciting.

Yusef-Zadeh: There's one question I have, that relates to the consequences of stellar winds from these cluster zones. You would expect to see stellar bubbles. If you have 1000 km s^{-1} winds, they should definitely shock to very high temperature gas, and you should be able to see some X-ray emission. For instance, this cluster near the Arches, if we could maybe take a look at the IRAS images, and I'd like to show one image here ...

Sellgren: Sure.

Yusef-Zadeh: ... because it may be interesting for the discussion. This is a 25μm image from IRAS. You can see the Galactic Center is somewhere here, and the Arches are here, and the non-thermal filaments which are here. If you look at this, it's somewhat difficult to see, but there is a very nice shell, a 25μm- shell, close to where the Arches Cluster is located. Now the contours here show the X-ray emission. This is an ASCA picture of the Galactic Center. I think it's interesting to consider that perhaps this X-ray gas is pushing the bubble, perhaps created by the cluster. In fact the distribution of the molecular cloud is very different here. You see very dense molecular cloud gas, on this side; I think the gas is pretty confined, because you have stellar winds from the cluster. Whereas on this side, there is not much dense gas, so I think perhaps this X-ray gas and the shell in the 25μm emission map are associated with this cluster, rather than with the Galactic Center.

Wardle: Do you know, what the density and temperature of the X-ray gas are?

Yusef-Zadeh: Yeah, that's the only problem. The temperature of the X-ray gas is about 10^8 K. This is an ASCA image.

Figer: To illustrate Farhad's point about the bubble, I'm putting up the MSX images. Can you see the bubble in the MSX image? Yeah, so that's in longitude left-right and latitude up-down, and you can see the Pistol and the Sickle.

Yusef-Zadeh: This is the Arches. The molecular cloud is basically distributed on this side, and the Galactic Center is here. What I was showing, in the 25μm IRAS image, is actually seen a lot better in the MSX picture. This bubble, and the X-ray emission, seem to be concentrated right at the center of this bubble. It's interesting. I think the ionized gas may be associated with the Arches Cluster, and on one side you don't have a lot of dense gas. It's quite possible that the gas is actually moving the slowest, in the direction where the density gradient is least. I think the problem is how do you get 10^8 K gas from a stellar cluster?

Sellgren: One thing we have to remember about OB associations, which is effectively what these three clusters are, is that you get some of the energy going into stellar winds, you also get stellar supernovae, and both of those are going to make big round things.

Figer: I want to follow up on that. In Paris [ISO conference, October 1998] we're presenting work on the bubble using ISO to look at the actual emission and excitation conditions, and I think we have some data on that region. I don't know much about that, Mark [Morris] is in charge of that.

Maeda: I'd like to comment on the X-ray observations. The soft X-ray emission is not that symmetric, but the hard X-ray emission is nearly symmetric. It is really unclear as whether this is intrinscially X-ray bright or not. It's a problem.

Unknown: Would you repeat that?

Maeda: Okay. The X-ray bright peak is seen in the ASCA image here [soft X-ray enhancement near the radio Arc]. And soft X-ray bright means it's not strongly absorbed. So, one possibility is there is no molecular cloud in front of the radio emission [because the soft X-ray emission correlates well with the radio].

Yusef-Zadeh: Well, I think we know that there is plenty of molecular gas to the sides where the Arches are, so if there is anything there, it's probably obscured a lot more.

Sellgren: Any comments?

Sakano: There is a compact dense molecular cloud at the X-ray peak.

Yusef-Zadeh: I think that's where we see the 6.4 keV line. This is a fluorescent iron line that we see in X-rays, and coincides with the peak of a molecular cloud. It's not a thermal X-ray gas, it's a reflection of X-rays from a dense molecular cloud.

Gezari: I was just going to make a comment about the MIR observations that I mentioned earlier. The luminosity derived from MIR observations is really very high, in the neighborhood of 10^4–10^5 $L_\odot$/arcsec^2. My feeling is that in fact, there may very well be multiple sources at these prominent infrared peaks. For example, the whole BNKL Orion infrared complex would be a 1$''$ diameter source at the distance of the Galactic Center, so it would be barely resolved by my observations, and possibly resolved by the Keck observations.

Cotera: Several of the classic MIR sources in the Galactic Center, such as IRS 9, can be resolved into several compact sources in the Keck images.

Gezari: Okay. So that's a very important thing to look for. There may be a mixture of types of objects other than single stars when we're thinking about the IRS sources such as 1, 10, 5, etc. We are used to thinking of these sources as being single stars of single types that are evolving, when it could be a mixture of stars. The same may be true in the tight clusters which are just a few tenths of an arcsecond apart. In Orion there's a real menagerie of objects, from O stars, which are unseen because they're so heavily extincted, to very young, forming objects. Many of these objects have individual luminosities of $10^3 L_\odot$, but not more than $10^3 L_\odot$. However, there's just a whole bunch of stuff in this type of cluster, and so there could be these much smaller clusters with separations of a few tenths of an arcsecond instead of several arcseconds, like the IRS 16 sources. Again it will be extremely interesting to see what the Keck data shows.

Sellgren: One thing that occurs to me, if there are little clusters—if there are objects like BNKL put at the distance of the Galactic Center—you should see signs characteristic of protostars. The bi-polar outflow has already been mentioned. A lot of times, even though they're incredibly embedded, you still see either the lines from a disk around the star, or sometimes you see photospheric absorption, and I'm wondering whether or not anyone's looked for, or ruled that out, for some of these compact MIR sources in the Galactic Center.

Gezari: I'm not aware of spectroscopy that shows that. Another part of the problem is there's either a lot of extended emission or a lot of unresolved or confused extended emission, and there's an enormous ionizing radiation field, and so there's a significant contribution to heating from all those other stars that we know about—all those other populations of stars. So, whatever embedded stars there are, are competing with very, very powerful additional processes that are going on. This is all related to the ionized gas and so it's a very complicated situation. I'd just like to be sure we keep in mind, that a source like the Trapezium in Orion would just be a couple of tenths of an arcsecond in size at the Galactic Center.

M. Rieke: About the comment on the stellar spectra. I've taken 2 μm spectra of sources like IRS 1 and those guys, none of them have ever showed the 2.3-micron CO emission like BN does. So, there's no evidence that way for a thick disk. The spectra are featureless and of course that really cannot tell you much. There's no smoking gun on any of those for something that could in anyway be identified as a protostellar disk.

Biermann: One more comment on X-rays. This is unpublished and therefore an advertisement for the Japanese ASCA people. The thermal X-ray emission, the 10 keV emission is very high in the Galaxy, and is extended around the Galactic disk. This is also found in other galaxies. So we see in the Galactic Center, I think, the deep thermal gas and those conditions which would presuppose a continuum radiation field. We probably have to think of shocks and supernovae and other such events to produce such a thermal gas which is quite common in galaxies, and there is no published explanation.

Mezger: In response to Dan Gezari's question about the ionizing radiation: I mentioned it already in my paper; based on observations— relatively old observations which could be improved or should be improved—about the heat flux density in the central 200 pc, which accounts for about 10% of the total free-free emission from our own Galactic disk. This implies a similar star formation rate for very massive stars. About 10% of the massive stars have been formed in the past several million years in the central 200 pc. On the other hand, and it's also an old story, people have looked for signs of very recent O star formation, like very compact HII regions, maser sources, and they are all not really there. It looks like star formation is not going on continuously, it goes in waves and at the moment we are over the crest of the wave. Star formation is already dying.

Roberts: I had made a point I'd like to reiterate about the difference between the stellar clusters seen in the Galactic Center versus average star formation you see in other places in the Galaxy. In the normal star forming regions, magnetic fields start with an ambient value of a few μG and increase to the core density of maybe a mG or two. It is a very slow process of ambipolar diffusion, allowing enough mass to concentrate to actually to start the process of star formation. Whereas here, if we have an ambient value which is near the value of the collapsed core to begin with ... you think about the environment of a cloud that's flying through at several hundred km s^{-1}, not the sort of quiescent environment you'd imagine the core of a dark cloud to be. Something else is going on to inject large densities to go over the critical mass density to allow stars to form.

Falcke: If we are over the crest of the wave, and star formation is dying down, was it the crest of the wave that has produced all three of the star clusters we've heard so much about? I mean, it's not only the center of the Galaxy but also the Arches Cluster and the Quintuplet Cluster. So than that crest, the peak of star formation, has to affect the entire region. What has actually produced that crest? What was the event that happened? Was it just a cloud coming in? Does it stop at various places in the Galactic Center or was is simply huge?

Mezger: That's something that is up to you to explain, what is the case. If you take the Lyman continuum photon production rate and you try to express it in terms of a star formation rate, even if you use a bimodal star formation rate, it means you have to cut off the IMF at about 3 $M_\odot$. Then you have to have a consumption rate of interstellar material going into forming stars which is about 0.4 $M_\odot$/yr. This is much, much more than you can account for by influx of interstellar material through the Galactic Plane. So that's already a sign that you have to have something like an oscillating star formation, because no way can you maintain such a high star formation rate just by infalling material. What people don't often realize is that all this is a situation which depends on the nuclear bulge, the inner few hundred parsecs. The physical state in the inner few hundred parsecs is so different from, for example, what's happening in the Galactic disk. You don't have the atomic intercloud gas here, only the molecular intercloud gas, because the density is so high that even in the intercloud space the atomic hydrogen is turned into molecular hydrogen. There are all kinds of differences. The radiation field is up by a factor of 100. We have to understand the physical state and the physics in the nuclear bulge in order to understand star formation in this region.

Sellgren: Okay, we're at the end of the discussion period. Thanks everyone.

Session VI

What is the Metallicity as Determined from Old and New Stars?

The Central Parsecs of the Galaxy
ASP Conference Series, Vol. 186, 1999
H. Falcke, A. Cotera, W.J. Duschl, F. Melia, M.J. Rieke, eds.

Stellar Abundances at the Galactic Center

S. V. Ramírez[1], K. Sellgren, D. M. Terndrup
The Ohio State University, Columbus OH 43210, USA

J. S. Carr
Naval Research Laboratory, Washington, DC 20375, USA

S. Balachandran
University of Maryland, College Park, MD 20742, USA

R. D. Blum
CTIO, La Serena, Chile

Abstract. We are measuring Fe abundances of cool, luminous stars within 30 pc of the center of the Milky Way. Our sample contains both AGB stars and M supergiants. Low-resolution ($\lambda/\Delta\lambda = 500$) H and K band spectra are used to estimate temperatures and gravities. Stellar Fe abundances are determined from high-resolution ($\lambda/\Delta\lambda$ = 40,000) K band spectra obtained on the IRTF using CSHELL. We find that Fe abundances of stars in the Galactic Center are consistent with the solar Fe abundance.

1. Introduction

We have measured iron abundances of cool luminous stars located within 30 pc of Galactic Center (GC), using $K-$band (2.2 μm) high-resolution spectra. Infrared techniques are required because of the large optical extinction towards the GC ($A_V \sim 30$ mag). Our sample includes both AGB stars and M supergiant stars.

Chemical abundances in the disks of spiral galaxies are observed to reach their highest values at the center (see Pagel & Edmunds 1981 and Shields 1990). H II regions in the Milky Way also show a radial metallicity gradient (Peimbert *et al.* 1978; Hawley 1978; Shaver *et al.* 1983; Pipher *et al.* 1984; Simpson & Rubin 1990; Simpson *et al.* 1995; Afflerbach *et al.* 1997). Chemical evolution models strive to explain these gradients by considering the relative star formation and gas infall/outflow rates, the star formation history, and the metal abundance of the gas compared to the stars (see Audouze & Tinsley 1976 and Rana 1991). The

[1]US Gemini Fellow.

gas metal abundance of the GC varies between 1 and 5 times solar depending on the element (Aitken *et al.* 1974, 1976; Willner *et al.* 1979; Lacy *et al.* 1980; Lester *et al.* 1981; Lester *et al.* 1987; Rubin *et al.* 1988; Shields & Ferland 1994; Simpson *et al.* 1995; Afflerbach *et al.* 1997), but Fe has not been measured in the gas phase. Our measurements of [Fe/H] are the first stellar abundance determinations in the GC.

2. Observations

We have observed cool luminous stars in the GC using the CSHELL spectrograph at the NASA Infrared Telescope Facility, a 3–m telescope located at Mauna Kea Observatory in Hawaii. The K–band spectra taken with this instrument have a spectral resolution of $\lambda/\Delta\lambda$ = 40,000 or 7 km s^{-1}. We have chosen iron lines which are free of molecular contamination, especially CN in M supergiants and water in M giant and AGB stars.

We have observed all cool stars that have low water absorption and that are brighter than $K = 9$, which is the limit set by the sensitivity of CSHELL. Our sample consists of seven M supergiants and three AGB stars. Nine of these stars, including the three AGB stars, are located in the central cluster within 2.5 pc of the GC. The remaining M supergiant belongs to the Quintuplet cluster, located 30 pc away from the central cluster.

3. Abundance Analysis

For our abundance analysis we are using model atmospheres for late type giants and supergiants from Kurucz (1993) and Plez (1992), with the spectral synthesis program MOOG (Sneden 1973). We have found a good agreement between abundances derived from both sets of models in the range of stellar effective temperature (T_{eff}) and surface gravity (log g) where they overlap. Our abundance analysis requires as input parameters: T_{eff}, log g, the microturbulent velocity of the star (ξ), and the oscillator strengths of the lines we are studying. We have verified our analysis technique for the GC stars by observing and analyzing cool luminous stars of known abundances (Smith & Lambert 1985, 1986, 1990; Luck & Bond 1989; McWilliam 1990).

Effective Temperature: For all our GC stars, we use the CO and H_2O absorption observed in the H and K bands at low resolution ($\lambda/\Delta\lambda$ = 500) to get T_{eff}. The CO strength increases with decreasing T_{eff} and decreasing log g. The H_2O strength has the same T_{eff} dependence but opposite surface gravity dependence. These two features together provide 2-dimensional spectral classification (Kleinmann & Hall 1986). Once we establish the luminosity class of each star, we use the CO strength to get T_{eff}. In addition, we can use lines of CO present in H–band (1.6 μm) high resolution spectra to get T_{eff}. Lines with different lower excitation potentials give the same carbon abundance when T_{eff} is correct. We have observed these CO lines only in IRS 7, which is the brightest star in the GC. Both methods give the same value of T_{eff} for IRS 7.

Surface Gravity: We derive the bolometric magnitude (M_{bol}) of the GC stars using the observed K magnitude, extinction, distance, and T_{eff}. We locate the GC stars in the HR diagram, using T_{eff} and M_{bol}. We estimate their masses by comparing their location in the HR diagram with evolutionary tracks for stars of different masses (Schaller *et al.* 1992; Schaerer *et al.* 1993*a*, *b*; Charbonnel *et al.* 1996). We compute the surface gravity using the estimated mass, luminosity and T_{eff}.

Microturbulent Velocity: For all our GC stars, we estimate ξ from log g (McWilliam *et al.* 1995). In addition, we can use lines of CO present in the H−band to get ξ spectroscopically. Lines of different equivalent widths give the same carbon abundance when ξ is correct. We have observed these CO lines only in IRS 7. These two techniques agree for IRS 7.

Oscillator strengths: We have derived empirical oscillator strengths for our abundance analysis. We use the spectrum of α Boo (Wallace & Hinkle 1996), the abundance measurements of Peterson *et al.* (1993) for α Boo, and the synthetic spectrum generated by MOOG to derive gf-values.

We have studied how errors in T_{eff} and log g would affect our Fe abundance determination. The Fe abundance is quite insensitive to errors in log g, since a typical error of ± 0.2 in log g corresponds to an error in [Fe/H] of ± 0.05 dex. The Fe abundance is more sensitive to errors in T_{eff}, since a typical error of ± 200 K represents an error in [Fe/H] of ± 0.1 dex.

4. Results

In the Table we present results for our sample of 10 stars within 30 pc of the GC. Note that IRS 7 is the brightest star in the GC, and VR 5-7 is the star located at the Quintuplet cluster.

A statistical analysis gives us an unweighted mean of [Fe/H] = +0.03 and a mean weighted by the errors of [Fe/H] = +0.13 . The estimate of the dispersion (σ) is 0.20 dex. This is not significantly different from the error in [Fe/H] for each star, given in the Table.

Star	# lines	[Fe/H]
IRS 7	4	+0.08 ± 0.18
IRS 11	3	+0.15 ± 0.20
IRS 19	3	+0.30 ± 0.09
IRS 22	4	+0.14 ± 0.23
RCS 72	2	−0.16 ± 0.34
RCS 114	2	−0.35 ± 0.27
RCS 124	2	+0.19 ± 0.09
RCS 129	2	+0.05 ± 0.23
RCS 140	2	+0.09 ± 0.43
VR 5-7	2	−0.18 ± 0.13

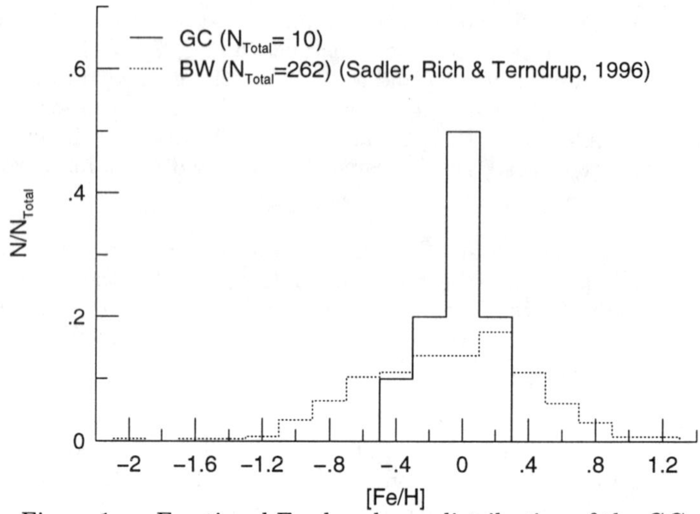

Figure 1. Fractional Fe abundance distribution of the GC compared with that of Baade's Window ($l, b = 1°, -4°$) in the Galactic bulge.

5. Discussion

In Figure 1 we compare our Fe abundance distribution in the GC with the Fe abundance distribution of Baade's window in the bulge (Sadler *et al.* 1996). Sadler *et al.* (1996) resolve the abundance spread of the bulge distribution. We have not detected a GC [Fe/H] distribution that is wider than the errors, in spite of having similar errors per star as Sadler *et al.* (1996). Thus, the dispersion of the GC distribution is significantly narrower than that of the bulge [Fe/H] distribution.

In Figure 2 we compare our GC Fe abundance distribution with the Fe abundance distribution of supergiants in the solar neighborhood (Luck & Bond 1989). We see that the GC abundance distribution is consistent with the abundance distribution of supergiants in the solar neighborhood.

We find that the average GC stellar Fe abundance is near solar. This result differs from the gas abundances in the GC, which vary between 1 and 5 times solar, depending on the element. In the future, we will measure stellar abundances of α–elements (Ca, Si, Ti) to address the issue of selective enrichment in the GC.

6. Conclusions

- Galactic Center Fe abundances are consistent with the solar value.
- Galactic Center Fe abundance distribution is consistent with the Fe abundance distribution of supergiants in the solar neighborhood.
- VR 5-7 (located 30 pc away from the center of the Galaxy) is not different in [Fe/H] from stars located in the central cluster.

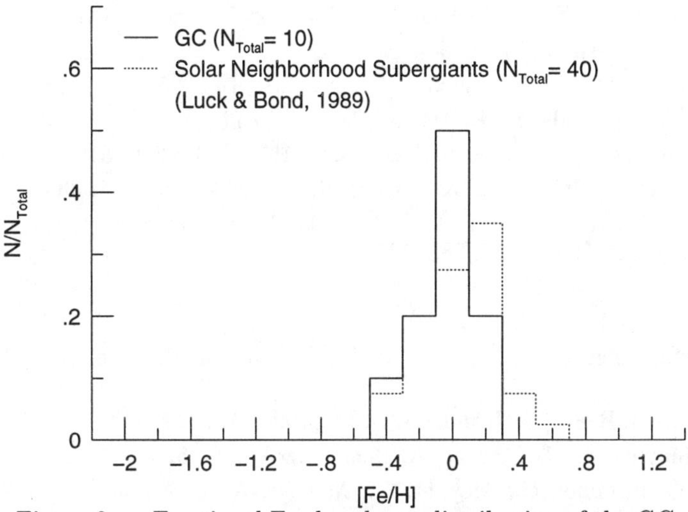

Figure 2. Fractional Fe abundance distribution of the GC compared with that of supergiants in the solar neighborhood.

Acknowledgments. Support for this work was provided by the National Science Foundation through the grant # GF-1003-97 from the Association of Universities for Research in Astronomy, Inc., under NSF cooperative agreement AST-8947990 and from Fundación Andes under project C-12984, and from NSF grant AST-9619230.

References

Afflerbach, A., Churchwell, E., & Werner, M. W. 1997, ApJ, 478, 190

Aitken, D. K., Jones, B., & Penman, J. M. 1974, MNRAS, 169, 35P

Aitken, D. K., Griffiths, J., Jones, B., & Penman, J. M. 1976, MNRAS, 174, 41P

Audouze, J. & Tinsley, B. M. 1976, ARA&A, 14, 43

Charbonnel, C., Meynet, G., Maeder, A., & Schaerer, D., 1996, A&AS, 115, 339

Hawley, S. A. 1978, ApJ, 224, 417

Kleinmann, S. G. & Hall, D. N. B. 1986, ApJS, 62, 501

Kurucz, R. L., Kurucz CD-ROM No. 13 "ATLAS9 Stellar Atmosphere Programs and the 2 km/s Grid",[Cambridge:Smithsonian Astrophysical Observatory]

Lacy, H. J., Townes, C. H., Geballe, T. R., & Hollenbach, D. J. 1980, ApJ, 241, 132

Lester, D. F., Bregman, J. D., Witteborn, F. C., Rank, D. M., & Dinerstein, H. L. 1981, ApJ, 248, 524

Lester, D. F., Dinerstein, H. L., Werner, M. W., Watson, D. M., Genzel, R., & Storey, J. W. V. 1987, ApJ, 320, 573

Luck, R. E. & Bond, H. E., 1989, ApJS, 71, 559

McWilliam, A. 1990, ApJS, 74, 1075

McWilliam, A., Preston, G. W., Sneden, C., & Searle, L. 1995, AJ, 109, 2757

Pagel, B. E. J. & Edmunds, M. G. 1981, ARA&A, 19 77

Peimbert, M., Torres-Peimbert, S., & Rayo, J. F. 1978, ApJ, 220, 516

Peterson, R. C., Dalle Ore, C. M., & Kurucz, R. L., 1993, ApJ, 404, 333

Pipher, J. L., Helfer, H. L., Herter, T., Briotta, D. A., Houck, J. R., Willner, S. P., & Jones, B. 1984, ApJ, 285, 174

Plez, B. 1992, A&AS, 94, 527

Rana, N. C. 1991, ARA&A, 29, 129

Rubin, R. H., Simpson, J. P., Erickson, E. F., & Haas, M. R. 1988, ApJ, 327, 377

Sadler, E. M., Rich, R. M., & Terndrup, D. M., 1996, AJ, 112, 171

Schaerer, D., Meynet, G., & Maeder, A. 1993*a*, A&AS, 98, 523

Schaerer, D., Charbonnel, C., Meynet, G., Maeder, A. & Schaller, G. 1993*b*, A&AS, 102, 339

Schaller, G., Schaerer, D., Meynet, G., & Maeder, A. 1992, A&AS, 96, 269

Shaver, P. A., McGee, R. X., Newton, L. M., Danks, A. C., & Pottash, S. R. 1983, MNRAS, 204, 53

Shields, G. A., 1990, ARA&A, 28 525

Shields, J. C. & Ferland, G. J. 1994, ApJ, 430, 236

Simpson, J. P. & Rubin, R. H. 1990, ApJ, 354, 165

Simpson, J. P., Colgan, S. W. J., Rubin, R. H., Erickson, E. F., & Haas, M. R., 1995, ApJ, 444, 721

Smith, V. V. & Lambert, D. L. 1985, ApJ, 294, 326

Smith, V. V. & Lambert, D. L. 1986, ApJ, 311, 843

Smith, V. V. & Lambert, D. L. 1990, ApJS, 72, 387

Sneden, C. 1973, Ph.D. thesis, University of Texas

Wallace, L. & Hinkle, K. 1996, ApJS, 107, 312

Willner, S. P., Russell, R. W., Puetter, R. C., Soifer, B. T., & Harvey, P. M. 1979, ApJ, 229, L65

The Central Parsecs of the Galaxy
ASP Conference Series, Vol. 186, 1999
H. Falcke, A. Cotera, W.J. Duschl, F. Melia, M.J. Rieke, eds.

OH/IR Stars as Signposts for Ancient Starburst Activity in the Galactic Center

L.O. Sjouwerman

Joint Institute for VLBI in Europe, P.O. Box 2, 7990 AA Dwingeloo, The Netherlands (sjouwerm@jive.nfra.nl)

H.J. Habing

Sterrewacht Leiden

M. Lindqvist

Onsala Space Observatory

H.J. van Langevelde

Joint Institute for VLBI in Europe

A. Winnberg

Onsala Space Observatory

Abstract. A recent survey has considerably increased the number of OH/IR stars known within 50 pc of the Galactic center (GC). The distribution of expansion velocities of the circumstellar shells of these stars shows an unexpectedly narrow peak around 19 km/s at the high end tail. From this, we conclude that a large fraction of stars was formed in an epoch of enhanced star formation at least one Gyr ago. Thus one can distinguish three different groups of stars in the direction of the GC: an old bulge-like population, a group of relatively young stars formed from continuously infalling gas, and a shrinking population of low-mass stars that formed in an isolated era—probably in a starburst.

1. Introduction

Although the Galactic center region (GC) is totally obscured in the optical, individual stars can be observed at infrared wavelengths (IR). Because the GC is located at about 8 kpc (Reid 1993), stars that are studied in general are the very IR-luminous massive objects; mostly red and blue supergiants. These stars have characteristic life-times of less than a few tens of million years. Conclusions derived from these studies, e.g. on stellar abundances and star formation history in the GC, thus only reflect the most recent developments in the GC region, i.e. within the last ≈ 100 Myr.

Stars of low and intermediate mass, with ages that are much larger than 100 Myr (up to $\sim$ 15 Gyr), should contain information on much longer time scales. However, not only are these stars intrinsically weak in the IR, the field in the GC is crowded with such objects. An exception to both constraints forms the sample of OH/IR stars. After passing the main-sequence and red giant branch stages, an average star ($\sim$ 1-7 $M_\odot$) may become an OH/IR star just before it transforms into a planetary nebula (Iben 1991). It is then in an evolutionary very short-lived stage ($\sim 10^5$ yr) of high luminosity ($L \gtrsim 5\,000\ L_\odot$) and heavy mass-loss ($10^{-6}$-$10^{-4}$ $M_\odot$/yr) on the asymptotic giant branch (AGB). The matter lost by the central star resides in an expanding circumstellar envelope that re-radiates the absorbed variable ($200 < P < 2000$ days) stellar emission in the IR. This IR radiation is easily observed. Moreover, many of the envelopes around AGB stars form maser emission - in particular the characteristic 1612 MHz OH maser line, which is bright and easily recognizable (e.g. Habing 1996).

We have followed up the pioneering 1612 MHz surveys in the GC, done by Habing et al. (1983), Lindqvist et al. (1992a) and van Langevelde et al. (1992), with a sensitive search for OH/IR stars within 50 pc of the GC (Sjouwerman et al. 1998a). During the analysis of the new sample of OH/IR stars, we were struck by the distribution of expansion velocities ($v_{\rm exp}$) of the circumstellar envelopes. It is unlike any other distribution of samples elsewhere in the Galaxy (Sect. 2, Fig. 1), and has led us to re-analyze the path of Lindqvist et al. (1992b; Sect. 3). We conclude that we have found the remnants of an epoch of enhanced star formation, a starburst, more than one Gyr ago (Sect. 4). Early reports on this subject appeared in Sjouwerman (1997) and Sjouwerman et al. (1998b).

2. OH/IR Stars in the Galaxy and the GC

Since the discovery of OH/IR stars in the late 1960's, many surveys have been carried out in all parts of the Galaxy and the Large Magellanic Cloud (LMC). In particular, surveys after 1983 exploited the highly successful predictive property of the IRAS two-color diagram (e.g., Olnon et al. 1984, Eder et al. 1988, te Lintel Hekkert et al. 1991, Blommaert et al. 1993). In regions where IRAS suffered from confusion (in the Galactic plane and bulge), the IRAS survey has been complemented by "blind" 1612 MHz surveys (e.g Lindqvist et al. 1992a, Blommaert et al. 1994, Sevenster et al. 1997). The line-of-sight velocity of each OH/IR star ($v_{\rm los}$) is easily measured as the *average* of the blue and red Doppler shifted 1612 MHz OH emission peaks from the expanding shell. In addition, the *difference* in velocity of the blue and red shifted peak ($2 \times v_{\rm exp}$) is a physical property related to the underlying star. Furthermore, a distance can be determined for some stars by measuring the angular size and the phase-lag between the blue and red shifted peak (e.g. Herman & Habing 1985). It is the combination of these properties that make OH/IR stars excellent probes to study the Galaxy.

In summary, the study of samples of OH/IR stars in the Galaxy have shown a statistical correlation between the $v_{\rm exp}$ and the age of the OH/IR stars (e.g. Baud et al. 1981). The age of the star, the time it takes to reach this very specific OH/IR stage in its evolution, is related to the main-sequence mass that determines the luminosity on the AGB, and its metallicity as reflected in the dust-to-gas ratio δ of the shell. The *average* age of a sample is also reflected statistically by the Galactic distribution and velocity dispersion of the samples: the older, the more spread out in phase-space.

Lindqvist et al. (1992b) have analyzed a sample of $\sim$ 150 OH/IR stars located within 200 pc of the GC (Lindqvist et al. 1992a). Statistically they found a difference in spatial and kinematic distribution between samples of OH/IR stars with low and high expansion velocities. The high expansion velocity objects ($v_{\rm exp} > 18$ km/s) are spatially much more concentrated to the GC (and plane) than the objects with $v_{\rm exp} < 18$ km/s. In addition, the high $v_{\rm exp}$ stars show a rapid rotation around the center, whereas the low $v_{\rm exp}$ group is dynamically very relaxed and scattered in both velocity and sky distribution. This division in groups of stars with high and low expansion velocities in the GC is analogous to other samples of OH/IR stars in the Galaxy (e.g. Baud et al. 1981). Also here, the stars with high $v_{\rm exp}$ are on average more massive and younger than the stars in the low $v_{\rm exp}$ group. Sevenster et al. (1995) drew the same conclusion from a dynamical analysis without explicitly splitting the sample of OH/IR stars into two according to $v_{\rm exp}$. From the tight rotation they even suggest that the high $v_{\rm exp}$ group formed in a distinct event. However, this view is contradicted by Serabyn & Morris (1996) who argue that the (young, massive OH/IR) stars were formed from separate infalling molecular clouds.

Apart from $v_{\rm exp}$, one would expect that other intrinsic physical properties also show statistical differences between both groups in the GC. However, recent attempts to find statistical differences, mainly in $M_{\rm bol}$, between groups with $v_{\rm exp} < 18$ km/s and $v_{\rm exp} > 18$ km/s have not been very convincing (e.g. Blommaert et al. 1998, Wood et al. 1998). The division in expansion velocities smaller and larger than 18 km/s, which has been used in almost every analysis of the sample in the GC, was chosen by Lindqvist et al. (1992b) simply to divide the sample into equally sized groups. In Sect. 3 we will show that this picture changes drastically if one does not require two roughly equal groups.

Qualitatively, the OH/IR star phenomena is generally well understood. The exact driving force of the mass-loss mechanism is still under investigation, but after matter escapes from the star it cools down, and dust and molecules start to form. The dust gets accelerated by the absorbed stellar radiation pressure, as quantified by the bolometric luminosity or magnitude of the star ($L_*, M_{\rm bol}$). The accelerated dust in turn is responsible for transferring momentum to the gas molecules. The particles reach a maximum velocity far away from the star: the expansion velocity of the envelope or shell. It is this (gas) velocity $v_{\rm exp}$ that we measure from the 1612 MHz OH masers at about 1000-10 000 AU from the central star. Clearly, strong radiation pressure results in large acceleration of the dust. Because the dust entrains the gas, the dust-to-gas ratio (δ), also influences the value of $v_{\rm exp}$. Typically one measures $v_{\rm exp} = 15$ (± 10) km/s. It can be shown that (Habing et al. 1994)

$$v_{\rm exp}^2 \propto \delta \cdot \sqrt{L_*} \; . \qquad (1)$$

Samples of OH/IR stars in general consist of a mix of old low mass-stars, and younger, more luminous massive stars. Also, stars of similar mass may have different metallicities due to different amounts of recycled stellar material in the star. The samples of OH/IR stars found in the Galactic bulge and

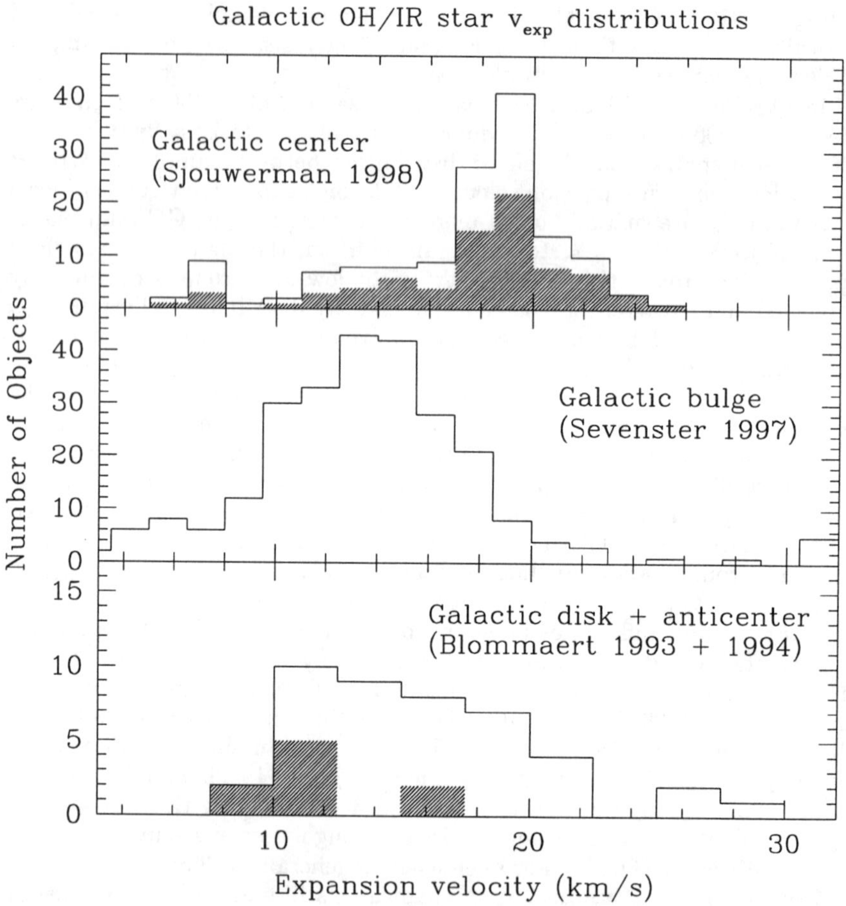

Figure 1. Expansion velocity distribution for OH/IR stars in the Galactic center (top frame, Sjouwerman et al. 1998a), the Galactic bulge (middle, Sevenster et al. 1997) and the combined Galactic disk and anticenter (bottom, Blommaert et al. 1993 and Blommaert et al 1994). In the bottom frame the shaded part shows the anticenter data, the histogram is for the combined sets. The shaded histogram in the top panel depicts the fraction of the stars in the bin for which a bolometric magnitude M_{bol} has been measured (Jones et al. 1994, Blommaert et al. 1998, Wood et al. 1998). For example, of the 41 stars in the bin with v_{exp} = 19 km/s, 22 stars have a measured M_{bol}. Note that this shaded histogram itself does not show any reference to the value of M_{bol} of the star, just that it is available. Because the shape of the shaded histogram in the top frame follows the histogram itself at about a 50% level, we may assume that the measured values and range in M_{bol} in each bin *roughly* represent the values and range in M_{bol} for all the stars with v_{exp} values of that bin.

disk+anticenter therefore show a wide distribution in v_{exp}. The tail toward higher v_{exp} in the disk distribution can be understood in terms of more young, massive, and relatively metal-rich OH/IR stars compared to the bulge. Furthermore, for equally bright anticenter stars (shaded in the bottom frame of Fig. 1), the low metallicity causes the average v_{exp} to be significantly lower than for disk and bulge stars (e.g. Blommaert et al. 1993, Blommaert et al. 1994, and Wood et al. 1992 for LMC stars).

However, the distribution of v_{exp} of the sample of $\sim$ 140 OH/IR stars within 50 pc of the GC (Sjouwerman et al. 1998a) shows an unexpectedly narrow and high peak around the relatively high value of $v_{exp} = 19$ km/s. We are not aware of any physical phenomenon, such as an interstellar pressure or density field, that can transform a distribution similar to that of the bulge into that of the GC without decreasing all values for v_{exp}. There must be relatively many stars with $v_{exp} \approx 19$ km/s. Because the OH/IR stars in the GC are not more luminous than bulge stars, on average the stars in the GC have high (but unknown) metallicity (Eq. 1, Habing et al. 1994). But more importantly, with a similar range in M_{bol} (or L_*) as the bulge, the OH/IR stars in the GC have a very narrow distribution on top of a broad and roughly constant distribution. We identify the broad and constant distribution with the general mix of stars as also found in the disk. We observe that there is an over-abundance, or excess of stars with $v_{exp} \approx 19$ km/s. We will refer to these stars as "the excess stars". Assuming Eq. 1 also holds for the GC, a narrow peak in v_{exp} for the excess stars can only exist if for a large fraction of the excess stars that form the peak, *either the effects of metallicity and luminosity cancel out, or the range in both δ and L_* is small.*

3. Orbital Angular Momentum and Physical Properties

One can almost immediately reject the explanation of cancelling effects of δ and L_*. It is true that δ and L_* are slightly anti-correlated (Vassiliadis & Wood 1993), but when one wants to explain the narrow peak by an overlap of the young, massive stars with the old, low mass stars, one has to assume that the low mass stars are considerably more metal-rich than the high mass stars (Eq. 1). This is in contradiction to what is generally found. Furthermore, the effect of cancelling δ and L_* should be *exactly the same* for all excess stars - otherwise the v_{exp} distribution would be broader due to the strong influence of δ on v_{exp}.

Both Lindqvist et al. (1992b) and Sevenster et al. (1995) found that a major difference between the two groups is the rotation around the GC: the rotation of the high v_{exp} group is much more tight than for the group with $v_{exp} < 18$ km/s. Actually, the slight rotation found in the low v_{exp} group is almost entirely due to including stars with $16.5 < v_{exp} < 18$ km/s. It is unfortunate that Lindqvist et al. (1992b) chose $v_{exp} = 18$ km/s to split the sample, because this is exactly where the excess stars also split into both the low and high v_{exp} groups. Thus, the "non-rotating group" has been polluted by the rotation of excess stars.

The increase in rotation around the GC with v_{exp} can be investigated by the contribution of each star to the cumulative sum of the total angular momentum of the sample. We show a measure of the angular momentum along the rotation axis of the Galaxy as function of v_{exp} in Fig. 2. It is the normalized sum over all stars, ranked according to increasing v_{exp}, of $l \times v_{los}$ (with l the Galactic longi-

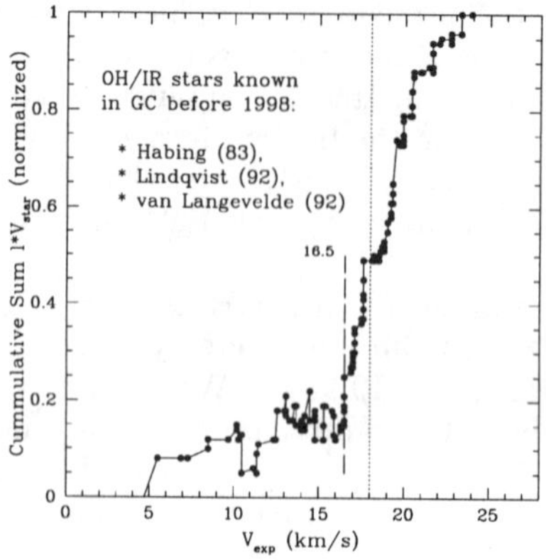

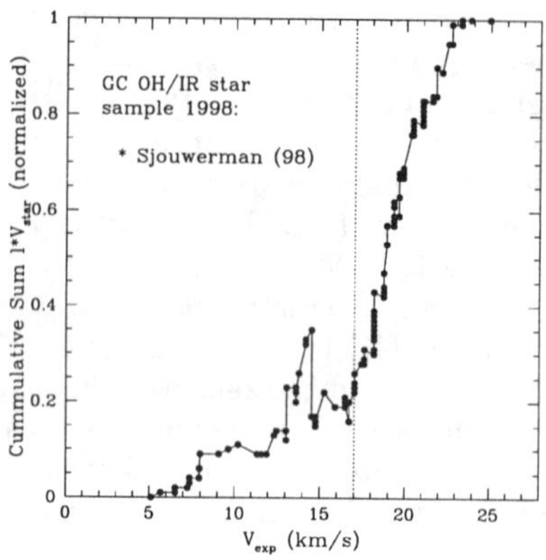

Figure 2. Normalized cumulative sum of $l \times v_{\rm los}$ as function of $v_{\rm exp}$. Left the OH/IR stars within 200 pc from the GC known from Habing et al. (1983), Lindqvist et al. (1992a) and van Langevelde et al. (1992). For most of these stars bolometric magnitudes and periods have been determined. Right the new sample of stars within 50 pc of the GC (Sjouwerman et al. 1998a). By dividing in two groups at $v_{\rm exp} = 18$ km/s, Lindqvist et al. (1992b) included part of the rotating high $v_{\rm exp}$ group into the non-rotating low $v_{\rm exp}$ group. A division at $v_{\rm exp} \approx 16.5$ km/s would have been more proper. In both figures, there is a clear transition at $v_{\rm exp} \approx 17$ km/s. Deviations from the general trend at 11 and 14 km/s are due to high velocity stars ($v_{\rm los} \ll -200$ km/s).

tude). For low $v_{\rm exp}$, the stars are dispersed over the $(l, v_{\rm los})$-diagram; individual contributions will be positive or negative (prograde or retrograde rotation) and no net rotation is expected. On the other hand, in the case of (prograde) rotation there is a strong correlation between l and $v_{\rm los}$. For the high $v_{\rm exp}$ one would expect the cumulative sum to accumulate.

It is clear from Fig. 2 that there is a sharp transition at $v_{\rm exp} \approx 17$ km/s. Below that value, the cumulative sum roughly hovers around a constant value indicating small or no net rotation. For comparison, for the IRAS based sample of te Lintel Hekkert et al. (1991), this value is $v_{\rm exp} = 13.5$ km/s. From both Fig. 1 and Fig. 2, we conclude that the low $v_{\rm exp}$ group should be characterized as stars with $v_{\rm exp} < 17$ km/s. Taking into account the steep drop-off at $v_{\rm exp} = 20$ km/s (cf. the rise at $v_{\rm exp} = 17$ km/s) in Fig. 1, we define the high $v_{\rm exp}$ group as stars with $v_{\rm exp} > 20$ km/s. The stars with intermediate $v_{\rm exp}$, the group with $17 < v_{\rm exp} < 20$ km/s, should be examined separately. From Fig. 2, a conclusion on kinematics for all three groups can be drawn immediately. Both the high and intermediate $v_{\rm exp}$ groups rotate rapidly, whereas the rotation of the sample of low $v_{\rm exp}$ stars is insignificant. With a slightly different definition of the high and low $v_{\rm exp}$ groups, we thus confirm the results of Lindqvist et al. (1992b) and Sevenster et al. (1995) on the dynamics of both groups.

We show histograms of several physical properties of the OH/IR stars in the GC in Fig. 3 (Blommaert et al., 1998, plus Wood et al., 1998, and Sjouwerman

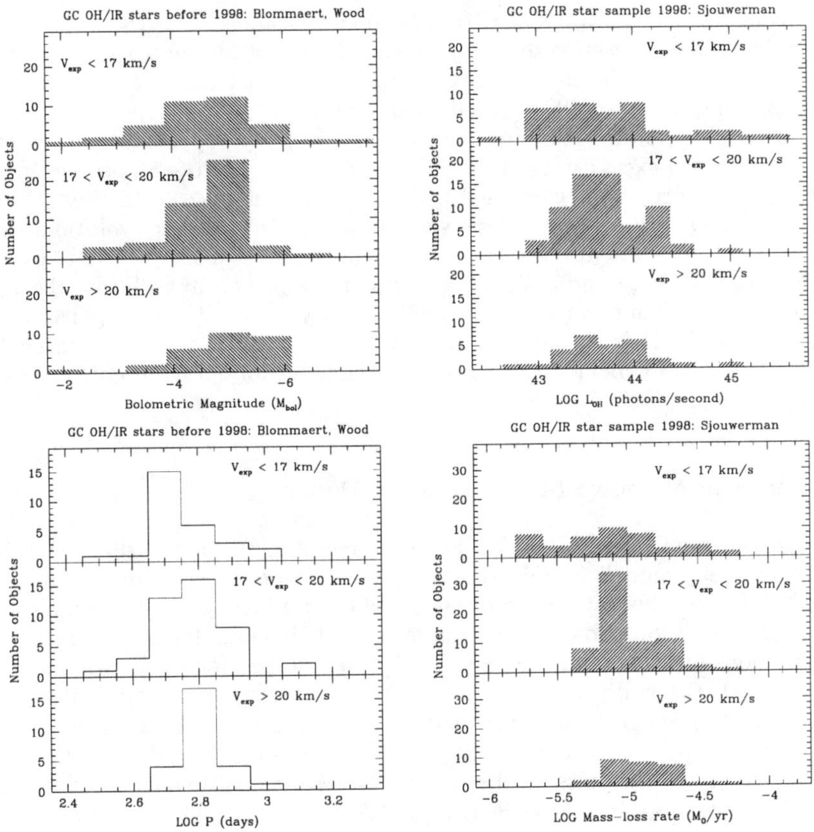

Figure 3. Physical properties of the central star ($M_{\rm bol}, P$) and the expanding OH masing shell ($\dot{\rm M}$, $L_{\rm OH}$) for each $v_{\rm exp}$-group. The latter properties are a function of δ (as well as L_*). See text.

et al. 1998a; see also Jones et al. 1994 and van Langevelde et al. 1993). For $M_{\rm bol}$, $L_{\rm OH}$ and $\dot{\rm M}$, the peaked distributions in the group with $17 < v_{\rm exp} < 20$ km/s *cannot* be explained by the small range of 3 km/s in $v_{\rm exp}$; it is almost as wide as for the high $v_{\rm exp}$ group ($\sim$ 4 km/s). Also, it cannot be explained by a combination of stars with the wide distributions from the low and/or high $v_{\rm exp}$ groups, in particular when the dynamics of both groups are taken into account. That is, the velocity *and* sky distribution of the stars in this intermediate group resembles the velocity *and* sky distributions of the stars in the high $v_{\rm exp}$ group. Thus the intermediate $v_{\rm exp}$ group, which includes *all* the excess stars, may also contain a number of stars from the high $v_{\rm exp}$ group. Because both differ from the low $v_{\rm exp}$ group, it is unlikely that this intermediate $v_{\rm exp}$ group has a large contribution of stars from the low $v_{\rm exp}$ group.

We therefore *observe* that the peaked distributions for the group with $17 < v_{\rm exp} < 20$ km/s is caused by an excess of stars with almost *identical* physical properties, both in central star ($M_{\rm bol}$) as in circumstellar shell ($\dot{\rm M}$, $L_{\rm OH}$). That the period P does not seem to correlate with the $v_{\rm exp}$ group in our view confirms that the Mira P-$M_{\rm bol}$-relation breaks down for OH/IR stars (e.g. Feast et al. 1989, Blommaert et al. 1998, Wood et al. 1998). From Fig. 1, we see that the measured $M_{\rm bol}$ values and range per bin are roughly representative for each $v_{\rm exp}$-bin. For the excess stars we then derive $M_{\rm bol} \approx -4.5$ to -5 (also depending on extinction correction) combined with roughly identical metallicities (Eq. 1). Because the OH/IR stage is a very short lived phase in stellar evolution, these "excess" OH/IR stars all have the same main-sequence mass (M) and age (T), i.e. just as $M_{\rm bol}$ (L_*) and δ also in a very narrow range. In addition, the $17 < v_{\rm exp} < 20$ km/s group is rotating strongly and dynamically young (with a high age, see below). This all indicates that the excess OH/IR stars were formed in a short time. We conclude that this is evidence for an epoch of enhanced star formation in the GC.

4. Implications for a Star Formation Epoch

Because the exact metallicity of the excess stars is not known, and due to a lack of high-metallicity models for OH/IR stars, we will use the models of Vassiliadis & Wood (1993) to illustrate the implications of our findings with some numbers in orders of magnitude. However, keep in mind that these models were constructed for luminous OH/IR stars in the LMC. We derive that for a star with $M_{\rm bol} \approx -5$ (or less), the main-sequence mass is 2 $M_\odot$ (or less), the metallicity of the originating molecular cloud ranges within $0.004 < Z < 0.008$, and the age T is $\gtrsim 1.1$ Gyr (or more). A much higher overall metallicity ($Z \gtrsim 0.03$), and thus a higher main-sequence mass for the excess OH/IR stars, is then inconsistent with the OH/IR star pulsation periods ($P \lesssim 1000^{\rm d}$, van Langevelde et al. 1993; Jones et al. 1994, Wood et al. 1998). This result is obviously one of the shortcomings of applying the models of Vassiliadis & Wood (1993) to the GC OH/IR stars.

Regardless of a precise knowledge of the metallicities, we can make a very simplified calculation of the total number of massive (OB) stars during the epoch of star formation outlined above. Although many factors are not well known, or might be different in a high metallicity environment as the GC, we will use as example the $Z = 0.008$ models of Vassiliadis & Wood (1993; see also e.g. Fagotto et al. 1994). We assume that an AGB star is observable as an OH/IR star only during the so-called 'superwind' phase on the AGB (9×10^4 yr for $M = 2$ $M_\odot$) and read in Table 2 of Vassiliadis & Wood (1993), that the total AGB phase lasts 1.5×10^7 yr. By comparing lifetimes of stars with small mass differences with the lifetime of the 'superwind' phase, we estimate an upper limit of the mass range for the OH/IR stars of 9×10^{-5} $M_\odot$ and infer that each OH/IR star represents 1.8×10^6 AGB stars in that mass range. Hence, because all OH/IR stars have the same mass and age, each OH/IR star observed can be converted to 1.8×10^6 stars with $M = 2$ $M_\odot$ during the formation of the stars. From our $v_{\rm exp}$ distribution, we estimate the minimum number N of OH/IR stars that is clearly in excess of what can be expected from the general mix of stars formed at other times, i.e. $N \gtrsim 20$. Without performing a detailed starburst calculation (see e.g. Rieke et al. 1993), we estimate the total number of OB stars ($10 < M < 60$ $M_\odot$) to be roughly 10^6 (or a few 10^7 over $1 < M < 60$ $M_\odot$) during this event.

To keep the metallicity of the cloud roughly constant, and the dynamics of the sample tight, the formation period must have been very short ($\ll$ 50 Myr). The luminosity produced in the nucleus during this event ($\sim 10^{10} L_\odot$) is comparable to the luminosity of the Galaxy as a whole. Accordingly, the Galaxy would have been a starburst galaxy during this epoch of enhanced star formation.

Note that the calculations are very sensitive to the AGB star mass, mass range, as well as the adopted stellar models. However, by selecting the lowest estimates for the excess of OH/IR stars and the conversion into AGB stars (the zero point of the initial main-sequence mass function; IMF), we believe that this is a lower limit for the number of OB stars. Furthermore, the formation of low-mass stars is probably suppressed (e.g. Morris 1993). The low-mass cut-off in the IMF is probably a few $M_\odot$. If this cut-off for the GC is higher than the mass of the observed OH/IR stars, there is a relative deficit of observable OH/IR stars (thus underestimating N). Moreover, we have used a relatively steep power of -3 in the IMF, close to the value for the high-mass end of the IMF (e.g. Rieke et al. 1993). The intermediate-mass part of the IMF is likely to be less steep, resulting in a severe underestimate of the number of OB stars when starting the calculations from $M = 2\ M_\odot$. When we continue, and estimate the mass of the molecular material from which our starburst resulted, we come to a cloud of a few times $10^8\ M_\odot$ (with a star formation efficiency of $\sim$ 20%; Genzel et al. 1995). We find our rough estimates consistent with the observations for starburst galaxies, for example the nearby starburst galaxies NGC 3504 and M 82 (Kenney et al. 1993; McLeod et al. 1993).

5. Conclusions

With a slightly different division in groups, we confirm the findings of Lindqvist et al. (1992b), Sevenster et al. (1995) and Serabyn & Morris (1996), i.e. one can statistically distinguish two groups of OH/IR stars with low and high $v_{\rm exp}$. The group with $v_{\rm exp} < 17$ km/s is in statistical sense "old" and is probably a bulge-like population. The group with $v_{\rm exp} \gtrsim 20$ km/s is statistically "young", and probably a result from sustained star formation. Furthermore, we have found a shrinking population of old stars (> 1 Gyr) with high (but unknown) metallicity (δ) and rapid rotation around the GC. These stars were formed in a short isolated era, for which the OH/IR stars with shell expansion velocities ($v_{\rm exp}$) between 17 and 20 km/s and $M \lesssim 2\ M_\odot$ are currently the most massive representatives. At the time of formation, the Galaxy would have been classified as a starburst galaxy.

References

Baud B., Habing H.J., Matthews H.E., Winnberg A. 1981, A&A 95, 156

Blommaert J.A.D.L., van der Veen W.E.C.J., Habing H.J. 1993, A&A 267, 39

Blommaert J.A.D.L., van Langevelde H.J., Michiels W.F.P. 1994, A&A 287, 479

Blommaert J.A.D.L., van der Veen W.E.C.J., van Langevelde H.J., Habing H.J., Sjouwerman L.O. 1998, A&A 329, 991

Eder J., Lewis B.M., Terzian Y. 1988, ApJS 66, 183

Fagotto F., Bressan A., Bertelli G., Chiosi C. 1994, A&AS 105, 29

Feast M.W., Glass I.S., Whitelock P.A., Catchpole R.M. 1989, MNRAS 241, 375

Genzel R., Weitzel L., Tacconi-Garman L.E., Blietz M., Cameron M., Krabbe A., Lutz D. 1995, ApJ 444, 129

Habing H.J. 1996, A&AR 7, 97

Habing H.J., Olnon F.M., Winnberg A., Matthews H.E., Baud B. 1983, A&A 128, 230

Habing H.J., Tignon J., Tielens A.G.G.M. 1994, A&A 286, 523

Herman J., Habing H.J. 1985, A&AS 59, 523

Iben I. 1991, ApJS 76, 55

Jones T.J., McGregor P.J., Gehrz R.D., Lawrence G.F. 1994, AJ 107, 1111

Kenney J.D.P., Carlstrom J.E., Young J.S. 1993, ApJ 418, 687

Lindqvist M., Winnberg A., Habing H.J., Matthews H.E. 1992a, A&AS 92, 43

Lindqvist M., Habing H.J., Winnberg A. 1992b, A&A 259, 118

McLeod, K.K.,Rieke G.H., Rieke M.J., Kelly D.M. 1993, ApJ 412, 111

Morris M. 1993, ApJ 408, 496

Olnon F.M., Baud B., Habing H.J., de Jong T., Harris S., Pottasch S.R. 1984, ApJ 278, L41

Reid M.J. 1993, ARAA 31, 345

Rieke G.H., Loken K., Rieke M.J., Tamblyn P. 1993, ApJ 412, 99

Serabyn E., Morris M. 1996, Nat 382, 602

Sevenster M.N., Dejonghe H., Habing H.J. 1995, A&A 299, 689

Sevenster M.N., Chapman J.M., Habing H.J., Killeen N.E.B., Lindqvist M. 1997, A&AS 122, 79

Sjouwerman L.O. 1997, PhD thesis, Chamlers University of Technology, Ch 6

Sjouwerman L.O., van Langevelde H.J., Winnberg A., Habing H.J. 1998a, A&AS 128, 35

Sjouwerman L.O., Habing H.J., van Langevelde H.J., Lindqvist M., Winnberg A. 1998b, in "The Central region of the Galaxy and Galaxies", Y. Sofue (Ed.), Kluwer Academic Publishers

te Lintel Hekkert P., Caswell J.L., Habing H.J., Haynes R.F., Norris R.P. 1991, A&AS 90, 327

van Langevelde H.J., Brown A.G.A., Lindqvist M., Habing H.J., de Zeeuw P.T. 1992, A&A 261, L17

van Langevelde H.J., Janssens A.M., Goss W.M., Habing H.J., Winnberg A. 1993, A&AS 101, 109

Vassiliadis E., Wood P.R. 1993, ApJ 413, 641

Wood P.R., Whiteoak J.B., Hughes S.M.G., Bessel M.S., Gardner F.F., Hyland R. 1992, ApJ 397, 552

Wood P.R., Habing H.J., McGregor P.J. 1998, A&A 336, 925

The Central Parsecs of the Galaxy
ASP Conference Series, Vol. 186, 1999
H. Falcke, A. Cotera, W.J. Duschl, F. Melia, M.J. Rieke, eds.

Physical Properties of OH/IR Stars Close to the Galactic Center

Anders Winnberg[1] and Michael Lindqvist

Onsala Space Observatory, S - 439 92 Onsala, Sweden

Harm J. Habing

Sterrewacht Leiden, P.O. Box 9513, NL - 2300 RA Leiden, The Netherlands

Abstract. Physical properties deduced from OH flux density and circumstellar envelope expansion velocity are studied in a statistical sense for a sample of about 140 OH/IR stars close to the Galactic Center (GC). The sample is confined to an area on the sky of about 1.14 square degrees roughly centered on the GC and the 3σ sensitivity of the survey is about 60 mJy. The distribution of the expansion velocities shows an asymmetric shape with a much steeper gradient on the high $v_{\rm exp}$ side than on the low $v_{\rm exp}$ side. It has a median value of about 18 km s^{-1} whereas OH/IR stars elsewhere in the Galaxy show significantly lower average expansion velocities. The flux density distributions for OH/IR stars with expansion velocities smaller than (Group I) and larger than 18 km s^{-1} (Group II) are significantly different. Group I contains stars with higher flux densities than Group II. Although Group I has a more extended volume density distribution than Group II this fact alone probably cannot explain the difference in their flux densities and a real difference in the OH luminosity functions has to be invoked.

A complete paper on this project will soon appear in Astronomy and Astrophysics.

1. Introduction

We discuss some physical properties of a sample of OH/IR stars associated with the Galactic Center (GC). These stars have been detected using the Very Large Array (Lindqvist et al. 1992a) and some of them using the 100-m radio telescope (Habing et al. 1983). These surveys have led to the first secure estimate of the mass density distribution in the central regions of the Galaxy between about 1 and 100 pc from the GC (Lindqvist et al. 1992b). The stars are very concentrated to the GC and thus have about the same distance from us.

[1]Guest Scientist, National Radio Astronomy Observatory, Array Operations Center, Socorro, New Mexico, U.S.A.

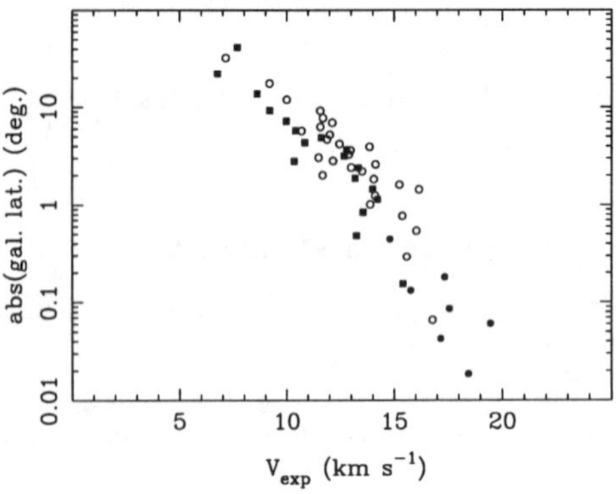

Figure 1. Absolute Galactic latitude versus expansion velocity for the sample close to the GC (filled circles; this paper), the te Lintel Hekkert sample (open circles; te Lintel Hekkert et al. 1991), and the Arecibo surveys (filled squares; Eder et al. 1988, Lewis et al. 1990, Chengalur et al. 1993). Each symbol represents the average of $\sim$ 20 OH/IR stars

They are therefore suitable for studies of distance-dependent quantities such as luminosity and mass-loss rate.

Recently a more sensitive survey for OH/IR stars in the central regions has been carried out by Sjouwerman et al. (1998a) using both the VLA and the Australia Telescope (AT). Therefore, today, about 200 OH/IR stars are known within 100 pc of the GC. Many of the stars have been identified with 43-GHz SiO masers and a few with 22-GHz H_2O masers (e.g. Sjouwerman 1997). A long-term project to measure the proper motions of these stars has been initiated (Sjouwerman et al. 1998b).

Several infrared studies of these stars have been carried out (Wood et al. 1998, and references therein) and some of them have been monitored in their OH 1612-MHz line (van Langevelde et al. 1993). In this way light-variation periods and bolometric luminosities have been determined for many of the stars.

2. Circumstellar Expansion Velocities

The distance between the two 1612-MHz line components of OH/IR stars is a measure of twice the circumstellar envelope expansion velocity. It has been shown in several studies, starting with Baud et al. (1981), that expansion velocity is inversely proportional to Galactic latitude and to radial velocity dispersion. Thus, OH/IR stars with large expansion velocity are younger than OH/IR stars with small expansion velocity. The anti-correlation with Galactic latitude is shown convincingly in Figure 1 for a large sample of OH/IR stars in the GC, in the Bulge, and in the Plane. There is a distance effect which works in the same direction, but we believe that it is unlikely that this effect is responsible for the major trend seen in Figure 1. The frequency function of expansion velocities in the GC is skewed to higher values (Figure 2). Whereas in the Bulge and in the

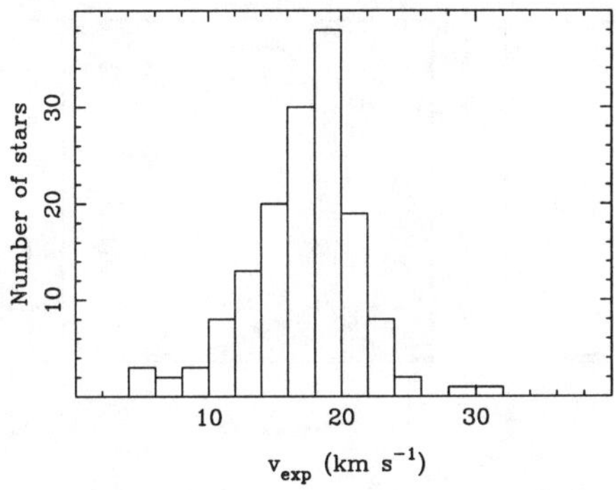

Figure 2. The distribution of circumstellar envelope expansion velocities for OH/IR stars close to the GC (this paper)

Disk the median value is about 15 km s^{-1}, in the GC it is about 18 km s^{-1}. And even more extreme, as shown by Sjouwerman et al. (1999) at this workshop, in the very center it shows a sharp peak at about 19 km s^{-1} hinting at a star burst which occurred about 1 Gyr ago. The higher expansion velocities in the center are probably due to younger stars with higher metal contents.

3. The OH Luminosity Function

The distances to AGB objects in general and to OH/IR stars in particular are poorly known. However, we know that the OH/IR stars in the GC sample are all at nearly the same distance from us. This fact allows us to deduce the correct shape of the OH luminosity function by merely constructing the flux density distribution. This can be turned into a luminosity function by application of the distance to the GC. If in the future this value is revised the effect on the luminosity function is just a shift without any change in the shape. The OH maser luminosity function and in particular its shape might be very important for a determination of the duration of the OH/IR phase (Baud and Habing 1983).

The distributions of the peak flux densities of the strongest OH line component, $S_{\rm max}$, are presented in Figure 3 for all stars (a), for stars with expansion velocities smaller than 18 km s^{-1} (b) and for stars with expansion velocities larger than 18 km s^{-1} (c).

Although the low-flux side of the distribution is strongly influenced by the detection probability, we believe that these histograms and curves indicate that the flux density distribution of OH/IR stars in the GC reaches a maximum slightly shortward of the average completeness limit. If the true distribution would have continued to grow as a power function towards lower flux densities, the corrected model, i.e. the dashed curve, would have reached down to the observed distribution. Likewise, in the same case, the corrected distribution, i.e. the open columns, would have reached up to the model (the solid curve). There is further evidence for this in the data of Sjouwerman et al. (1998a).

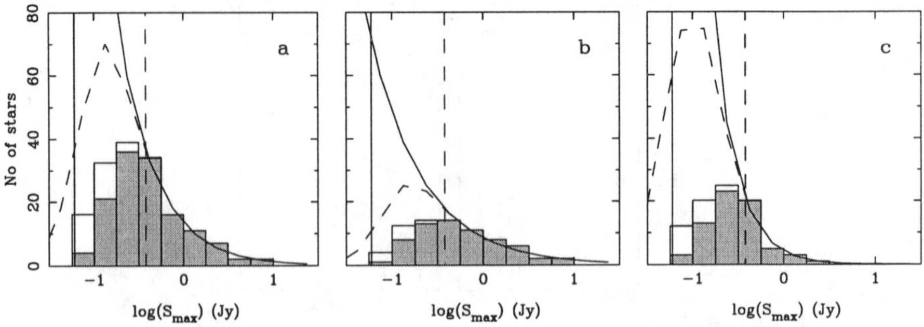

Figure 3. The distribution functions of the peak flux densities of the strongest OH line component for (a) all stars, (b) for stars with $v_{\rm exp} <$ 18 km s^{-1}, and (c) for stars with $v_{\rm exp} >$ 18 km s^{-1}. The solid columns depict the observed distributions and the open columns show the distributions corrected for the average probability of detection according to a mathematical model. The solid lines are least-square fits of power functions to the high-flux side of the distributions and the dashed lines are these fitted functions multiplied by the detection probabilities. The solid vertical line marks the 3-σ noise level (60 mJy) and the dashed vertical line marks the average completeness limit (380 mJy)

We have fitted a power-law function to the high-flux side of the flux density distributions according to:

$$N(\log_{10} S) = AS^{-\alpha} \tag{1}$$

where S is the OH flux density and A and α are constants. The number of stars per unit of logarithmic S can be related to the number of stars per unit S according to:

$$N(S) = N(\log_{10} S) \frac{\mathrm{d}(\log_{10} S)}{\mathrm{d}S} \tag{2}$$

Thus, if the slope of $N(\log_{10} S)$ is α, the slope of $N(S)$ is $\alpha + 1$. For the flux distribution of all stars we get $\alpha = 1.0 \pm 0.3$ corresponding to $N(S) \propto S^{-2.0 \pm 0.3}$. For stars in the group with $v_{\rm exp} < 18$ km s^{-1} we get $\alpha = 0.75 \pm 0.3$ and for stars with $v_{\rm exp} > 18$ km s^{-1} we get $\alpha = 1.6 \pm 0.6$. The errors are 3σ. Thus we may conclude that the exponents are different for the two sub-samples. Moreover the strongest OH emitters are found in the group with low expansion velocities. This group has a more extended sky distribution than the group with high expansion velocities. One would therefore suspect that the stronger masers could be due to more foreground stars in this group. However, Monte Carlo calculations show that this is very unlikely. The flux density distributions therefore indicate different underlying luminosity functions.

4. Conclusions

The OH/IR stars within 100 pc from the GC have a median circumstellar envelope expansion velocity of $\sim$ 18 km s^{-1} significantly different from that of OH/IR stars in the Bulge and in the Disk ($\sim$ 15 km s^{-1}). This is interpreted as the presence of many stars with a high metallicity close to the GC.

The OH flux density distribution of the OH/IR stars close to the GC mimics the OH luminosity function of these stars since their distances are all close to the distance of the GC. Although the sensitivity of the finding survey was not sufficient to detect OH/IR stars on the low-luminosity side of the luminosity function we believe that the maximum of the luminosity function has been reached. We base this conclusion on a consideration of the probability of detection.

The flux density distributions for OH/IR stars with $v_{\rm exp} < 18$ km s^{-1} and with $v_{\rm exp} > 18$ km s^{-1} are significantly different. The slope of the high-flux side of the distribution for stars with $v_{\rm exp} > 18$ km s^{-1} is higher (1.6 ± 0.6) than that for stars with $v_{\rm exp} < 18$ km s^{-1} (0.75 ± 0.3). Thus, the most luminous stars are found in the latter group. The physical reason for this is not clear.

Acknowledgments. Part of this work was carried out while AW was a Guest Scientist at National Radio Astronomy Observatory in Socorro, New Mexico, U.S.A. This project is supported by the Swedish Natural Science Research Council.

References

Baud, B., Habing, H.J., Matthews, H. E., Winnberg, A. 1981, A&A, 95, 171

Baud, B., Habing, H.J. 1983, A&A, 127, 73

Chengalur, J.N., Eder, J., Lewis, B.M., Terzian, Y. 1993, ApJS, 89, 189

Eder, J., Lewis, B.M., Terzian, Y. 1988, ApJS, 66, 183

Habing, H.J., Olnon, F.M., Winnberg, A., Matthews, H.E., Baud, B. 1983, A&A, 128, 230

Lewis, B.M., Eder, J., Terzian, Y. 1990, ApJ, 362, 634

Lindqvist, M., Winnberg, A., Habing, H.J., Matthews, H.E. 1992a, A&AS, 92, 143

Lindqvist, M., Habing, H.J., Winnberg, A. 1992b, A&A, 259, 118

Sjouwerman, L.O. 1997, PhD Thesis, Technical Report No. 316, Chalmers University of Technology, Göteborg

Sjouwerman, L.O., et al., 1999, this volume, p. 379

Sjouwerman, L.O., van Langevelde, H.J., Winnberg, A., Habing, H.J. 1998a, A&AS, 128, 35

Sjouwerman, L.O., van Langevelde, H.J., Diamond, P.J. 1998b, A&A, 339, 897

te Lintel Hekkert, P., Caswell, J.L., Habing, H.J., Haynes, R.F., Norris, R.P. 1991, A&AS, 90, 327

van Langevelde, H.J., Janssens, A.M., Goss, W.M., Habing, H.J., Winnberg, A. 1993, A&AS, 101, 109

Wood, P.R., Habing, H.J., McGregor, P.J. 1998, A&A, 336, 925

The Central Parsecs of the Galaxy
ASP Conference Series, Vol. 186, 1999
H. Falcke, A. Cotera, W.J. Duschl, F. Melia, M.J. Rieke, eds.

Discussion Session VI: What is the Metallicity as Determined from Old and New Stars?

W. J. Duschl[1]

Institut für Theoretische Astrophysik der Universität Heidelberg
Tiergartenstr. 15 D-69121 Heidelberg

R. F. Coker[2], A. Cotera[2]

Steward Observatory, The University of Arizona, Tucson, AZ 85721

Abstract. This is an edited transcript of a discussion session about the metallicity of stars in the Galactic Center that took place during the Galactic Center conference "The Central Parsecs" in Tucson, Arizona, Sep. 7-11, 1998. The session was chaired by the moderator as indicated in the author list. The discussions were taped, transcribed, and finally edited by editors who are listed as co-authors. The contributions of the conference participants are preceded by their names and thus references to specific aspects and ideas of this discussion session must also include a reference to the respective speaker(s), e.g., in a footnote.

Discussion

Duschl: OK, let's get the discussion going. How do we reconcile the differences in metallicity presented by Lorant Sjouwerman and Solange Ramierz?

Ramírez: Acutally, I have some questions about the hot gas. For example, the X-ray sources I have seen in the Galactic Center, could they force the gas out somehow, such that we can get all the high metallicity gas out of the interstellar medium where the new stars are formed, so those stars have a lower metallicity?

Cotera: But if you get all of the gas out of the region, what do you form all the stars out of?

Ramierz: The bar perhaps?

Cotera: Perhaps the bar drives low metallicity gas in, then starts the process that makes the high metallicity gases?

[1]Moderator of the discussion session

[2]Editor of the discussion session

Sellgren: That is one explanation. But there is also the question: are the elements that they measure in the gas phase the same as the elements that are measured in the stars? Because we are measuring iron, and we have data that we have not reduced yet on other elements like silicon and titanium. But the gas phase material is argon, neon, sulfur, nitrogen. None of those have been measured.

Cotera: Is there enough silicon in the gas phase?

Sellgren: Silicon has the problem that it gets trapped in the dust grains. And so does iron.

Cotera: But how about you, Paco? You were saying that from iron you were getting higher than solar metallicities.

Najarro: Well, I do not know. I am very confused with most of the results on this. It may be just a matter of relaxing all the assumptions. You are using LTE, and you say, I have to deal with very strange guys, lots of winds and many other possibilities, and I think we have got to be very careful when we just give a number. My feeling is that it does not look like an iron ridge, but when I see the final results I may believe it. We see also nitrogen from the HII regions enhanced so there must be some enrichment.

Sellgren: We've seen nitrogen enriched at IRS 7 but we believe that's due to dredge up. For example, we see carbon and oxygen low and nitrogen high, but we think that IRS 7 is bringing stuff up to its surface that it has processed in its interior rather than reflecting initial abundances in that case. Iron does not get processed as much.

Najarro: One of the tests you should do also is to relax the LTE assumption, which is very important when you have winds. It has been proven that this otherwise can really lead to a false abundance. We have done differential analyses that show that. The other models should be run as well. So that is still open. I do not know.

Ramírez: What is the effect of not considering LTE? Are you going to get higher abundances or lower abundances?

Najarro: I think if you have strong winds, you will fill up your lines. So I guess you would need higher abundances to compensate for that and then get down to what you are seeing right now with your LTE.

Sellgren: The thing I want to emphasize, though, is that we are doing a differential analysis, and we compare it to α Ori (in the solar neighborhood) which is the same spectral type as IRS 7.

Najarro: Yeah, but you need an absolute calibrator, and if α Ori is such a kind of star, you need at least one non-LTE analysis for α Ori with winds taken into account.

Sellgren: Well, I wouldn't say that the Galactic Center has the same metallicity as α Ori. You are right, we can not be sure about the absolute value, but we can certainly be sure of a comparison between the solar neighborhood stars and Galactic Center stars.

Najarro: Most of the stars show some growth—you are seeing the wind there. You have to consider that.

Ramírez: That is just the case of IRS 7. We have 9 other stars, and we do not see those broad lines in the other 9 stars.

Najarro: What is your resolution?

Ramírez: 7 km s^{-1}. You also have other effects, such as the microturbulence that makes the lines broader.

Najarro: But microturbulence is a patch that you put in to make the profile match. That is not physics anymore. There is no physics in microturbulence. Microturblence was something that was invented for fitting the profiles to the observations. So you grow them. You have the thermal upper winds of the ions and you grow them. That would include a different temperature, and that is what makes a problem for your profile. Microturbulence is a patch. There is no physics in it.

Mezger: Lorant [Sjouwerman], what accounts for the abundance of early stars and the abundance of the interstellar metal? There are several quantitative indications that the metal abundance in the Galactic Center is at least twice that in the solar vicinity. For example, Solange Ramírez has shown the abundances of the interstellar metal, wasn't it? I thought you had shown an abundance gradient which goes from the sun towards the Galactic Center, which constantly increases, is this correct?

Ramírez: Yes, that was made from HII regions.

Mezger: It was an HII region? One can then state that the metallicity of the gas of interstellar metal in the Galactic Center is very high?

Ramírez: Yes, maybe twice or three times solar.

Sanders: Even out here in this part of the Galaxy, the metallicity varies, doesn't it? It has been recently discovered that Orion is very different from solar. Could some of the same things be going on in the Galactic Center, with these different metal analyses we are getting?

Sjouwerman: Oh yes, I think. The ages we are talking about are very large. It could be that 2 Gyrs ago there was still no bar which drives your gas inwards. So, the metallicity would have been enriched in this region [around the center] until the bar was created, and after that you get solar.

Anantharamaiah: So, I take the message home with me, that if you look at the Galactic Center today, it is not like a starburst region, but if you looked at it 10 million years ago, 100 million years ago, it would look like a starburst—a typical starburst that we see in other galaxies?

Sjouwerman: I am saying that you should have looked between 1-4 Gyrs ago. I cannot talk about the recent past since my stars are much older than that.

Anatharamaiah: So that is on the basis of the OH/IR stars. Is there a concensus on that?

Novak: But where are all the stars? Can we get a handle on any of the other stars that were in this burst?

Sjouwerman: Well, they should show up as G-type main sequence stars.

Najarro: The other ones are already on the AGB.

Cotera: You are using the stars to trace metallicity. Have there been studies using the same sort of technique in other galaxies? Is there some way to check that if you do this with a star, you also get a metallicity gradient?

Unknown: It has been done for A supergiants in a couple of cases.

Cotera: And do they see metallicity gradient using stars?

Unknown: They are starting to see it now in K band.

Sellgren: There are certainly line strength gradients seen in the optical in elliptical galaxies. I do not know if that is true for spirals. Most of the extragalactic metallicity gradient observations are from metals in the gas phase, such as oxygen.

Cotera: Yes, that is what I am thinking. We are saying that there should be a metallicity gradient in our Galaxy but that is from gas phases. And now you are saying in the stellar phase we do not see that in these lines. And my question is, would we expect to?

Sellgren: I do not know. To summarize, we actually have several conflicting lines of evidence, and I think one of the things that all of us are interested in is that our work on the M supergiants is showing solar metallicity, but we have not published it because we have been checking and rechecking and rechecking everything we can think of, because it is such a weird result. The line from the OH/IR stars is that they have very high gas to dust ratios, which infers to me high metallicity for the OH/IR stars to explain the expansion velocity. And Paco is finding very high iron abundances.

Najarro: But it is one star, so . . .

Sellgren: One star.

Figer: Maybe two more.

Sellgren: Yeah, right. Two months from now you will be able to tell us whether or not there are high metal abundances in more than one star. And there are gas phase measurements which show things like argon and nitrogen and sulfur to be high in the gas phase. So we have conflicting lines of evidence, and I think there are two things to do. One is to work on understanding the systematic uncertainties to make sure that there is not something going on where one of us or all of us have some factor of two floating around we have not accounted for. And the other is that we need to start thinking about what is expected in the Galactic Center. We need to build a model of how much gas bubbles in, how much is retained, and how many metals are produced, averaged over a very long time, and then have some form of evolutionary model to actually predict something for the Galactic Center to test against. Would you expect the gas and the stars to be the same? So we need to look for systematic effects and make some theoretical predictions to test against.

Yusef-Zadeh: There is one place that would be very interesting to study: the metallicities of IRS 7. There is ionized gas around IRS 7 and we actually see neon lines from the tail and from the bowshock region. We could make a comparison to the stellar abundance and the gas abundance, and see if we get the same results.

Sellgren: We have done the Minispiral itself. It can go higher but neon is potentially solar, so it is the other elements that are higher and this may be at least part of the discrepancy is comparing different elements. That does not explain the difference between our results and Paco's [Najarro] results.

Najarro: That is a good point. Can you get that information on the LMC? That would be a nice test.

Sellgren: You can't observe the LMC with our instrument. We do not have high enough resolution.

Unknown: With a starbust population you get the right metallicities, but they assume that you have a steady inflow as you require for the distance corrections. But now we know that merging is an important ingredient in Galactic evolution, and we just discovered a galaxy which is merging into the Galactic central region. And so that means that in addition to the steady enriched gas that you can not get from the outside—which has been true for all the star formation and regeneration—suddenly you get fresh gas from outside. Bob Sanders explained yesterday, how you get star formation very close to the center, and I checked it. This mechanism, this pulling in and shocking of the gas works beautifully, but it requires an initial gas density of 10^5, and so you could have fresh material coming from the outside of the central region, from across the Galaxy—which has been speculated ages ago—come in there, and then you have star formation for a short period with much different abundances. Then you have all these Wolf-Rayet stars that just pump heavy elements into the medium like crazy. So you have extreme inhomegenities, and it seems to me this fits everything we see very nicely. It is just a big mess.

Sofue: Well, how about the starburst. In 10 Gyr how many of these stars will become supernovae?

Unknown: OB stars? I think, with the IMF I used, it is roughly 10^6.

Sofue: One million supernovae within how many years?

Unknown: Ten million years.

Unknown: It would be a nice galaxy.

Unknown: So there is no trace of it left anywhere?

Unknown: No.

Duschl: Let us take that home from this discussion: we live a couple of Gyrs too late for living in a starburst galaxy. Perhaps the Galaxy will try it again. Okay, thank you very much.

Session VII

How Do the Various Components of the Inner 100 Parsecs Interact With Each Other?

The Central Parsecs of the Galaxy
ASP Conference Series, Vol. 186, 1999
H. Falcke, A. Cotera, W.J. Duschl, F. Melia, M.J. Rieke, eds.

Wide Field Radio Imaging of the Galactic Center

Namir E. Kassim

Naval Research Laboratory, Code 7213, Washington, DC 20375-5351; kassim@rsd.nrl.navy.mil

Theodore N. LaRosa

Department of Bio and Physical Sciences, Kennesaw State University, Kennesaw, Georgia 30144; ted@avatar.kennesaw.edu

T. Joseph W. Lazio

National Research Council and Naval Research Laboratory, Code 7210, Washington, DC 20375-5351; lazio@rsd.nrl.navy.mil

Scott D. Hyman

Department of Physics, Sweet Briar College, Sweet Briar, VA, 24595; hyman@sbc.edu

Abstract. We present a large and sensitive sub-arcminute resolution meter wavelength image of the Galactic Center. Archival 330 MHz (λ0.9 m) VLA observations were reprocessed utilizing wide-field imaging algorithms developed at the National Radio Astronomy Laboratory and the Naval Research Laboratory. The resulting image is a significant improvement over previous ones, both providing higher fidelity images of known sources and revealing many new sources. We have also used "targeted faceting" to generate images at the full resolution of the data for the first time for regions well away from Sgr A.

At 330 MHz the VLA is sensitive to both thermal and non-thermal sources, and this image provides an excellent overview of the large-scale structure ($2^\circ \times 2^\circ$) in the Galactic Center. Several new sources have been found, and a catalogue of over 100 sources is being prepared. Subsequent observations of one of these new features has led to the discovery of a new non-thermal filament with a unique orientation along the plane; observations of other new features are in progress.

Improved imaging and sensitivity has allowed for better spectral index determinations. In particular, the spectral index (between 330 and 1400 MHz) as a function of length along several non-thermal filaments has been measured. The spectral index was found to be constant, indicating that these may be young features.

A second epoch 330 MHz Galactic Center imaging campaign, an efficient monitor for source variability and transients, is in progress as is a first epoch campaign with the new 74 MHz VLA observing system.

1. Introduction

Low frequency radio observations offer unique insight into the rich phenomenology present in the Galactic Center. At meter wavelengths, non-thermal emission, from sources such as supernova remnants (SNRs), pulsars, and the non-thermal filaments (NTFs), is enhanced. Thermal emission from H II regions is generally optically thick, but thermal sources are still fairly bright. Thus, at meter wavelengths sources with different emission mechanisms can be studied simultaneously. Moreover, the intrinsically large field of view at long wavelengths provides an extremely efficient means of mapping the large regions, and even moderate angular resolution ($\sim 1'$) observations can provide excellent surface brightness sensitivity to extended emission.

The 330 MHz VLA system, with its $156'$ (FWHP) primary beam, is well suited for studying the Galactic Center in a single pointing. In addition, the subarcminute angular resolution provided by the VLA in its extended configurations is capable of resolving structures with a wide range of scale sizes. The first epoch of 330 MHz VLA Galactic Center imaging was published by Pedlar et al. (1989) and Anantharamaiah et al. (1991). Those observations generated a number of important results, particularly for regions of emission near the pointing center. However the so-called "wide-field imaging problem," associated with imaging large fields at long wavelengths with non-coplanar arrays (such as the VLA), limited the dynamic range of the original 330 MHz images, which were processed with conventional "2-D" imaging algorithms.

Since those initial 330 MHz VLA Galactic Center observations, Kassim et al. (1998) have expanded the capabilities of a 3-D wide-field imaging algorithm (Cornwell & Perley 1992) and have ported it to powerful supercomputing platforms at the Naval Research Laboratory. This program, `dragon`, is now utilized routinely to generate 74 and 330 MHz wide-field images of VLA data with unprecedented fields of view and sensitivities near the thermal noise limit. Kassim & Frail (1996) employed it to re-image the archival Galactic Center data set (Pedlar et al. 1989). Here we present the full wide-field image which emerged from that effort. Additional wide-field observations of the Galactic Center at 74 MHz are currently being acquired as well as a second epoch 330 MHz data set that will be used to study source variability.

2. Motivation: The Unidentified 80 MHz Clark Lake Source

The initial effort to re-image the Galactic Center was motivated by an unidentified non-thermal source first imaged at 80 MHz (Fig. 1) with the Clark Lake array (LaRosa & Kassim 1985; Kassim et al. 1986). Its appearance on the reprocessed 330 MHz image as as a limb-brightened shell just north of the Arc (Fig. 2) together with its non-thermal spectrum, $\alpha = -0.56$ ($S_\nu \propto \nu^\alpha$), led Kassim & Frail (1996) to identify it as SNR 0.33+0.04. These are the same criteria used to identify Sgr A East as an SNR by Ekers et al. (1983), and, if correct, make G0.33+0.04 the second closest known SNR to the Galactic Center.

The orientation of the major axis of SNR 0.33+0.04 along the Galactic plane gives it a striking similarity to Sgr A East, though it is much more extended and, thus, presumably older. There are, however, well-known problems with the

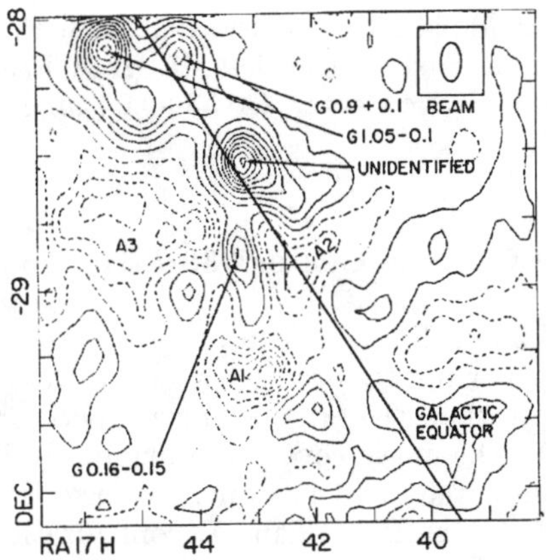

Figure 1. Clark Lake image of the Galactic Center at 80 MHz (LaRosa & Kassim 1985). The source labeled "Unidentified" was later identified as SNR 0.33+0.04 (Kassim & Frail 1996). The angular resolution is approximately 8′.

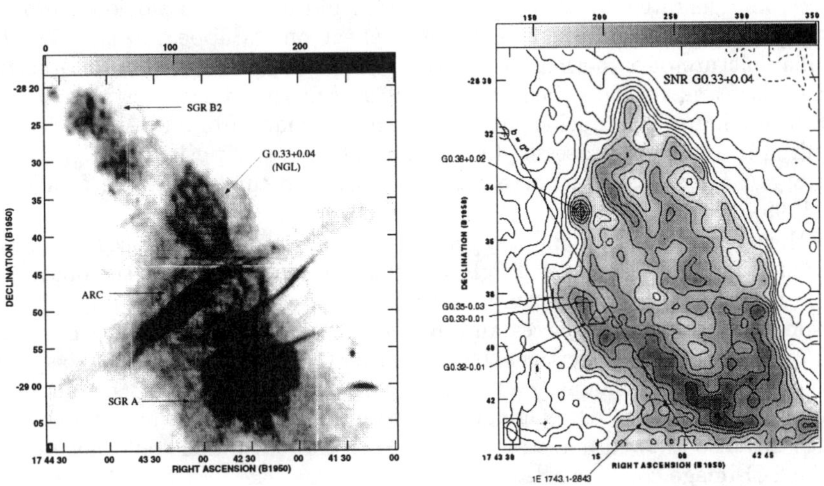

Figure 2. *(Left)* Sub-image of the wide-field VLA 330 MHz image showing SNR 0.33+0.04 and its immediate surroundings. *(Right)* Sub-image of wide-field VLA 330 MHz image showing detailed structure of SNR 0.33+0.04 (Kassim & Frail 1996). The angular resolution is approximately 45″.

interpretation of Sgr A East as a normal SNR, such as its unusually high radio surface brightness and its large inferred explosion energy (Mezger et al. 1989). Alternate scenarios, such as tidal disruption of a massive star by the central black hole (Khokhlov & Melia 1996) or the debris from a gamma-ray burst, also exist to explain Sgr A East. Regardless, if Sgr A East is an indication of outburst activity at the Galactic Center, G0.33+0.04 might represent an earlier episode of similar activity.

3. New Wide-Field Galactic Center 330 MHz Image

Figure 3 shows the full 330 MHz wide-field image of the Galactic Center. It is one of the most spectacular images ever generated with the VLA, and is both larger and more sensitive than the original images (Pedlar et al. 1989; Anantharamaiah et al. 1991). The image covers approximately $2^\circ \times 2^\circ$ at an angular resolution of approximately $45''$ and has an rms noise level of approximately 5 mJy beam^{-1}. Data from the B-, C-, and D configurations of the VLA were combined to generate it. Figure 4 shows a schematic of the Galactic Center, based on this new wide-field image.

At this frequency and over this large field of view, the VLA can no longer be assumed to be a co-planar array. Neglecting the non-coplanar array geometry, as Pedlar et al. (1989) and Anantharamaiah et al. (1991) did, does not degrade the image fidelity significantly near the center of the image, for two reasons. First, the image distortions caused by the non-coplanar geometry increase with distance from the center of the image. Second, Sgr A East is so bright that the errors generated by the improper deconvolution of the sidelobe response to (angularly) distant sources have little effect on its appearance. Thus the familiar picture of a thermal Sgr A West in absorption against a non-thermal Sgr A East and its halo (Pedlar et al. 1989; Anantharamaiah et al. 1991), as shown in Figure 5, is unchanged. Figure 5 was made from VLA observations acquired in 1998 and processed by `dragon`. The relative position of Sgr A East and Sgr A West based on these data are examined further by Yusef-Zadeh et al. (1999).

However, neglecting the array's non-coplanar nature produces significant errors for moderate to weaker brightness sources located far from the phase center of the image. Properly compensating for the array geometry improves the response to both point sources and extended features dramatically. Figures 6 and 7 illustrate these distortions, by comparing the distorted response to sources in the original (2-D imaging) 330 MHz images and the undistorted sources revealed by the proper 3-D imaging. The reduction in the noise floor on the new image has also allowed detection of additional sources which could not be seen in the 2-D image counterparts.

Finally, the computational power available to Pedlar et al. (1989) and Anantharamaiah et al. (1991) limited the extent to which they could image the data at the intrinsic angular resolution of the data ($\approx 12''$ for the A-configuration data). Only the immediate region of Sgr A (e.g., Fig. 5) could be imaged at that resolution. Sufficient computational power now exists to implement "targeted faceting" (Cornwell & Perley 1992) over the entire field of view. In this scheme, the fact that the radio sky is mostly dark is exploited; only a small region sur-

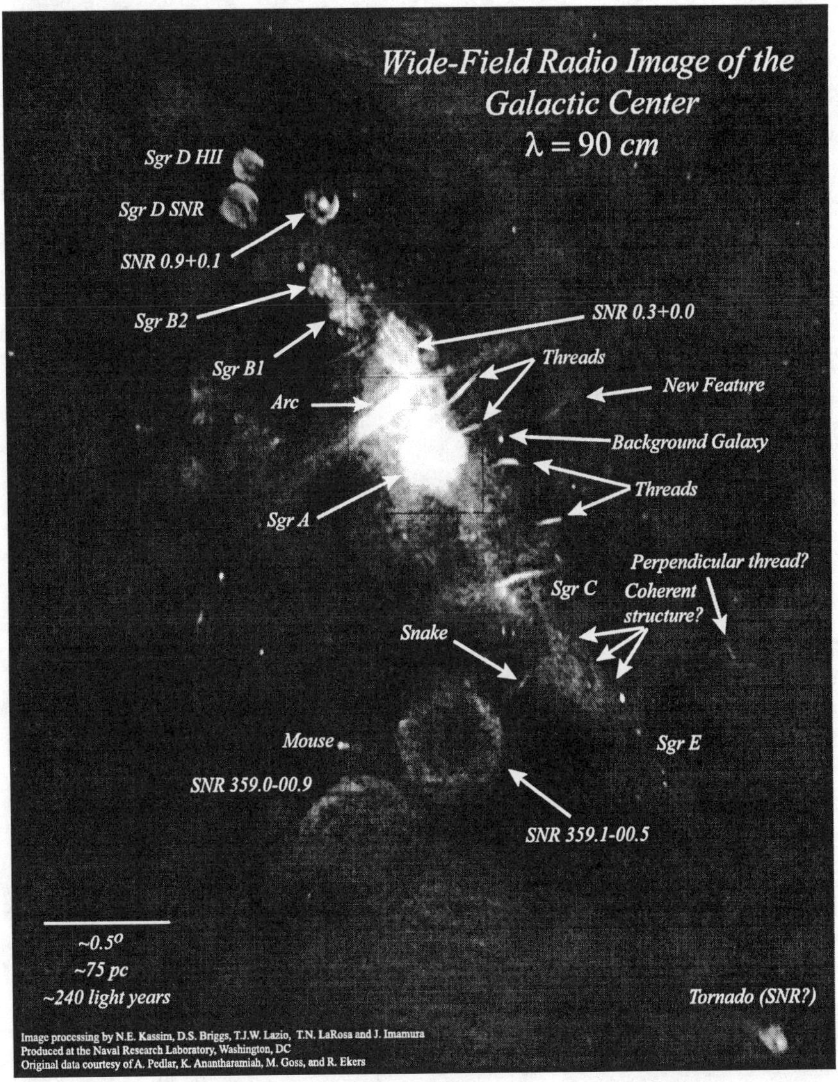

Figure 3. *(See Plate IV for color version)* Full 330 MHz wide-field image of the Galactic Center at an angular resolution of approximately 45″. The original data were obtained by Pedlar et al. (1989) and Anantharamaiah et al. (1991) using the VLA in the B-, C-, and D configurations. The rms noise in the image is approximately 5 mJy beam^{-1}. The image has not been corrected for primary beam attenuation.

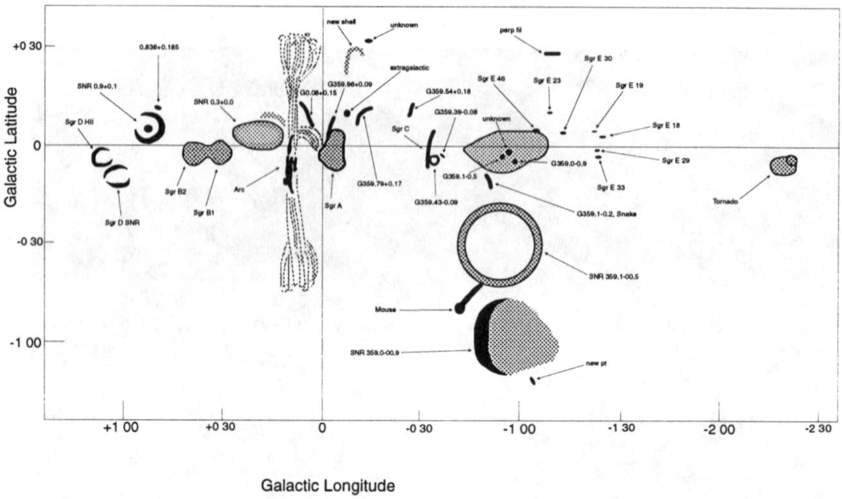

Figure 4. Schematic of the Galactic Center based on Figure 3.

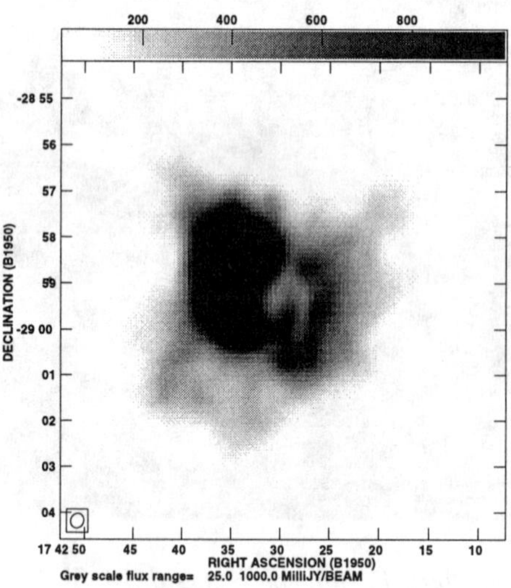

Figure 5. Sgr A West in absorption against Sgr A East and its halo at 330 MHz. This image is from a recent observation with the A-configuration VLA (Yusef-Zadeh et al. 1999) and was produced using **dragon**. It is similar to that found by Pedlar et al. (1989) since their lack of 3-D processing produced minimal distortions at the center of their image. The angular resolution is approximately 20″.

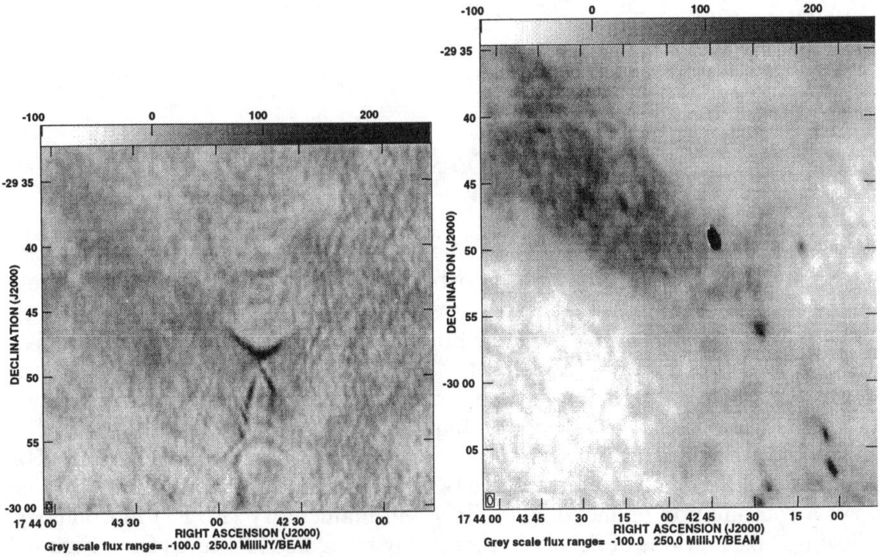

Figure 6. *(Left)* Distorted response to a point source located far (> 1°) from the field center from the original 2-D processed 330 MHz image. *(Right)* Undistorted sub-image of Figure 3 made to the same scale and of approximately the same region as shown to the left. The 3-D image reveals many more sources, possibly including an extended emission feature labeled "Coherent structure?" on Figure 3.

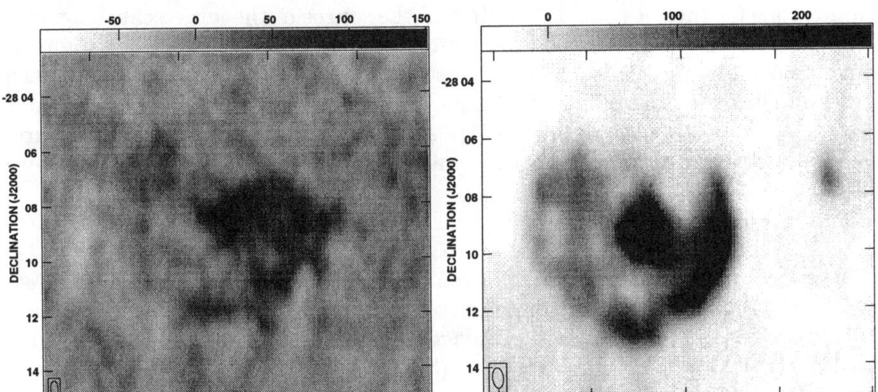

Figure 7. *(Left)* Distorted response to SNR 0.9+0.1 located far (> 1°) from the field center from the original 2-D processed 330 MHz image. *(Right)* Sub-image of Figure 3 made to the same scale and of approximately the same region as shown to the left. The undistorted image reveals the accurate shell and plerionic emission of this composite SNR.

rounding each source is imaged at the highest possible resolution (Kassim et al. 1998).

4. New Sources

As a result of its larger size, undistorted response to emission far from the phase center, and lower noise level, we have been able to identify over 100 discrete emission features in Figure 3 (LaRosa et al. 1999).

Newly recognized sources include both compact and extended emission features, in addition to SNR 0.33+0.04. The former may represent pulsars or background radio galaxies and are important, among other reasons, for scattering studies (Lazio et al. 1999). The extended features may represent new SNRs, or new and unusual structures unique to the Galactic Center environment. For example, Figure 8 shows a new feature G359.85+0.39, named the "Cane," which shows a combination of shell-like (SNR?) and filamentary (NTF?) morphology. Follow-up observations of this and other newly recognized features from Figure 3 are in progress.

The first new feature recognized on this image, and the one for which the largest number of follow-up observations have been conducted, is G358.85+0.47, labeled "Perpendicular Filament?" on Figure 3. Anantharamaiah et al. (1999) present high-resolution observations of this feature and show that it appears to be a previously-unrecognized NTF. As such, it is the NTF most (angularly) distant from Sgr A and the only NTF oriented along the Galactic plane. The implications for theories of NTF origin and the source of the large-scale magnetic field in which they are presumed to be immersed are clearly important. Figure 8 shows an overlay of the 1400 MHz (grey-scale) and the 330 MHz (contour) image of the "Pelican." The 330 MHz image was processed at the highest angular resolution intrinsic to the data ($\approx 12''$) using the "targeted faceting" procedure. Its filamentary appearance at that resolution motivated the 1400 MHz follow-up observations.

Figure 3 also shows hints of tantalizing structure on even larger scales ($> 30'$). For example, the Arc appears to extend on either side of the Galactic plane into plume-like lobe structures reminiscent of extragalactic sources. Also recognizable on Figure 3 are previously noted connections between moderate (1–10$'$) scale radio features at the Galactic Center and even larger scale ($> 1^\circ$) structures such as the "Omega Lobe" (Sofue & Handa 1984). Unfortunately, much of this largest scale emission is presently at too low a level to accept as real. We hope that the "second epoch" VLA Galactic Center campaign, with the expanded and improved 330 MHz system, will shed light on the reality of this emission. The most severe limit on the sensitivity of Figure 3 to the largest scales and weakest emission comes from the lack of short interferometric baselines (the "zero-spacing" problem). We are exploring the possibility of adding single-dish data, thereby eliminating this problem. A second epoch 330 MHz observation will also serve as a means of monitoring the Galactic Center for source variability and transients.

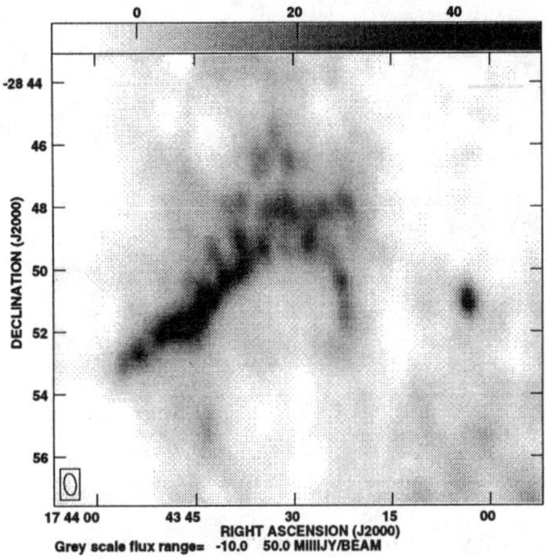

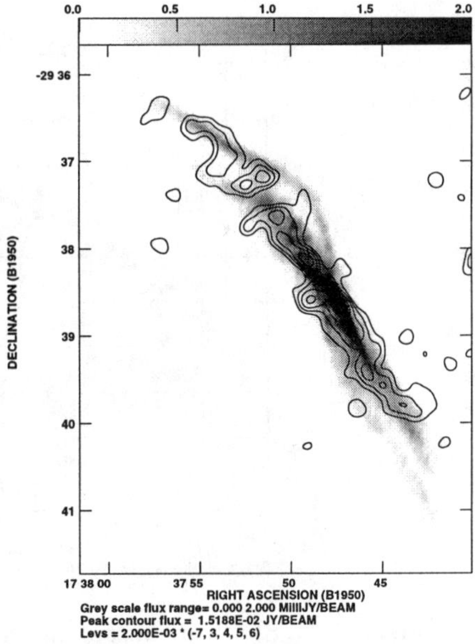

Figure 8. New sources in the Galactic Center. *(Left)* G359.85+0.39, the "Cane." A subimage from Figure 3 showing this new shell-like feature. *(Right)* Overlay of 1400 MHz (grey-scale) and 330 MHz (contour) of the newly discovered NTF "the Pelican." The 330 MHz image has a resolution of approximately 12″ and was produced using "targeted faceting." The 1400 MHz image has a resolution of approximately 6″ and is discussed further by Anantharamaiah et al. (1999).

5. Revisiting NTF Spectra

The improved sensitivity of our new image motivated us to revisit the question of NTF continuum spectral indices. We have acquired the 1400 MHz VLA data from Liszt & Spiker (1995) of one of the brightest NTFs, Sgr C. Figure 9 shows the spectral index between 330 and 1400 MHz along the length of the Sgr C filament. In contrast to previous studies, which find significant variations in NTF spectral index (Anantharamaiah et al. 1991), we see no significant variation in spectral index along the Sgr C filament. Previous suggestions of a spectral-index flattening along the filament, toward its eastern terminus, may have been caused by the superposition of an H II region across the eastern end of the NTF.

The lack of significant variations in spectral index along this filament are consistent with simple synchrotron aging lifetime estimates of about 1.4×10^6 yrs. A minimum energy calculation yields a magnetic field of 0.12 mG. For an assumed ambient density of 10 cm^{-3}, the Alfvén velocity is $v_A \sim 80$ km s^{-1}, and the corresponding streaming length is 120 pc. The length of the Sgr C filament, at an assumed distance of 8.5 kpc, is only 25 pc. Thus, we would not have

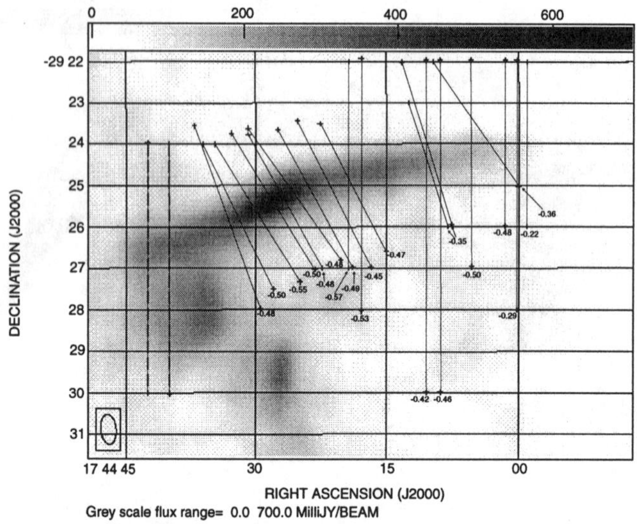

Figure 9. Spectral index along the Sgr C filament based on 330 MHz (this work) and 1400 MHz (Liszt & Spiker 1995) VLA data. The locations of the cross-cuts are shown superposed on a subimage of Figure 3.

expected to see synchrotron aging along its length. Conversely, the restricted length of this NTF may imply that it is relatively young.

Finally, the qualitative appearance of many of the NTFs on our higher resolution images (e.g., Fig. 8) is that of twisted strands which tend to unravel near their ends. This appearance is somewhat at odds with most NTF generation mechanisms which typically assume relativistic electrons are accelerated at one end of the NTF (Morris & Serabyn 1996). Measurements of NTFs from the newly processed 330 MHz data and their implications are examined more closely in LaRosa et al. (1999).

6. Summary and Future Plans

We have successfully reprocessed archival 330 MHz VLA Galactic Center observations using wide-field imaging algorithms, thereby compensating for the distortions produced by the non-coplanar array geometry. The resulting image is a significant improvement over previous images and provides an excellent overview of Galactic Center structure. We have also used "targeted faceting" to generate images at the full resolution of the data for the first time for regions well away from Sgr A.

An extensive catalogue of over 100 sources is being prepared. Many of the sources have not been previously detected, and include a variety of unresolved and extended features. Follow-up observations of one of these has already led to the discovery of G358.85+0.47, the "Pelican," a new NTF located (angularly) more distant from the Sgr A and at a significantly different orientation than all previously known NTFs. Follow-up observations of other new features are already underway.

Preliminary examination of the spectral index variations along the NTFs, made from measurements based on the new 330 MHz image, indicates that their spectra are relatively constant and that they are young. Moreover their morphology would seem inconsistent with present models of NTF formation and may suggest that a new theory is required.

A second epoch 330 MHz VLA Galactic Center campaign is currently in progress. The significantly improved hardware and data reduction techniques of the modern VLA should produce and even more sensitive wide-field 330 MHz image than was made from the decade-old archival data. Moreover, a second epoch map can be used in conjunction with the first epoch map presented here as a monitor of variable sources and transient phenomena associated with the Galactic Center. A first epoch 74 MHz campaign is also in progress.

Acknowledgments. We thank A. Pedlar for sharing his calibrated visibility data with us. We thank J. Imamura and S. Bollinger for assistance with the identification of sources. TNL was supported by a Navy-ASEE summer faculty fellowship and a NASA-JOVE grant to KSU. TJWL is supported by an NRC-NRL Research Associateship. SDH is supported by a grant from the Jeffress Memorial Trust. Basic research in radio astronomy at the NRL is supported by the Office of Naval Research.

References

Anantharamaiah, K. R., Lang, C. C., Kassim, N. E., Lazio, T. J. W., & Goss, W. M. 1999, this volume, p. 507

Anantharamaiah, K. R., Pedlar, A., Ekers, R. D., & Goss, W. M. 1991, MNRAS, 249, 262

Cornwell, T. J. & Perley, R. 1992, A&A, 261, 353

Ekers, R. D., van Gorkom, J. H., Schwarz, U. J., & Goss, W. M. 1983, A&A, 122, 143

Kassim, N. E.., Briggs, D. S., & Foster, R. S. 1998, in New Horizons in Multi-Wavelength Sky Surveys, eds. B. J. McLean, D. A. Golombek, J. J. E. Hayes, & H. E. Payne (Kluwer: Dordrecht) p. 89

Kassim, N. E. & Frail, D. A. 1996, MNRAS, 283, L51

Kassim, N. E., LaRosa, T. N., & Erickson, W. C. 1986, Nature, 322, 522

Khokhlov, A. & Melia, F. 1996, ApJ, 457, L61

LaRosa, T. N., Kassim, N. E., Lazio, T. J. W., & Hyman, S. D. 1999, AJ, in preparation

LaRosa, T. N. & Kassim, N. E. 1985, ApJ, 299, L112

Lazio, T. J. W., Cordes, J. M., Anantharamaiah, K. R., Goss, W. M., & Kassim, N. E. 1999, this volume, p. 441

Liszt, H. S. & Spiker, R. W. 1995, ApJS, 98, 259

Mezger, P. G., Zylka, R., Salter, C. J., Wink, J. E., Chini, R., Kreysa, E., & Tuffs, R. 1989, A&A, 209, 337

Morris, M. & Serabyn, E. 1996, ARA&A, 34, 645

Pedlar, A., Anantharamaiah, K. R., Ekers, R. D., Goss, W. M., van Gorkom, J. H., Schwarz, U. J., & Zhao, J. H. 1989, ApJ, 342, 769

Sofue, Y. & Handa, T. 1984, Nature, 310, 568

Yusef-Zadeh, F., et al. 1999, this volume, p. 197

The Central Parsecs of the Galaxy
ASP Conference Series, Vol. 186, 1999
H. Falcke, A. Cotera, W.J. Duschl, F. Melia, M.J. Rieke, eds.

Sgr A East and its Interaction With Surrounding Gas

Robert Zylka[1,2], Rolf Güsten[2], Sabine Philipp[2,1], Hans Ungerechts[3], Peter G. Mezger[2], Wolfgang J. Duschl[1,2]

Abstract. We have started a spectral line on-the-fly mapping project at the IRAM 30m Telescope in the gas volume density and column density tracing transitions of CS(2–1), CS(3–2) and ^{13}CO(2–1). In the continuum 1.2 mm emission, a $\sim 15'$ wide band along the Galactic plane between Sgr C and Sgr B2 has been mosaiced. The spectroscopic data obtained so far cover the central $\sim 21 \times 20'$. We present some preliminary results revealing the kinematics of the compact GMCs labeled M–0.13–0.08 and M–0.02–0.07, as well as of the Circumnuclear Disc, and we derive insights for the physical interaction with the non-thermal source Sgr A East and the associated gas. In addition a possible scenario for the distribution of the gas along the line of sight is derived.

1. Introduction

The giant molecular clouds associated with Sgr A were among the first detected by radio astronomy, but the kinematics of the neutral gas in the Galactic Center region still exhibits many unresolved questions. The situation becomes most uncertain for the central ~ 50 pc, where many features – often quite isolated in velocity-space – are observed with peculiar kinematics (revealing high velocity gradients, some at "forbidden" velocities). The location of the gas within the nuclear region and the interaction of the clouds are still not very well understood. One particular example is the interaction of the non-thermal source Sgr A East with the surrounding material.

Only very limited sets of spectroscopy data with an angular resolution sufficient to study the fine kinematic details of the gas (better than $25'' \approx 1$ parsec) have been obtained there so far. Serabyn and Güsten (1987) mapped the Bridge region (north of Sgr A*) in CS(2-1), Zylka et al. (1990) the Sgr A Complex ($\sim 7 \times 6'$) in the 1–0 and 2–1 transitions of ^{13}CO and C^{18}O with the IRAM 30-m telescope. Only with recent progress in advanced observing and data acquistion techniques, more complete mapping of the extended structures became possible, with good sensitivity and high angular resolution.

[1]Institut für Theoretische Astrophysik, Tiergartenstraße 15, 69121 Heidelberg, Germany

[2]Max-Planck-Institut für Radioastronomie, Auf dem Hügel 69, 53121 Bonn, Germany

[3]Instituto de Radioastronomía Milimétrica, Avenida Divina Pastora 7, 18012 Granada, Spain

2. Observations

We present first results of ongoing molecular line and dust continuum surveys, performed with the IRAM 30m-telescope. To complement continuum observations with the MPIfR 19- and 37-element bolometer array receivers, we have started an on-the-fly mapping project in the gas volume density and column density tracing transitions of CS(2–1), CS(3–2) and ^{13}CO(2–1). In the continuum 1.2 mm emission, a $\sim 15'$ wide band along the Galactic plane between Sgr C and Sgr B2 has been mosaiced. The spectroscopic data obtained so far cover the central $\sim 21 \times 20'$ (see Fig. 1).

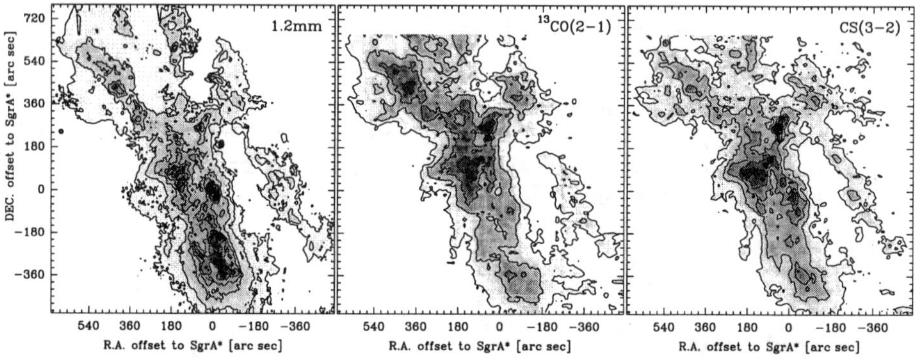

Figure 1. The central $\sim 15'$ of the Galaxy, as seen in the 1.2 mm continuum and velocity integrated molecular emission. The angular resolution is 11, 12 and 17″, respectively. The data were obtained during several observing sessions with the IRAM 30 m telescope in the on-the-fly mapping mode.

3. The Main Components of the Sgr A Radio and Cloud Complex

The distribution of molecular clouds as observed in the 1.2mm continuum, and in the ^{13}CO(2–1) and CS(3–2) emission is shown in Fig. 1. In the central 40″ the observed 1.2 mm continuum is dominated by the synchrotron radiation of Sgr A*, centered at (0,0), and the free-free emission of Sgr A West. The cloud located to the south is generally referred to as M–0.13–0.08 or the "20 km s^{-1} cloud", the one on the eastern edge of Sgr A East is M–0.02–0.07 or the "50 km s^{-1} cloud". – One should be aware, however, that these names are misleading, because the gas motion in this region shows very large velocity gradients and, as we will show below, these two clouds form one cloud complex. To this cloud complex belongs also the Circumnuclear Disc (CND), the only known structure within the central parsecs with kinematics clearly determined by the gravitational potential of the central stellar cluster.

The location of these giant molecular clouds along the line-of-sight is uncertain. The fact that they are responsible for the extinction of NIR emission from stars in the Nuclear Bulge (see Philipp et al. 1999) clearly shows that they are located in the Galactic Center region and embedded within the Nulcear Bulge. If these clouds were much behind the Nuclear Bulge, we would not see any extinction. If they were at large galactocentric distances from the solar system, given the clouds huge column densities corresponding to $\tau_K \sim 0.3$, we would expect much deeper absorption than is actually observed.

Comparison between the 1.2mm continuum and molecular line emission reveals some notable differences. Outside of the "Bridge" region, where the bolometer observations loose a significant fraction of this very extended emission, the low intensity emission compares rather well. The denser cores of these clouds, however, appear significantly different. Towards Sgr A free-free and synchrotron emission add to the underlying lower-level dust emission. This can explain the observed difference of this region. Towards other cloud cores, effects due to radiative transport and excitation of the molecular emission have to be studied in more detail.

4. Kinematics of the Sgr A Complex

The kinematics of the gas within the central $\sim 10'$ changes rapidly across Sgr A East. The velocity gradients are generally larger along Galactic latitude

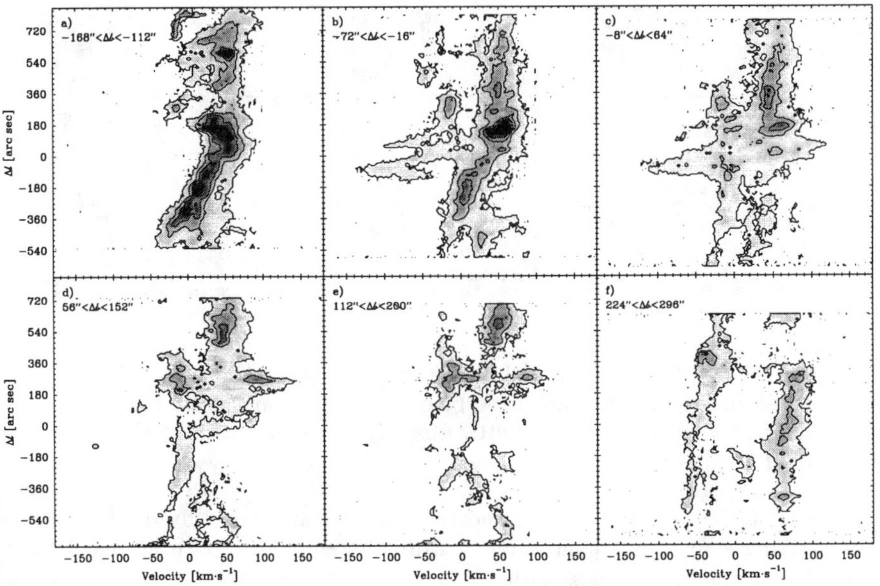

Figure 2. Representative Galactic longitude-velocity plots (v$-\Delta l$) of the CS(3$-$2) emission. In order to emphasize the main kinematic features, the data have been averaged over selected ranges in Galactic latitude, Δb.

than along Galactic longitude. Within the telescope beam of 12″, some features cover a velocity range of more than 100 km s^{-1}. Along the eastern edge of Sgr A East typically 1–2 very strong lines are visible, with $T_{MB} > 20$ K in the $^{13}CO(2-1)$ line. Along its western edge, as many as 6–7 components do exist (Zylka 1996).

To demonstrate the complex kinematics of the gas, in Figs. 2 and 3 we present representative position-velocity diagrams along Galactic longitude and latitude, respectively. In Fig. 2a, the continuity of emission between M−0.13 − 0.08 and M−0.02 − 0.07 is shown, with a band of emission extending smoothly from ~ -30 km s^{-1} at the most negative longitude offset to $\sim$40 km s^{-1} at $\Delta l \sim 90''$, at the position of M−0.02 − 0.07. Here a bow-shaped velocity feature is visible, with a strong velocity gradient. This gas is located exactly at the edge of the 6 cm continuum emission from Sgr A East (Yusef-Zadeh & Morris 1987), suggestive of intimate physical interaction. The extension of the emission band to higher longitude offsets is unclear. In Fig. 2b, high negative-velocity gas can be seen, both at positive and negative longitudes. The feature at negative longitude offsets is the southern lobe of the circumnuclear disk. The feature at positive offsets, revealing a rather similar velocity gradient, seems to be associated with Sgr A East (Figs. 3c,d).

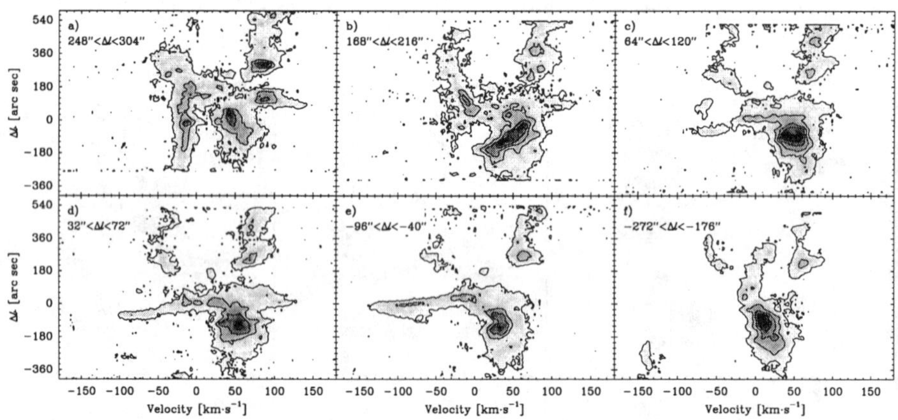

Figure 3. Galactic latitude-velocity plots (v−Δb) of the CS(3−2) emission. Again, the data have been averaged over selected ranges in Galactic longitude, Δl, to better display the kinematics of the Sgr A cloud complex.

In plots 2d) – f) the emission north of Sgr A East is displayed. Here, where the striking stream-like continuum features extend outward from the bulk of the 6 cm emission (compare with Fig. 4), the gas shows very peculiar kinematics. This is nicely visible in the longitude-averaged data in Fig. 3, showing a cloud complex whose components seem to point radially to a common center in velocity-latitude space. At smaller longitudes (Fig. 3b) the gas bordering the northern edge of Sgr A East, defined by the 6 cm emission, is shown. In Figs. 3b-d several features with an enormous velocity gradient (in latitude) are displayed.

At negative Galactic longitude offsets (Fig. 3e), gas from the northern part of M−0.13 − 0.08 bends towards negative velocities and seems to combine with the southern lobe of the CND. Finally in Fig. 3f the kinematics of M−0.13 − 0.08 are presented.

In all velocity-Δb plots a paraboloid structure above the Galactic plane is visible, which opens towards positive latitudes. This structure was also seen by Tsuboi et al. (1985).

5. Gas Compressed by Sgr A East

The morphology and velocity structure of the gas shell compressed by Sgr A East is shown in Fig. 4. Because of the high opacity of the $^{13}CO(2-1)$ transition, the

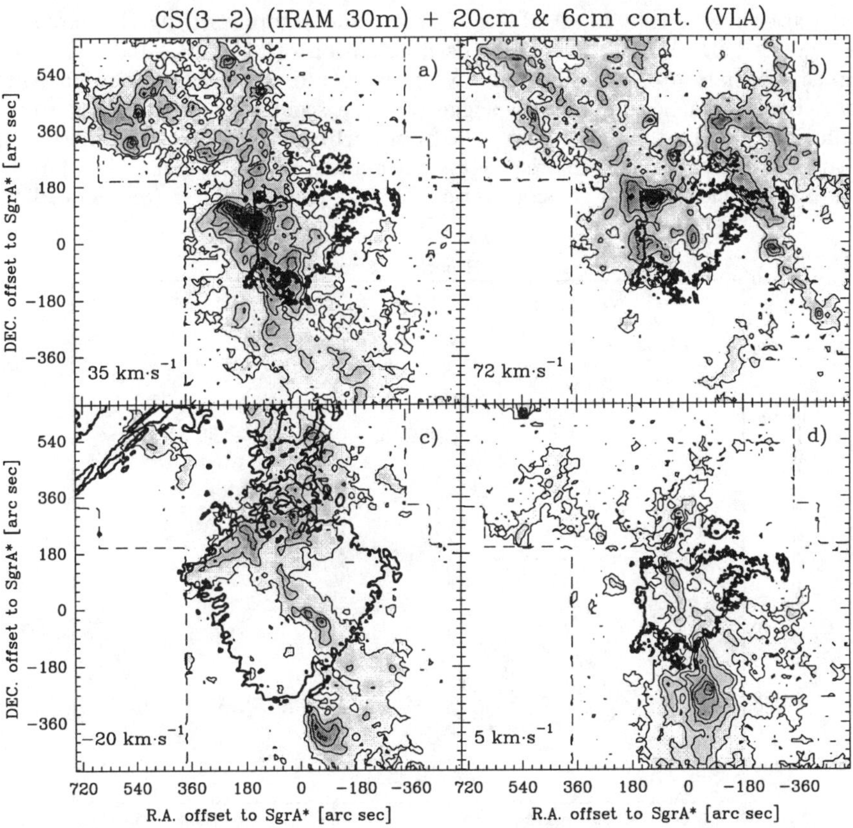

Figure 4. Four representative velocity channel maps of the CS(3−2) emission, showing the gas towards Sgr A East proper (panel a and b), towards the Bridge (panel c) and an example of disrupted shell (panel d) associated with Sgr A East. The overlaid heavy lines correspond to low-level contours of the 6 cm (panels a, b & d; Yusef-Zadeh & Morris 1987) and 20 cm (panel c; Yusef-Zadeh 1989) continuum emission.

kinematic details are best visible in the CS(3−2) transition. Therefore we use this CS(3−2) transition to discuss the properties of the gas associated with Sgr A East. Between ∼ 0 and ∼ 80 km·s^{-1} the gas with the highest column density follows closely the envelope of Sgr A East. Figs. 4a,b show two representative velocity channels at 35 and 72 km·s^{-1}. At v ∼ −20 km·s^{-1} (Fig. 4c) the gas in the "Bridge" closely correlates with the northern edge of Sgr A East.

At lower velocities, shell-like structures become visible that open in the same direction as the fountain-like streamers visible at 6 and 20 cm. An example is shown in panel d of Fig.4 (CS(3−2) at v∼5 km s^{-1}).

6. Possible Scenario for the Gas Distribution

Our data suggest that M−0.02 − 0.07, the cloud compressed by the explosion that created Sgr A East, is the northern extension of M−0.13 − 0.08. The apparent association of the high-velocity gradient and negative velocity gas with Sgr A East (Figs. 2b and 3c,d), is suggestive of a scenario in which – due to the explosion that created Sgr A East – the front part of the cloud became disrupted, creating tidally unstable debris. Sgr A East and the debris must therefore be located very close to the nucleus and might even feed the CND (Sanders 1999).

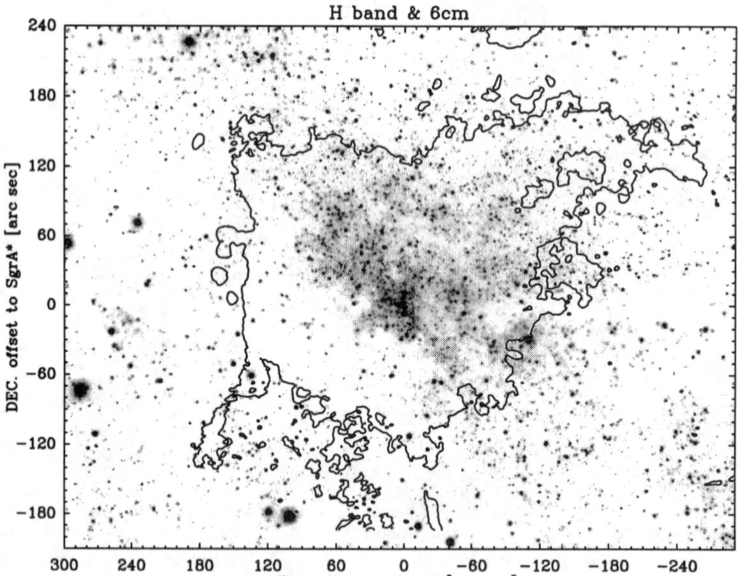

Figure 5. H band (λ1.65 μm) mosaic of the central ∼ 600″ × 500″. The overlaid contour corresponds to the 6 cm continuum emission (Yusef-Zadeh & Morris 1987) of Sgr A East. Clearly one sees the the most luminous stars and the highest star density in the H mosaic within the central part of Sgr A East.

The explosion removed much of the gas along the line-of-sight towards the center of Sgr A East, giving us a much deeper insight into the central stellar cluster. In such a scenario we can understand why outside of the Central Cavity (i.e. outside of the central $\sim 30''$) we find the most luminous stars and the highest star density in the central part of Sgr A East (Fig. 5; see also Philipp et al. 1999).

References

Philipp S., Mezger P.G., Zylka R., 1998, this volume, p. 300

Sanders, R. 1999, this volume, p. 250

Serabyn E., Güsten R., 1987, A&A 184, 133

Tsuboi M., Inoue M., Handa T., Tabara H., Kato T. et al., 1986, AJ 92, 818

Yusef-Zadeh F., Morris M, 1987, ApJ 320, 545

Yusef-Zadeh F., 1989, IAU-Symp 139, 243

Zylka R., Mezger P.G., Wink J.E., 1990, A&A 234, 133

Zylka R., 1996, ASPC Series 102, 'The Galactic Center', 77

The Central Parsecs of the Galaxy
ASP Conference Series, Vol. 186, 1999
H. Falcke, A. Cotera, W.J. Duschl, F. Melia, M.J. Rieke, eds.

Ionized Gas in the Sgr A Complex – VLA Observations of H168α and H270α Recombination Lines

K.R. Anantharamaiah[1]

National Radio Astronomy Observatory, Socorro, NM 87801, USA

A. Pedlar

Nuffield Radio Astronomy Laboratories, Jodrell Bank, UK

W.M. Goss

National Radio Astronomy Observatory, Socorro, NM 87801, USA

Abstract. Radio recombination lines at $\lambda = 20$ cm reveal the presence of an extended component of ionized gas of lower density (~ 100 cm^{-3}) in the Sgr A complex. This component extends well beyond the thermal 'mini-spiral' Sgr A-West which seem to be embedded in it. The low-density component is present over the entire extent of Sgr A East and possibly beyond and it is responsible for the turnover in the spectrum of Sgr A East and the halo observed by Pedlar et al (1989) at $\lambda = 90$ cm. The radial velocity of the extended ionized gas ranges from +50 km s^{-1} to –200 km s^{-1} with minima of emission near –50 and –150 km s^{-1}. The total mass of the low-density component is estimated to be $\sim 10^4$ M$_\odot$. A possible extended feature at a high velocity of 470 km s^{-1} is also detected, but needs further observations to confirm it.

1. Introduction

At centimeter wavelengths, most of the radio continuum emission from the Sgr A complex arises from four distinct components: (1) the ultra-compact non-thermal radio source Sgr A* which may be at the very center of the Galaxy (Lo et al 1998)), (2) the 'mini-spiral' Sgr A West which is a rotating ionized ring around Sgr A* (Ekers et al 1983, Roberts & Goss 1993), (3) the shell-type, non-thermal source Sgr A East which is possibly an energetic supernova remnant, located off-centered on Sgr A* (Ekers et al 1983), and (4) a triangular shaped halo which seem to surround all the other three components (Pedlar et al 1989). In addition to these four main components, the Sgr A complex also contains many weaker but distinct thermal and non-thermal radio features which have been classified and cataloged by Yusef-Zadeh and Morris (1987)

[1]Raman Research Institute, Bangalore 560 080, India

Using high resolution observations at 90 cm and 20 cm, Pedlar et al (1989) obtained constraints on the relative location along the line of sight of the main components of the Sgr A complex. They showed clearly that much of the thermal gas associated with Sgr A West lies in front of Sgr A* and Sgr A East. Figure 1a shows a 20 cm image of the Sgr A region. The large shell like structure is Sgr A East. The bright compact source within the shell and the bright extended region around it consists of Sgr A* and Sgr A West. At this resolution ($10'' \times 8''$), Sgr A* is not separated from Sgr A West and the mini-spiral-like structure of Sgr A West is not resolved. Figure 1b shows an image of the same region at 90 cm with the same angular resolution. The difference between the structures in Figures 1a and 1b is remarkable. The change in the structure is caused by the different emission mechanisms for Sgr A West and Sgr A East (thermal and non-thermal) and their relative location along the line of sight. The thermal source Sgr A West is seen as a depression in Figure 1b whereas it appears as a bright emission feature superposed over the non-thermal source Sgr A East in Figure 1a. The conclusion is obvious - Sgr A West is in front of Sgr A East and the former has become optically thick at $\lambda = 90$ cm and thus absorbs the radiation from the non-thermal source Sgr A East.

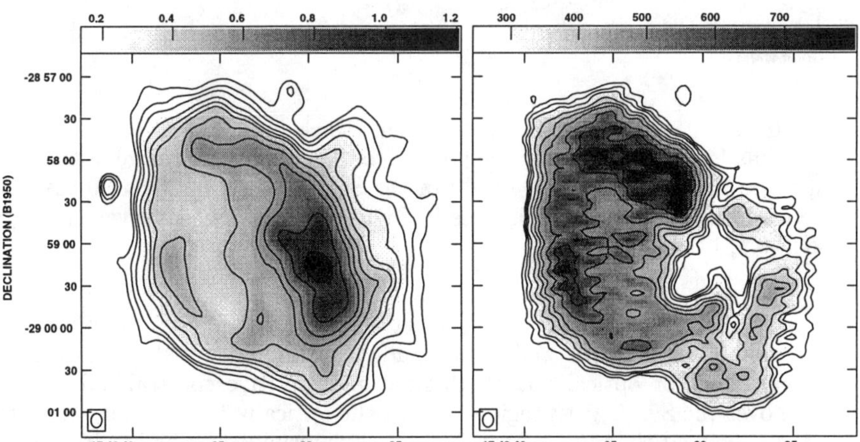

Figure 1. (a) A 20 cm image of the Sgr A region. Contour levels are 150, 175, 200, 250, 300, 350, 450 ... 1250 mJy/beam. (b) A 90 cm image of the same region. Contour levels are 250, 275, 300, 350, 400, 450, 550, 650, 750 mJy/beam. The images are taken from Anantharamaiah et al (1991) and convolved to a resolution of $10'' \times 8''$.

The difference between the emissions at 20 cm and 90 cm is seen quantitatively in Figure 2a, which displays EW cross-cuts through the images in Figures 1a and 1b at the position of Sgr A*. At 20 cm (thick line) the minor peak on the east is due to the shell of Sgr A East and the strong peak is due to the combination of Sgr A* and Sgr A West. The strong peak at 20 cm turns to an apparent absorption-like feature at 90 cm (thin line) indicating that both Sgr A* and Sgr A East are absorbed by the optically thick Sgr A West. An important implication of this result is that *it is not possible to obtain any measurement of the flux density of Sgr A* at frequencies below about 500 MHz* and thus no constraints on theoretical models which try to explain the overall spectrum of Sgr A* can be placed at these frequencies (e.g. Mahadevan 1998). At these frequencies, Sgr A* is simply hidden behind an absorbing screen.

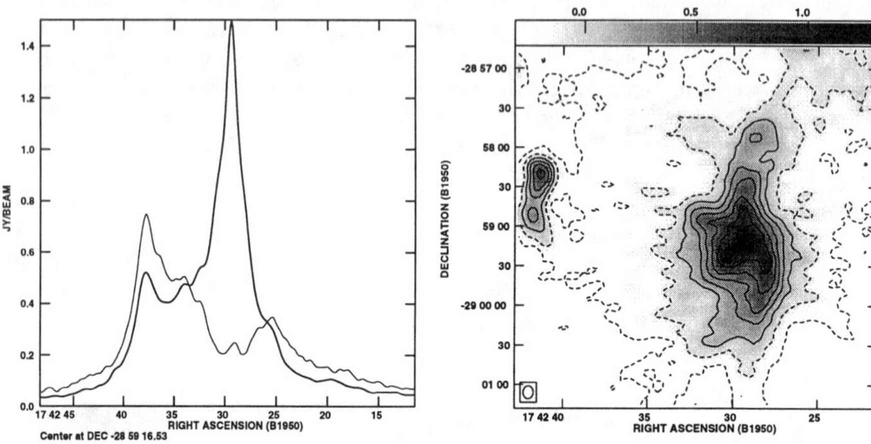

Figure 2. (a) EW cross-cuts at 20 cm (thick line) and 90 cm (thin line) made from Figure 1 at the position of Sgr A* (b) An image of the spectral index α (where $S_\nu \propto \nu^\alpha$) of the Sgr A region between 90 cm and 20 cm made from the images in Figure 1. The contour levels are -0.3,-0.1,.1,0.3 ... 1.3.

Figure 2b shows an image of the spectral index α (where $S_\nu \propto \nu^\alpha$) of the Sgr A region between 20 cm and 90 cm made from the images shown in Figure 1. As expected for optically thick thermal gas, positive spectral indices are observed over the Sgr A West region. In fact the optically thick region seems to extend well beyond the range of Sgr A West as determined by higher frequency observations (Ekers et al 1983). As noted by Pedlar et al (1989), the mini-spiral Sgr A West seems to be embedded in a somewhat lower density extended ionized gas. Furthermore, the spectral index of Sgr A East and the halo between 20 cm and 90 cm is 'flatter' ($\alpha \sim -0.3$ to 0.0) when compared to the spectral index between 20 cm and 6 cm ($\alpha \sim -1.1$ to -0.9: Ekers et al 1983, Pedlar et al 1989) indicating that free-free absorption by thermal gas causes the turnover in their non-thermal spectrum. A quantitative estimate of the amount of thermal gas in front of Sgr A East and also in the halo was made by Pedlar et al (1989) using high resolution images at 2 cm, 6 cm, 20 cm and 90 cm. They showed: (1) the mini-spiral of Sgr A West is embedded in a thermal-halo, about 1.5′ in angular extent and a ionized gas mass of $\sim 500 M_\odot$, (2) ionized gas with an emission measures $\sim 10^5$ pc cm^{-6} and $\tau_{90cm} \sim 1$ is present over the extent of Sgr A East ($\sim 3'$) and possibly beyond, and (3) the triangular-shaped halo surrounding both Sgr A East and Sgr A West is possibly a mixture of thermal and non-thermal gas and may be responsible for part or all of the spectral turnover of Sgr A East between 20 cm and 90 cm.

In this paper, we report observations of recombination lines at 20 cm and 90 cm using the VLA. The 20 cm data clearly shows the presence of non-zero velocity ionized gas present over the entire angular extent of Sgr A East and also beyond. The 90 cm data shows zero-velocity lower-density ionized gas present along the line of sight to the Galactic Center.

2. Observations of Recombination Lines Near 1374 and 332 MHz

Recombination lines (RL) at lower frequencies (e.g. $\nu \leq 1.4$ GHz) are sensitive to lower density extended ionized gas due to enhanced stimulated emission in such gas at higher quantum levels (Shaver 1975, Pedlar et al 1978, Anantharamaiah 1985). On the other hand, at higher frequencies ($\nu \geq 5$ GHz), it is easier to observe higher density, compact ionized regions. High resolution recombination line observations of the Galactic center region has so far been made only at frequencies above 5 GHz. (Van Gorkom et al 1985, Roberts & Goss 1993, Yusef-Zadeh et al 1995) and they have been useful in studying relatively high density ($n_e > 10^3$ cm^{-3}) ionized gas in Sgr A West, in the Arched Filaments and in other compact HII regions in that area. However, the extended lower density ionized gas in the Sgr A complex, for which evidence was presented in the previous section, has not been detected in high frequency recombination lines. We therefore undertook observations of the H168α line at $\nu_{rest} = 1374.6$ MHz and the H270α line at $\nu_{rest} = 332.25$ MHz using the VLA.

Observations at 1374 MHz were made in 1996 in both C and D configurations of the VLA using a bandwidth of 6.25 MHz and 64 channels. The velocity resolution was 21.3 km s^{-1} and the velocity coverage was $\pm$650 km s^{-1}. The data were edited, calibrated and imaged using standard procedures in AIPS software. The angular resolution of the final line and continuum images is $56.7'' \times 24.6''$ and PA = 35°. For the H270α line at 332 MHz, we used the data obtained in the D configuration of the VLA in 1987 as a part of the continuum study published by Pedlar et al (1989). The velocity resolution of the data is 11.0 km s^{-1} and covers a range of $\pm$300 km s^{-1}. The angular resolution is $8.2' \times 3.7'$ with PA = 10°.

Figure 3 shows the spatially integrated profiles of the H168α and the H270α lines. The two line profiles are different which illustrate that recombination lines at different frequencies are sensitivity to different components of ionized gas. The higher frequency line (i.e. H168α) extends over a large velocity range ($\sim$ 200 km s^{-1}) and arises in relatively higher density gas ($n_e > 100$ cm^{-3}) close to the Galactic Center whereas the lower frequency line (i.e. H270α) is narrow ($\sim$30 km s^{-1}) and arises in a lower density gas ($n_e < 10$ cm^{-3}) which may be anywhere along the line of sight (Anantharamaiah & Bhattacharya 1985). Note that the expected radial velocity for the line-of-sight gas towards the Galactic Center is 0 km s^{-1}. The narrow feature near -150 km s^{-1} in the 90 cm profile is the C270α line. The H270α and the C270α lines are dominated by stimulated emission (Pedlar et al 1978, Anish Roshi & Anantharamaiah 1997) and thus their spatial distribution follows that of the continuum. With the sensitivity of the present observations, the lines are clearly detected only near the continuum

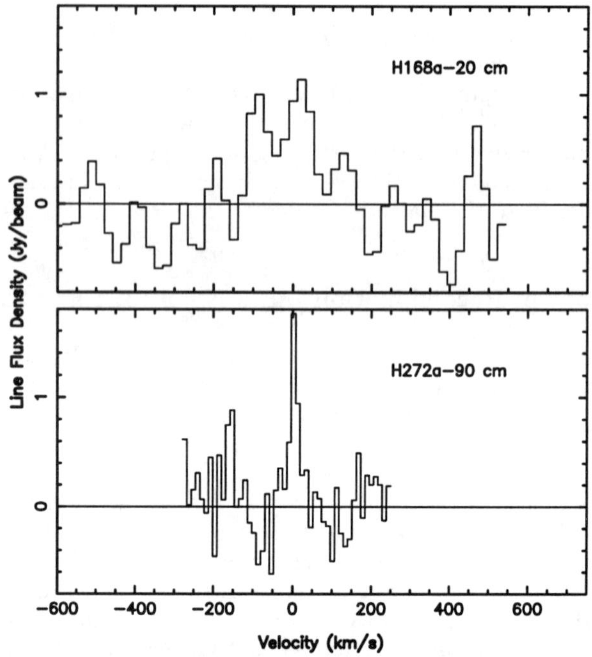

Figure 3. H168α (*top*) and H270α (*bottom*) recombination line profiles integrated over the Sgr A complex. The narrow feature near -150 km s^{-1} in the lower profile is the C272α line.

peak. The lower frequency line is not discussed further here since it does not arise close to the Galactic center.

The H168α line profile in Figure 3 is consistent with single-dish measurements at this position by Kestevan and Pedlar (1977). A slight uncertainty in the baseline (i.e. the zero-level) is still visible in the spectrum which is due to the residual 3-MHz ripple in the VLA waveguide system. A significant narrow feature at a velocity of 470 km s^{-1} is present in the H168α spectrum which needs to be confirmed by further observations.

Figure 4 shows the continuum image of the Sgr A complex at 20 cm (1374 MHz) in contours with H168α line emission integrated over the velocity range -200 km s^{-1} to $+42$ km s^{-1} (moment 0) superposed in grey scale. The angular resolution is $56.7'' \times 24.6''$ with PA = 35°. The continuum image shows the three main components of the Sgr A complex: the triangular shaped outer halo which is about 10′ in size, the shell-like structure of Sgr A East and the bright central region, Sgr A West. At this angular resolution, Sgr A* is not separated from Sgr A West. The integrated line emission in grey scale shows that the H168α line emission extends over the whole of Sgr A East. There can be little doubt that this extended emission arises in ionized gas associated with the Galactic Center since the velocities are non-zero and the velocity extent of this gas is much larger than that of the line-of-sight gas detected in the H270α line. The detection of an extended region of H168α line emission is consistent with the

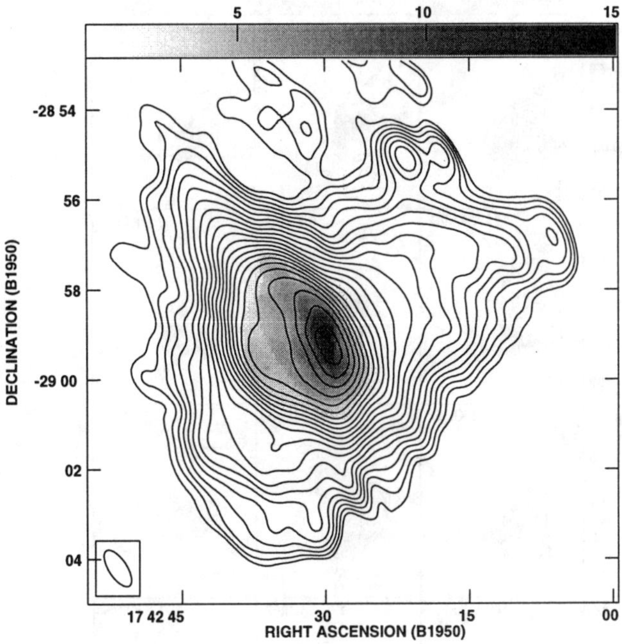

Figure 4. Continuum image of Sgr A complex at 1.37 GHz in contours with H168α line emission integrated over the range +50 to –200 km s^{-1} in grey scale. The first contour level is 0.3 mJy/beam and the subsequent ones are successively higher by a factor of 1.2. The grey scale range is 1 to 15 Jy/beam km s^{-1}. The beam is 56.7″ × 24.6″ and PA = 35°.

conclusions of Pedlar et al (1989) who attributed the turnover in the spectrum of Sgr A East to free-free absorption by thermal gas. There is also some weak H168α line emission to the north of Sgr A East where the base of the arched filaments meet the Sgr A halo at (l, b) = (0,0).

The velocities and spatial distribution of the H168α emission can be seen in the channel maps shown in Figure 5. The H168α emission can be identified in Figure 5 in three distinct velocity ranges. Taking into account the channel width of 21.3 km s^{-1}, these three velocity ranges are (1) +50 to –10 km s^{-1}, (2) –75 to –115 km s^{-1} and (3) –180 to –200 km s^{-1}. In addition to these emissions, a narrow high velocity feature with V_{lsr} = 470km s^{-1} was detected in the spectrum shown in Figure 3. The channel image at V_{lsr} = 470km s^{-1} is shown in Figure 6 in contours superposed over the continuum image in grey scale. If this line emission is confirmed by further observations, then it represents in-falling ionized gas at a high velocity and extended over a few parsecs.

In the velocity ranges +50 to -10 km s^{-1} and –75 to –115 km s^{-1}, the line emission is spread over the entire extent of Sgr A East with the peak of emission in front of Sgr A West. Comparing the observed velocities of the H168α line at

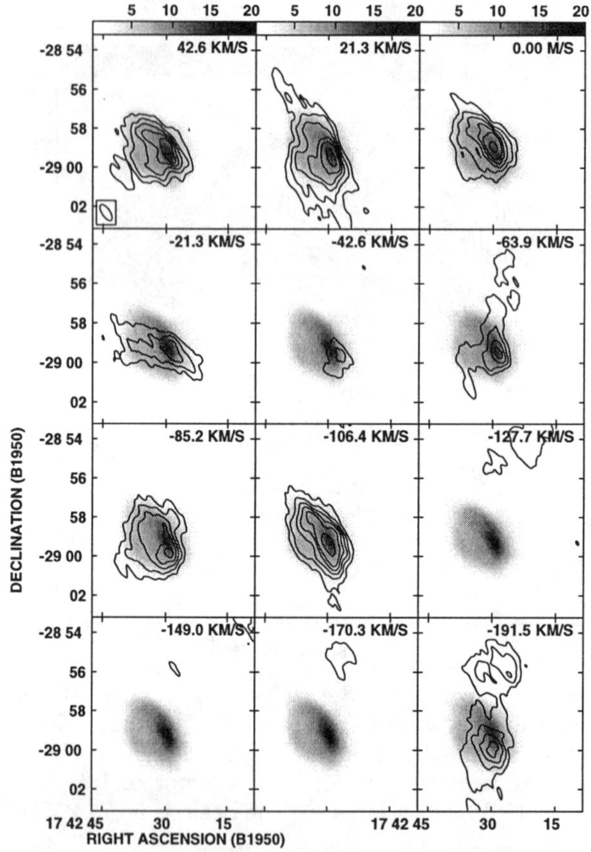

Figure 5. Channel maps of H168α emission in contours superposed over a grey scale continuum image at 20 cm. The LSR velocities are indicated in each frame. Contour levels are 20,30,40,50,60,80 mJy/beam. The beam size is as in Figure 3.

the position of Sgr A West with the velocities observed at higher frequencies by Roberts & Goss (1993) and Yusef-Zadeh et al (1995), it appears that the H168α line emission does not arise in the ionized gas that is directly associated with Sgr A West. The ionized gas detected in the H168α line is clearly extended and possibly lies in front of both Sgr A East and West sources. The similarity of the spatial structure of line emission (which is from thermal gas) and that of Sgr A East (which is a non-thermal source) strongly suggests that the H168α line is dominated by stimulated emission. It is therefore possible that the line emission extends even beyond Sgr A East which may not be detected here due to lack of sensitivity and diminished background continuum. Due to limited angular and velocity resolution of these observations, no systematic motion of this gas can be discerned. Since there are three distinct velocity ranges where line emission occurs, the three regions are likely to be separated along the line of sight. Some of this ionized gas may be associated with the halo and mixed with the non-thermal gas as suggested by Pedlar et al (1989).

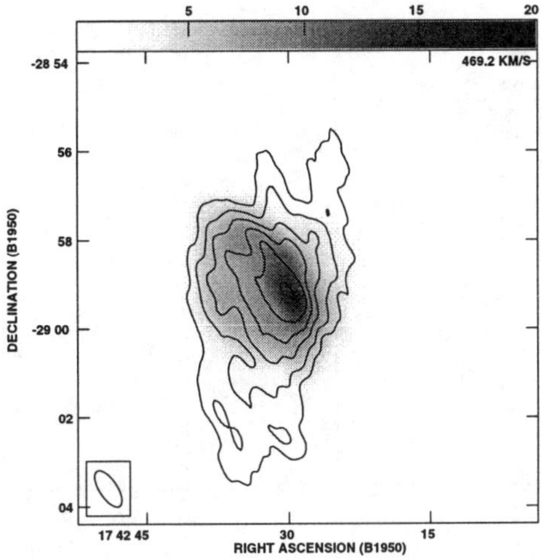

Figure 6. Channel image of the H168α emission at V = 470 km s^{-1} (in contours) superposed over the continuum image at 20 cm. The contour levels are 20, 30, 40, 50, 60, 80 mJy/beam. The beam size is as in Figure 4.

The peak of emission near −190 km s^{-1} occurs south of the continuum peak. Some weak emission at this velocity also occurs slightly north of Sgr A East. Although negative velocities are observed in the higher frequency H92α lines (Roberts & Goss 1993) in the southern arm of Sgr A West, the H168α line emission at this velocity is more extended than the southern arm and thus may represent a distinct component.

3. Constraints on the Density and Mass of Ionized gas

The extended ionized gas observed here in the H168α line is not observed in the lower frequency H270α line nor in the higher frequency H110α and H92α lines. High resolution H110α and H92α VLA observations may also have missed detecting extended gas due to lack of sensitivity to such structures. As explained in Section 1, lower frequency continuum observations indicate the gas has a free-free optical depth $\tau_{ff} \sim 1$ at $\lambda = 90$ cm. These facts allow us to place some constraints on the density and emission measure of the ionized gas.

Figure 7a shows a H168α line profile spatially integrated over a $54'' \times 135''$ region of Sgr A East centered at RA = $17^h42^m35.2^s$ and DEC = $-28°59'18.0''$. This region is to the east of Sgr A West and thus the line profile has no contribution from the latter. As discussed in Section 1, this region shows a turnover in the spectrum of Sgr A East which is attributed to free-free absorption by thermal gas. It is reasonable to suppose that the H168α line in Figure 7a arises in this gas. Although the line profile in Figure 7a is complex consisting of at least 3 components, for the purpose of illustration of the model, we approximate

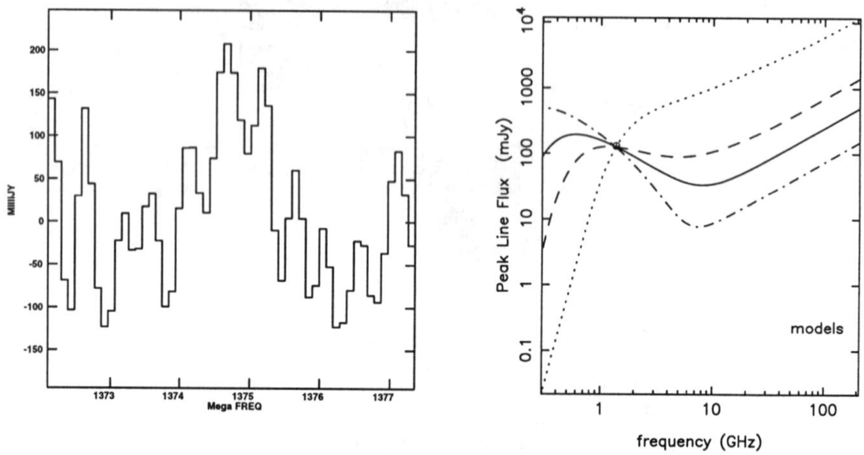

Figure 7. (a) H168α line profile integrated over a $54'' \times 135''$ region on Sgr A East at the position RA = $17^h42^m35.2^s$ and DEC = $-28°59'18.0''$. At this position there is no contribution to the line from Sgr A West. (b) Possible models to account for the line emission. See text for parameters of the models

it by a Gaussian with a peak of 130 mJy/beam, FWHM = 220 km s^{-1} and a centroid near 0 km s^{-1}. The integrated line flux is 1.3×10^{-21} W m^{-2}. The continuum flux density of Sgr A East over this region is $\sim$ 36 Jy/beam at 1.37 GHz. We assume an intrinsic spectral index of -0.9 based on higher frequency measurements (Ekers et al 1983).

To illustrate possible models which can account for the observed line emission, we assume a homogeneous slab of ionized gas at a temperature of 10^4 K to be in front of Sgr A East. Figure 7b shows four models with different densities. The densities are 10 cm^{-3} for the dash-dot-dash line, 100 cm^{-3} for the solid line, 500 cm^{-3} for the dashed line and 5000cm^{-3} for the dotted line. The emission measure in each model is adjusted to fit the observed H168α line. It is clear from Figure 7b that the models with the highest (5000 cm^{-3}) and lowest (10cm^{-3}) densities do not produce the observed behavior. The high density model predicts a sharp increase in line strength towards higher frequencies, whereas the lower density model predicts an increase towards lower frequencies. Only the intermediate density models predict lower line strengths at both higher and lower frequencies as observed. We favor the model represented by the solid line which has n_e = 100 cm^{-3}, EM = 3.3×10^5 pc cm^{-6}. This model predicts a free-free optical depth of $\sim$1.1 at 90 cm consistent with the observed turnover in the spectrum of Sgr A East (Pedlar et al 1989). Stimulated emission accounts for 90% of the line strength. The predicted line strengths in Figure 1 (solid line) are consistent with non-detection at 330 MHz and between 5 GHz and 10 GHz.

The model also predicts that recombination lines at higher frequencies ($\nu > 10$ GHz) may be detectable. It is possible that some of the H66α line emission near 22 GHz observed by Mezger and Wink (1986) from an 'extended' component is from this gas. In the model represented by the solid line in Figure 1b, the total mass of the ionized gas is about 10^3 $M_\odot$. Ionization of this gas can be maintained by a single O5 star. Note that this model, which is for illustration, is only for the ionized gas which is in front of a part of Sgr A East and which is responsible for the line emission shown in Figure 7a. To account for the total line emission shown in Figures 3 and 5, it is necessary to separate the possible contribution from Sgr A West. This separation is not attempted here. If we assume that ionized gas with the same properties extends over whole of Sgr A East (i.e. about 4 arminutes), then the total ionized mass is $\sim 8 \times 10^4$ $M_\odot$.

The National Radio Astronomy Observatory is a facility of the National Science Foundation operated under cooperative agreement by Associated Universities, Inc.

References

Anantharamaiah, K.R. 1985, J. Ap. Astr., 6, 203

Anantharamaiah, K.R. & Bhattacharya, D. 1986, J. Ap. Astr. 7, 141

Anantharamaiah, K.R., Pedlar, A., Ekers, R.D., & Goss, W.M. 1991, MNRAS, 249, 262

Anish Roshi, D., & Anantharamaiah, K.R. 1997, MNRAS, 292,63

Ekers, R.D., van Gorkom, J.H., Schwartz, U.J. & Goss, W.M. 1983, A&A, 122, 143

Kestevan, M.J. & Pedlar, A. 1977, MNRAS, 180, 731

Lo, K.Y., Shen, Z.Q., Zhao, J.H. & Ho, P.T.P 1998 ApJ, 508, L61

Mahadevan, R., 1998, Nature, 394, 651

Mezger, P.G. & Wink, J.E. 1986, A&A, 157, 252

Pedlar, A., Davis, R.D., Hart, L. & Shaver, P.A. 1978, MNRAS182, 473

Pedlar, A., Anantharamaiah, K.R., Ekers, R.D., Goss, W.M., Van Gorkom, J.H., Schawrtz, U.J., & Zhao, J.H. 1989, ApJ, 342, 769

Roberts, D. & Goss, W.M. 1993, ApJS, 86, 133

Shaver, P.A. 1975, Pramana, 5, 1

van Gorkom, J.H., Schwartz, U.J, & Bregman, J.D. 1985, IAU symp 106, The Milky Way Galaxy, ed. H. von Woerden, R.J. Allen & W.B. Burton (Dordrect:Kluwer), 443

Yusef-Zadeh, F. & Morris, M. 1987, ApJ, 322, 721

Yusef-Zadeh, F., Zhao, J.H & Goss, W.M. 1995, ApJ, 442, 646

The Central Parsecs of the Galaxy
ASP Conference Series, Vol. 186, 1999
H. Falcke, A. Cotera, W.J. Duschl, F. Melia, M.J. Rieke, eds.

OH(1720 MHz) Masers at the Galactic Center

Mark Wardle

Special Research Centre for Theoretical Astrophysics, University of Sydney, NSW 2006, Australia

Farhad Yusef-Zadeh

Department of Physics & Astronomy, Northwestern University, 2131 N. Sheridan Rd., Evanston, IL 60208, USA

T.R. Geballe

Joint Astronomy Centre, 660 N. A'ohoku Pl., Hilo, HI 96720, USA

Abstract. OH(1720 MHz) masers permit direct measurements of the magnetic field strength at the Galactic Center, and their angular broadening constrains models for the medium that scatters radio waves in the central 100 pc of the Galaxy. As the 1720 MHz masers are unaccompanied by the main-line masers at 1665 and 1667 MHz, they must be pumped by collisions in molecular gas that is cooling after being overrun by a non-dissociative (i.e. C-type) shock wave. In particular, this confirms that the "SNR" Sgr A East is driving a shock into the M-0.02-0.07 molecular cloud. The intensity of the v=1–0 S(1) line of H_2 is consistent with the shock strength expected to be driven into the molecular gas by the pressure within Sgr A East.

1. Introduction

OH masers are "point" probes of the gas in molecular clouds and star forming regions, allowing precise velocity measurements and Zeeman determinations of the magnetic field strength. At the Galactic Center, angular broadening of maser spots is also of interest for understanding the location and nature of the ionized medium that scatters radio waves. If the conditions for level inversion are understood then in principle temperatures and densities in the gas, and some information about the radiation field can be inferred. It turns out that the inversion of the 1720 MHz transition of OH requires a restricted set of physical conditions that can only be achieved in molecular gas that is cooling after being overrun by a non-dissociative (C-type) shock. Therefore the masers mark a subset of the locations at the Galactic Center where molecular gas is being violently disturbed.

Several OH(1720 MHz) masers have been detected in Sgr A (Yusef-Zadeh et al. 1996); their locations are shown in Fig. 1. The gross dynamics of molecular gas associated with the Sgr A region have been studied in CO, HCN and CS

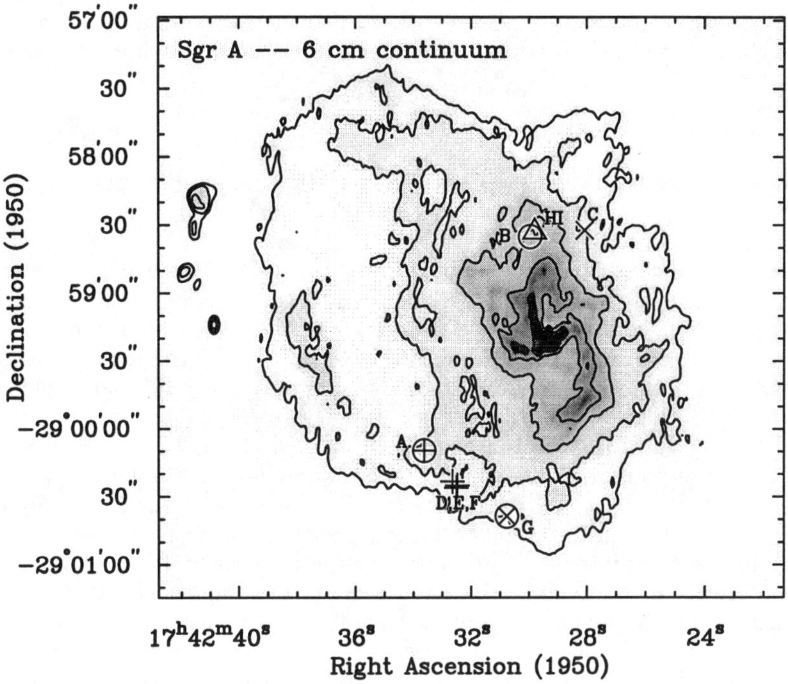

Figure 1. OH(1720 MHz) masers at the Galactic center (Yusef-Zadeh et al. 1996). The maser positions are labeled A–G, and the location of an HI Zeeman measurement by Plante et al. (1995) is labeled 'HI'

(Serabyn et al. 1986; Güsten et al. 1987; Serabyn, Lacy & Achtermann 1992) so the maser velocities unambiguously associate them with the Galactic Center. The features A and D–G at velocities of +50–60 km/s are located within the M-0.02-0.07 ("50 km/s") cloud, and B (+130 km/s) is located within the northern part of the circumnuclear ring. Feature 'C' at +43 km/s may also be physically associated with the northwest side of Sgr A East.

The masers allow an accurate determination of the magnetic field strength associated with the molecular gas at the Galactic Center. Previous Zeeman absorption measurements, in HI (Schwarz et al. 1990; Plante, Lo & Crutcher 1995) and in OH (Killeen, Lo & Crutcher 1992), detected milliGauss field strengths, but the weak signal had to be combined over large regions. These measurements are confused by spatial gradients in line profile and probably by changes in the magnetic field along the line of sight (e.g., through the circumnuclear ring, Wardle & Königl 1990). The compactness of the masers avoids these problems, and Zeeman measurements by Yusef Zadeh et al. (1996) yield line-of-sight fields of +3mG in the M-0.02-0.07 sources, −(3–4) mG for source B, and +2 mG for source C. In principle, the inferred line-of-sight field strength is overestimated if the masers are extremely saturated (Elitzur 1998), but this is not expected for OH(1720 MHz) masers in general (Lockett, Gauthier & Elitzur 1998), and does not appear to be the case at the Galactic Center (Yusef-Zadeh et al. 1998).

2. Interstellar Scattering

Studies of radio-wave scattering at the Galactic Center have previously been restricted to Sgr A* (e.g., Backer 1988; Yusef-Zadeh et al. 1994) and OH/IR stars (van Langevelde et al. 1992; Frail, Diamond et al. 1994). These studies indicate that the scattering medium probably lies within 100 pc of the center. As the nearest OH/IR star is about 3 arcminutes away from Sgr A*, the OH(1720 MHz) masers usefully sample the properties of the screen along lines of sight that are closer in projection on the plane of the sky. The broadening of a source depends on proximity to the scattering medium, so it is helpful that the location of the masers along the line of sight is well-determined, unlike that of individual OH/IR stars which may lie 100 pc in front or behind Sgr A*.

The masers suffer an angular broadening similar to that of Sgr A*, about an arcsecond at these frequencies (Yusef-Zadeh et al. 1998). The scattering disks (see Fig. 2) differ in both ellipticity and position angle because of variations in the line-of-sight-integrated properties of the medium. The conclusions that can be drawn are limited by the small number of samples available, but it is significant that the shape and orientation of the broadening differs between lines of sight separated by less than a parsec – if the intervening scattering medium were thicker than a few parsecs the integration along the line of sight would tend to wash these differences out. This favors "interface" models (Lazio & Cordes 1998) over, for example, the 10^8 K gas that fills the inner half a degree of the Galaxy (Yamauchi et al. 1990). Further, maser C does not show the same degree of broadening as the other sources, implying that if it is located at the Galactic Center then the scattering screen must be patchy.

3. OH(1720 MHz) Masers and Shock Waves

OH masers have traditionally been associated with HII regions and evolved stars. However, it has recently become apparent that 1720 MHz OH masers unaccompanied by 1665 and 1667 MHz masers are produced where supernova remnants (SNRs) interact with molecular clouds (e.g., Frail, Goss & Slysh 1994; Claussen et al. 1997; Koralesky et al. 1998). This is supported by studies of the pumping of the OH maser lines (Elitzur 1976; Pavlakis & Kylafis 1996a,b; Lockett et al. 1998), which show that the OH(1720 MHz) maser is collisionally pumped in molecular gas at temperatures and densities between 50-125 K and $10^5 - 10^6$ cm^{-3} respectively, whereas the 1665 and 1667 MHz masers are generally pumped by far-infrared photons. Thus the 1665/7 MHz masers are associated with HII regions and evolved stars. Occasionally 1720 MHz masers are also associated with main-line masers, but in their absence are located in cooling, shocked molecular gas under specific physical conditions.

Although the pumping conditions for the 1720 MHz OH masers suggest that they are associated with shock waves, OH itself is not formed directly by shock chemistry (e.g., Draine, Roberge & Dalgarno 1983; Hollenbach & McKee 1989;

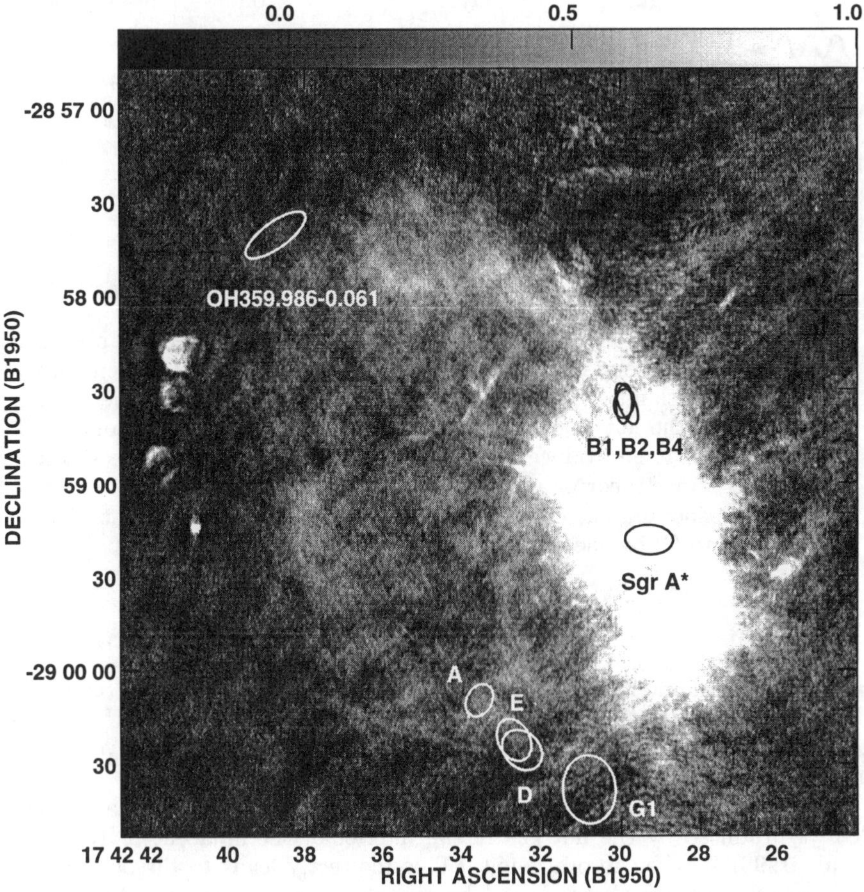

Figure 2. A λ6cm radiograph of Sgr A showing the scattering disks (enlarged by a factor of 15) of Sgr A*, an OH/IR star, and the OH(1720 MHz) masers (Yusef-Zadeh et al. 1998).

Kaufman & Neufeld 1996). Instead, the OH must be produced by dissociation of the copious amounts of water formed within a C-type shock (Wardle et al. 1998; Lockett et al. 1998). OH masers adjacent to compact HII regions are produced by photodissociation of water (Elitzur & de Jong 1978; Hartquist & Sternberg 1991; Hartquist et al. 1995), but in that case there is a strong dissociating FUV flux from the central star. This dissociating flux is largely absorbed and re-radiated by grains, providing sufficient FIR photons to also pump the 1665/7 MHz transitions (e.g., Pavlakis & Kylafis 1996b). This cannot be the case for the unaccompanied 1720 MHz masers associated with SNR-

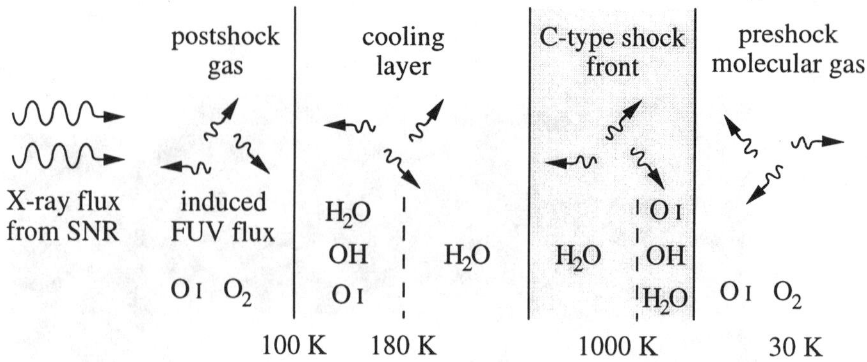

Figure 3. A model for the production of OH in molecular clouds associated with supernova remnants. A C-type shock is driven into a molecular cloud adjacent to a SNR. The X-ray flux from the SNR interior (to the left) permeates the cloud, inducing a weak secondary FUV flux that is produced locally throughout the cloud. The shock efficiently wave incorporates atomic and molecular oxygen into water. Once the shocked gas cools, water is subsequently dissociated to OH and then to O I by the secondary UV flux.

molecular cloud interactions. Although only a weak UV field is required to dissociate the water (Lockett et al. 1998), it is difficult to deliver UV photons to the postshock gas, as the attenuation of an external flux by intervening dust generates a FIR continuum that inverts the main line transitions.

This problem is circumvented if the irradiation of the molecular cloud by X-rays produced by the hot gas in the interior of the adjacent SNR (Wardle et al. 1998) is taken into account. This indirectly leads to dissociation: energetic electrons are photo-ejected by the X-rays and thermalize by collisions with molecules. Some of the collisions excite the Werner Ly-α band of H_2, and the subsequent radiative decay contributes a photon to the local FUV radiation field. Because X-rays are much more penetrating than UV and the ratio of the grain to molecule absorption cross-sections for X-rays is much lower than for UV photons (e.g., Maloney, Hollenbach & Tielens 1996) very little FIR is generated by this mechanism.

A sketch of the model is shown in Fig. 3, which shows a C-shock driven into a molecular cloud. The pre-shock gas at the right of Fig. 3 is at rest, and the oxygen that is not bound in CO exists as O I or O_2. As the gas is accelerated, compressed, and heated within the shock front the atomic and molecular oxygen is rapidly converted to water. The shocked, water-rich gas cools as it drifts behind the shock front. The entire shock structure is subject to the weak dissociating UV flux produced by the X-rays permeating the cloud. The dissociations cannot compete with the rapid reformation of water until the temperature drops to below about 200 K, at which point water is dissociated to OH and subsequently to O I.

4. H_2 2μm Emission Associated with Sgr A East

The masers to the southeast of Sgr A East, at positions B and D–F in Fig. 1, arise in shocked gas in the M-0.02-0.07 cloud, where the morphology and velocity gradients have already been interpreted in terms of an interaction with Sgr A East (Serabyn et al. 1992). The maser at position C on the northwest boundary of Sgr A East is likely to be indicative of a similar interaction. At position A the circumnuclear ring appears to be disturbed by gas associated with the Northern Arm (Jackson et al. 1993; Yusef-Zadeh et al. 1996).

In the case of Sgr A East there is sufficient information to think about the interaction in more detail. The hot interior of Sgr A East, which supplies the overpressure driving the shock wave into the M-0.02-0.07 cloud, should be in rough pressure equilibrium with the magnetic pressure inferred from Zeeman measurements, $4 \times 10^{-7}\,\mathrm{erg\,cm^{-3}}$, which dominates the pressure in the shocked molecular gas. Indeed, ASCA observations of the X-ray emitting gas filling Sgr A East (Koyama et al. 1996) imply $n_e \approx 6\,\mathrm{cm^{-3}}$ and $T \approx 10\,\mathrm{keV}$, or a pressure of $2 \times 10^{-7}\mathrm{erg\,cm^{-3}}$. Further, the X-ray luminosity from Sgr A East, $10^{36}\,\mathrm{erg\,s^{-1}}$, is sufficient to produce OH behind the shock front. Equating the magnetic pressure, $4 \times 10^{-7}\,\mathrm{erg\,cm^{-3}}$, to ρv_s^2, where ρ is the preshock density and v_s is the shock speed, and adopting a preshock density of $n_{\mathrm{H}} = 2 \times 10^4\,\mathrm{cm^{-3}}$ (Mezger et al. 1989), we infer $v_s \approx 25$–$30\,\mathrm{km\,s^{-1}}$. A C-type shock at this speed produces an intensity in the 1-0 S(1) line of 10^{-4}–$10^{-3}\,\mathrm{erg\,s^{-1}\,cm^{-2}}$ (Draine et al. 1983; Kaufman & Neufeld 1996).

This should be observable. Recently we used the long slit 1-5μm spectrometer, CGS4, at UKIRT to search for shocked H_2 emission associated with the OH(1720 MHz) masers in the 50 km/s cloud (covering positions A and D–G in Fig. 1). The spectra we obtained are shown in Fig. 4. The 1-0 S(1) (2.122 μm) line can be seen in the individual spectra, but other lines of H_2 are too weak for them to be securely identified. The 1-0 S(0) (2.223 μm) and 2-1 S(1) (2.248 μm) lines of H_2, and H I Br γ (2.166 μm) are, however, evident in the averaged spectrum.

The intensity of the 1-0 S(1) line, assuming that $A_K \approx 2.5$, is $1.5 \times 10^{-4}\,\mathrm{erg\,s^{-1}\,cm^{-2}}$, consistent with the levels "predicted" above. However, the the 2-1 S(1) / 1-0 S(1) and 1-0 S(0) / 1-0 S(1) line ratios in the average spectrum are about 1/7 and 1/3 respectively, somewhere between those expected for collisional excitation and for pure fluorescence. These ratios could be produced by intense UV irradiation of molecular gas (Sternberg & Dalgarno 1989), but the UV radiation field associated with Sgr A East is likely to be much less than in the inner parsec of the Galaxy, and the maser spots are distributed in an area free of the HII regions that would be associated with massive star formation. Given the weak detections of the 2-1 S(1) and 1-0 S(0) lines, further observations are required to determine the line ratios accurately.

Preliminary results from AAT and HST observations of shocked H_2 associated with the masers are discussed elsewhere in these proceedings (Yusef-Zadeh et al. 1999).

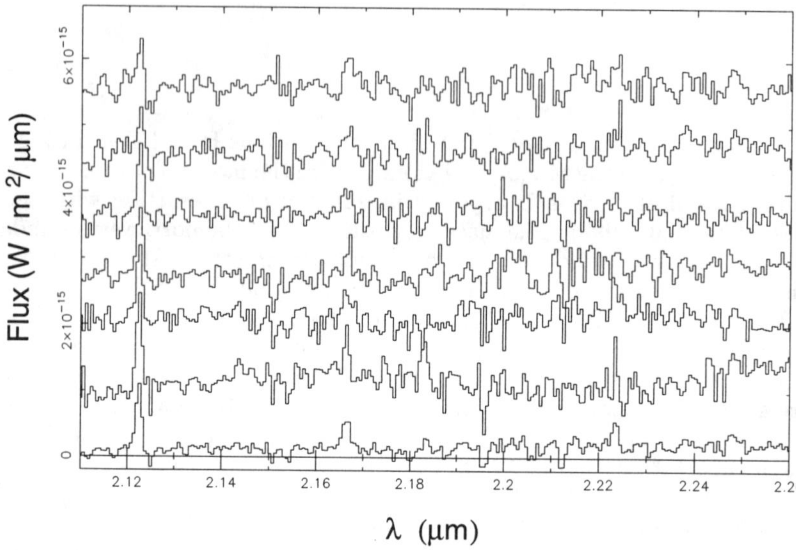

Figure 4. Six adjacent spectra, with vertical scales offset from one another, extracted from a CGS4 observation of the OH(1720 MHz) maser spots in Sgr A East and (at the bottom) the averaged spectrum. Each spectrum is of an area 1.2″ × 2.5″. The second row from the top coincides with position A, $\alpha(1950) = 17^h42^m33.6^s, \delta(1950) = -29^000'09.6''$. The top row is 1.2 arcsec SW, and the 3rd - 6th spectra are 1.2, 2.5, 3.7, and 4.9 arcsec to the NE of the above coordinates.

5. Discussion

In principle a low-energy cosmic-ray flux enhanced by a factor of 100 over the solar neighbourhood value has a similar effect on the heating and chemistry of molecular gas as X-ray irradiation from adjacent SNR (e.g., Maloney et al. 1996). In this case, energetic electrons produced by cosmic ray ionisation play an analogous role to the electrons photoejected by X-rays. This could well apply to many galactic SNRs – for example, the high-energy cosmic-ray flux in the W44, W28 and IC 443 SNRs that is required to produce the associated EGRET sources is roughly two orders of magnitude larger than the solar neighbourhood value (Esposito et al. 1996), so a naive extrapolation down to the cosmic-ray energies important for ionisation suggests that the cosmic-ray and X-ray contributions to the secondary FUV flux may be similar. In the case of Sgr A East, Melia et al. (1998) have argued that the Galactic Center EGRET source 2EG J1746-2852 can be explained similarly (see also Fatuzzo et al. 1999).

Acknowledgments. The CGS4 spectra were obtained as part of the UKIRT Service Programme. UKIRT is operated by the Joint Astronomy Center on behalf of the U.K. Particle Physics and Astronomy Research Council. The Special

Research Center for Theoretical Astrophysics is funded by the Australian Research Council under the Special Research Centres programme. F. Yusef-Zadeh's work was supported in part by NASA grant NAGW-2518.

References

Backer, D. C. 1988, in AIP Conf Proc 174: Radio Wave Scattering in the Interstellar Medium, ed. J. Cordes, B. Rickett & D. Backer (New York: AIP), 111

Claussen, M. J., Frail, D. A., Goss, W. M. & Gaume, R. A. 1997, ApJ, 498, 143

Draine, B. T., Roberge, W. G. & Dalgarno, A. 1983, ApJ, 264, 485

Elitzur, M. 1976, ApJ, 203, 124

Elitzur, M. 1998, ApJ, 504, 390

Elitzur, M. & de Jong, T. 1978, A&A, 67, 323

Esposito, J. A., Hunter, S. D., Kanbach, G. & Sreekumar, P. 1996, ApJ, 461, 820

Fatuzzo, M., et al. 1999, this volume, p. 560

Frail, D. A., Diamond, P. J., Cordes, J. M. & van Langevelde, H. J. 1994, ApJ, 427, L43

Frail, D.A., Goss, W. M. & Slysh, V.I. 1994, ApJ, 424, L111

Güsten, R., Genzel, R., Wright, M. C. H., Jaffe, D. T., Stutzki, J. & Harris, A. I. 1987, ApJ, 318, 124

Hartquist, T. W., Menten, K. M., Lepp, S. & Dalgarno, A. 1995, MNRAS, 272, 184

Hartquist, T. W. & Sternberg, A. 1991, MNRAS, 248, 48

Hollenbach, D. J. & McKee, C. F. 1989, ApJ, 342, 306

Jackson, J. M., Geis, N., Genzel, R., Harris, A. I., Madden,S., Poglitsch, A., Stacey, G. J. & Townes, C. H. 1993, ApJ, 402, 173

Kaufman, M. J. & Neufeld, D. A. 1996, ApJ, 456, 611

Killeen, N. E. B., Lo, K. Y. & Crutcher, R. 1992, ApJ, 385, 585

Koralesky, B., Frail, D. A., Goss, W. M., Claussen, M. J. & Green, A. J. 1998, AJ, in press

Koyama, K., Maeda, Y., Sonobe, T., Takeshima, T., Tanaka, Y. & Yamauchi, S. 1996, PASJ, 48, 249

Lazio, T. J. W. & Cordes, J. M. 1998, ApJ, in press (astro-ph/9804157)

Lockett, P., Gauthier, E. & Elitzur, M. 1998, ApJ, in press

Maloney, P. R., Hollenbach, D. J. & Tielens, A. G. G. M. 1996, ApJ, 466, 561

Melia, F., Fatuzzo, M., Yusef-Zadeh, F. & Markoff, S. 1998, ApJ, 508, L65

Mezger, P. G., Zylka, R., Salter, C. J., Wink, J. E., Chini, R., Kreysa, E. & Tuffs, R. 1989, A&A, 209, 337

Pavlakis, K. G. & Kylafis, N. D. 1996a, ApJ, 467, 300

Pavlakis, K. G. & Kylafis, N. D. 1996b, ApJ, 467, 309

Plante, R. L., Lo, K. Y. & Crutcher, R. M. 1995, ApJ, 445, L113

Schwarz, U. J., Lasenby, J., Kronberg, P. P. & Wielebinski, R. 1990, in Galactic and Intergalactic Magnetic Fields, ed. R. Beck (Dordrecht : Kluwer), 383

Serabyn, E., Güsten, R., Walmsley, C. M., Wink, J. E. & Zylka, R. 1986, A&A, 169, 85

Serabyn, E., Lacy, J. H. & Achtermann, J. M. 1992, ApJ, 395, 166

Sternberg, A. & Dalgarno, A. 1989, ApJ, 338, 197

van Langevelde, H. J., Frail, D. A., Cordes, J. M. & Diamond, P. J. 1992, ApJ, 396, 686

Wardle, M., & Königl, A. 1990, ApJ, 362, 120

Wardle, M., Yusef-Zadeh, F. & Geballe, T. R. 1998 ApJ, submitted (astro-ph/9804146)

Yamauchi, S., Kawada, M., Koyama, K., Kunieda, H. & Tawara, Y. 1990, ApJ, 365, 532

Yusef-Zadeh, F., Cotton, W., Wardle, M., Melia, F. & Roberts, D. A. 1994, ApJ, 434, L33

Yusef-Zadeh, F., Roberts, D. A., Goss, W. M., Frail, D. A. & Green, A. J. 1998,ApJ, in press (astro-ph/9809279)

Yusef-Zadeh, F., Roberts, D.A., Goss, W.M., Frail, D. & Green, A. 1996, ApJ, 466, L25 & Green, A. J. 1998,ApJ, in press (astro-ph/9809279)

Yusef-Zadeh, F., et al. 1999, this volume, p. 197

The Central Parsecs of the Galaxy
ASP Conference Series, Vol. 186, 1999
H. Falcke, A. Cotera, W.J. Duschl, F. Melia, M.J. Rieke, eds.

Hyperstrong Radio-Wave Scattering and Free Electrons in the Galactic Center

T. Joseph W. Lazio

National Research Council and Naval Research Laboratory, Code 7210, Washington, DC 20375-5351; lazio@rsd.nrl.navy.mil

James M. Cordes

Department of Astronomy, Cornell University and NAIC, Ithaca, NY, 14853-6801; cordes@spacenet.tn.cornell.edu

K. R. Anantharamaiah and W. M. Goss

NRAO, P. O. Box 0, Socorro, NM 87801; anantha@aoc.nrao.edu; mgoss@nrao.edu

Namir E. Kassim

Naval Research Laboratory, Code 7213, Washington, DC, 20375-5351; nkassim@rsd.nrl.navy.mil

Abstract. The scattering diameters of Sgr A* and several nearby OH masers ($\approx 1''$ at 1 GHz) indicate that a region of enhanced scattering is along the line of sight to the Galactic center. We have combined radio-wave scattering data and free-free emission measurements in a likelihood analysis that constrains the GC-scattering region separation to be $\Delta_{\rm GC} = 133^{+200}_{-80}$ pc and the angular extent of the region to be $0.5^\circ \leq \psi_\ell \lesssim 1^\circ$. The close correspondence between $\Delta_{\rm GC}$ and $\psi_\ell D_{\rm GC}$ suggests that the scattering region encloses the GC.

A uniform scattering region would cover G359.87+0.18, a Fanaroff-Riley II radio galaxy $15'$ from Sgr A*, producing a scattering diameter of at least $600''$ at 0.33 GHz. Its observed diameter is $20''$. We therefore conclude that the GC scattering region is patchy, with a typical separation between scattering clumps of approximately $5'$ (≈ 10 pc at a distance of 8.5 kpc). This separation is comparable to the size scale of GC molecular clouds. The similarity between these two length scales is an indication that the scattering region is linked to the GC molecular clouds. We favor a model in which the scattering arises in the interfaces between molecular clouds and the 10^7 K ambient gas; in this model the scattering medium is hot ($T_e \sim 10^6$ K) and dense ($n_e \sim 10$ cm^{-3}).

Pulsars seen through this region will have a dispersion measure of approximately 2000 pc cm^{-3}, of which approximately 1500 pc cm^{-3} arises from the GC, and will have a pulse broadening time of $350\nu_{\rm GHz}^{-4}$ *seconds*. We stress the uniqueness of the GC scattering region, probably resulting from the high-pressure environment in the GC.

1. Introduction

The observed diameter of Sgr A*, the compact source in the Galactic center (GC), scales as λ^2 from 30 cm to 3 mm and is anisotropic at least over the wavelength range 21 cm to 7 mm (e.g., Yusef-Zadeh et al. 1994; Bower & Backer 1998), as expected if interstellar scattering from electron density micro-structure determines the observed diameter. Maser spots in OH/IR stars within 25′ of Sgr A* also show enhanced, anisotropic broadening (van Langevelde et al. 1992; Frail et al. 1994).

These observations do not constrain the *radial* location of the scattering region for the following reason: The sources are in or near the GC, and a region of moderate scattering located far from the GC can produce angular broadening equivalent to that from a region of intense scattering associated with the GC. Although a scattering region associated with the GC is attractive for phenomenological reasons, other sites of enhanced interstellar scattering are found throughout the Galaxy (e.g., NGC 6634, Moran et al. 1990; Cyg X-3, Molnar et al. 1995) and the mean free path for encountering such a region is approximately 8 kpc (Cordes et al. 1991).

Associating the scattering with a specific region may elucidate the mechanism for the generation of the density fluctuations responsible for the scattering. The currently favored mechanism is that velocity or magnetic field fluctuations generate the density fluctuations (Higdon 1984, 1986; Spangler 1991; Sridhar & Goldreich 1994; Goldreich & Sridhar 1995). If correct, scattering observations provide a sub-parsec probe of the velocity or magnetic field within the region. Velocity or magnetic field fluctuations are also a natural means for inducing anisotropy in the density fluctuations and thereby in the scattering disks.

The location of the scattering region also has implications for radio pulsar searches toward the GC. Even if the scattering region is far from the GC, pulsars seen through it will suffer pulse broadening of at least 5 s at 1 GHz (Cordes & Lazio 1997). If the region is associated with the GC, only at frequencies above 10 GHz will pulsations be detectable (because pulse broadening $\tau_d \propto \nu^{-4}$).

We describe a series of observations and a likelihood analysis that indicate the scattering region is associated with the GC. A survey for extragalactic sources seen through the GC is described in §2., §3. describes the likelihood analysis, §§4. and 5. report what we have inferred about the GC scattering region, §6. summarizes the implications for the detection of radio pulsars in the GC, and §7. presents our conclusions.

2. A Survey for Extragalactic Sources Seen Through the Galactic Center

The scattering diameter of a compact extragalactic source viewed through the scattering region is (van Langevelde et al. 1992)

$$\theta_{\rm xgal} = \frac{D_{\rm GC}}{\Delta_{\rm GC}}\theta_{\rm Gal}, \tag{1}$$

where $\theta_{\rm Gal}$ is the characteristic diameter of a GC source—the diameter of Sgr A* at 1 GHz is $1\overset{\prime\prime}{.}3$, we take the GC-Sun distance to be $D_{\rm GC} = 8.5$ kpc ($1' = 2.5$ pc),

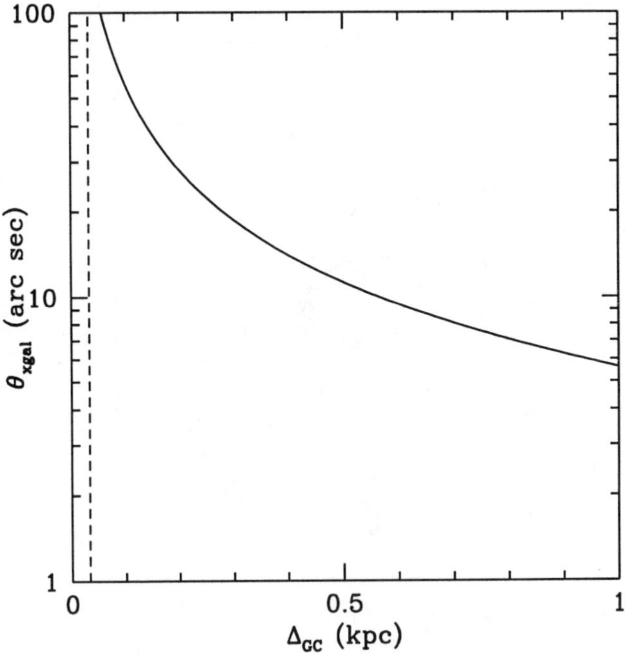

Figure 1. The diameter of an extragalactic source at 1.4 GHz seen through the scattering region in front of Sgr A* as a function of the distance of the scattering region from the GC, Δ_{GC}. The dotted line indicates an extreme lower limit on Δ_{GC} as derived from the lack of free-free absorption toward Sgr A* at centimeter wavelengths. At 1.4 GHz, the scattering diameter of Sgr A* is 0.″7. The scattering diameter scales as $\theta_s \propto \nu^{-2}$.

and Δ_{GC} is a characteristic separation between the GC and the scattering region. Figure 1 shows θ_{xgal} as a function of Δ_{GC}. If $\Delta_{GC} \gtrsim 1$ kpc—the scattering region is not associated with the GC—we expect extragalactic source diameters to be a few arcseconds; if $\Delta_{GC} \approx 100$ pc—the scattering region is associated with the GC—extragalactic source diameters could exceed 1 *arc min*. However, few extragalactic sources have been identified toward the GC. The two closest sources are B1739−298 (Dickey et al. 1983) and GPSR 0.539+0.263 (Bartel 1994, private communication), which are 48′ and 40′ from Sgr A*, respectively. Neither of these is within the region of enhanced scattering defined by the OH masers.

Motivated by the prediction of large angular diameters illustrated in Fig. 1, we used the VLA to survey 10 fields at 20 cm between 7′ and 137′ from Sgr A* (Lazio & Cordes 1998a). Our objective was to identify extragalactic sources and measure their scattering diameters so as to constrain the GC-scattering region separation. In order to find sources within these fields, we employed pdfCLEAN, a source detection algorithm in which sources in an image are identified by comparing the intensity histogram of the image to that expected from a noise-only image. We found over 100 sources, with the faintest sources having a

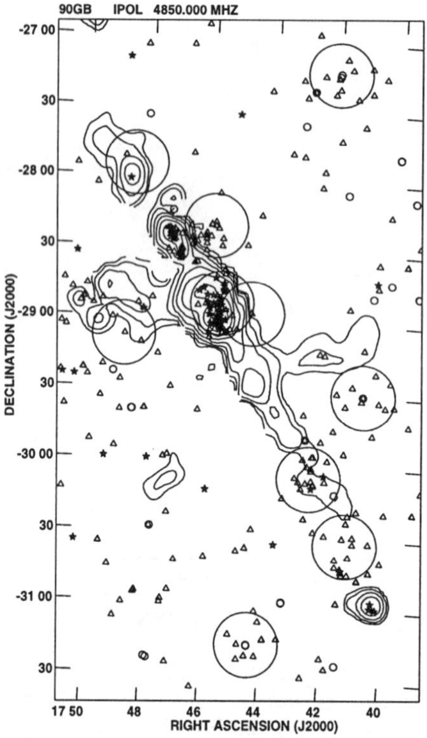

Figure 2. Radio sources detected in the surveys of Lazio & Cordes (1998a), Zoonematkermani et al. (1990), and Helfand et al. (1992). The large circles are the fields observed by Lazio & Cordes (1998a). The contours are from the 5 GHz survey with the NRAO 91 m telescope (Condon et al. 1991).

flux density of 3 mJy (Fig. 2). The average number of sources per field is approximately 10, though fields close to Sgr A* tend to contain fewer sources. This apparent deficit of sources within 1° of Sgr A* is suggestive of extragalactic sources being broadened enough to be resolved out by our observations.

3. A Likelihood Analysis of Galactic Center Scattering

The measurements available for constraining scattering in the GC consist of broadening observations of OH masers, source counts, and free-free emission and absorption. The joint likelihood function for these scattering measurements is (Lazio & Cordes 1998b)

$$\begin{aligned} \mathcal{L} &= \mathcal{L}_{\rm ff}\mathcal{L}_{\rm counts}\mathcal{L}_{\rm broaden} \\ &= \mathcal{L}(\mathbf{T}_{\rm ff}, \tau_{\rm ff}|\theta_{\rm OH}, \mathbf{N}; \mathbf{\Delta}_{\rm GC}, \psi_\ell, \psi_{\mathbf{b}}, \mathbf{l}_0, \mathbf{T}_{\mathbf{e}})\mathcal{L}(\mathbf{N}|\theta_{\rm OH}; \mathbf{\Delta}_{\rm GC}, \psi_\ell, \psi_{\mathbf{b}}) \times \\ &\quad \mathcal{L}(\theta_{\rm OH}|\mathbf{\Delta}_{\rm GC}, \psi_\ell, \psi_{\mathbf{b}}). \end{aligned} \qquad (2)$$

$\mathcal{L}_{\rm broaden} = \mathcal{L}(\theta_{\rm OH}|\mathbf{\Delta}_{\rm GC}, \psi_\ell, \psi_{\rm b})$: The likelihood function for angular broadening measurements of OH masers associated with OH/IR stars. The three-dimensional spatial distribution of this stellar population in the GC has been inferred (Lindqvist et al. 1992), allowing us transform this spatial distribution into a distribution of scattering diameters. There are two radio transients with measured scattering diameters (Davies et al. 1976; Zhao et al. 1992), but both are within 1′ of Sgr A*, and their contribution to $\mathcal{L}_{\rm broaden}$ would not be substantial. Although no H_2O masers or extragalactic sources with measured angular diameters have been seen through the screen, these classes of sources can still be used to place limits on the angular extent of the screen.

$\mathcal{L}_{\rm counts} = \mathcal{L}(\mathbf{N}|\theta_{\rm OH}; \mathbf{\Delta}_{\rm GC}, \psi_\ell, \psi_{\rm b})$: The likelihood function for counts of individual sources in VLA surveys (Lazio & Cordes 1998a; Zoonemetkermani et al. 1990; Helfand et al. 1992). Our likelihood analysis takes into account that (1) the sources detected are drawn both from Galactic and extragalactic populations; (2) the VLA configuration used limits the detection of sources larger than 30–60″ in diameter; (3) extragalactic sources show a distribution of intrinsic angular diameters; and (4) a large scattering diameter diminishes the detectability of sources.

$\mathcal{L}_{\rm ff} = \mathcal{L}(\mathbf{T}_{\rm ff}, \tau_{\rm ff}|\theta_{\rm OH}, \mathbf{N}; \mathbf{\Delta}_{\rm GC}, \psi_\ell, \psi_{\rm b}, \mathbf{l}_0, \mathbf{T}_{\rm e})$: The likelihood function for free-free measurements. The density fluctuations responsible for scattering also contribute to the emission measure. The EM corresponding to a source of diameter θ_s is $\mathrm{EM} = 10^{2.74}\,\mathrm{pc\,cm^{-6}}(\mathrm{l_0}/1\,\mathrm{pc})^{2/3}(\mathrm{D_{GC}}/\Delta_{\rm GC})^2(\theta_{\rm s}/133\,\mathrm{mas})^2$ (van Langevelde et al. 1992). Of particular interest are five masers that have angular diameters comparable to or greater than that of Sgr A*, show elliptical scattering disks, and are at least 15′ from Sgr A* (Frail et al. 1994). As the free-free emission is highly concentrated toward Sgr A*, these masers provide a stringent constraint on the amount of emission contributed by the density fluctuations responsible for scattering.

Figure 3 shows the global likelihood, which is a function of (1) $\Delta_{\rm GC}$; (2) ψ_ℓ and ψ_b, the angular extent of the scattering screen; (3) $l_0^{2/3}T_e^{-1/2}$, the outer scale of electron density fluctuations in the region and the temperature of the gas responsible for the scattering. We found that ψ_b is poorly constrained and fixed it at $\psi_b = 0^\circ\!.5$.

The likelihood function is maximized at $\Delta_{\rm GC} = 133^{+200}_{-80}$ pc and over the range $0^\circ\!.5 \le \psi_\ell \le 1^\circ\!.2$. *Scattering toward the GC is dominated by a scattering region local to the GC.* We adopt $\Delta_{\rm GC} = 150$ pc. The maximum likelihood estimate for $\Delta_{\rm GC}$ is determined primarily by the $\mathcal{L}_{\rm broaden}$ factor. $\mathcal{L}_{\rm counts}$ is constant for $\Delta_{\rm GC} \lesssim 500$ pc while $\mathcal{L}_{\rm ff}$ cannot constrain $\Delta_{\rm GC}$.

The angular extent of the region is $0^\circ\!.5 \le \psi_\ell \le 1^\circ$ and is further constrained by combining $\mathcal{L}$ with angular broadening measurements for other classes of sources. At positive longitudes, H_2O masers in Sgr B ($\approx 45'$ from Sgr A*) have diameters nearly five times smaller than that of Sgr A* (Gwinn et al. 1988), and the upper limit on the diameter of the extragalactic source GPSR 0.539+0.263 requires $\psi_\ell \lesssim 40'$. Toward negative longitudes the maser OH 359.517+0.001 is $0^\circ\!.5$ from Sgr A*, yet has a diameter 1.5 times larger than that of Sgr A*,

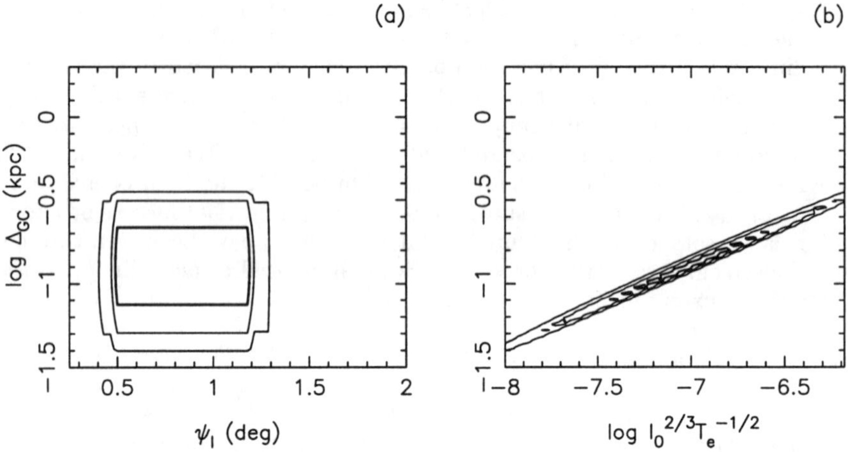

Figure 3. The global likelihood for GC scattering. Contours show the 67%, 90%, and 99% confidence regions. We assume $\psi_b = 0\overset{\circ}{.}5$. (*a*) $\mathcal{L}$ as a function of $\Delta_{\rm GC}$ and ψ_ℓ. The sharp edges in the contours reflect the underlying granularity of the data. (*b*) $\mathcal{L}$ as a function of $\Delta_{\rm GC}$ and $l_0^{2/3}T_e^{-1/2}$, with l_0 in pc and T_e in K. The maximum likelihood occurs at $\log l_0^{2/3}T_e^{-1/2} = -7.0$ and $\log \Delta_{\rm GC} = -0.88$.

while the extragalactic source B1739−298 ($\ell = 358.9$, $b = 0.0$) has an angular diameter smaller than that of Sgr A*. We conclude that there is no evidence for an asymmetric distribution in ℓ of scattering, though, a larger angular extent is possible if the scattering screen is patchy.

At the distance of the GC, 150 pc $\approx 1^\circ$. The close correspondence between the radial, $\Delta_{\rm GC}$, and transverse sizes, $\psi_\ell D_{\rm GC}$, indicates that the scattering region probably encloses or fills the GC, rather than being a simple screen along the line of sight. The scattering diameter for a background source seen through the GC is at least 90″ and may exceed 180″, if the scattering region encloses the GC. This large scattering diameter makes the GC the most extreme site of scattering known in the Galaxy. The large scattering diameter is almost certainly a reflection of the unique conditions in the GC.

The likelihood also has a single maximum at $l_0^{2/3}T_e^{-1/2} = 10^{-7.0\pm0.8}$. This value is considerably smaller than that in the solar neighborhood, $l_0^{2/3}T_e^{-1/2} \sim 10^{-2}$ (Spangler 1991). The absence of free-free absorption toward Sgr A* above 1 GHz constrains $\Delta_{\rm GC}/D_{\rm GC}$ to be $0.1 \le \Delta_{\rm GC}/D_{\rm GC} \le 0.32$, for nominal values of $l_0 = 1$ pc and $T_e = 10^4$ K (van Langevelde et al. 1992). Since $\Delta_{\rm GC}/D_{\rm GC} = 0.018$, we anticipate that the scattering region has $l_0 < 1$ pc, $T_e > 10^4$ K, or both.

4. G359.87+0.18 and the Covering Factor of the Scattering Region

Heretofore we have been describing the scattering region as if it covered the GC uniformly. The assumption of a uniform screen is also implicit in our likelihood analysis of extragalactic source counts.

We have conducted observations between 0.33 and 15 GHz of G359.87+0.18 (Lazio et al. 1998), a source that is only 15′ from Sgr A* and has been variously classified as extragalactic, a Galactic jet source, and a young supernova remnant. We classify G359.87+0.18 as extragalactic, most likely an FR II radio galaxy:

Morphology Images at 1.5 and 8.5 GHz (Fig. 4) show two components (A and C) surrounding an unresolved central component (B). Component C is elongated in the direction toward component B, and component B has a steep-spectrum, jet-like feature pointed toward component A. At our highest resolution, 15 GHz, component A is resolved into two sub-components, one of which is possibly an edge-brightened hot spot.

Spectrum At frequencies higher than 1.5 GHz, all components have steep spectra, $\alpha \gtrsim -1$ (Fig. 4). The spectrum flattens between 0.33 and 1.5 GHz with $\alpha \approx -0.8$, probably due to free-free absorption. An optical depth $\tau_{0.33\,\mathrm{GHz}} \approx 0.3$ would account for the difference between the observed flux density and that expected from an extrapolation of its high-frequency spectrum. Optical depths $\tau_{0.33\,\mathrm{GHz}} \gtrsim 1$ are seen elsewhere in the GC, and an extended, low-density H II region covers the inner degree of the GC (Mezger & Pauls 1979; Anantharamaiah et al. 1991; Anish Roshi & Anantharamaiah 1997).

Variability On time scales of a few days to a month, the source shows no structural variations, and the flux density is essentially constant. Since 1986 the angular separation between components A and B has remained constant, and the flux density of the source has not varied by more than 10% at any frequency over this this time span.

Distance The H I absorption spectrum of G359.87+0.18 is similar to that seen toward the extragalactic source B1739−298 (Dickey et al. 1983), with both sources showing absorption at 0 km s^{-1}. This absorption may be a blend of two components—one near 0 km s^{-1}, the other near −20 km s^{-1}. The gas at −20 km s^{-1} does not appear in the spectrum of Sgr A* and presumably lies beyond Sgr A*. In order that its power output be consistent with that of other FR II galaxies, G359.87+0.18 must be at a redshift $z > 0.1$. By comparison with other distant radio galaxies (Röttgering et al. 1994), a redshift $z \gtrsim 2$ is indicated.

G359.87+0.18 has other characteristics typical of distant radio galaxies (Röttgering et al. 1994): The majority of such sources consist of more than one component—10% are triples—and the 12″ separation between components A and C is comparable to the median angular size of distant radio galaxies.

The observed diameter of G359.87+0.18 at 0.33 GHz is 19″. Accounting for the intrinsic diameter implies a scattering diameter $14'' \lesssim \theta_s \leq 19''$. The range results from uncertainty about the intrinsic diameter below 1 GHz. The equivalent 1 GHz scattering diameter is 1.″5–2″. While large, such a scattering diameter is not unusually so, nor is it unique toward the GC. The scattering diameters of the extragalactic sources B1739−298 and B1741−312 ($\ell = 357.^\circ8$, $b = -1^\circ$), scaled to 0.33 GHz, are only a factor of 2–3 smaller than that in-

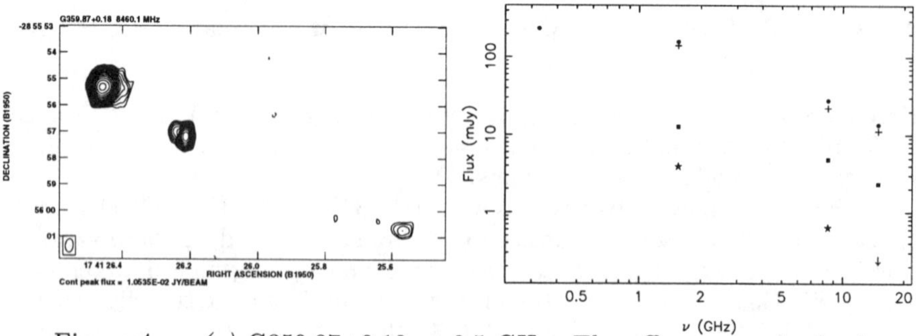

Figure 4. *(a)* G359.87+0.18 at 8.5 GHz. The off-source noise level in the image is 28 μJy beam^{-1}, and contours are -0.1, 0.1, 0.141, 0.2, 0.283, 0.4, ... mJy beam^{-1}. The beam is shown in the lower left. Component A is to the northeast; component C is to the southwest. *(b)* Spectra of components of G359.87+0.18. Crosses, component A; filled squares, component B; and stars, component C. At 15 GHz only an upper limit can be placed on the flux density of component C. Filled circles show the combined flux density for all three components; at 0.33 GHz, the individual components cannot be distinguished.

ferred for G359.87+0.18. While heavily scattered, they are not affected by the hyperstrong region in front of Sgr A* (Lazio & Cordes 1998b).

Our likelihood analysis implies that the scattering diameter at 0.33 GHz is at least 600″ for an extragalactic source. How can G359.87+0.18 be both extragalactic and not heavily scattered?

One possibility is that the hyperstrong scattering region does not extend to the Galactic latitude of G359.87+0.18. However, within 15′ are three sources with comparable latitudes whose sizes are affected by scattering: The maser OH 359.762+0.120, 7.′4 south, the non-thermal thread G0.08+0.15, 13′ northeast, and the non-thermal thread G359.79+0.17, 4.′7 south.

An alternate option is that the scattering region is inhomogeneous, with "holes" in it. The size of the clumps would have to be smaller than the 4.′7 (12 pc) separation between G359.87+0.18 and the thread G359.79+0.17. A "clumpy" scattering region would also explain the contrast between the diameters of OH 0.190+0.036 and OH 0.319−0.040 (van Langevelde & Diamond 1991; Frail et al. 1994) and of Sgr A OH1720:C and Sgr A OH1720:B masers (Yusef-Zadeh et al. 1998b). In both cases, order of magnitude changes in the size of scattering disks are seen over scales of 10′ or less.

5. Physical Conditions in the Scattering Region

The diameters of Sgr A* and the OH masers require a density within the scattering medium of $\overline{n}_e \sim 10^3\,\mathrm{cm}^{-3}(1/\varepsilon\sqrt{\mathrm{f}})(\mathrm{l}_0/1\,\mathrm{pc})^{1/3}(\Delta_{\mathrm{GC}}/150\,\mathrm{pc})^{-3/2}$ (Lazio & Cordes 1998b), where $\varepsilon \equiv \delta n_e/\overline{n}_e$ and f is the volume filling factor of the density fluctuations δn_e. Though $l_0 \ll 1$ pc or $f \ll 1$ could alter our estimate by an order of magnitude, nonetheless, the scattering medium must be dense,

$n_e \gtrsim 10^2$ cm^{-3}. For comparison, $n_e \sim 1$ cm^{-3} for scattering regions in the Galactic disk (Spangler 1991).

The presence of small-scale (≈ 0.1 pc) magnetoionic cloudlets in the GC has already been inferred to explain large changes in Faraday rotation measures (G359.1−00.2, the "Snake," Gray et al. 1995; G359.54+0.18, the non-thermal filaments, Yusef-Zadeh et al. 1997), though the inferred density in these cloudlets, 0.3–10 cm^{-3}, is less than our nominal estimate.

We now consider two models for the host medium, using our maximum likelihood estimates (§3. and Fig. 3). In both models, the scattering arises on the surfaces of molecular clouds.

5.1. Photoionized Surfaces of Molecular Clouds

Over the region $|\ell| \lesssim 1^\circ\!.5$ and $|b| \lesssim 0^\circ\!.5$, $n_e \sim 10$ cm^{-3} (Mezger & Pauls 1979), within which are H II regions with $n_e \sim 10^3$–10^5 cm^{-3}, primarily within Sgr A West and Sgr B2 (e.g., Mehringer et al. 1993). Some of these high density regions are the photoionized surfaces (size $\sim 10^{-4}$ pc) of molecular clouds ($n \gtrsim 10^4$ cm^{-3}) irradiated by the ambient radiation field (effective temperature $\approx$ 35 000 K). Yusef-Zadeh et al. (1994) identified these molecular skins as the source of the scattering and associated their thicknesses with the outer scale, l_0.

Our constraint on $l_0^{2/3}T_e^{-1/2}$ and the effective temperature of the radiation field imply $l_0 \sim 10^{-7.1}$ pc. This value is considerably smaller than that derived by Yusef-Zadeh et al. (1994) for the ionized molecular skins. However, Yusef-Zadeh et al. (1994) used a value of the ionizing flux appropriate to the inner few parsecs. Outside the inner few parsecs, the ionizing flux would be lower than that assumed, bringing their estimate of the skin depth and our estimate of l_0 into better agreement.

5.2. "Warm" Interfaces

The central $1^\circ\!.8 \times 0^\circ\!.9$ of the GC contains a diffuse X-ray source (Yamauchi et al. 1990). Frail et al. (1994) identified this X-ray emitting gas as responsible for the scattering.

The density and temperature of this region are 0.05 cm^{-3} and 10^7–10^8 K. This region cannot itself be the host of the density fluctuations. However, this gas appears spatially coincident with the central zone of intense molecular emission and presumably abuts cooler gas. We identify the interfaces where the GC molecular clouds are exposed to this ambient hot medium as the source of the scattering. The X-ray emitting gas and molecular clouds appear to be in rough pressure equilibrium with $P \sim 5 \times 10^6$ K cm^{-3} (Blitz & Spergel 1991). Taking pressure balance to extend throughout the interface region, we find a density $n_e \sim 5$–50 cm^{-3} for $T_e \sim 10^5$–10^6 K. If the size of the interface region is set by thermal conduction, the portion of the interface with $T_e < 10^6$ K has a size $\lesssim 10^{-1}$ pc. Clearly if not all of the interface contributes to the scattering better agreement would be obtained between l_0 and the interface size.

One point in favor of this model is the distribution of the X-ray emitting gas. The size of the X-ray emitting region is similar to the extent of the scattering region, approximately 1°. In contrast, the molecular cloud distribution extends over the range $-1^\circ \lesssim \ell \lesssim 2^\circ$. If the scattering traced massive stars within these molecular clouds, as the photoionized molecular cloud skins model suggests, the scattering should extend further in longitude than it does. In this respect, the lack of enhanced scattering for H_2O masers in Sgr B is particularly problematic.

We stress the importance, and probable uniqueness, of the high density in the GC. In the Galactic disk density fluctuations cannot be supported in media with $T_e \sim 10^6$ K (Spangler 1991). However, densities in the Galactic disk are a factor of 10^2–10^3 smaller than those in the GC.

6. Radio Pulsar Searches Toward the Galactic Center

We predict a dispersion measure DM $\approx$ 2000 pc cm^{-3} for a radio pulsar in the GC, with approximately 1500 pc cm^{-3} arising from the GC itself. For comparison, the largest DM known is for PSR B1758−23 with DM = 1074 pc cm^{-3} (Kaspi et al. 1993). Furthermore, the temporal broadening of pulses from pulsars seen through this region will be $350\,\nu_{\rm GHz}^{-4}$ *seconds*, requiring high-frequency ($\nu \approx 10$ GHz) periodicity searches to detect pulsations (Cordes & Lazio 1997).

7. Conclusions

We have used a likelihood analysis to constrain the following parameters of the GC scattering region (Fig. 3): The GC-scattering region separation, $\Delta_{\rm GC}$ = 150 pc; the angular extent of the region, $0.5° \leq \psi_\ell \lesssim 1°$; and the outer scale of the density fluctuations and the gas temperature, $l_0^{2/3} T_e^{-1/2} = 10^{-7}$ with l_0 in pc and T_e in K. The close correspondence between $\Delta_{\rm GC}$ and $\psi_\ell D_{\rm GC}$ suggests that the scattering region encloses the GC.

The GC scattering region produces a 1 GHz scattering diameter of 90″, if the region is a single screen, or 180″, if the region wraps around the GC. Pulsars seen through this region will have a dispersion measure of approximately 2000 pc cm^{-3}, of which approximately 1500 pc cm^{-3} arises from the GC, and suffer pulse broadening of $350\,\nu_{\rm GHz}^{-4}$ *seconds*; pulsations will be detected only for frequencies above 10 GHz (Cordes & Lazio 1997).

Observations of the source G359.87+0.18 between 0.33 and 15 GHz and an H I absorption spectrum show that this source is a Fanaroff-Riley II radio galaxy. A lower limit to its redshift is 0.1, and its steep spectrum suggests $z \gtrsim 2$.

Though G359.87+0.18 is only 15′ from Sgr A*, it is not affected by the hyperstrong scattering region that is responsible for the large diameter of Sgr A*. We attribute the lack of hyperstrong scattering toward G359.87+0.18 to an inhomogeneous scattering region. An inhomogeneous region would also explain the small diameters of the masers OH 0.190+0.036 and Sgr A OH1720:C. The scattering region would have to be inhomogeneous on 5–10′ scales (10–20 pc at the Galactic center) to explain the small diameters of these sources.

We favor a model in which the scattering medium arises on interfaces between molecular clouds and the 10^7 K ambient gas; the scattering medium is hot ($T_e \sim 10^6$ K) and dense ($n_e \sim 10$ cm^{-3}). The GC scattering region is unique in the Galaxy, probably because it is a high-pressure environment and can sustain densities and temperatures much higher than in the Galactic disk.

Acknowledgments. We thank D. Frail, N. Bartel, C. Carilli, C. Lang, H. van Langevelde, and F. Yusef-Zadeh for assistance with various aspects of this work. This research made use of Simbad, operated at the CDS, Strasbourg, France. TJWL holds a National Research Council-NRL Research Associateship. Basic research in astronomy at the NRL is supported by the Office of Naval Research. This research was supported by the National Science Foundation under grants AST 92-18075 and AST 95-28394. The National Astronomy & Ionosphere Center is operated by Cornell University under a cooperative agreement with the NSF. The National Radio Astronomy Observatory is a facility of the NSF operated under cooperative agreement by Associated Universities, Inc.

References

Anantharamaiah, K. R., Pedlar, A., Ekers, R. D., & Goss, W. M. 1991, MNRAS, 249, 262

Anish Roshi, D. & Anantharamaiah, K. R. 1997, MNRAS, 292, 63

Blitz, L. & Spergel, D. N. 1991, Nature, 379, 631

Bower, G. C. & Backer, D. C. 1998, ApJ, 496, L97

Condon, J. J., Broderick, J. J., & Seielstad, G. A. 1991, AJ, 102, 2041

Cordes, J. M. & Lazio, T. J. W. 1997, ApJ, 475, 557

Cordes, J. M., Weisberg, J. M., Frail, D. A., Spangler, S. R., & Ryan, M. 1991, Nature, 354, 121

Davies, R. D., Walsh, D., Browne, I. W. A., Edwards, M. R., & Noble, R. G. 1976, Nature, 261, 476

Dickey, J. M., Kulkarni, S. R., van Gorkom, J. H., & Heiles, C. 1983, ApJS, 53, 591

Frail, D. A., Diamond, P. J., Cordes, J. M., & van Langevelde, H. J. 1994, ApJ, 427, L43

Goldreich, P. & Sridhar, S. 1995, ApJ, 438, 763

Gray, A. D., Nicholls, J., Ekers, R. D., & Cram, L. E. 1995, ApJ, 448, 164

Gwinn, C. R., Moran, J. M., Reid, M. J., & Schneps, M. H. 1988, ApJ, 330, 817

Helfand, D. J., Zoonematkermani, S., Becker, R. H., & White, R. L. 1992, ApJS, 80, 211

Higdon, 1984, ApJ, 285, 109

Higdon, 1986, ApJ, 309, 342

Kaspi, V. M., Lyne, A. G., Manchester, R. N., Johnston, S., D'Amico, N., & Shemar, S. L. 1993, ApJ, 409, L57

Lazio, T. J. W., Anantharamaiah, K. R., Goss, W. M., Kassim, N. E., & Cordes, J. M. 1998, ApJ, submitted

Lazio, T. J. W. & Cordes, J. M. 1998a, ApJS, 118, 201

Lazio, T. J. W. & Cordes, J. M. 1998b, ApJ, 505, 715

Lindqvist, M., Habing, H. J., & Winnberg, A. 1992, A&A, 259, 118

Mehringer, D. M., Palmer, P., Goss, W. M., & Yusef-Zadeh, F. 1993, ApJ, 412, 684

Mezger, P. G. & Pauls, T. 1979, in The Large-Scale Characteristics of the Galaxy, ed. W. B. Burton (Dordrecht: Reidel) p. 357

Molnar, L. A., Mutel, R. L., Reid, M. J., Johnston, K. J., 1995, ApJ, 438, 708

Moran, J. M., Rodríquez, L. F., Greene, B., & Backer, D. C. 1990, ApJ, 348, 150

Röttgering, H. J. A., Lacy, M., Miley, G. K., Chambers, K. C., & Saunders, R. 1994, A&AS, 108, 79

Spangler, S. R. 1991, ApJ, 376, 540

Sridhar, S. & Goldreich, P. 1994, ApJ, 432, 612

van Langevelde, H. J. & Diamond, P. J. 1991, MNRAS, 249, 7

van Langevelde, H. J., Frail, D. A., Cordes, J. M., & Diamond, P. J. 1992, ApJ, 396, 686

Yamauchi, S., Kawada, M., Koyama, K., Kunieda, H., & Tawara, Y. 1990, ApJ, 365, 532

Yusef-Zadeh, F., Cotton, W., Wardle, M., Melia, F., & Roberts, D. A. 1994, ApJ, 434, L63

Yusef-Zadeh, F., Roberts, D. A., Goss, W. M., Frail, D. A., & Green, A. J. 1998b, ApJ, submitted

Yusef-Zadeh, F., Wardle, M., Parastaran, P. 1997, ApJ, 475, L119

Zhao, J., et al. 1992, Science, 255, 1538

Zoonematkermani, S., Helfand, D. J., Becker, R. H., White, R. L., & Perley, R. A. 1990, ApJS, 74, 181

The Central Parsecs of the Galaxy
ASP Conference Series, Vol. 186, 1999
H. Falcke, A. Cotera, W.J. Duschl, F. Melia, M.J. Rieke, eds.

SHARC 350 μm Mapping of the Galactic Center from the Caltech Submillimeter Observatory

C. D. Dowell, D. C. Lis, E. Serabyn, M. Gardner, A. Kovacs, & S. Yamashita

Dept. Physics, Caltech, Mail Code 320-47, Pasadena, CA 91125
cdd@submm.caltech.edu

Abstract.
Using the 20-pixel camera SHARC, we have surveyed the 350 μm emission from a 60′ by 12′ region in the Galactic center. A comparable region has been observed at 800 μm by Lis & Carlstrom (1994); the SHARC map has better spatial resolution and sensitivity to extended emission, however. This paper introduces several features not visible in prior maps of dust emission. We have detected a 10+ pc band of emission peaking at $l = -0.14°, b = +0.02°$, which is probably associated with the 'negative velocity arc' observed in ^{13}CO (Bally et al. 1988). We suggest an association of a rounded 350 μm feature near the Radio Arc with an expanding shell observed in CS (Tsuboi, Ukita, & Handa 1997). Faint emission at the same Galactic longitude as the Dust Ridge (Lis & Carlstrom 1994) but at opposite latitude is visible as well as a compact source at the base of a thermal radio filament at negative Galactic latitude. Diffuse dust emission within a few arc-minutes of the Sgr D core is detected for the first time. 350 μm sources are associated with both H_2O masers in Sgr D, but only one out of four OH masers (Mehringer et al. 1998). The mean 350 μm/800 μm flux ratio for the Galactic center is approximately 17 ($\beta \sim 2.0$) over the map but is higher ($\beta \sim 2.5$) in parts of the Dust Ridge and lower in Sgr B2 (N).

1. Introduction

Submillimeter continuum observations of the Galactic center provide a measure of the amount and structure of dense material on scales of 0.5 - 200 pc. Broadband observations at 350 μm suffer very little from contamination by molecular line, free-free, and synchrotron emission, and, except toward the Sgr B2 core, the emission is optically thin.

Lis & Carlstrom (1994) mapped a large portion of the Galactic center at 800 μm using a single bolometer and an 'on-the-fly' mapping technique; many sources were detected for the first time, including cold clouds not seen in the far-IR ($\lambda < 100$ μm). However, the beam size of the experiment (30″) was larger than can be obtained at shorter wavelengths, and the observing time per unit solid angle was small. These shortcomings have been addressed by mapping the region at 350 μm with SHARC, a multi-pixel bolometer camera.

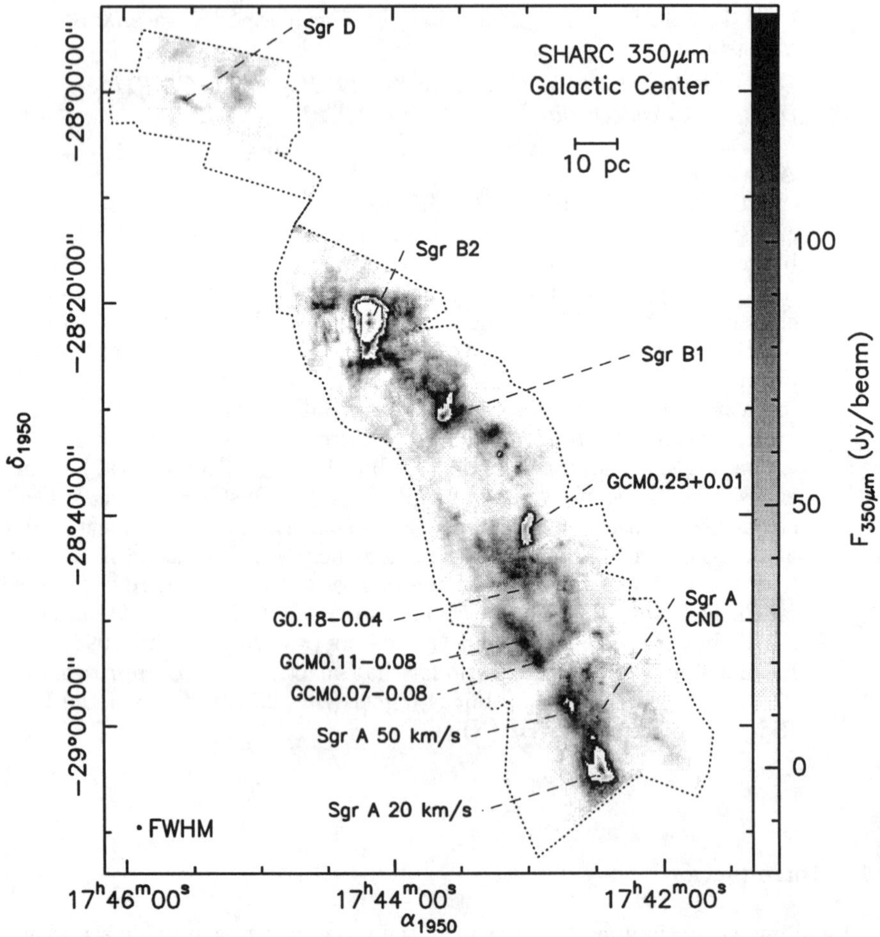

Figure 1. 350 μm map of the Galactic center. The beam size is 15″. The intensity scale wraps around and becomes nonlinear above 145 Jy/beam. For position reference, Sgr A* is located inside the Circum-Nuclear Disk ('CND'), and the non-thermal radio filaments (not shown) cross the Galactic plane near the Sickle (G0.18–0.04).

2. SHARC Observations

SHARC (Wang et al. 1996) uses a linear array of 20 bolometers. Each pixel subtends 5″ by 10″. The diffraction beam and choice of smoothing algorithm lead to an effective resolution of 15″ FWHM in our map.

The data were acquired by 'on-the-fly' mapping (chopping and continuous scanning) and restored with the NOD2 algorithm (Haslam 1974). The observations took place during April-September 1998. During the very dry conditions in April, the point-source sensitivity per pixel was ~2 Jy $s^{0.5}$ at Galactic center transit. However, due to map reconstruction noise and correlated sky noise (causing striping), the sensitivity in the raw resultant map is degraded. Fourier filtering was applied to the map to remove the sky noise stripes on rapid time scales (spatial wavelengths $\leq$ 30″); the filtering was quite satisfactory in that source structure was preserved. After filtering, the typical small-scale noise was 6 Jy RMS over most of the map.

The SHARC map is shown in Figure 1. The coverage is approximately 0.3 square degrees. Some imperfections are evident in the map. There are reconstruction artifacts to the east and west of the bright peaks of Sgr B2, a few regions of negative flux caused by emission at the ends of the scans, and signal-to-noise variations.

There is very good correspondence between the features in our 350 μm map, the 800 μm map of Lis & Carlstrom (1994), and the 1.3 mm map of Mezger, Duschl, & Zylka (1996) where coverage and sensitivity permit comparison. In general, we confirm the poor association of 350 μm dust emission with mid- and far-infrared (λ < 100 μm) and radio continuum emission. The 350 μm map resembles maps of some molecular transitions, however (NH_3 – Güsten, Walmsley, & Pauls 1981; CS – Serabyn & Güsten 1987).

Calibration was accomplished by observing Uranus, assuming a temperature of 64 K (Griffin & Orton 1993). The map calibration assumes point-like sources, and map fluxes are given as Jy/15″ beam. The calibration uncertainty is 25%.

3. Discussion

3.1. Source Fluxes

The source fluxes measured in our 350 μm map are given in Table 1. Assuming a dust temperature of 30 K (Lis & Carlstrom 1994), a grain emissivity of Q(350 μm) = 1.9 x 10^{-4}, a grain density of 3 g/cm^3, a grain radius of 0.1 μm, a dust-to-gas density ratio of 0.01, and a distance of 8.5 kpc, the conversion from flux to gas mass is approximately 20 $M_\odot$/Jy (Hildebrand 1983). The total mass detected in our map is then 4.3 x 10^6 $M_\odot$. By using an average temperature of 30 K we underestimate the mass in cold regions (GCM0.25+0.01, T = 18 K; Lis & Menten 1988) and overestimate the mass in hot regions (Sgr A CND, T $\geq$ 40K; Davidson et al. 1992).

3.2. Sgr A

The (sub-)millimeter appearance of the molecular clouds in the central 10′ of the Galaxy is well known. The SHARC image (Figure 2) of the bright sources GCM–

Table 1. Measured 350 μm Fluxes (Jy)

Source	15″ beam	30″ beam	total
20 km/s cloud (GCM–0.13–0.08)	294	640	16,000
50 km/s cloud (GCM–0.02–0.07)	166	370	8000
CSO–0.14+0.02	73	148	3100
Sgr A C.N.D.	97	220	1500
GCM0.07–0.08/GCM0.11–0.08	142	319	13,000
GCM0.25+0.01	236	523	5900
Sgr B1	290	625	9300
Sgr B2 (M)	2740	4740	49,000
Sgr B2 (N)	2350	3810	
Sgr D	106	197	7600
Total			210,000

0.13–0.08 (+20 km/s cloud) and GCM–0.02–0.07 (+50 km/s cloud) reproduces many features observed before at longer wavelengths (Mezger et al. 1996; Lis & Carlstrom 1994; Dent et al. 1993).

The emission from the vicinity of Sgr A* is relatively weak at 350 μm. Sgr A* itself, having a relatively flat submillimeter spectrum (Serabyn et al. 1997), is swamped by dust emission from the 2 pc Circum-Nuclear Disk (CND). The CND, the dominant emitter in the central 10 pc at $\lambda \sim 100\mu$m (Odenwald & Fazio 1984; Davidson et al. 1992), is not nearly as prominent in the submillimeter, indicating a lack of cold dust along the line of sight.

An elongated emission feature is visible to the west of GCM–0.13–0.08 running approximately parallel to the Galactic plane. We have labeled the peak of this band as CSO–0.14+0.02. The location and extent of the feature suggest an association with the 'negative velocity arc' observed in ^{13}CO at v = –100 to –40 km/s (Bally et al. 1988), although the dust emission appears to be at a slightly more negative Galactic latitude. A far-IR source (Odenwald & Fazio 1984) lies along the arc, so star formation may be occurring in the cloud.

Oka et al. (1999) have recently identified a compact cloud (CO 0.02–0.02) about 5′ from Sgr A* with a large velocity width ($\geq$100 km/s). Dust emission in the vicinity of this source is seen in the SHARC map (Figure 2) as well as the maps by Dent et al. (1993), Lis & Carlstrom (1994), and Mezger et al. (1996).

3.3. Radio Arc

Among the prominent 20 cm features in the Galactic center are the Thermal Arched Filaments at $l = 0.10, b = 0.05$ (Yusef-Zadeh 1986). The SHARC map of this region is shown in Figure 3. The eastern filament has much more 350 μm emission associated with it than the western filament. The majority of the 350 μm flux is east of the eastern filament, displaced further from the radio features than is the far-IR (50 - 90 μm) emission (Morris, Davidson, & Werner 1995), implying an east (cold) to west (hot) temperature gradient. The 'Arches' star cluster (Cotera et al. 1996; Serabyn, Shupe, & Figer 1998) does not seem well

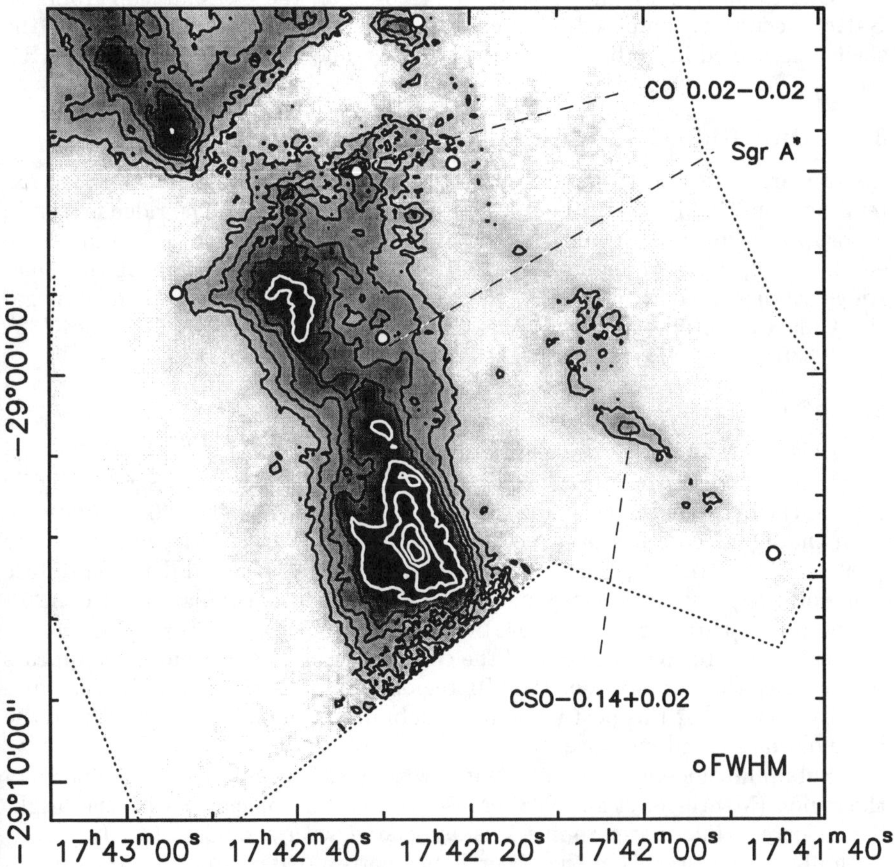

Figure 2. 350 μm map of Sagittarius A. The contour levels are 40, 60, 80, 100, 140, 180, 220, and 260 Jy/beam. The brightest submillimeter emission is from the 20 km/s and 50 km/s clouds. The fainter Circum-Nuclear Disk is visible partially surrounding Sgr A*. Far-IR sources (Odenwald & Fazio 1984) are labeled with circles.

placed to produce this temperature gradient; a heating source further to the west seems to be required.

A fainter thermal radio filament is located on the opposite side of the Galactic plane, possibly associated with a compact radio feature at $l = 0.100, b = -0.168$. A compact 350 μm source is detected in this region with a flux of 49 Jy/15″ beam, suggesting recent or ongoing star formation.

Oka et al. (1998) find that at 17″ resolution, the CO emission from the Galactic center is composed of numerous filaments, arcs and shells. One of the shells, discovered in CS by Tsuboi et al. (1997), may also be seen in the SHARC map.

3.4. Dust Ridge

Lis & Carlstrom (1994) discovered a cold ridge of submillimeter emission extending from GCM0.25+0.10 near the Radio Arc to Sgr B1. The ridge is seen in absorption against the general Galactic center emission at $\lambda \lesssim 70$ μm (Egan et al. 1998; Lis & Menten 1998). The SHARC 350 μm map of the Dust Ridge is shown in Figure 4. A fainter ridge of emission is located at negative Galactic latitude, including an extended source at $17^h43^m31.7^s$, $-28°40'32''$ ($l = 0.330, b = -0.076$) with a peak flux of 77 Jy/15″ beam.

3.5. Sgr D

Sgr D is a site of active star formation located at $l = 1.13, b = -0.10$. The projected distance from the Galactic center is 170 pc; however, Sgr D may lie beyond the Galactic center region (Mehringer et al. 1998). The 800 μm emission from the Sgr D core has previously been mapped by Lis, Carlstrom, & Keene (1991). With SHARC, we have mapped a much larger area and detected diffuse emission over much of the region. The 350 μm map is compared with the 18 cm map of Mehringer et al. (1998) in Figure 5. The peak flux at the Sgr D core is 106 Jy. To the northwest of the core, the submillimeter emission appears to lie along the boundary of the HII region. Perhaps the radio HII region is density bounded at this position and the submillimeter emission comes from the HII-molecular cloud interface.

Submillimeter emission is associated with both 1.3 cm H_2O masers known in the region (Mehringer et al. 1998). Maser source 1 is coincident with the bright Sgr D core, while maser source 2 is associated with a modest (34 Jy/beam) submillimeter source. On the other hand, correspondence of 350 μm sources with 18 cm OH maser sources is lacking. Of the 4 OH maser sources falling within our map (E, G, H, and I – Mehringer et al. 1998), only one of them (I, coincident with H_2O maser 2) has associated submillimeter emission. This supports the suspicion by Mehringer et al. (1998) that sources E, G, and H are evolved stars rather than young stellar objects.

3.6. 350 μm to 800 μm Flux Ratio

The large coverage of the 350 μm SHARC map and the 800 μm map of Lis & Carlstrom (1994) allows an investigation of the spectral index over the Galactic center region (Figure 6). The mean value of $F_{350\mu m}/F_{800\mu m}$ is approximately 17, which corresponds to a spectral index of $\beta = 2.0$ for 30 K dust. The flux ratio is low (12) at Sgr B2 (N) due to its high optical depth. The flux ratio is

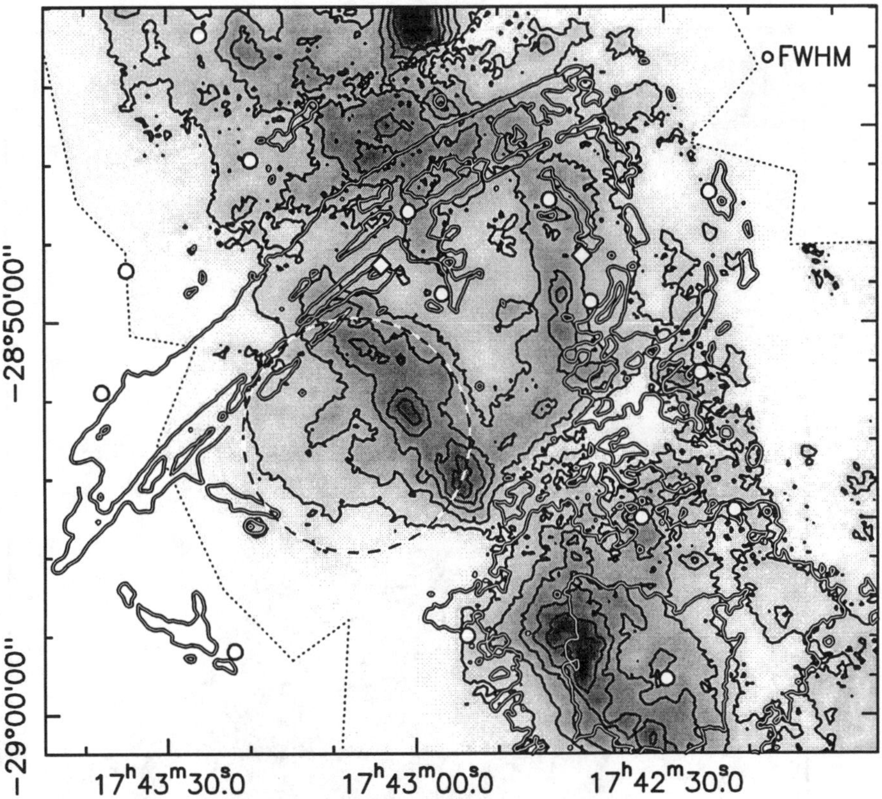

Figure 3. SHARC map of the Radio Arc region. The gray scale and single contours show the 350 μm emission with levels of 25, 50, 75, 100, and 125 Jy/beam. The double contours show 20 cm continuum emission (Yusef-Zadeh 1986). The system of Thermal Arched Filaments – located at $17^h42.5^m$, $-28°49'$ – lies to the west of a ridge of dust emission while the radio emission from the Sickle (G0.18–0.04) at $17^h43.0^m$, $-28°47'$ is southeast of a dense cloud. The diamonds mark known clusters of hot stars that may ionize the Filaments and Sickle (Cotera et al. 1996). The partial radio contour at lower left shows two thermal radio filaments at negative Galactic latitude. One of the filaments lies near a compact radio and 350 μm continuum source at $17^h43^m19.2^s$, $-28°55'16''$. Its association with the filament is unknown. The dashed circle shows the rough location of an expanding CS shell discovered by Tsuboi et al. (1997). The shell may also be detected at 350 μm, as suggested by the rounded outer contour, superposed on the ridge formed by GCM0.07–0.08 and GCM0.11–0.08. Circles mark far-IR sources (Odenwald & Fazio 1984).

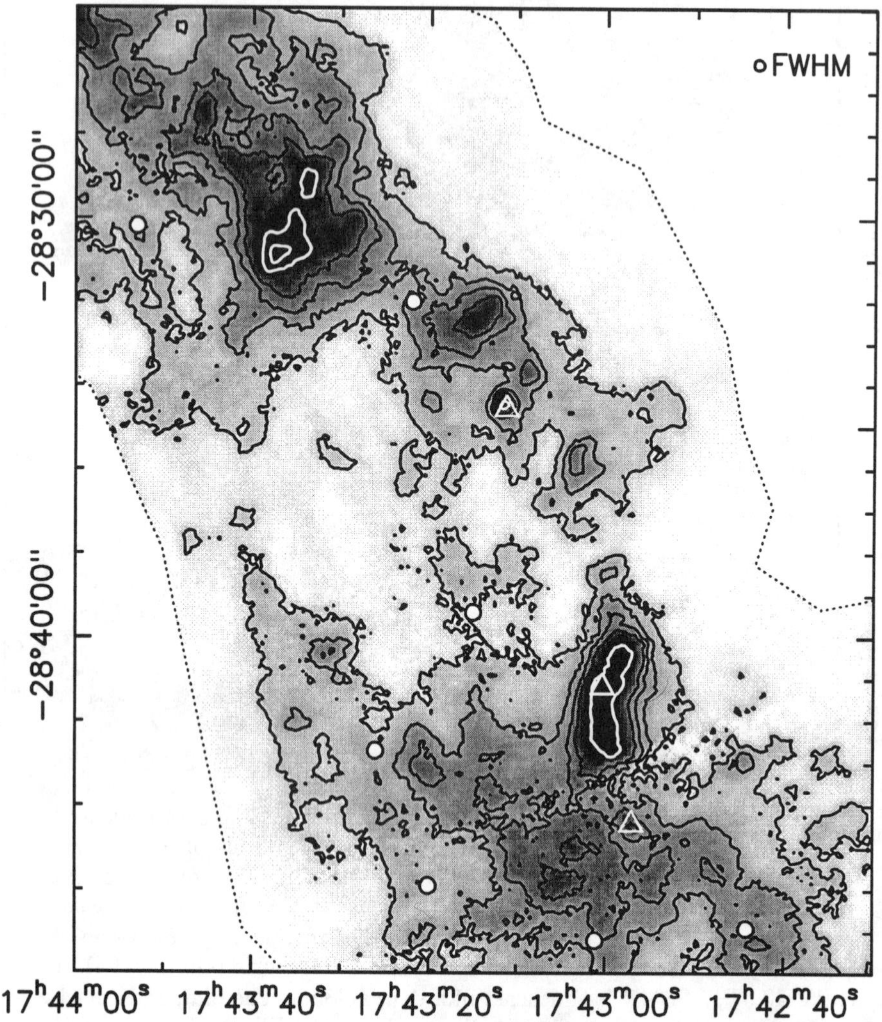

Figure 4. 350 μm map of Dust Ridge. Contour levels are 25, 50, 75, 100, 175, and 250 Jy/beam. The Dust Ridge lies outside the Radio Arc and consists of a chain of sources extending from the bright elongated cloud GCM0.25+0.01 to Sgr B1. A fainter ridge of sources is located at negative Galactic latitude (eastern part of the map). Three H_2O masers (triangles) have been detected by Lis & Menten (1995); however, the star formation rate in the Dust Ridge as a whole is low. FIR sources (Odenwald & Fazio 1984) are labeled with circles.

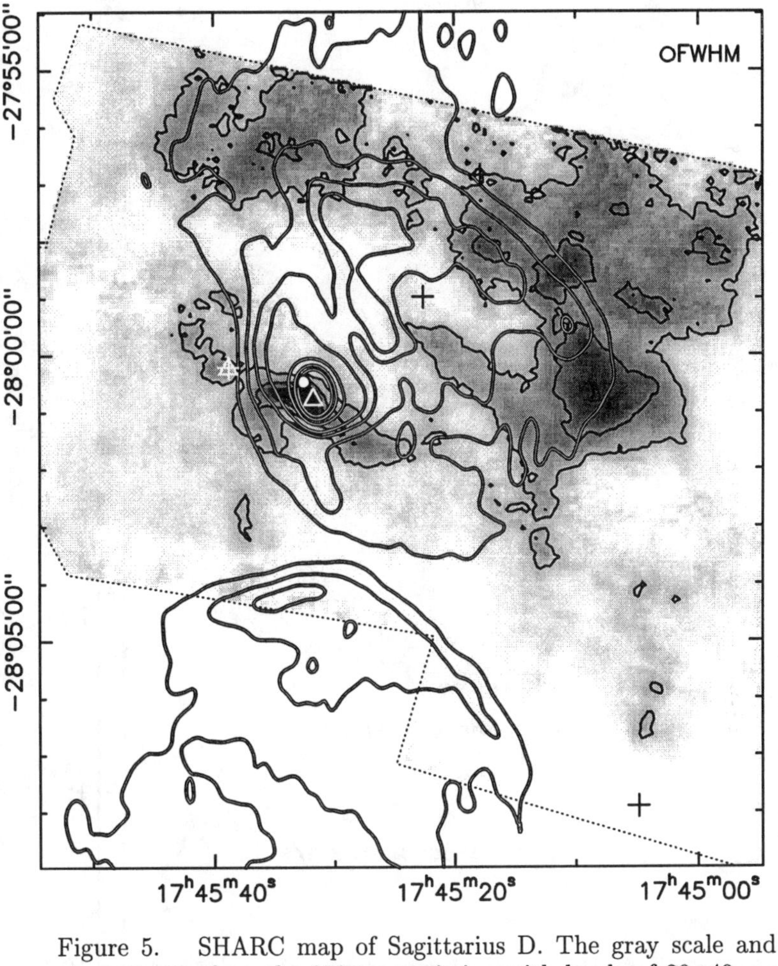

Figure 5. SHARC map of Sagittarius D. The gray scale and single contours show the 350 μm emission with levels of 20, 40, and 60 Jy/beam. The double contours show 18 cm continuum measured by Mehringer et al. (1998). The triangles show the positions of H_2O masers, and the plus symbols show OH masers (Mehringer et al. 1998). The H_2O masers correspond more closely with regions of submillimeter emission. The diffuse emission to the northwest of the Sgr D core lies along the boundary of the HII region. The FIR peak from Odenwald & Fazio (1984) is marked with a circle.

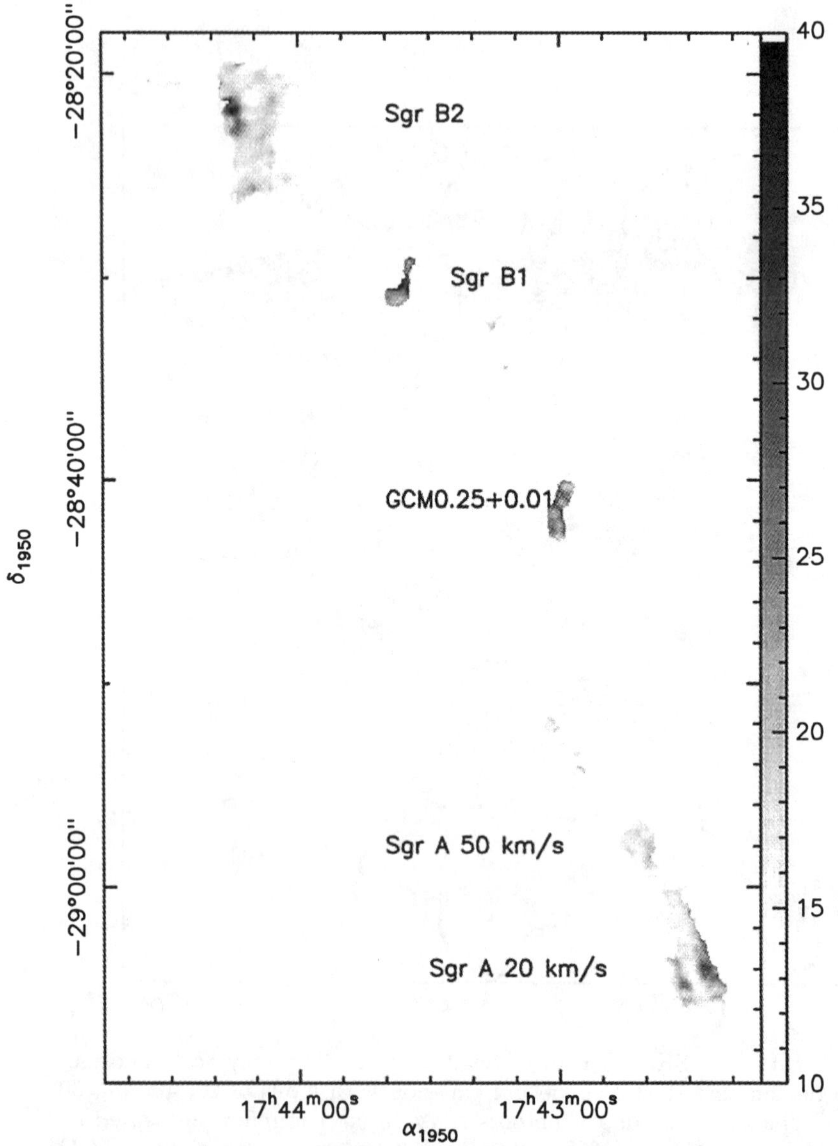

Figure 6. Flux ratio $F_{350\mu m}/F_{800\mu m}$. The 800 μm map is by Lis & Carlstrom (1994). The 350 μm map has been smoothed to 30″ resolution to match the 800 μm map. The flux ratio is approximately 17 over most of the map. A low spectral index is seen at the cold, optically thick source Sgr B2 (N) at $17^h44^m10.3^s$, $-28°21'17''$, and a high spectral index is observed at M0.25+0.01 and Sgr B1 in the Dust Ridge. The high spectral index to the east of Sgr B2 and to the west of Sgr A is likely an artifact.

high (27) at GCM0.25+0.01 and Sgr B1 in the Dust Ridge; the large implied $\beta = 2.5$ may be a signature of the lack of high-mass star formation (Lis & Menten 1998; Lis et al. 1998).

Acknowledgments. We thank T. Hunter for assistance in acquiring the data and F. Yusef-Zadeh and D. Mehringer for providing the radio continuum maps. The Caltech Submillimeter Observatory is supported by grant AST 96-15025 from the NSF.

References

Bally, J., Stark, A.A., Wilson, R.W., & Henkel, C. 1988, ApJ, 324, 223

Cotera, A.S., Erickson, E.F., Colgan, S.W.J., Simpson, J.P., Allen, D.A., & Burton, M.G. 1996, ApJ, 461, 750

Davidson, J.A., Werner, M.W., Wu, X., Lester, D.F., Harvey, P.M., Joy, M., & Morris, M. 1992, ApJ, 387, 189

Dent, W.R.F., Matthews, H.E., Wade, R., & Duncan, W.D. 1993, ApJ, 410, 650

Egan, M.P., Shipman, R.F., Price, S.D., Carey, S.J., Clark, F.O., & Cohen, M. 1998, ApJ, 494, L199

Griffin, M.J., & Orton, G.S. 1993, Icarus, 105, 537

Güsten, R., Walmsley, C.M., & Pauls, T. 1981, A&A, 103, 197

Haslam, G. 1974, A&AS, 15, 333

Hildebrand, R.H. 1983, QJRAS, 24, 267

Lis, D.C., Carlstrom, J.E., & Keene, J. 1991, ApJ, 380, 429

Lis, D.C., & Carlstrom, J.E. 1994, ApJ, 424, 189

Lis, D.C., & Menten, K. 1995, in Airborne Astronomy Symp. on the Galactic Ecosystem, ed. M.R. Haas, J.A. Davidson, & E.F. Erickson (San Francisco: ASP), 499

Lis, D.C., & Menten, K. 1998, ApJ, accepted

Lis, D.C., Serabyn, E., Keene, J., Dowell, C.D., Benford, D.J., Phillips, T.G., Hunter, T.R., & Wang, N. 1998, ApJ, accepted

Mehringer, D.M., Goss, W.M., Lis, D.C., Palmer, P., & Menten, K. 1998, ApJ, 493, 274

Mezger, P.G., Duschl, W.J., & Zylka, R. 1996, A&A Rev., 7, 289

Morris, M., Davidson, J.A., & Werner, M.W. 1995, in Airborne Astronomy Symp. on the Galactic Ecosystem, ed. M.R. Haas, J.A. Davidson, & E.F. Erickson (San Francisco: ASP), 477

Odenwald, S.F., & Fazio, G.G. 1984, ApJ, 283, 601

Oka, T., Hasegawa, T., Sato, F., Tsuboi, M., & Miyazaki, A. 1998, ApJS, in press

Oka, T., White, G.J., Hasegawa, T., Sato, F., Tsuboi, M., & Miyazaki, A. 1999, ApJ, accepted

Serabyn, E., Carlstrom, J.E., Lay, O., Lis, D.C., Hunter, T.R., Lacy, J.H., & Hills, R.E. 1997, ApJ, 490, L77

Serabyn, E., & Güsten 1987, A&A, 184, 133

Serabyn, E., Shupe, D., & Figer, D.F. 1998, Nature, 394, 448

Tsuboi, M., Ukita, N., Handa, T. 1997, ApJ, 481, 263

Wang, N., et al. 1996, Appl.Optics, 35, 6629

Yusef-Zadeh, F. 1986, in The Center of the Galaxy, ed. M. Morris (Dordrecht: Kluwer), 243

The Central Parsecs of the Galaxy
ASP Conference Series, Vol. 186, 1999
H. Falcke, A. Cotera, W.J. Duschl, F. Melia, M.J. Rieke, eds.

Discussion Session VII: How Do the Various Components of the Inner 100 Parsecs Interact With Each Other?

W.M. Goss[1]

National Radio Astronomy Observatory, P.O. Box O, Socorro, New Mexico 87801

R. Coker[2], A. Cotera[2]

Steward Observatory, The University of Arizona, Tucson, AZ 85721

Abstract. This is an edited transcript of a discussion session about the various components (Sgr A East & West, CND, ionized gas, large-scale radio structure etc.) filling the inner 10 to 100 parsecs of the Galactic Center that took place during the Galactic Center conference "The Central Parsecs" in Tucson, Arizona, Sep. 7-11, 1998. The session was chaired by the moderator as indicated in the author list. The discussions were taped, transcribed, and finally edited by editors who are listed as co-authors. The contributions of the conference participants are preceded by their names and thus references to specific aspects and ideas of this discussion session must also include a reference to the respective speaker(s), e.g., in a footnote.

Discussion

Goss: I think Namir [Kassim], with his 90 cm map, has shown yet again that with an image of the Galactic Center, at a new wavelength and a new resolution, you will often discover something new. Few regions in the Milky Way, at least for the radio astronomer, have such a wide diversity of spectral indices and angular scales. We know that there are regions that are completely optically thick with spectral indices of +2 and there are non-thermal sources. What this new 90 cm image has shown is how important it is to have a large dynamic range in spatial frequencies, especially when imaging objects on the scale of a degree in size. Of course we also know that there are small-scale structures at the level of milli-arcseconds from VLBI, and certainly at the arcsecond and tenth of a arcsecond level—even in Sgr A West. There are of course many explanations for this, one of these is fairly trivial, since there is free-free absorption going on. Of course, one trivial explanation of this is free-free absorption. We can see in Namir's image and in the paper by Anantharamaiah, that if we

[1]Moderator of the discussion session

[2]Editor of the discussion session

have the absorbing gases external to the synchrotron emitter there is a very dramatic cut-off that is almost like a threshold. But in many cases there is a mixture of the non-thermal gas and the gas that is doing the absorption—then you get an intermediate case. Questions that occurred to me, and I will take your comments: can we do interesting statistics on the number of filaments, supernova remnants, HII regions, etc., that really are in the Galactic Center within a hundred parsecs? Do we have a complete census? How would we do this? Something that has occurred to Anantha and myself is that in the next couple of years the GMRT in India will be coming on line. It is a large, sensitive radio telescope with baselines up to 25 km and 30 antennas, each of 45 meter diameter that are optimized for low frequencies. It will be optimized to work at frequencies of 1.4 GHz and below. A 600 MHz image with a resolution of a few seconds of arc is going to be fascinating because we are going to see through the electrons in many cases and see new structures.

Kassim: I will point out that Andrew Gray's work was at 843 MHz with the MOST (Molonglo Observatory Synthesis Telescope) of Sydney University. It showed that there was definitely an excess in the number of supernova remnants in the Galactic Center. There are 14 remnants over what you would have expected. That was at the 3σ level, and there are at least two or three new remnants since then.

Goss: Yes, he also found lots of new features in the Galactic Center, which I am still impressed with because it was done with a fairly low dynamic range radio telescope. The main reason is that the resolution was about a minute of arc. Certainly, Andrew obtained a number of fascinating new results from that data.

Kassim: I completely agree. I think the 600 MHz observations will be extremely interesting. It takes you up to a level where the opacity from the thermal gas becomes much less of an observational constraint.

Yusef-Zadeh: What is missing is the spectral index measurements in these images. I think you are getting a lot of very beautiful images at many different frequencies. The problem here is that we would really like to know what these spectral indices are, to really model and be able to discuss what is happening in the Galactic Center region. So I think the emphasis should be on finding a way to find the spectral index with good accuracy. I do not know if there is any prospect for doing that.

Reid: Oh there certainly is. If you are interested, Namir [Kassim] has this image with a resolution of ... $25''$?

Kassim: Yes, $25''$ but we could easily go to $6''$.

Reid: You could easily go to 20 centimeters with the VLA and have a deconvolved resolution of $25''$—you could also easily get the maximum resolution.

Yusef-Zadeh: Yes, maximum resolution is one thing but the field of view is another. I think that one of the problems is that the extended structures really depress the emission, and it is very difficult to get an accurate flux density. One thing that I may suggest is to observe at 20 cm, separating the IFs by a very large frequency, and get the spectral index based on two IFs separated by wide frequencies, let us say 18 and probably 20 cm.

Goss: ... bandwidth synthesis, essentially.

Yusef-Zadeh: Yes that is really the way to do it, because then you have the same field of view and the same resolution, you more or less eliminate all these uncertainties.

Roberts: I have a question that was similar to what Miller asked: with a lower noise level—i.e. with images better than Namir's 90 cm image—are we going to see more sources popping up, or have we seen all there is to see?

Kassim: The indication is that we have hardly even scraped the surface. As soon as you just scrape off a little more we would get a much better picture of the Galactic Center. Right now we are just fighting for signal-to-noise and resolution. But I would say what you would really like to do is just have the VLA for a year and ...

Goss: Look, you are a young guy Namir, and I am going to retire [as VLA director] one of these soon days, so please wait. [Laughter]

Kassim: If you could do a deep 20 cm wide-field image you would find all sorts of things.

Novak: I do not know how feasible this is, but there has been some work on Faraday rotation using the radio. I think it is very valuable for probing the extent of this polodial field that we seem to see and to find out anything about the line-of-sight magnetic field.

Goss: Yes, we will be hearing more about that in Cornelia's [Lang] talk later on. And of course Farhad [Yusef-Zadeh], through the years, has done quite a lot of Faraday rotation work.

Townes: I just had a specific question for Kassim. He said that the synchrotron lifetime was something like 1.4 million years. How do you derive that?

Kassim: It is a simple back-of-the-envelope calculation. That is just from assuming there is a synchrotron spectrum, and integrating to get the luminosity and assuming equipartition. And you calculate the amount of energy in the particles. You can assume the distance, calculate the luminosity and the amount of energy in the magnetic field and the particles.

Townes: This gives the decay time of the electrons, which is 1.4 million years?

Kassim: Right.

Wardle: Do you get some estimate of the magnetic field from the equipartition?

Kassim: We have only done it for Sgr C, and we were getting around 115 μG. This is about a tenth of the milligauss fields that you see a little bit closer to the center.

Townes: It is a very weak field [compared to the inner regions].

Kassim: That is right.

Anantharamaiah: In response to the question about new features, when we first made an image of a smaller field with the same data, one of the goals we had was to look for new features, and I remember looking over the images and not finding anything that was new. But since then this image has been improved. After Andrew Gray found this feature called the Snake, we went back and looked at the 90cm image and saw evidence for it. Similarly, there is another small feature that Cornelia has looked at that has become a nice beautiful filament now. If it comes to whether we have exhausted the known features or not, I would say it's definitely not so. There are many other features right in this image, which I suspect will all turn out to be filaments.

Goss: I wanted to make one slight modification to the statement that Namir made. The background sources that you see in these images here, and especially near W28, those are extragalactic sources. They are non-thermal and they have the right source density. Most of these sources are not Galactic.

Kassim: Especially the point sources, you mean?

Goss: Yes.

Kassim: I think there are lots of little filaments in the image.

Goss: Yes, but what about thermal sources – you can see M8, of course, in your image and also W28 – but you are not going to see little planetary nebulae and similar object that are all optically thick at 327 MHz in the continuum. It is just the confusion of the quasar and the radio galaxy population that is of course everywhere. Each one of these images has about 10 Jy of confusion with probably a few hundred or a thousand background sources.

Melia: This is another question for Namir: the image that you showed with the spectral index cross cuts in one of those filaments, how did you do the averaging in the spectral index?

Kassim: We just did the peaks.

Melia: Oh, just the peaks along the [filament]?

Kassim: Because at that time it was probably the safest way. We just convolved the two images to the same beam size.

Melia: I wondered how the particles are loaded in that particular filament? Is this due to magnetic field dissipation, where you get the twisting of the filaments? Would that be consistent with what you see? Because in that case the particles would be accelerated in the middle, where you get the twisting flux tubes and the particles diffuse outwards in both directions from the middle. But then it would not look like it was loaded from one end to the other, but rather the particles were generated in the middle of the filament. I wonder if that is common? Do you see that in any of the other filaments, or is that the only one?

Kassim: No, a lot of the earlier work suggested the filaments did not have a good peak brightness in the middle. When you look at them more carefully, and try to have more of a census of them, they really all look very similar. They all give the impression of being bright in the middle with strands that come apart at the ends.

Melia: Do they look like they are twisted in the middle? Because that is what you would need in order to dissipate the field.

Yusef-Zadeh: You do not know for sure if it is twisted, because you can always argue that it is the juxtaposition of two filaments. You do not really know, though you could interpret it. We just do not have the evidence for it now.

Goss: We will probably hear about that later today. Let us move on. We have had Robert Zylka's presentation. I was really struck by several things when I saw these fantastic images of CS and ^{13}CO and the continuum at 1.2 mm. Fantastic progress has been made in the last few years. I remember that in the La Serena GC meeting in 1996, there were a number of groups that were working on millimeter and even sub-millimeter imaging, and it seems to me that a great deal of progress has occurred in the last several years. I certainly had the impression that there were a couple of PhD theses in trying to interpret the maps—all that fantastic velocity data—and trying to make sense of the kinematics. One of the most interesting things that I picked up was this issue of whether the Sgr A East so-called supernova event did something to the +50 km s $^{-1}$ cloud, and created the circumnuclear disk? Some of us were talking about this at lunch and I found it remarkable and so interesting that I would like to hear comments.

Roberts: What struck me about that is that throughout the week it seems like there has been more and more evidence that there is a connection between Sgr A East and Sgr A West. One of the things was that you have the circumnuclear disk (CND) which appears to have some sort of circular velocities. We could have this Sgr A East explosion happening after the CND had already formed, but I have a hard time believing that it would not have disrupted the CND in some way. But this seems to be the much cleaner way to connect the two: an explosion forms the disk, so you do not have to worry about the energetics having disrupted the disk. You push the cloud in, and then gravity takes over after that.

Zylka: Yes, he is right. The crucial point in these models is the time scale. If we knew how much time was necessary, maybe we could really calculate what is going on with the disk. But the dynamical time scale for the CND is something like 10^4 yrs.

Sanders: The dynamical time scale of the CND is more like 10^5 yr. That is the rotation time. I think the rotation period about 3 pc is around 1.5×10^5 yr.

Zylka: Well I think that the CND is not really just gas wrapping around the center. If the time scale for the CND is even larger, then we do not have a problem. If the CND was there when Sgr A East's explosion happened, it could have affected the CND or even destroyed it. The point is that we see that the CND is not isotropic. It is warped and makes an S pattern. The central part of the CND is more or less in the Galactic core, an ellipse with a long axis along the Galactic Plane. But the open edges go north-south. So this is dynamically an extremely complicated feature. And in principle we need something like this to explain all the dynamical or kinematical features that we see in the disk. If you put the explosion in first – I am not sure of the numbers – but I think it can be explained.

Anantharamaiah: We also know from various other evidence that Sgr A* seems to be quite embedded inside Sgr A West, if we look at all the models or the pictures, Sgr A* is inside (and that is the candidate for the black hole). But now if you look at the shape of Sgr A East carefully, it looks like a shell-like remnant, and the shell is over the edge of Sgr A West; it goes past Sgr A West. It looks as if the black hole is inside the supernova, at least in 2-dimensional projection. Now the supernova remnant, Sgr A East, obviously went off recently, because we can still see there is a nice remnant of the shell-like structure. And it is amazing to me that such an explosion should develop such a nice shell-like structure while a 2 million solar mass black hole was sitting next to it.

Goss: I think it's a question that Fulvio [Melia] already asked earlier this morning. Something that's bothered me for years: is this a proper supernova remnant? Or is this, as Peter [Mezger] and Fulvio [Melia] have said, not the standard "supernova remnant" that we know and love so well.

Mezger: That is why I am not very happy that we call Sgr A East a supernova remnant. The argument is that this is something which exploded in a cloud. Then you can model it using someone else's computations, and you find that you need an energy which is about 50 times that of a supernova.

Unknown: So what is it?

Mezger: I do not know. I think Morris was suggesting it is a cloud that drifted over the black hole, and fed the black hole for a while, days or weeks and so on.

Falcke: I think Fulvio's idea of this—it might not even be his idea in the first place, but he calculated that—is intriguing. This could have been one of those stellar disruptions Martin Rees was always talking about. He was always saying that it should happen. The problem with there being a black hole is that on average every ten thousand years or so a star should be falling into the hole and make a big explosion. Well, there is one. That seems to be very consistent.

Mezger: The interesting thing is that one of your colleagues in Japan who made these X-ray observations found out that from the X-ray background emission, if one assumes that the thing is expanding at the velocity of sound, one gets approximately the same age which we get from modelling it as a supernova. It is some kind of explosive event but where it comes from, I do not know.

Townes: If it were a star falling in, why has it drifted off so far from Sgr A West?

Mezger: If you take the age of 4×10^4 years then you get a projected drift velocity of about 50 km s^{-1}, which is not terribly much. I mean, everything in the vicinity of the Galactic Center has a higher velocity.

Townes: But an explosion you would think would tend to average that out.

Mezger: It is an explosion which may have happened inside a very massive molecular cloud.

Melia: If it is due to a tidally disrupted star where the explosion works, the explosion will occur when the star is coming out of its closest approach to the black hole. So that naturally creates an asymmetry in the efflux. In fact, if you want to calculate the solid angle into which the efflux goes, it corresponds to 30°which is enough to create the observed structure. I would like to add that in some of tomorrow's talks we will hear about this energetics question a little bit more. It turns out that if the EGRET source is associated with Sgr A East, and if we compare the supernova remnants that EGRET has observed, then the gamma-ray power is about 50 times what one would expect from a supernova remnant. It is interesting to see that this over-abundance of power has a value not very different from that expected from Peter's work.

Yusef-Zadeh: When you look at OH observations you do see an absorption only toward Sgr A East. You do not see any molecular gas in absorption against Sgr A West. On the other hand, if you look at HCN, HCO^+ or other molecular lines, you do not see any 50 km s^{-1} gas towards Sgr A West. In other words there is really no molecular gas in absorption or emission towards Sgr A West.

Zhao: There is a kind of finger-like OH absorption feature towards Sgr A West.

Yusef-Zadeh: That is a new one outside the CND. I am talking right within Sgr A West, the CND sort of surrounds it. We do not really see any emission or absorption of any molecular lines that we know. The point is that if you are associating the 50 km s^{-1} cloud with the Galactic Center at Sgr A West, there has to be a way to remove that 50 km s^{-1} cloud at the position of Sgr A West. Somehow you would have to remove that, because we do not see the 50 km s^{-1} feature behind Sgr A East or in front. Somehow the 50 km s^{-1} gas must have been removed towards Sgr A West.

Goss: Is that little finger at 50 km s^{-1}? I thought it was at a higher velocity.

Zhao: It is at 70 km s^{-1}. It could also be gas that is ionized and moving into the Galactic Center region.

Coil: Could we discuss all of the arguments for it being a supernova to begin with? Was it just the shell shape and spectral index?

Kassim: Shell-like morphology and non-thermal spectral index.

Coil: Maybe we are holding on to that. It is not a perfect sphere; it is slightly elongated.

Goss: Well, at least slightly to our credit, we did appreciate in our early papers that there were lots of problems – the surface brightness was wrong, and there were other things that did not fit in with the conventional SNR.

Coil: I would also like to ask Mark [Wardle]: the OH masers – those to the north appear to be associated with the CND – but the ones to the south, how far away are they from the CND? They could be associated with the CND and not Sgr A East depending on whether or not they are far away.

Wardle: I think they are too far away to be within the CND, based on the CS maps.

Coil: Those are old CS numbers with poor resolution. I have a poster upstairs which shows a streamer gas around 25 km s^{-1}.

Wardle: The velocity is 50 to 60 km s^{-1}.

Coil: Alright, that is a little high.

Falcke: I want to come back to Doug Roberts' talk for a minute. I still can't get it quite straight in my mind, the geometry of those clouds and their belonging to the CND. So, the 50 km s^{-1} cloud – does it actually touch the CND? Is this all just one complex, or are they completely separated?

Zylka: Not very far separated. They are along the line of sight, yes. We believe that Sgr A West is within Sgr A East on the front part of Sgr A East, so they are closely related. You can not have gas rotating at 100 km s^{-1} within this cloud.

Coil: I think most people agree that the 25 km s^{-1} cloud is connected to the 50 kms^{-1} cloud, so there is a larger gas stream.

Goss: Anantharamaiah stressed the importance of ionized gas, which is *not* associated with Sgr A West. Again, it's a question of finding the right filter in terms of resolution, observing frequency, and the right recombination line to see something that is sensitive to stimulated emission. But how do we confirm these ideas? When we talk about ionized gas we want to see them in other transitions: are there other recombination line observations we can carry out? Obviously in order to do that we need to know the densities, path lengths and the temperatures. You might be able to use the infrared (NICMOS - Paschen lines). And don't forget that it is possible to observe the winds of IRS7 in both the radio and IR using recombination lines.

M. Rieke: Let me quantify the trend of the IRS 7 tails. That is right at the limit of what NICMOS can do at Paα. However, this emission measurement is a good estimate of the things which envelope IRS 7. There is also another complication, which is that you are observing against an intense continuum. In the near-infrared, we always show these lines with black background but that is actually on top of quite a bright continuum from, for example, stars in the Galactic bulge and so put all these together and we are not winning by as much as you think.

Mezger: We looked for that extended, diffuse ionized gas about 15 years ago. Tom Pauls and I later on published it together with Ed Churchwell's [radio recombination] emission line data. The separation [into thermal and non-thermal emission] was done on the basis of different spectral indices and that is where we found out that the very strong extended component extends about $\pm$100-150 pc in the Galactic Plane of this very narrow region, what we call nowadays the nuclear bulge. What makes the observation of this separation somewhat uncertain is the fact that there is flat-spectrum non-thermal emission from the Galactic Center. We found that the extended emission could be approximated by two ellipses here. The free-free flux density which comes from this area is about one-tenth of the total free- free emission of our Galaxy, and that is where my statement came from that about 10% of the O stars through the past million years have formed very close to the central region. The problem is that part of the non-thermal emission that we observe today was not known earlier. The Arc has a flat spectrum which is not due to free-free emission and so we may have over-estimated the extended free-free emission. Tom Pauls and I searched and found the recombination line emission in this region. But, there are observations clearly indicating that there is extended emission. Our estimate was that only 10% of the free-free emission comes from very compact HII regions, and 90% comes from the extended low density stuff of the type we have heard about today. Certainly this measurement should be repeated with much better data available today, and probably the best way to separate thermal from non-thermal emission is to acquire recombination line observations.

Anantharamaiah: There are very low frequency recombination line observations with the right angular resolution of the entire Galactic Plane now, starting from somewhat negative longitudes. This data comes from Anish Roshi's Phd thesis using the Ooty telescope in India at 327 MHz. The paper will appear soon in ApJ. We see evidence for extended ionized gas, not just in the central regions but all through the disk. There is no dramatic difference in that low-density component near the Galactic Center as opposed to the gas you find if you move away by a few degrees in longitude. Along the line of sight, they are all at velocities near zero km s^{-1} near the center, and it extends for a good part of 2° from the center. When you separate the regions, the way you did, probably you are picking up radiation from that gas. I do not know whether we could actually associate that with the Galactic Center region. So this would be several kpc away, as an educated guess.

Goss: Here's a question for some of you: are there possibilities for spectroscopy in the mid-infrared or even the far infrared, to look for forbidden lines or recombination lines?

M. Rieke: I think there is going to be a big improvement in what one can do in the 10 μm window because there are now finally good arrays, some 10 μm cameras and spectrometers being put on the 4m-10m class telescopes, and there are other interesting forbidden lines with a variety of excitations that one can do.

Simpson: I am going to show some spectra on Ar^+ tomorrow that do extend fairly far from the Galactic Center.

Townes: There is certainly a marked improvement in 10 μm and in the far infrared. When SOFIA goes up we will have an order of magnitude improvement in the far infrared. The Keck instruments have been recognized as being very powerful. Whether that will detect these or not, I do not know, but we will have to see.

Yusef-Zadeh: The Green Bank telescope, when it comes up, I think could be very useful to do recombination lines, surveying the Galactic Center region. When we had the 43-meter telescope many years ago, Anantha and I did a survey of the Galactic Center, and we found a lot of line emission from that region. Now, with a much better sensitivity, one could do a survey of the region.

Goss: And with modern spectrometers you can do many recombination lines at once.

Goss: For years I was frustrated at not being able to have a good, clear picture of where the scattering was occurring and how to make a coherent picture. I think Joe Lazio (in this thesis work at Cornell) and Jim Cordes have gone a long way in resolving a lot of these problems and making a model that fits in the OH masers, scattering in the direction of Sgr A*, and the (on the sky) nearby extragalactic sources. But it is like a lot of things, when you answer a question you raise some new ones, and certainly I found that to be the case with this whole issue of the connection with the X-ray heating of the edges of molecular clouds, and how this might be related to the 10^4 K HII gas that is around.

Lazio: One of the questions after my talk was: could it still really be the 10^4 K gas? I said earlier that you could make a relation between the observed scatter broadened size of the masers, and the HII or the free-free emission. You then get an emission measure from the observed size. Now the observed sizes are around $1.5''$. These observed diameters imply an emission measure of almost 10^5 pc cm^{-6}. If you now say that my likelihood analysis is wrong, and that the scattering screen is really closer, say, one kiloparsec away, so it is really not associated with the Galactic Center, we are still talking about emission measures of 10^7 pc cm^{-6}. That was one of the things that really drove me to try to find an alternative to the 10^4 K gas, because certainly my bias was for a 10^4 K gas. With an emission measure of at least 10^6 pc cm^{-6}, and probably 10^8 pc cm^{-6}, you really must go through contortions to avoid the free-free emission constraints.

Anantharamaiah?: How does the connection come about with an emission measure of 10^8 pc cm^{-6} ?

Lazio: Well, the scattering is caused by density fluctuations. So, the density fluctuations imply there is some mean density, and once you have a mean density, then you have an emission measure. So there are various parameters in here. This is in some sense a minimal estimate since I have assumed that the RMS density fluctuations are comparable to the mean density, which is certainly the case for what we see elsewhere in the Galaxy. If the RMS density fluctuations are a hundredth of the mean density, then I think this problem becomes even more extreme.

Mahadevan: You must be making some assumptions here, because the scattering is done by very small scale fluctuations.

Lazio: Yes.

Mahadevan: Mean density and RMS may be matching on some other scale.

Lazio: Which is at the L_0 parameter—the outer scale of the fluctuations—right there.

Narayan: Right. What do you know about L_0?

Lazio: Very little. Well, actually, that is not entirely true. I have parametrized that as 1 pc. You can play around with that a little bit; in other directions in the Galaxy you see maybe a 0.01 pc for some very heavy scattering images. So you can bring this down a little bit, but again, this is only at two-thirds power. So I certainly think that the L_0 is going to be fairly small. I am a bit uncomfortable with the fact that I still have to force this L_0 to be very small in order to avoid violating the free-free emission observations.

Anantharamaiah: Many years ago, Ramesh [Narayan] and I did a very simplistic kind of calculation to try to see whether we can explain the scattering, not near the Galactic Center, but generally in the Galactic Plane, the inner Galaxy where the scattering is much larger, as you know. Just scaling from the ionized gas that we know has been detected in thermal emission and recombination lines – 10^4 K gas – making some simple assumption that the outer scale is like the sizes of the regions that we detect, and then extrapolating it all the way down. We were able to say, if we believe the turbulence went that way, then we can account, quantitatively, for all of the increased scattering that you see in the inner region, using this 10^4 K gas, and an emission measure of only 10^3 pc cm $^{-6}$ or less. Now, if you go from there to the Galactic Center region, you need a bit more, right? That sort of ionized gas is already seen, in the center. So why do not believe in the 10^4 K gas and why does the emission measure really becomes so much larger? There must be something in the assumption that we made which was not right.

Lazio: Maybe we need to sit down and actually work through the derivation, but, really, the emission measure is going to be huge, unless these things are very small, and L_0 is really small. I think the other thing to keep in mind—we have not heard much about it—are the pressures in the Galactic Center. They are at least factors of 10–100 higher than what they are in the disk. So at some level it is perhaps not that surprising that the properties of the scattering medium are much different from what we are accustomed to in the disk.

Goss: There are many lines of sight in the Galaxy. We know about scattering in Cyg X-3 and GRS 1915+105. Does your model work in those?

Lazio: No, because I think in those cases, the pressure is low, but there is a good connection between Hα emitting gas and the scattering gas.

Lazio: There are still some questions as to exactly how it arises in the HII regions, but it's not as unusual in the regions near Cyg X-3 for example.

Goss: Other comments on this?

Yusef-Zadeh: There are so many OH/IR stars near the Galactic Center, why is it that we only see three or four of them being scattered?

Lazio: It is not measured for the other ones.

Yusef-Zadeh: Oh, so nobody has looked at them?

Unknown: No. The resolution we have is just too high.

Yusef-Zadeh: So it would be really nice if somebody could do VLBI observations of these.

Unknown: It is 100-200 stars.

Yusef-Zadeh: Because to test the idea of the scattering medium being very lumpy and patchy, these OH masers should be very useful.

Goss: It is unfortunate, but the 1720 MHz OH masers we know about—even outside of scattering regions—are resolved out, i.e. they are too large. This is the case for at least the ones we have done so far with the VLBA. The strongest masers that we have been working on in W28 (with Mark Claussen, Dale Frail and Ketan Desai) have an angular size of about 50 mas. We can just do it with the VLBA, but we are not sure if it is scatter-broadened or if it is intrinsic. It is certainly not going to be a good object for scattering measurements, based on a sample of one, but it was certainly a typical one.

Wardle: So what is the physical size corresponding to 25 mas?

Goss: A few hundred AU.

Wardle: So it is 10^{15} cm, which is comparable with what's going on in the shells.

Goss: We know those supernovae have a fair bit of ionized gas near them.

M. Rieke: I just wondered if anyone had actually added up all of the radio emission that you might see from the central say 10-30 pc to compare with what we see from the nuclei of starburst galaxies where we can not see the thermal emission, but can only see the non-thermal emission, and I wondered if that would actually be true if we looked at our own Galactic Center from a distance.

Goss: I am sure that has been done. I have heard people talk about that through the years. That is one of Ron Ekers' favorite topics, does anybody know the answer?

Eckart: I did not quite understand why we are not seeing the thermal emission? I thought in the Galactic Center, we have a mix of thermal and non-thermal emissions.

M. Rieke: But say in M82, all you see is non-thermal. It's very hard to see the thermal emission.

Eckart: Why is that?

M. Rieke: Because the non-thermal is so strong.

Reid: But mostly it's in a bright strip.

Goss: We know it is a few hundred Janskys, isn't it? Spread out over about 100 pc?

Kassim: When we add the zero-spacing flux maybe we'll have a very good answer.

Yusef-Zadeh: I think this idea that half of the continuum emission is thermal and half non-thermal, may not be really correct any more because of the flat spectrum radio sources.

M. Rieke: But the question is, would it scale and look like a very weak starburst as we heard yesterday. Would it scale up to a strong starburst? Would the thermal and non-thermal emission go up in the same way, or would there be a different scaling?

Mezger: Within the central 200 pc the crossover, where non-thermal and thermal emission are approximately equal, occurs at 5 to 6 GHz. So, only if you observe at centimeter wavelengths would you see predominantly thermal emission and not too many observations of extragalactic sources with this high resolution have been made.

Goss: Finally, let us turn to Mark Wardle's discussion of 1720 MHz OH masers in the Galactic Center: can we use this as a diagnostic to tell us something about the relevant SNR model. This is surely a question for Mark and Farhad [Yusef-Zadeh]. Do you think this model is going to be required for all the SNR associated with 1720 MHz OH masers?

Yusef-Zadeh: I think all the ones that we have observed seem to have an X-ray source within them.

Goss: Yes, that is the point. And what is the main difference between your model and Phil Lockett's?

Wardle: Well, they did the new version of the pumping calculations, and they had some arguments against J-type versus C-type shocks that involved the cooling time. They basically ruled out J-type shocks using different arguments. Well, you have to dissociate the water to make the maser, and they leaned towards a weak UV field. They were the ones who pointed out the problem in the far infrared.

Goss: That the far infrared would lead to main line masers [at 1665/1667 MHz], is that the issue?

Wardle: Yes. My impression is, the pumping calculations are more uncertain when the far infrared radiation field is around. I think you have to be a little careful. As far as the nature of Sgr A East is concerned, we already know there are X-rays. The pressure you need is consistent. It tells you the interaction really is occurring in the molecular clouds, but I do not know if it tells you much about its nature directly.

Roberts: Pressure does not give any help for understanding the nature of the explosion?

Wardle: Well, it does, but you can get the same number from looking at the X-rays.

Yusef-Zadeh: One question I have is whether we expect to see any variability of these masers? Are they going to have the same type of variability as the ones we see in HII region OH masers?

Goss: We know the observational evidence is pretty meager.

Yusef-Zadeh: Yes, but is there any theoretical idea that could predict some kind of variability in these shocks?

Goss: I should point out that there have been a couple of groups at the VLA, not involving any of us here today, that have made a very concerted effort to look for these 1720 MHz OH masers in extragalactic SNRs. These groups are David Frayer, Dale Frail, and Ernie Seaquist and Huib van Langevelde et al who have observed M82, NGC 253, and Cen A. In these galaxies they find 1720 MHz masers, but it is not clear to me that they are the same type of objects. The luminosities are much larger compared to the galactic sources such as W28 and W44. I am sure Farhad [Yusef-Zadeh] has some comments about this.

Yusef-Zadeh: Well, Doug Roberts knows about the LMC source.

Roberts: We (Anne Green, Farhad [Yusef-Zadeh], and I) found one 1720 MHz maser in the LMC.

Goss: Oh, you found one in the LMC. And this source is possibly associated with a SNR close to the prominent HII region 30 Dor ?

Roberts: I have heard of a weird source observed with the VLA recently—the HII region S106. It has an OH 1720 MHz maser which does not have any main lines [1665 and 1667 MHz], and what is interesting about that is, independently through OH abundance analysis, we suggested that OH was over-abundant. The weird thing is that there is no supernova remnant associated with it, and we checked: there is no observed X-ray object in the region either.

Wardle: One thing I would like people to talk about, who know more about the observations, is the possibility of finding 1612 MHz OH masers that are sort of analogous to the 1720 MHz masers.

Goss: Farhad [Yusef-Zadeh], Doug Roberts, and I have looked at about five of these 1720 MHz sources at 1612 MHz. As you know, you hit confusion everywhere in the Galactic Plane when you observe 1612 MHz OH, i.e., there are many Type II OH/IR stars in your primary VLA beam (even when observing with high angular resolution). But even if you eliminate those OH/IR stars, you still do not observe any 1612 MHz masers at the positions and velocities associated with the SNRs.

Sjouwerman: But I think I can see these 1612 MHz masers, at a low level, about 17 mJy or so. They are on top of the continuum, but you have to subtract the continuum to see the masers. It is published.

Goss: Is there something you can show?

Sjouwerman: I was looking for OH/IR stars, and when you hit the Galactic Center, you hit continuum issues. And what you try to do with this continuum emission is you try to subtract it; however, if you subtract the continuum emission, you still have some residuals. I must say this is not the actual map you get. I took the cube, subtracted the continuum emission, and then for every position in the sky, I looked at the velocities corresponding to this position, and I just remembered the maximum value for it. So this is what I call a max map. It shows you where the emission is, but it does not tell you what the velocity is. And as you can see, the OH/IR stars are very prominent in this picture. What you have here is a sort of residual map. And these residuals clearly fit into the 6 cm radio maps. So I think this supports the picture of Mark [Wardle]. I do not know how to interpret this, but I just published it as it is.

Goss: Thank you very much. Thanks very much to all of the speakers.

Acknowledgments. Alison Coil, Cornelia Lang, and Anantharamaiah provided additional input for editing this session.

Session VIII

What is the Strength and Influence of the Magnetic Field in the Galactic Center?

The Central Parsecs of the Galaxy
ASP Conference Series, Vol. 186, 1999
H. Falcke, A. Cotera, W.J. Duschl, F. Melia, M.J. Rieke, eds.

Magnetic Fields in the Galactic Center

D.A. Roberts
Astronomy Department, University of Illinois, 1002 W. Green St., Urbana, IL 61801 USA

Abstract. Early models of the magnetic field structure of the Galactic Center suggested very strong field strengths. In the past decade or so, there have been many projects to investigate the morphology and strength of the magnetic fields in the inner parsec of the Galactic Center. Magnetic field observations include primarily near and far-infrared polarization and Zeeman measurements. The infrared polarization measurements along with a model for the polarization mechanism provide the magnetic field morphology in the plane of the sky.

The only direct quantitative measurement of the magnetic field strength in the Galactic Center is obtained from observing the Zeeman effect in radio frequency spectral lines. Centimeter-wavelength Zeeman observations of Sgr A have been carried out in several gas components: neutral hydrogen (HI), hydroxyl (OH), and hydrogen recombination lines (H92α). Line-of-sight fields have been measured at the level of a few milligauss. These fields are much stronger than those in the ambient interstellar medium (whose average is a few microgauss) and are similar to fields in the cores of dense molecular clouds. This talk will discuss new observations of the Zeeman effect in OH(1720MHz) line observed in a maser in the Circumnuclear disk (CND) and the correlation with the HI Zeeman measurements at the same location. New OH (1720MHz) Zeeman results are interpreted in the context of previous Zeeman and polarization observations.

1. Background

Since linear radio filaments were first observed in the Galactic Center, it had been proposed that strong magnetic must exist in the center of our Galaxy. More recent low-frequency radio imaging of the inner 100 parsecs (Kassim et al. 1999, this volume, p. 403) show many linear features throughout the region. Despite the implications of low frequency morphology, it has been difficult to measure the magnetic field in the inner few parsecs of the Galaxy. In the past decade, magnetic field studies have been made using the following techniques.

Infrared polarization from rotating dust grains has been used to investigate magnetic fields in the Galactic Center. These observations can be used to determine the orientation of $B_{\perp}$. However, because the polarization mechanisms and grain characteristics are poorly understood, infrared polarization measurements can not be used to determine the *magnitude* of the magnetic field.

The only technique to directly measure the magnetic field in the Galactic Center is by use of the radio frequency Zeeman effect. Zeeman observations detect the splitting of a radio frequency spectral line (in either emission or absorption) into components with opposite circular polarizations. The amount of the splitting can be measured directly or in the case of a weak field the derivative spectrum of Stokes I spectrum is fitted to the Stokes V spectrum. The coefficient of the fit is the product of the line-of-sight magnetic field times the Zeeman splitting factor. For the Galactic Center and most galactic HII regions, the most sensitive spectral lines are the 1420 MHz line of HI and the various components of the hydroxyl ion (OH) at 1665, 1667, and 1720 MHz. In the most favorable cases (*e.g.*, sources with significant fields and spectral lines with high Zeeman splitting factors), interferometers can be used to create images of the line-of-sight magnetic field.

2. Previous Magnetic Field Observations of the CND

Previous IR polarization observations of the circumnuclear disk (CND) at 100 μm carried out by Hildebrand & Davidson (1994) using the Kupier Airborne Observatory (KAO) provide a view of the magnetic field in the plane of the sky, assuming polarization was due to absorption by rotating dust grains. The IR polarization results were modeled by Wardle & Königel (1990) as an accretion disk with axial symmetry. Desch & Roberge (1997) extended the Wardle & Königel model by adding a large random component and ambipolar diffusion. However, the IR results do not provide any measurement of the magnitude of the field.

Previous Zeeman observations have been carried out in the CND using the HI and thermal OH lines. Killeen, Lo, & Crutcher (1992) carried out Zeeman observations of the the CND using the OH (1667MHz) absorption line with the VLA. These observations covered a velocity range of 57 to 123 kms s^{-1}. Through a technique to detect a magnetic fields over an extended area (spatial summing), they detected a field of $-2.0\pm$ 0.5 mG and $-3.4\pm$ 1.0 mG in the northern part of the CND. A similar value was obtained by Plante, Lo, & Crutcher (1995), who observed the Galactic Center region in HI absorption with the VLA. The HI absorption suffers from extensive blends from multiple components across the band, however, a Zeeman signal was detected in the HI data from a blend of lines at 130 km s^{-1}. A field of $-3.0\pm$ 0.5 mG is measured for this component.

In the past few years there has been growing interest in observing OH (1720MHz) masers to probe regions that contain slow shocks (usually from supernovae remnants) and moderate density ($\sim 10^6$ cm^{-3}) molecular gas (for an early review see, Frail, Goss & Slysh 1994). Observations of OH (1720MHz) masers in the CND, using the VLA D-array by Yusef-Zadeh et al. (1996) first showed Zeeman splitting in the OH (1720MHz) line toward the Galactic Center. At the very position of the HI Zeeman detection by Plante, Lo, & Crutcher, a Zeeman splitting from multiple OH components is observed. Because of the low spatial resolution of the observations, the emission from all components appears to originate from the same location with three spectral components from 132 to 136 km s^{-1}. The fitted line-of-sight magnetic field appears to be -4.0 to $-3.0\pm$ 0.3 mG from these components.

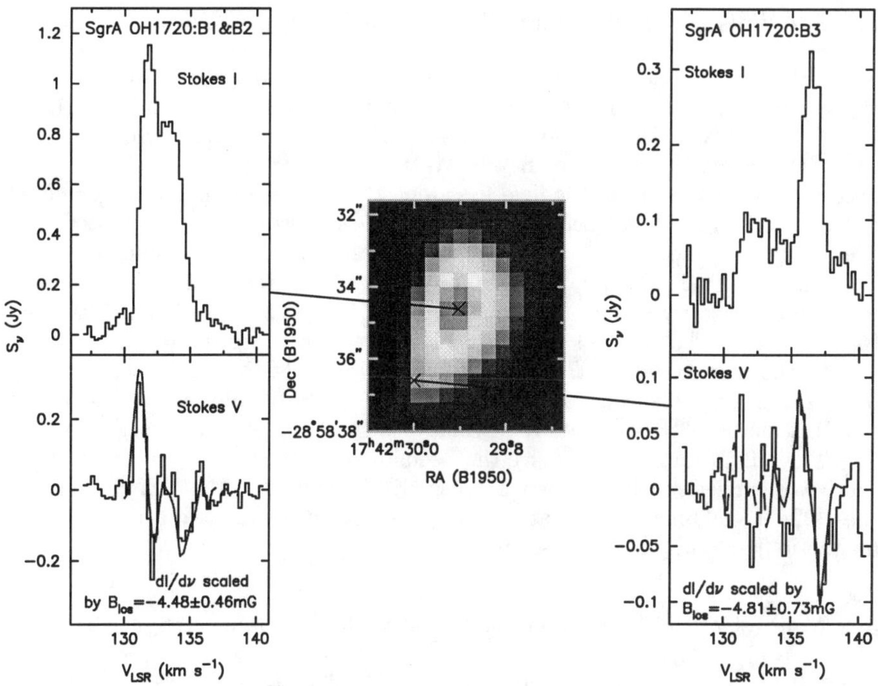

Figure 1. Profiles of Stokes I (top histogram) and V (bottom histogram) in the OH (1720 MHz) line for our maser detections for SgrA OH (1720MHz) maser sources in the northern part of the CND (masers B1-B3). The dashed line superposed on the V spectrum is the derivative of the I spectrum scaled by the indicated magnetic fields.

3. New OH (1720MHz) Maser Observations of the CND

The masers in Sgr A detected in the low resolution data of Yusef-Zadeh et al. (1996) were recently re-observed with the VLA A-array with spatial and spectral resolutions of $2''.48 \times 1''.28$ and 0.27 km s^{-1}, respectively. The important maser associated with the CND (which was determined spectrally to have three components) is observed to break into two spatial components. The spectral identification is improved due to the finer spectral resolution.

Figure 1 shows the I and V spectra from the masers in the northern part of the CND. The image at the center of the figure shows the velocity integrated OH (1720MHz) line emission between 129.7 and 138.5 km s^{-1}. The brightest sources B1 and B2 are partially blended spectrally, though their peak emission with different line widths are displaced from each other. The highest velocity feature at 136.5 km s^{-1} feature is displaced to the southwest by about $2''$ from the peaks of lower velocity components at 131.7 and 133.4 km s^{-1}.

4. Previous Magnetic Field Observations of Sgr A West

Several far infrared polarization observations have been carried out toward Sgr A West to determine the magnetic field morphology in the plane of the sky. Hildebrand & Davidson (1994) have carried out infrared polarimetry observations of the northern northern arm of Sgr A West at 100 μm using the KAO. Their observations suggests that the field lies along the northern arm and the detected polarization follows the morphology of the 50 μm emission. It is suggested that the hot gas detected at 50 μm is coincident with the ionized streamer.

The NIMPOL imaging polarimeter on the AAT was used to observe mid infrared polarization of the northern arm of Sgr A West at 12.5 μm (Aitken 1996, Roche 1996). The inferred magnetic field direction is strikingly uniform and follows the morphology of the northern arm, similar to that of the far infrared polarization of Hildebrand & Davidson (1994). Additionally, a high degree of polarization of 7-8% was observed, which is near the maximum observed value.

The only attempt to measure the Zeeman effect in the ionized gas of Sgr A West was carried out by Roberts & Goss (1993), who used the VLA to observe the H92α recombination line at 8.4 GHz. Their observations did not detect a magnetic field and yielded a 3-σ upper limit of 8 mG.

5. Magnetic Fields Farther from the Galactic Center: Sgr B2

There have been several attempts to investigate the magnetic fields in the massive star forming region of Sgr B2. Novak et al. (1997) have carried out polarization observations of Sgr B2 at 115 μm. They suggest that the global polarization properties are a combination of polarization by absorption in the core and emission in the envelope. The suggested magnetic field structure is uniformly north-south with small spatial variations ($< 30°$).

VLA HI Zeeman observations were carried out by Crutcher et al. (1996). The HI absorption line profiles toward Sgr B2 showed wide lines that were a result of blending from narrow, unresolved velocity components. The field was detected in four locations with values about 0.5 mG with small variations. The suggested strength is near the critical value to support the cloud against gravitational collapse.

6. General Discussion and Conclusions

Magnetic field measurements in the Galactic center are important and have been extensively pursued. Although the observations are difficult and strength of the field has only been determined in a few locations, there are a few general points concerning the results so far. Sensitivity limitations place a limit on the positions where the magnetic field can be measured using the Zeeman effect. However, where the effect is measured, the field is strong. In the CND, it is as high as 4 mG. One of the major questions concerning the magnetic field structure is "what scales do large fields exist in the Galactic Center?" At an intermediate distance to the Galactic Center, Sgr B2 has magnetic fields of a few hundred microgauss and magnetic flux to mass ratio near the critical value. These properties are similar to other Galactic HII regions far away from the center such as W3 (Roberts et al. 1993) and S106 (Roberts et al. 1995). Within 1-2 pc, gravity and organized mas motions dominate the dynamics of the gas. This is the case for the CND and the northern arm. Outside 1-2 pc, if magnetic fields are as strong as the detected few mG they would dominate, or at the least, significantly influence the gas dynamics. In the inner 100 pc, strong magnetic fields would influence star formation.

References

Aitken, D.K. 1996, in Polarimetry in the Interstellar Medium, ASP Conf. Ser. Vol. 97, eds. W.G. Roberge & D.C.B. Whittet, (San Francisco:PASP) p 225

Crutcher, R.M., Roberts, D.A., Mehringer, D.M., & Troland, T.H. 1996, ApJ462, L79

Desch, S.J. & Roberge, W.G. 1997, ApJ475, L115

Frail, D.A., Goss, W.M. & Slysh, V.I. 1994, ApJ424, L111

Kassim et al. this volume

Hildebrand, R. H. & Davidson, J. A. 1994, in The Nuclei of Normal Galaxies: Lessons from the Galactic Center, eds. R. Genzel and A. I. Harris, (Dordrecht:Kluwer) p 199

Killeen, N.E.B., Lo, K.Y., & Crutcher, R. 1992, ApJ385, 585

Novak, G., Dotson, J.L., Dowell, C.D., Goldsmith, P.F. Hildebrand, R.H., Platt, S.R., Schleuning, D.A. 1997, ApJ487, 320

Plante, R.L., Lo, K.Y., & Crutcher, R.M. 1995 ApJ445, L113

Roberts, D.A. & Goss, W.M. 1993, ApJS86, 133

Roberts, D.A., Crutcher, R.M., Troland, T.H., & Goss, W.M. 1993, ApJ412, 675

Roberts, D.A., Crutcher, R.M., & Troland, T.H. 1995, ApJ442, 208

Roche, P.F. 1996, in Polarimetry in the Interstellar Medium, ASP Conf. Ser. Vol. 97, eds. W.G. Roberge & D.C.B. Whittet, (San Francisco:PASP) p 551

Wardle, M. & Königel, A. 1990, ApJ362, 120

Yusef-Zadeh, F., Roberts, D.A., Goss, W.M., Frail, D.A., & Green A.J. 1996, ApJ466, L25

The Central Parsecs of the Galaxy
ASP Conference Series, Vol. 186, 1999
H. Falcke, A. Cotera, W.J. Duschl, F. Melia, M.J. Rieke, eds.

Far-infrared/Submillimeter Polarimetric Observations of the Galactic Center

G. Novak

Department of Physics and Astronomy, Northwestern University, Evanston, Illinois 60208

Abstract. Measurements of the linear polarization of far-infrared and submillimeter thermal emission from interstellar dust grains provide information on the configuration of the magnetic field lines that thread molecular clouds. During the last ten years, far-infrared and/or submillimeter polarization has been measured for six distinct regions within the Galactic Center. We compare recent results, including the first detection of submillimeter polarization from clouds in Sgr A, with earlier work, including far-infrared polarimetry of the 3-pc circumnuclear disk, the arched filaments, and Sgr B2. We argue that (1) Within the Galactic Center neutral gas layer, the measured magnetic field directions appear to be more consistent with a globally azimuthal field geometry than with one that is globally poloidal, (2) the energy available from magnetic reconnection between this postulated large-scale azimuthal field and the poloidal field traced by the non-thermal filaments is sufficient to power the filaments, and (3) recent submillimeter polarimetric results for the "50 km/s cloud" in Sgr A support the hypothesis that the formation of massive stars in this cloud was triggered by the expansion of Sgr A (East).

1. Large-Scale Field Geometry: Poloidal or Azimuthal?

The discovery of filamentary structure in the non-thermal Radio Arc (Yusef-Zadeh, Morris, and Chance 1984) provided early evidence for large-scale magnetic fields in the Galactic Center. These non-thermal filaments, that trace magnetic field lines, are orthogonal to the Galactic plane. This result, combined with the later discovery of more non-thermal filaments (and other radio structures) of similar orientation, has led some observers to infer that the central 100-200 pc of the Galaxy are permeated by a large-scale field that has a "poloidal" or "vertical" configuration (e.g. see Morris 1994 and Wielebinski 1994). The field lines in the disk of the Galaxy, in contrast, are known to run along the spiral arms (on average), and thus have a global geometry that is nearly azimuthal.

The basically azimuthal geometry of the Galactic disk's field can be understood by comparing the energy density of the magnetic field with the kinetic energy density of the gas. If we consider only local kinetic energy, then the two energy densities are comparable in the disk, but if we consider kinetic energy due to rotation of the disk, then the magnetic energy is overwhelmed by the kinetic energy. On the largest scales, the field must therefore get sheared out by the differential rotation of the gas and it will end up following the gas streamlines, that run parallel to the spiral arms.

It is important to remember, however, that the morphology of the non-thermal emission is selectively tracing the field in hot, diffuse, ionized regions. We cannot necessarily draw conclusions about the magnetic field in the denser, neutral gas based on morphology and orientation of non-thermal filaments. Most of the gas in the Galactic Center, in fact, is in this neutral form: it lies in the "Galactic Center neutral gas layer" (Güsten 1989), and here the field must be probed using other methods. Due to the magnetic alignment of interstellar dust grains, magnetic fields can be mapped using far-IR/submillimeter linear polarization measurements, that give the direction (but not the strength) of the interstellar magnetic field (Hildebrand 1988; Hildebrand et al. 1995; Roberge 1996). (Zeeman radio measurements provide another method for observing fields in neutral regions within the Galactic Center, but we will not discuss these here.)

The Galactic Center neutral gas layer extends about 500 pc along the Galactic plane and has a width in the perpendicular (Galactic latitude) direction of observations of several regions within this layer have been carried out using the (now decommissioned) Kuiper Airborne Observatory (KAO). Close to 100 separate sky positions were observed, and the results are very different from the poloidal configuration seen in the diffuse ionized gas: A recent review (Davidson 1996) concluded that "the magnetic fields observed in the diffuse gas in the inner 200 pc of the Galaxy seem to lie perpendicular to the Galactic plane, whereas the fields measured in the dense gas of this region ... seem to lie almost parallel to the Galactic plane". The Galactic Center may thus have a globally azimuthal field in the neutral gas and a globally poloidal field in the hot ionized regions.

Next, we will briefly review these far-IR KAO observations, and then discuss more recent results obtained at submillimeter wavelengths using the Caltech Submillimeter Observatory (CSO).

The first Galactic Center source to be observed polarimetrically in the far-IR was the 3-pc circumnuclear disk (CND), where Werner et al. (1988) and Hildebrand et al. (1990, 1993) found a magnetic field running approximately parallel to the Galactic plane. The simplest explanation was that the field geometry was azimuthal with respect to the Galaxy. This discovery did not lead to a reevaluation of the common assumption of a poloidal field on 100 pc scales, because the effects of gravity could be expected to be dominant within the central few pc, due to the large central mass of $2 - 3 \cdot 10^6 M_\odot$. In fact, the model that was developed to fit the far-IR polarimetry of the CND involved poloidal fields outside the CND and azimuthal fields inside it. (In this model, the fields frozen into the CND become wound up into an azimuthal configuration by differential rotation.)

Morris et al. (1992) made a far-IR polarimetric map of the thermal arched filament region, located 20-30 pc away from Sgr A* in the Galactic East direction, and here too found fields running parallel to the Galactic plane. They noted that a large shear had been observed in the neutral gas near the arched filaments, and on this basis argued that the general Galactic Center field, presumed to be poloidal, was, at this location, being sheared out along a direction parallel to the Galactic plane.

The only other Galactic Center cloud for which far-IR polarimetry has been published in the refereed literature is Sgr B2, located at a projected distance of 110 pc from Sgr A* (Dowell 1997, Novak et al. 1997). Millimeter/submillimeter polarimetry is also available for this source (Kane et al. 1993, Greaves et al. 1995). The largest spatial coverage was obtained by Novak et al. who concluded "At positions in the envelope that are far from the star-forming core, where disruptive forces are at work, the field is nearly parallel to the Galactic plane". The pattern noted by Davidson was becoming clear: far-IR observations provide no direct evidence for a pervasive large-scale poloidal field in the Galactic Center.

Still more far-IR polarimetry has been published in conference proceedings and review articles (Morris, Davidson, and Werner 1995; Morris and Serabyn 1996). These results do not change the overall pattern. Individual sky positions often show a wide range of position angles for the inferred magnetic field, and even when one averages the position angle over a large region, the alignment with the Galactic plane is not always very good. But the alignment with the direction perpendicular to the Galactic plane is in all cases worse, usually much worse.

My collaborators and I were able to further explore this issue in 1997, when we used the CSO to make the first submillimeter polarimetric map of Sgr A (Novak et al. 1999). The University of Chicago submillimeter array polarimeter (Schleuning et al. 1997, Dowell et al. 1998) was used for these observations. We observed the CND and two massive cool clouds lying within 20 pc (in projection) of Sgr A*: the "50 km/s" cloud (M-0.02-0.07), and the "20 km/s" cloud (M-0.13-0.08).

The results are shown in Figure 1. In the CND region, we see field directions that are similar to those seen previously here in the far-IR. The 50 km/s cloud is known to be interacting with the expanding non-thermal shell source Sgr A (East), and, as expected, we observe a strong imprint of the expansion of Sgr A (East) into this cloud. Our results for the 50 km/s cloud are discussed in section 3. The 20 km/s cloud is not interacting with any sources comparable in energy to Sgr A (East), so its field may be more representative of the large-scale field in the neutral ISM at the Galactic Center. It shows an average magnetic field direction that is close to the direction of the Galactic plane. To first order, then, we can see that the CSO results support the picture of a globally azimuthal field in the neutral gas at the Galactic Center.

Certainly much more work is needed before drawing firm conclusions, but the results of far-IR and submillimeter polarimetry do suggest that a large-scale azimuthal field may coexist with the poloidal field, with the former dominating the neutral gas while the latter dominates the ionized gas. This would be reminiscent of the Earth's field that is (to first order) azimuthal inside and poloidal outside. At a minimum, caution is indicated when making general statements about the magnetic field in our Galaxy's nucleus. For example, we now know that Figure 3 of the 1994 review by Wielebinski, that shows a picture of the Milky Way with fields running along the spiral arms in the disk and a dipole field in the center, is probably an oversimplification.

We cannot eliminate the possibility that there may be large volumes of space where the poloidal field dominates even in dense clouds. For example, it may be that the poloidal fields are confined to Galactocentric radii smaller

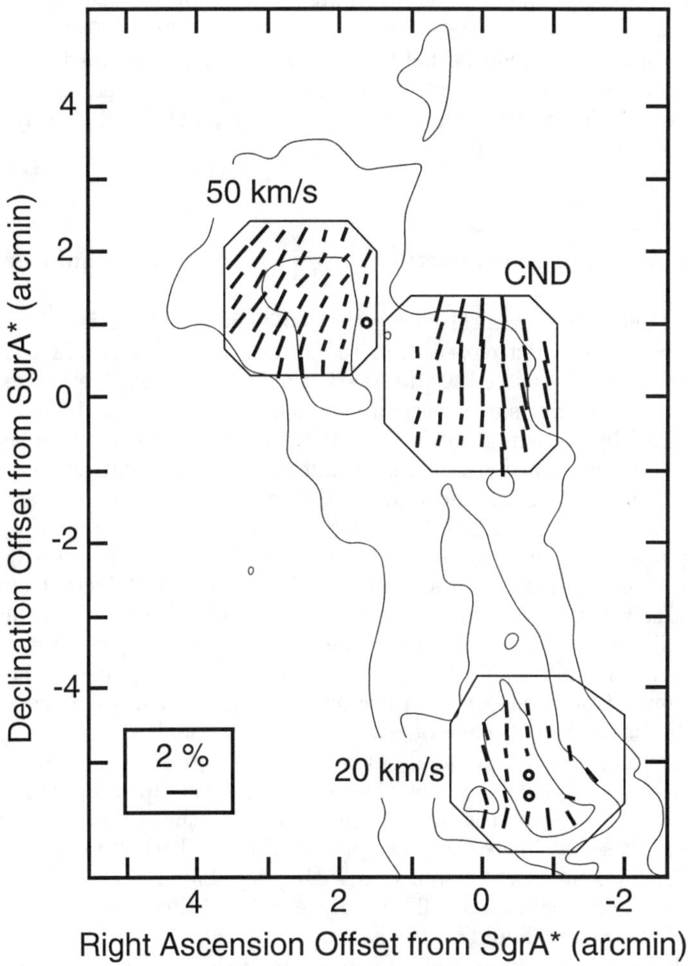

Figure 1. 350 micron polarization results for Sgr A (Novak et al. 1999), shown together with contours of 800 micron emission (Lis and Carlstrom 1994). Three distinct regions were observed. These are outlined in the figure by three irregular boxes, each enclosing a group of bars. Each bar indicates a sky position for which we obtained a 3-sigma detection of polarization. The orientations of the bars indicate the inferred magnetic field directions. (They are orthogonal to the measured angles of polarization.) The length of each bar is proportional to the degree of polarization (see key at lower left of figure). Sky positions for which Novak et al. did not detect polarization but were able to set low upper limits on the magnitude of the polarization are shown as open circles.

than 70 pc, while the azimuthal fields dominate beyond this radius. Because of uncertainty in the line-of-sight positions of the Galactic Center clouds that we have mapped polarimetrically, this cannot be ruled out. But it does seem likely that a globally azimuthal field exists over some significant fraction of the Galactic Center neutral gas layer, and it may therefore be interesting to explore a possible interaction between this field and the poloidal field. This is the subject of the next section.

2. Does Magnetic Reconnection Power the Non-Thermal Filaments?

Stark et al. (1991) show that the clouds that make up the Galactic Center neutral gas layer have a finite lifetime: Within 10^9 yr, they will all spiral into the center due to dynamical friction. Other dissipation mechanisms work even faster.

Let us assume that some reasonable fraction of the neutral gas layer is in fact permeated by an azimuthal field. Due to its finite lifetime, this layer must inevitably migrate inwards, with its azimuthal field lines entrained. It seems likely that the poloidal fields traced by the Galactic Center non-thermal filaments are strong (≈ 1 mG, Morris 1994), and we will further assume that they are not a transient phenomenon. E.g. , we will assume that one billion years from now, similar non-thermal filamentary structures will exist in the Galactic Center. The topology of this situation is then such that the circular field lines in the neutral gas layer will not be able to cross the poloidal fields that they encounter as they drift inwards. This is because the poloidal fields cross the plane of the Galaxy, are many tens of pc in length, and are open at both ends. (This can be seen, for example, in the radio image of Anantharamaiah et al. 1991). Note that we need not assume that the poloidal fields occupy a significant volume of space. For the purposes of our argument they could, for example, be confined to a cylindrical sheet. The point is that because the azimuthal field cannot cross the poloidal field, it seems likely that the (presumably weaker) azimuthal fields will get destroyed by magnetic reconnection with the poloidal fields. (Alternatively, the trapped poloidal field lines will be compressed and therefore will get stronger - in the end the result should be the same.)

It is plausible that a non-negligible fraction of the energy contained in these azimuthal fields will be liberated in this process. In an analogy with the physics of solar flares, magnetic reconnection has been suggested by several authors as the mechanism that provides the relativistic electrons that light up the the non-thermal filaments (e.g. see Lesch and Reich 1992, and Serabyn and Morris 1994). Here we make an estimate of the rate at which the infalling neutral clouds provide magnetic energy, and compare this with the luminosity of the filaments.

Novak et al. (1997) used a combination of Zeeman splitting observations and far-IR polarimetry data to set a conservative lower limit of 150 μGauss on the strength of the (apparently azimuthal) large-scale field they observed in Sgr B2. Using the model of Lis and Goldsmith (1989) to obtain mass and radius estimates for this cloud of $6 \cdot 10^6 M_\odot$ and 22.5 pc, and assuming it to be spherical, we can estimate a lower limit on the the ratio of magnetic energy (in the large-scale field) to cloud mass. Assuming that this ratio applies to the neutral gas layer as a whole, and using a total layer mass of $2 - 610^7 M_\odot$ (for a

400 pc radius; Sodroski et al. 1995), we then obtain a lower limit on the total magnetic energy of the neutral gas layer of $\approx 8 \cdot 10^{51} erg$.

If the azimuthal global geometry of the field in the neutral gas layer can be taken to indicate that magnetic forces are smaller than gravitational forces (e.g. see Werner et al. 1988), then we can also derive an upper limit on the magnetic energy by equating it to the orbital kinetic energy. Using the above estimate of Sodroski et al. for the mass of the layer and an orbital velocity of 180 km/s (Genzel and Townes 1987), we obtain an upper limit of $13 \cdot 10^{54} erg$.

Dividing the above energies by 10^9 yr, we obtain an available luminosity of between $\approx 10^2$ and $\approx 10^5 L_{\odot}$. The estimated total luminosity of the non-thermal filaments in the Galactic Center is $1 - 2 \cdot 10^2 L_{\odot}$ (Yusef-Zadeh and Morris 1987, Lesch and Reich 1992). We conclude that the energy provided by the inevitable reconnection between infalling azimuthal fields and strong poloidal fields is probably sufficient to power the filaments.

3. Star Formation in the 50 km/s Cloud - the Role of Sgr A (East)

CS 5-4 observations of the 50 km/s cloud by Serabyn, Lacy, and Achterman (1992) show two high-density components. These are referred to as the "core", and the "curved ridge". (In Figure 1, they are essentially blended: together they form a single feature as outlined by the highest contour.)

Figure 2 shows our results superposed on a 1.3 mm continuum map by Mezger et al. (1989). This map traces dust emission, but with much better angular resolution than was obtained for the 800 micron map of Figure 1. The 1.3 mm map has been processed to remove the flux contributions from Sgr A* and Sgr A West. It is important to note that, for reasons explained by Mezger et al. , their map is relatively insensitive to extended emission. The shaded region in Figure 2 is the Sgr A (East) radio source.

The "curved ridge" component of the 50 km/s cloud is clearly visible in the 1.3 mm map. It is the brighter, northeastern section of what appears to be a nearly complete ring of emission surrounding Sgr A (East). The "core" component is not visible, partly due to the suppression of low spatial frequencies, and partly because the 1.3 mm map has limited spatial coverage. The core lies about an arcminute to the East of the curved ridge. The ring of emission seen in the 1.3 mm map is believed to correspond to a shell of dust and gas that has been swept up by the expanding non-thermal shell source Sgr A (East).

The filled circles in Figure 2 show the locations of five compact H II regions that lie within the 50 km/s cloud (Ekers et al. 1983). These contain young massive stars. Based on the curvature evident in the line of H II regions, and on their location close to Sgr A (East), Ho et al. (1985) suggested that the formation of the stars in these H II regions was triggered by the expansion of Sgr A (East). However, Serabyn, Lacy, and Achterman (1992) noted the approximately round shape of the contours of CS emission from the core component, and concluded that the core has not yet been affected by the expansion of Sgr A (East). Furthermore, these authors argued that the H II regions are associated with the core rather than with the curved ridge, and that the expansion of Sgr A (East) has thus not yet affected the H II regions, and hence could not be the star formation trigger. Serabyn et al. also argue this last point on the grounds

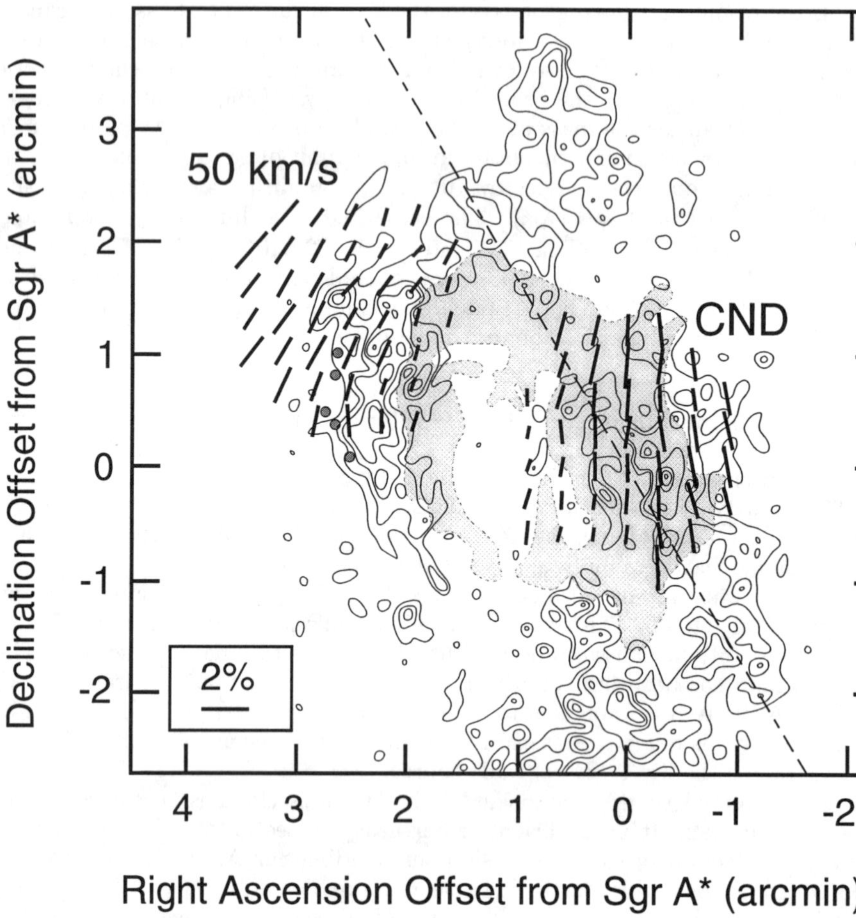

Figure 2. 350 micron polarization results shown together with a high-resolution 1.3 mm dust emission map of Mezger et al. (1989; solid contours) and non-thermal emission from Sgr A East (shaded area; corresponding to the second lowest contour from the 6 cm map of Ekers et al. 1983). The orientations of the bars indicate the inferred magnetic field directions. (They are orthogonal to the measured angle of polarization.) As in Figure 1, the length of each bar is proportional to the degree of polarization. Filled (grey) circles show the positions of compact HII regions in the 50 km/s cloud. A dashed line indicates the orientation of the Galactic plane. We note that the superposition of the 6 cm and 1.3 mm maps shown here is identical to that of Fig. 3b in Mezger et al. (1989).

of inconsistency between the ages of the stars in the H II regions and the age of Sgr A (East).

When a thin slab of gas is formed by compression of an initially roundish cloud, the magnetic field in the slab should end up nearly parallel to the plane of the slab, due to flux-freezing. It can be seen in Figure 2 that for sight-lines passing through the curved ridge, both the mean field direction and the observed field curvature are consistent with this expectation. Furthermore, note also that the curved field overlaps both the line of H II regions and the curved ridge. This supports the suggestion that Sgr A (East) has triggered the formation of the stars. Also, Novak et al. (1999) argue that there is no serious inconsistency between the ages of the stars and that of Sgr A (East).

The inferred field directions for points lying at the northeastern extreme of our polarization map for the 50 km/s cloud are nearly perpendicular to the Galactic plane (see Figure 2). Could this be evidence for poloidal fields in neutral regions? A more complete set of observations are needed to answer this question.

4. Future Observations

Galactic Center magnetic fields can play important roles in star formation, cosmic ray transport, and also gas infall and angular momentum transport on a variety of spatial scales. In addition, much of the energetic activity associated with the thermal and non-thermal radio filaments appears to have a magnetohydrodynamic explanation, judging from the models and theories that have been advanced (e.g. Benford 1988, Shibata and Uchida 1987).

In the near term, new data will come from continued observations with the Hertz instrument at the CSO, and larger-scale mapping using the Submillimeter Polarimeter for Antarctic Remote Observations (SPARO; Novak et al. 1996, Dotson et al. 1998). Over the next five to ten years, the technique of far-IR/submillimeter polarimetry can play an increasingly important role, due to major technical advances such as bolometer arrays with many more pixels, higher bandwidth for continuum interferometry, and major new facilities. Two very promising future developments are multi-wavelengths polarimetry using arrays with hundreds of pixels on the 2.5 meter Stratospheric Observatory for Infrared Astronomy (SOFIA), and continued development of the very new field of submillimeter/millimeter interferometric polarimetry (e.g. see Rao et al. 1998).

Acknowledgments. This work has benefitted greatly from the hard work and talents of my collaborators in the CSO polarimetry of Sgr A: J. Dotson, C.D. Dowell, R. Hildebrand, T. Renbarger, and D. Schleuning. I also wish to thank M. Houde and the staff of the CSO for assistance with the observations, and J. Davidson, M. Morris, and F. Yusef-Zadeh for fruitful discussions. This work was made possible by an NSF Faculty Early Career Development Award (OPP-9618319).

References

Anantharamaiah, K.R., Pedlar, A., Ekers, R.D. and Goss, W.M. 1991, MNRAS, 249, 262

Benford, G. 1988, ApJ, 333, 735

Davidson, J.A. 1996, in ASP Conf. Ser. 97, Polarimetry of the Interstellar Medium, ed. D.C.B. Whittet and W.G. Roberge (San Francisco: Astron. Soc. of the Pacific), 504

Dotson, J.,L., Novak, G., Renbarger, T., Pernic, D., Sundwall, J. 1998, in Proceedings of SPIE International Symposium on Astromomical Telescopes and Instrumentation, held in Kona, Hawaii, 20-28 March, 1998.

Dowell, C.D. 1997, ApJ, 487, 237

Dowell, C.D., Hildebrand, R.H., Schleuning, D.A., Vaillancourt, J., Dotson, J.L., Novak, G., Renbarger, T., and Houde, M. 1998, ApJ, 504, 589

Ekers, R.D., van Gorkom, J.H., Schwarz, U.J., and Goss, W.M. 1983, Astron. and Astrophys. 122, 143

Genzel, R. and Townes, C.H. 1987, Ann. Rev. Astron. Astrophys., 25, 377

Greaves, J. S., Holland, W.S., Murray, A.G., and Nartallo, R. 1995, MNRAS, 272, L1.

Güsten, R. 1989, in Proceedings of I.A.U. Symposium No. 136, The Center of the Galaxy, ed. M. Morris (Boston:Kluwer), 89

Hildebrand, R.H. 1988, QJRAS, 29, 327

Hildebrand, R.H., Davidson, J.A., Dotson, J., Figer, D.F., Novak, G., Platt, S.R., and Tao, L. 1993, ApJ, 417, 565

Hildebrand, R.H., Dotson, J., Dowell, C.D., Platt, S.R., Schleuning, D., Davidson, J.A., and Novak, G. 1995, in ASP Conf. Ser. 73, Airborne Astronomy Conference on the Galactic Ecosystem: From Gas to Stars to Dust, ed. M.R. Haas, J. A. Davidson, and E.F. Erickson (San Francisco, ASP), 97

Hildebrand, R.H., Gonatas, D.P., Platt, S.R., Wu, X.D., Davidson, J.A., Werner, M.W., Novak, G., and Morris, M. 1990, ApJ, 362, 114

Ho, P.T.P., Jackson, J.M., Barrett, A.H., and Armstrong, J.T. 1985, ApJ, 288, 575

Kane, B.D., Clemens, D.P., Barvainis, R., and Leach, R.W. 1993, ApJ, 411, 708

Lesch, H., and Reich, W. 1992, Astron. and Astrophys. 264, 493

Lis, D.C., and Carlstrom, J.E. 1994, ApJ, 424, 189

Lis, D.C., and Goldsmith, P.F. 1989, ApJ, 337, 704

Mezger, P.G., Zylka, R., Salter, C.J., Wink, J.E., Chini, R., Kreysa, E., and Tuffs, R. 1989, Astron. and Astrophys. 209, 337

Morris, M. 1994, in Normal Galactic Nuclei: Lessons from the Galactic Center, ed. R. Genzel and A. I. Harris (Dordrecht, Kluwer), 185

Morris, M., Davidson, J.A., and Werner, M. 1995, in ASP Conf. Ser. 73, Airborne Astronomy Conference on the Galactic Ecosystem: From Gas to Stars to Dust, ed. M.R. Haas, J. A. Davidson, and E.F. Erickson (San Francisco, ASP), 477

Morris, M., Davidson, J.A., Werner, M., Dotson, J., Figer, D.F., Hildebrand, R.H., Novak, G., and Platt, S.R. 1992, ApJ, 399, L63

Morris, M., and Serabyn, E. 1996, Ann. Rev. Astron. Astrophys. 34, 645.

Novak, G., Dotson, J.L., Dowell, C.D., Goldsmith, P.F., Hildebrand, R.H., Platt, S.R. and Schleuning, D.A. 1997, ApJ, 487, 320

Novak, G., Dotson, J.L., Dowell, C.D., Hildebrand, R.H., Renbarger, T., and Schleuning, D. A. 1999, ApJ, submitted

Novak, G., Dotson, J., Renbarger, T., Platt, S.R., and Dragovan, M. 1996, in ASP Conf. Ser. 97, Polarimetry of the Interstellar Medium, ed. D.C.B. Whittet and W.G. Roberge (San Francisco: Astron. Soc. of the Pacific), 111

Rao, R., Crutcher, R.M., Plambeck, R.L., and Wright, M.C.H. 1998, ApJ, 502, L75

Roberge, W.G. 1996, in ASP Conf. Ser. 97, Polarimetry of the Interstellar Medium, ed. D.C.B. Whittet and W.G. Roberge (San Francisco: Astron. Soc. of the Pacific), 401

Schleuning, D.A., Dowell, C.D., Hildebrand, R.H., Platt, S.R., and Novak, G. 1997, PASP, 109, 307

Serabyn, E., Lacy, J. H., and Achterman, J.M. 1992, ApJ, 395, 166

Serabyn, E., and Morris, M. 1994, ApJ, 424, L91

Shibata, K., and Uchida, Y. 1987, Pub. Astron. Soc. Japan, 39, 559

Sodroski, T.J., Odegard, N., Dwek, E., Hauser, M.G., Franz, B.A., Freedman, I., Kelsall, T., Wall, W.F., Berriman, G.B., Odenwald, S.F., Bennett, C., Reach, W.T., and Weiland, J.L. 1995, ApJ, 452, 262

Stark, A.A., Gerhard, O.E., Binney, J., and Bally, J. 1991, MNRAS, 248, 14P

Werner, M.W., Davidson, J.A., Morris, M., Novak, G., Platt, S.R., and Hildebrand, R.H. 1988, ApJ, 333, 729

Wielebinski, R. 1994, Proceedings of NATO Advanced Research Workshop on Cosmical Magnetism, ed. D. Lynden-Bell (Dordrecht:Kluwer), 143

Yusef-Zadeh, F., and Morris, M. 1987, ApJ, 320, 545

Yusef-Zadeh, F., Morris, M., and Chance, D. 1984, Nature, 310, 557

The Central Parsecs of the Galaxy
ASP Conference Series, Vol. 186, 1999
H. Falcke, A. Cotera, W.J. Duschl, F. Melia, M.J. Rieke, eds.

Radio Polarmetric Studies of the Galactic Center Threads

C. C. Lang

National Radio Astronomy Observatory, Socorro NM

Department of Astronomy & Astrophysics, UCLA, Los Angeles CA

Abstract. The Galactic Center non-thermal filaments (NTFs) are a unique class of objects found only in the inner 150 pc of the Galaxy. These NTFs are oriented essentially perpendicular to the Galactic plane and extend for 10's of pc in length. Polarimetric data indicate that the emission is synchrotron in nature, and estimations of the field strength are in the range of a few mG. The NTFs therefore trace the vertical magnetic field component in the center of the Galaxy. Multifrequency, polarimetric observations made with the VLA of the Northern and Southern Threads and the Snake NTFs are presented here. Spectral index, rotation measure, and intrinstic magnetic field information will be discussed in each case. Future observations of several newly identified NTFs are discussed.

1. Introduction

VLA observations during the last 15 years have revealed a number of unusual filamentary structures in the inner 150 pc of the Galaxy (Yusef-Zadeh, Morris, & Chance 1984, Morris & Yusef-Zadeh 1985, Bally & Yusef-Zadeh 1989, Lizst & Spiker 1995, Gray et al. 1995). There are now seven well-studied filaments that share a large number of common properties and are known as the non-thermal filaments (NTFs). Figure 1 shows the concentration of radio emission in the inner 2 degrees of the Galaxy, with the known NTFs labelled. In addition to showing elongated, linear structure perpendicular to the Galactic Plane, the NTFs share these characteristics (see also review by Morris 1996): (I) the NTFs are unique to the Galactic Center and are found only within a projected distance of 150 pc from SgrA (we assume a distance of 8.0 kpc to the Galactic Center), (II) lengths of 10's pc and widths <0.5 pc, (III) strong linear polarizaton ($\sim$50%), (IV) spectral indicies which are falling with frequency (V) rotation measures are $>$ 1000 rad m^{-2}, and (VI) after correction for Faraday rotation, intrinsic magnetic fields are aligned along the filament long axes.

Recent multi-wavelength studies have shown that every NTF is associated and potentially interacting with both ionized and molecular gas. However, there are several unresolved issues related to this unusual class of objects. The first question is: what is the origin of the NTFs, and in particular, the acceleration of relativistic particles? Several theories of NTF generation have been proposed and are summarized by Morris (1996), yet existing observations have been unable to distinguish among the suggestions.

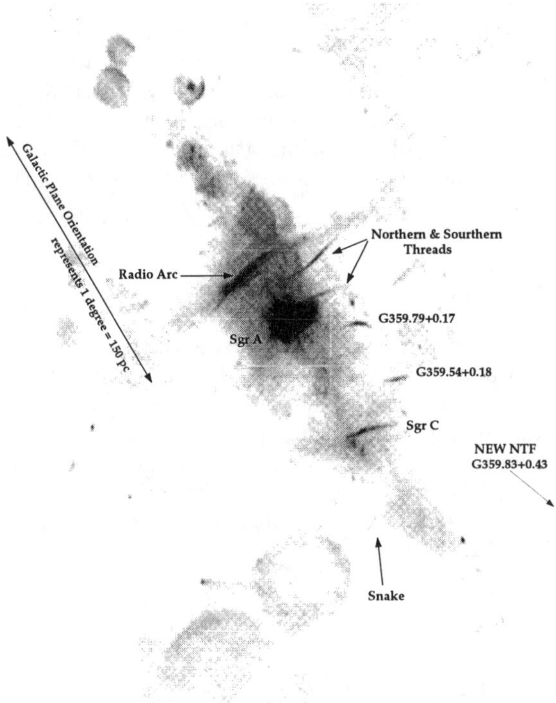

Figure 1. VLA λ=90 cm continuum image showing the inner few degrees with a resolution of 43″. The data originally taken by Pedlar et al. (1989) have been reprocessed and are discussed by Kassim et al. (1999).

Recent observations suggest that all of the NTFs are associated with adjacent ionized and molecular gas. The second issue is: are these observed associations related to and perhaps a crucial element for NTF generation? Serabyn & Morris (1994) have proposed a model based on observations of the molecular gas in the vicinity of the Radio Arc NTFs, in which magnetic reconnection occurs at the interface betweeen a molecular cloud, its ionized edge, and each magnetic NTF. This model has been applied to observations of several other NTF systems, including G359.54+0.18 and the adjacent molecular gas (Staghun et al. 1998) and the HII and molecular complex at one end of the Snake (Uchida et al. 1996).

Finally, studies of the NTFs reveal valuable information about the nature of the magnetic field configuration at the Galactic Center. One of the key questions is: do the NTFs represent a ubiquitous magnetic field concentrated in the inner

100 pc of the Galaxy, or do they represent isolated concentrations of magnetic field (flux tubes)? (Morris 1994).

Here we present high resolution, polarimetric data of the Northern and Southern Threads and the Snake NTF. This study is the first detailed, multi-wavelength study of the Threads since they were discovered by Morris & Yusef-Zadeh (1985) and detected at λ=90 cm by Anantharamaiah et al. (1991). The data on the Snake are a continuation of the polarization study by Gray et al. (1995), with the aim of finding the intrinsic magnetic field in this unusually kinked NTF.

2. Observations

The Northern and Southern Threads have been observed with the VLA at λ=2, 3.6, 6 and 20 cm. Data from a number of different epochs and configurations of the VLA have been combined to achieve the highest resolution images (for details see Lang, Morris & Echevarria, in prep.). λ=3.6 cm data of the Snake were taken with the VLA in January 1996, in the CnB configuration. In all cases, standard calibrators (3C286, NRAO530, 1748-253) and data reduction techniques were employed. All images were made using the deconvolution procedure IMAGR in AIPS.

3. Results

3.1. The Northern and Southern Threads

The Northern Threads are prominent sources in Figure 2, each extending for nearly 40 pc. In addition, the Radio Arc NTFs and the curved filament known as G359.79+0.17 are also detected (see Figure 1 for locations). All NTFs systems show sub-filamentation in this image: the Southern Thread has the most dramatic splitting along its length, the Northern Thread appears to have a parallel filament at its W diffuse edge. G359.79+0.17 has the appearance of multiple filaments along its broad and difuse ends. Similar sub-structure is detected in all of the well-studied NTFs, including G359.54+0.18 (Yusef-Zadeh et al. 1997), the Snake (Gray et al. 1995) and Sgr C (Lizst & Spiker 1995). In addition, a number of shorter features with brightnesses comparable to those of the known NTFs and elongation perpendicular to the Galactic plane are apparent in Figure 2, particularly surrounding the SgrA complex and S of the Arched Filaments. We refer to these features as "Streaks".

The 2″ resolution of the λ=6 and $\bar{3}$.6 cm images of the Northern Thread shows that the width of the filament is only 4″ (0.15 pc) at its E extent, but it broadens along its length. At its most diffuse, W extent, the Northern Thread has a width of 12″ (0.5 pc).

The spectrum of the Northern Thread between λ=90 and λ=2 cm is shown in Figure 5. The spectrum between λ90 cm and λ6 cm is α=−0.5, and it steepens sharply at λ6 cm to α=−2.0 at higher frequencies.

The Northern Thread has a patchy distribution of polarized emission as shown in Figure 4, but fractional polarizations at both 4585 MHz and 8085 MHz are as high as 70%. No potential sources of depolarization are apparent in the

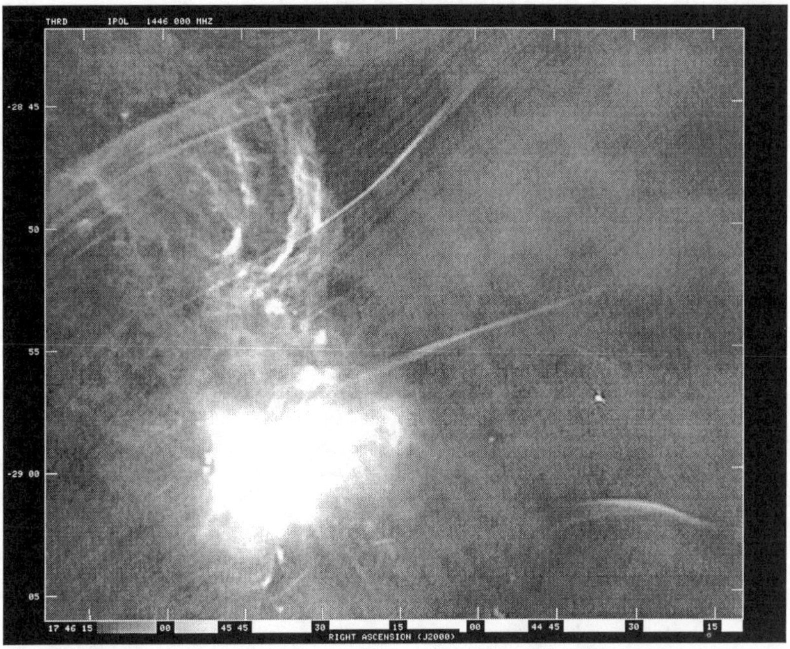

Figure 2. VLA λ=20 cm image which shows the inner 50 pc of the Galaxy. There are four NTF systems obvious in this figure from upper left to lower right: the Northern and Southern Threads, the Radio Arc, and G359.79+0.17. This image has a resolution of $\sim 6''$.

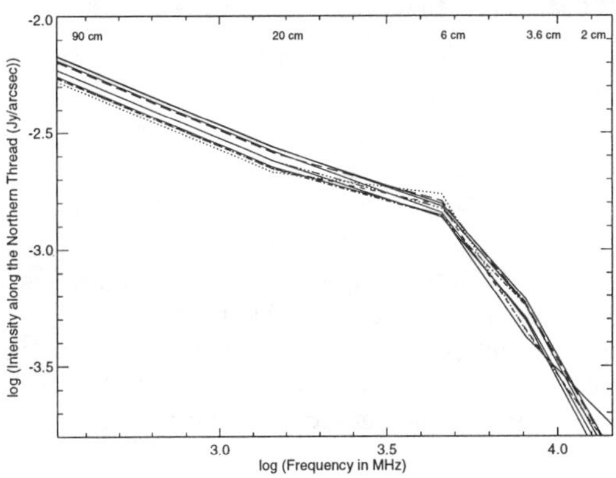

Figure 3. The spectrum of the Northern Thread between λ=90 cm and λ=2 cm. Linestyles correspond to positions along the NTF where the intensity was sampled.

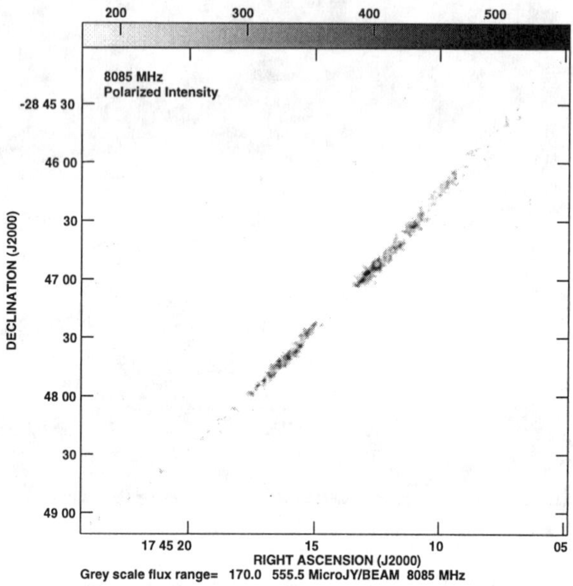

Figure 4. Polarized emission at 8085 MHz arising from the Northern Thread. The data were taken with the VLA and have a resolution of 2.″5.

total intensity images to account for the observed patchiness. The RMs toward the Northern Thread range from 300–3000 rad m^{-2} and vary across the NTF itself without any obvious patterns. The intrinsic magnetic field orientation determined after correction for Faraday rotation is shown in Figure 5 and is aligned predominately along the long axis of the filament, as in all of the other NTFs.

The Southern Thread data were less conclusive: the Southern Thread did not stand out against the halo of the SgrA complex in total λ=6 or λ=3.6 cm intensity. A weak detection of polarized emission at λ=3.6 was made, which shows the same bifurcation along the length of the filament, as is detected in total intensity.

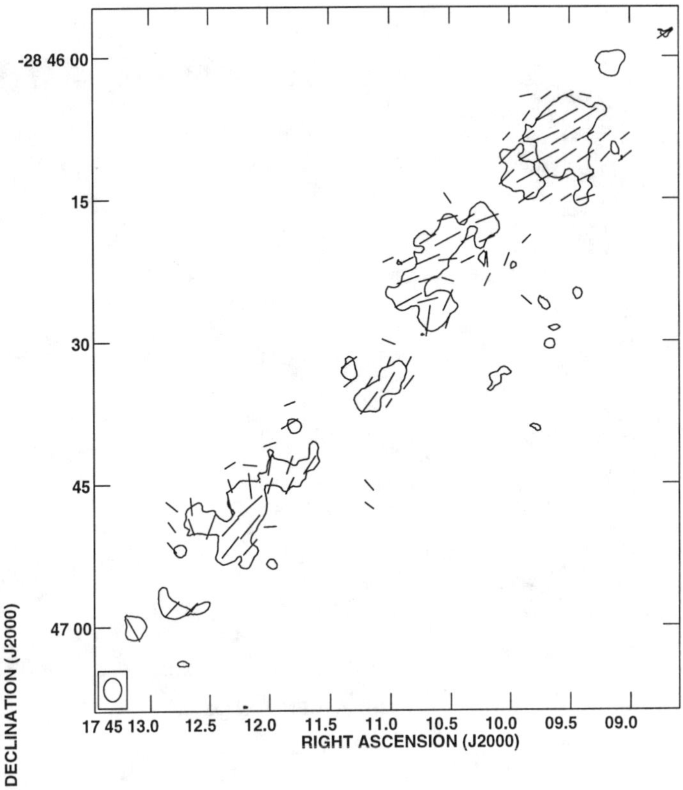

Figure 5. Vectors representing the intrinsic orientation of the magnetic field are shown here superimposed on the contours of VLA polarized intensity at 4585 MHz for the NW region of the Northern Thread, with a resolution of 2."5.

3.2. The Snake

The Snake (G359.1-0.2) has several features which set it apart from the rest of the NTFs: (1) it is the longest NTF detected (~60 pc), (2) it is located the farthest from SgrA (~120 pc in projection), and (3) it shows a number of striking kinks along its length (Gray et al. 1995). A study of the magnetic field orientation in this unusual NTF may provide insight as to the mechanism that is organizing the intrinsic magnetic field, and perhaps causing the kinks along the Snake's length.

Figure 6 shows the 8085 MHz polarized emission arising from the portion of the Snake that was imaged by mosaicking together two fields, each with a primary beam of 5′. The polarized emission decreases noticably in the regions where the Snake shows a kink. RM calculations were done, using three closely

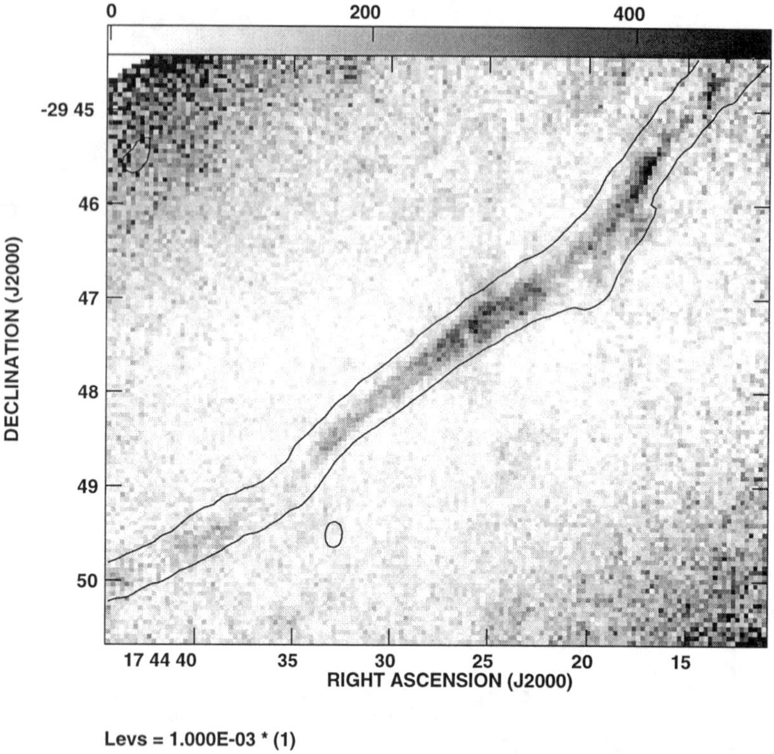

Figure 6. 8085 MHz polarized emission arising from the Snake with a low level 4585 MHz total intensity contour from Gray et al. (1995) overlaid. The resolution of the 8085 MHz polarized emission is $\sim 3''$.

spaced frequencies near $\lambda 3.6$ cm and one at $\lambda 6$ cm, supplied by Gray et al. (1995). Values of the RM are in the range of 2000–5000 rad m^{-2}, similar to values toward the other NTFs detected at the Galactic Center. There is a gradient in the distribution of RM, with values increasing toward the kink region from both sides. Presumably, the depolarization in this region is due to the large RM gradient; it is thus difficult to determine the intrinsic field orientation in the kinked region. Preliminary results suggest that the field is aligned along the filament over most of the Snake, as in the other NTFs. However, in the region where the Snake kinks, the field appears to deviate from this orientation.

4. Discussion

4.1. Northern Thread

The multi-frequency and polarization study that was carried out for the Northern and Southern Threads shows that these sources share the same characteristics as the other well-studied Galactic Center NTFs. Radio recombination line observations have shown that the Arched Filaments H II regions (Yusef-Zadeh 1986) lie on the edge of the -30 km s^{-1} molecular cloud (Serabyn & Morris 1987), both of which the Northern Thread intersects. The presence of these three factors fits the criteria for the model of magnetic reconnection that was proposed by Serabyn & Morris (1994). According this model, the electrons in the Radio Arc are injected at the intersection between the magnetic field and the ionized and molecular gas. In an attempt to detect evidence that acceleration of the electrons occurs at the interface between the NTF and the molecular cloud, we looked for variations of the spectral index along the length of the Northern Thread. We would expect a steepening of the spectral index westward from the intersection of the Northern Thread with the Arched Filaments. However, the 20/6 cm spectral index remained constant for $3'$ along the Northern Thread. A possible explanation is that the electrons may be diffusing along the filament more quickly than they are losing energy via synchrotron radiation. It is therefore difficult to draw conclusions about the location of acceleration or reacceleration of the electrons along the Northern Thread. Kassim et al. (1999) also find no variation of the 90/20 cm spectral index along the SgrC filament.

4.2. Magnetic Field Configuration in the Galactic Center

The detection of several prominent Streaks in Figure 2 (see §3.1) suggests that the multiplicity of such NTFs may far surpass what is currently detected. It also indicates that magnetic field may in fact, pervade this region with only selected field lines illuminated by synchrotron emission. Morris & Serabyn (1996) have pointed out that the presence of the NTFs is the best evidence for a pervasive field which occurs on scales of a few hundred parsecs (a scale comparable to the central concentration of molecular gas). This type of magnetic field configuration can provide the support via magnetic pressure to account for the narrowly confined NTFs, since magnetic pressure acts perpendicular to the field orientation. Models which rely on localized field enhancements (i.e. flux tubes or isolated magnetic field systems) have difficulties providing a mechanism for keeping the NTFs so highly collimated and from preventing expansion of these features (Morris 1994).

CCL thanks Anantharamaiah for supplying the λ90 cm data. The National Radio Astronomy Observatory is a facility of the National Science Foundation, operated under a cooperative agreement with the Associated Universities, Inc.

References

Anantharamaiah, K.R., Pedlar, A., Ekers, R.D., and Goss, W.M. 1991, MNRAS, 249, 262

Bally, J. & Yusef-Zadeh, F. 1989, ApJ, 336, 173

Gray, A., Nicholls, J., Ekers, R.D., and Cram, L. 1995, ApJ, 448, 164
Kassim, N.E., LaRosa, T.N., Lazio, T.J.W., and Hyman, S., this volume, p. 403
Lang, C. C., Morris, M., & Echevarria, L., in prep.
Liszt, H.S., Spiker, R.W., 1995, ApJS, 98, 259
Morris, M. 1994, in *The Nuclei of Normal Galaxies:Lessons from the Galactic Center*, eds. R. Genzel, A. Harris, pp. 185-98
Morris, M. 1996, in *Unsolved Problems of the Milky Way*, Blitz, L. and Teuben, P., ed. Netherlands: IAU, p. 247
Morris, M., and Serabyn, E. 1996, A&AReview, 34, 645
Morris, M., and Yusef-Zadeh, F. 1985, AJ, 90 (12), 2511
Pedlar et al. 1989, ApJ, 342, 769
Serabyn, E., Güsten, R., 1987, A&A 184, 133
Serabyn, E., and Morris, M. 1994, ApJ, 424, L91-94
Staghun, J., Stutzki, J., Uchida, K.I., Yusef-Zadeh, F., 1998, A&A
Uchida, K.I., Morris, M., Serabyn, E.. Güsten, R., 1996, ApJ 462, 768
Yusef-Zadeh, F, 1986, Ph.D. Thesis, Columbia University
Yusef-Zadeh, F., Morris, M., Chance, D., 1984, Nature 310, 557
Yusef-Zadeh, F., Wardle, M., and Parastaran, P., 1997, ApJ, 475, L119

The Central Parsecs of the Galaxy
ASP Conference Series, Vol. 186, 1999
H. Falcke, A. Cotera, W.J. Duschl, F. Melia, M.J. Rieke, eds.

Discovery of a New Filamentary Structure G358.85+0.47 – The Pelican

K.R. Anantharamaiah[1], C.C. Lang[2]

National Radio Astronomy Observatory, Socorro, NM 87801, USA

N.E. Kassim, T.J.W. Lazio

Naval Research Laboratory, Washington D.C., USA

W.M. Goss

National Radio Astronomy Observatory, Socorro, NM 87801, USA

Abstract. We report the discovery of a new filamentary structure, G358.85+0.47, consisting of at least three mutually parallel but bent 'strands', located about 1.5° SW of the Sgr A complex. Unlike all the other known Galactic center filaments, G358.85+0.47 is oriented parallel to the galactic plane. Based on its appearance in a 20 cm image, we give it the name 'the Pelican' for further reference.

1. Introduction

The region within about 150 pc of the Galactic Center contains many unique radio structures. Particularly interesting among these are several linear, non-thermal structures termed "arcs", "threads", and "filaments". The Radio Arc, is a bundle of about a dozen parallel filamentary structures, was discovered by Yusef-Zadeh, Morris and Chance (1984). Two isolated linear structures known as "threads" (G0.08+0.15 and G359.96+0.09) were first reported by Morris and Yusef-Zadeh (1985). The other known linear structures near the Galactic center are G359.79+0.17, Sgr C (Liszt 1985), G359.54+0.18 (Bally & Yusef-Zadeh 1989) and G359.1-0.2 or the "snake" (Gray et al 1991). The linear structures are tens of parsecs long and only a fraction of a parsec wide. Their origin is still not understood. Some of the known properties of these filaments are (Morris 1996, Lang 1999): (1) most of them seem to be oriented perpendicular to the Galactic plane and lie at positive Galactic latitudes, (2) the radio emission is non-thermal and strongly linearly polarized, (3) the spectra of the filaments, where they have been measured, is steep ($S \propto \nu^{-0.6}$), flat, or even inverted (Anantharamaiah et al 1991), (4) there are no obvious radio point sources associated with the fila-

[1]Raman Research Institute, Bangalore 560 080, India

[2]Dept of Astronomy, UCLA, Los Angeles, CA, USA

ments and (5) in every well studied case, there appears to be an association with a molecular cloud.

Although the origin of these filaments is still enigmatic, there is general agreement that there is an intimate connection between these filaments and the large scale magnetic field structure in the Galactic center region. In fact, most of the models that are proposed to explain these structures, make use of some type of magnetic phenomenon including generation of magnetic loops (Heyvaerts, Norman & Pudritz 1988), induced electric field through the $v \times B$ interaction between a moving molecular cloud and a uniform field (Benford 1988), particle acceleration by field line annihilation around current pinches (Lesch & Reich 1992), and field annihilation at the surface of molecular clumps (Serabyn and Morris 1996).

In most of the models, the filaments are thus thought to define the magnetic field direction. Polarization measurements of some of the filaments do support this idea (Lang 1999). As these threads and filaments are oriented essentially perpendicular to the Galactic plane, these radio structures are often considered as strong evidence for the presence of an intense dipolar poloidal magnetic field in the galactic center region (Morris 1996).

In this paper we report the discovery of a new filamentary structure which is about 1.5° SW of the Sgr A complex. This filamentary structure is unique in that it is oriented parallel to the galactic plane whereas all the other previously known filaments are oriented roughly perpendicular to the galactic plane. The orientation of this filament may have implications for the large scale structure of the magnetic filed in the galactic center region.

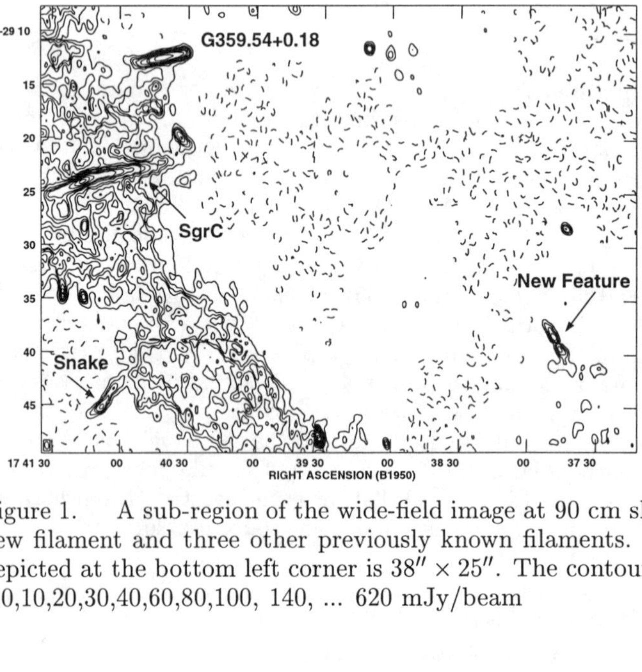

Figure 1. A sub-region of the wide-field image at 90 cm showing the new filament and three other previously known filaments. The beam depicted at the bottom left corner is 38″ × 25″. The contour levels are -10,10,20,30,40,60,80,100, 140, ... 620 mJy/beam

2. Identification of the New Filament in a Wide-field 90 cm Image

Using the data at $\lambda = 90$ cm from all the four VLA configurations obtained by Pedlar et al (1989) and Anantharamaiah et al (1991), a new wide-field ($\sim 4°$) radio image of the Galactic center region was recently made by Kassim et al (1999). Fig 1 shows a sub-region of this image. The new image was made using a 3D-imaging algorithm developed by Cornwell and Perley (1992) which corrects for the non-coplanarity of the VLA and allows sensitive, distortion-free imaging of sources far from the field center. In a careful examination of the image made using the B+C+D configuration data with a resolution of $38'' \times 25''$, we identified a new linear structure which is located about 1.5°south-west of SgrA. In Fig 1, the new linear feature is located in the bottom right corner. The band of emission on the left hand side of Fig 1 is the emission from the Galactic plane. Fig 1 also shows three other previously known filamentary structures G359.54+0.18, Sgr C and the G359.1-0.2 (the Snake). The new linear feature is located at the position $\alpha_{1950} = 17^h37^m47^s.6$ and $\delta_{1950} = -29°38'36''$ and is similar in appearance to the other linear structures but is oriented parallel to the Galactic plane. We designate this feature as G358.85+0.47 based on its galactic coordinates.

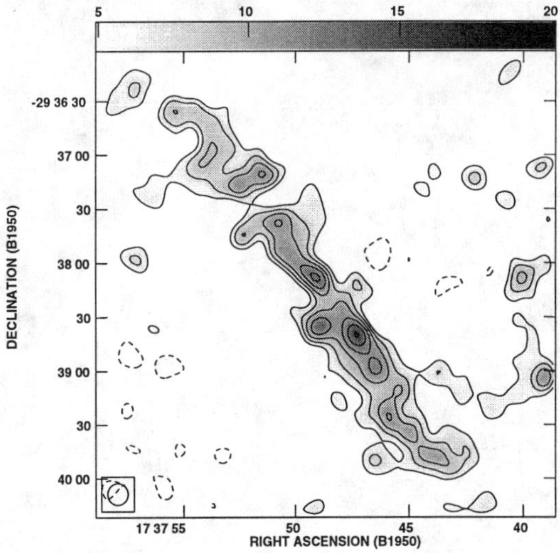

Figure 2. Higher resolution (12″) image of G358.85+0.47 at 90 cm made using only the A&B configuration data. The contour levels are -5,5, 7,9,11,13,15,17 mJy/beam

This newly found linear feature was examined further at higher angular resolution using an image made with only the A and B configuration data. Fig 2 shows a 90 cm image of G358.85+0.47 at a resolution of 12″. G358.85+0.47 has a linear extent of $\sim 4'$ and appears barely resolved along its width. There are a few spots of enhanced emission along its length. The brightness of G358.85+0.47 ranges from 10 to 15 mJy/beam at λ=90 cm.

3. Follow up Observations at $\lambda = 20$ cm

Follow up observations of G358.85+0.47 were made on Aug 24, 1998 using the B-configuration of the VLA at $\lambda = 20$ cm. Data was acquired in two frequency bands each of width 50 MHz and two orthogonal (RR and LL) polarizations. The amplitudes were calibrated using the source 3C286 and initial phase calibration was performed using frequent observations of the source 1748-253. The final image made after self-calibration and combining both the frequency bands is shown in Fig 3 in grey scale and in Fig 4 in contours.

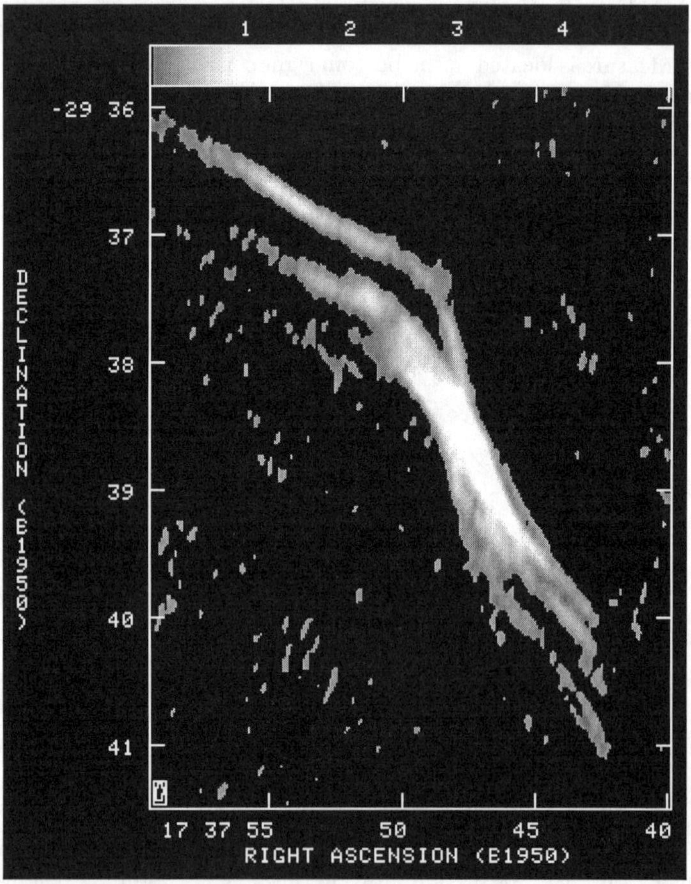

Figure 3. 20 cm image of the new filament G358.85+0.47 in Gray scale. The beam size is 8.4″ × 3.4″, PA = -7°.

The filamentary nature of the new feature is obvious from Figs 3 and 4. There could be little doubt that this feature belongs to the same category of radio filaments such as Sgr C, G359.54+0.18, G359.79+0.17, the Snake and the 'threads' G0.08+0.15 and G359.96+0.09. The distinguishing feature of the new filament is that it is oriented parallel to the galactic plane. Although we do not yet have any radio polarization data for the new filament, based on what is known about the other filaments (Lang 1999), it is reasonable to expect that the magnetic field lines will be aligned along the length of the filament or in other words parallel to the galactic plane. Such a orientation of the field lines will have to be reconciled with the strong poloidal magnetic field that has been postulated based on the orientation of the previously known filaments. It is possible that the location and the orientation of the new filament may set limits on the distance from the galactic center up to which a dipolar field may be extended. For example, the new filament may indicate the position where the field lines are turning around to close the loop.

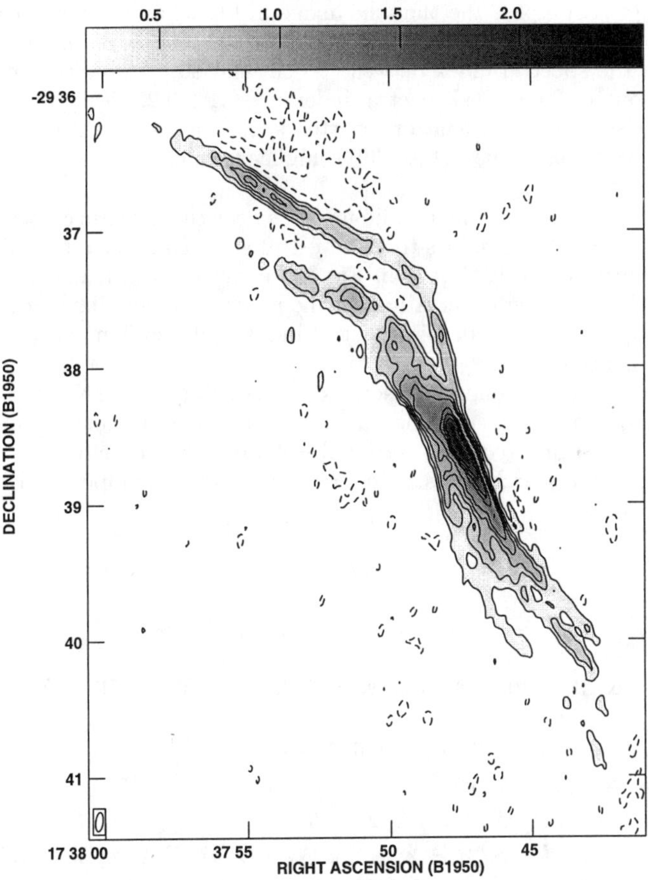

Figure 4. 20 cm image of the new filament G358.85+0.47 in contours. Contour levels are -0.2,0.2,0.4, 0.6 ... 2.2 mJy/beam. The beam is depicted at the bottom left corner (8.4″ × 3.4″, PA = -7°)

The overall structure of the new filament is also unique. It consist of at least three sub-filaments which are bent in an inverted S-shape. The bend in the northern most filament is particularly pronounced. The distortion in the filaments implies either that the field lines in this regions are themselves distorted or that the structure of the filament is not dominated by the magnetic field. The minimum magnetic field in the northern sub-filament implied by assuming that radio emission at 1.4 GHz is due to synchrotron emission is 70 μG.

It is not clear why there are significant differences in the structure of the filament at 90 cm and 20 cm (Figs 2 and 4). The differences persist even when the images are compared at the same angular resolution (12″). While the filamentary structure is obvious in the 20 cm image, at 90 cm there are several peaks along the length of the filament and the northern most bent filament is absent in Fig 2. The 90 cm image is limited by the dynamic range of the wide-field image. Note that at 90 cm, the field of view of the VLA includes the emission from the entire Sgr A complex and thus the image in Fig 2 had to be made after subtracting strong emission from other regions. Because of these differences, the determination of the spectral index between 90 cm and 20 cm is uncertain. At the peak of the source the derived spectral index is -0.1 ± 0.2. While this spectral index is consistent with the flat or inverted spectrum exhibited by several other filaments (Anantharamaiah et al 1991), higher frequency observations are required for a better estimate.

As there are a number of unique radio features near the galactic center, it is essential to assign identifying names to these features for easy reference. Based on the visual appearance of its image in Fig 3, we suggest the name 'Pelican' for the new filament. Further observations of the pelican-shaped filament in Fig 3 are underway to determine its radio spectral index, polarization and possible association with molecular gas.

The National Radio Astronomy Observatory is a facility of the National Science Foundation, operated under a cooperative agreement by Associated Universities, Inc. TJWL is supported by a National Research Council-NRL Research Associateship. Basic research in astronomy at the NRL is supported by the Office of Naval Research.

References

Anantharamaiah, K.R., Pedlar, A., Ekers, R.D. & Goss, W.M. 1991, MNRAS, 249, 262

Bally, J. & Yusef-Zadeh, F. 1989, ApJ, 336, 173

Benford, G. 1988, ApJ, 333, 735

Cornwell, T.J. & Perley, R. 1992, A&A, 261, 353

Gray, A.D., Cram, L.E., Ekers, R.D. & Goss, W.M. 1991, Nature, 353, 237

Kassim, N.E., LaRosa, T.N., Lazio, T.J.W & Hyman, S.D.1999 (this book)

Lang, C.C. 1999, this volume, p. 498

Heyvaerts, J., Norman, C. & Pudritz, R. 1988, ApJ, 330, 718

Lesch, H. & Reich, W. 1992, A&A, 264, 493

Liszt, H.S., 1985, ApJ, 203, L65

Pedlar, A., Anantharamaiah, K.R., Ekers, R.D., Goss, W.M., Van Gorkom, J.H., Schawarz, U.J., & Zhao, J.H. 1989, ApJ, 342, 769

Morris, M. 1996, *Unsolved Problems of the Milky Way*, ed. L. Blitz & P. Teuben, (Kluwer) p247

Morris, M. & Yusef-Zadeh, F. 1985, AJ, 90, 2511

Serabyn, E. & Morris, M. 1994, ApJ, 424, L91

Yusef-Zadeh, F., Morris, M., & Chance, D. 1984, Nature, 310, 557

The Central Parsecs of the Galaxy
ASP Conference Series, Vol. 186, 1999
H. Falcke, A. Cotera, W.J. Duschl, F. Melia, M.J. Rieke, eds.

The Age of the Galactic Center Arc Determined by its mm-Wave Spectral Cut-Off

Yoshiaki Sofue[1], Wolfgang Reich, Patricia Reich, and Richard Wielebinski

Max-Planck-Institut für Radioastronomie, Auf dem Hügel 69, D-53121, Bonn, Germany

Abstract. High-resolution radio continuum observations at 43 GHz of the Radio Arc near the Galactic Center were made using the Effelsberg 100-m telescope. The straight non-thermal filaments are hardly seen at 43 GHz, while they are clearly seen at 32 GHz and lower frequencies. The spectral index between 32 GHz and 43 GHz of the brightest straight filament is as steep as $\alpha = -1, (S \propto \nu^{\alpha})$, indicating a sharp turnover of the spectrum at these frequencies. Assuming a magnetic field strength of about 1 mG, we estimate the life time of cosmic-ray electrons to be about 4000 years. We conclude that the sharpest non-thermal radio filaments are very short-lived structures. On the other hand, the "sickle" is clearly visible at 43 GHz and has a flat spectrum confirming its thermal characteristics.

1. Introduction

The Galactic Center Arc comprises a bunch of highly aligned vertical magnetic fields, emitting highly polarized synchrotron radio emissions (Yusef-Zadeh et al. 1984, Yusef-Zadeh 1986; Tsuboi et al. 1986; Sofue et al. 1987; Reich 1989). Although the radio emission is clearly non-thermal, its radio spectrum is flat, or even inverted, over a wide range of frequencies from decimeter to millimeter wave ranges (Reich et al. 1988), which implies that the cosmic-ray electrons are young and have a peculiar energy spectrum. A mono-energetic electron spectrum would explain the inverted radio spectrum, while its acceleration mechanism is open to question (e.g., Lesch and Reich 1992, Morris 1996).

In order to clarify the origin of the filaments, it is important to discriminate the location and epoch of acceleration. In particular, the age of the filaments can be obtained by a high-resolution map of the Arc at high frequencies where the emission might have a spectral turnover. In our earlier 43 GHz observations of the bright non-thermal arc crossing the "sickle" region at G0.18–0.04 at 43 GHz using the Nobeyama mm-wave Array (Sofue et al. 1992), we have found that the non-thermal filaments are hardly visible, although the whole Arc is well seen at the lower resolution single-dish map at the same frequency (Sofue et al. 1986). The "sickle" is clearly seen, indicating a flat spectrum for its thermal

[1]Institute of Astronomy, The University of Tokyo, Mitaka, Tokyo 181

origin (Serabyn and Güsten 1991). The spectral turnover of the non-thermal filaments at 43 GHz suggests that the thin filaments near the "sickle" would be a short-lived feature with a life time of about 4000 years.

In order to perform a more detailed, and better calibrated spectral variation in a wider area of the Arc, we have performed high-resolution, single-dish observations of the Arc in radio continuum at 43 GHz using the Effelsberg 100-m telescope.

2. Observations and Data Reduction

The observations were made using the Effelsberg 100-m telescope on 1998 May 15, 16, 18, and 21 under exceptionally fine weather conditions. We used a HEMT receiver at the center frequency of 43.0 GHz and a bandwidth of 2.8 GHz. Two circular polarizations have been observed, while we use only one channel which had a much lower noise temperature of about 120 K. The beam width was measured to be 20.0$''$ using the nearby quasar NRAO 530. The pointing was calibrated by mapping Sgr A*. Including the effect of residual pointing errors of the individual coverages, the effective beam width was 21.5$''$ for the final maps. We have mapped the following three regions, each with $6' \times 6'$ area: (1) Centered at RA_{1950}=17h 43m 05s, $Dec_{1950} = -28°48'00''$, (2) Centered at RA_{1950}=17h 43m 20s, $Dec_{1950} = -28°51'30''$, (3) Centered close to Sgr A* at RA_{1950}=17h 42m 30s, $Dec_{1950} = -28°59'25''$.

The mapping was done by a raster-scan mode at orthogonal directions along the right ascension and declination with a scan interval of 10$''$. Each of the arc fields (area 1 and 2) was covered 8 times, and the mapping of Sgr A was inserted every two maps. A RMS noise of about 30 mJy per 21.5$''$ beam was measured from the final map.

3. Results

The obtained 43 GHz map of the Arc region is shown in the right panel of Figure 1. We also show a 32 GHz map of the same region taken with the same telescope (Lesch and Reich 1992), and a VLA 20 cm map for the same region taken from Yusef-Zadeh et al. (1984). All the maps have been smoothed to a 27$''$ beam, what is the 32 GHz original resolution.

3.1. Non-thermal Filaments

The VLA non-thermal filaments are well recognized in the 32 GHz map. However, the bright filaments seen at lower frequencies hardly recognized in our 43 GHz map. This can be more clearly seen by comparing the 5 GHz VLA map (taken from Yusef-Zadeh 1986) and the 43 GHz map from Nobeyama mm-Array (NMA: Sofue et al. 1992) as shown in Figure 2. This fact indicates that the spectrum of the brightest filament is nearly flat between 1.4 GHz and 32 GHz, while it turns to drop steeply between 32 GHz and 43 GHz. Namely, the filament has a steep spectral cutoff at $\sim$ 40 GHz. This is also clearly recognized by comparing the intensity ratio maps between 1.4 GHz and 32 GHz and 32 GHz to 43 GHz, respectively, in Figure 3.

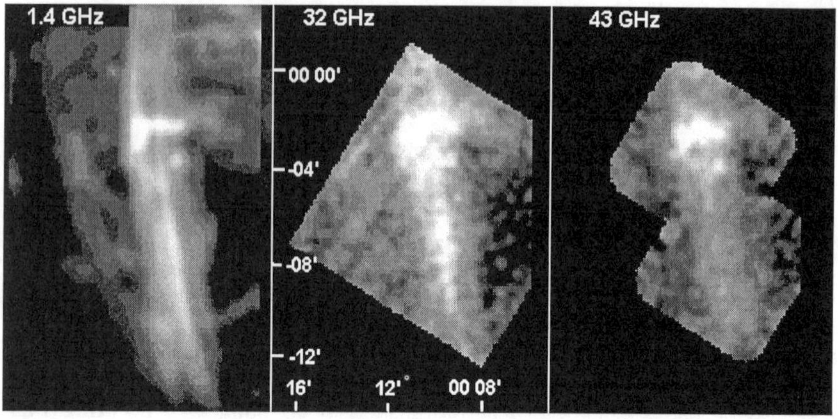

Fig. 1. (a) 1.4 GHz VLA map of the Radio Arc convolved to a beam of 27″ (Yusef-Zadeh et al. 1984)(left). (b) 32 GHz map observed with the 100-m telescope at the original resolution of 27″ (middle). (c) 43 GHz map observed with the 100-m telescope, convolved to 27″ (right). [Coordinates are (l, b). The map centers are at $(l, b) = (00°10', -00°06')$.

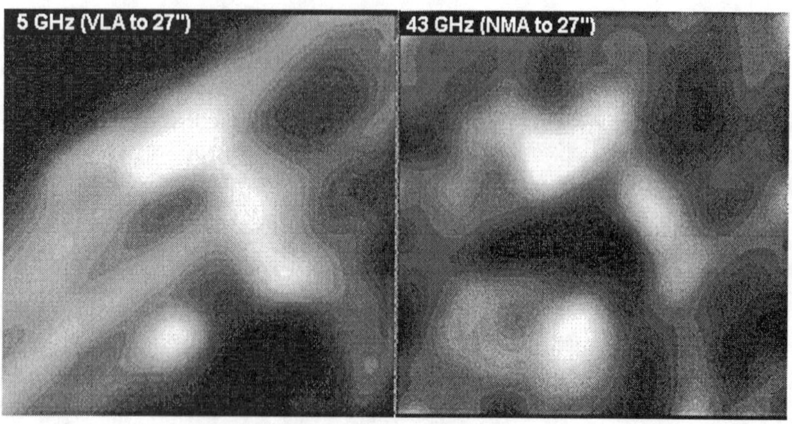

Fig. 2. (a) 5 GHz VLA map of the Radio Arc convolved to a beam of 27″ (Yusef-Zadeh 1986)(left). (b) 43 GHz map observed with the NMA convolved to a 27″ beam (right). [Coordinates are (Ra, Dec).]

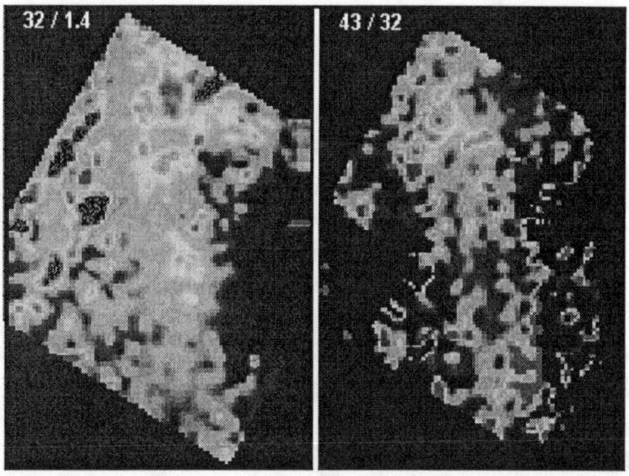

Fig. 3. (a) Intensity ratio of the 32 GHz to 1.4 GHz maps (left) for the same region in Fig. 1. (b)The same but 43 GHz to 32 GHz maps (right).

3.2. The "Sickle" and "Pistol"

The "sickle" is clearly seen as a loose association of clumps at 5 GHz, 32 GHz and 43 GHz, having flat spectra, corresponding to their thermal characteristics. The "Pistol" is also clearly seen at 43 GHz. It appears to be brighter at 43 GHz than at 32 GHz and 1.4 GHz, suggesting an inverted spectrum above 32 GHz. The "Pistol" might have more than one component or is partly optically thick.

4. Discussion

The straight Arc-filaments are hardly seen in our 43 GHz map. They have a sharp spectral turnover between 32 GHz and 43 GHz. This has been already found in a filament near the "Sickle" in our 43 GHz interferometer observations (Sofue et al. 1992). On the other hand, it is known that the entire Arc structure, when observed at low resolution with single dish telescopes, has a flat, or rather inverted, spectrum (Reich et al. 1988). This indicates that high-energy cosmic-ray electrons emitting at 43 GHz in the filaments have already decayed away, while they are widely spread in the entire Arc region. On the other hand, lower-energy electrons, which emit at lower frequencies (32 GHz and 1.4 GHz), are still alive and confined in the filaments. These facts give constraints on the epoch of the acceleration of cosmic-ray electrons in the filaments.

The magnetic field strength in the filaments is of the order of 1 mG (e.g., Yusef-Zadeh et al. 1984). The energy of synchrotron electrons which emit at 43 GHz in such a magnetic field is approximately

$$E \sim \sqrt{\frac{2\pi m^3 c^5 \nu}{eB}} \sim 2 \text{ GeV}.$$

The lifetime of an electron of this energy is then

$$\tau \sim \frac{3m^4c^7}{2e^4B^2E} \sim 4 \times 10^3 \text{ years.}$$

Here, ν, m, e, E, B and c are the frequency, electrons mass, electric charge, energy, magnetic field strength, and light velocity, respectively. Thus, the non-thermal filaments in the Arc would be magnetic tubes, temporarily illuminated by high-energy electron fluxes, whose acceleration would have taken place about 4000 years ago. The Larmor radius of such electrons is small enough compared to the thickness of a filament resolved with the VLA. Hence, the electrons are localized in the filaments, and cannot escape across the field lines.

We point out the similarity of the filaments to the "threads", which, however, have steep spectra (Anantharamaiah et al. 1991; Lang 1999). The threads may be magnetic tubes, which were illuminated by high-energy electrons accelerated earlier than the Arc electrons. The high-energy electrons have decayed meanwhile and the spectrum has steepened.

References

Anantharamaiah, K. R., Pedlar, A., Ekers, R. D., and Goss, W. M. 1991, MNRAS, 249, 262.

Lang, C. 1999, this volume, p. 498.

Lesch, H. and Reich, W. 1992, AA, 264, 493.

Morris, M. 1996, in *Unsolved Problems of the Milky Way, Proc. IAU Symp. 169*, eds. L. Blitz, P. Teuben, (Kluwer Academic Publishers, Dordrecht) p.247.

Reich, W. 1989, in *The Center of the Galaxy, Proc. IAU Symp. 136*, ed. M. Morris, (Kluwer Academic Publishers, Dordrecht) p.265.

Reich, W., Sofue, Y., Wielebinski, R. and Seiradakis, J. H. 1988, AA, 191, 303.

Serabyn, E., and Güsten, R. 1991, AA, **242**, 376.

Sofue, Y., Inoue, M., Handa, T., Tsuboi, M., Hirabayashi, H., Morimoto, M., Akabane, K. 1986, PASJ, 38, 483

Sofue, Y., Murata, Y., and Reich, W. 1992, PASJ, 44, 367

Sofue, Y., Reich, W., Inoue, M., Seiradakis, J. H. 1987, PASJ, 39, 359.

Tsuboi, M., Inoue, M., Handa, T., Tabara, H., Kato, T., Sofue, Y., and Kaifu, N. 1986, AJ, 92, 818.

Yusef-Zadeh, F., Morris, M., Chance, D. 1984, Nature, 310, 557.

Yusef-Zadeh, F., 1986, PhD Thesis, Columbia University.

The Central Parsecs of the Galaxy
ASP Conference Series, Vol. 186, 1999
H. Falcke, A. Cotera, W.J. Duschl, F. Melia, M.J. Rieke, eds.

Discussion Session VIII: The Magnetic Field in the Galactic Center

Y. Sofue[1]

Institute of Astronomy, The University of Tokyo, Mitaka, Tokyo 181

C. C. Lang[2]

National Radio Astronomy Observatory, Box 0, Socorro, NM 87801

Abstract. This is an edited transcript of a discussion session about the magnetic field in the Galactic Center that took place during the Galactic Center conference "The Central Parsecs" in Tucson, Arizona, Sep. 7-11, 1998. The session was chaired by the moderator as indicated in the author list. The discussions were taped, transcribed, and finally edited by an editor who is listed as co-author. The contributions of the conference participants are preceded by their names and thus references to specific aspects and ideas of this discussion session must also include a reference to the respective speaker(s), e.g., in a footnote.

Discussion

Sofue: Before we go on I'd like to summarize the ways in which we can learn about the magnetic field at the Galactic Center. I think that people here would agree that we have several ways to do this: (1) the Zeeman effect in maser and radio recombination lines (summarized by Mark Wardle and Doug Roberts), (2) far infrared polarization of dust grains in molecular clouds (reported on by Giles Novak), and (3) radio synchrotron emission and polarization (as Cornelia Lang talked about). From Zeeman measurements, we can derive the magnetic field strength and orientation. Far infrared observations give us the orientation of the plane-of-sky component, but no information on the strength. And, finally, the radio component gives a lot of information: intensity, morphology, and linear polarization of synchrotron emission. Also, there is the rotation measure, which gives information on the line of sight magnetic field. But for the rotation measure, we need information on the thermal electron density, for which Anantha[ramaiah] gave us some idea. I think people have also reached some consensus about the strength of the magnetic field in the Galactic Center region from Zeeman splitting—we probably get a field strength of 1–10 mG, right?

[1]Moderator of the discussion session

[2]Editor of the discussion session

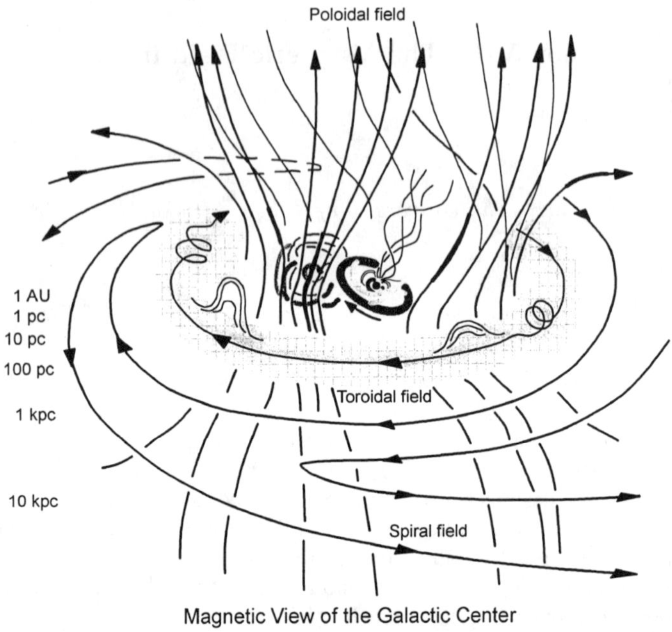

Figure 1. Sketch of the magnetic field configuration in the Galactic Center (from Y. Sofue; not to scale)

Roberts: 8 mG is the upper limit for the magnetic field strength from Zeeman splitting; 5 mG was the highest actual detection.

Sofue: What can you tell us about the direction of these detections?

Roberts: In all of the cases we have observed, the direction has been away from the observer, away from the Earth, except for the Northern Arm, in which case the orientation is towards us. There's one other measurement I didn't mention, but that is in the Southern Arm. There's another potential HI Zeeman detection which is also away from us, but that is a marginal result.

Sofue: It is extremely important to see if the magnetic field is toroidal or poloidal in the circumnuclear region.

Roberts: If it was a toroidal field, we expect to see a sign reversal across the symmetric structure, and so far we have not detected that.

Sofue: Probably we can then agree that the magnetic field strength in some clouds is as high as 1–10 mG.

Yusef-Zadeh: We may have a potential to measure the total field by doing Faraday rotation towards these masers, i.e. by converting the linear polarization of these masers into a field strength. This is something that has been done for W44 and W28.

Roberts: What are the results that are found using this method?

Goss: We got very messy results in W28 that we could not interpret with the linear polarization. Some of the problem was the lack of angular resolution, and it ended up being completely ambiguous.

Sofue: Now they've given me the strength of the magnetic field, we can make some comparisons between the equipartition magnetic fields and the interstellar medium: ionized gas, cosmic ray energy density, and molecular gas. On size scales of 10–100 pc,the magnetic field dominates. But, of course, the gravity must dominate everything close in near the black hole. From the question about the energy ratio between the magnetic field and the gas, we learned that the gas farther out may be dominated by the magnetic field.

Cotera: Actually, I had a general question. It does not seem like there is any systematic magnetic field: there appears to be a large-scale magnetic field, but when you start getting closer and closer to the center, the magnetic field seems to be very tangled. So I'm wondering how, if you have tangled magnetic fields everywhere, does that affect star formation? Do we have the situation that Doug was talking about, where you have to have massive clumps already to overcome magnetic pressure along field lines? If the magnetic field is tangled, do you still have a magnetic field pressure to overcome?

Unknown: It makes it more difficult, because you can't have free fall, and if it's even more tangled, the ability to collapse is even worse.

Unknown: I've got a follow-up question. If a field is really poloidal as the filaments seem to suggest, and if they really originate from this sort of H II/molecular cloud interaction, why do we only see them in the Northern hemisphere? Take the Northern Thread for example. There is an H II/molecular gas complex there, the filament seems to only extend in one direction away from this complex, to positive latitudes, and not the other, right?

Lang: But we do see something [in the other direction]. If you look closely at the λ20 cm image, you see hints of parallel features arising at negative latitudes, too. However, they are just faint hints of filaments. But, if you actually look at the whole length of the Northern Thread, it extends over 50 parsecs, and there is a 10 pc spur of the filament that is on the other side of the Arched Filaments [at negative latitudes]. And there is also the Snake filament, located at negative latitudes.

Roberts: I have a follow-up question to that: I was just asking Giles if at the position where the Northern Thread intersects the Arched Filaments, is there any disturbance in the orientation of the plane of sky vectors?

Novak: It appears to follow the Arched Filaments over its entire length.

Sofue: What about the field strength in the Arched Filaments?

Novak: Well, Mark Morris estimated a field strength of 8 mG based on using the Chandrasekhar and Fermi method, but you can easily get into problems when you use that method. You have conflicts of superposition along the line of sight in the clumps, so I think that it is, as Doug said, very difficult to derive a field strength here.

Yusef-Zadeh: I guess I should say that, if you look at all the magnetic field measurements of thermal gas, you really don't see any poloidal magnetic field. Everything seems to be azimuthal. That is the field inside of the [molecular] clouds, isn't it? And there is nothing poloidal about it.

Figer: Why should the magnetic field inside of a cloud match the magnetic field in the Galactic Center, with the cloud rotating, and presumably getting ripped apart?

Yusef-Zadeh: Right. I agree, there is no strong evidence for poloidal magnetic fields in the molecular gas. The only evidence we have are the magnetized filaments, that appear to trace field lines perpendicular to the plane of the Galaxy. So I think it's important to find more evidence for this poloidal field.

Figer: There's evidence for a poloidal field, and no evidence against it. Because in every case where it's azimuthal, it's *inside* a cloud.

Yusef-Zadeh: Yeah, the thing is that everybody talks about it, and we came up with it, the suggestion that there is a poloidal magnetic field pervasive throughout the Galactic Center, but it is surprising that you do not find any evidence for it. [Laughter]

Sofue: I'm confused, but that's okay. Here is a sketch (Fig. 1) of the different scales of the magnetic field in the Galactic Center, and its strength and orientation. As Cornelia [Lang] suggested in her discussion today, there may be a transition at the position of the new filament, and from the dust polarization that Giles [Novak] talked about, there is a suggestion that the magnetic field near the center (i.e. the circumnuclear disk) is toroidal.

Unknown: I have a question: does that say that the circumnuclear disk has had to be turned around several times to turn on the magnetic field?

Novak: I don't know if it is several times, but it has to turn a little bit at least, because if one relies on differential rotation ... I guess you would have to calculate how extreme the differential rotation is. Can Mark Wardle comment on this?

Wardle: If you stretch the cloud out first before it gets in there, then you pull the magnetic field along too, and then wrap it around once. I'm thinking about the Northern Arm where you have the magnetic field running along that, for example. Our model, the Wardle and Königl model, which is essentially an accretion disk model, is assuming very different initial conditions.

Sanders: This is based on very recent work, about twenty minutes ago ... and probably the numbers should be checked again. I want to make a remark on the problems of star formation in the presence of a very strong gravitational field—that is, very near the black hole, very near the point mass. So, it would be star formation within a few tenths of a parsec from Sgr A*. And basically, these are very general arguments, which you could find in Spitzer's first book. For a collapse, if a magnetic field is not is going to prevent collapse, the gravitational potential energy has to be greater than the magnetic field energy. That means, roughly, that $\frac{GM^2}{R} > B^2R^3$, which gives you a limit on the mass, M > $\frac{BR^2}{\sqrt{G}}$. And you can use this similar relationship between mass, density, and radius [$M \propto \rho R^3$] to eliminate the radius. And what you find is that you can get gravitational collapse in the presence of a magnetic field, as long as the mass of this spherically collapsing clump is greater than a critical value, which is given roughly by $\frac{B^3}{\rho^2}$. Now I went through some numbers, and if you have a magnetic field of 1 mG, let's say, normal interstellar conditions with a density of 10 cm^{-3}, then this critical mass is about 10^6 $M_\odot$.

Wardle: A magnetic field of 1 mG in normal interstellar conditions corresponds to much higher densities than that, about 10^4 or 10^6 cm^{-3}.

Sanders: I'm just trying this out as sort of a numerical example. Certainly if you increase the density then you get into less trouble.

Sofue: So you have no chance to make a star?

Sanders: No chance to make stars unless you can get rid of the magnetic field by experienced techniques people talk about, but these are on the long-term, long time scales. In the Galactic Center, though, if you want star formation to start near the black hole, you have to exceed the Roche limit. And the Roche limit, as we heard yesterday, is something like 10^{10} cm^{-3}. That gives you a critical mass for a magnetic field of 1 mG of 10^{-12} $M_\odot$. So in other words, as long as the densities are high enough anyway to satisfy the Roche limits, magnetic fields should be no problem for star formation near the black hole. Now there's one caveat here, and that is, if you start with the 1 mG magnetic field, and you also compress it in a shock—up to maybe 1–10 G—then you have a problem. But you need fields in excess of 1 G to really hurt star formation, if the density exceeds the Roche limit.

Unknown: If you now use the argument that you made yesterday, with an estimated density of 10^6, then you are just right in the middle where you want to be.

Sanders: That's good to hear. I suspect actually, angular momentum might be more of a problem for star formation than magnetic fields. Because you have these streams of gas going by the black hole, you might expect a lot of shear and a lot of turbulence, a lot of vorticity. I haven't worked it out, but that could be a more serious problem.

Wardle: The magnetic fields will usually kill that, if you hang on to them long enough.

Sofue: As for the future, clearly more MHD simulations should be made, and include instabilities like the one Bob Sanders was just talking about. I must also mention that we need to get more observations of the magnetic fields near to the accretion disk.

Coker: I believe it was Geoff Bower in his talk, who said that the polarization is very small in the radio for Sgr A*, which suggests that at least very close to the black hole, the field is very tangled. Especially if it is synchrotron emission, it has to be tangled, because otherwise it had to be polarized, so there is a little bit of an argument that on a small scale you have a very tangled field, while only on the large scale you have an ordered field.

Zhao: That limit is still having problems, because of large rotation measures.

Bower: In the case of the millimeter data, though, we don't have a problem with large rotation measures. The rotation measures there are 10^7 rad m^{-2} and the fractional polarization is one plus or minus one, which is quite small and indicative of a tangled field.

Sofue: I think that today's discussion is now closed.

Session IX

What Would the Galactic Center Look Like from 1 Mpc?

The Central Parsecs of the Galaxy
ASP Conference Series, Vol. 186, 1999
H. Falcke, A. Cotera, W.J. Duschl, F. Melia, M.J. Rieke, eds.

The Mid-Infrared Spectrum of the Galactic Center – a Starburst Nucleus

J. P. Simpson[1] and F. C. Witteborn[1]

NASA/Ames Research Center, MS 245-6, Moffett Field, CA 94035-1000

M. Cohen

Radio Astronomy Laboratory, University of California, Berkeley, CA 94720-3411

S. D. Price

Air Force Research Laboratory, VSBC, Hanscom AFB, MA 01731-3010

Abstract. Using the Michelson interferometer on the *Midcourse Space Experiment (MSX)*, we have taken spectra of many positions in the central 25′ of the Galactic Center (GC) with a 6′ × 9′ FOV. The spectral coverage was 380 to 1700 cm^{-1} (6 to 26 μm) and the resolution was $\sim$ 21 cm^{-1}. The spectra exhibit strong UIR/PAH features at 6.2, 7.7, 8.6, and 11.3 μm, in addition to the ionic lines of [Ne II] at 12.8 μm, [S III] 18.7 μm, and [Ar II] 6.98 μm. There are deep silicate absorption features at 10 and 18 μm and a cold continuum increasing at the longest wavelengths. Additional weak features are present in the spectra. We discuss the variation in the extinction at 10 μm as a function of location in the GC. Compared to the MSX spectrum of the Orion Nebula (Simpson et al. 1998), smoothed to the same resolution and multiplied by the estimated GC extinction, the GC spectra have similar PAH features, but the Orion Nebula also has strong lines of [Ne III] 15.6 μm, [S IV] 10.5 μm, and [Ar III] 8.99 μm and its 25 μm continuum is stronger (colder). Thus, the GC exhibits the mid-IR spectrum of a low excitation H II region and a nearby molecular cloud with a surface photodissociation region (PDR). This is in excellent agreement with the canonical model of a starburst nucleus in which the hot stars and molecular clouds are randomly distributed. The outer surfaces of the clouds are photodissociated and ionized by the photons from the stars located outside the clouds. The PAH molecules are transiently heated by the stellar photons. Since the exciting stars are located well outside the clouds, the radiation field is dilute compared to a newly-formed blister H II region like Orion; this dilute radiation field causes the relatively low excitation of the ionic lines.

[1] SETI Institute

1. Introduction

Galaxy nuclei are normally classified for their dominant sources of excitation by their optical spectra. Galaxies whose optical spectra exhibit featureless continuum emission (thought to arise from matter falling into black holes) and/or emission from broad lines of a very high range of ionization are known as Active Galactic Nuclei (AGN) or even as QSOs if the normal stellar regions of the galaxy can hardly be detected. Other common sources of excitation of galaxy nuclei are evolved stars or the hot young stars that ionize the interstellar gas and produce optical spectra resembling H II regions.

The different sources of excitation are also distinguishable in the infrared spectra of the galaxies. This is especially useful for those galactic nuclei that are obscured optically by extinction. Infrared spectra have the advantage over the radio spectra in that there are a number of elements that have emission lines from both low and high levels of ionization. Moreover, the dust that is the source of the infrared continuum also has features that can be used in identification of the dust composition and the source of the excitation. Roche et al. (1991) compared the 10 μm spectra of a number of galaxies of known classification from optical spectra. They found that galaxies with H II region nuclei have spectra dominated by the unidentified infrared features (UIR) that are often attributed to PAHs plus emission lines of [Ne II] at 12.8 μm. On the other hand, they found that galaxies regarded as AGNs have either featureless spectra or spectra showing the 10 μm silicate feature in absorption but no PAHs or atomic forbidden lines.

Spoon et al. (1998) and Genzel et al. (1998) used *ISO* to measure mid-IR spectra of a large number of galaxies to try to identify the source of the excitation of the class known as Ultraluminous *IRAS* Galaxies. Genzel et al. discovered that a good criterion is found on a plot of the ratio of the high to low excitation forbidden lines versus the ratio of the 7.7 μm PAH feature to the adjacent continuum. The H II region (also known as starburst) galaxies have low excitation forbidden lines and strong 7.7 μm PAH features whereas the AGNs have much higher excitation forbidden lines and essentially no 7.7 μm PAH feature. Some of the Ultraluminous *IRAS* Galaxies have ratios more like the starburst galaxies and some have ratios more like the AGNs; thus Genzel et al. concluded that these ratios can be used for galaxy classification from infrared spectra.

Some examples are M82 and the Circinus Galaxy. M82 (Lutz et al. 1998a) is a good example of Roche et al.'s H II region galaxy; it is also called a starburst galaxy because of the dominance of the spectrum by hot young stars and their effects on the surrounding molecular clouds. On the other hand, Circinus Galaxy (Moorwood et al. 1996) has some features of the H II region spectrum — PAHs and the low excitation molecular and ionic lines — but it also has high excitation lines like [Ne V], [O IV], and [Ne VI], which require photons that are much more energetic than found in any H II region and which consequently indicate a high energy central source. The weaker PAH feature could be due to additional continuum from dust heated by the central AGN or the PAH molecules could be dissociated by the high energy photons (e.g., Lutz et al. 1998b).

Lutz et al. (1996) published a spectrum of the Galactic Center taken in a $14''$ by $20''$ aperture with SWS on *ISO*. The spectrum is dominated by the deep 10

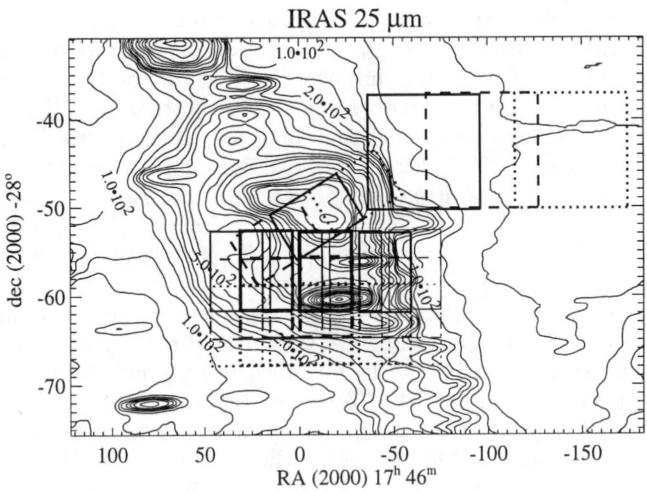

Figure 1. The positions measured in the Galactic Center are plotted on a 25 μm contour map from *IRAS*.

and 20 μm silicate absorption features; also present are hydrogen recombination lines and the forbidden lines typically found in H II regions. Although there is a very small amount of [O IV] in the spectrum, for the most part the excitation is low. Lutz et al. attributed the [O IV] to hot old stars like Wolf-Rayet stars (known to exist in the Galactic Center) rather than to hard photons from an AGN. Genzel et al. (1998) put the Galactic Center into the starburst group. On the other hand, there is no evidence for PAHs in the *ISO* spectrum of Lutz et al. (1996); Genzel et al. suggested that the PAH molecules are destroyed or transformed in the radiation field of the region around Sgr A*.

The Galactic Center clearly contains indications of both AGN-like excitation sources (Sgr A* and the lack of PAH features in the $14'' \times 18''$ region surrounding it) and young hot stars (the central cluster, AFGL 2004/Quintuplet cluster, and the Arches Cluster and their H II regions). In order to compare the Galactic Center with the nuclei of external galaxies, one really should measure the GC spectrum as integrated over comparable physical sizes. This means either integrating over a number of different locations or measuring the spectrum in a much larger field of view. This paper presents mid-IR spectra of the Galactic Center taken in September and October, 1996, with the *Midcourse Space Experiment (MSX)* Michelson interferometer.

2. Observations

The *MSX* Michelson interferometer had 6 Si:As impurity-band conduction detectors with fields of view of 6′ by 9′ and 11′ by 11′. The *MSX* instruments are further described by Mill et al. 1994 and Egan et al. 1998. We used the 6′ × 9′ detector with the widest wavelength coverage to map the Galactic Center in a 7 by 3 position grid (RA by Dec). At the same time, spectra from 6 to 9 μm were measured in an 11′ × 11′ detector; however, because of the lower signal, these spectra were averaged, resulting in only 3 positions plotted. On a different orbit, the telescope scanned repeatedly back and forth over the Galactic Center. Here, the spectra from the 6′ × 9′ detector were averaged for 3 positions in Galactic longitude, with the center position being the base of the Arched Filaments. For all spectra, the resolution was $\sim$ 21 cm^{-1}. Figure 1 shows these positions overlaid on the *IRAS* 25 μm contours[1] and Figures 2 – 4 show the spectra obtained at these positions.

The features immediately apparent in all these spectra are the 10 and 18 μm silicate absorption features, the strong PAH features at 6.2, 7.7, and 11.3 μm (the 8.6 μm is partly submerged by the 10 μm absorption), and the low excitation lines of [Ar II] 6.98 μm and [Ne II] 12.8 μm. The 18.7 μm [S III] line is also visible as is the 17.0 μm line of H_2 (S1) at some positions. What are *not* apparent are any strong high excitation forbidden lines, although a weak [Ne III] 15.6 μm line could be present. These spectra are very similar (albeit with lower resolution and more extinction) to those of M82, the prototypical starburst galaxy.

Would we see the high excitation lines at this low resolution? The *MSX* interferometer also measured the spectrum of the Orion Nebula with 2 cm^{-1} resolution (Simpson et al. 1998). The higher excitation lines of [Ar III] 8.99 μm, [S IV] 10.5 μm, and [Ne III] 15.6 μm are clearly present and strong; in addition, the PAH features from the Orion photodissociation region (PDR) plus molecular hydrogen lines from the Orion molecular cloud are present.

Figure 5 shows the Orion Nebula spectrum smoothed to the same resolution as the Galactic Center spectra and multiplied by $\exp(-\tau_{9.7} * \kappa_\nu)$ where $\tau_{9.7} = 3.2$, 3.6, and 4.0. The extinction coefficients, κ_ν, as a function of frequency, ν, were estimated from the *ISO* Galactic Center spectrum of Lutz et al. (1996) (fit by the sum of 3000 K and 250 K blackbodies multiplied by the derived extinction coefficients times $\tau_{9.7} = 3.6$) and footnote 7 of Genzel et al. (1998) and normalized to unity at 9.7 μm. The chief differences with the Galactic Center spectra are that the Orion Nebula has a much stronger cold continuum and that the higher excitation lines are easily visible at the lower resolution. Even the [Ar III] and [S IV] lines are visible, the [Ar II] line is much weaker, and the ratio of [Ne III]/[Ne II] is much larger in Orion than in the Galactic Center (the two neon lines have very similar extinctions).

Thus we can conclusively say that the Galactic Center in a large field of view exhibits strong PAH features and low ionization; its mid-IR spectrum can best be described as that of a starburst galaxy.

---

[1]The IRAS image was obtained using the facilities of the IPAC, which is funded by NASA as part of the IRAS extended mission program under contract to JPL.

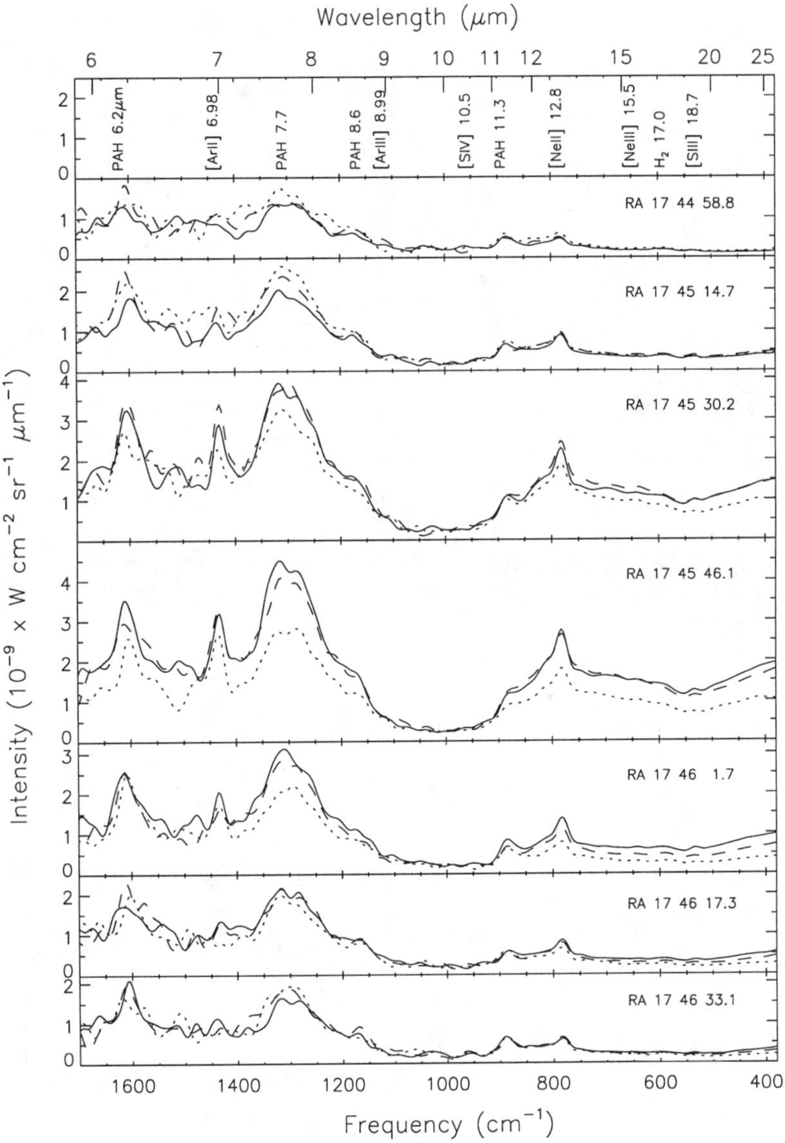

Figure 2. The 21 spectra of the Galactic Center are plotted. The 7 right ascensions (J2000) are plotted separately; the 3 declinations are plotted as north (solid), center (dashed), and south (dotted).

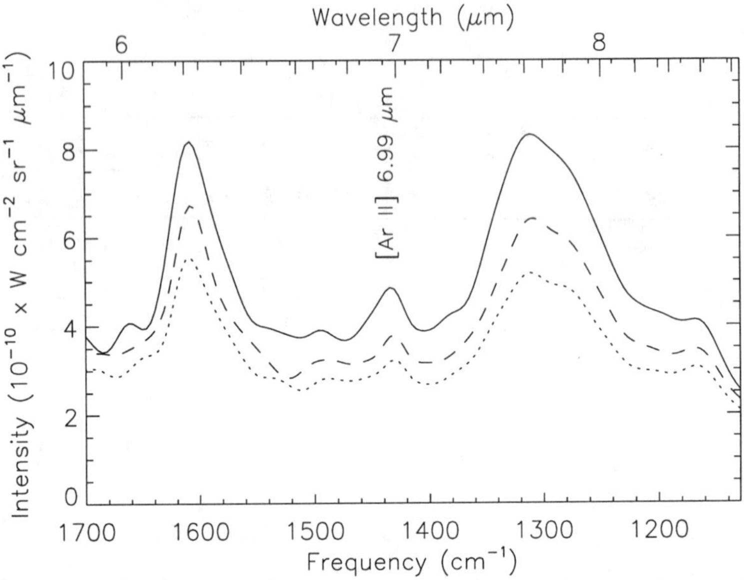

Figure 3. Spectra from the 11′ × 11′ detector are plotted. The 3 right ascensions are plotted as east (solid), center (dashed), and west (dotted).

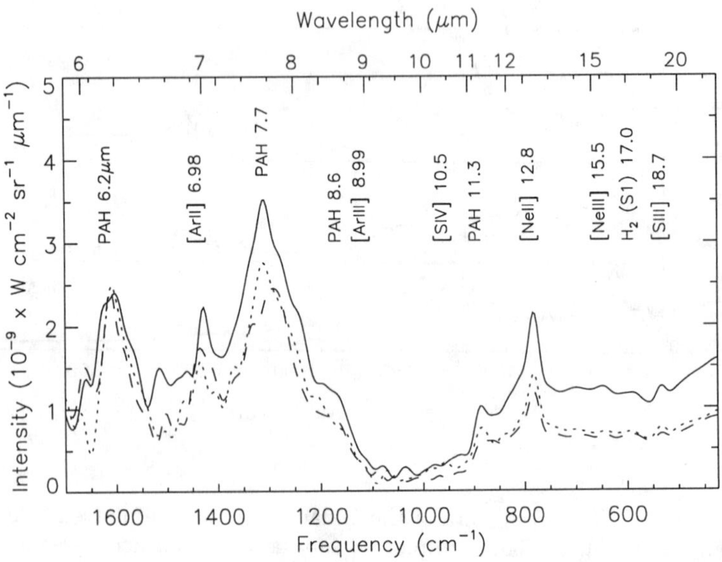

Figure 4. Three spectra of the Arched Filaments region are plotted.

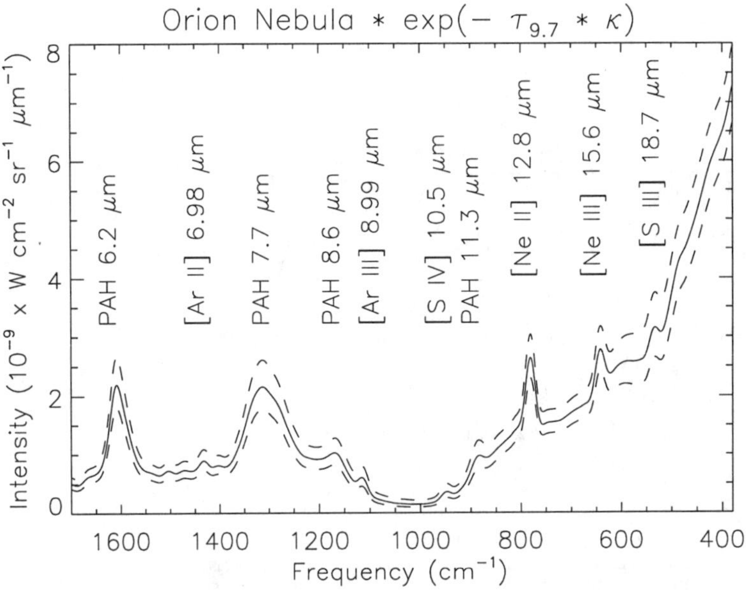

Figure 5. The spectrum of the Orion Nebula, smoothed to the same resolution as the Galactic Center spectra and multiplied by extinctions corresponding to $\tau_{9.7} = 3.2$, 3.6, and 4.0.

3. Discussion

It appears from Figure 2 that the extinction is not the same for all the spectra. For example, the 8.6 and 11.3 μm PAH features are much more distinct in the spectra of the outer positions than in the spectra for the center. We have attempted to find the values of $\tau_{9.7}$ such that when each spectrum is multiplied by exp(+ $\tau_{9.7} * \kappa$), the resulting spectrum is flat through the 10 μm silicate feature. (The estimated contribution of the zodiacal emission (Leinert et al. 1998) for solar elongation $\sim$ 73 deg and ecliptic latitude $\sim$ 6 deg was removed first.) Figure 6 shows the variation in the extinction as determined by this method. For future work we will use these spectra to estimate the relative contributions from the different components. For example, the ratio of PAHs to 25 μm continuum is smaller closer to Sgr A than it is further out.

Wolfire, Tielens, and Hollenbach (1990) discussed the application of PDR theory to H II regions and galactic nuclei. Their Figure 7 showed 3 different possible scenarios: a) hot stars closely associated with their natal molecular clouds such that the far-UV radiation field on the clouds is high, as occurs in an H II region and its PDR, b) hot stars and molecular clouds randomly distributed such that the radiation field is the average interstellar field, and c) molecular clouds heated by an AGN. Scenario b may be the most applicable to the Galactic Center. The low ionization would be due to the stars being located, on the average, at some distance from the molecular clouds. We must note, on the other hand, that all the H II regions in the inner Galaxy are of low

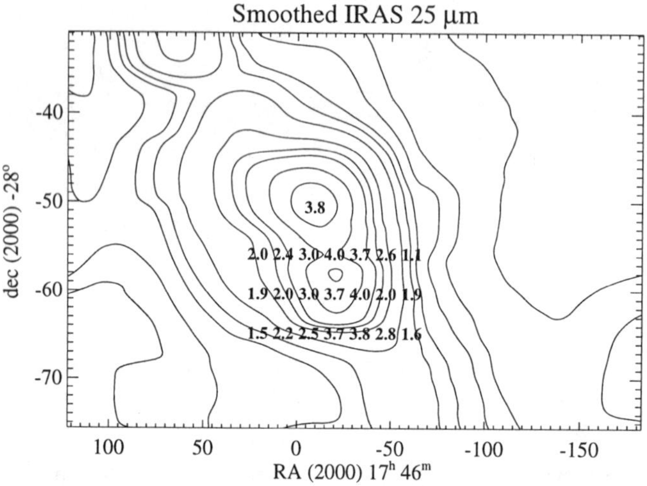

Figure 6. The estimated optical depth, $\tau_{9.7}$, is plotted on the *IRAS* 25 μm contour map smoothed to a resolution of 6′, approximately corresponding to FOV in the *MSX* interferometer.

excitation; the other effect that could be important here is that high abundance H II regions are of low excitation because of their high UV opacity (Simpson et al. 1995).

Scenario b is compatible with the FIR observations of the H II regions in the Galactic Center - low gas density, low ionization, strong PDR lines and continuum indicating excitation by the known clusters plus other random stars (Erickson et al. 1991; Colgan et al. 1995; Colgan et al. 1996; Simpson et al. 1997; Timmermann et al. 1997).

4. Conclusion

When observed in a large FOV, the Galactic Center shows strong PAH features, low excitation lines (H_2, [Ar II], [Ne II], [S III]), essentially no higher excitation lines ([Ne III]/[Ne II] $<<$ 1) even compared to the Orion Nebula, and deep silicate absorption at 10 and 18 μm, especially right on the Galactic plane. These are the same characteristics that are seen in starburst galaxies and that are not apparent in AGNs, which have high excitation lines, flatter continua, and/or weak PAH features. In fact, the Galactic Center appears to be of even lower excitation than some starburst galaxies, such as M 82. The very low excitation in the Galactic Center could be due to either high abundances in the exciting stars and H II regions or to the ionizing radiation field being more diffuse than is found in the young H II regions surrounding newly formed stars. This diffuse radiation field could be the result of a starburst occurring within the last few million years but long enough ago that the hot stars are no longer physically associated with the molecular clouds of their birth.

Acknowledgments. *MSX* is sponsored by the Ballistic Missile Defense Organization. We gratefully acknowledge the steadfast support of the BMDO. JPS and FCW acknowledge support from NASA/Ames Research Center Research Interchange grant NCC2-900.

References

Colgan, S. W. J., Erickson, E. F., Simpson, J. P., Haas, M. R., & Morris, M. 1996, ApJ, 470, 882

Colgan, S. W. J., Simpson, J. P., Erickson, E. F., Haas, M. R., Rubin, R. H., & Morris, M. 1995, BAAS, 27, 884

Egan, M. P., Shipman, R. F., Price, S. D., Carey, S. J., Clark, F. O., & Cohen, M. 1998, ApJ, 494, L199

Erickson, E. F., Colgan, S. W. J., Simpson, J. P., Rubin, R. H., Morris, M., & Haas, M. R. 1991, ApJ, 370, L69

Genzel, R. et al. 1998, ApJ, 498, 579

Leinert, C. et al. 1998, A&AS, 127, 1

Lutz, D. et al. 1996, A&A, 315, L269

Lutz, D. et al. 1998a, in Star Formation with the *Infrared Space Observatory*, ASP Conf. Ser. 132, ed. J. L. Yun & R. Liseau (San Francisco: ASP), 89

Lutz, D., Spoon, H. W. W., Rigopoulou, D., Moorwood, A. F. M., & Genzel, R. 1998b, ApJ, 505, L103

Mill, J. D. et al. 1994, J. Spacecraft & Rockets, 31, 900

Moorwood, A. F. M. et al. 1996, A&A, 315, L109

Roche, P. F., Aitken, D. K., Smith, C. H., & Ward, M. J. 1991, MNRAS, 248, 606

Simpson, J. P., Colgan, S. W. J., Cotera, A. S., Erickson, E. F., Haas, M. R., Morris, M., & Rubin, R. H. 1997, ApJ, 487, 689

Simpson, J. P., Colgan, S. W. J., Rubin, R. H., Erickson, E. F., & Haas, M. R. 1995, ApJ, 444, 721

Simpson. J. P., Witteborn, F. C., Price, S. D., & Cohen, M. 1998, ApJ, 508, in press

Spoon, H. W. W., Genzel, R. & Lutz, D. 1998, in Star Formation with the *Infrared Space Observatory*, ASP Conf. Ser. 132, ed. J. L. Yun & R. Liseau (San Francisco: ASP), 414

Timmermann, R. et al. 1996, ApJ, 466, 242

Wolfire, M. F., Tielens, A. G. G. M., & Hollenbach, D. 1990, ApJ, 358, 116

The Central Parsecs of the Galaxy
ASP Conference Series, Vol. 186, 1999
H. Falcke, A. Cotera, W.J. Duschl, F. Melia, M.J. Rieke, eds.

Discussion Session IX: How "Standard" is the Central Region of Our Galaxy? What Would the GC Look Like from One Mpc Distance?

H. Falcke[1]

Max-Planck-Institut für Radioastronomie, Auf dem Hügel 69, D-53121 Bonn, Germany

S. Kern[2], A. Cotera[2]

Steward Observatory, The University of Arizona, Tucson, AZ 85721

Abstract. This is an edited transcript of a discussion session about the comparison of the Galactic Center with other galaxies and Active Galactic Nuclei (AGN) that took place during the Galactic Center conference "The Central Parsecs" in Tucson, Arizona, Sep. 7-11, 1998. The session was chaired by the moderator as indicated in the author list. The discussions were taped, transcribed, and finally edited by editors who are listed as co-authors. The contributions of the conference participants are preceded by their names and thus references to specific aspects and ideas of this discussion session must also include a reference to the respective speaker(s), e.g., in a footnote.

Discussion

Falcke: I would like to start our discussion by considering some things you're never going to see in any other galaxy—only in our own Galaxy. For example, I was just wondering: all the magnetic reconnection going in the Galactic Center (GC) with perpendicular and parallel magnetic fields, is that what is actually happening in an accretion disk somewhere else? Is that an accretion disk at some level which we actually see at work? Especially, Sofue's view graph yesterday was very stimulating. I was wondering what is going to happen if we have these poloidal and azimuthal magnetic fields accreting towards the center. I was always seeing in my mind a jet coming out ...

Moreover, if 'we' were a real AGN with a high accretion rate what would Sgr A West and Sgr A East be? Do they have anything to do with, for example, what we see in Seyferts? In fact, we do see strong, complex, steep spectrum radio cores in Seyfert galaxies—is that Sgr A East? We do

[1] Moderator of the discussion session

[2] Editor of the discussion session

see a narrow line region of hot gas—Ski showed some of this in NGC 1068 for example—is that Sgr A West? Is the CND related to the obscuring torus in Seyferts? No doubt, however, Sgr A* is something which is clearly related to an AGN.

We heard a lot about star formation and there seems to be an agreement that the Galactic Center is some sort of starburst or post-starburst galaxy—a mini starburst galaxy. In fact, I've written down something Ski might like, perhaps this is actually a Starburst 2 galaxy because the GC is actually obscured from our view. I don't think this classification has ever been used in the context of starbursts, since there are no broad lines, but the GC is really at a very unfortunate inclination.

We can probably exclude that the GC is currently a Seyfert galaxy, but if we look at it from face on, could we actually call it a LINER galaxy to some degree because LINER galaxies are probably a mix of active and star forming galaxies? There are a number of questions and I'm wondering if someone wants to comment on some of them.

Unknown: I think the most important thing is to have high mass accretion rates, to bring the mass inside. That is where the difference in star formation between barred spirals and AGN's comes into play.

Antonucci: Yeah there have been studies of that, and really people haven't found anything. There didn't seem to be a correlation at all, which was really surprising.

Falcke: Yesterday, there was speculation about a connection between the bar and the metallicity. Has that been actually found in other galaxies?

Antonucci: I'd like to ask the question a different way: what are people's thoughts on why the CND is tipped relative to the Milky Way's plane, are there good ideas about that?

Unknown: I think in the Galactic Center, in the region where the molecular clouds are, they can do anything they want: they are not bound to the Galactic Bulge. I think the disk has a large proportion of molecular clouds which cannot be disrupted because of the angular momentum.

Doeleman: I think we realize the CND is still gas. There are links between the CND and tori in galaxies. The gas dynamics seem to be completely the same, and they change on a short time scale.

Falcke: Well I thought that was something that could be stable over quite a long time, which was what Sanders showed. That was something that was not clear in my mind in the end whether the streamers we see now are really short lived features or are features that have to survive a number of million years. So has that been actually settled? I'm not quite so sure about this.

Novak: You were asking what causes the CND to be different? I can't remember who it was, but somebody had mentioned that the interaction of Sgr

A East might have some relation with the tipping of the CND. I think that's possible.

Falcke: What is the rate of precession of the CND? It's a bit longer than what Shep [Doeleman] is talking about I think—not on the time scale of 10^4 yr.

Sanders: I think as long as the gravitational potential is more or less spherically symmetric you could have any orientation you want for as long as you want and so indeed, as you were saying, at a random angle it will tidally stretch and make a disk or annulus if its tilted. I find it hard to understand how structures on a larger scale, say out to several hundred parsecs, can stay tilted for a long period of time—that is for more than a few rotation periods of differential precession. There you are getting away from a spherically symmetric potential and you would expect differential precession. Certainly near the center you could have any orientation, due to different orbits of individual clouds.

Falcke: I'm quite sure there are actually two processes. You have certainly normal rotation in the disk, especially at large scales, which is more or less aligned with the galaxy disk, but within the disk and close to the nucleus you can probably have all kinds of orientations.

Anatharamaiah: Many years ago there was a report of an Seyfert line that was more than 1000 km s^{-1} wide, seen in the nucleus, similar to the broad line region in Seyferts. Does anyone remember that?

M. Rieke: Those were HeI lines. Those lines still exist, and they are coming from the IRS16 cluster of hot stars.

Antonucci: Then can we argue that our Galactic Center looks like a Seyfert, a mini-Seyfert, or an AGN?

Biermann: This might then also apply to a starburst galaxy like M82 and the question whether you have a black hole—a starved AGN—despite the emission lines in M82 not coming from an AGN. We know from our little black hole that some of them are starving to such a degree that they don't do anything except disturb the kinematics—when we have enough spatial resolution and spectroscopy. In our Galaxy is this tiny little bit of radio emission that we see, Sgr A*, and if we didn't have this emission we wouldn't bother at all; there would be no such conference.

Falcke: Certainly I wouldn't be here ...

Sanders: I have a question for Ski about NGC 1068, is the polarized light still due to scattering?

Antonucci: That was the original argument and there have been corroborating arguments. There is, for example, significant evidence that the X-rays are scattered. The iron K_α line has an enormous equivalent width. The X-rays are thought to ionize the iron K_α line. Its equivalent width is so enormous that based on that alone I think people would conclude we are not seeing the line directly but it is scattered; that would imply that electrons rather than dust grains are scattering X-rays the same way as in the optical.

Sanders: Are there any cases where it is dust scattering and not electrons in the optical?

Antonucci: I think that there are many cases where there's not enough knowledge to say and I think that there may well be some cases for dust scattering. There are some IRAS galaxies where studies show that the polarized flux distribution is bluer than in a typical AGN. Similarly in the X-rays in NGC 1068, you see that high ionization K_α line which is probably scattered and indicative of electron scattering. In other Seyfert 2's you can see the same. Some of them also have low energy K_α lines.

Lazio: I'm just curious: Is it possible to do spectropolarimetry around the Galaxy, or do we have too much resolution?

Antonucci: I think I may have attempted that many years ago, we looked at some things which were high latitude reflection nebulae, and they all turned out to be HII regions. So if you tell me that there are some bona fide high latitude reflection nebulae that may be worth trying again, I will try. Bill Keel got some similar results by looking in Baade's Window region at [NII]+Hα regions to see if we were a LINER galaxy. I don't think he published the results because he considered it marginal. But he thought he was seeing a high ratio of [NII] to Hα, but it wasn't publication quality. Perhaps that should be attempted again, because he did it on a 40-inch telescope.

Melia: How closely have you been following the high energy observations of Seyferts? Could you summarize what the high energy data from OSSE are telling us about the fluorescence in Seyfert I's versus Seyfert II's? Is it consistent with our current picture?

Antonucci: It's a really powerful statement. In Seyfert I's you see the direct X-ray continuum and you also see the so called Compton hump in the K_α line. The strength of these features depends on the covering factor on the sky of matter which has a column density of the order of 10^{24} [cm−2]. Some of the K_α-lines and some of the onsets of the Compton hump are very broadened, and that's the evidence for a relativistic potential. In many other cases you see a narrow K_α-line and you know that there's a large coupling factor between very high density column material that's not in the potential well. There's is one wonderful case that Kim Weaver published who did reverberation mapping of the torus. They looked at the K_α-line and the Compton hump in NGC 2992, and they could see the high energy X-rays penetrate the torus and see them vary, just like the Seyfert I's. Then with a delay of a few years, she saw the K_α-line and the Compton hump vary. So, it was definitely from a few parsec scale outside the sublimation radius. What limits a little bit the power of the method is that many Seyfert II's have column densities of 10^{23} in which case you wouldn't see much reflection. But many have 10^{24}, in which case you would.

Falcke: I have a hard time stopping the discussion here, but we have a busy day ahead. So, thanks to everyone involved.

Session X

What Can We Learn from High-Energy Emission in the Galactic Center?

The Central Parsecs of the Galaxy
ASP Conference Series, Vol. 186, 1999
H. Falcke, A. Cotera, W.J. Duschl, F. Melia, M.J. Rieke, eds.

Cosmic Rays in the Galactic Center Region

W. Rhode[1]

Physics Dept., University of California, Berkeley, CA 94720, USA

T.A. Enßlin, P.L. Biermann[2]

Max Planck Institute for Radioastronomy, D-53010 Bonn, Germany

Abstract. EGRET data on the Gamma ray emission from the inner Galaxy have shown a rather flat spectrum. This spectrum extends to about 50 GeV in photon energy. It is usually assumed that these gamma-rays arise from the interactions of cosmic ray nuclei with ambient matter. Cosmic Ray particles have been observed up to $3\,10^{20}$ eV, with many arguments suggesting, that up to about $3\,10^{18}$ eV they are of Galactic origin. Cosmic ray particles get injected by their sources, presumably supernova explosions. Their injected spectrum is steepened by diffusive losses from the Galaxy to yield the observed spectrum. As cosmic ray particles roam around in the Galactic disk, and finally depart, they encounter molecular clouds and through p-p collisions produce gamma rays from pion decay. The flux and spectrum of these gamma rays is then a clear signature of cosmic rays throughout the Galaxy. Star formation activity peaks in the central region of the Galaxy, around the Galactic Center, the focus of this meeting. Looking then at the gamma ray spectrum of the central region of our Galaxy yields clues as to where the cosmic ray particles interact, and with what spectrum. Using the FLUKA Monte-Carlo, we have modeled this spectrum, and find a best fit for a powerlaw spectrum of cosmic rays with a spectrum of 2.34, rather close to the suggested injection spectrum for supernovae which explode into their own winds. This suggests that most cosmic ray interaction happens near the sources of injection; it has already been shown elsewhere that this is consistent with the spectrum of cosmic ray nuclei derived from spallation. One important consequence is that cosmic ray heating and ionization should be strong in the Galactic Center region.

1. Introduction

The spectrum of Galactic cosmic ray particles extends to probably $3\,10^{18}$ eV; the various contributions have been reviewed extensively by Wiebel-Sooth &

---

[1]Physics Dept., Univ. of Wuppertal, D-42097 Wuppertal, Germany

[2]*Speaker at the conference*

Biermann (1998), and the basic fits to the data for the various chemical elements have also been given in Wiebel-Sooth, Biermann & Meyer (1998). These cosmic ray particles interact with interstellar matter, and so spallate to produce the secondary nuclei (see, e.g., Garcia-Munoz et al. 1987), as well as the gamma ray emission above GeV photon energy (see, e.g., Stecker 1971).

There is a new AGASA paper (Hayashida et al. 1998) suggesting that neutrons near 10^{18} eV are seen coming from both the Galactic Center region as well as the Cygnus region, detected by a correlation in arrival direction. The data suggest neutrons, because protons at such an energy cannot get through the Galaxy on a straight line path with its magnetic field, since at that energy the Larmor radius is about 1 kpc. The data cannot also easily be explained as gamma ray photons, since then the correlation would be stronger at lower energy even, where nothing is seen. Of course, neutrons are produced by isospin flip in p-p collisions. It is worth remembering, that the Galactic Center region, as well as the Cygnus region are the prime candidate regions for star formation and supernova activity in the Galaxy, as clearly shown in radio, far-infrared and gamma ray data. Neutrons also can get here within their life time from the Galactic Center, but not at much lower energy. This leaves neutrons as the most probable origin of these events. If this result is accepted with the interpretation as neutrons, then it strongly supports the argument that cosmic rays are indeed Galactic up to 10^{18} eV. However, it is also immediately obvious, that the flux of cosmic ray particles required in the source regions to produce so many neutrons as suggested by the AGASA data, needs to be rather high relative to the flux observed at Earth. This then leads to considerable ionization and heating by low energy cosmic rays in the Galactic Center region.

In this paper we wish to review first the interpretations of the cosmic ray particles in the Galactic Center region, and then show that a simple concept may be sufficient to explain the new data from EGRET, which show a rather flat gamma ray spectrum. This is a severe test for any theory of cosmic ray origin.

2. Origin of High Energy Cosmic Rays

Our Galactic Center harbors a black hole, which probably went through many activity episodes during its growth. Therefore we want to ask first whether this activity could possibly explain high energy cosmic rays, and as a consequence gamma rays.

Biermann & Strittmatter (1987) have shown that radio galaxy hot spots can accelerate protons to about 10^{21} eV. Scaling this result with the power of the underlying source and using the jet/disk-symbiosis picture developed by Falcke et al. (1995 and later papers) we obtain for the maximum proton energy

$$E_{p,max} \;=\; 6.7\,10^{20}\,Q_{jet,46}^{1/2}\,\mathrm{eV} \tag{1}$$

where $Q_{jet,46}$ is the power of the jet in units of 10^{46} erg/s. The most extreme inferred jet luminosity is about $3\,10^{47}$ erg/s, and so energies up to

$$E_{p,max} \;=\; 4.\,10^{21}\,\mathrm{eV} \tag{2}$$

appear possible (Biermann 1998a). Therefore radio galaxies and their various counterparts such as compact radio quasars (see also Farrar & Biermann 1998) are clearly a suitable source for high energy cosmic rays. The jet-disk symbiosis does seem to work down to stellar size black holes (Falcke & Biermann 1999), and so we may be permitted to use it for intermediate powers of a proposed source.

The jet power of our Galactic Center, assuming that the compact radio source does signify the existence of a jet, is roughly

$$Q_{jet} = 5\,10^{38}\,\mathrm{erg/s} \tag{3}$$

and so $E_{p,max} = 1.5\,10^{17}$ eV. Therefore the activity of our central black hole is insufficient to produce neutrons at 10^{18} eV, and so is unlikely to help to explain the correlations in arrival directions in the data.

The Galactic Center region does harbor many interesting binary systems, some of which are referred to as mini-quasars; however, there again, their power is just not sufficient to explain particles near 10^{18} eV.

There is a new hot disk model, where weakly relativistic protons produce various secondaries in their interaction in the disk (Mahadevan 1998), but this model also cannot explain any particles at 10^{18} eV.

Therefore we propose to explore in the following the activity and cosmic ray injection properties of supernovae focusing on those supernovae that explode into their own stellar winds (Biermann 1997).

3. Galactic Cosmic Rays

In a series of papers Biermann et al. (1993 and later) have proposed that cosmic rays get injected from three sites predominantly:

- Supernovae that explode into the interstellar medium.
- Supernovae that explode into their own stellar wind.
- Radio galaxies and compact radio quasars.

The predictions of these models have been given in various reviews, and we summarize here briefly:

The cosmic ray particles which interact the most derive from the wind-supernovae. This happens since massive stars explode close to their birthplace, where the original material is still around from which they formed (Biermann & Tinsley 1974). Their source spectrum has been predicted to be $E^{-2.33-0.02\pm0.02}$ below the knee at $5\,10^{15}$ eV particle energy. For particles above the knee the corresponding prediction is $E^{-2.74-0.07\pm0.07}$. This peculiar way of writing the expected theoretical error range signifies an asymmetric error distribution, extending here from the most probable value of 2.33 to 2.37 in the first case. The bend (to explain the knee) has been predicted to be at 600 Z TeV, where Z is the charge of the chemical element nucleus under consideration, and the cutoff is near 100 Z PeV. Because the energy of the bend depends on the charge, the element abundance gets heavier at the knee, as noted already by Peters (1959, 1961).

To understand the concept of wind-supernovae we must remember that massive stars come in a variety of clothes, which are their different wind shells: First, stars above 8 solar masses, but below about 15 solar masses explode as supernovae, but do so directly into the interstellar medium. This is the classical case. Then stars above 15 but below about 25 solar masses explode into their wind, and that wind may be powerful enough to sweep up interstellar material into a shell mixed with shocked wind material, but the shell is still rather thin. The material in this shell is enriched in Helium from the nuclear reactions inside the star. Finally, above about 25 solar masses the winds get very powerful, leading to Wolf-Rayet stars, and is heavily enriched. In this case the wind-shell may be rather thick.

The transport through the Galaxy is described by a diffusion coefficient, which depends on the wave field that derives from a Kolmogorov spectrum of interstellar turbulence, and so the cosmic ray spectrum is steepened by 1/3 (see, e.g., Biermann 1995), to yield $E^{-2.67-0.02\pm0.02}$ below the knee.

This result can be directly compared with the data for Helium through Iron, which give a best fit of $E^{-2.64\pm0.04}$ (Wiebel-Sooth, Biermann & Meyer 1998).

However, where the interaction really happens is not clear. There are many possible points of view on this question, but two conceptually simple notions are documented in the literature:

First, there is the CR-standard model (see, e.g., Garcia-Munoz et al., 1987) that the average cosmic rays interact with the interstellar matter. In such a picture the gamma rays should have a spectrum that nicely fits the average cosmic ray spectrum, near $E^{-2.7}$. In such a picture the secondary to primary ratio of spallation products such as boron derived from carbon spallation gives the spectrum of interstellar irregularities with an implied energy dependence of the leakage time scale as $E^{-0.6}$. One problem with this argument is that there is little evidence for such a spectrum of irregularities (Biermann 1995), but it does give a good fit.

Second, there is the notion that most cosmic ray interaction happens near the source (Biermann 1998b). In such a picture the spallation leading to secondary nuclei production happens in the shell around the stellar wind, when the supernova induced shock smashes through that shell. This leads to an energy dependence for the local leakage time scale of $E^{-0.55}$ (Biermann 1998b); however, this happens only when the shell is thick enough to allow diffusive interaction to be dominant over convective losses. This latter process is likely to dominate for the more abundant, but thinner shells around slightly lower mass stars. Therefore, in such a picture we expect that the more common stars in the range 15 to 25 solar masses would produce most gamma rays. And as a corollary we expect that the gamma rays should correspond to the injection spectrum.

3.1. The Failure of the CR-Standard Model

The standard model has been explored in two papers recently, using the best EGRET data (Hunter et al. 1997, Mori 1997). The standard model fails by a wide margin.

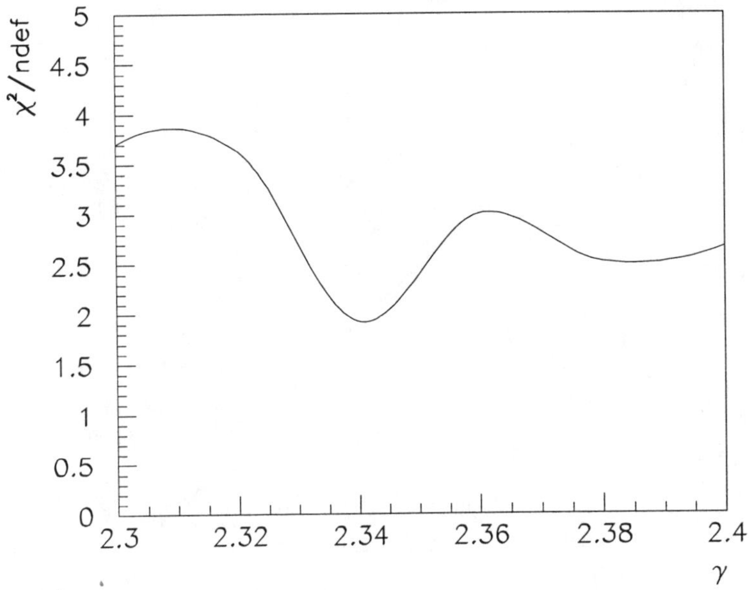

Figure 1. χ^2_{red} versus spectral index.

The failure is due to the spectrum. The observed gamma ray spectrum is just too flat in order to be produced by a cosmic ray spectrum near $E^{-2.7}$.

There are several ways out of this conundrum.

First, one might argue that the Monte-Carlo codes used to predict the gamma rays are not good enough. This is what Mori (1997) has tried. The uncertainties in the Monte-Carlos are not sufficient to explain the flat spectrum.

Second, one might argue, that pion decay does not explain the data. This has been tried by Pohl & Esposito (1998). They suggest that the spectrum can be partially derived from inverse Compton scattering off a population of energetic electrons produced by supernova remnants. In the progressive leakage of the electrons as a function of energy and time from injection the observed spectrum can be matched. If the new AGASA data are correctly interpreted with arising from energetic neutrons, then cosmic ray nucleon interaction is about as high as can possibly be, and so it is difficult to see how to avoid gamma ray production from pion decay being a strong contributor.

Conversely, a success of an alternative model that also explains other data would be very helpful, and this is what we have tried.

3.2. A Fit to the Data

Therefore we have adopted a simple powerlaw model for the Bremsstrahlung and inverse Compton contribution and then fitted the data with one main parameter in mind: the power law spectrum of the cosmic rays; this law is given as a strict power law in momentum from many MeV to many GeV and beyond in energy.

We have tried three different Monte-Carlo codes from CERN and Fermi-Lab to do this analysis, and we have adopted for this work the code FLUKA

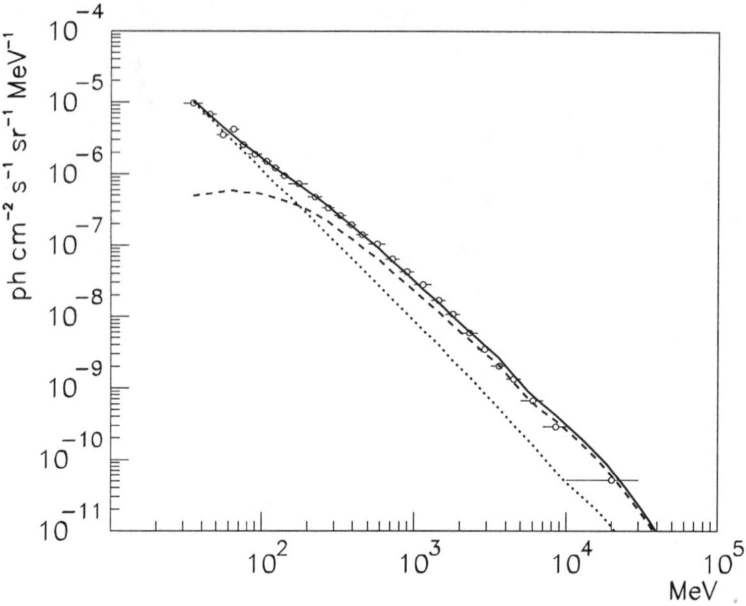

Figure 2. The fit to the gamma ray data of the inner Galaxy.

(specifically the version GEANT3.21/FLUKA from the CERN library). This code could readily be adapted to include the subtle effects of Helium for instance.

Fig. 1 gives the χ^2_{red} fit as a function of spectral index. The wiggle at spectral index 2.36 is due to a systematic pattern in the data from the primary data analysis. In the plot shown here, a clear minimum is visible at spectral index 2.34. This minimum is at a level of $\chi^2_{red} \simeq 1.9$. Taking into account the uncertainties of the IC and Bremsstrahlung contribution this is quite acceptable. Fig. 2 gives the resulting fit, which still shows some model dependent waves from the finite resolution of the Monte-Carlo used.

The next step will be to check how high in energy we can push this model of cosmic ray interaction near the source; there are severe limits now on the inner Galaxy from the CASA-MIA experiment (Ong 1998).

4. Consequences

First of all, what the agreeable fit demonstrates is that there is a spectrum for the cosmic rays that fits their gamma ray emission. This spectrum is consistent with a power law, and is in fact quite close to the original prediction of a source spectrum for wind-supernovae.

Second, it shows that source related interaction may be worth pursuing in detail. What has not yet been done here, is a fit to the detailed isotope abundances (see the recent discussion of this point by Westphal et al. 1998). Together with the paper presented at the Hirschegg conference (Biermann 1998b) this means that there is a viable proposal how to explain a) the cosmic ray spectrum itself, b) the gamma ray spectrum, and c) the spectrum of spallation secondaries.

If this scenario could be confirmed there are some consequences also for the stars with strong winds:

- Wolf-Rayet and OB stars have shock waves running through their winds.
- These shocks accelerate electrons and produce observed radio emission (Biermann & Cassinelli 1993).
- These shocks also accelerate protons, resulting in a steep pion decay spectrum; this spectrum is steep because the Alfvénic Mach number of these shocks is low.
- These shocks also accelerate nuclei, which, in collisions, can give rise to spallation products in an excited nuclear state, then explaining gamma ray lines (Nath & Biermann 1994b) from active regions of star formation.

Finally, to summarize the essential idea again, there are consequences of this scenario as well for exploding stars with strong winds:

- The supernova shock races through the wind.
- The shock accelerates particles.
- Cosmic ray injection of elements such as helium and most heavier elements originates from this acceleration.
- Once outside their site of origin the protons (and other nuclei) at energies below about 50 MeV use up their ionization and heating power near their origin (Nath & Biermann 1994a).
- In the Galactic Center region the cosmic ray induced ionization and heating should be high.

Acknowledgments. We acknowledge numerous discussions on these matters with many friends and colleagues, especially John Bieging, Alina Donea, Tom Gaisser, Stan Hunter, Bob Kinzer, Norbert Langer, Karl Mannheim, Sera Markoff, Jim Matthews, Hinrich Meyer, Rene Ong, Buford Price, Ray Protheroe, Jörg Rachen, Eun-Suk Seo, Todor Stanev, and Dave Thompson. WR would like to express his gratefulness for the warm hospitality he has received at Berkeley in the group of Buford Price.

References

Biermann, P.L. 1993, A&A 271, 649; astro-ph/9301008

Biermann, P.L. 1995, in the Proc. Gamow Jubilee Seminar, St. Peterburg, Sept. 1994, Space Science Rev. 74, 385; astro-ph/9501003

Biermann, P.L. 1997, chapter in "Cosmic winds and the Heliosphere", Eds. J. R. Jokipii et al., Univ. of Arizona Press, p. 887; astro-ph/9501030

Biermann, P.L. 1998a, in the NASA Workshop on "Observing Giant Cosmic Ray Air Showers for $E > 10^{20}$ eV Particles from Space", Eds: John F. Krizmanic, Jonathan F. Ormes, and Robert E. Streitmatter, in AIP conf. proc. 433, book, p. 22

Biermann, P.L. 1998b, in Proc. Hirschegg Conference on Nuclear Astrophysics, GSI, Darmstadt, p. 211

Biermann, P.L., Cassinelli, J.P. 1993, A&A 277, 691; astro-ph/9305003

Biermann, P.L., Strittmatter, P.A. 1987, ApJ 322, 643

Biermann, P.L., Tinsley, B.M. 1974, PASP 86, 791

CERN GEANT3.21/FLUKA: CERN Program library write-up W5013 and references therein

Falcke, H., Biermann, P.L. 1995, A&A 293, 665; astro-ph/9411096

Falcke, H., Biermann, P.L. 1998, A&A 342, 49; astro-ph/9810226

Farrar, G.R., Biermann, P.L 1998, PRL 81, 3579; astro-ph/9806242

Garcia-Munoz, M., Simpson, J.A., Guzik, T.G., Wefel, J.P., Margolis, S.H. 1987, ApJS 64, 269

Hayashida *et al.* 1998, preprint; astro-ph/9807045

Hunter, S.D. *et al.* 1997, ApJ 481, 205

Mahadevan, R. 1998, Nature 394, 651

Mori, M. 1997, ApJ 478, 225

Nath, B.B., Biermann, P.L. 1994a, MNRAS 267, 447; astro-ph/9311048

Nath, B.B., Biermann, P.L. 1994b, MNRAS 270, L33; astro-ph/9407001

Ong, R. 1998, Physics Reports 305, 93

Peters, B. 1959, Nuovo Cimento Suppl. XIV, ser. X, p. 436 - 456.

Peters, B. 1961, Nuovo Cimento XXII, p. 800 - 819

Pohl, M., Esposito, J.A. 1998, ApJ 507, 327; astro-ph/9806160

Stecker, F.W. 1971, "Cosmic Gamma Rays", NASA SP-249, NASA, Washington, DC

Westphal, A.J., et al. 1998, Nature 396, 50

Wiebel-Sooth, B., Peter L. Biermann, P.L. 1988, "Cosmic Rays" in Landolt-Börnstein, Handbook of Physics, Springer Publ. Comp., (in press)

Wiebel-Sooth, B. Biermann, P.L., Meyer, H. 1998, A&A 330, 389; astro-ph/950925

The Central Parsecs of the Galaxy
ASP Conference Series, Vol. 186, 1999
H. Falcke, A. Cotera, W.J. Duschl, F. Melia, M.J. Rieke, eds.

Gamma-Rays from the Galactic Center: A Point-Like or Diffuse Source?

Sera Markoff[1]

Physics Department, The University of Arizona, Tucson, AZ 85721

Fulvio Melia[2]

Physics Department and Steward Observatory, The University of Arizona, Tucson, AZ 85721

Abstract. EGRET has detected a γ-ray flux from the Galactic center region, but its resolution is not sufficient to determine the exact source. While the lack of variation in the flux down to detector sensitivity suggests a diffuse emitter, a compact source (possibly coincident with the massive black hole candidate Sgr A*) cannot be ruled out on this basis: recent hydrodynamical simulations of the accretion process indicate that a lower level of fluctuation is consistent with this amplitude on a timescale comparable to the current observation timeline. Until instruments with the capability to resolve the details of this region at extremely high-energies are available, we are left to determine the source by identifying the most likely candidates and attempting to model their spectra based on what we know of their physical environments. We present some details of the general model here, and its particular application to Sgr A*. A detailed analysis of the photon production from hadronic collisions and other processes, using the most current particle physics data and theoretical models, indicates that an accreting black hole environment is inconsistent with the broad-band data.

1. Introduction

EGRET, on-board the Compton GRO, has detected a γ-ray excess between 30 MeV$-$10 GeV, located within an error circle of 0.2° for photons with $E_\gamma > 1$ GeV (Mayer-Hasselwander et al. 1998). This error circle includes the Galactic center (GC) but is not necessarily centered on it. However, the absence of any other significant sources of γ-rays within $\sim 15^\circ$ makes it unlikely that the emission is not associated with the GC.

The shape of the detected γ-ray spectrum is not unique: it shows a pronounced bump around 100 MeV that is similar to other EGRET detections now associated with either supernova remnants or pulsars (Esposito et al. 1996), and

[1]NSF Graduate Student Fellow

[2]Presidential Young Investigator

to the diffuse Galactic γ-ray spectrum (Hunter et al. 1997). Assuming 8.5 kpc to the GC, the observation error circle for the emission at energies greater than 1 GeV corresponds to $\sim$ 30 pc, an area which could easily accommodate many sources. One possibility is that there is no single source, and that the γ-rays are the result of Galactic cosmic rays interacting throughout the GC region. This idea is discussed in another paper in this volume by Rhode et al. (1999). The power integrated over $E_\gamma > 100$ MeV is $2.2 \pm 0.2 \times 10^{37}$ (Mayer-Hasselwander et al. 1998), which is sufficiently high that if the γ-rays are associated instead with a single source, this object must be rather unique in the GC. Since Sgr A*, the massive black hole candidate at the dynamical center of the GC, is also unique, it is appealing to consider this as a choice for a compact source of the emission. The lack of variability has led others to also consider diffuse sources for the γ-rays, such as Pohl (1997) and Fatuzzo et al. (1999), in this volume.

The idea of associating accretion onto a central compact object with the EGRET detection was first explored by Mastichiadis & Ozernoy (1994). However, they over-simplified some of the particle physics, and concluded that the presence of a $\sim 10^3\ M_\odot$ black hole was consistent with the EGRET data. Mahadevan et al. (1997) considered thermal and non-thermal protons in an ADAF scenario around Sgr A, and concluded that they cannot simultaneously explain the low- and high-energy data. In this article, we will discuss in general our model for producing the γ-rays, utilizing the most current data for the energy-dependent cross sections, inelasticity and pion multiplicity, allowing for a better calculation of the relevant cooling processes and spectral components, together with a self-consistent treatment of the relativistic particle distributions. Then we will discuss the application of this physics to Sgr A* in particular.

We do not attempt to account for the radio through infrared spectrum via the processes invoked here (instead see, e.g., Melia 1994; Beckert & Duschl 1997; Falcke et al. 1993; Narayan et al. 1998). However, these low energy data act as crucial constraints on the secondary particle emissivity.

2. Physical Picture

Discovered first as a non-thermal radio source, Sgr A* lies at the dynamical center of the Galaxy (Menten et al. 1997), and has a mass of $\sim 2.6 \times 10^6 M_\odot$ (Haller et al. 1996; Genzel et al. 1997; Ghez et al. 1998) concentrated within $\sim$ 0.02 pc. This high density is suggestive of a massive black hole, although other possibilities are still being considered. The nearby IRS 16 complex of hot giant stars is most likely the main contributor to a Galactic center wind of terminal velocity $v_{gw} \approx 500 - 700$ km s^{-1} and a mass-loss rate $\dot{M}_{gw} \approx 3 - 4 \times 10^{-3}\ M_\odot$ yr^{-1}. As the wind flows past the central region, a portion is captured by the gravitational field of Sgr A* and begins to fall inward towards the central object. To simplify the calculation we neglect any residual angular momentum in the infalling plasma, resulting in a quasi-spherical accretion flow with free fall velocity v_{ff}. The gas eventually becomes supersonic, and presumably results in a shock between $40 - 120 r_g$ (Babul et al. 1989), where r_g is the Schwarzschild radius.

Not much is known about the structure of the magnetic field at the shock, but since we assume collisionless processes, the acceleration will be enhanced by interactions with the perpendicular (to the flow) component of the magnetic field (see Jokipii 1987). We assume that the shock accelerates both electrons and protons to relativistic energies, but the increased sensitivity of the electrons to synchrotron and inverse Compton cooling compared to the protons limits their maximum Lorentz factor by several orders of magnitude, and so for reasons we discuss in more detail below, we find their contribution to be negligible. We assume that the accelerated protons are injected by the shock at a rate of $\dot{\rho}_p(E_p) = \rho_o E_p^{-x}$ cm^{-3} s^{-1} GeV^{-1}. The steady-state proton spectral index z is thought to lie in the range of $\sim 2.0 - 2.4$ (Jones & Ellison 1991), and is determined from a diffusion loss treatment. We choose to define the injection in terms of two free parameters, the normalization ρ_o and index x, so that we can restrain our results by the shock efficiency η and test for self-consistency. These three parameters are related by:

$$\int \rho_o E_p^{-x} E_p \, dE_p = \eta L_{grav} \equiv \frac{\eta G M_h \dot{M}}{r_{sh}} , \tag{1}$$

where r_{sh} is the location of the shock.

The relativistic particles will interact with the ambient particles and the magnetic field, producing photons via synchrotron, inverse Compton scatterings, the decay of mesons created during *p*-*p* collisions and via the interactions of secondary electrons. Because the protons can be ultra-relativistic, the leading order protons produced will suffer multiple collisions, contributing to the spectrum repeatedly as they cascade down the distribution in energy space, eventually giving up their entire kinetic energy to secondary particles. The secondaries consist of pions, which will decay either to photons ($\pi^0 \rightarrow \gamma\gamma$) or leptons ($\pi^\pm \rightarrow \mu^\pm \nu_\mu$, with $\mu^\pm \rightarrow e^\pm \nu_e \nu_\mu$).

We take our model of the gas dynamics from Melia (1994), the relevant parameters from which are contained in his paper in this volume (Melia 1999).

3. The Relativistic Particle Interactions

The gravitational energy liberated during the infall can be converted into high-energy radiation through the interactions of the relativistic protons. These include

$$pp \rightarrow pp \quad \& \quad pN \rightarrow pN\, M_\pi , \tag{2}$$

where N is either a proton or a neutron and M_π (known as the pion multiplicity) represents the energy-dependent distribution of pions created during the inelastic collision. The first (elastic) interaction serves to redistribute the energy among the protons.

Other possible interactions are:

$$p\gamma \rightarrow p\pi^0\gamma \quad p\gamma \rightarrow n\pi^+\gamma \quad p\gamma \rightarrow e^+e^-p \quad pe \rightarrow eNM_\pi . \tag{3}$$

However, the extremely low density and energy of photons near Sgr A* suppresses the *p*-γ interactions compared to those in more luminous galactic nuclei. Similarly, there are not enough extremely energetic e^-'s present to make the last interaction indicated above significant—it is analogous to *p*-γ interactions because the electron interacts with the proton via a virtual photon.

For each inelastic p-p collision at a particular energy, a multiplicity M_π of pions is produced, with a ratio of charged to neutral pions of roughly 2:1 (exactly for the case when there are two leading protons, but there will be a surplus π^+ when there is a leading neutron so charge is always conserved). These have a distribution in transverse (to the beam in experiments; in our case, to the direction of the boost back to the laboratory frame) momentum and rapidity y $dN_\pi/dp_\perp dy$, or in terms of the pseudo-rapidity η, $dN_\pi/dp_\perp d\eta$, which is measured as a function of center of mass (CM) energy at collider experiments. The rapidity is related to the parallel component of the momentum, $p_{||}*$, and energy of the pion in the CM frame, E_p^*, by $y = (1/2)\ln([E_\pi^* + p_{||}^*]/[E_\pi^* - p_{||}^*])$, and $y \approx -\ln\tan(\theta^*/2) \equiv \eta$ for relativistic energies, where $\cos\theta^* = p_{||}^*/ \mid p^* \mid$.

There have been many p-p and p-$\bar{p}$ collider experiments over the past 25 years, and each produces a data set based on a different apparatus and different region of phase space. In many cases, two experiments at the same energy have somewhat inconsistent results due to differences in procedure, trigger specifications and normalization. We sifted through the multitude of data and chose the largest set of published data available that are self-consistent. In practical terms, this means the $dN_\pi/dp_\perp dy$ distributions for each CM energy should not cross, so that interpolation between distributions is possible. We calculate $dN_\pi/dp_\perp dy$ from the inclusive cross section, usually measured in the form $Ed^3\sigma/dp^3$. In this case, $dN_\pi/dp_\perp dy = (2\pi p_\perp/\sigma)Ed^3\sigma/dp^3$.

3.1. The Steady State Particle Distributions

Assuming the system is in equilibrium, we can calculate the steady-state proton distribution resulting by solving the space-independent diffusion loss equation over the entire energy range. Specifically, we include terms for losses due to p-p and p-γ (where relevant) collisions, synchrotron and inverse Compton cooling, and for injection due to the shock and the influx of proton secondaries from p-p collisions at higher energy. We also include the probability for $p \to n$ charge exchange. Note that in order to calculate the inverse Compton scattering of relativistic protons correctly, one must know the fraction of the total p-γ cross section that comprises only elastic collisions. Unfortunately, there is no measurement, about which we know, for this alone. It is possible to calculate using Regge theory, and we feel this is worthwhile in the case of more luminous galactic nuclei, where the energetic photon population often dominates the interactions. However, in the case of Sgr A*, an upper limit will suffice, since it is a relatively weak component. We therefore calculate the proton inverse Compton spectrum using $\sigma_{p\gamma,tot}$, which is an overestimation by possible over an order of magnitude, but as we shall show, still not the major contributor to the spectrum.

The collisions of these relativistic protons with the ambient ones create electrons and positrons via charged pion decay. Following the same procedure, we solve the diffusion loss equation for the electrons and positrons for the steady-state distribution, which we then use to calculate the synchrotron and inverse Compton spectra from the cascade $e^\pm$s. We include Klein-Nishina corrections at high-energies.

4. The Calculational Procedure

The first step is to identify which photon-producing processes dominate in which energy ranges. Following Begelman et al. (1990) and Chodorowski et al. (1992), we do this by comparing the cooling rates of relativistic protons from synchrotron, inverse Compton scattering, and the collisions as functions of energy. Then, comparing the total loss rate from all processes combined to the proton acceleration rate, it is possible to find the maximum Lorentz factor attained by the protons. This sets the high-energy cutoff for the injected power-law distribution.

Once we have $\rho_p(E_p)$ at a given energy, we find the distribution in the CM frame of pions $dN_\pi/dp_\perp dy(\sqrt{s})$ for each *p-p* collision at any particular CM energy. Because we can relate y to $p_{||}^*$, and because in the CM frame the pion energy $E_\pi^* = \sqrt{m_\pi^2 + p_{||}^{*2} + p_\perp^{*2}}$, when we multiply $dN_\pi/dp_\perp dy dp_\perp dy$ by the number of collisions per unit volume, we find the distribution of pions with energy E_π^* per unit volume, per unit time, per *p-p* collision at E_p, in the *p-p* CM frame.

The next step is to find the distributions of the resulting decay products and then transform them to the observer's frame. The details of this can be found in Markoff et al. (1998). By summing over all particle contributions, we find the spectral components due to hadronic interactions.

5. Results and Discussion

Using the parameters discussed in Melia (1994) and his article in this volume, we find that the maximum attainable Lorentz factor for the relativistic protons is $\gamma_{p,max} \approx 4 \times 10^8$, which is fairly insensitive to changes in these parameter values. This is a much higher limit than one would expect for a typical AGN, where the higher photon density would lead to more efficient cooling by inverse Compton and *p*-γ interactions. By comparison, the Lorentz factor for shock-accelerated leptons is $\gamma_{e,max} \approx 10^5$, corresponding to an energy of $\approx 5 \times 10^4$ MeV. Assuming the same total injection rate of relativistic electrons and protons, the total energy contained in accelerated electrons is significantly lower than that in cascade leptons, which will therefore dominate the leptonic contribution to the spectrum. The gyroradius is $\approx 4 \times 10^{12}$ cm for the highest energy protons. This is roughly a seventh of the shock radius for $r_{sh} = 40 r_g$, which is a reasonable size for the acceleration region.

As we can see in Figure 1, *p-p* processes dominate the cooling for most of the physical energy range by several orders of magnitude. In an AGN, *p*-γ processes typically begin to dominate at $\gamma_p \approx 10^5 - 10^6$ (see, e.g. Begelman et al. 1990), but the extreme dearth of photons around Sgr A*, particularly at high energies, suppresses this channel until the thresholds for pair and pion production can be crossed. Once crossed, these rates will respectively dominate, but only at such high proton energy that their effect on the overall spectrum will be negligible. For this reason we can also neglect *p-e* interactions, whose cross section is even smaller. The rate for Compton cooling is quite an overestimation due to the use of the total cross section, which in addition we fix at the lowest energy datum for CM energies less than 1.1 GeV even though the data indicate it is in fact

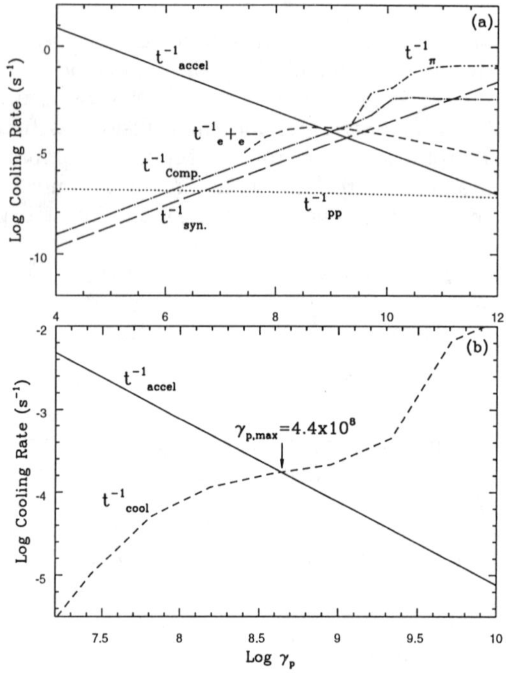

Figure 1. Cooling rates for radiation processes in Sgr A*. (a) Solid: proton acceleration rate, dotted: rate of cooling by p-p collisions, short dash: rate of cooling by p-$\gamma \rightarrow pe^+e^-$, short dash-dot: rate of cooling by p-$\gamma \rightarrow NM_\pi$, long dash: rate of cooling by synchrotron, long dash-dot: rate of cooling by elastic p-γ scattering ("p inverse Compton"). (b) Solid: proton acceleration rate, dashed: sum of all cooling rates in (a).

decreasing quite rapidly at this point. For Sgr A*, where a collision of the highest energy photon and proton will still fall under this limit, the overestimation is thus compounded. Beyond emission from p-p collision by-products, synchrotron is the only other component which assuredly contributes significantly to the γ-ray spectrum.

We do not include interactions with relativistic neutrons produced during the cascade, because only those with $\gamma_n \gtrsim 10^8$ will remain in the system long enough to interact (Sikora et al. 1989), and such a population is negligible. High-energy photons will also escape unhindered because of the dearth of ambient matter, and so we assume that what is observed is very close to what is produced at the shock.

Figure 2a gives an example of a set of self-consistent, reasonable physical parameters that results in a reasonable fit to the γ-ray data, but unfortunately this fit does not satisfy the radio and IR constraints. Although the new low energy EGRET data points are most likely contaminated by nearby sources (Mayer-Hasselwander et al. 1998), they serve as upper limits of a probable rise in the flux towards lower energy.

For comparison, we show in Figure 2b the spectrum resulting from the same parameters as in Figure 2a, except for the magnetic field, which has been reduced until the curve is consistent with the low-frequency data. This reduced magnetic field, while sub-equipartition, is not necessarily unrealistic. Hydrodynamical simulations indicate the field may be significantly lower than the equipartition value (Coker & Melia 1999). However, even in this case, the proton synchrotron cannot adequately account for the low energy EGRET data.

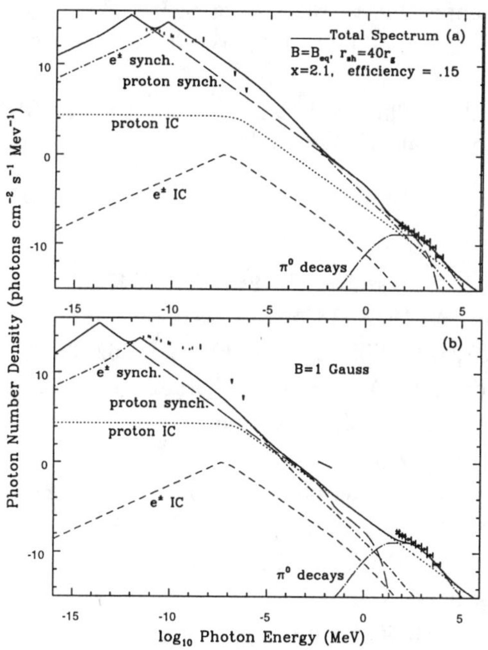

Figure 2. (a) Five spectral components (as labeled) resulting from accelerated protons and secondary $e^{\pm}$'s. The shock is here located at $40r_g$, and the injected proton spectral index is $x = 2.09$, yielding a steady state proton index $z = 2.10$. The inferred efficiency in this case is $\eta = 15\%$. With an equipartition field, the synchrotron components from secondaries overwhelm the lower energy data. (b) For a 1 Gauss field, the synchrotron is lowered enough to fall below the data, but now the lower energy EGRET points can no longer be accounted for. The data points or upper limits (arrows) are from: (radio) Lo (1987), Zylka et al. (1993); (IR) Menten et al. (1997), Stolovy (1996); (X-rays) Pavlinskii et al. (1992); (γ-rays) Mayer-Hasselwander et al. (1998).

We conclude that a reasonable fit to the EGRET data cannot produce a spectrum consistent with both the low and high-energy data. The emissivity of the cascade particles is simply too high to satisfy the lower energy constraints.

6. Concluding Remarks

In this paper, we have used our improved treatment of the proton-proton cascade within the accreting environment of Sgr A* to show that shock acceleration near the central engine does not appear to be the mechanism producing the EGRET γ-ray flux from the Galactic center region. This seems to be consistent with growing evidence that a diffuse source is at work to produce the γ-ray photons. Conclusive evidence may come when GLAST (Gamma-ray Large Area Space Telescope) is launched. This mission may be able to resolve the source if it is diffuse on a scale comparable to the size of Sgr A East. GLAST's projected spatial resolution is $30''$ to 5 arcmin, compared to the ~ 4 arcmin size of the Sgr A East shell.

Acknowledgments. This work was supported by an NSF Graduate Fellowship and the NASA grant NAGW-2518.

References

Babul, A., Ostriker, J. & Mészáros, P. 1989, ApJ, 347, 59.

Beckert, T. & Duschl, W. 1997, A&A, 328, 95.

Begelman, M., Rudak, B. & Sikora, M. 1990, ApJ, 362, 38.

Chodorowski, M., Zdziarski, A. & Sikora, M. 1992, ApJ, 400, 181.

Coker, R. & Melia, F. 1999, ApJ, submitted.

Esposito, J.A. et al. 1996, ApJ, 461, 820.

Falcke, H., Mannheim, K., & Biermann, P.L. 1993, A&A, 278, L1.

Fatuzzo, M., et al. 1999, this volume, p. 560.

Genzel, R. et al. 1997, MNRAS, 291, 219.

Ghez, A. et al. 1998, ApJ, in press.

Haller, J. et al. 1996, ApJ, 468, 955.

Hunter, S.D. et al. 1997, ApJ, 481, 205.

Jokipii, J. R. 1987, ApJ, 313, 842.

Jones, F. & Ellison, D. 1991, SSRv, 58, 259.

Lo, K.Y. 1987, in AIP Proc. 155, ed. D.C. Backer, (AIP: New York), 155, 30.

Mahadevan, R., Narayan, R. & Krolik, J. 1997, ApJ, 486, 268.

Markoff, S., Melia, F. & Sarcevic, I. 1998, ApJ, submitted.

Mastichiadis, A. & Ozernoy, L. 1994 ApJ, 426, 599.

Mayer-Hasselwander, H.A. et al. 1998, A&A, 335, 161.

Melia, F. 1994, ApJ, 426, 577.

Melia, F. 1999, this volume, p. 127.

Menten, K. et al. 1997, ApJ, 475, L111.

Narayan, R. et al. 1998, ApJ, 492, 554.
Pavlinskii, M., et al. 1992, Soviet Astr. Lett., 18, 291P.
Pohl, M. 1997, A&A, 317, 441.
Rhode, W., Enßlin, T. & Biermann, P. 1999, this volume, p. 543.
Sikora, M. Begelman, M. & Rudak, B. 1989, ApJ, 341, L33.
Stolovy, S.R., Hayward, T.L. & Herter, T. 1996, ApJ, 470, L45.
Zylka, R., Mezger, P.G. & Lesch, H. 1993, A&A, 261, 119.

The Central Parsecs of the Galaxy
ASP Conference Series, Vol. 186, 1999
H. Falcke, A. Cotera, W.J. Duschl, F. Melia, M.J. Rieke, eds.

The Galactic Center EGRET Source and the Broad-band Spectrum of Sgr A East

Marco Fatuzzo

Physics Department, The University of Arizona, Tucson AZ 85721

Fulvio Melia

Physics Department and Steward Observatory, The University of Arizona, Tucson AZ 85721

Farhad Yusef-Zadeh

Department of Physics and Astronomy, Northwestern University, Evanston, IL 60208

Sera Markoff

Physics Department, The University of Arizona, Tucson AZ 85721

Abstract. High energy (30 MeV - 10 GeV) emission associated with a Galactic Center source has recently been observed by EGRET on board the Compton Gamma-ray Observatory. Continuum low-frequency radio observations of the Galactic Center reveal the presence of several viable source candidates, one of which is a prominent elliptical shell structure known as Sgr A East. Although morphologically similar to supernova remnants (SNRs), its inferred energetics and size appear to be extreme and have generated some uncertainty regarding this interpretation. Additionally, the radio spectral index determined from 6 cm and 20 cm observations of Sgr A East (~ 1) differs from the indices normally found for SNRs (~ 0.5). This difference is quite significant given that the corresponding index for the electron distribution for Sgr A East is then ~ 3 (versus ~ 2 for SNRs) and therefore unlike that thought to be produced directly from the shock acceleration processes believed to occur in such an environment. Regardless, recent observations of this region at 1720 MHz (the transition frequency of OH maser emission) have revealed the presence of several maser spots, suggesting a kinship between Sgr A East and three EGRET SNRs also observed at 1720 MHz: IC443, W28 and W44. This possible kinship suggests that Sgr A East could be the source of the EGRET emission. We show here that the pion-production and decay mechanism invoked to account for gamma-ray emission from SNRs can also naturally account for the radio through TeV spectrum of Sgr A East (including the EGRET data). The differences between the spectral features of Sgr A East and the other SNRs then results from the differences between the underlying energetics and the properties of the surrounding interstellar medium.

1. Introduction

EGRET on board the Compton Gamma-ray Observatory has identified a central $\sim$ 30 MeV - 10 GeV continuum source (designated as 2EG J1746-2852) with luminosity $\approx 2 \times 10^{37}$ ergs s^{-1} (Mayer-Hasselwander et al. 1998). The highest-energy (> 1 GeV) component of the emission is peaked within 0.2 degrees of the Galactic Center, and the lack of other strong γ-ray sources within 15 degrees of this position makes a chance coincidence highly unlikely. Two important characteristics of the observed emission are a strong bump-like feature starting at around 100 MeV and a lack of temporal variability down to the instrument sensitivity (roughly a 20% amplitude), the latter characteristic suggesting an extended source.

There are several known processes by which MeV to GeV photons can be produced by an underlying nonthermal distribution of particles. Nonthermal radio structures within the inner tens of parsecs of the Galactic center must therefore be considered as viable candidates for the source of the EGRET emission. Of course, the Galactic center harbors several such structures, including Sgr A*, the 7 arcminute halo, the radio arc, and Sgr A East. The work of Markoff, Melia and Sarcevic (1998) seem to rule out Sgr A* as the source of the EGRET emission. Melia, Yusef-Zadeh and Fatuzzo (1998) found that inverse Compton scattering of the UV and IR fields produced within the central parsecs of the Galactic Center by the relativistic electron population inferred from radio data of the 7 arcminute halo can account for the EGRET emission, but only if the electron distribution has a low-energy cutoff at around 1 GeV. It remains unclear how such a particle distribution would arise. Similarly, Pohl (1997) found that the inverse Compton scattering of ambient far IR photons by monoenergetic electrons injected into the radio arc can also account for the EGRET emission. However, recent observations of the arc indicate a centrally brightened morphology (Lang 1999) which seems to be in conflict with such a scenario. We consider in this work the possibility that Sgr A East is the source of the EGRET emission.

Sgr A East is an elliptical structure with nonthermal emission peaked at its periphery. Geometrically, the structure is elongated along the Galactic plane with a major axis of length 10.5 pc and a center displaced from the apparent dynamical nucleus, Sgr A West, by 2.5 pc in projection toward negative Galactic latitudes. The latter is a ($\sim 1-2$ pc) three-arm spiral configuration of ionized gas (Ekers et al. 1983; Lo & Claussen 1983) that engulfs the compact nonthermal radio source, Sgr A*, having a dimension between 1.1 AU and 0.1 AU (Morris & Serabyn 1996). The bulk of the thermally emitting Sgr A West appears to be located in front of Sgr A East (Yusef-Zadeh & Morris 1987; Pedlar et al. 1989), whose exact distance behind the nucleus is not known, but a number of arguments suggest that it is positioned very close to the Galactic center (e.g., Güsten & Downes 1980; Goss et al. 1989; Yusef-Zadeh et al. 1998). Thus, if present, the massive black hole (presumably associated with Sgr A*) lies within the Sgr A East shell, but at one extreme end of its elongated structure.

Sgr A East may be a supernova remnant (SNR), but its inferred energetics and size appear to be extreme and have generated some uncertainty regarding this interpretation. For example, Mezger et al. (1989) deduced that a lower limit of $\sim 4 \times 10^{52}$ ergs of energy is required to produce a remnant in a dense cloud with the observed diameter of the radio synchrotron source. With a typical energy of $\sim 10^{51}$ ergs in its ejecta, a supernova may not have had the power to produce this remnant, but an explosion of this magnitude can be generated by the tidal disruption of a main sequence star whose trajectory has taken it within ten or so Schwarzschild radii of the central object (Khokhlov & Melia 1996). These events should occur roughly once every $10^4 - 10^5$ years; the most recent one may have produced Sgr A East.

Regardless of what actually caused the initial explosion, recent observations of this region at 1720 MHz (the transition frequency of OH maser emission) have revealed the presence of several maser spots at the SE boundary of Sgr A East with a velocity of ≈ 50 km s^{-1}, and one near the Northern arm of Sgr A West at a velocity of 134 km s^{-1} (Yusef-Zadeh et al. 1996). This is important because OH masers can be effective tracers of shocks produced at the interface with molecular clouds, particularly for supernova remnants (Frail et al. 1994; Wardle, Yusef-Zadeh & Geballe 1998). It has therefore been argued that the expansion of an explosion's ejecta (possibly a supernova remnant) into the Sgr A East molecular cloud (i.e., the 50 km s^{-1} cloud surrounding the shell) is responsible for producing the shocked OH maser emission at ≈ 50 and 134 km s^{-1}. It is not surprising, therefore, that radio continuum observations of this object at $\lambda\lambda 20$ and 6 cm (Pedlar et al. 1989) reveal the presence of nonthermal emission.

The possible kinship between Sgr A East and SNRs is particularly tantalizing in that SNRs are natural environments for gamma-ray production. Indeed, a comparison of 14 radio bright SNRs and 38 EGRET pointlike sources at latitudes less than 10 degrees by Esposito et al. (1997) revealed five coincidences. It is interesting to note that 3 of these EGRET SNRs (IC443, W28 and W44) have also been observed at 1720 MHz (Claussen et al. 1997). On the other hand, if the radio emission of Sgr A East is due to synchrotron, then a relativistic electron distribution with $dN/dE \sim E^{-\alpha}$ is required, where $\alpha \sim 2.5 - 3.3$. This is rather peculiar for several reasons. First, such a particle spectrum is unlike that thought to be produced directly in shock acceleration, which instead results in steady-state spectral indices $\alpha \sim 2.0 - 2.4$ (Jones & Ellison 1991). The particles may have been cooled, though one would need to show this self-consistently. Second, this situation appears to be different from that of other SNRs in which $\alpha \sim 2.2 - 2.4$ (e.g., Gaisser et al. 1998). However, one must keep in mind the different energy budgets available to Sgr A East and SNRs. An observationally significant pion production rate in Sgr A East offers the interesting possibility that the e^+e^- pion decay byproducts may produce the $\alpha \sim 3$ power-law distribution inferred from Sgr A East's radio spectrum, and these particles then also self-consistently account for the underlying γ-ray spectrum via either bremsstrahlung, or the IC scattering of the central IR and UV photons.

Owing to the uniqueness of the Galactic Center, differences between Sgr A East and the EGRET SNRs may also be due to different properties of the surrounding medium. In their detailed calculations of the relative importance of pion production, bremsstrahlung, and inverse Compton (IC) scattering (including the contribution to IC from microwave background radiation, the Galactic IR/optical background radiation and the infrared emission resulting from shock-heated dust) within the SNR shells of IC 443 and γ Cygni, Gaisser, Protheroe and Stanev (1998) showed that, at least in these SNRs, the dominant contributions to the γ-ray spectrum come from bremsstrahlung (by relativistic electrons colliding with the ambient proton distribution) and pion decay. In contrast, the presence of relatively intense UV and IR photon fields at the Galactic center suggests that IC scattering may play a dominant in the spectrum of Sgr A East.

2. Calculation

We work with the hypothesis that the EGRET γ-ray bump is due primarily to π^0 decays ($\pi^0 \rightarrow \gamma\gamma$), these being produced in *pp* scatterings between ambient protons and a relativistic population accelerated by shocks within the Sgr A East shell. In this process, a multiplicity of pions is produced, the charged members of which decay leptonically ($\pi^\pm \rightarrow \mu^\pm \nu_\mu$, with $\mu^\pm \rightarrow e^\pm \nu_e \nu_\mu$). We first find the physical conditions required to produce this γ-ray bump, and then use the associated electrons and positrons produced in the pion decays to calculate the radio spectrum. Fitting the synchrotron emission to the VLA data (see Fig. 3) is then a matter of adjusting just one unknown, viz. the average magnetic field B in the gyration region. This same population of e^+e^- also radiates by bremsstrahlung, and by inverse Compton scattering the infrared (IR) and ultraviolet (UV) photons from the nucleus.

A full description of our *pp* scattering algorithm is given elsewhere (Markoff, Melia & Sarcevic 1997, 1998). Briefly, we assume that the accelerated protons are injected into the shock at a rate of $\dot{\rho}_p(E_p) = \rho_o E_p^{-\alpha_p}$ cm^{-3} s^{-1} GeV^{-1}. The steady-state proton spectral index is thought to lie in the range of $2.0 - 2.4$ (Jones & Ellison 1991), and is determined from a diffusion loss treatment. We choose to define the injection in terms of two free parameters, the normalization ρ_o and index α_p, so that we can restrain our results by the shock efficiency and test for self-consistency. Because the protons are ultrarelativistic, the leading order nucleons produced will continue contributing to the spectrum via multiple collisions in an ensuing cascade, until they lose enough of their energy to rejoin the ambient plasma. The main products of these collisions are pions, which decay either to photons or leptons, as described above. Although the procedure for calculating the pion production rate is rather straightforward, it is complicated by the energy dependence of the pion multiplicity and the *pp* scattering cross section. There have been many *pp* and $p\bar{p}$ collider experiments over the past 25 years, and each produces a data set based on a different apparatus and different region of phase space. In many cases, two experiments at the same energy have somewhat inconsistent results due to differences in procedure, trigger specifications and normalization. We have sifted through the multitude of data sets and have chosen the largest compilation of published data available that are self-consistent (see Markoff, Melia & Sarcevic 1997, 1998, and references cited therein). With these functional forms, we calculate the steady-state proton distribution resulting from the balance of injection from the shock, cascade collisions producing secondary particles which feed back into the system, and synchrotron and inverse Compton losses. Following a similar procedure, we also solve the diffusion loss equation for the decay e^+e^-, taking into account their creation rate from $\pi^\pm$ decays, and their energy loss from radiative processes.

In Figure 1, we show the γ-ray spectrum produced from π^0 decays for two different cases, characterized by their respective values of α_p and the ambient proton density n_p in the region where the relativistic protons engage in the particle cascade described above. Of course, the full spectrum will be the sum of this component and those arising from bremsstrahlung and inverse Compton scattering, which we consider below. However, we can already see that the fit to the γ-ray data between ~ 100 MeV and 10 GeV is rather robust over a range of conditions. What does change from case to case is the energy content in the relativistic proton population, which is related to the efficiency introduced above. So, for example, when the ambient proton density n_p is low (Case 2), the removal rate of protons from the relativistic distribution is smaller than that in Case 1, resulting in a higher equilibrium density of relativistic protons. This is why the energy content of these particles is much higher in Case 2 than in Case 1, accounting for roughly 77% and 1% of a typical SN burst of 10^{51} ergs for these two situations, respectively.

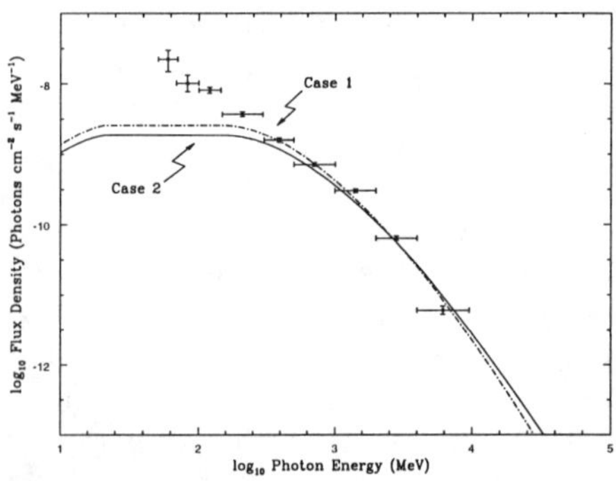

Figure 1. The γ-ray spectrum resulting solely from pion decays in the Sgr A East shell. The two cases shown here represent the range of physical conditions within which a good fit may be obtained simultaneously for the EGRET and VLA data (see text). Case 1: $n_p = 10^3$ cm^{-3}, $\alpha_p = 2.2$. Case 2: $n_p = 10$ cm^{-3}, $\alpha_p = 2.0$. In both cases, the shell is assumed to have a radius of 5 pc and a width of 1 pc. In steady state, the relativistic (shock accelerated) protons constitute a total energy of 8.8×10^{48} ergs for Case 1, and 7.7×10^{50} ergs for Case 2.

The electron distribution resulting from these two Cases is shown in Figure 2. The main result to infer from this plot is that the e^- and e^+ distributions can mimic a power-law with a spectral index ~ 3 very well, consistent with the value implied by the radio spectrum of Sgr A East. We shall see below that this single process of *pp* scatterings may therefore self-consistently account for both the EGRET γ-ray and the VLA GHz spectral components of this source, and may also produce a detectable flux at TeV energies. Apparently, the leptons produced in $\pi^\pm$ decays in this source are numerically dominant over those accelerated directly within shocks.

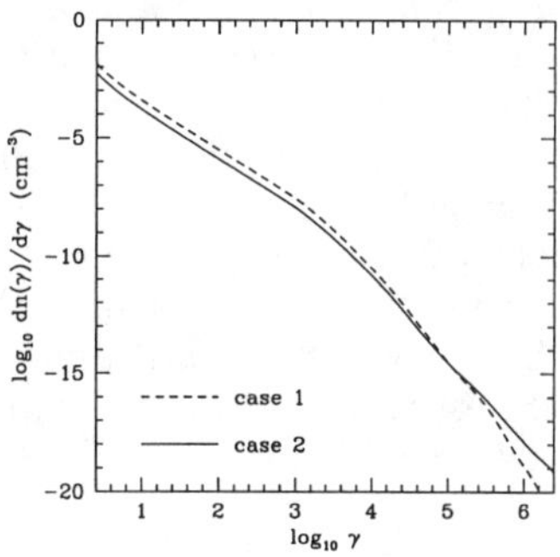

Figure 2. The steady state electron distribution resulting from pion decays for the same population that produces the spectra shown in Fig. 1. Cases 1 and 2 correspond to those defined in that figure. The positron population is the same as that shown here for the electrons.

These relativistic electrons are immersed in an intense source of UV and IR radiation from the central $1-2$ parsecs. The ring of molecular gas (also known as the Circumnuclear Disk, or CND) is rotating, and is heated by a centrally concentrated source of UV radiation (predominantly the IRS 16 cluster of bright, blue stars). The CND is a powerful source ($\approx 10^7\ L_\odot$) of mid to far-infrared continuum emission with a dust temperature of $\approx$ 100 K (e.g., Telesco et al. 1996; Davidson et al. 1992). This radiation is due to reprocessing by warm dust that has absorbed the same power in the UV (Becklin, Gatley and Werner 1982; Davidson et al. 1992). Models of the photodissociation regions in the CND require an incident dissociating flux (6 eV $< h\nu <$ 13.6 eV) of 10^2–10^3 erg cm^{-2} s^{-1} (Wolfire, Tielens & Hollenbach, 1990), implying a total UV luminosity of about $2\times10^7\ L_\odot$, consistent with the radio continuum emission from Sgr A West (Genzel et al. 1985). This intensity is also suggested by the detection of radio continuum emission from the outer envelope of IRS 7, a cool supergiant being photoionized by the UV radiation field (e.g., Yusef-Zadeh and Melia 1992), and is consistent with the inferred ionizing flux in Sgr A West, corresponding to a centrally concentrated source of 2×10^{50} ionizing photons per second (Lacy, et al. 1982; Ekers, et al. 1983; Mezger and Wink 1986).

The calculation of the inverse Compton (X-ray and γ-ray) emissivity for this known e^+e^- population bathed by the IR and UV radiation is carried out according to what is now a standard procedure (see, for example, Melia, Yusef-Zadeh & Fatuzzo 1998). For simplicity, we assume that the electron distribution is locally isotropic, which then also implies that the upscattered radiation field is emitted isotropically from within a volume $V\sim250$ pc^3 in Sgr A East, corresponding to a shell with radius $R\approx5$ pc and thickness $\Delta R\approx1$ pc. The centroid of this structure is assumed to lie at 7 pc behind the nucleus. We note, however, that unlike our previous treatment of inverse Compton scattering in this source (Melia, Yusef-Zadeh and Fatuzzo, 1998), we do not here need to invoke a γ_{min} for the leptons, using instead the full distribution displayed in Figure 2.

The remaining continuum component of importance is bremsstrahlung radiation resulting from the interaction between the relativistic leptons and the ambient (fixed target) nuclei. This component is calculated using the bremsstrahlung differential cross-section with no screening (formula 3BN in Koch & Motz 1959) for $Z = 1$ and then multiplying by a factor of 1.4 to account for a 10:1 ratio of hydrogen to helium. Thus, the overall spectrum from Sgr A East is a superposition of the γ-rays from π^0 decays, synchrotron radiation by the relativistic leptons produced during the decay of the charged pions, bremsstrahlung emission by these electrons and positrons, and their Comptonization of the IR and UV radiation from the central $1 - 2$ pc.

3. Discussion

The key result of our calculations is displayed in Figure 3, which shows the complete spectrum corresponding to Case 2 above. The magnetic field inferred from the fit through the VLA points is $\approx 10^{-5}$ Gauss. This is within a factor of two of the equipartition value of B in Sgr A East, assuming a radio luminosity of 1.8×10^{35} ergs s^{-1} arising from relativistic leptons gyrating in a magnetic field over a volume of 250 pc^3. The ASCA measurement is an upper limit, since it does not distinguish between thermal and nonthermal emission from this region. The spectrum we show here is strictly the contribution from the nonthermal particles in the Sgr A East Shell. Our predicted hard X-ray spectrum is also consistent with Ginga observations, which show evidence of a hard tail in the diffuse X-ray spectrum from the inner region of the Galaxy (Yamasaki et al. 1996). At the highest energy measurement to date, a 270 minute exposure by the Whipple observatory produced a small excess (2.5σ) of > 2.0 TeV emission from within the EGRET error box of 2EG J1746-2852, yielding an integrated flux upper limit of 0.45×10^{-11} cm^{-2} s^{-1} (Buckley et al. 1997). Although this signal is at best only 2.5σ above the noise, our calculated spectrum would be consistent with a detection at this level. We are therefore in favor of a deeper integration at this frequency to help verify the extension of Sgr A East's spectrum to the TeV domain.

Case 1 would not produce an acceptable spectrum because the larger ambient proton density ($n_p = 10^3$ cm^{-3}) results in a significantly higher lepton bremsstrahlung flux, greatly in excess of the ASCA and EGRET measurements. The spectral shape would then also not be a good fit to the data since the γ-rays from π^0 decays would be swamped by the other radiative components. In fact, we found it very difficult to fit this broadband spectrum with any ambient density in excess of ~ 10 cm^{-3}. This raises an interesting issue, since the energy content in the relativistic protons (see above) is then a large fraction ($\sim 0.5 - 0.8$) of the energy in a "standard" SN burst ($\approx 10^{51}$ ergs). It is difficult to envisage the efficiency of shock acceleration being this high. On the other hand, this relativistic energy content would constitute only $\sim 1 - 2\%$ of the available energy if the burst that produced the Sgr A East remnant was $\sim 4 \times 10^{52}$ ergs, as suggested by other observations (see above). Our results may therefore be somewhat consistent with the supposition that Sgr A East was produced not by a single, standard supernova explosion, but rather by several such events, or a single more catastrophic incident, such as the tidal disruption of a star.

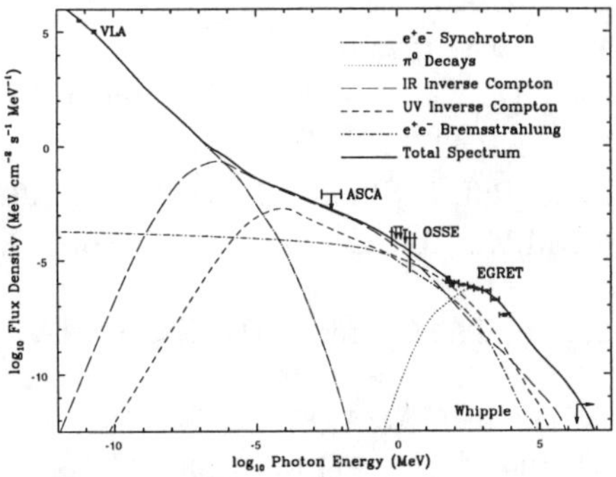

Figure 3. The broadband spectrum calculated self-consistently using the particle decay products from the fit of Case 2 in Fig. 1. The data points for this plot are from Pedlar et al. (1989) (VLA), Koyama et al. (1996) (ASCA), Purcell et al. (1997) (OSSE), Mayer-Hasselwander et al. (1998) (EGRET), and Buckley et al. (1997).

If the γ-rays detected by EGRET are indeed produced in this fashion, it seems inevitable to us that the upscattered IR and UV photons should result in a significant intensity at intermediate (i.e., $\sim 10 - 100$ keV) energies, as indicated in Figure 3. This flux density ($\sim 10^{-3}$ photons cm^{-2} s^{-1} MeV^{-1}) may be above the sensitivity limit (at $\sim 10^{-7}$ photons cm^{-2} s^{-1} MeV^{-1} for a point source) of UNEX, a proposed balloon flight instrument (Rothschild 1998). With its expected spatial resolution of several arcmin or less, it should therefore have little trouble imaging the hard X-ray counterpart to 2EGJ1746-2852, if this source is extended and is associated with Sgr A East.

Other than the overall energetics, it appears that the main distinction between Sgr A East and the EGRET SNRs is the intensity of the external radiation upscattered by the relativistic electrons and positrons. Whereas the ambient soft photon intensity in the latter is too low to produce an observationally significant spectral component compared to bremsstrahlung (Gaisser, Protheroe & Stanev 1998), the IR and UV radiation from the Galactic center does produce a dominant broadband continuum when scattering with the relativistic leptons in Sgr A East.

Acknowledgments. This research was partially supported by NASA under grant NAGW-2518 and by an NSF Graduate Fellowship, and has made use of NASA's Astrophysics Data System Abstract Service. We are also indebted to Pasquale Blasi for helpful discussions.

References

Becklin, E.E., Gatley, I. & Werner, M.W. 1982, ApJ, 258, 134.

Buckley, J.H. et al. 1997, The 25th International Cosmic Ray Conference.

Claussen, M. J., Frail, D. A., Goss, W. M. & Gaume, R. A. 1997, ApJ, 489, 143.

Davidson, J.A., et al. 1992, ApJ, 387, 189.

Ekers, R.D. et al. 1983, A&A, 122, 143.

Esposito, J.A., Hunter, S.D., Kanback, G. & Sreekumar, P. 1996, ApJ, 461, 820.

Frail, D.A. et al. 1994, ApJ, 427, L43.

Gaisser, T.K., Protheroe, R.J. & Stanev, T. 1998, ApJ, 492, 219.

Genzel, R., Crawford, M.K., Townes, C.H. & Watson, D.M. 1985, ApJ, 297, 766.

Goss, W.M. et al. 1989, in The Center of the Galaxy, IAU 136, ed. M. Morris, p345.

Güsten, R. & Downes, D. 1980, A&A, 87, 6.

Jones, F. & Ellison, D. 1991, Space Sci.Rev., 58, 259.

Khokhlov, A. & Melia, F. 1996, ApJ, 457, L61.

Koch, H. W. & Motz, J. W. 1959, Rev. Mod. Phys., 31, 920.

Koyama, K., Maeda, Y., Sonobe, T. et al. 1996, PASJ, 48, 249

Lacy, J.H., Townes, C.H. & Hollenbach, D.J. 1982, ApJ, 262, 120.

Lang, C. 1999, this volume, p. 498

Lo, K.Y. & Claussen, M.J. 1983, Nature, 306, 647.

Markoff, S., Melia, F. & Sarcevic, I. 1997, ApJ, 489, L47.

Markoff, S., Melia, F. & Sarcevic, I. 1998, ApJ, submitted.

Mayer-Hasselwander, H.A., et al. 1998, A&A, in press.

Melia, F., Yusef-Zadeh, F. & Fatuzzo, M. 1998, ApJ, in press.

Mezger, P.G. & Wink, J.E. 1986, A&A, 157, 252.

Mezger, P. et al. 1989, A&A, 209, 337.

Morris, M. & Serabyn, E. 1996, ARA&A, 34, 645.

Pedlar, A. et al. 1989, ApJ, 342, 769.

Purcell, W. R., et al. 1997, ApJ, 491, L725.

Pohl, M. 1997, A&A, 317, 441.

Rothschild, R. 1998, private communication.

Telesco, C.M., Davidson, J.A. & Werner, M.W. 1996, ApJ, 456, 541.

Wardle, M., Yusef-Zadeh, F. & Geballe, T. 1998, ApJ, submitted.

Wolfire, M.G., Tielens, A. & Hollenbach, D. 1990, ApJ, 358, 116.

Yamasaki, N.Y., et al. 1996, AAS, 120, 393.

Yusef-Zadeh, F. & Melia, F. 1992, ApJ, 385, L41.

Yusef-Zadeh, F. & Morris, M. 1987, 320, 545.

Yusef-Zadeh, F., Roberts, D.A., Goss, W.M., Frail, D.A. & Green, A.J. 1996, ApJ, 466, L25

Yusef-Zadeh, F., Roberts, D. A., Goss, W.M., Frail, D.A. & Green, A.J. 1998, ApJ, submitted.

The Central Parsecs of the Galaxy
ASP Conference Series, Vol. 186, 1999
H. Falcke, A. Cotera, W.J. Duschl, F. Melia, M.J. Rieke, eds.

Cold Matter in the Galactic Center Region Measured with the X-ray Satellite ASCA

M. Sakano[1], M. Nishiuchi[1], K. Koyama[2] and J. Yokogawa

Department of Physics, Kyoto University Sakyo, Kyoto 606-8502, Japan

Y. Maeda[1]

Department of Astronomy and Astrophysics, Pennsylvania State University, University Park, PA 16802-6305, U.S.A.

Abstract. We report the cold matter distribution in the Galactic Center region. We have measured the column density of X-ray binaries in the Galactic Center using the X-ray satellite ASCA, and derive the total mass and distribution of cold matter in this region. The column densities are given by a smooth function of angular distance from the Galactic Plane. Assuming a disk-like mass distribution with 500 pc radius, we estimate the total mass to be (1–several)$\times 10^7 M_\odot$. Together with results at other wavelengths, we discuss the physical conditions of the cold matter in the Galactic Center region.

1. Introduction

The interstellar matter (ISM) in the Galactic Center region is known to be quite unusual and complicated. Radio molecular line and far-infrared observations have revealed that there exist many dense molecular clouds, which have a rather high velocity dispersion in the Galactic Center region (e.g., Morris & Serabyn 1996). On the other hand, hard X-ray observations have revealed that this region is filled with unusually high temperature plasma ($kT \sim$ 10keV), of which the pressure is also quite high $nT \sim 10^7$ K cm^{-3} (Koyama et al. 1989, Yamauchi et al. 1991, Koyama et al. 1996). Moreover, places in which the magnetic strength is measured all show high magnetic strength of B ~mG, hence the whole Galactic Center region may be possibly filled with a strong magnetic field.

To approach the physical state in this region, intensive studies at all available wavelengths are essential. However, our knowledge about the ISM in this region is still limited, probably because the Galactic Center region is veiled with dense dust grains. The recent developments of hard X-ray detectors have made X-ray studies possible for which obscuration in the Galactic Plane is not a problem.

---

[1]Research Fellow of the Japan Society for the Promotion of Science.

[2]CREST: Japan Science and Technology Corporation (JST)

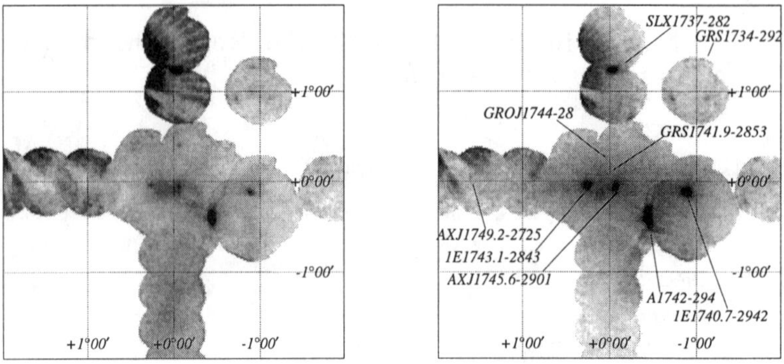

Figure 1. Global *ASCA* images in the Galactic Center region of $4° \times 4°$ with the energy bands of (left) 0.7–3 keV and (right) 3–10 keV. The coordinates are galactic; the central horizontal line is the Galactic Plane. The images are corrected for exposure, vignetting and grid efficiency with GIS after subtraction of the non-X-ray background. The color levels are logarithmically spaced. Some pointings including too bright transient sources are excluded from this image, for example, the pointings including the transient GRO J1744−28 (they also include the transient GRS 1741.9−2853), or the variable source AX J1749.2−2725. Since GRS 1734−292 was detected closely near the edge of the detectors, it cannot be seen in the mosaic image, in which the most outer regions of the detectors are cut.

X-ray measurements of the cold ISM in the line of sight to a certain source is based on the low energy cut-off in the spectrum. With this method, one can directly measure the total matter quantity in the line of sight to the source because the X-ray measurement is independent of the physical or chemical conditions in the ISM, such as, density, temperature, or composition of the matter (atomic, molecular, or dust).

We here apply this new method of X-ray measurements of the cold matter distribution to the Galactic Center region. Then we discuss the physical condition of the ISM in this region, comparing our result with results at other wavelengths. We assume the distance to the Galactic Center to be 8.5 kpc.

2. Observations

The X-ray satellite *ASCA* has observed the Galactic Center region recurrently since 1993 fall. We used almost all the pointing data available: 33 pointings where $-2° < l_{\rm II} < 2°$ and $-2° < b_{\rm II} < +2.5°$ with total exposure of about 1 Msec. The exposure time of each pointing varied highly from pointing to pointing (8–1000 ksec).

ASCA is equipped with two Gas Imaging Spectrometers (GISs) and two Solid-State Imaging Spectrometers (SISs) both located separately at foci of four identical thin-foil X-ray telescopes (XRTs). Details of this instrumentation are found in Tanaka et al. (1994).

3. Results

3.1. Images

Figure 1 shows the mosaic image of the Galactic Center region. The most prominent features in the images are X-ray stars, probably X-ray binaries. For the bright X-ray stars, the spectra are not much influenced by fluctuations of the diffuse background plasma, which spreads over the entire field of the Galactic Center region. In addition, the X-ray spectra of X-ray stars are known to be generally featureless and simple. Hence we can extract the column density to each X-ray star rather accurately.

3.2. X-ray Measurement of the Column Density to Each Source

We picked out bright X-ray stars with hard spectra, and measured their X-ray absorption, *i.e.*, column density of cold matter. Since hard X-rays are nearly transparent to the Galactic Center, the column densities to these hard X-ray sources probably reflect the cold ISM mass distribution in the Galactic Center region if these sources are distributed symmetrically around the Galactic Center.

We extracted the spectra from sources for which no reports with *ASCA* had been published. Since most sources are located out of SIS field of views, we used GIS data only. Those spectra were found to be featureless with no emission line. Thus, we fitted each spectrum with an absorbed power-law model. Figure 2 shows the spectra with the best-fit model, whereas Table 1 shows the fitting results.

Table 1. Fitting results of the X-ray stars

Sources	Γ^a	N_H (10^{22} H cm^{-2})	$F_X{}^b$ (10^{-11} erg s^{-1} cm^{-2})
1E 1743.1−2843	1.65±0.15	19±1	12
GRS 1741.9−2853	2.3±0.2	11.4±0.9	10
A1742−294	1.8±0.05	5.7±0.5	50
SLX 1737−282	2.1±0.1	1.86±0.16	6
GRS 1734−292	1.3±0.15	1.7±0.3	4

[a]Photon index.

[b]Observed flux within 2–10 keV. Since all the sources are variable, we only show the flux measured in one of the observations.

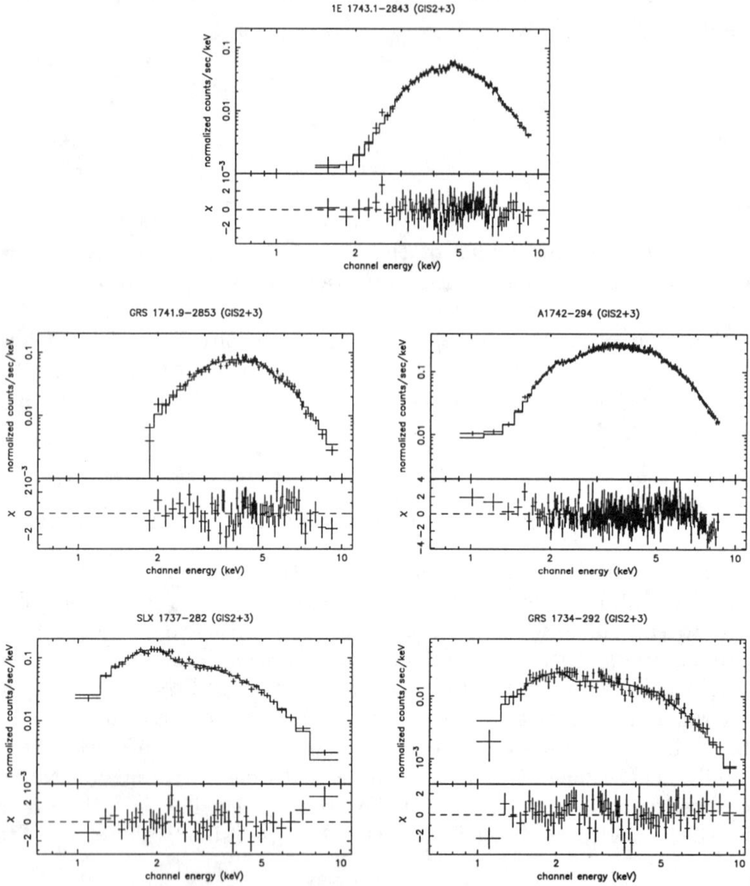

Figure 2. Spectrum of each X-ray star taken with GIS2+3: (Top-left) 1E 1743.1−2843; (top-right) GRS 1741.9−2853; (middle-left) A 1742−294; (middle-right) SLX 1737−282; (bottom) GRS 1734−292. The best-fit model is overlaid when we fitted each spectrum with absorbed power-law model.

Table 2. Column densities of X-ray stars ($|l_{II}| < 2°$)

Sources	l_{II} (deg.)	b_{II} (deg.)	$\|b'\|$ (deg.)	N_H (10^{22} H cm^{-2})	Ref.[a]
AX J1745.6−2901[b]	359.92	−0.04	0.002	26^{+1}_{-2}	(2)
1E 1743.1−2843	0.25	−0.03	0.02	19^{+1}_{-1}	(1)
1E 1740.7−2942	359.14	−0.12	0.08	10^{+1}_{-2}	(3)
AX J1749.2−2725	1.70	0.12	0.16	10^{+3}_{-3}	(4)
GRS 1741.9−2853	359.96	0.13	0.17	$11.4^{+0.9}_{-0.9}$	(1)
SAX J1747.0−2853[c]	0.21	−0.23	0.19	9^{+1}_{-1}	(5)
A1742−294	359.58	−0.38	0.33	$5.7^{+0.5}_{-0.5}$	(1)
GRO J1744−28	0.04	0.30	0.35	$6.1^{+0.9}_{-0.9}$	(6)
SLX 1737−282	359.97	1.25	1.30	$1.86^{+0.16}_{-0.16}$	(1)
GRS 1734−292	358.87	1.40	1.45	$1.7^{+0.3}_{-0.3}$	(1)
SLX 1735−269	0.79	2.40	2.45	$1.3^{+0.15}_{-0.15}$	(7)

[a] (1)this work; (2)Maeda et al. 1996; (3)Sakano et al. 1998; (4)Torii et al. 1998; (5)Sidoli et al. 1998 (6)Nishiuchi et al. 1998; (7)David et al. 1997;

[b] It was identified as A1742−289 in reference (2). However, Kennea and Skinner (1996) indicated that the identification is not correct and it is actually a newly discovered source with *ASCA*, AX J1745.6−2901.

[c] There is no *ASCA* observation for this source, hence the result with SAX is referenced.

3.3. Estimate of the Cold Mass Distribution in the Galactic Center Region

Table 2 summarizes the column densities for the hard X-ray stars in the Galactic Center region. For sources with published *ASCA* data, we adopted the column density from the references.

We plot the column density to the angular distance from the Galactic Plane, $|b'|$, in Figure 3, where $b' \equiv b_{II} + 0.0460°$ because the galactic latitude of Sgr A* is $-0.0460°$. The smooth profile in Figure 3 suggests that these X-ray sources are located actually in the Galactic Center region.

In order to derive the mass distribution in the Galactic Center region, we need to subtract the disk mass component in front of the Galactic Center. We adopted the Galactic interstellar mass model from Olling & Merrifield (1998), as mentioned below.

According to the disk mass model, the hydrogen density in the disk, $n_H(r, z)$, is a function of the Galactocentric radius (r) and the vertical distance from the Plane (z). The radial dependence is described as,

$$\begin{aligned} n_H(r) = & \ 20\rho(R_0) && (0 < r \leq \frac{R_0}{10}) \\ = & \ 2\rho(R_0)R_0 \cdot \tfrac{1}{r} && (\frac{R_0}{10} \leq r \leq \frac{R_0}{2}) \end{aligned}$$

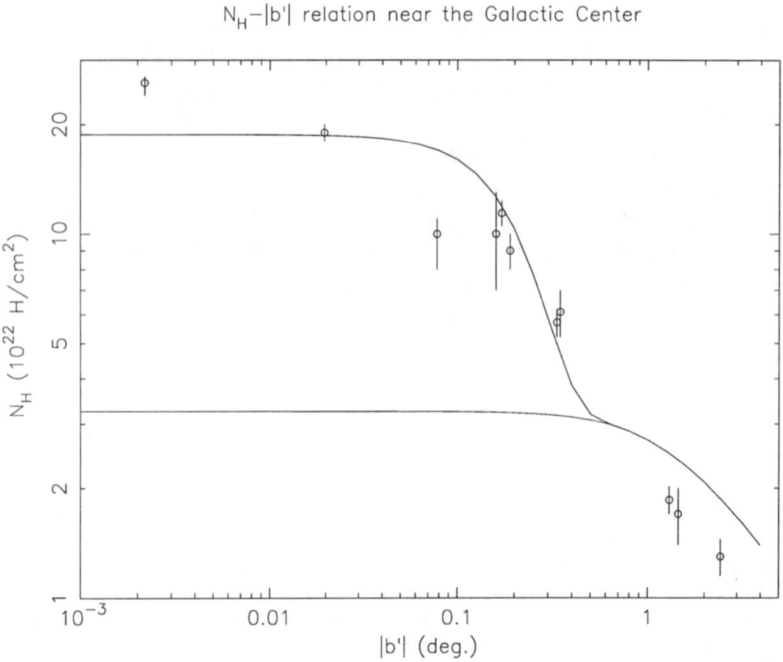

Figure 3. The column densities of X-ray stars as a function of the angular distance from the Galactic Plane ($b' \equiv b_{II} + 0.0460°$). The upper solid line shows the best-fit model. The lower solid line shows the profile of the assumed disk component.

$$= \quad 2\rho(R_0)R_0{}^2 \cdot \tfrac{1}{r^2} \quad (\frac{R_0}{2} \leq r \leq R_0), \qquad (1)$$

where R_0 is the distance to the Galactic Center from the Sun and $\rho(R_0)$ is the hydrogen density near the Sun. We adopted $\rho(R_0)$ to be 0.2 H cm^{-3} according to Dame et al. (1987). The dependence of the density for the vertical direction to the disk was assumed to be a Gaussian distribution with $\sigma = 74$pc, *i,e.*, half width of half maximum is 87pc, according to Dame et al. (1987). The lower solid line in Figure 3 represents the profile of the assumed disk mass model.

Then we fitted the profile in Figure 3 with an exponential profile in addition to the Galactic interstellar mass model. The exponential profile represents the ISM distribution in the Galactic Center region on the assumption that all these sources are located at the same distance as the Galactic Center.

We overlaid the best-fit model profile in Figure 3. The best-fit peak hydrogen column density was found to be about 1.9×10^{23}H cm^{-2}, and the scale width, 0.16° (1σ), corresponding to 24pc. This width is consistent with that derived from dense molecular observations with radio (e.g., Tsuboi et al. 1998).

Then we estimated the total ISM in the Galactic Center region, assuming a 500-pc radius disk, in which the mass distribution depends only on latitude. The derived total mass from the best-fit parameters is $4\times10^7 M_\odot$. The statistical uncertainty of this estimate is rather small because the estimated mass is not sensitive to the fitting parameters, but is directly derived from the profile if we only give the mass distribution shape for an azimuthal distribution. Hence, we conclude that the total ISM in the Galactic Center region is (a few – several)$\times10^7 M_\odot$. Note that we have implicitly assumed a solar abundance ISM. The possibility of an abundance anomaly is discussed later on in Section 4.2.

In Figure 3, one can see the slight inconsistency between the data and the model for the outer three data points, which implies that the assumed disk mass model should be slightly reduced. However, such a change would cause no serious influence on our estimate of the mass profile in the Galactic Center region.

4. Discussion

4.1. Past X-ray Observations in this Region

Past X-ray observations have suggested that the averaged column density to X-ray sources in this region is about several $\times10^{22}$H cm^{-2} (e.g., Pavlinsky et al. 1994). However, our result revealed that the column density distribution follows a certain function depending on the angular distance from the Plane, and accordingly expects that the sources with a few tens arc-minutes offset from the Plane have $N_{\rm H} \sim$ several $\times10^{22}$H cm^{-2} as previously suggested, however the sources near the Plane have about $N_{\rm H} \sim \times10^{23}$H cm^{-2} or larger. This formalization may be a good indicator whether a source in this region is located actually in or before this region.

4.2. Abundance Problem

X-ray measurements of column density depend on the abundance. In the above analysis, we assumed the abundance to be the same as the solar value. Therefore, if the abundance in the Galactic Center region is significantly different from the solar value, our result should be modified.

Many measurements of the galactic abundance found an abundance gradient along the radial direction from the Galactic Center to the outer part. Thus the abundance in the Galactic Center region is expected to be at least a few times larger than the solar value (e.g., Majewski 1993).

Recent X-ray measurement by Murakami et al. (1998) of the X-ray reflection nebula Sgr B2 suggested the abundance of iron and lighter elements in the Galactic region to be two times of the solar value or larger. On the contrary, recent infrared observations of giant stars in the Galactic Center region by Sellgren et al. (1997) and Ramirez et al. (1998) derived the iron abundance to be consistent with the solar value.

Thus, the abundance in the Galactic Center region is still uncertain. However, it may be at least consistent with, or larger than the solar value. The column density measured with X-ray absorption is basically anti-proportional to the abundance, *i.e.*, if the abundance in the Galactic Center region is larger than the solar value, which may be the real case, the derived column density profile and accordingly the estimated total ISM mass in the Galactic Center region will decrease. Consequently, if we take into account the possibility of an abundance anomaly, the derived total mass is (1–several)$\times 10^7 M_\odot$.

4.3. Comparison with Other Wavelengths

Many efforts have been made to measure the total ISM in the Galactic region at various wavelengths. Blitz et al. (1985) estimated at upper-limit to the mass of $6 \times 10^7 M_\odot$ with γ-ray observations. Infrared observations by Cox & Laureijs (1989) also support this result. However, classical CO observations (Bania 1986) estimated it at a few $\times 10^8 M_\odot$ or larger, which is apparently inconsistent with γ-ray or infrared estimates. Recently, Oka et al. (1998) and Dahmen et al. (1998) derived the total mass with molecular observations to be (2–6)$\times 10^7 M_\odot$, assuming that the molecular clouds in the Galactic Center region are in pressure equilibrium, not in self-gravitational equilibrium as is the usual assumption. Their result is consistent with the γ-ray and infrared results, but inconsistent with previous radio results.

We have estimated the total ISM in the Galactic Center region with X-rays, which is a fully independent method compared to those at other wavelengths. Moreover, being independent of the physical condition of the ISM is a major advantage of this method. The total mass that we estimated is (a few – several)$\times 10^7 M_\odot$. It strongly supports the value from the γ-ray, infrared, and recent molecular observations. Therefore, these results all together implies that the cold ISM in the Galactic Center region is not in self-gravitational, but in pressure equilibrium. This is probably due to the high pressure of a surrounding high-temperature plasma or due to a strong magnetic field, or possibly both.

Acknowledgments. We are grateful to the *ASCA* team, especially to the *ASCA* Galactic Plane Survey team whose effort made this study possible. We also thank Drs. T. Oka and H. Sogawa for discussion. M. S., M. N. and Y. M. acknowledge support from the Japan Society for the Promotion of Science for Young Scientists.

References

Bania, T. M 1986, ApJ, 308, 868

Blitz, L., Bloemen, J. B. G. M., Hermsen, W., & Bania, T. M. 1985, A&A, 143, 267

Cox, P. & Laureijs, R. 1989, IAU symp.136: The Center of the Galaxy, p.121

Dahmen, G., Huttemeister, S., Wilson, T. L., & Mauersberger, R. 1998, A&A, 331, 959

Dame, T. M., Ungerechts, H., Cohen, R. S., de Geus, E. J., Grenier, I. A. et al. 1987, ApJ, 322, 706

David, P., Goldwurm, A., Murakami, T., Paul, J., Laurent, P. et al. 1997, A&A, 322, 229

Kennea, J. A. & Skinner, G. K. 1996, PASJ, 48L, 117

Koyama, K., Awaki, H., Kunieda, H., Takano, S. et al. 1989, Nature, 339, 603

Koyama, K., Maeda, Y., Sonobe, T, Takeshima, T. et al. 1996, PASJ, 48, 249

Maeda, Y., Koyama, K., Sakano, M., Takeshima, T., & Yamauchi, S. 1996, PASJ, 48, 417

Majewski, S. R. 1993, ARA&A, 31, 575

Morris, M. & Serabyn, E. 1996, ARA&A, 34, 645

Murakami, H., Koyama, K., Sakano, M., Tsujimoto, M., & Maeda, Y. 1998, ApJ, submitted

Nishiuchi, M., Koyama, K., Dotani, T., Nagase, F. et al. 1998, ApJ, submitted

Oka, T., Hasegawa, T., Hayashi, M., Handa, T., & Sakamoto, S. 1998, ApJ, 493, 730

Olling, R. P. & Merrifield, M. R. 1998, MNRAS, 297, 943

Pavlinsky, M. N., Grebenev, S. A., & Sunyaev, R. A. 1994, ApJ, 425, 110

Ramirez, S. V., Sellgren, K., Carr, J. S., Balachandran, S. C., Blum, R. D., & Terndrup, D. 1998, in this proceeding

Sakano, M., Koyama, K., Imanishi, K., Tsujimoto, M., & Maeda, Y. 1998, ApJ, submitted

Sellgren, K., Carr, J. S., & Balachandran, S. C. 1997, IAU Symp. 184, The Central regions of the Galaxy and Galaxies, ed. Y. Sofue (London: Kluwer Academic Publishers), 21

Sidoli, L., Mereghetti, S., Israel, G. L., Cusumano, G., Chiappetti, L., & Treves A. 1998, A&A, 336L, 81

Tanaka, Y., Inoue, H., & Holt, S. S. 1994, PASJ, 46, L37

Torii, K., Kinugasa, K., Katayama, K., Kohmura, T. et al. 1998, ApJ, in press

Tsuboi, M., Handa, T. & Ukita, N. 1998, ApJS, in press

Yamauchi, S., Kawada, M., Koyama, K., Kunieda, H., & Tawara, Y. 1990, ApJ, 365, 532

The Central Parsecs of the Galaxy
ASP Conference Series, Vol. 186, 1999
H. Falcke, A. Cotera, W.J. Duschl, F. Melia, M.J. Rieke, eds.

ASCA Observations of the Sgr A Region

Y. Maeda

Department of Astronomy and Astrophysics, 525 Davey Laboratory, The Pennsylvania State University, University Park, PA 16802-6305, U.S.A.

K. Koyama[1], H. Murakami and M. Sakano

Department of Physics, Graduate School of Science, Kyoto University, Sakyo-ku, Kyoto, 606-8502, Japan

Abstract. *ASCA* observed the Galactic Center Sgr A in 1993 September, 1994 September and 1997 March. We detected one bright spot whose peak coincided with the position of Sgr A*. However this spot was dominated by diffuse emission. The flux and spectral shape of this peak were the same through the three observations. Thus we could derive that the upper limit of absorption-corrected luminosity $L_{\mathrm{x,SgrA*}}$ is 10^{35} erg s^{-1}. Since AGN activity usually accompanies intense hard X-rays and variability, we conclude that present activity of Sgr A* has been quiescent for the past several years.

1. Introduction

Sgr A is the complex radio source at the central region of our Galaxy. The compact non-thermal radio source SgrA* is located at the dynamical center of the Sgr A region. Since dynamical measurements indicate a dark mass of $\sim 3\times10^6$ $M_\odot$ within $\sim$0.1 pc centered at Sgr A*, Sgr A* is a candidate of a nuclear massive blackhole (Eckart & Genzel 1997 and reference therein). However, in spite of the extensive observations, direct evidence for AGN activity, i.e., compact emission exceeding the Eddington limit of a neutron star, is still lacking from Sgr A* (for a recent review, Morris & Serabyn 1996).

AGN activity usually accompanies intense hard X-rays (ex. Awaki et al. 1991; Ueno 1996). In the hard X-ray wavelength, many observations near Sgr A* had been carried out before *ASCA* observations. To compare the flux with these observations, we assumed that Sgr A* has a 10 keV thermal bremsstrahlung with $N_{\mathrm{H}} = 1\times10^{23}$ H cm^{-2}, and converted the reported X-ray fluxes into the 3–10 keV band fluxes. The Spacelab-2 observations in 1985 July–August gave a 3–10 keV flux 3×10^{-11} erg s^{-1} cm^{-2} (Skinner et al. 1987). The SPARTAN observation, made about 6 weeks earlier than the Spacelab-2 mission, detected no significant flux with an upper limit of 1×10^{-11} erg s^{-1} cm^{-2} (Kawai et al. 1988). In

[1]CREST, Japan Science and Technology Corporation (JST), 4-1-8 Honmachi, Kawaguchi, Saitama, 332-0012, Japan

1990 March and April, Syunyaev et al. (1991) reported a source within $1'$ of the nucleus, whose 3–10 keV fluxes corresponded to 10 $\times 10^{-11}$ erg s^{-1} cm^{-2} and 16 $\times 10^{-11}$ erg s^{-1} cm^{-2}, respectively. In 1991 September–October, the emission from this region was at the same level in the 3–10 keV band. Therefore, the previous observations suggest a long-term time variability of Sgr A* with the 3–10 keV flux ranging (1–16) $\times 10^{-11}$ erg s^{-1} cm^{-2}, which corresponds to an absorption-corrected luminosity of (1–20) $\times 10^{35}$ erg s^{-1} at a distance of 8.5 kpc[1].

The *Ariel 5* sky survey made clear that a significant fraction of Seyfert 1 shows a large amplitude ($\delta I/I > 1$) variability on time scales of days (Marshall, Warwick & Pounds 1981). Recent re-analysis of the *EXOSAT* data (Grandi et al. 1992) revealed that $\sim$ 40% of AGNs in a hard X-ray selected sample show variability on a time scale of less than one day. On a longer time scale (typically weeks to months) 97% of the same sample showed significant variability. The reported X-ray variability from Sgr A* was thus believed to be a phenomenon of currently weak AGN activity.

The first *ASCA* observation near to Sgr A* was made in 1993 autumn. In the SIS/*ASCA* hard-band image, Koyama et al. (1996) detected the two bright spots within the inner 2 arc min of the Sgr A region. One spot was $1'.3$ away from Sgr A*. During the first follow-up observation in 1994 autumn, Maeda et al. (1996) discovered an X-ray burst and an 8.4 hr period dip from this spot, establishing that this spot is a new eclipsing low-mass X-ray binary AX J1745.6−2901 (Kennea & Skinner 1996). The peak of the other spot coincided the position of Sgr A*. However, this spot was dominated by diffuse emission (hereafter "Sgr A-diffuse"). No time variability from Sgr A-diffuse was detected through the two observations.

Since AX J1745.6−2901 is only $1'.3$ away from Sgr A*, previously reported X-ray fluxes of Sgr A* with non-imaging instruments might have suffered contamination due to AX J1745.6−2901. Therefore the variability reported by the previous X-ray observations should be tested with repeated *ASCA* monitoring observations.

On the contrary to these *ASCA* results against current AGN activity of Sgr A*, some possibilities of past AGN activities were reported through the analysis of the diffuse X-ray emission detected with *Ginga* and *ASCA*.

One is the large-scale hot-plasma of about 100 pc diameter, which is associated with prominent $K\alpha$ lines of He- or H-like irons (Koyama et al. 1989; Yamauchi et al. 1990; Koyama et al. 1996). The total energy of the large-scale plasma is as large as 10^{54} erg, with a dynamical age of $\sim 10^5$ yr. A possible origin is an energetic explosion of about 10^{54} erg which took place in the Galactic Center region, possibly Sgr A*, during the last 10^5 yr.

With *ASCA*, Koyama et al. (1996) found another small diffuse structure near Sgr A "Sgr A-diffuse", which exhibits K-lines from highly ionized atoms. We found a high temperature plasma inside the Sgr A shell (an oval region of $\sim 2' \times 3'$ across), with a total energy and a dynamical age of 10^{50} erg and 10^3 yr, respectively.

---

[1]Hereafter we adopted the distance of the Galactic Center Sgr A to be 8.5 kpc.

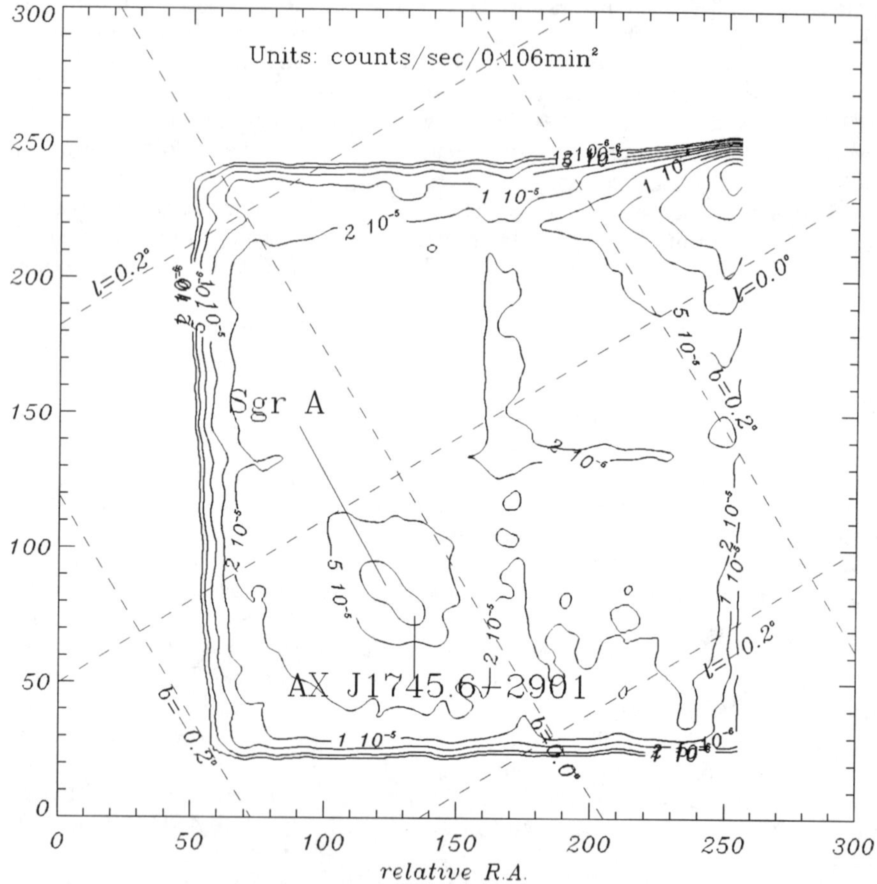

Figure 1. Contour map of the smoothed SIS image observed in 1997 spring. The energy band covers 3–10 keV. No corrections for vignetting and background-subtraction were applied.

ASCA further found extended emissions of the 6.4-keV line, the brightest region of which is located near the giant molecular cloud Sgr B2, with the shifted X-ray peak to the Sgr A side from the peak of the molecular gas of Sgr B2 (Koyama et al. 1996; Murakami et al. 1998). The 6.4-keV line is attributable to fluorescent Kα lines of neutral or low ionized iron. The shifted X-ray peak indicates that an irradiating X-ray source should be in the direction of Sgr A. Nevertheless, no bright source is found in this direction.

These facts led Koyama et al. (1996) to suspect that Sgr B2 was irradiated by the Sgr A* X-rays, which is presently dim but was bright some hundred years ago, the light traveling time between Sgr B2 and Sgr A*. This putative activity in the past may also explain Sgr A-diffuse, because the dynamical age of the plasma (about 10^3 yr) is estimated to be close to the light traveling time (Koyama et al. 1996).

These possibilities of the past AGN activity of Sgr A* imply upcoming high-energy release from Sgr A*, which encourages *ASCA* monitoring observation. Based on the above motivation, we performed a third Sgr A observation (second follow-up observation) in 1997 spring. Although we have already published a part of the results for the first two observations (Koyama et al. 1996; Maeda et al. 1996), we report our results with them.

2. Observations

The *ASCA* observations of the Sgr A region were performed three times in 1993 autumn, 1994 autumn and 1997 spring. *ASCA* has on board two *Solid-state Imaging Spectrometers* (SIS) and two *Gas Imaging Spectrometers* (GIS) instruments which simultaneously view the target (Tanaka, Inoue & Holt 1994).

3. SIS Image

Figure 1 shows a SIS image taken in 1997 spring. Sgr A is located at the center of the south-east chip. Figure 1 shows two bright spots in the Sgr A region. One is AX J1745.6−2901. The other is Sgr A-diffuse, which appears as a diffuse structure with a peak at the position of Sgr A*. The bright source GRO J1744−28 is located at the north-west corner. We can see the spill-over X-rays from GRO J1744−28 over the entire CCD chips. In this paper, we report *ASCA* results of Sgr A-diffuse. The result for GRO J1744−28 is given by Nishiuchi et al. (1998).

4. SIS Spectra

4.1. Data Reduction

For the three observations, we made the SIS spectra from a "source region" of $2'.5$ radius circle centered on Sgr A*, excluding an area of $0'.8$ radius from AX J1745.6−2901. We made the background spectra from a $2'.5$–$4'.0$ radius annular "background region" centered on Sgr A*. We subtracted the background by normalizing the integrated area.

We estimated the spill-over X-rays to the source region from nearby bright sources AX J1745.6−2901 and GRO J1744−28 with the angular response of XRT. Using this background, we found that we can remove the spill-over X-rays from GRO J1744−28. However, about 1/20 of the total X-ray photons of AX J1745.6−2901 still remained in the background-subtracted spectra.

The flux of AX J1745.6−2901 in the high state (1994 autumn) was an order of magnitude larger than that of Sgr A-diffuse. The background-subtracted spectrum during the persistent phase of AX J1745.6−2901 in the high state included a large amount of the spill-over X-rays from AX J1745.6−2901 which became comparable to the events of Sgr A-diffuse itself. In order to reduce the remaining contamination, we used the spectrum during the eclipse phase of AX J1745.6−2901 for the 1994 autumn observation. Since AX J1745.6−2901 was in the low state in 1993 autumn and 1997 spring, we used the data both in the eclipse and persistent phases. However the spill-over X-rays from AX J1745.6−2901 still remained in the background-subtracted spectra. Therefore we further subtracted these remained spill-over X-rays using the best-fit value of AX J1745.6−2901.

Inversely, the spill-over X-rays from Sgr A-diffuse to the background region are subtracted from the source flux. However, the over-subtraction by these spill-over X-rays is roughly 15%. Therefore, we ignore this over-subtraction. The background-subtracted spectra are shown in Figure 2.

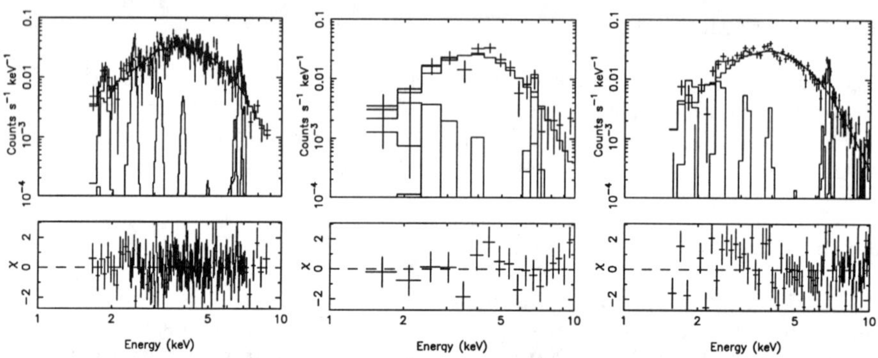

Figure 2. The SIS spectra of Sgr A-diffuse. Top, middle and bottom panels show the spectra obtained with the 1993 autumn, 1994 autumn and 1997 spring observations, respectively. Error bars are 1 σ. The solid line corresponds to the best-fit value with the multi-line fits as summarized in Tables 1 & 2. Fit residuals are shown at the bottom of each panel.

4.2. Model Fitting

We first analyze the spectrum in 1993 autumn because this spectrum was least contaminated by the spill-over X-rays. As Koyama et al. (1996) have already reported, several emission lines, which can provide key information for the origin of the X-rays, were found in the spectrum (Figure 2). We determined the line energy, flux and equivalent width, which are summarized in Table 1. We confirmed Kα lines from helium-like ions of silicon, sulphur, argon, calcium and iron atoms. We also confirmed that the 6.4 keV line from neutral iron atoms was not detected in the spectrum. The X-rays from Sgr A-diffuse would come from a hot gas filling the interstellar space with thermal energy of about 2×10^{50} ergs. A possible origin of this plasma is a remain of the past activity at the Galactic Center. Since detailed discussion has been given by Koyama et al. 1996, we move to study the time variability of Sgr A-diffuse.

4.3. Long-term Time Variability

In order to study the flux variations of Sgr A-diffuse, we fitted the spectra obtained by the follow-up observations (1994 autumn & 1997 spring) with the same model given in Table 1. For simplicity, we fixed the model parameters to

Table 1. Multi-line fits to the Sgr A-diffuse spectrum in 1993 autumn

——Continuum——			
N_H (H cm^{-2})		$8.1(6.9–9.4)\times10^{22}$	
kT (keV)		9.5(6.0–20.3)	
Normalization[a]		$8.7(7.3–11.4)\times10^{-3}$	
——Emission lines——			
(Element)	(Line name)	(Line center) (keV)	(Line flux) (ph s^{-1} cm^{-2})
Si	He-like Kα	1.86(fix[b])	$1.7(0.7–2.7)\times10^{-5}$
S	He-like Kα	2.45(fix[b])	$3.8(2.2–5.4)\times10^{-5}$
Ar	He-like Kα	3.14(fix[b])	$1.6(0.1–3.0)\times10^{-5}$
Ca	He-like Kα	3.90(fix[b])	$<2.2\times10^{-5}$
Fe	Neutral Kα	6.4(fix[b])	$<3.4\times10^{-5}$
	He-like Kα	6.7(fix[b])	$1.3(1.0–1.7)\times10^{-4}$
	H-like Kα	6.97(fix[b])	$<5.6\times10^{-5}$
——X-ray flux & luminosity——			
F_x(2–10 keV) (erg s^{-1} cm^{-2})		1.9×10^{-11}	
L_x(2–10 keV) (erg s^{-1})		2.9×10^{35}	
χ^2(d.o.f.)		135.5(129)	

The observed line flux was determined by fitting a model of 7 Gaussian lines and an absorbed thermal bremsstrahlung to the data in 1.5 - 10 keV band. Line width was fixed to zero because of the limited statistics. Errors are 90% level.

a: $3.02 \times 10^{-15} \int n_e n_i dV \ / \ (4\pi D^2)$, where n_e is the electron density (cm^{-3}), n_i is the ion density (cm^{-3}), and D is the distance to th e source (cm). $n_e = 1.08 \times n_i$.

b: Line energy was fixed to the theoretically predicted value because of the limited statistics.

the best-fit values given in Table 1 but allowed the global normalization to vary. The best-fit parameters are given in Table 2 & Figure 2. These models reproduce the spectral shapes of the two spectra obtained with the follow-up observations, which have the characteristics of a flat continuum and many emission lines with low-energy absorption.

The best-fit fluxes of the three observations are almost the same within statistical errors. Therefore we can safely conclude that the flux and spectral shape of Sgr A-diffuse are nearly constant with the flux level around 2×10^{-11} erg s^{-1} cm^{-2} (2–10 keV) among the three observations.

5. X-ray Activity of Sgr A*

The peak position of Sgr A-diffuse coincided with Sgr A* and exhibited the diffuse structure. Therefore we can estimate the upper limit flux of Sgr A* using the X-ray flux of Sgr A-diffuse, i.e.

$$F_{\rm x,SgrA*} \lesssim 2 \times 10^{-11} {\rm erg\ s^{-1}\ cm^{-2}}\ (2 - 10{\rm keV}). \qquad (1)$$

Table 2. Fluxes of Sgr A-diffuse for the three observations

Observation date	$\mathrm{Flux}_{2-10\mathrm{keV}}{}^{a}$ (erg s^{-1} cm^{-2})	χ^2(d.o.f.)
1993 autumn	$2.09(\pm 0.09)\times 10^{-11}$	135.5(138)
1994 autumn	$1.92(\pm 0.27)\times 10^{-11}$	15.95(17)
1997 spring	$1.90(\pm 0.06)\times 10^{-11}$	136.1(71)

Errors are 90% level. All the errors include only the statistical errors.
a: The systematic decrease of the detection-efficiency in the 4-CCD mode was corrected. The decreasing rates are 8% and 22% in 1993 autumn and 1994 autumn, respectively. For detail, see Maeda (1998).

The X-ray luminosity $L_{\mathrm{x,SgrA*}}$ at a distance of 8.5 kpc corrected for interstellar absorption of 8×10^{22} H cm^{-2} (Table 1) is given by

$$L_{\mathrm{x,SgrA*}} \lesssim 10^{35} \mathrm{erg\ s^{-1}}\ (2 - 10\mathrm{keV}). \tag{2}$$

The upper limit of the X-ray luminosity from Sgr A* is much less than the Eddington luminosity 10^{38} erg s^{-1} of a stellar-mass ($\sim 1\ M_{\odot}$) black hole. We confirmed that the present activity of Sgr A* is much quieter than that of AGNs.

We also detected no flux variations larger than its flux-level (2×10^{-11} erg s^{-1} cm^{-2}) from Sgr A-diffuse with various time scale. Therefore, we find no flux variability with various time scale from Sgr A*. Since we find neither luminous X-rays nor X-ray flux-variability from Sgr A*, we do not find any direct evidence for observational similarity between Sgr A* and AGNs. We can conclude that Sgr A* has been quiescent, at least, during the past several years.

Although *ASCA* detected no direct hard X-rays from Sgr A*, the soft X-ray telescopes *Einstein* and *ROSAT* with higher angular resolution found a point-like source which coincides with the position of Sgr A* (Watson et al. 1981; Predehl & Trümper 1994). Since soft X-rays suffer large interstellar absorption toward the Galactic Center, it must be confirmed in the hard X-ray band. *AXAF* with an angular resolution of 0.5 arcsec will be able to spatially resolve Sgr A-diffuse in the hard X-ray band and detect the hard X-rays from Sgr A*.

Acknowledgments. We express our thanks to the *ASCA* team for many efforts on the fabrication of the satellite, launch, daily operation, software development and calibration. Y. M. and M. S. are financially supported by *the Japan Society for the Promotion of Science (JSPS)*.

References

Awaki, H., Koyama, K., Inoue, H., & Halpern, J. P., 1991, PASJ 43, 195

Eckart, A., & Genzel, R. 1997, MNRAS 284, 576

Grandi, P., Tagliaferri, G., Giommi, P., Barr, P., & Paslumbo, G., 1992 ApJS, 82, 93

Kawai, N., Fenimore, E. E., Middleditch, J., Cruddace, R. G., Fritz G. G., Snyder W. A., Ulmer M. P. 1988, ApJ 330, 130

Kennea J. A., & Skinner G. K. 1996, PASJ 48, L117

Koyama, K. Awaki, H. Kunieda, H. Takano, S. Tawara, Y. Yamauchi, S. Hatsukade, I., & Nagase, F. 1989, Nature 339, 603

Koyama, K., Maeda, Y., Sonobe, T., Takeshima, T., Tanaka, Y., Yamauchi, S. 1996, PASJ 48, 249

Maeda, Y., 1998 Doctor Thesis (Kyoto University)

Maeda, Y., Koyama, K., Sakano, M., Takeshima, Y., & Yamauchi, S. 1996, PASJ 48, 417

Marshall, N., Warwick, R. S., & Pounds, K. A., 1981, MNRAS, 194, 987

Morris,M., & Serabyn,E. 1996, Ann. Rev. Astron. Astrophys. 34, 645

Morrison, R., & McCammon, D. 1983, ApJ 270, 119

Murakami, H. Koyama, K., Sakano, M., Tsujimoto, M. & Maeda, Y. 1998, ApJ, submitted

Nishiuchi, M., Koyama, K., Maeda, Y., Asai, K., Dotani, T., Inoue, H., Mitsuda, K., Nagase, F., Ueda, Y., & Kouveliotou, C. 1998, ApJ, submitted

Predehl, P., & Trümper, T. 1994, A&A 290, L29

Skinner G. K., Willmore A. P., Eyles C. J., Bertram D., Church M. J., Harper P. K. S., Herring J. R. H., Peden J. C. M. et al. 1987, Nature, 330, 544

Syunyaev R., Pavlinskii M., Gil'fanov M., Churazov.E., Grebenev S., Markevitch M., Dekhanov I. Yamburenko N., & Babalyan G. 1991, *Sov .Astron.* 17, L42

Tanaka, Y., Inoue, H., & Holt, S. S. 1994, PASJ 46, L37

Ueno, S., 1996, Doctor thesis, Kyoto University

Watson, M. G., Willingale, R., Grindlay, J.E., & Hertz, P. 1981, ApJ, 250, 142

Yamauchi, S., Kawada, M., Koyama, K., Kunieda, H., Tawara, Y., & Hatsukade, I. 1990, ApJ, 365, 532

The Central Parsecs of the Galaxy
ASP Conference Series, Vol. 186, 1999
H. Falcke, A. Cotera, W.J. Duschl, F. Melia, M.J. Rieke, eds.

Is the Sgr B2 Cloud an X-ray Reflection Nebula?

Hiroshi Murakami, Katsuji Koyama[1], Masaaki Sakano, and Masahiro Tsujimoto

Department of Physics, Faculty of Science, Kyoto University, Sakyo-ku, Kyoto 606-8502, Japan

Yoshitomo Maeda

Department of Astronomy and Astrophysics, The Pennsylvania State University, University Park, PA 16802-6305, U.S.A.

Abstract. We present the *ASCA* results of imaging spectroscopy of the giant molecular cloud Sgr B2. We found very peculiar features in this cloud; a strong 6.4-keV line emission, a low energy cutoff below 4 keV and a pronounced edge-structure at 7.1 keV. The X-ray image is extended and its peak position is shifted from the core of the molecular cloud toward the Galactic Center by about 1–2′. The X-ray spectrum and the morphology are well reproduced by a scenario where X-rays from an external source located in the Galactic Center direction are scattered by the molecular cloud Sgr B2, and come into our line of sight. Thus Sgr B2 is a new type of X-ray object, which may be called an "X-ray reflection nebula". The luminosity of the irradiating source is required to be $\geq 10^{38}$ erg s^{-1}, but no source has enough brightness. One possibility is that within the recent past, perhaps a few hundred years ago, Sgr A* was very luminous, and we are observing the "reflected" X-rays at present.

1. Introduction

X-rays from the Galactic center (GC) region have been repeatedly observed with past X-ray instruments. One of the remarkable discoveries is a large-scale hot plasma of about 100-pc in diameter, which is associated with prominent Kα lines of He- or H-like irons (Koyama et al. 1989; Yamauchi et al. 1990). The total energy of the large-scale plasma is about 10^{54} erg, with the dynamical age of $\sim 10^5$ yr. The *ASCA* satellite, with imaging capability in the wide energy band (0.5–10 keV) and reasonable energy resolution, has further found a high temperature plasma inside the Sgr A shell (an oval region of $\sim 2' \times 3'$ across), with the total energy of 10^{50} erg and the dynamical age of 10^3 yr.

ASCA also found diffuse emission from the 6.4-keV line of iron; the brightest region is located over the giant molecular cloud Sgr B2, and its X-ray peak is

[1]CREST, Japan Science and Technology Corporation (JST), 4-1-8 Honmachi, Kawaguchi, Saitama, 332-0012, Japan

shifted toward the GC from that of the molecular gas (Koyama et al. 1996). Koyama et al. (1996) speculated that Sgr A* had been bright until some hundred years ago, the light travel time between Sgr B2 and Sgr A*, but is currently dim.

The 6.4-keV line regions, together with the hot plasmas surrounding the Sgr A*, may thus provide a challenging scenario for the past activity of the GC: however because the previous report was based on limited data sets, their interpretations were rather preliminary and qualitative. We therefore have analyzed the X-ray spectrum and morphology of the Sgr B2 cloud in further detail with all the available data combined, and try to give more quantitative results and implications.

We assume the distance to Sgr B2 is the same as that to the GC or 8.5 kpc nominally, since Sgr B2 is located a few hundred pc from the GC, and the distance is estimated to be 7.1 ± 1.5 kpc (Reid et al. 1988). Then the distance between Sgr B2 and Sgr A* is 100 pc.

2. Observations

Two observations of Sgr B2 were made with *ASCA* on 1993 October 1, and on 1994 September 22–24. In both observations, all four detectors, two Solid-state Imaging Spectrometers (SIS0, SIS1) and two Gas Imaging Spectrometers (GIS2, GIS3) were operated in parallel. Details of the instruments are found in Burke et al. (1991), Tanaka, Inoue, & Holt (1994), Serlemitsos et al. (1995), Ohashi et al. (1996), Makishima et al. (1996), and Gotthelf (1996). Each of the GIS was operated in the PH mode with the standard bit-assignment. The SIS data were obtained in the 4-CCD bright mode. The net observing time was 95 ksec for GIS and 85 ksec for SIS.

3. Results and Analysis

3.1. X-Ray Images of Sgr B2

We made the X-ray images in the narrow energy band with center energy of 6.4 keV and width of twice the energy resolution (FWHM): 5.8–7.0 keV for GIS and 6.2–6.6 keV for SIS. Figure 1a shows the narrow band SIS image laid over the radio intensity contours of the CH_3CN line (Bally et al. 1988). Figure 1b shows the GIS image.

From the X-ray images, we determined the X-ray peak positions of the Sgr B2 cloud in the GIS and SIS images to be RA(2000) = $17^h\ 47^m\ 20^s.5$, Dec(2000) = $-28^\circ\ 24'\ 20''$ and RA(2000) = $17^h\ 47^m\ 21^s.3$, Dec(2000) = $-28^\circ\ 24'\ 13''$, respectively. The statistical errors are $10''$ for GIS and $15''$ for SIS. There is an additional uncertainty of the *ASCA* attitude determination of less than $40''$ (Gotthelf 1996). The radio peak of Sgr B2 is RA(2000) = $17^\mathrm{h}\ 47^\mathrm{m}\ 20^s.1$, Dec(2000) = $-28^\circ\ 23'\ 6''$ in $4''.5 \times 3''.7$ beam size from the HC_3N observation by Lis et al. (1993). Thus the X-ray peak is shifted from the cloud core to the south by $1'.2$, which is significantly larger than the X-ray position errors. In other words, the X-ray peak is shifted towards the Galactic Center (to the right in Figure 1).

We then compared the radial profiles with the point spread function (PSF) in 6.4-keV line (6.2–6.6keV) and continuum band (4.0–10.0 keV, excluding the

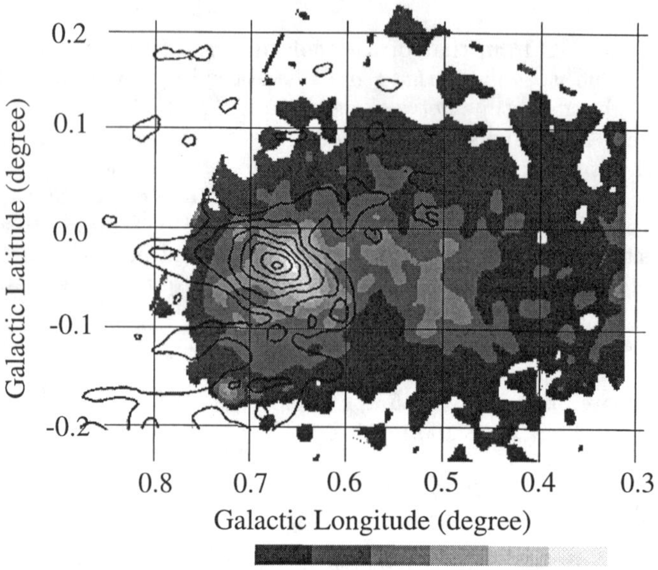

(a) SIS image

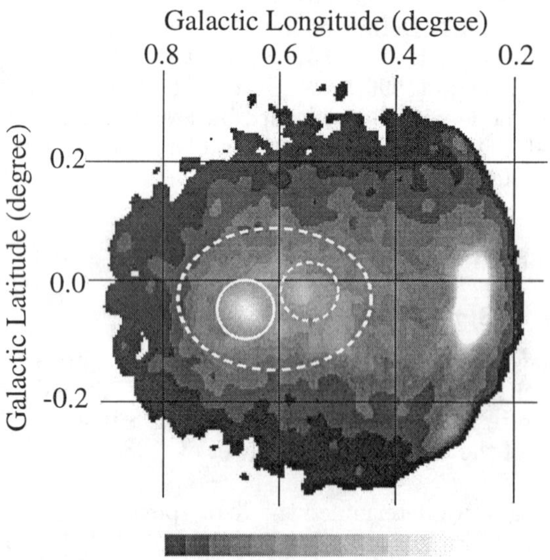

(b) GIS image

Figure 1. (a) The 6.4-keV line image around the Sgr B2 cloud obtained with SIS, laid over the CH_3CN line contours (Bally et al. 1988). (b) The 6.4-keV line image with GIS. The source and the background regions are shown by the solid circle and the dotted ellipse, respectively. The bright source at the boundary is a galactic binary X-ray source (1E 1743.1−2843).

6.2–6.6 keV). We used SIS data for this analysis, because spatial resolution of SIS is better than that of GIS. The χ^2 fittings of the PSF to the observed radial profiles lead to the best-fit reduced χ^2 to be 1.75 (40 d.o.f.) for the 6.4-keV line and 3.46 (40 d.o.f.) for the continuum band, which reject the point source assumption.

3.2. Spectrum of Sgr B2

Figure 2 shows the GIS spectrum. The source and the background regions are shown in Figure 1b with the solid circle and dotted ellipse, respectively. We fitted the spectrum to a model of two narrow Gaussian lines at 6.4 keV and 6.7 keV, which are K_α line from neutral and He-like iron respectively, and a power-law of fixed photon index 2.0, all with same absorption. For the low energy absorption, we used Bałucińska-Church & McCammon (1992) cross-sections with the solar abundances (Feldman 1992). The iron column density, since it is sensitive to the edge depth, was allowed to vary.

The best-fit parameters are given in Table 1. The 6.4-keV line, as we expected, is very strong with the equivalent width of 1.9 keV. The 6.7-keV line flux is less than 10% of the 6.4-keV line. The iron column density is $\sim 4 \times 10^{19}$ Fe cm^{-2}. This value is converted to a hydrogen column $N_{\rm H}$ of 1.3×10^{24} H cm^{-2} if the abundance is solar. The hydrogen column density determined from the low energy cutoff is $N_{\rm H} \sim 8 \times 10^{23}$ H cm^{-2}. Both of these $N_{\rm H}$ values are at least 5 times larger than that of interstellar gas to the GC region (Sakano et al. 1997). Therefore it is concluded that large absorption column is present near or at the Sgr B2 cloud.

Table 1. Results of Spectral Fitting of Sgr B2

Model Components	Parameters	Unit	Best-Fit Values
Absorption	$N_{\rm H}^{\rm a}$	(H cm^{-2})	$8.5^{+2.4}_{-2.1} \times 10^{23}$
	$N_{\rm Fe}^{\rm b}$	(Fe cm^{-2})	$4.1 \pm 2.0 \times 10^{19}$
Continuum	Photon Index		2.0 (fixed)
	Flux$^{\rm c}$ (4–10 keV)	(ph s^{-1} cm^{-2})	1.4×10^{-4}
Fe 6.4-keV Line	Flux$^{\rm c}$	(ph s^{-1} cm^{-2})	8.6×10^{-5}
	Equivalent Width	(keV)	$1.9^{+0.8}_{-0.5}$
Fe 6.7-keV Line	Flux$^{\rm c}$	(ph s^{-1} cm^{-2})	$< 5 \times 10^{-6}$
	Equivalent Width	(keV)	< 0.03
Total Luminosity	$L^{\rm d}_{4-10{\rm keV}}$	(erg s^{-1})	1.8×10^{35}
Reduced χ^2 (d.o.f.)			0.846 (32)

The errors are at 90% confidence level.
[a] The equivalent hydrogen column density for the solar abundance material except for iron.
[b] The iron column density. For solar materials, the iron abundance is 3.24×10^{-5} of the hydrogen (Feldman 1992).
[c] The fluxes are not corrected for absorption.
[d] Here and elsewhere, all X-ray luminosities are corrected for absorption.

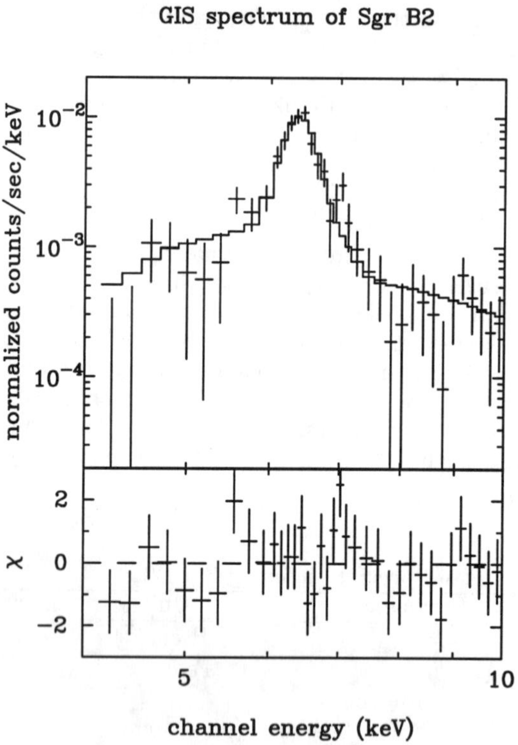

Figure 2. The GIS (GIS2 + 3) spectrum of Sgr B2. The solid line shows the simulated spectrum of an *X-ray reflection nebula.*

4. X-Ray Reflection Model and Simulation

With the fitting of the spectrum of Sgr B2, we confirmed the very peculiar features; very strong emission line at 6.4 keV, a large low-energy cutoff and a deep absorption edge at 7.1 keV. The X-ray emitting region is extended with its peak position about 1–2 arcminute offset from the cloud center to the GC side. These facts lead us to suspect that the X-rays from Sgr B2 are due to Thomson scattering (continuum emissions) or fluorescence (6.4 keV) after photo-electric absorption of neutral iron atoms by irradiation by an external X-ray source.

Then we simulated the secondary X-ray emission from Sgr B2 when the cloud is irradiated by an external X-ray source. For simplicity, we assume that the gas distribution in Sgr B2 is cylindrically symmetric with the axis parallel to the line of sight. The mass distribution of the cloud is taken from the result of ^{13}CO and C^{18}O observations by Lis & Goldsmith (1989), and is given by;

$$\left(\frac{n}{1\mathrm{H_2\ cm^{-3}}}\right) = 5.5\times10^4\left(\frac{r}{1.25\mathrm{pc}}\right)^{-2} + 2.2\times10^3, \tag{1}$$

where n is the number density of hydrogen molecules, and r is the projected distance from the center of the cloud. This formula gives the total molecular mass to be 6.3×10^6 $M_\odot$ in the 45-pc diameter region.

We divided the model cloud into $225\times225\times113$ cells (each cell is 0.2 pc in size), and calculated absorbed, reflected, and fluorescent X-rays in each cell as a function of incident X-ray energy.

4.1. Spectrum

Assuming the Galactic absorption to be 1 $\times10^{23}$H cm^{-3} (Sakano et al. 1997), we firstly made simulated X-ray spectrum for the case of solar abundances. This model, however, failed to fit the observed spectrum with a reduced χ^2 of 2.35 (36 d.o.f.), where the free parameter is only the normalization of the flux.

Large residuals are found in the flux of the 6.4 keV line, depth of the K-edge and low-energy absorption. We, accordingly, fitted the spectrum allowing the abundances to be free but with the fixed solar ratio while the iron abundance is independently free. Then we got an acceptable fit with reduced-χ^2 of 0.78 (34 d.o.f.). The best-fit parameters of both abundances are 2.4 for iron and 1.6 for the other elements. The model spectrum is given in Figure 2 by the solid line.

4.2. Morphology

With the same procedure for the spectral model, we simulated images of the 6.4-keV fluorescent line (Figure 3) and the continuum, then fitted to the observed profiles in the same way as section 3.1. Free parameters for the fitting are only the normalization and the background level. The best-fit reduced-χ^2 values are 1.25 for 40 degrees of freedom (d.o.f.) for the 6.4 keV line and 1.06 (40 d.o.f.) for the continuum X-rays.

We thus found acceptable fits to the radial profiles even for solar abundances. We note that no significant constraint of the abundances is obtained from the radial profile alone. However, we can independently constrain the abundances from the separation angle between the 6.4-keV X-ray peak and the

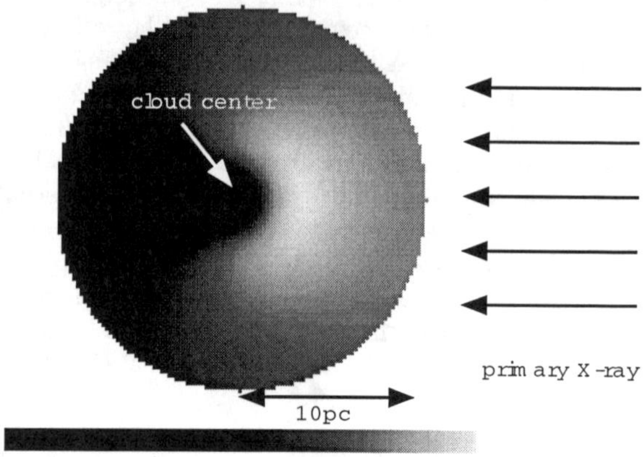

Figure 3. The simulated image of the 6.4-keV line expected with the *X-ray reflection nebula*. The brightness distribution is shifted from the cloud center to the side of primary X-ray source.

molecular core, because larger absorption in the molecular cloud leads to a further shift of the 6.4 keV peak to the irradiating source side.

The observed 6.4 keV peak is 3 ± 2 pc away from the radio peak. We thus made constant-separation lines of 1 pc and 5 pc by the reflection model simulations allowing abundances of iron and the other elements to be free parameters. The results are given in Figure 4a with solid lines. The 90% confidence contour from the spectral analysis is also given with the solid line ellipse. We found no overlapping region in the abundances of iron and the other elements between the simulated results for the morphology and spectrum.

4.3. Mass and Abundance

The disagreement of abundances between the imaging and spectral simulations may be attributable to an improper assumption on the gas distribution for the Sgr B2 cloud, either in the shape, total gas mass, or probably both. However, for simplicity, we allowed the total gas mass to be free and simulated the spectrum and morphology with variable abundances. Consistent results are obtained by reducing the total gas mass to be $< 1/2$ of the initial gas assumption of Lis & Goldsmith (1989). For example, we show the case of total gas mass of 3/4, 1/2, 1/4 of Lis & Goldsmith (1989). The allowable regions for both the imaging and spectral analysis are obtained as given by solid lines and ellipse of each graph in Figure 4. Total gas mass of Sgr B2 is constraint to be $< 3 \times 10^6$ $M_\odot$, consistent abundances both for the spectrum and image simulations are > 2 solar.

Thus the *X-ray reflection nebula* scenario has the potentiality for determination of the total gas mass and chemical abundance. However, since the observed value of the peak separation still has a large error, further detail along

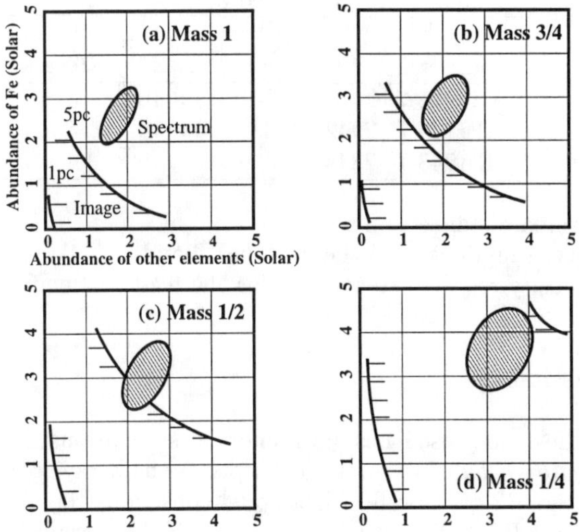

Figure 4. The 90% confidence contours between the abundances of iron and other elements for the spectrum analysis in four cases of the assumptions of the total mass (a: 6.3 $\times 10^6$, b: 4.7 $\times 10^6$, c: 3.2 $\times 10^6$, d: 1.6 $\times 10^6$ in the unit of solar mass). The constant-separation lines for the 1-pc and 5-pc separations between the 6.4-keV X-ray peak and the molecular core are also indicated in each case.

this approach should be postponed until high quality data in the morphology and the spectrum become available.

4.4. Irradiating Source

Comparing the simulated luminosity and observed value of Sgr B2, we can estimate the required luminosity of the external source to be,

$$L_{2-10\mathrm{keV}} \sim 5 \times 10^{39} \left(\frac{d}{100\mathrm{pc}}\right)^2 \left(\frac{D}{8.5\mathrm{kpc}}\right)^2 \mathrm{erg\ s^{-1}},$$

where d and D are the distances from the putative X-ray source to Sgr B2 and to the sun, respectively. Since the reflected X-rays are extending to a large fraction of Sgr B2, the required X-ray flux should be the value averaged over a time of about 100 yr. This requirement may exclude galactic binary transient sources, whose duty cycle of flaring may be very small. Therefore, we compared the required flux with the observed luminosities of rather stable sources found in the catalogue near the GC in Table 2. The observed luminosities of catalogued bright X-rays sources near the GC are 2–3 $\times 10^{36}$ erg s^{-1}, are much too low to explain the reflected luminosity of Sgr B2.

Table 2. Comparison of Required and Observed Luminosities

X-ray Binary Sources	d^{a} (pc)	L^{b}_{obs} (erg s^{-1})	L^{c}_{req} (erg s^{-1})
A 1742−294	170	3×10^{36}	3×10^{39}
1E 1740.7−2942	233	3×10^{36}	5×10^{39}
1E 1743.1−2843	63	2×10^{36}	4×10^{38}

[a] The projected distance to Sgr B2.
[b] The observed luminosity in the 2–10 keV band.
[c] The luminosity required to account for the 6.4-keV line from Sgr B2.

5. Discussion

Although the model used for the numerical simulations is very simplified, an *X-ray reflection nebula* interpretation explains the X-ray spectrum and the morphology of Sgr B2 quite well. This model constrains the abundances and total gas mass of Sgr B2 to be > 2 solar and < 3×10^{6} $M_{\odot}$, respectively. These are the first X-ray determinations of the abundances and the mass of molecular clouds. Since Sgr B2 is located near the GC, the chemical abundances would be similar to those of the GC region. It has been debatable whether the GC region is overabundant or not (Binette et al. 1982; Luck 1982; Shaver et al. 1983; Ratag et al. 1992; Sellgren, Carr, & Balachandran 1997; Ramirez et al. 1997). Our results reinforce the overabundance argument on the GC region.

Total gas mass of Sgr B2 has been poorly known. Lis & Carlstrom (1994) estimated the mass to be 2×10^{6} $M_{\odot}$ based on the observation of the 800 μm dust emission. Oka et al. (1998) estimated the mass to be 7.1×10^{6} $M_{\odot}$ in a 42 pc diameter region. This is almost the same value as Lis & Goldsmith (1989); however, they also pointed out that the molecular clouds near the GC may be in pressure equilibrium with the hot gas and/or magnetic field in this region, in which case, the mass of Sgr B2 is estimated to be 2.0 $\times10^{5}$ $M_{\odot}$. Our results favor a lower value, which may be conceivable if the central dense core of the Sgr B2 is in self-gravitational equilibrium and the outer part is in pressure equilibrium.

We found no X-ray source which exhibits enough X-rays to produce the flux of the *X-ray reflection nebula* Sgr B2. Koyama et al. (1996) pointed out that the hot plasma prevailing in the GC region can explain only 10 % of the Sgr B2 flux. Thus, one plausible possibility is, as already proposed by Koyama et al. (1996), that the Galactic nucleus Sgr A* was bright some hundred years ago, the light travel time between Sgr A* and Sgr B2, and is dim at present.

Sunyaev & Churazov (1998) simulated the morphology of the surface brightness distribution, the equivalent width and the shape of the fluorescent line from Sgr B2 cloud as a function of time and relative position of the *X-ray reflection nebula* and the irradiating source. They examined both in the cases of a flare-like and a step-function (X-ray flux is constant but is suddenly decreased to zero) irradiation by an X-ray source. As they demonstrated, a short time variability is the key to investigate the nature of the irradiating source, whether it is the

Galactic nuclei or a binary X-ray source. For a comparison with data, observations with higher angular resolution are necessary. The *AXAF* resolution of 0.″5, for example, can attain the time resolution of about 1 month, which is typical duration of a transient binary. The future observations may also reveal more realistic 2-dimensional gas distribution.

If Sgr A* was a bright X-ray source in the past, we would also expect molecular clouds other than Sgr B2 to be 6.4-keV sources. The X-ray bright spot near the radio arc is such a candidate. High quality X-ray and radio data of the cloud are crucial to judge the reflection nebula scenario, to search for past activity of the GC region and Sgr A*, or to investigate the mass distribution and abundances of the molecular cloud.

6. Summary

The *ASCA* observations revealed that the giant molecular cloud Sgr B2 is a new type of X-ray source, which may be called an *X-ray reflection nebula.* The observational facts and its implications are the following:

1. We found a strong emission line at 6.4 keV from cold iron with an equivalent width of about 1.9 keV. No significant emission line from He-like iron, which is commonly observed near the GC region, was found from Sgr B2.

2. The spectrum exhibits a large absorption at energies lower than 4 keV, equivalent to 8×10^{23} H cm^{-2} (with the solar abundances) and a pronounced edge-structure at 7.1 keV with equivalent N_{Fe} of 4×10^{19} Fe cm^{-2}.

3. The X-ray image of the 6.4-keV line as well as the continuum band is found to be extended with the peak shifted toward the GC by about 1′–2′.

4. The spectrum and the X-ray morphology are well reproduced by an *X-ray reflection nebula* model, in which a strong X-ray source is irradiating the molecular cloud Sgr B2 from the direction of the GC. Then X-ray photons below 4 keV and those above the iron K-edge energy are highly absorbed. The latter process is followed by the fluorescent iron line at 6.4 keV. Some fractions of the continuum X-rays are also reflected by the Thomson scattering process.

5. The *X-ray reflection nebula* model favors an overabundance and smaller total gas mass of the Sgr B2 cloud.

6. No luminous X-ray source to produce the reflection nebula is found. One possibility is that past activity of the GC is responsible for the presence of reflected X-rays and absence of the irradiating source.

Acknowledgments. The authors express their thanks to all the members of the *ASCA* team. M.S. and Y.M. are financially supported by the Japan Society for the Promotion of Science. The authors also thank Dr. T. Oka for his useful discussion.

References

Bally, J., Stark, A. A., Wilson, R. W., & Henkel, C. 1988, ApJ, 324, 223

Bałucińska-Church, M., & McCammon, D. 1992, ApJ, 400, 699

Binette, L., Dopita, M. A., D'Odorico, S., & Benvenuti, P. 1982, A&A, 115, 315

Burke, B. E., Mountain, R. W., Harrison, D. C., Bautz, M. W., Doty, J. P., Ricker, G. R., & Daniels, P. J. 1991, IEEE Trans., ED-38, 1069

Feldman, U. 1992, Phys. Scr, 46, 202

Gotthelf, E. 1996, The *ASCA* news (Greenbelt: NASA GSFC), 4, 31

Koyama, K., Awaki, H., Kunieda, H., Takano, S., Tawara, Y., Yamauchi, S., & Nagase, F. 1989, nature, 339, 603

Koyama, K., Maeda, Y., Sonobe, T., Takeshima, T., Tanaka, Y., & Yamauchi, S. 1996, PASJ, 48, 249

Lis, D. C., & Goldsmith, P. F. 1989, ApJ, 337, 704

Lis, D. C., Goldsmith, P. F., Carlstrom, J. E., & Scoville N. Z. 1993, ApJ, 402, 238

Lis, D. C., & Carlstrom, J. E. 1994, ApJ, 424, 189

Luck, R. E. 1982, ApJ, 256, 177

Makishima, K., et al. 1996, PASJ, 48, 171

Ohashi, T., et al. 1996, PASJ, 48, 157

Oka, T., Hasegawa, T., Hayashi, M., Hanada, T., & Sakamoto, S. 1998, ApJ, 493, 730

Ramirez, S. V., Sellgren, K., Terndrup, D. M., Carr, J. S., Balachadran, S., & Blum, R. D. 1997, IAU Symp. 184, The Central Regions of the Galaxy and Galaxies, ed. Y.Sofue (London: Kluwer Academic Publishers), 67

Ratag, M. A., Pottasch, S. R., Dennefeld, M., & Menzies, J. W. 1992, A&A, 255, 255

Reid, M. J., Schneps, M. H., Moran, J. M., Gwinn, C. R., Genzel, R., Downes, D., & Rönnäng, B. 1988, ApJ, 330, 809

Sakano, M., Nishiuchi, M., Maeda, Y., Koyama, K., & Yokogawa, J. 1997, IAU Symp. 184, The Central Regions of the Galaxy and Galaxies, ed. Y.Sofue (London: Kluwer Academic Publishers), 443

Sellgren, K., Carr, J. S., & Balachandran, S. C. 1997, IAU Symp. 184, The Central Regions of the Galaxy and Galaxies, ed. Y.Sofue (London: Kluwer Academic Publishers), 21

Serlemitsos, P. J., et al. 1995, PASJ, 47, 105

Shaver, P. A., McGee, R. X., Newton, L. M., Danks, A. C., & Pottasch S. R. 1983, MNRAS, 204, 53

Sunyaev, R., & Churazov, E. 1998, MNRAS, 297,1279

Tanaka, Y., Inoue, H., & Holt S. S. 1994, PASJ, 46, L37

Yamauchi, S., Kawada, M., Koyama, K., Kunieda, H., Tawara, Y., & Hatsukade, I. 1990, ApJ, 365, 532

The Central Parsecs of the Galaxy
ASP Conference Series, Vol. 186, 1999
H. Falcke, A. Cotera, W.J. Duschl, F. Melia, M.J. Rieke, eds.

Discussion Session X: What Can We Learn from High-Energy Emission in the Galactic Center?

F. Melia[1]

Physics Department and Steward Observatory, The University of Arizona, Tucson, AZ 85721

S. Markoff[2]

Physics Department, The University of Arizona, Tucson, AZ 85721

Abstract. This is an edited transcript of a discussion session about high energy phenomena (cosmic rays, X-rays, gamma-rays, Sgr A East) in the Galactic Center that took place during the Galactic Center conference "The Central Parsecs" in Tucson, Arizona, Sep. 7-11, 1998. The session was chaired by the moderator as indicated in the author list. The discussions were taped, transcribed, and finally edited by an editor who is listed as co-author. The contributions of the conference participants are preceded by their names and thus references to specific aspects and ideas of this discussion session must also include a reference to the respective speaker(s), e.g., in a footnote.

Discussion

Melia: I've written down a couple of notes as the speakers were going through their presentations, and we probably shouldn't try to stick to any order in this, particularly since some people have to leave pretty soon. So I'd be happy to entertain anyone at the beginning of the discussion with any of these points. The energetics of Sagittarius A East may be of common interest; perhaps we could start with that, but if anyone else would like to start with any of the other ones, we could do that just as well. Does anyone have any suggestions, or should we just start with Sagittarius A East? Let's do that. Okay, Ramesh, you had an interesting...

Narayan: I just had the question, if it's a hundred times more powerful than a normal supernova, then what is it?

Melia: It might actually help for many of us who are newer to the field if we could get a historical perspective, and I don't want to dump this on Miller, but . . .

---

[1]Moderator of the discussion session

[2]Editor of the discussion session

Goss: I don't need to worry, because every time I forget how old I am, I am immediately reminded ...

Melia: So maybe if we could have a little summary about what the issues were from the very beginning? I guess it started in the late 80s, and it was obvious from that time that there was a problem with the energetics.

Goss: Of course, historically there were even problems with the morphology, because we were plagued by the low declination of Sagittarius A–the fact that high-resolution radio telescopes were in the northern hemisphere, and Sagittarius A is a southern object. Even though the recognition of the association of strong radio emission with the Galactic Center was made by McGee and Bolton (1954) in the early 1950s, from low-resolution radio telescopes in Australia. A lot of people were working on this. Dennis Downes was worried about it, and I was worried about it. We struggled from the Netherlands to make images of Sgr A, where it was a few degrees above the horizon, and we started seeing evidence for the shell structure. In the 70s and in the early 80s when the VLA came on line, we did finally, for the first time, see the shell, about the same time that Farhad did, and other people. And immediately we realized that the surface brightness was all wrong for being a conventional supernova remnant. That's just another way, after all, of saying that it was too energetic for its size and presumed distance. And so in the first papers that we wrote, with Ekers, Pedlar, and Anantha[ramaiah], we already pointed out that if we assumed it was a normal SNR, the energy was going be wrong by factors of 10 or more. Then we realized that the confining mechanism was probably a dense molecular cloud around there. And the other thing that influenced our thinking in those early years were the HII regions. Those A, B, C, and D HII regions that we have—we knew they were HII regions—and maybe they were related to the confining high-density gas.

Melia: I was looking for Peter Mezger, I don't see him–

Kassim: Yeah, I'll just mention that the number of 4×10^{52} erg/sec, that people have been floating around, comes out of the Mezger et al. paper, 1993 or something like that. It's just the back-of-the-envelope Sedov calculation: most of the time when you do the Sedov calculation for remnants, you assume an explosion energy, and you calculate the density of the external medium based on the size and assumed shock velocities. In their case they actually knew the density from radio observations. They used that in turn to constrain the explosion energy. So it's the extremely high density into which it was exploding that drives the explosion energy up to 50 times what we expect from a normal remnant. It's kind of a very simple back-of-the-envelope calculation.

Townes: A few years ago there was a general supposition that perhaps this is just a large number of supernovae all added together. Does anybody take that seriously now? How probable is that?

Narayan: The SNR is not very old.

Townes: Not very old, no.

Narayan: So you need 30 or 40 explosions in ...

Townes: About 50 of them in a very short time. That seems very unlikely to me. Does anybody like that model?

Biermann: I think it might be worth mentioning that probably from Berkeley, Karl Heilas has found these big supershells, both in our Galaxy and in other galaxies. You can take exactly the same type of analysis and you'll find the same type of energy. So there are many sources in the Galaxy and other galaxies like that, where you find the requirement to have 50 to 100 supernovae in one region which, because of lower density, is very much larger. To me this suggests that indeed there could be 50 to 100 supernovae that would explode in close sequence. And I don't know of any other model for the other supershells outside the Galactic Center, which easily does all these things. But you have a model ... ?

Narayan: There is a model that my colleague published recently—Avi Loeb and his student. They said that this could be a gamma-ray burst, because the energy that's involved in a gamma-ray burst is just about this order, maybe 10^{53} erg. And according to him, the numbers of supershells coincides with the numbers of gamma-ray bursts that you typically expect in a galaxy. Based on some lifetime and the ages of shells. I don't know the details.

Biermann: Yah, yah, I didn't mention this model because as I understand gamma-ray burst statistics, they are sort of floating around by 2 to 4 powers of ten right now. So these things are possible—clearly possible—but extremely uncertain.

Narayan: I think the energetics are now pretty much pinned down—now that we know the redshifts.

Biermann: I don't agree with that. I think that depends on the model you use, even so.

Narayan: No, this is just what you observe.

Biermann: Yeah, that's a problem, I agree. But that's a real problem.

Melia: Maybe what Peter's getting at is whether the energy is beamed or not.

Biermann: Yes, of course. For example, we are currently working on a jet model for gamma-ray bursts.

Melia: That would be the distinction. But if you see isotropic emission, then ...

Biermann: ... then this is perfect.

Melia: Anantha, you had your hand up.

Anantharamaiah: I brought up this question the other day, but I'll bring it up again. It's related to Sagittarius A East. Now we did some work to find out the actual shape of Sagittarius A East, because it seems to be mixed up with Sgr A West on one side. So by taking optically thin radiation at centimeter wavelengths, we would like to subtract away Sagittarius A West from the total emission to find the shape of Sgr A East. And it does look like a shell structure with the edge enhanced like in supernova remnants. So just from morphology, it does look like something exploded in the center and remained in a shell. If there were multiple explosions, they all had to happen at the center. If it is a supernova remnant, many people now realize that it is something that is less than 100,000 years old, and may be only a few tens of thousands of years old. And it is offset from the location of Sgr A West, which we know, at the center of it must be Sgr A*, because the kinematics, the dynamics, and everything that goes with Sgr A* looks like it must be a center of activity. Now you take this supernova-like object, that is maybe just one parsec away from there, and if we make Sgr A West and Sgr A East actually connected to each of them, and if we assume Sgr A East to be a supernova remnant, then everybody is suggesting that this black hole and Sgr A West is inside the supernova remnant, and that I find really hard to sort of realize. How can you have an explosion that went off only about a few tens of thousands of years ago and looks like a nice, symmetrical shell-like spectrum that we expect from an explosion. Yet inside it is this spectacular object, a black hole, plus all the other mumbo-jumbo that goes with it.

Melia: Unless the object that exploded was near Sgr A* and Sgr A West. Because it could be that the shell just passed through that region and now looks like it's enveloping the center.

Cotera: Wouldn't you expect it to be much more disturbed if it had passed through, I mean, wouldn't you expect the expanding shell to be slightly affected by passing by a black hole?

Melia: Well, unless the explosion happened very close to the black hole, or within a couple of arcseconds of Sgr A*. I don't think the shell is that perfect, that you'd be able to exclude that.

Townes: Isn't the most natural explanation stellar infall into the black hole?

Melia: That's one possibility.

Townes: It seems to me that's a very natural explanation.

Melia: That would be one possibility. That would be a good reason why the shell now envelops the Sgr A* region, because if it's true there was a tidally disrupted star, that would have happened within about 20 Schwarzschild radii, so it's right at the location of the black hole.

Anantharamaiah: Would it still produce a symmetric looking shell?

Melia: No, actually, it wouldn't. It would produce an extended feature because most of the efflux comes out within a relatively small solid angle. Because when the explosion occurs, it occurs as the star comes out of its point of closest approach along its trajectory, so most of the efflux actually comes out into a relatively narrow cone. It doesn't explode as a sphere. The tidal force is still compressing the star in one direction, and the efflux escapes into the outbound trajectory direction. That could be the explanation for it.

Narayan: Is the energetics okay for tidal disruption?

Melia: Yeah. If it comes within ten Schwarzschild radii, it is. That's how Alexei Khoklov and I estimated that number, because it's how much work the tidal field does on the star in compressing it that determines how much energy is released when it explodes. The way this works, by the way, which is something that I never really appreciated before we started thinking about this about a year and a half ago, is that when the tidal force compresses the star, the work that it does gets dissipated into internal energy because the time scale for dissipation is shorter than the transit time. So if this picture is right, the reason it happens is that the work done by the tidal force is converted into internal energy, and when the star is freed from the tidal field, it explodes with that stored energy. This work corresponds to about 10^{53} erg, which is then the energy of the explosion.

Falcke: I just want to mention one point that I was discussing with Charles Townes over coffee either yesterday or two days ago when we talked about exactly that point. You expect, based on these arguments, that every 10,000 years or say 10^5 years such a thing to happen: stars are falling in, they explode, BOOM! Sgr A East is there. But of course then it has to happen in the next 10 to 100,000 years, then another 100,000 years, and that matter gets injected into the ISM in the central region. I'm just wondering what would happen to this continuously injected mass. Would it form a diffuse, very hot gas in the center? Do we have any observational evidence for this, or can we even link it to something we've already observed?

Melia: Right. The morphological question is, of course, a very important one. Farhad would be able to address the large scale structure. Is he here?

Goss: They left.

Melia: The problem that we have in the Galactic Center is that there's a lot of complexity, of course, and the 50 km/s cloud is dominating many of the features that we see. I forget what the transit time is. Let's see, for 50 parsecs, moving at 50 km/s, how long would that have taken?

Biermann: One million years.

Melia: One million years. Okay, so maybe one might have expected to see two such remnants built on top of each other, and I don't know if the morphology of Sagittarius A East is sharp enough, clean enough, that we can tell whether it's a one shell feature or two shells, one catching up with the other one.

Biermann: But we have 10^8 K gas; we just heard that. We do have evidence for something energetic.

Narayan: At least it's some continuous injection, so to speak, on the time scale of a few million years...

Melia: Yes, but Heino, I think, was asking about the morphology ...

Falcke: There could be both effects. If you have every 100,000 years a star falling in, and injecting energy into the ISM, it should have an effect on what we see there, either in a diffuse form or in distinct shells.

Novak: Related to your question, there is this big structure called the Omega. I'm sure the radio observers know it. Would that be related at all to what Heino's looking for? Maybe it's not hot enough, I don't know.

Kassim: Well, Janet, you showed one of these old low-resolution radio single dish images. It's on there. Remember that one you put on? The beginnings of it are on there.

Goss: Why don't we get Sofue's image on, since that's what it is. He's one of the discoverers of this.

Kassim: See how the arc kind of goes way out and forms a second Omega at the top of that.

Sofue: Right.

Novak: I have a nice image of it here.

Kassim: Yeah, we've seen several pictures of it, actually.

Biermann: It was the last explosion, I guess. [Laughter]

Melia: So what's the age of that feature? It's about one degree in size, right?

Sofue: Yeah, about 300 parsecs.

Melia: What would the age be for that?

Sofue: I'd say maybe 10^7 yrs, if it is an explosion. The problem is if it is associated with a strong magnetic field.

Biermann: Are we now coming to the point where we need 100 gamma-ray bursts to do this?

Novak: How do we know the age of a few million years? Is that known?

Sofue: Not very well known, no.

Novak: Can you put a limit on the age of this structure? At all?

Sofue: Yeah. If you simply take the input energy, size of the shell, and the expansion velocity you can get an age.

Novak: And where do you get the expansion velocity from?

Sofue: From the associated molecular cloud. It's possible, but not certain.

Melia: Go ahead, Susan.

Stolovy: Fulvio, in your model doesn't it matter what kind of star you have?

Melia: Oh, those numbers were based on a one solar-mass star.

Stolovy: I see. Well what if you have the stars that are in the inner Sagittarius IR cluster, that are perhaps O stars ...

Melia: The more massive ones?

Stolovy: Right.

Melia: The probability of those coming in would be much smaller, though, because there are far fewer of them. That's why we picked one solar mass, because that would be the most likely star coming in. If the mass is bigger, the frequency of course would go down, but you also win out because the star wouldn't have to go as close to the black hole in order to be compressed with enough energy to explode. So you can play with those numbers, but those don't really matter too much. The time scale is always between 10 and 100,000 years. But let me add that this model is only a possibility, and it will probably be difficult to ascertain whether it is correct. Maybe we should go on a bit, because we only have about 10 or 15 minutes left. Does anyone have a preferred topic to go to, or should I just pick one?

Why don't we address the X-rays, this issue of the recurrence time scale for producing the X-rays—the X-ray illumination from Sagittarius A*—I thought was interesting. Masaki, can you say something about what the expected recurrence time scale is?

Sakano: The recurrence time scale ...

Melia: ... assuming it's coming from Sagittarius A* ...

Sakano: It is difficult to say, but the most recent activity is maybe about several hundred years ago, or one thousand years ago.

Melia: This is a relatively short time scale.

Sakano: Yes.

Melia: What type of X-ray flux is required for that? And also connected with that was the issue of the location of the peak in the 6.4 keV emission. Why should it be at Sagittarius B2? Is that a density effect, or is it the geometry of the emitter that's causing that?

Sakano: If the plasma is widespread in the central region and was heated by a recent phase of Galactic Center activity, the required luminosity is about 10^{42} erg per second, which is quite large.

Melia: And the duration of that phase is?

Sakano: Duration?

Melia: Yes, I'm trying to get an idea of what the total energy released should be.

Sakano: The total energy is more than 10^{54} erg.

Biermann: Wow, we *do* need a gamma-ray burst ... [Laughter]

Melia: And why is it that Sagittarius B2 is the peak of the emission? Is there a clear sense of that? Is it the density distribution, or is it the geometry of the X-ray source, being at the location of Sagittarius A*?

Sakano: It appears to be the X-ray source.

Melia: And that's consistent with the X-rays coming from Sagittarius A*?

Sakano: It's really consistent, yes.

Melia: Did anyone else have any comments or questions about the X-ray image from the Galactic Center? Can we spend a few minutes talking about the possibility that Sagittarius A* is the source of the gamma-rays? I guess the issues here are whether changing some of the physical assumptions, for example, if the electron temperature is not equal to the proton temperature, whether that might change the relative fluxes in the radio compared to the gamma-rays. Because this would impact the number density of relativistic protons and decay products that come from those. Ramesh, do you have a sense of how that would go?

Narayan: Within the ADAF model, taking the $\dot{M}$ that we fixed by looking at the radio spectrum, for instance, we could not produce enough gamma-rays to fit the data. We missed by about an order of magnitude in flux, which corresponds to something like a factor of 3 or 4 in mass accretion rate.

Melia: This is fixing the radio . . .

Narayan: Fixing the radio and X-ray, and then looking at what the protons will do, we didn't get enough gamma-rays. You could do the opposite, which is more or less what you and Sera did, namely, just fit the gamma-rays, forget the rest, and then see whether it will fit the rest of the spectrum. You and Sera couldn't do it. It's an open question whether Rohan's charged pion decay would fit. Because the properties of the ADAF, the self-absorption, etc., are a little different from your model. Maybe it would work.

Melia: Does it matter what you assume from the proton distribution, whether you take it to be thermal or ...

Narayan: Oh, thermal, no hope at all. It has to be highly non-thermal.

Biermann: All the models require non-thermal proton distributions.

Narayan: Absolutely, yes. If it's thermal, you miss by orders of magnitude.

Melia: So it is fair to say, right now, that it doesn't look likely that even if you have a two-temperature plasma you'd be able to fit the gamma-rays and the radio data together, right?

Narayan: Probably not. I think one thing to talk about is, you have developed some really accurate ways of computing the gamma-ray flux. It may be interesting to take the ADAF model and apply it to your machinery to see what it gives.

Melia: Sera, that sounds like a good application to work on.

Mahadevan: There's one thing about the gamma-ray flux that the ADAF model does: it produces this turnover at 70 MeV because of π^0 production. So now, since these EGRET observations have shown that this 30 MeV detection is close to an upper limit, I don't know how you could actually push it down to a lower energy.

Melia: You may also be able to fill part of that spectrum in with continuum emission, like inverse Compton and bremsstrahlung.

Narayan: Not at 30 MeV. All of that cuts off at a few hundred keV.

Melia: And Wolfgang, what about the quasi-monoenergetic distribution, could that produce the gamma-rays from Sgr A*?

Duschl: I can't tell you right now all I can promise is to check it.

Melia: Alright, so I guess one conclusion we can make from this discussion is that Sagittarius A* probably is not the source of gamma-rays, which means that looking at viable alternatives is very worthwhile. I think right now Sagittarius A East is probably the most viable candidate. When it comes to cosmic rays, that's again a slightly different issue, because the cosmic rays might or might not be associated with the gamma-rays. Just to start off that discussion, it wasn't clear to me whether the proton distribution is really unique, given the neutrons that we see from the Galactic Center. Can one reliably argue that the proton distribution has to be fixed fairly strongly by the neutron flux that we see, or is there some possible variance?

Biermann: Variability with time?

Melia: I was thinking about some variant possible fit based on what we see in the neutrons.

Biermann: Well, we haven't checked the details, but as far as I can tell from just crude estimates based on the relevant Japanese paper, it is consistent—and actually I was quite surprised that it is consistent—with a cosmic ray injection spectrum [i.e., injection spectrum in cosmic ray sources] all the way out to the high-energy. I found this very surprising, I had not expected that, because at a high-energy meeting recently we talked about what we could expect there, and I expected the flux would be lower. And this is an interesting result. Basically you need the injection spectrum all the way up to 10^{18} eV, and then it seems to work. We will check this now with our big Monte Carlo code. In the big Monte Carlo code, I might mention, you get the π^0 decay spectrum much more accurately than what we can do with these normal simple codes. You get all these details with very high accuracy. These big codes are not perfect, they do have their problems, but they are the best you can do.

Townes: I haven't heard anyone bring up the gamma-rays from aluminium and magnesium. And it seems to me that has some pertinence. It leads me to believe it has to have been a very compact explosion.

Biermann: Through nuclear decay right?

Townes: Yeah.

Biermann: I mentioned those.

Townes: Oh, did you? Sorry. Doesn't that show that there's been a compact explosion?

Biermann: I mentioned it very quickly in my talk, but you can show—and we showed this for the Orion region then—in the normal shocks that were close to a Sedov solution—as we've seen in UV and in the radio—accelerate particles and heavy nuclei. In fact you can directly reproduce almost trivially those nuclear gamma-ray lines.

Townes: You can?

Biermann: Yah. Sure. From the shock size you need a normal star without any supernova, and this is one of the many models of this kind, what I would call the trivial model. Because you don't even need a supernova to do it. That was published four years ago. There are many other models, but this is sort of the easy way. Because we have so many Wolf-Rayet stars in the Galactic Center region, and they have powerful shocks in their wind—that's why we see the non-thermal radio emission which was originally detected by people in Berkeley. And this also underlines the points that in these powerful winds you have strong magnetic fields. Right there, before they blow up. And so it is my feeling that this assembly of Wolf-Rayet stars and He stars, this is enough to do almost anything you want. You don't need anything special, that's at least my suspicion. And I may be over-reacting, but I have the feeling that Wolf-Rayet stars and massive OB stars they can do lots of things for you.

Kassim: What's the spectrum of the emission?

Biermann: The spectrum is normal to a little steep, as far as I remember. You have absorption, that is a classical problem in stellar winds, thermal absorption of non-thermal radiation, and in our first paper we went a little bit through that argument, but I was hoping that there would be more data sometime soon on these things. But I think it's enough to tell you that there is a fairly strong magnetic field. There's a fairly normal shock acceleration but in a different stellar wind.

Melia: It's too bad that we've only gotten part of the way through. This is a very interesting discussion, but somebody just pointed out to me that we're running out of time, so I ask if anyone has any last comment or question about this session? No? In that case, we'll have Miller Goss express a few thoughts on the organization of this conference, followed with some closing remarks by Charlie Townes to terminate the meeting. Thank you all for your enthusiastic participation in this very useful meeting.

Summary

The Central Parsecs of the Galaxy
ASP Conference Series, Vol. 186, 1999
H. Falcke, A. Cotera, W.J. Duschl, F. Melia, M.J. Rieke, eds.

Conference Summary: The Central Parsecs

C. H. Townes[1]
University of California, Department of Physics, 366 Le Conte Hall #7300, Berkeley, CA 94720-7300

S. Stolovy[2]
Steward Observatory, The University of Arizona, Tucson, AZ 85721

H. Falcke[2]
Max-Planck-Institut für Radioastronomie, Auf dem Hügel 69, D-53121 Bonn, Germany

Abstract. This is an edited transcript of the summary talk of the Galactic Center conference "The Central Parsecs" in Tucson, Arizona, Sep. 7-11, 1998, given by C. Townes. The talk was taped, transcribed, and finally edited by editors who are listed as co-authors.

Talk

Thank you. I have an impossible task, to summarize this workshop. But there is one point that I feel sure we would all agree on as the most obvious conclusion: that this has been wonderful conference. Miller Goss has already made that point, with thanks to the people who have organized it and helped us. I know we're all very appreciative. Now, what I have to say will not be a precise summary. It is more of a personal perspective, and I will make a few specific points. I hope that some of you may have additional points that I have missed, but I'm going ahead with my own thoughts.

In the first place, the Galactic Center of course turns out to be a wonderful laboratory. It's obviously the closest galactic center to us. Furthermore, a great deal is happening there. In a way, too many things are happening, which then complicates the situation for us because they're all interacting. To isolate them all is part of the difficulty. By now, with the wealth of measurements available, we really know an enormous amount about what's there. How much we understand is another question. I'll try to outline those things I think we more or less understand and things which we don't.

Before doing that, however, let me go back in history a bit and see how we've gotten where we are, because the information has been building up very

[1]Speaker

[2]Editor of the talk

rapidly, almost exponentially. I'm sure you know that in the early part of this century we didn't even know that we were in a galaxy. Well, that was finally found out and recognized, but the first real observation of the Galactic Center was Jansky's work in the radio region. Astronomers were, of course, kind of floored, because in the optical region nothing could be seen, and that was part of the initial misunderstanding of where we were. Radio waves came along, and Jansky's discovery was the first observation. That did not excite astronomers; interpretation of it wasn't clear. Although it didn't excite astronomers for a while, nevertheless eventually it did, and has become very important.

I believe the next important observation, of a different character, was in 1968: Becklin and Neugebauer in the infrared. Intense infrared radiation was detected! It's of course the waves which go through dust that we have to use: radio, infrared, X-rays, gamma rays, and so on. In 1974 the first spectral work was done by Aitken, Jones and Penman, on Neon II. And that's what excited our group at Berkeley. Also by then Balick and Brown had detected Sgr A*, a rather exciting discovery. At Berkeley, the first of my students to work on the Galactic Center was Wollman, whose task was to do some spectroscopy on Ne II and see what was going on. And from the Ne II spectra, the first thing we found was that a large part of it was going away from us fairly fast. And we were immediately bombarded by an astronomer in England who said "Well, I said this all the time; we are moving away from the Galactic Center and rather fast, and this proves it". He had already written a paper, and our evidence he believed was the confirmation. This is just how the interpretation of things can go a bit wrong, when we have incomplete information. And of course our information is never absolutely complete.

We proceeded to make more measurements, and what was then enormously helpful and revealing was the high-resolution work of radio astronomy: Ekers and Goss et al., Lo and others, Brown and so on, with the VLA. We began to get the picture. We had been measuring random points around the Galactic Center, to see what the velocity, density, and temperature of the gas was. Suddenly we realized there was structure. The radio people had found structure of curved streamers of gas which made systematic measurements much more practical for us. Finally we could see overall what was happening.

Then OH stars came along, observed by Lindqvist and others. And in addition to gas velocities, stellar velocities began to be measured by Sellgren, Becklin, Hall, Haller and the two Riekes. So it seemed fairly clear from the gaseous and stellar measurements that, yes, in fact, there was a concentrated mass. But the very recent work on stellar motions, by Eckart and Genzel and by Ghez and Becklin and others—the two groups—has been the linchpin which has really firmed up the case. Not because it wasn't pretty clear that there was a concentrated mass of the order of 3 million solar masses, but this brought us much closer in, so the mass had to be enormously more concentrated. And in addition, the statistics were much better. So now the situation is much clearer, and this conference started off with some of those discussions, which were very important ones.

Now, so much for some of the history and the build-up. I haven't mentioned things like the gamma-ray and X-ray work, that came along. With the big group of people who are present with such a wide variety of approaches, interests, measurements, and theory, the field is building up very rapidly.

What do we know now, from this conference? I would say first, we have a rather good measurement of the mass; a concentrated mass in the center of 2.6 $\pm$0.2 million solar masses. Nevertheless, we don't know it quite that accurately. That's the statistical error in the measurements themselves. But another thing we have to know is the distance to the Galactic Center. It's the uncertainty in that distance, measured well first by Mark Reid, which dominates the uncertainty in the mass itself. The mass interpretation is very dependent on the distance. In fact the actual uncertainty is almost half a million solar masses, if we accept the standard uncertainty in the distance. The statistics of velocity measurements are much better, but there's a systematic error involved, and thank goodness Mark Reid is going to measure the distance still more accurately for us soon.

Next we also know the spectrum of Sgr A*. We have only upper limits at infrared and X-ray radiation, but a lot of the spectrum we know pretty well. Nevertheless, there's still a lot to be done, and hopefully new satellite measurements will give us much more detailed information about the X-rays. Somehow we have to find ways of getting adequate infrared measurements of Sgr A* .

On the motion of Sgr A*, we also have pretty good information. That's been followed for some years by Backer and by Reid. We have reports that Sgr A* is stationary within about 15 kilometers per second, but one measurement is +15 and the other one is -15, and so we still have an uncertainty. We hope that will be improved.

The position of Sgr A* compared to the center of mass we're closing in on to perhaps a fraction of an arcsecond. That of course is the position perpendicular to our line of sight. We don't know the position along the line of sight. It's presumably within a few arcseconds; that's something we don't know, but I suggest that the two are closely coincident.

The size of Sgr A* we also know remarkably well, with all the radio measurements, particularly the millimeter wave measurements. Krichbaum has said the best measurement of a diameter at one millimeter wavelength is 0.11 $\pm$ 0.06 milliarcseconds, and it is somewhat elongated according to Lo and Zhao. But there's still a lot of important work to be done there; we think we know the mass, but how big it is, and what it is, are different questions. To see the black hole, or to see at least closer and closer to it, the millimeter wave measurements and maybe the far infrared measurements are probably very important. To get very high angular resolution, good quality images, and good sensitivity, interferometry is very important. What else can we do to really see that object and understand more about what is there? Well, I think perhaps additional X-ray work is one of the most crucial things, because we don't know for sure just what kind of an object it is. The natural interpretation is, for sure, that it's a black hole. We know what black holes are like, and essentially it's so concentrated there's nothing else we know that it could be. On the other hand, it isn't behaving quite the way it should, in terms of generating energy. It's easy to dismiss the fact that we don't see much energy now, by saying: well, this is a temporary affair, it happens there's not much material falling in right now, and so it's not very intense at the moment. On the other hand, we know that things have to have fallen in from time to time. Calculation says that a star should fall in about every 10^4 years, in addition to the gas we know is there. If we can go back in time, it would be very important to know has it ever emitted anything? And that's what the Russian work in X-rays attempts to do, looking at the scattering of X-rays at various distances from the center. Hopefully AXAF can give us not only much better measurements astrometrically, more sensitive measurements of what's coming out of the central region, but maybe also tell whether there's been anything that's happened in the past that would fit what we generally expect.

There are many other remarkable things that have been developed and that we've seen at this meeting. There are striking new, precise, detailed maps in the radio frequency region, and in the infrared region from the mid-infrared camera results which Cotera was showing us. And those will tell us many more things, I'm sure. We also have detailed maps of molecular and ionic densities, temperatures, and velocities. And I am impressed by the detailed information we now have about magnetic fields. Maybe we don't understand very much about magnetic fields and what they're doing dynamically, but we are beginning to get more and more detailed, good maps of magnetic fields in the Galactic Center. There seems to be a fairly uniform agreement that the magnetic field, at least close in, is in the plane of the gas itself, whereas elsewhere in many cases it's perpendicular. And now we also think of quarks and strings.

I realize now, I should have put in my history the discovery by Yusef-Zadeh and Mark Morris of the radio arc. That was a big event and really opened up the picture for us. And now we have additional strings and snakes and things of this type, discovered in the radio. What has not been discussed very much is the exact origin of the ionized particles that are producing the synchrotron radiation. We can imagine, yes, they were generated by magnetic fields, but I don't think we know very much about how.

Stellar populations have also been mapped and discussed thoroughly, and I think we now have a very good picture of the stellar populations. We also have a pretty good picture of the radio scattering, and the kind of precision with which we can expect to do radio mapping to locate activity precisely.

Now, let me try to face some of the questions: what are the things that puzzle us? There are indeed puzzles, and I'm grateful that there are, because if everything had fitted our understanding would not have been challenged. But it happens that this particular system isn't so simple; it turns out not to fit our previous ideas very well. And I'd like to lay out some challenges. The challenges require a lot of additional theoretical work, which has already been important. Maybe in the long run we'll learn something absolutely brand new and quite unexpected as a result.

One of the big questions, of course, concerns the radiation flux that's being emitted. And we have a number of interesting suggestions as to what might limit that flux. One is that the infall has very little angular momentum, and the other one is advection. There could also be jet and wind phenomena. Still another suggestion that hasn't been very much discussed here is a heavy dust blanket in front of Sgr A*, which might be cutting down the radiation an order of magnitude, and could make our understanding a little simpler. But we haven't a clear-cut fit that's very satisfying. We do see temporal fluctuations; I think those are important, and I hope they will be continuously watched. But nevertheless, maybe *the* most puzzling phenomenon is why hasn't there been more radiation? I'm hopeful that this AXAF work and other X-ray work will help us understand that, yes, there has been in the past some radiation. Or if it doesn't, then maybe it'll be still more exciting to see what in the world it all means.

Another puzzle that strikes me, though, is why don't we see more phenomena connected with the infall of stars? Because if calculations are right, and a star will fall in about every 10^4 years, and I think that's a fairly conservative calculation, why don't we see more remnants of these big explosions? One can say that maybe we see at least a couple of them: perhaps Sgr A East is one of them. Perhaps also the distribution of gas immediately around Sgr A West is another, because that gas certainly is not in equilibrium, and calculated time scales say something disturbed it within the last 50,000 years or so. So maybe there are a couple of things, but why aren't there more, and more evident? That seems to me another important puzzle.

Do we already understand this concentrated mass, which according to the gravitational data has to be there? Reinhard [Genzel] raised the question, do we have to have an event horizon or can it be something else—not a black hole? People have proposed other things. But we were shown that it couldn't be a simple bunch of rocks or such things; this couldn't produce the kind of spectrum we see, but maybe there could be some combination of a black hole and other materials, such as the new suggestion of massive neutrinos, plus a black hole in the center or something like that. That is a wild possibility but I think we need to examine such things and either disprove them if they're not possible, or perhaps recognize there is something different from what we presently believe. Perhaps lensing is still another way that we can get very close to the Galactic Center to see what's there and how concentrated it really is.

Let me move on quickly to the stars. Another puzzle appears to be that we don't see any accelerations in the proper motion of the stars near the Galactic Center. It's not highly probable we would see a particular star accelerated, but it's rather improbable that we have not seen any accelerations. In a few years, clearly we'll know more, and perhaps we'll see curvature in the orbits, giving evidence for acceleration.

Another puzzle is the low metal abundance in the ten red giants which were analyzed. We know that the metal abundance in the gas observed is fairly high. We're told that the metal abundance of the OH stars seen through Baade's window is high, but ten red giants near the Galactic Center have metal abundances comparable with what we have around the Sun, and that is very strange. Also the absence of red giants very close to the center is strange. We know the population of red giants dips as we get very close to the Galactic Center. Why? The old explanation was that they get hit by other stars and so their shells are ripped off and they no longer look like red giants. Recent calculation claims that's not true. The probability of the shells being ripped off is too low. But for some reason, they're not there, and we don't know why.

We have thought for some time that the very bright collections of hot Wolf-Rayet stars—hot, but not very very hot ones—is very strange. Closeness to the Galactic Center might have seemed an explanation for exceptional star formation, but we know they're up in the Pistol region, and now we have a new one in the Arches. It seems common in that general region that recent stellar formation has occurred. That's not surprising in view of the dynamics and the density of material, but why these particular types are formed isn't understood. These star forming regions give us a new kind of laboratory, the Pistol star is claimed to be 200 solar masses, maybe the biggest one that we know. These stars clearly deserve additional study, as well as the rapidly burning stars close to the center.

Finally, let me touch on the magnetic field again. I think the magnetic field clearly deserves a lot more study. It's complex. I think many of us have avoided it because of this complexity. We now have many good measurements. Clearly magnetic fields have an important dynamic role, and we need to understand this, despite the complications, better than we do.

We've seen maps of the whole area, with all the different phenomena. The time scales of formation and ages of these things, the connections between them, the different structures and all the interactions, make a very complex picture. But it's important for us to understand these complexities if we're going to understand each individual phenomenon better, and I hope we can.

Author Index

Subject Index

DUNELM UNIV

DURHAM UNIVERSITY LIBRARY
3 0104 01160677 3